COMPREHENSIVE CHEMICAL KINETICS

COMPREHENSIVE

Section 1. THE PRACTICE AND THEORY OF KINETICS

Volume 1 The Practice of Kinetics

Volume 2 The Theory of Kinetics

Volume 3 The Formation and Decay of Excited Species

Section 2. HOMOGENEOUS DECOMPOSITION AND ISOMERISATION REACTIONS

Volume 4 Decomposition of Inorganic and Organometallic Compounds

Volume 5 Decomposition and Isomerisation of Organic Compounds

Section 3. INORGANIC REACTIONS

Volume 6 Reactions of Non-metallic Inorganic Compounds

Volume 7 Reactions of Metallic Salts and Complexes, and Organometallic Compounds

Section 4. ORGANIC REACTIONS (6 volumes)

Volume 9 Addition and Elimination Reactions of Aliphatic Compounds

Volume 10 Ester Formation and Hydrolysis and Related Reactions

Volume 12 Electrophilic Substitution at a Saturated Carbon Atom

Volume 13 Reactions of Aromatic Compounds

Section 5. POLYMERISATION REACTIONS (2 volumes)

Section 6. OXIDATION AND COMBUSTION REACTIONS (2 volumes)

Section 7. SELECTED ELEMENTARY REACTIONS (2 volumes)

Additional Sections

HETEROGENEOUS REACTIONS

SOLID STATE REACTIONS

KINETICS AND TECHNOLOGICAL PROCESSES

CHEMICAL KINETICS

EDITED BY

C. H. BAMFORD

M.A., Ph.D., Sc.D. (Cantab.), F.R.I.C., F.R.S.
Campbell-Brown Professor of Industrial Chemistry,
University of Liverpool

AND

C. F. H. TIPPER

Ph.D. (Bristol), D.Sc. (Edinburgh)
Senior Lecturer in Physical Chemistry,
University of Liverpool

VOLUME 9

ADDITION AND ELIMINATION REACTIONS OF ALIPHATIC COMPOUNDS

ELSEVIER SCIENTIFIC PUBLISHING COMPANY
AMSTERDAM – LONDON – NEW YORK
1973

ELSEVIER SCIENTIFIC PUBLISHING COMPANY
335 JAN VAN GALENSTRAAT
P. O. BOX 1270, AMSTERDAM, THE NETHERLANDS

AMERICAN ELSEVIER PUBLISHING COMPANY, INC.
52 VANDERBILT AVENUE
NEW YORK, NEW YORK 10017

LIBRARY OF CONGRESS CARD NUMBER 72-83195
ISBN 0-444-41051-1

WITH 13 ILLUSTRATIONS AND 86 TABLES

PRINTED IN THE NETHERLANDS

COMPREHENSIVE CHEMICAL KINETICS

Contributors to Volume 9

P. Beltrame	Università di Cagliani, Istituto Chimica Policattedra, Cagliari, Italy
R. Bolton	Department of Chemistry, Bedford College, University of London, London, England
A. F. Cockerill	University of East Anglia, Norwich, England (*Now* Lilly Research Centre, Erl Wood Manor, Windlesham, Surrey, England)
R. H. DeWolfe	Department of Chemistry, University of California, Santa Barbara, Calif., U.S.A.
W. Kirmse	Abteilung für Chemie, Ruhr-Universität Bochum, Bochum, West Germany

Preface

Section 4 deals almost exclusively with reactions recognised as organic in a traditional sense, but excluding (unless very relevant) those already considered in Sections 2 and 3 and biochemical systems. Also oxidations, *e.g.* of hydrocarbons by molecular oxygen, polymerization reactions and fully heterogeneous processes are considered later. The relationships of mechanism and kinetics, *e.g.* effects of structure of reactants and solvent, isotope effects, are fully discussed. Rate parameters of individual elementary steps, as well as of overall processes, are given if available. We have endeavoured, in conformity with our earlier policy, to organise this section according to the types of chemical transformation and with the minimum of recourse to mechanistic classification. Nevertheless it seemed desirable to divide up certain general processes on the basis of their nucleophilic or electrophilic character.

Volume 9 deals with the majority of addition and elimination reactions involving aliphatic compounds. Chapter 1 covers electrophilic addition processes, mainly of water, acids and halogens to olefins and acetylenes, and Chapter 2 the addition of unsaturated compounds to each other (the Diels–Alder reaction and other cycloadditions). This is followed by a full discussion of α-, β-, γ- and δ-eliminations (mainly olefin and alkyne forming) and fragmentation reactions. In Chapter 4 carbene and carbenoid reactions, and in Chapter 5 alkene isomerisation (including prototropic and anionotropic, and Cope and Claisen rearrangements), are discussed.

The editors are very grateful for the continuing support and advice from members of the editorial board, and also for much invaluable assistance from D. Bethell of the Department of Organic Chemistry.

C. H. Bamford
C. F. H. Tipper

Liverpool
November, 1972

Contents

Chapter 1

ELECTROPHILIC ADDITIONS TO UNSATURATED SYSTEMS

R. BOLTON

Submitted June 1969

1. Introduction

1.1 GENERAL PRINCIPLES

One of the most characteristic reactions of an unsaturated system is addition, in which an entity X–Y attacks and saturates the multiple bond. The means by which this reaction occurs depends upon a number of factors which will be discussed as a background to this article.

The bond X–Y may break in two distinct ways; firstly, each fragment may take with it one of the bonding electrons to give two free-radicals, $\dot{X}$ and $\dot{Y}$ (homolysis) which may then subsequently add to the unsaturated system. Such a situation occurs in the peroxide-catalysed addition of HBr to olefins[1], *viz.*

$$RCH{=}CH_2 + HBr \xrightarrow{\text{peroxide}} RCH_2CH_2Br \tag{1.1}$$

The second means of breaking the X–Y bond involves one of the fragments retaining both of the bonding electrons. In this case X and Y are left with formal charges (*e.g.* X^+, Y^-) and the subsequent attack upon the multiple-bond system by one or other of these fragments involves an ionic or heterolytic mechanism. The "normal" reaction of olefins with hydrogen bromide demonstrates this form, *viz.*

$$RCH{=}CH_2 + HBr \rightarrow RCHBrCH_3 \tag{1.2}$$

Within this latter category, two new possibilities occur. The initial attack upon the unsaturated system may be due (*i*) to a positively charged species (X^+) or to the positively charged end of the X–Y dipolar molecule, when the addition is initiated by an electrophilic attack, or it may be due (*ii*) to a negatively charged species (Y^-) or to the negatively charged end of the X–Y dipole, when a nucleophilic addition is said to take place. The first mode of reaction is exemplified by the acid-catalysed hydration of an olefin, *viz.*

$$RCH{=}CH_2 + H_2O \xrightarrow{H^+} RCH(OH)CH_3 \qquad (1.3)$$

and the second by the addition of malonic ester anion to ethyl crotonate, *viz.*

$$CH_3CH{=}CHCO_2Et + \bar{C}H(CO_2Et)_2 \rightarrow CH_3\underset{\displaystyle CH(CO_2Et)_2}{\underset{|}{C}}H\bar{C}HCO_2Et \qquad (1.4)$$

We are particularly concerned with electrophilic addition mechanisms, but some general statements, covering both heterolytic reaction mechanisms, can be made. In contrast with the homolytic mechanism, partial or complete ionisation takes place at some stage in the heterolytic reaction—charges are either formed or destroyed or dispersed. It would therefore be expected (and is indeed found) that a heterolytic reaction would be profoundly affected by the solvating ability of the environment. The formation of ionic species from covalent molecules requires considerable energy which is mostly supplied by the heat of solvation of the ions; where this energy is small, ionisation is unlikely to occur. The Hughes–Ingold theory of solvent effects[2] develops along these lines; a reaction which involves the formation of charge in its rate-determining stage will be inhibited in a "poor" solvent and assisted in a "good" solvent, and *vice versa.* However, the bulk medium need not supply all or even much of the energy stabilising the incipient ions; these may be helped to form by the intervention of some other species in solution which can coordinate with either of the fragments of the X–Y dipole. In doing this, and making the rate-determining stage of the reaction feasible, terms involving the concentration of the catalyst may appear in the rate expression where they have thermodynamic rather than kinetic significance. An example of "solvation" occurring through an entity other than the bulk medium is found in the Friedel–Crafts reaction of benzene with benzoyl chloride. Aluminium chloride is essential for the success of the reaction, and appears in the rate expression[3]

$$\text{Rate} = k_2[\text{benzene}][\text{AlCl}_3] \qquad (1.5)$$

(benzoyl chloride, as the solvent, does not appear in the expression) although its function is to polarise the acid chloride molecule, probably through the interaction shown in I[3], *viz.*

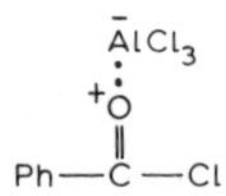

I

Other examples of this will be mentioned subsequently.

The products of the reaction may have some influence upon the observed kinetics, usually by complicating the rate expression when there is some interaction between a molecule of product and one of reactant, or when the reaction studied has an equilibrium position far from the extreme as in the acid-catalysed hydration of some $\alpha\beta$-unsaturated carbonyl compounds[4]. Such mathematical complications, however, can usually be allowed for, and kinetic studies are usually made under conditions to minimise such effects.

1.2 STRUCTURAL EFFECTS

Since the rate-determining stage in an electrophilic addition reaction often involves the attack of the electrophile upon the unsaturated system, factors which affect the electronegativity of the atom being attacked will influence the rate of the reaction. In the acid-catalysed hydration of olefins, which in dilute solutions follows the simple kinetic form

$$\text{Rate} = k[\text{olefin}][\text{H}^+] \qquad (1.6)$$

alkyl groups have a considerable effect (Table 1)[5].

TABLE 1

RELATIVE RATES OF HYDRATION OF OLEFINES, $CH_2{=}C(Me)R$, IN 29.6% PERCHLORIC ACID AT 38°C

R	H	CH_2Cl	CH_3	C_2H_5	$C(CH_3)_3$	C_6H_5
k_{rel}	1	1	8×10^3	10×10^3	8×10^3	5×10^3

Here the reaction is assisted by conjugative and hyperconjugative effects, and is retarded by the inductive effect which withdraws electrons from the reaction site. A similar effect is found in the addition of bromine to olefins ($RCH{=}CH_2$) in acetic acid at 25°, *viz.*

R	C_4H_9	H	CH_2OAc	CH_2Cl	CH_2CN	CO_2Et
k_{rel}	26.2	1.0	0.21	0.02	0.0027	5×10^{-5}

when any electronic effect which removes electrons from the reaction site strongly decreases the observed reaction rate[6].

Structural effects in the reactant also have an influence upon the reaction rate. In the formation of the X–Y dipole, when this is the true attacking species, the extent of charge separation will reflect the electronegativities of the two fragments (or rather, the atoms whose common bond is broken in the transition state) and the stabilities of the incipient ionic fragments. Thus, the order of reactivity

$$\text{Cl–OAc} > \text{Cl–OCl} > \text{Cl–Cl} > \text{ClOH}$$

has been provisionally established for addition to olefins, and this sequence reflects neither the ease of heterolysis of the Cl–X bond nor its bond strength[7].

The mechanisms of electrophilic addition reactions to unsaturated systems has been the subject of a recent monograph[7]; for this reason it is not proposed to discuss the criteria for adopting a mechanism in any detail in the present work, although it must not be thought that all such proposed mechanisms have been demonstrated incontravertibly.

2. Acid-catalysed hydration of olefins

2.1 ACIDITY CONCEPTS (see also Volume 2, Chapter 4)

The addition of the elements of water to an olefin is an acid-catalysed process. In dilute aqueous solution the reaction has the kinetic form[8]

$$\text{Rate} = k[\text{olefin}][\text{H}^+] \tag{2.1}$$

in which the hydrogen ion concentration is taken to be a measure of the acidity of the solvent. At higher acid concentrations, however, this simple rate expression is not always valid[9]. The rate-determining stage in the reaction is, however, the conversion of the olefin to a positively-charged intermediate having carbonium ionic character; although its exact structure is still a source of controversy, it may be thought of as a solvated carbonium ion and we can write a simplified reaction scheme such as Scheme 1.

$$\text{RCH=CH}_2 + \text{H}^+ \rightleftharpoons \text{R}\overset{+}{\text{C}}\text{HCH}_3 \underset{-\text{H}_2\text{O}}{\overset{+\text{H}_2\text{O}}{\rightleftharpoons}} \text{RCH(OH)CH}_3$$

Scheme 1 Simplified mechanism of the hydration of olefins in acid.

It is not relevant here to discuss whether the carbonium ion is "encumbered" (with solvent molecules) or "unencumbered", but to point out the analogy between the first stage of the reaction and the protonation of a simple base, *viz.*

$$B + H^+ \rightleftharpoons BH^+ \tag{2.2}$$

In more concentrated solutions of acids the extent of such protonation far exceeds that predicted from the hydrogen ion concentration and the Hammett acidity function, H_0, was the first of a number of attempts to describe the proton-donating properties of such solutions[10]. The definition of this function[11], *viz.*

$$H_0 = pK_{BH^+} - \log \frac{[BH^+]}{[B]} \tag{2.3}$$

was made in terms of a number of indicator bases whose equilibrium constants could be measured, usually spectrophotometrically, by measuring the concentrations of the free base and the protonated form. By using successively weaker bases a scale could be built up which ranged from dilute aqueous solution ($[H^+] = 0.1\,M$) to highly concentrated acids. The limitations of this concept, and the assumptions made in putting it forward, need not be discussed here[12]. The function is, however, only a measure of acidity towards neutral bases whose conjugate acids undergo no further reaction. Allied functions (H_+, H_-) can be similarly defined for bases bearing positive or negative charge, and for bases where the sequence

$$XOH \xrightarrow{H^+} [XOH_2]^+ \rightarrow X^+ + H_2O \tag{2.4}$$

occurs (*e.g.* nitric acid, triarylcarbinols), another function (J_0, C_R) is defined[12,13], *viz.*

$$-J_0 = pK_{ROH} + \log \frac{[R^+]}{[ROH]} \tag{2.5}$$

$$J_0 = H_0 + \log (a_{H_2O}) - \log \frac{f_{BH^+}}{f_B} \cdot \frac{f_{ROH}}{f_{R^+}} \tag{2.6}$$

The H_0 function was measured using mainly substituted anilines as bases. It became evident that, using other bases, other H_0 scales could be defined and measured[14], so that it was possible to report[15] that the general order of the acidity functions was

$J_0 > H_R'$ (olefins) $> H_0'''$ (tertiary amines) $> H_0$ (primary amines)

although each of the functions was intended to measure the acidity of a solution irrespective of the base employed. Bunnett and Olsen[16] have published a linear free-energy treatment of the situation. It succeeds the earlier w and w^*

References pp. 80–86

treatments in that no direct correlation with H_0 need be assumed, but allows a discussion of the hydration changes and the role of water in the reaction from considerations of the correlation parameter (ϕ), and seems to hold promise both as a means of describing reactions in acidic media and as a diagnostic tool.

Much of the reported work on the hydration of olefins has tried to fit the acid-dependence of the reaction either to the stoichiometric hydrogen ion concentration or to the origin Hammett function. In the latter case, even when the reaction shows linear dependence upon H_0, the slope of the plot is often greater than unity. It has been suggested that this discrepancy is due to changes in the activity coefficient terms[17] similar to the variations found between the different acidity functions[15], *viz.*

$$\log k = -H_0 + \log \frac{f_S}{f^*} \cdot \frac{f_{BH^+}}{f_B} + \text{constant} \qquad (2.7)$$

and indeed values for the $\log k$ *vs.* H_0 slope are closer to unity after such allowances have been made[17]. In the present work, recognising the diversity of the problems, we have reported correlations found in the original papers without further comment. This does not imply that more recent acidity functions might not be more appropriate.

2.2 EQUILIBRIA IN HYDRATION

The acid-catalysed hydration of olefins is a reversible process (Scheme 1, p. 4) for which the rate-determining stage involves the protonation of the olefin. As a result, the kinetics of the reaction may be complicated if hydration is not substantially complete under the reaction conditions. In dilute solution, however, this is generally not the case, and some equilibrium constants for the reaction

$$\rangle C{=}C\langle \; + H_2O \rightleftharpoons \; \rangle CH{-}C(OH)\langle \qquad (2.8)$$

are recorded in Table 2.

The unexpectedly low K value of crotonaldehyde can be easily explained on the basis of electronic effects, excepting for the observation that its rate of hydration is considerably greater than those of crotonic acid or of propene. A mechanism involving proton attack upon the ethylenic bond would require the sequence of reactivity

TABLE 2

EQUILIBRIUM CONSTANTS MEASURED FOR THE HYDRATION OF OLEFINS

Olefin	*Acid*	*K*	*Temp. (°C)*	*Ref.*
Acrolein	$HClO_4$	12.0[a]	19.98	18
	$HClO_4$	8.4[a]	29.7	18
	$HClO_4$	6.3[a]	39.93	18
Acrylic acid	$HClO_4$	11.3[a]	110.6	18
	$HClO_4$	9.5[a]	119.8	18
	$HClO_4$	6.7[a]	134.7	18
Crotonic acid	$HClO_4$	5.0[a]	90.03	19
	$HClO_4$	3.3[a]	111.3	19
Crotonaldehyde	$HClO_4$	0.8[a]	25.0	19
2-Methylpropene	HNO_3	1.0[c]	30.0	20
Methylenecyclobutane	HNO_3	200[c]	30.0	20
1-Methylcyclobutene	HNO_3	30[c]	30.0	20
1-Methylcyclopentene	HNO_3	0·007[c]	30·0	20
Styrene	$HClO_4$	41[b]	25.0	21
α-Deuterostyrene	$HClO_4$	55[b]	25.0	21
p-Methoxystyrene	$HClO_4$	34[b]	25.0	21
p-Methylstyrene	$HClO_4$	43[b]	25.0	21
p-Chlorostyrene	$HClO_4$	40[b]	25.0	21
p-Nitrostyrene	$HClO_4$	10[b]	25.0	21
2-Phenylpropene	H_2SO_4	14[d]	25.0	22
2-(*p*-Methoxyphenyl)propene	H_2SO_4	8[d]	25.0	22
2-(*p*-Nitrophenyl)propene	H_2SO_4	11[d]	25.0	22
2-(*p*-Chlorophenyl)propene	H_2SO_4	13[d]	25.0	22
2-(*p*-Methylphenyl)propene	H_2SO_4	6[d]	25.0	22
2-(*m*-Chlorophenyl)propene	H_2SO_4	5[d]	25.0	22
4-Isopropenylbenzoic acid	H_2SO_4	7[d]	25.0	22

[a] Specific acid catalysis (by H^+) was found [*F. G. Ciapetta and M. Kilpatrick, J. Am. Chem. Soc.*, **70** (1948) 639] in the hydration of isobutene; general acid-catalysis was not, and for this reason it was thought that the transition state comprised only a proton and the olefin. Schubert *et al.*[21], however, showed that formic acid itself could protonate *p*-methoxy-α-methylstyrene, and hence that general acid catalysis occurred.

[b] $K = k_1/(k_{-1})$.

[c] K relative to isobutene (2-methylpropene).

[d] $K = k_1/(k_{-1}.a_{H_2O})$; values quoted are averaged from the data.

$$\mathrm{RCH{=}CHCHO < RCH{=}CHCOOH \ll RCH{=}CH_2}$$

The most reasonable explanation of the situation is to postulate a pre-equilibrium in which the carbonyl group of the aldehyde is protonated, followed by a rate-determining attack by water upon the conjugate acid, *viz.*

$$CH_3CH{=}CHCHO \overset{H^+}{\rightleftharpoons} [CH_3CHCHCHOH]^+ \text{ (fast)}$$

$$[CH_3CHCHCHOH]^+ \xrightarrow{H_2O} \text{Product} + H^+ \text{ (slow)}$$

In the hydration of mesityl oxide[23], the slow step in the sequence appears to be the uptake of a proton by the 2,4-dihydroxy-2-methyl-3-pentene which results from a fast pre-equilibrium hydration of the mesityl oxide. In this situation, however, the rate of the overall reaction does not vary linearly with either $[H^+]$ or h_0 at higher acidities; this is in keeping with the requisites of the reaction mechanism which is neither strictly A-1 or A-2[12].

In an A-1 mechanism, only the protonated substrate, SH^+, is involved in the rate-determining stage, and the effective concentration of this ion, and hence the observed reaction rate, ought to depend upon h_0. In an A-2 mechanism, SH^+ undergoes attack by a nucleophile in the rate-determining stage. For the special, but common, case when this nucleophile is water, it has been shown[24] that the rates of some acid-catalysed reactions are dependent upon $[H^+]$ (see ref. 12 for the derivation of this relationship for an A-2 process). The proposal that such a relationship is *diagnostic* of the A-2 mechanism has been criticised[25].

2.3 TABULATION OF KINETIC RESULTS

Table 3 contains kinetic results for the acid-catalysed hydration of olefins. In the second column of the table, rate coefficients are quoted in units of 10^{-4} $l.mole^{-1}.sec^{-1}$. k_1 is defined by the expression

$$\text{Rate} = k_1[\text{olefin}]$$

and is therefore specific to a definite acidity; k_2 is defined by the expression

$$\text{Rate} = k_2[\text{olefin}][\text{acidity}]$$

where "acidity" is defined either as $[H^+]$ or by a function such as H_0 and its negative antilogarithm h_0. In the latter case a note will indicate the use of an acidity function; in all other cases the acidity will be defined as the stoichiometric hydrogen-ion concentration. The temperature, and the solvent (when not water), are quoted in parentheses after each rate coefficient; energies and entropies of activation, where available, are quoted in the third column of the table, together with any further results or comments. Separate entries to the table deal with other relevant material, such as kinetic isotope effects.

TABLE 3

ACID-CATALYSED HYDRATION OF OLEFINS

Acid	$10^4 \times k$	*Comments*	*Ref.*
Propene			
D_2SO_4	$k_{(\text{anti-Mark.})}/k_{(\text{Mark.})} = ca.\ 4 \times 10^{-5}$[a]		26
Acrolein			
$HClO_4$	$k_2 = 0.37$ (19.98)		18
	$= 0.97$ (29.70)		18
H_2SO_4	$= 1.23h_0$ (20.0)		27
	$= 3{\cdot}89h_0$ (30.0)		27
	$= 6.92h_0$ (40.0)		27
	$= 17.78h_0$ (50.0)		27
Acrylic acid			
$HClO_4$	$k_2 = 0.44$ (110.6)		18
	$= 0.82$ (119.8)		18
2-Methylpropene (isobutene)			
HNO_3			
0.1 *M*	$k_1 = 0.39$ (25.0)		28
	$= 0.033$ (25.0; 50% dioxan)		28
	$= 0.0014$ (25.0; 95% dioxan)		28
0.05 *M*	$k_1 = 0.185$ (25.0)	Salt effects also reported	29
0.1005 *M*	$= 0.39$ (25.0)		29
0.248 *M*	$= 1.02$ (25.0)		29
0.498 *M*	$= 2.30$ (25.0)		29
1.000 *M*	$= 5.95$ (25.0)		29
0.0909 *M*	$k_1 = 0.114$ (25.00)[b]	$\Delta H^{\ddagger} = 17.3$ kcal.mole^{-1}	9
	$= 0.302$ (34.93)[b]		30
	$= 0.754$ (44.90)[b]	$\Delta S^{\ddagger} = -31.2$ eu	30
	$= 1.80$ (54.94)[b]		30
0.506 *M*	$= 0.834$ (25.00)[b]		9
0.973 *M*	$= 0.788$ (14.96)[b]	$\Delta H^{\ddagger} = 16.7$ kcal.mole^{-1}	30
	$= 2.19$ (24.89)[b]		30
	$= 2.21$ (25.00)[b]	$\Delta S^{\ddagger} = -27.3$ eu	9
	$= 5.55$ (34.84)[b]		30
	$= 14.2$ (45.03)[b]		30
1.515 *M*	$k_1 = 4.78$ (25.00)[b]		9
2.525 *M*	$= 14.35$ (25.00)[b]		9
4.03 *M*	$= 48.4$ (25.00)[b]		9
5.04 *M*	$= 110$ (25.00)[b]		9
H_2SO_4			
1.242 *M*	$k_1 = 3.83$ (25.00)[b]	H_0 *vs.* $\log k$ profile	9
$HClO_4$	$k_{H_2O}/k_{D_2O} = 1.45$ (25.0)	Solvent effect	31
Crotonaldehyde			
HNO_3			
0.5–2.00 *M*	$k_1 = 0.066$ (25.0)[c]		32
	$= 0.179$ (35.0)[c]	$\Delta H^{\ddagger} = 24.5$ kcal.mole^{-1}	32
	$k_1 = 0.60$ (25.0)		33
$HClO_4$			
0.5–2.00 *M*	$k_1 = 0.062$ (25.0)[c]		32
	0.168 (25.0)[c]	$\Delta H^{\ddagger} = 24.5$ kcal.mole^{-1}	32
H_2SO_4	$k_1 = 1.0$ (25.0)[d]		33

References pp. 80–86

TABLE 3 (*Contd.*)

Acid	$10^4 \times k$	*Comments*	*Ref.*
Crotonic acid			
HNO_3			
	k_1 very small (25.0)		19, 32
$HClO_4$	$k_1 = 0.0195$ (90.0)		34
	$= 0.0914$ (111.3)		34
1-Methylcyclobutene			
HNO_3			
2.523 *M*	$k_1 = 4.64$ (25.0)[b]	Competition with methylenecyclobutane	35
Methylenecyclobutane			
HNO_3			
0.972 *M*	$k_1 = 3.97$ (25.17)[b]	$k_1 = 2.95 \times 10^{-4}.h_0^{1.29}$	35
	$= 9.47$ (34.58)[b]		35
	$= 22.75$ (45.15)[b]		35
2.011 *M*	$= 16.14$ (25.17)		36
2.523 *M*	$k_1 = 32$ (25.0)[b]		35
2.995 *M*	$= 45.0$ (25.17)[b]		36
4.000 *M*	$= 109.2$ (25.17)[b]		36
4.941 *M*	$= 209$ (25.17)[b]		36
2-Methyl-1-butene			
HNO_3			
0.973 *M*	$k_1 = 7.58$ (10.00)[b]	$\Delta H^{\ddagger} = -15.2$ kcal.mole^{-1}	37
	$= 19.8$ (20.00)[b]	$\Delta S^{\ddagger} = -32$ eu	37
	$= 48.4$ (30.00)[b]		37
		$\Delta H^{\ddagger} = 21$ kcal.mole^{-1}	38
2-Methyl-2-butene			
HNO_3			
0.053 *N*[e]	$k_1 = 2.1$ (25.0)	Rates with other acids	39
0.107 *N*[e]	$= 2.3$ (25.0)	also reported	39
0.516 *N*[e]	$= 2.4$ (25.0)	$\Delta H^{\ddagger} = 18.9$ kcal.mole^{-1}	39
1.000 *N*[e]	$= 3.7$ (25.0)		39
0.973 *M*	$= 0.142$ (0.00)[b]		30
	$= 0.387$ (9.70)[b]	$\Delta H^{\ddagger} = 15.4$ kcal.mole^{-1}	30
	$= 1.021$ (20.00)[b]		30
	$= 2.58$ (30.00)[b]	$\Delta S^{\ddagger} = 32.8$ eu	30
	$= 5.94$ (40.00)[b]		30
0.973 *M*	$k_1 = 0.662$ (15.00)[b]	$\Delta H^{\ddagger} = 15.2$ kcal.mole^{-1}	37
	$= 1.69$ (25.00)[b]		37
	$= 4.84$ (35.00)[b]	$\Delta S^{\ddagger} = -33$ eu	37
1.500 *M*	$k_1 = 5.43$ (30.00)[b]	$k_1 = 2.04 \times 10^{-4}.h_0^{1.20}$	35
2.011 *M*	$= 10.04$ (30.00)[b]		35
2.995 *M*	$= 25.7$ (30.00)[b]		35
4.000 *M*	$= 56.5$ (30.00)[b]		35
4.941 *M*	$= 99.0$ (30.00)[b]		35
HNO_3 in D_2O–H_2O			
0.973 *M*	$k_1 = 2.58$ (30.00; %D = 0.00)[b]		40
	$= 2.50$ (30.00; %D = 25.2)[b]		40
	$= 2.60$ (30.00; %D = 50.6)[b]		40
	$= 2.61$ (30.00; %D = 58.9)[b]		40
	$= 2.23$ (30.00; %D = 95.7)[b]		40

TABLE 3 (*Contd.*)

Acid	$10^4 \times k$	*Comments*	*Ref.*
1-Methylcyclopentene			
HNO_3			
0.0909 *M*	$k_1 = 0.741\ (29.89)^b$	$\Delta H^{\ddagger} = 13.8$ kcal.mole^{-1}	41
	$= 1.11\ (34.84)^b$		41
	$= 2.31\ (44.92)^b$	$\Delta S^{\ddagger} = -40$ eu	41
HNO_3 in D_2O–H_2O^f			
0.0909 *M*	$k_1 = 1.176$ (35.20; %D = 0.00)^b		40
	$= 1.169$ (35.20; %D = 29.9)^b		40
	$= 1.229$ (35.20; %D = 42.3)^b		40
	$= 1.276$ (35.20; %D = 51.4)^b		40
	$= 1.298$ (35.20; %D = 65.0)^b		40
	$= 1.360$ (35.20; %D = 74.8)^b		40
	$= 1.350$ (35.20; %D = 87.2)^b		40
	$= 1.340$ (35.20; %D = 90.3)^b		40
	$= 1.301$ (35.20; %D = 93.7)^b		40
Mesityl oxide (4-Methyl-3-penten-2-one)			
HNO_3			
1.00 N^e	$k_1 = 0.75\ (25.00)^d$		42
	$= 0.58$ (25.00)		33
H_2SO_4			
	$= 0.80\ (25.00)^d$		33
$HClO_4$			
1.00 N^e	$= 0.83$ (25.00)		42
2.00 N^e	$= 0.95$ (25.00)		42
	$= 0.53\ (25.00)^d$		33
HCl	$= 0.60\ (25.00)^d$		33
H_3PO_4	$= 1.3\ (25.00)^d$		33
D_2SO_4	$= 0.26\ (25.00)^d$		33
DCl	$= 0.20\ (25.00)^d$		33
2,2-Dimethyl-3-pentene			
HNO_3			
0.973 *M*	$k_1 = 1.22\ (25.12)^b$	$k_1 = 10^{-4}.h_0^{1.30}$	35
	$= 15.1\ (25.12)^b$		35
	$= 74.2\ (25.12)^b$		35
Styrene (phenylethylene)^g			
$HClO_4$			
4.564 *M*	$k_1 = 0.533$ (25.0)		21
4.58 *M*–7.41 *M*	$k_2 = 1.6 \times 10^{-3}.h_0^{1.26}$ (25.0)	$\Delta H^{\ddagger} = 19.5$ kcal.mole^{-1}	21, 43
6.54 *M*–7.71 *M*	$= 1.0 \times 10^{-4}.h_0^{1.39}$ (10.0)	$\Delta S^{\ddagger} = -5$ eu	43
5.72 *M*	$k_2 = 3.95$ (25.0; β,β-D_2)	$k_H/k_D = 1.02 \pm 0.02$	21
5.72 *M*	$k_2 = 4.15$ (25.0; α-D)	$k_H/k_D = 1.02 \pm 0.02$	21
0.10 *M*–3.53 *M*	$k_2 = 8 \times 10^{-1}.h_0^{1.27}$ (25.0; *p*-OMe)		21
3.53 *M*–5.50 *M*	$k_2 = 3 \times 10^{-2}.h_0^{1.33}$ (25.0; *p*-Me)		21
4.60 *M*–7.41 *M*	$k_2 = 5 \times 10^{-4}.h_0^{1.24}$ (25.0; *p*-Cl)		21
9.39 *M*	$k_2 = 6 \times 10^{-6}.h_0^{1.23}$ (50.0; *p*-NO_2)		21
0.102 *M*	$k_2 = 9.69$ (25.0; *p*-OMe, α-Me)		21
H_2SO_4			
2.95 *M*–7.75 *M*	$k_2 = 3.7 \times 10^{-3}.h_0^{1.17}\ (25.0)^h$		44
1.72 *M*–4.00 *M*	$k_2 = 6.0 \times 10^{-1}.h_0^{1.27}$ (25.0; *p*-OMe)		44
2.00 *M*–6.12 *M*	$k_2 = 3.4 \times 10^{-2}.h_0^{1.21}$ (25.0; *p*-Me)		44

References pp. 80–86

TABLE 3 (*Contd.*)

Acid	$10^4 \times k$	*Comments*	*Ref.*
2.95 *M*–7.86 *M*	$k_2 = 9.8 \times 10^{-3}.h_0^{1.12}$ (25.0; *m*-Me)[h]		44
5.40 *M*–9.22 *M*	$k_2 = 1.6 \times 10^{-3}.h_0^{1.14}$ (25.0; *p*-Cl)[h]		44
4.05 *M*–8.65 *M*	$k_2 = 2.0 \times 10^{-3}.h_0^{1.08}$ (25.0; *p*-Br)		44
5.60 *M*–10.8 *M*	$k_2 = 2.3 \times 10^{-4}.h_0^{1.10}$ (25.0; *m*-Cl)[h]		44
6.90 *M*–10.8 *M*	$k_2 = 1.8 \times 10^{-4}.h_0^{1.11}$ (25.0; *m*-Br)		44
8.55 *M*–12.8 *M*	$k_2 = 5.3 \times 10^{-5}.h_0^{1.16}$ (25.0; *m*-NO_2)[h]		44
7.68 *M*–10.6 *M*	$k_2 = 2.7 \times 10^{-6}.h_0^{1.28}$ (25.0; *m*-NH_3^+)		44
2.34 *M*	$k_2 = 27.0$ (25.0; α-Me)		45
0.5 *M*–5.3 *M*	$k_2 = 3.2h_0$ (25.0; α-Me)		46
2.34 *M*	$k_2 = 1600$ (25.0; *p*-OMe, α-Me)		45
1.5 *M*–3.5 *M*	$= 126h_0$ (25.0; *p*-OMe, α-Me)		46
2.30 *M*	$k_2 = 117$ (25.0; *p*-Me, α-Me)		45
2.34 *M*	$= 120$ (25.0; *p*-Me, α-Me)		46
2.34 *M*	$k_2 = 29.5$ (25.0; *m*-Me, α-Me)		45
0.21 *M*–3.10 *M*	$= 2.9h_0^{1.02}$ (25.0; *m*-Me, α-Me)		47
2.3 *M*	$k_2 = 6.7$ (25.0; *p*-Cl, α-Me)		46
2.34 *M*	$= 9.3$ (25.0; *p*-Cl, α-Me)		45
2.3 *M*	$k_2 = 1.53$ (25.0; *m*-Cl, α-Me)		46
2.34 *M*	$= 2.55$ (25.0; *m*-Cl, α-Me)		45
2.3 *M*	$k_2 = 0.058$ (25.0; *p*-NO_2, α-Me)		45
3.7 *M*–9.2 *M*	$= 0.008h_0$ (25.0; *p*-NO_2, α-Me)		46
2.3 *M*–3.5 *M*	$k_2 = 0.014h_0^{1.27}$ (25.0; *p*-CO_2H, α-Me)		46
HCO_2H			
	$k_2 = 95.5[H_3O^+]+6[HCO_2H]$ (50.0; *p*-OMe, α-Me)		21
2,6-Dimethylhepta-2,5-dien-4-one (phorone)			
HCl		$\Delta H^{\ddagger} = 7.6$ kcal.mole^{-1}	48
		$\Delta S^{\ddagger} = 18.5$ eu	49
2,6-Dimethyl-2-hydroxyhept-5-en-4-one (semiphorone)			
HCl			
1.0 *M*	$k_2 = 9.84$ (15.6)	$\Delta H^{\ddagger} = 13.7$ kcal.mole^{-1}	49
	$= 18.8$ (22.9)	$\Delta S^{\ddagger} = 33$ eu	49
	$= 30.2$ (30.0)		49
	$= 65.7$ (40.5)		49
d-3-p-Menthene			
H_2SO_4 in EtOH–H_2O (56 w% EtOH)			
2.99 *N*	$k_1 = 0.195$[j], 0,199[k] (44.0)		50
3.24 *N*	0.260[j], 0.257[k] (44.0)		51
3.48 *N*	$= 0.328$[j], 0.325[k] (44.0)		50
H_2SO_4			
5.0 *M*–7.0 *M*	$k_2 = 0.0028h_0$		51
trans-β-Methylcinnamic acid			
H_2SO_4			
6.4 *M*–9.0 *M*	$k_2 = 0.00025h_0$		51
4-Methoxy-3-buten-2-one			
	$k_2 = 4700$ (22.0)	$\Delta H^{\ddagger} = 10.1$ kcal.mole^{-1}	52
	$= 7200$ (30.0)	$\Delta S^{\ddagger} = -26$ eu	52
	$= 15800$ (30.0; D_2O)		52
	$= 10300$ (35.0)		
	$= 13000$ (40.0)		52

TABLE 3 (*Contd.*)

Acid	$10^4 \times k$	*Comments*	*Ref.*
1-Phenyl-1,3-butadiene			
H_2SO_4			
3.91 *M*	$k_1 = 11.1$ (20.0)	d(log *k*)/d$H_0 = -1.22$	53
	= 16.8 (25.0)		53
	= 55 (35.0)		53
	= 84 (40.0)		53
3.14 *M*	$k_1 = 3.6$ (20.0)		53
	= 6.5 (25.0)		53
	= 20.5 (35.0)		53
	= 34.1 (40.0)		53
$HClO_4$			
2.924 *M*	$k_1 = 0.71$ (15.0; 40% aq. dioxan)	d(log *k*)/d$H_0 = -1.49$	53
	= 2.17 (25.0; 40% aq. dioxan)		53
	= 6.4 (35.0; 40% aq. dioxan)		53
3.37 *M*	$k_1 = 1.90$ (15.0; 40% aq. dioxan)		53
	= 6.3 (25.0; 40% aq. dioxan)		53
	= 17.0 (35.0; 40% aq. dioxan)		53

[a] Relative rates of anti-Markownikov addition and Markownikov addition, *i.e.* $k_{(\text{propan-1-ol})}/k_{(\text{propan-2-ol})}$.
[b] $10^4\,k$ quoted in units of l.mole^{-1}.atm^{-1}.min^{-1} referring to gaseous rather than dissolved olefin.
[c] Rate coefficients show some dependence upon acidity other than a linear relation with $[H^+]$.
[d] At H_0 = zero; *k* does not vary linearly with either $[H^+]$ or with h_0.
[e] Total ionic strength.
[f] %D = [D]100/([H]+[D]).
[g] Substituents shown in parentheses after temperature.
[h] $\Delta H^\ddagger$ (kcal.mole^{-1}) and $\Delta S^\ddagger$ (eu) are: H, 17.5, 13.1; *m*-Me, 17.3, 12.5; *p*-Cl, 18.0, 13.6; *m*-Cl, 19.8, 11.8; and *m*-NO_2, 21.4, 13.1. (ref. 44).
[j] Racemisation of olefin.
[k] Hydration of olefin.

2.3.1 *Efficiencies of catalysing acids*

The acids used for kinetic studies of hydration of olefins are generally those with anions of low nucleophilicity. Within this series, however, there are appreciable differences in the efficiencies of various acids in promoting the hydration reaction. The sequence

$$H_2SO_4 > HClO_4 \sim HNO_3$$

reflects the reaction rates in the presence of acids at the same formal concentration of hydrogen ion[9,27,32,33,42]; this order is not altered if the Hammett acidity function is used in preference to the stoichiometric hydrogen ion concentration, when the sequence becomes[33]

$$H_3PO_4 > H_2SO_4 > HCl > HNO_3 > HClO_4 > D_2SO_4 > DCl$$

at H_0 = zero. It seems, therefore, that the requirements for adding a proton to an olefin, and those for protonating an aromatic base, are not identical; this might be better expressed by the Bunnett relationship[16] in which some more appropriate protonation reaction could be used for the linear free-energy relation. This treatment would also permit a more narrow definition of the reaction mechanism if the correlation coefficients (ϕ) are considered to be influenced by the degree of protonation of the olefin in the transition states.

3. The addition of acids to olefins through electrophilic hydrogen

3.1 INTRODUCTION

The addition of weak acids (MeOH, CH_3COOH) and strong acids (HBr, H_2SO_4) to olefins can proceed in a similar manner to the acid-catalysed hydration reactions. A similar reaction scheme involving initial attack on the double bond by a proton can be proposed, *viz.*

$$RCH{=}CH_2 + H^+ \rightleftharpoons R\overset{+}{C}HCH_3 \xrightarrow{X^-} RCHXCH_3 \qquad (3.1)$$

and is, in fact, observed under certain conditions. In the case of a weak acid, however, there is another possible mechanism. Because the conjugate base of a weak acid is often a powerful nucleophile (*e.g.* azide ion, thiophenoxide ion), nucleophilic addition can become the favoured mode of reaction; the addition is now catalysed by base and involves the rate-determining attack of X^- upon the olefin[54]. In order to ensure the electrophilic addition mechanism and to cause faster rates, catalytic amounts of strong acids are often added.

3.2 REACTION ORDER

The simple reaction (3.1) does not show the function of the solvent molecules, although, as we have discussed in an earlier section (p. 2), their role in assisting the course of an ionic mechanism is vital. In a hydroxylic solvent, the solvation of incipient ionic species is readily achieved by the bulk medium; if the solvent is also the reactant, the rate law has the simple kinetic form

$$\text{Rate} = k[\text{olefin}] \qquad (3.2)$$

as in the case of trifluoroacetic acid[55], or the form

$$\text{Rate} = k[\text{olefin}][\text{strong acid}] \qquad (3.3)$$

as in the case of methanol[56]. In other situations, the absence of an effective solvent is evident. The addition of hydrogen bromide to olefins is a good example. In nitrobenzene the reaction appears to obey the kinetics

$$\text{Rate} = k[\text{olefin}][\text{HBr}] \tag{3.4}$$

for the experimental results fitted a second-order plot well, although no detailed proof of the reaction order was carried out[57]. In heptane, however, an extremely complex kinetic situation was found in which the reaction was of up to the third order with respect to hydrogen bromide, and was extremely sensitive to adventitious catalysts such as water, which lowered the order with respect to hydrogen bromide and accelerated the rate[58]. In this situation, it seems possible that molecules of hydrogen bromide are acting as solvent molecules. However, other interpretations include a termolecular collision process, involving a transition state such as I (ref. 59)

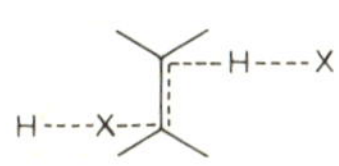

I

or the formation of polymers such as H_2Br_2 (3.5) which are thought to be more acidic species[60].

$$\text{HBr} + \text{HBr} \rightleftharpoons \text{H}^+[\text{HBr}_2]^- \tag{3.5}$$

3.3 SIDE REACTIONS

The addition of acids to olefins, since it involves the formation of a carbonium ionic species, will be subject to a number of side reactions which divert the course of the reaction by providing other routes through which the carbonium ion may be attacked. In a hydroxylic solvent, solvolysis of the carbonium may occur preferentially to the reaction with X^- (Scheme 3.), or polymerisation or cyclization may take place.

While all of these reactions will not affect the kinetic order, since they occur after the rate-determining protonation of the unsaturated system, they must be considered if the reactions are to be used preparatively, or if detailed speculation about the mechanism is to be made.

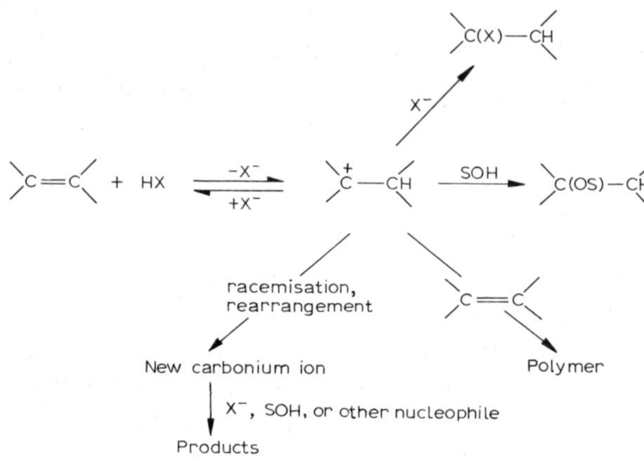

Scheme 3. The addition of acids to olefins.

3.4 KINETIC DATA

The kinetic data on addition of acids to olefins are summarised in Table 4. Within the reactions covered by this Table there is a wide variety of reaction order and also some differences between the ways in which reaction velocity has been reported. Rate coefficients are quoted in the appropriate units (sec^{-1}; $l.mole^{-1}.sec^{-1}$, *etc.*) and the order of the reaction is indicated by the subscript (*e.g.* k_1, k_4) at each entry. Olefins are arranged in increasing carbon content.

TABLE 4

ADDITION OF ACIDS TO OLEFINS

$$\text{Rate} = k_2[\text{olefin}][\text{HX}] = k_3[\text{olefin}][\text{HX}]^2 = k_4[\text{olefin}][\text{HX}]^3$$

$$\text{Rate} = k[\text{cyclohexene}][\text{HX}]^2 = k_2'[\text{HX}]^2$$

Olefin	*k*	*T* (°*C*)	*Solvent*	*Catalyst*	*% Reaction*	*Time (sec)*	$\Delta H^\ddagger$ (*kcal.mole*$^{-1}$)	$-\Delta S^\ddagger$ (*eu*)	*Ref.*
Hydrogen chloride									
Propene	0.00016% sec^{-1} (*sic*)	0							61
	0.00026% sec^{-1}	20							61
2-Methylpropene	$k_4 = 10^{-3}$	0	Heptane						62
Styrene	$k_3 = 1.6$	0	CCl_4	$SnCl_4$			9		63
	$= 7$	25	CCl_4	$SnCl_4$			9		63
α-Pinene		25	$CHCl_3$		99	600			64
		25	Xylene		95	600			64
		25	$PhNO_2$		87	600			64
		25	MeOH		3.8	600			64
		25	EtOBu, Dioxan		1.0	600			64
3-Methyl-2--cholestene		20	23% w/v HCl–dioxan		50	840			65
2-Methyl-1-butene	$k_3 = 1.85$	25	$MeNO_2$						66
2-Methyl-2-butene	$k_3 = 0.55$	25	$MeNO_2$						66
Isoprene	$k_3 = 0.16$	25	$MeNO_2$						66
1-Methyl-cyclopentene	$k_3 = 0.82$	25	$MeNO_2$						67
	$= 1.32$	25	$MeNO_2$	DCl used					67
3,3-Dimethyl--1-butene	$k_2 = 1.5 \times 10^{-9}$	25	HOAc				16.4		68
	$= 1.8 \times 10^{-8}$	50	HOAc						68
	$= 4.0 \times 10^{-6}$	125	HOAc						68
Styrene	$k_3 = 2.5 \times 10^{-6}$	25	HOAc				18.6		68
	$= 2.0 \times 10^{-5}$	50	HOAc						68
	$= 1.0 \times 10^{-6}$	16.5	HOAc						68

TABLE 4 (*Contd.*)

Olefin	*k*	*T* (°C)	*Solvent*	*Catalyst*	% *Reaction*	*Time* (*sec*)	$\Delta H^{\ddagger}$ (*kcal.mole*$^{-1}$)	$-\Delta S^{\ddagger}$ (*eu*)	*Ref.*
Hydrogen bromide									
Propene	$k_4 = 2.7 \times 10^{-4}$	0							58
Isopropenyl acetate	$k_2 = 7.7 \times 10^{-3}$	25	PhCl						69
Cyclohexene	$k_2' = 7.8 \times 10^{-4}$	20	cyclohexene				6.9[a]		70
	26.7×10^{-4}	40							70
	7.8×10^{-3}	60					14.6[b]		70
1-Phenylpropene	$k_{cis}/k_{trans\,\text{addn.}} = 1.18 \pm 0.10$		CH_2Cl_2						71
n-1-Undecene		20	$PhNO_2$		50	7320			72
		36	$PhNO_2$		50	5160			72
trans-*n*-2-Undecene		20	$PhNO_2$		50	16080			72
		36	$PhNO_2$		50	11040			72
trans-*n*-3-Undecene		20	$PhNO_2$		50	12720			72
		36	$PhNO_2$		50	8880			72
trans-*n*-4-Undecene		20	$PhNO_2$		50	14460			72
		36	$PhNO_2$		50	10200			72
trans-*n*-5-Undecene		20	$PhNO_2$		50	17520			72
		36	$PhNO_2$		50	12060			72
cis-*n*-2-Undecene		20	$PhNO_2$		50	10080			72
		36	$PhNO_2$		50	6960			72
cis-*n*-3-Undecene		20	$PhNO_2$		50	2280			72
		36	$PhNO_2$		50	1740			72
cis-*n*-4-Undecene		20	$PhNO_2$		50	8880			72
		36	$PhNO_2$		50	6600			72
cis-*n*-5-Undecene		20	$PhNO_2$		50	12600			72
		36	$PhNO_2$		50	9180			72
Hydrogen iodide									
Propene	$k = 1.14 \times k_{3,3\ \text{dimethylbutene}}$	−80							73
Cyclohexene	Iodine catalysis of HI addition								73

Trifluoroacetic acid						
4-X-1-Butenes						
X = H	$k_1 = 7.71 \times 10^{-6}$	0	CF_3COOH	14	28	74
	8.06×10^{-5}	20	CF_3COOH			74
	4.07×10^{-4}	45	CF_3COOH			74
X = Cl	$k_1 = 2.85 \times 10^{-6}$	60	CF_3COOH			74, 75
X = Br	$k_1 = 2.80 \times 10^{-6}$	60	CF_3COOH			74
X = OMe	$k_1 = 4.57 \times 10^{-6}$	60	CF_3COOH			74
X = $OCOCF_3$	$k_1 = 2.63 \times 10^{-7}$	60	CF_3COOH			74
X = $OCOCH_3$	$k_1 = 2.62 \times 10^{-6}$	60	CF_3COOH			74
5-X-1-Pentenes						
X = H	$k_1 = 1.14 \times 10^{-3}$	60	CF_3COOH			74
X = Cl	$k_1 = 2.17 \times 10^{-4}$	60	CF_3COOH			74, 75
X = Br	$k_1 = 1.77 \times 10^{-4}$	60	CF_3COOH			74
X = CN	$k_1 = 5.33 \times 10^{-6}$	60	CF_3COOH			74
X = OMe	$k_1 = 1.85 \times 10^{-4}$	60	CF_3COOH			74
X = $OCOCF_3$	$k_1 = 7.30 \times 10^{-6}$	60	CF_3COOH			74
X = $OCOCH_3$	$k_1 = 2.91 \times 10^{-5}$	60	CF_3COOH			74
X = $OCOC_6H_4$-*p*-NO_2	$k_1 = 1.16 \times 10^{-5}$	60	CF_3COOH			74
5-X-1-Hexenes						
X = H	$k_1 = 5.1 \times 10^{-5}$	25	CF_3COOH	14.3	29	76
	$= 2.35 \times 10^{-4}$	35	CF_3COOH			77
	$= 1.47 \times 10^{-3}$	60	CF_3COOH			77
X = Cl	$k_1 = 6.03 \times 10^{-5}$	35	CF_3COOH	13.9	33	77
	$= 3.56 \times 10^{-4}$	60	CF_3COOH			77
X = Br	$k_1 = 7.33 \times 10^{-5}$	35	CF_3COOH	14.0	32	77
	$= 4.39 \times 10^{-4}$	60	CF_3COOH			77
X = I	$k_1 = 7.22 \times 10^{-5}$	35	CF_3COOH	13.8	33	77
	$= 4.20 \times 10^{-4}$	60	CF_3COOH			77
X = OMe	$k_1 = 6.30 \times 10^{-5}$	35	CF_3COOH	16.3	25	77
	$= 5.02 \times 10^{-4}$	60	CF_3COOH			77
X = $OCOCF_3$	$k_1 = 1.12 \times 10^{-6}$	35	CF_3COOH	15.5	35	77
	$= 8.79 \times 10^{-6}$	60	CF_3COOH			77
X = $OCOCH_3$	$k_1 = 5.60 \times 10^{-6}$	35	CF_3COOH	16.4	29	77
	$= 4.51 \times 10^{-5}$	60	CF_3COOH			77

TABLE 4 (*Contd.*)

Olefin	*k*	*T* (°C)	*Solvent*	*Catalyst*	% *Reaction*	*Time* (*sec*)	$\Delta H^{\ddagger}$ (*kcal.mole*$^{-1}$)	$-\Delta S^{\ddagger}$ (*eu*)	*Ref.*
X = CN	$k = 3.9 \times 10^{-7}$	35	CF_3COOH				19.7	24	77
	$= 4.72 \times 10^{-6}$	60	CF_3COOH						77
6-X-1-Hexenes									
X = H	See under 5-X-hexenes								
X = Cl	$k_1 = 1.60 \times 10^{-4}$	60	CF_3COOH						75
	$= 1.75 \times 10^{-4}$	60	CF_3COOH						74
X = Br	$k_1 = 1.56 \times 10^{-4}$	60	CF_3COOH						74
X = CN	$k_1 = 2.38 \times 10^{-5}$	60	CF_3COOH						74
X = OMe	$k_1 = 2.05 \times 10^{-4}$	60	CF_3COOH						74
X = $OCOCF_3$	$k_1 = 4.56 \times 10^{-5}$	60	CF_3COOH						74
X = $OCOCH_3$	$k_1 = 9.00 \times 10^{-5}$	60	CF_3COOH						74
X = $OCOC_6H_4$-*p*-NO_2	$k_1 = 5.98 \times 10^{-5}$	60	CF_3COOH						74
1-Heptene	$k_1 = 2.75 \times 10^{-4}$	35	CF_3COOH						77
	$= 1.54 \times 10^{-3}$	60	CF_3COOH				13.4	32	74, 77
4,4-Me_2-1-pentene	$k_1 = 8.44 \times 10^{-5}$	35	CF_3COOH						77
7-Cl-1-heptene	$k_1 = 3.08 \times 10^{-4}$	60	CF_3COOH						75
6-Heptenyltrifluoro-acetate	$k_1 = 6.42 \times 10^{-6}$	35	CF_3COOH				13.6	38	77
	$= 3.64 \times 10^{-5}$	60	CF_3COOH						77
7-X-1-Heptenes									
X = H	$k_1 = 2.75 \times 10^{-4}$	35	CF_3COOH				13.4	32	77
	1.54×10^{-3}	60	CF_3COOH						74, 71
X = Cl	$k_1 = 3.08 \times 10^{-4}$	60	CF_3COOH						75, 84
X = CN	$k_1 = 1.10 \times 10^{-4}$	60	CF_3COOH						74
X = OMe	$k_1 = 2.93 \times 10^{-4}$	60	CF_3COOH						74
X = $OCOCF_3$	$k_1 = 1.66 \times 10^{-4}$	60	CF_3COOH						74
X = $OCOCH_3$	$k_1 = 2.49 \times 10^{-4}$	60	CF_3COOH						74
6-Hepten-2-yl-trifluoroacetate	$k_1 = 6.42 \times 10^{-6}$	35	CF_3COOH				13.6	38	77
	3.64×10^{-5}	60	CF_3COOH						77

8-X-1-Octenes						
X = H	$k_1 = 2.87 \times 10^{-4}$	35	CF_3COOH	13.8	30	77
	1.68×10^{-3}	60	CF_3COOH			74, 77
X = Cl	$k_1 = 7.08 \times 10^{-4}$	60	CF_3COOH			74, 75
X = CN	$k_1 = 2.98 \times 10^{-4}$	60	CF_3COOH			74
X = OMe	$k_1 = 5.68 \times 10^{-4}$	60	CF_3COOH			74
X = $OCOCF_3$	$k_1 = 3.77 \times 10^{-4}$	60	CF_3COOH			74
X = $OCOCH_3$	$k_1 = 5.82 \times 10^{-4}$	60	CF_3COOH			74
7-Octen-2-yl-trifluoroacetate	$k_1 = 2.63 \times 10^{-5}$	35	CF_3COOH	15.9	28	77
	2.00×10^{-4}	60	CF_3COOH			77
9-X-1-Nonenes						
X = H	$k_1 = 1.64 \times 10^{-3}$	60	CF_3COOH			74
X = Cl	$k_1 = 8.53 \times 10^{-4}$	60	CF_3COOH			74
	$= 8.68 \times 10^{-4}$	60	CF_3COOH			75
X = OMe	$k_1 = 9.51 \times 10^{-4}$	60	CF_3COOH			74
X = $OCOCF_3$	$k_1 = 6.55 \times 10^{-4}$	60	CF_3COOH			74
4-CH_2X-1-vinylcyclo-hexanes						
X = Cl	$k_1 = 6.41 \times 10^{-4}$	60	CF_3COOH			74
X = OMe	$k_1 = 5.50 \times 10^{-4}$	60	CF_3COOH			74
X = $OCOCF_3$	$k_1 = 2.52 \times 10^{-4}$	60	CF_3COOH			74
X = $OCOC_6H_4$-$p(NO_2)$	$k_1 = 3.01 \times 10^{-4}$	60	CF_3COOH			74
8-X-*trans*-bicyclo-[4,3,0]-3-nonene						
X = CN	$k_1 = 1.54 \times 10^{-5}$	60	CF_3COOH			74
X = COOH	$k_1 = 7.80 \times 10^{-5}$	60	CF_3COOH			74
X = $(COOEt)_2$	$k_1 = 1.92 \times 10^{-5}$	60	CF_3COOH			74
Cyclohexene	$k_1 = 2.34 \times 10^{-3}$	60	CF_3COOH			74
Vinylcyclohexane	$k_1 = 3.12 \times 10^{-3}$	60	CF_3COOH			74
6,6-Dicarboxyethyl-1-hexene	$k_1 = 4.07 \times 10^{-5}$	60	CF_3COOH			74
1-Decene	$k_1 = 1.84 \times 10^{-3}$	60	CF_3COOH			74
9-Decen-1-yl trifluoroacetate	$k_1 = 1.03 \times 10^{-3}$	60	CF_3COOH			74

TABLE 4 (*Contd.*)

Olefin	*k*	*T* (°C)	*Solvent*	*Catalyst*	*% Reaction*	*Time (sec)*	$\Delta H^{\ddagger}$ (*kcal.mole*$^{-1}$)	$-\Delta S^{\ddagger}$ (*eu*)	*Ref.*
1-Undecene	$k_1 = 1.59 \times 10^{-3}$	60	CF_3COOH						74
11-Chloro-1- -undecene	$k_1 = 1.25 \times 10^{-3}$	60	CF_3COOH						74
	1.28×10^{-3}	60	CF_3COOH						75
10-Undecen-1-yl- -trifluoroacetate	$k_1 = 1.06 \times 10^{-3}$	60	CF_3COOH						74
11-Cyano-1- -undecene	$k_1 = 9.83 \times 10^{-4}$	60	CF_3COOH						74
Acetic acid									
Norbornadiene	$k_1 = 1.5 \times 10^{-6}$	20	HOAc	H_2SO_4, $HClO_4$					78
Styrenes									
X = H	$10^8 k_1 = 1.74 h_0^{1.265}$	25	HOAc	H_2SO_4					79
	$10^4 k_1 = 0.55$	12	HOAc	H_2SO_4	($H_0 = -3.27$				79
	$= 1.38$	20	HOAc	H_2SO_4	over the				79
	$= 4.87$	32	HOAc	H_2SO_4	temperature				79
	$= 11.8$	40	HOAc	H_2SO_4	range)				79
	$10^9 k_1 = 8.7 h_0^{1.27}$	25	HOAc	$HClO_4$					79
X = *p*-OMe	$10^5 k_1 = 1.55 h_0^{1.19}$	25	HOAc	H_2SO_4					79
	$= 0.45 h_0^{1.19}$	12	HOAc	H_2SO_4					79
	$10^3 k_1 = 1.85$	22	HOAc	H_2SO_4	($H_0 = -1.83$)				79
	$= 3.06$	28	HOAc	H_2SO_4	($H_0 = -1.83$)				79
	$= 0.60$	32	HOAc	H_2SO_4	($H_0 = -1.11$)				79
	$= 6.20$	36	HOAc	H_2SO_4	($H_0 = -1.83$)		16.0		79
	$= 1.21$	40	HOAc	H_2SO_4	($H_0 = -1.11$)		16.0		79
X = *p*-Me	$10^7 k_1 = 2.51 h_0^{1.26}$	25	HOAc	H_2SO_4					79
	$= 0.32 h_0^{1.37}$	12	HOAc	H_2SO_4					79
	$= 1.13 h_0^{1.31}$	20	HOAc	H_2SO_4					79
	$= 6.53 h_0^{1.30}$	32	HOAc	H_2SO_4			18.3		79
	$= 1.18 h_0^{1.26}$	25	HOAc	$HClO_4$					79

X = *m*-Me	$10^8k_1 = 3.80h_0^{1.28}$	25	HOAc	H_2SO_4			79
	$10^3k_1 = 0.135$	12	HOAc	H_2SO_4	$(H_0 = -3.27)$		79
	$= 0.323$	20	HOAc	H_2SO_4	$(H_0 = -3.27)$		79
	$= 1.167$	32	HOAc	H_2SO_4	$(H_0 = -3.27)$	18.4	79
	$10^8k_1 = 2.80h_0^{1.23}$	25	HOAc	$HClO_4$			79
X = *p*-Cl	$10^4k_1 = 0.325$	20	HOAc	H_2SO_4	$(H_0 = -3.27)$		79
	$= 1.35$	32	HOAc	H_2SO_4	$(H_0 = -3.27)$		79
	$= 3.33$	40.5	HOAc	H_2SO_4	$(H_0 = -3.27)$	20.6	79
	$10^9k_1 = 3.4h_0^{1.29}$	25	HOAc	H_2SO_4			79
X = *p*-Br	$10^9k_1 = 7.2h_0^{1.17}$	25	HOAc	H_2SO_4			79
	$10^4k_1 = 0.267$	20	HOAc	H_2SO_4	$(H_0 = -3.27)$		79
	$= 0.98$	32	HOAc	H_2SO_4	$(H_0 = -3.27)$		79
	$= 2.30$	39.8	HOAc	H_2SO_4	$(H_0 = -3.27)$	20.1	79
X = *m*-Cl	$10^4k_1 = 8.17$	12	HOAc	H_2SO_4	$(H_0 = -4.87)$		79
	$10^9k_1 = 2.80h_0^{1.01}$	20	HOAc	H_2SO_4			79
	$= 4.92h_0^{1.01}$	25	HOAc	H_2SO_4			79[c]
	$= 22.4h_0^{0.94}$	32	HOAc	H_2SO_4			79
	$= 31.1h_0^{1.00}$	40	HOAc	H_2SO_4		21.4	79
X = *m*-Br	$10^4k_1 = 0.68$	12	HOAc	H_2SO_4	$(H_0 = -4.81)$		79
	$= 0.43$	20	HOAc	H_2SO_4	$(H_0 = -4.27)$		79
	$10^9k_1 = 0.47h_0^{1.2}$	25	HOAc	H_2SO_4			79
	$= 2.16h_0^{1.16}$	32	HOAc	H_2SO_4			79
	$= 5.07h_0^{1.16}$	40	HOAc	H_2SO_4		21.4	79
X = *m*-NO_2	$10^{11}k_1 = 1.4h_0^{1.26}$	25	HOAc	H_2SO_4			79
	$10^4k_1 = 1.16$	12	HOAc	H_2SO_4	$(H_0 = -6.16)$		79
	$= 1.25$	20	HOAc	H_2SO_4	$(H_0 = -5.77)$		79
	$= 13.5$	30	HOAc	H_2SO_4	$(H_0 = -6.16)$		79
	$= 6.50$	32.4	HOAc	H_2SO_4	$(H_0 = -5.77)$		79
	$= 28.3$	35.5	HOAc	H_2SO_4	$(H_0 = -6.16)$	23.7	79
	$= 15.8$	39.7	HOAc	H_2SO_4	$(H_0 = -5.77)$	23.8	79
X = *p*-NO_2	$10^{12}k_1 = 6.0h_0^{1.26}$	25	HOAc	H_2SO_4			79
	$10^4k_1 = 0.73$	20	HOAc	H_2SO_4	$(H_0 = -5.83)$		79
	$= 3.43$	32	HOAc	H_2SO_4	$(H_0 = -5.83)$		79
	$= 9.16$	40	HOAc	H_2SO_4	$(H_0 = -5.83)$	24.0	79

TABLE 4 (*Contd.*)

Olefin	*k*	*T* (°C)	*Solvent*	*Catalyst*	% *Reaction*	*Time* (*sec*)	$\Delta H^{\ddagger}$ (*kcal.mole*$^{-1}$)	$-\Delta S^{\ddagger}$ (*eu*)	*Ref.*
Vinylferrocene	$k_1 = 4.5 \times 10^{-4}$	40	HOAc						80
Vinylruthenocene	$k_1 = 5.4 \times 10^{-4}$	40	HOAc						80
Vinylosmocene	$k_1 = 2.08 \times 10^{-3}$	40	HOAc						80
Methanol									
2-Methylpropene	$k_1 = 2.44 \times 10^{-5}$	25	MeOH	$C_7H_7SO_3H$			23.2	1.7	81
	$= 9.23 \times 10^{-5}$	35	MeOH	$C_7H_7SO_3H$					81
2-Methyl-2-butene	$k_1 = 3.54 \times 10^{-5}$	25	MeOH	$C_7H_7SO_3H$			22.6	3.2	81
2-Methylpropene	$k_1 = 2.44 \times 10^{-5}$	25	MeOH	$C_7H_7SO_3H$					81
	$= 9.23 \times 10^{-5}$	35	MeOH	$C_7H_7SO_3H$			23.2	1.7	81
2-Methyl-2-butene	$k_1 = 1.08 \times 10^{-5}$	25	MeOH	$C_7H_7SO_3H$					81
	$= 3.55 \times 10^{-5}$	35	MeOH	$C_7H_7SO_3H$					81
	$= 10.96 \times 10^{-5}$	45	MeOH	$C_7H_7SO_3H$			21.2	10.2	81
2-Ethyl-1-butene	$k_1 = 3.54 \times 10^{-5}$	25	MeOH	$C_7H_7SO_3H$					81
	$= 11.99 \times 10^{-5}$	35	MeOH	$C_7H_7SO_3H$					81
	$= 43.0 \times 10^{-5}$	45.4	MeOH	$C_7H_7SO_3H$			22.6	3.2	81
2-Methyl-1-pentene	$k_1 = 2.16 \times 10^{-5}$	25	MeOH	$C_7H_7SO_3H$					81
	$= 7.35 \times 10^{-5}$	35	MeOH	$C_7H_7SO_3H$					81
	$= 25.8 \times 10^{-5}$	45.4	MeOH	$C_7H_7SO_3H$			22.6	4.1	81
2,4-Me_2-1-pentene	$k_1 = 1.80 \times 10^{-5}$	25	MeOH	$C_7H_7SO_3H$					81
	$= 6.74 \times 10^{-5}$	35	MeOH	$C_7H_7SO_3H$					81
	$= 24.2 \times 10^{-5}$	45	MeOH	$C_7H_7SO_3H$			23.4	1.8	81
2,4,4-Me_3-2-pentene	$k_1 = 1.68 \times 10^{-5}$	25	MeOH	$C_7H_7SO_3H$					81
	$= 6.47 \times 10^{-5}$	35	MeOH	$C_7H_7SO_3H$					81
	$= 22.6 \times 10^{-5}$	45	MeOH	$C_7H_7SO_3H$			23.5	1.6	81
2-Ethyl-1-hexene	$k_1 = 3.39 \times 10^{-5}$	25	MeOH	$C_7H_7SO_3H$					81
	$= 11.99 \times 10^{-5}$	35	MeOH	$C_7H_7SO_3H$					81
	$= 43.9 \times 10^{-5}$	45	MeOH	$C_7H_7SO_3H$			23.1	1.5	81
Styrenes and substituted styrenes									
X = H	$k_1 = 2.95 \times 10^{-4}$	115	MeOH	$C_7H_7SO_3H$			29	0.4	81
	$= 12.0 \times 10^{-4}$	130	MeOH	$C_7H_7SO_3H$					81
	$= 35.3 \times 10^{-4}$	142	MeOH	$C_7H_7SO_3H$					81

X = α-Me	$k_1 = 0.42 \times 10^{-4}$	25	MeOH	$C_7H_7SO_3H$			81
	$= 1.54 \times 10^{-4}$	35	MeOH	$C_7H_7SO_3H$			
	$= 4.76 \times 10^{-4}$	44.5	MeOH	$C_7H_7SO_3H$			
	$= 5.03 \times 10^{-4}$	45.4	MeOH	$C_7H_7SO_3H$	22.6	2.7	81
X = p-Me, α-Me	$k_1 = 3.42 \times 10^{-4}$	25	MeOH	$C_7H_7SO_3H$			
	$= 12.0 \times 10^{-4}$	35	MeOH	$C_7H_7SO_3H$			
	$= 39.0 \times 10^{-4}$	45	MeOH	$C_7H_7SO_3H$	21.8	1.4	81
X = p-C_6H_5	$k_1 = 5.61 \times 10^{-4}$	115	MeOH	$C_7H_7SO_3H$			
	$= 2.14 \times 10^{-3}$	130	MeOH	$C_7H_7SO_3H$	27	3.6	81
Sulphuric acid							
Ethylene[d]	$k_1 = 0.96 \times 10^{-4}$	30	H_2SO_4 aq.	(1.65 *M*)			82
	$= 1.10 \times 10^{-4}$	30	H_2SO_4 aq.	(1.70 *M*)			82
	$= 1.83 \times 10^{-4}$	30	H_2SO_4 aq.	(3.04 *M*)			82
	$= 2.28 \times 10^{-4}$	30	H_2SO_4 aq.	(3.14 *M*)			82
	$= 3.33 \times 10^{-4}$	30	H_2SO_4 aq.	(3.70 *M*)			82
	$= 3.66 \times 10^{-4}$	30	H_2SO_4 aq.	(3.75 *M*)			82

[a] For *trans* addition of DBr.

[b] For *cis* addition of DBr.

[c] Derived from quoted rate coefficients. From the parameters in Table III, ref. 79, $10^9 k_1 = 0.99 h_0^{1.18}$ at 25°C.

[d] $-\mathrm{d}[C_2H_4]/\mathrm{d}t = k_1[C_2H_4] + \text{constant}$. k_1 does not vary linearly with any acidity function; see however, refs. 83 and 84.

The first- and second-order rate coefficients are defined by the equations

$$\text{Rate} = k_1[\text{olefin}] \tag{3.2}$$

which will be specific to a defined acidity and concentration of weak acid, and by the equation

$$\text{Rate} = k_2[\text{olefin}][\text{acidity function}] \tag{3.6}$$

For higher reaction orders, the overall order will be indicated by the subscript to the rate coefficient, and the detailed kinetic equation will be shown in a footnote.

4. Electrophilic addition of halogens

4.1 INTRODUCTION

The reaction of halogens with olefins is one of the most studied reactions in organic chemistry, as well as one of the oldest[85]. The susceptibility of an olefin to attack by halogenating agents (ICl, Br_2) has been used to determine the presence and concentration of olefins, and the "iodine number" has been important in the estimation of unsaturated oils.

In general, the rate of addition of halogens to olefins is increased by electron-donating groups (alkyl, OH, OMe, NR_2) attached to the multiple-bond system, and decreased by electron-withdrawing groups (NO_2, CN, halogen, NMe_3), in the same way as in the acid-catalysed addition of HX. This implies an electrophilic attack of the halogen molecule or of some fragment formed in pre-equilibrium with it, giving a carbonium ionic species or the products. Since halogenation by molecular bromine in the presence of salts yields products in which the anions of the salts have been incorporated[86], the reaction sequence cannot involve

$$X_2 + \rangle C{=}C\langle \xrightarrow{\text{slow}} \rangle CX{-}CX\langle \tag{4.1}$$

where X = halogen, as the rate-determining step, but must require a two-stage mechanism in which first one part of the reactant adding attacks the olefin, and then the reaction is completed by a fast stage involving whatever nucleophiles are available. The simplest mechanism which is consistent with the facts is shown in Scheme 4.1.

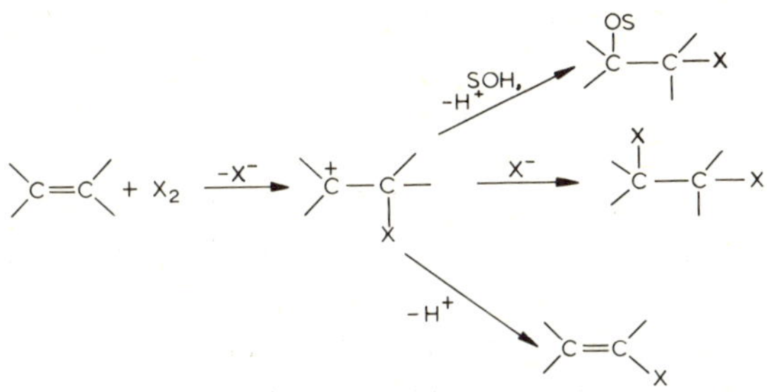

Scheme 4.1. Mechanism of halogen addition to olefins.

This mechanism assumes that all necessary solvation of the intermediates can be obtained from the solvent, and that neither molecules of the olefin nor molecules of the halogen (or interhalogen) assist in stabilising the charges formed. Such a situation is found in polar solvents (*e.g.* nitromethane, aqueous media, nitrobenzene) where the order of the reaction is simple, *viz.*

$$\text{Rate} = k[\text{olefin}][X_2] \tag{4.2}$$

However, a more complex kinetic expression occurs when the addition of the less reactive halogens is studied in solvents of low dielectric constant. The reaction of chlorine with olefins in most solvents shows second-order kinetics, although in carbon tetrachloride there seems to be a number of significant mechanisms which are effected in poorly defined kinetic orders[87]. In acetic acid, however, the simple kinetic equation (4.2) is observed for chlorine and for bromine additions at low concentrations. At higher concentrations of bromine, a term showing second-order dependence upon the halogen concentrations is found, *viz.*

$$\text{Rate} = k[\text{olefin}][X_2] + k_3[\text{olefin}][X_2]^2 \tag{4.3}$$

and such behaviour also occurs with the interhalogens ICl and IBr[88]. The explanation of this third-order reaction term is uncertain; the hypothesis that two molecules of halogen form a charged species ($Br^+Br_3^-$) in pre-equilibrium to the rate-determining stage is ruled out by the observed salt effects[89]. The possible formation of Br_4 in pre-equilibrium cannot be eliminated, but it requires the hypothesis that such species are formed in sufficiently large concentration to be kinetically significant and are appreciably more reactive than Br_2. The most simple explanation, which is strengthened by spectroscopic evidence, is that the decomposition of an initially formed complex (olefin, Br_2) is the rate-determining stage, and that the removal of bromide ion can be achieved[88] either by the solvent or by molecular halogen, due to the stability of Br_3^-.

The addition of iodine to olefins is unique among these halogen reactions in its reversibility. As a result, the appropriate, more complex kinetic equation is required to allow for the equilibrium position of the reaction, or else a large excess of the olefin is used to drive the reaction substantially to completion. In solvents of low polarity (carbon disulphide, carbon tetrachloride) the reaction is of high order with respect to iodine. Groh *et al.*[90] believed I_6 to be an important polymer of iodine, since their investigations showed the term

$$\text{Rate} = k_4[\text{olefin}][I_2]^3 \tag{4.4}$$

to be dominant. In nitrobenzene or acetic acid, however, the reaction showed second-order dependence upon the halogen concentration[91], *viz.*

$$\text{Rate} = k_3[\text{olefin}][I_2]^2 \tag{4.5}$$

Much attention has been given to the donor properties of the solvent towards iodine in explaining these results[92,93]. The present trend is towards a mechanism involving solvent polarisation of the halogen molecule prior to reaction (either by bulk solvent or by other halogen molecules) but the mechanism is by no means firmly decided.

4.2 CATALYSED HALOGEN ADDITION

Higher kinetic orders with respect to halogen (X_2) have been discussed in the previous section; it is not surprising that other species can take the place of halogen in the kinetic equation. Chloride ion and bromide ion can assist in the addition reactions of bromine[94] and, possibly, iodine[93]. This gives rise to a new kinetic contribution

$$\text{Rate} = k[\text{olefin}][X_2][X^-] \tag{4.6}$$

The function of the halide ion is not clearly understood; its efficiency depends upon the olefin under attack, and as a salt effect of unknown magnitude is simultaneously exerted upon the reaction, the separation of the overall rate into its components is not easy.

As the kinetic form (eqn. (4.6)) is equivalent to

$$\text{Rate} = k[\text{olefin}][X_3^-] \tag{4.7}$$

an investigation was made to determine whether trihalide ions could function as electrophiles, notwithstanding their formal negative charge. Robertson *et al.*[95] believe that they have evidence of nucleophilic attack upon less reactive compounds (*e.g.* methyl vinyl sulphone). A further mode of reaction involves acid-catalysed nucleophilic addition, in which the deactivating group (*e.g.* CHO) has appreciable basicity and, since the extent of the catalysis depends upon (*i*) the presence of hydrogen ion (since added base represses the usual autocatalysis), (*ii*) the strength of the added acid (since perchloric acid is more effective than sulphuric acid but nitric acid is almost ineffective, paralleling their acidities in the solvent, acetic acid), and (*iii*) the basicity of the olefin (since cinnamaldehyde is more affected than ω-nitrostyrene although the latter possesses greater electron-withdrawing ability), the rate-determining stage is thought to involve nucleophilic attack upon the conjugate acid of the olefin, *viz.*

$$\text{RCH=CHX} \xrightleftharpoons{\text{H}^+} [\text{RCH=CHXH}]^+ \xrightarrow{\text{X}_2} \text{Products} \tag{4.8}$$

The greater efficiency of the acids HCl and HBr than that expected from their acidities in acetic acid can be explained by postulating attack upon the protonated olefin by the trihalide ion. Such acid-catalysed halogenation is observed with both bromine and, to a lesser extent, chlorine.

In the presence of "halogen-carriers", such as iodine or pyridine, further kinetic complexities result. Buckles and Yuk[96] found that the bromination of crotonic acid was a kinetically simple reaction only in the presence of tetramethylammonium tribromide, the overall kinetic form in ethylene dichloride being

$$\text{Rate} = k[\text{olefin}][\text{Br}_2][\text{Br}_3^-] \tag{4.9}$$

Iodine-catalysed halogenation is apparently complex; the kinetic forms

$$\text{Rate} = k[\text{olefin}][\text{Cl}_2][\text{ICl}] \tag{4.10}$$

and

$$\text{Rate} = k[\text{olefin}][\text{Br}_2]^2[\text{IBr}] \tag{4.11}$$

can be recognised[87].

References pp. 80–86

4.3 INTERHALOGEN ADDITION

The two interhalogens, ICl and IBr, have been used extensively to measure unsaturation, but the kinetics of their reaction with olefins has received very little attention. White and Robertson[88] have established the third-order kinetics of addition, where the simple form is

$$\text{Rate} = k[\text{olefin}][\text{IX}]^2 \tag{4.12}$$

They also found the order of reactivity

$$I_2 \ll \text{IBr} < Br_2 < \text{ICl} < \text{BrCl}$$

and showed that hydrogen chloride *decreased* the rate of addition of iodine monochloride to allyl acetate, presumably due to the formation of HCl_2. Although other kinetic forms could occur these do not seem to have been observed.

The high instability of bromine monochloride (the heat of formation of which has been estimated as 0.3 kcal.mole^{-1})[97] provides some difficulty in studying its reactions. Nonetheless, the rate of addition of mixtures of chlorine and bromine (with the total concentration of halogen constant) to *cis*-cinnamic acid in carbon tetrachloride–acetic acid mixtures was greatest when $[Br_2]/[Cl_2]$ was unity[88]. The kinetic form of the reaction was the same as that for the other interhalogens and for bromine, but not for chlorine. The possible intervention of bromine monochloride in the reaction of allyl trimethylammonium perchlorate with hypobromous acid in aqueous acid has also been reported; here, the kinetics were[98],

$$\text{Rate} = k[\text{olefin}][\text{BrOH}][\text{H}^+] + k'[\text{olefin}][\text{BrOH}][\text{H}^+][\text{X}^-] \tag{4.13}$$

and the second term of the equation was interpreted to imply attack by BrX, since the relative efficiencies of the anions (X^-) was in the order $NO_3^- <$ $HSO_4^- <$ OAc < Cl.

4.4 REACTION OF HYPOHALOUS ACIDS AND THEIR DERIVATIVES

The reactions of the hypohalous acids with olefins are kinetically complex, but some features can be discerned. In each case, a kinetic term

$$\text{Rate} = k[\text{olefin}][\text{HOX}] \tag{4.14}$$

involving electrophilic attack by HOX can be found[99]. At higher concentrations of hypochlorous acid, the addition to olefins is of order greater than one with respect to the electrophile. This could be interpreted as being due to chlorine monoxide produced by

$$2HOCl = Cl_2O + H_2O \tag{4.15}$$

and for sufficiently unreactive olefins the kinetic term

$$\text{Rate} = k[\text{olefin}][HOCl]^2 \tag{4.16}$$

was clearly recognisable. With the more reactive allyl alcohol, however, the formation of chlorine monoxide is partially rate-determining, and hence the order with respect to the olefin falls below unity. Such intervention clearly shows chlorine monoxide to be an electrophilic chlorinating agent as well. Nucleophilic anions in general probably participate in processes such as

$$OAc^- + ClOH + H^+ \rightarrow ClOAc + H_2O \tag{4.17}$$

and would give rise to kinetics similar to eqn. (4.13).

One of the complications in these studies results from adventitious acid formed by substitution reactions. In this case a new chlorinating species may be introduced, *viz.*

$$HOCl + HCl = Cl_2 + H_2O \tag{4.18}$$

However, acid catalysis is observed (by perchloric acid) in conditions where no halide ion is present and where the conjugate base of the acid is non-nucleophilic. This provides the rate term

$$\text{Rate} = k[\text{olefin}][HOX][H^+] \tag{4.19}$$

The significance of this kinetic form is the basis of considerable discussion, since it point towards the formation of species such as $ClOH_2^+$ and Cl^+, whose reality has been held in question.

4.5 KINETIC DATA

The kinetic data on addition of halogen, interhalogens, and hypohalous acids are collected in Table 5. The Table is subdivided according to the reagents initiating the addition; free halogens appear first, followed by hypohalous acids. The kinetic form is shown by reference to equations quoted previously in this section, and to the following supplementary kinetic equations.

$$\text{Rate} = k_2[\text{Br}_2][\text{olefin}] + k_3[\text{Br}_3^-][\text{olefin}] \quad (4.20)$$

$$\text{Rate} = k_3[\text{Br}_2]^2[\text{olefin}] \quad (4.21)$$

$$\text{Rate} = k_1[\text{Br}_2][\text{olefin}] + k_2[\text{Br}_2]^2[\text{olefin}] + k_3[\text{Br}_3^-][\text{olefin}] \quad (4.22)$$

$$\text{Rate} = k_2^{\text{I}}[\text{HOCl}][\text{olefin}] + k_2^{\text{II}}[\text{HOCl}]^2 \quad (4.23)$$

$$\text{Rate} = k_2^{\text{I}}[\text{HOCl}][\text{olefin}] + (k_2^{\text{II}} + k_3^{\text{II}}[\text{HOAc}])[\text{HOCl}]^2 + k_2^{\text{III}}[\text{HOCl}][\text{HOAc}] \quad (4.24)$$

$$\text{Rate} = k_2^{\text{I}}[\text{HOCl}]([\text{RCO}_2\text{H}] + [\text{RCO}_2^-]) + (k_2^{\text{II}} + k_3^{\text{II}}[\text{RCO}_2\text{H}])[\text{HOCl}]^2 \quad (4.25)$$

The rate coefficients are of the appropriate order in mole, l and sec units.

TABLE 5

ELECTROPHILIC ADDITION OF HALOGENS, INTERHALOGENS AND HYPOHALOUS ACIDS TO OLEFINS

Olefin	*Kinetic form*	*k(°C)*	*Solvent*	*Ref.*
Chlorine				
CH_2=CHBr	(4.2)	4.7×10^{-3} (24)	HOAc	100, 103
CH_2=$CHSO_3^-H^+$	(4.2)	1.8×10^{-3} (24)	HOAc	100
CH_2=$CHSO_3^-Na^+$	(4.2)	1.2×10^{-2} (24)	HOAc	100
CH_2=$CHSO_2Me$	(4.2)	1.6×10^{-6} (24)	HOAc	95
CH_2=CHCN	(4.2)	1.6×10^{-6} (24)	HOAc	100
CH_2=$CHCO_2H$	(4.2)	2.8×10^{-4} (24)	HOAc	104
		3.0×10^{-4} (25)	HOAc	105
CH_2=$CHCO_2Et$	(4.2)	9.3×10^{-4} (25)	HOAc	106
CHCl=CHCl (*cis*)	(4.2)	2.7×10^{-4} (25)	80% HOAc[a]	100
		7.0×10^{-3} (25)	60% HOAc[a]	100
CHCl=CHCl (*trans*)	(4.2)	8.8×10^{-5} (25)	80% HOAc[a]	100
CHCl=CCl_2	(4.2)	2.1×10^{-5} (25)	60% HOAc[a]	100
CH_2=$CHCHCl_2$	(4.2)	1.1×10^{-2} (25)	HOAc	100
CH_2=$CHCCl_3$	(4.2)	1.0×10^{-4} (25)	HOAc	100
CHCl=$CHCH_2Cl$ (*cis*)	(4.2)	9.1×10^{-4} (25)	HOAc	101
		1.6×10^{-2} (25)	90% HOAc[a]	102
CHCl=$CHCH_2Cl$ (*trans*)	(4.2)	8.3×10^{-4} (25)	HOAc	101
		1.4×10^{-2} (25)	90% HOAc[a]	102
		8.5×10^{-2} (25)	80% HOAc[a]	102
CHCl=$CHCHCl_2$	(4.2)	1.6×10^{-5} (25)	HOAc	101
		1.2×10^{-2} (25)	80% HOAc[a]	102
		6.2×10^{-2} (25)	60% HOAc[a]	102
CCl_2=$CHCH_2Cl$	(4.2)	1.46×10^{-2} (25)	60% HOAc[a]	102
Maleic acid	(4.2)	1.83×10^{-6} (24)	HOAc	104
Crotonic acid	(4.2)	1.03×10^{-2} (24)	HOAc	104
Crotonaldehyde	(4.2)	6.8×10^{-3} (24)	HOAc[b]	107

TABLE 5 (*Contd.*)

Olefin	*Kinetic form*	*k(°C)*	*Solvent*	*Ref.*
Tiglic acid	(4.2)	6.0×10^{-1} (24)	HOAc	106
$Me_2C{=}CHCO_2H$	(4.2)	8.3×10^{-1} (24)	HOAc	104
MeCH=CHCN	(4.2)	2.8×10^{-5} (24)	HOAc	108
PhCH=CHCN	(4.2)	3.7×10^{-4} (24)	HOAc	108
PhCH=CHBr	(4.2)	5.0×10^{-1} (24)	HOAc	100
PhCH=CHCHO	(4.2)	3.0×10^{-2} (24)	HOAc	107, 109
$PhCH{=}CHNO_2$	(4.2)	3.3×10^{-4} (24)	HOAc	107, 110
$PhCH{=}CHCO_2H$ (*cis*)	(4.2)	3.7×10^{-1} (25)	HOAc	88
$PhCH{=}CHCO_2H$ (*trans*)	(4.2)	8.2×10^{-2} (25)	HOAc	103
		1.03×10^{-1} (25)	HOAc	111
p-Me-cinnamic acid	(4.2)	1.72 (24)	HOAc	111
o-Cl-cinnamic acid	(4.2)	4.5×10^{-3} (25)	HOAc	106
p-Cl-cinnamic acid	(4.2)	4.2×10^{-2} (25)	HOAc	106
m-NO_2-cinnamic acid	(4.2)	1.8×10^{-4} (25)	HOAc	106
p-NO_2-cinnamic acid	(4.2)	6.2×10^{-5} (25)	HOAc	111
$PhCH{=}CHCO_2Me$	(4.2)	1.6×10^{-1} (24)	HOAc	111
$PhCH{=}CHCO_2Et$	(4.2)	1.6×10^{-1} (24)	HOAc	107, 109
o-NO_2-$C_6H_4CH{=}CHCO_2Me$	(4.2)	1.8×10^{-5} (24)	HOAc	111
m-NO_2-$C_6H_4CH{=}CHCO_2Me$	(4.2)	1.8×10^{-4} (24)	HOAc	109, 111
p-NO_2-$C_6H_4CH{=}CHCO_2Me$	(4.2)	6.1×10^{-5} (24)	HOAc	111
$PhCH{=}CMeCO_2H$	(4.2)	5.0×10^{-2} (25)	HOAc	106
$PhCH{=}CPhCO_2H$	(4.2)	6.0×10^{-3} (25)	HOAc	106
$PhCBr{=}CHCO_2H$	(4.2)	3.5×10^{-3} (25)	HOAc	106
$PhCH{=}CBrCO_2H$	(4.2)	1.3×10^{-4} (25)	HOAc	106
2,4-$(NO_2)_2$-stilbene	(4.2)	2.1×10^{-1} (24)	HOAc	110
Coumarin	(4.2)	5.0×10^{-5} (24)	HOAc	111
$PhCH{=}CHCOCMe_3$	(4.2)	6.0×10^{-1} (24)	HOAc	109
PhCH=CHCOPh	(4.2)	1.03 (24)	HOAc	109
p-$MeC_6H_4CH{=}CHCOPh$	(4.2)	*ca.* 23 (24)	HOAc	109
o-$ClC_6H_4CH{=}CHCOPh$	(4.2)	3.0×10^{-2} (24)	HOAc	109
p-$ClC_6H_4CH{=}CHCOPh$	(4.2)	3.8×10^{-1} (24)	HOAc	109
m-$NO_2C_6H_4CH{=}CHCOPh$	(4.2)	3.8×10^{-3} (24)	HOAc	109
$PhCH{=}CHCOC_6H_4$*p*-Me	(4.2)	1.03 (24)	HOAc	109
$PhCH{=}CHCOC_6H_4$*m*-NO_2	(4.2)	2.5×10^{-1} (24)	HOAc	109
Bromine				
C_2H_4	(4.2)	5.05×10^{-1} (25)	MeOH	112
	(4.2)?	$k_{rel} = 1.0$ (−78)	CH_2Cl_2	113
	(4.2)?	$\ln k = 22.9028 - 13{,}750/RT$	"non-hydroxylic solvents"	114
	(4.2)	1×10^{-4} (25)	CCl_4	115
	(4.2)	*ca.* 1.4 (25)	HOAc	103
	(4.20)	$k_2 = 3.94 \times 10^{5}$ (25) $k_3 = 2.0 \times 10^{4}$ (25)	H_2O	116
$CH_2{=}CHBr$	(4.7)	$k = 2.67 \times 10^{-4}$ (25)[c] $= 9.17 \times 10^{-4}$ (40)	HOAc	117
		$= 4.58 \times 10^{3}$ (40)[d]		118
	(4.21)	$k_3 = 7.0 \times 10^{-3}$ (25)	HOAc	100
$CH_2{=}CHCO_2Et$	(4.20)	$k_2 = 1.06 \times 10^{-1}$ (25) $k_3 = 6.7 \times 10^{-2}$ (25)	H_2O	116
	(4.2)	$k_2 = 6.6 \times 10^{-5}$ (25)	HOAc	103
$CH_2{=}CHOCOMe$	(4.2)	$k_2 = 1.22$ (25)	HOAc	119

TABLE 5 (*Contd.*)

Olefin	*Kinetic form*	*k(°C)*	*Solvent*	*Ref.*
CH_2=$CHSO_3^-K^+$	(4.22)	k_1 = 1.05 (0)[e]	H_2O	121
		= 4.1 (10)	H_2O	121
		= 13.7 (20)	H_2O	121
		k_2 = 35 (0)[e]	H_2O	121
		= 85 (10)	H_2O	121
		= 123 (20)	H_2O	121
		k_3 = 0.29 (10)[e]	H_2O	121
		= 0.81 (20)	H_2O	121
CH_2=$CHSO_2Me$	(4.2)	$k_2 = 1.8 \times 10^{-5}$ (25)	HOAc	122
CH_2=$CHSO_2$*n*-C_4H_9	(4.2)	$k_2 = 1.3 \times 10^{-5}$ (25)	HOAc	122
CH_2=CHMe	(4.2)	$k_2 = 3.3 \times 10^{-3}$ (25)	CCl_4	115
	(4.2)?	k_{rel} = 2.0 (−78)[f]	CH_2Cl_2	113
	(4.20)	$k_2 = 4.5 \times 10^6$ (25)	H_2O	116
		$k_3 = 3.2 \times 10^6$ (25)	H_2O	116
MeCH=$CHCO_2H$	(4.2)?	k_{rel} = 0.26 (−78)[f]	CH_2Cl_2	113
	(4.2)	$k_2 = 1.03 \times 10^{-2}$ (25)	75% HOAc	106
	(4.2)	k_2 = 4.0 (25)	37.8% MeOH	119
		= 0.7 (25)	48.8% MeOH	119
		= 0.23 (25)	59.8% MeOH	119
MeCH=$CHCO_2Et$	(4.20)	k_2 = 2.76 (25)	H_2O	116
		k_3 = 1.04 (25)	H_2O	116
MeCH=CHCN	(4.2)	k_2 = 6.5 (25)	37.8% MeOH	119
		= 2.2 (25)	48.8% MeOH	119
		= 0.45 (25)	59.8% MeOH	119
		= 0.17 (25)	61.3% MeOH	119
		= 0.08 (25)	70.4% MeOH	119
		= 0.03 (25)	79.8% MeOH	119
		= 0.003 (25)	MeOH	119
Allyl alcohol	(4.20)	$k_2 = 6.7 \times 10^5$ (25)	H_2O	116
		$k_3 = 6.9 \times 10^4$ (25)	H_2O	116
Allyl acetate	(4.21)	$k_3 = 8.1 \times 10$ (25)[g]	HOAc	120
		$= 1.15 \times 10^2$ (50)	HOAc	120
Allyl chloroacetate	(4.21)	3.5×10 (25)[h]	HOAc	120
		4.3×10 (50)	HOAc	120
Allyl benzoate	(4.21)	7.01×10 (25)[j]	HOAc	120
		1.18×10^2 (50)	HOAc	120
Allyl phenylacetate	(4.21)	7.6×10 (25)[k]	HOAc	120
		1.22×10^2 (50)	HOAc	120
Allyl propionate	(4.2)	3.8 (25)	HCOOH	119
		0.71 (25)	MeOH	119
		0.35 (25)	Ac_2O	119
		0.05 (25)	HOAc	119
Allyl fluoride	(4.2)	4.7×10^{-2} (25)	HOAc	122
Allyl chloride	(4.2)	2.7×10^{-2} (25)	HOAc	122
	(4.2)	2.0×10^{-1} (25)[l]	HOAc	94
Allyl bromide	(4.2)	1.7×10^{-2} (25)	HOAc	122
Allyl cyanide	(4.20)	k_2 = 440 (25)	H_2O	116
		k_3 = 100 (25)	H_2O	116
2-Methylpropene (isobutene)	(4.20)	$k_3 = 6 \times 10^6$ (25)	H_2O	116
	(4.20)	k_2 = 1.0 (25)	CCl_4	115
	(4.2)?	k_{rel} = 5.5 (−78)	CH_2Cl_2	113
1-Butene	(4.2)	$k_2 = 3.3 \times 10^{-2}$ (25)	CCl_4	115

TABLE 5 (*Contd.*)

Olefin	*Kinetic form*	*k(°C)*	*Solvent*	*Ref.*
	(4.2)	$k_2 = 4.9 \times 10$ (25)	MeOH	123
2-Butene	(4.2)	$k_2 = 1.0$ (25)	CCl_4	115
	(*cis*) (4.2)	$k_2 = 1.31 \times 10^3$ (25)	MeOH	123
	(*trans*) (4.2)	$k_2 = 8.5 \times 10^2$ (25)	MeOH	123
2-Methyl-1-chloro-1-propene	(4.21)	$k_3 = 3.66$ (25)	HOAc	
β-Methylcrotonic acid	(4.2)	$k_2 = 4.7 \times 10^{-3}$ (25)	HOAc	106
	(4.2)	$k_2 = 4.7 \times 10^{-1}$ (25)	79.8% MeOH	119
		$= 2.5 \times 10^{-1}$ (25)	90.0% MeOH	119
		$= 6.5 \times 10^{-2}$ (25)	MeOH	119
β-Ethylacrylic acid	(4.2)	$k_2 = 1.8 \times 10^{-2}$ (25)	HOAc	106
Methyl α-ethylacrylate	(4.2)	$k_2 = 6.0 \times 10^{-2}$ (25)	50% HOAc	124
1,3-Dichloro-2-methylenepropane ($CH_2{=}C(CH_2Cl)_2$)	(4.2)	$k_2 = 3.3 \times 10^{-4}$ (25)	HOAc	101
Maleic acid	(4.7)	$k_3 = 5.9 \times 10^{-4}$ (30)[m]	HOAc	117
		$= 3.06 \times 10^{-3}$ (50)	HOAc	117
		$= 1.45 \times 10^{-2}$ (70.5)	HOAc	117
Fumaric acid	(4.7)	$k_3 = 1.15 \times 10^{-4}$ (30)	HOAc	117
		$= 6.1 \times 10^{-4}$ (50)	HOAc	117
		$= 3.06 \times 10^{-3}$ (70.5)	HOAc	117
Diethyl fumarate	(4.20)	$k_2 = 1.00 \times 10^{-4}$ (25)	H_2O	116
		$k_3 = 3.9 \times 10^{-4}$ (25)	H_2O	116
1-Pentene	(4.2)	$k_2 = 34.8$ (25)	MeOH	112
2-Pentene	(*cis*) (4.2)	$k_2 = 2.1 \times 10^3$ (25)	MeOH	112
	(*trans*) (4.2)	$k_2 = 1.33 \times 10^3$ (25)	MeOH	123
2-Methyl-1-butene	(4.2)	$k_2 = 500$ (25)	HOAc	122
2-Methyl-2-butene	(4.2)	$k_2 = 4.2$ (25)	CCl_4	115
	(4.2)?	$k_{rel} = 10.4$ (−78)[f]	CH_2Cl_2	113
Cyclopentene	(4.2)	$k_2 = 660$ (25)	HOAc	122
1-Hexene	(4.2)	$k_2 = 33$ (25)	MeOH	112
	(4.2)	$k_2 = 17.5$ (25)	HOAc	122
2-Hexene	(*cis*) (4.2)	$k_2 = 1.46 \times 10^3$ (25)	MeOH	123
	(*trans*) (4.2)	$k_2 = 0.86 \times 10^3$ (25)	MeOH	123
3-Hexene	(*cis*) (4.2)	$k_2 = 3.25 \times 10^3$ (25)	MeOH	123
	(*trans*) (4.2)	$k_2 = 1.85 \times 10^3$ (25)	MeOH	123
4-Methyl-1-pentene	(4.2)	$k_2 = 1.54 \times 10$ (25)	MeOH	112
2-Methyl-3-pentene	(*cis*) (4.2)	$k_2 = 7.7 \times 10^2$ (25)	MeOH	123
	(*trans*) (4.2)	$k_2 = 6.0 \times 10^2$ (25)	MeOH	123
3-Methyl-2-pentene	(*cis*) (4.2)	$k_2 = 6.0 \times 10^4$ (25)	MeOH	123
	(*trans*) (4.2)	$k_2 = 5.95 \times 10^4$ (25)	MeOH	123
3,3-Dimethyl-1-butene	(4.2)	$k_2 = 11.6$ (25)	HOAc	122
2,3-Dimethyl-2-butene	(4.2)?	$k_{rel} = 14$ (−78)[f]	CH_2Cl_2	113
	(4.2)	$k_2 = 4.6 \times 10^5$ (25)	MeOH	112
Cyclopentylidene acetic acid	(4.2)	$k_2 = 3.26 \times 10^{-2}$ (25)	HOAc	106
2-Heptene	(*cis*) (4.2)	$k_2 = 1.64 \times 10^3$ (25)	MeOH	123
	(*trans*) (4.2)	$k_2 = 9.4 \times 10^2$ (25)	MeOH	123
3-Heptene	(*cis*) (4.2)	$k_2 = 2.16 \times 10^3$ (25)	MeOH	123
	(*trans*) (4.2)	$k_2 = 1.19 \times 10^3$ (25)	MeOH	123
2,2-Dimethyl-3-pentene	(*cis*) (4.2)	$k_2 = 6.65 \times 10^2$ (25)	MeOH	123
	(*trans*) (4.2)	$k_2 = 79.5$ (25)	MeOH	123
2,2-Dimethyl-4-pentene	(4.2)	$k_2 = 5.8$ (25)	MeOH	112
	(4.2)	$k_2 = 1.7$ (25)	HOAc	122
Cyclohexylidene acetic acid	(4.2)	$k_2 = 2.75 \times 10^{-2}$ (25)	HOAc	106

TABLE 5 (*Contd.*)

Olefin	*Kinetic form*	*k(°C)*	*Solvent*	*Ref.*
Di-isobutylene[p]	(4.2)	$k_2 = 0.33$ (25)	CCl_4	115
2,2-Dimethyl-4-hexene	(4.2)	$k_2 = 650$ (25)	MeOH	112
2,2,4-Trimethyl-3-pentene	(4.2)	$k_2 = 1.37 \times 10^3$ (25)	MeOH	112
Styrene and substituted styrenes				
X = H	(4.2)?	$k_{rel} = 3.35$ (−78)[f]	CH_2Cl_2	113
	(4.2)	$k_2 = 120$ (25)	HOAc	124
	(4.2)	$k_2 = 670$ (25)	MeOH	125
X = *p*-F	(4.2)	$k_2 = 755$ (25)	MeOH	125
X = *m*-F	(4.2)[q]	$k_2 = 0.455$ (25.4)	HOAc	126
X = *m*-Cl	(4.2)[q]	$k_2 = 0.292$ (25.4)	HOAc	126
X = *m*-Br	(4.2)[q]	$k_2 = 0.278$ (25.4)	HOAc	126
X = 3,4-Cl_2	(4.2)[q]	$k_2 = 0.123$ (25.4)	HOAc	126
		= 0.67 (25)	HOAc	103
X = 2,6-Cl_2	(4.2)	$k_2 = 1.3 \times 10^{-2}$ (25)	HOAc	103
X = *p*-OMe	(4.2)	$k_2 = 1.06 \times 10^5$ (25)	MeOH	125
X = *o*-Me	(4.2)	$k_2 = 254$ (25)	MeOH	125
X = *m*-Me	(4.2)	$k_2 = 159$ (25)	MeOH	125
X = *p*-Me	(4.2)	$k_2 = 1.07 \times 10^3$ (25)	MeOH	125
X = 2,4-Me_2	(4.2)	$k_2 = 4.4 \times 10^3$ (25)	MeOH	125
X = 2,5-Me_2	(4.2)	$k_2 = 671$ (25)	MeOH	125
X = 2,6-Me_2	(4.2)	$k_2 = 42.5$ (25)	MeOH	125
X = 2,4,6-Me_3	(4.2)	$k_2 = 985$ (25)	MeOH	125
X = *m*-NO_2	(4.2)	$k_2 = 0.038$ (18)[r]	HOAc	126
		= 0.066 (25.4)	HOAc	126
		= 0.121 (35.3)	HOAc	126
		= 0.192 (44.8)	HOAc	126
X = *p*-NO_2	(4.2)	$k_2 = 0.021$ (18.0)[s]	HOAc	126
		= 0.037 (25.4)	HOAc	126
		= 0.071 (35.3)	HOAc	126
		= 0.118 (44.8)	HOAc	126
X = β-Br	(4.21)	$k_3 = 0.866$ (25)	HOAc	100
X = α-Me	(4.2)	$k_2 = 500$ (25)	HOAc	122
Cinnamic acid and substituted cinnamic acids				
X = H	(*trans*) (4.2)	$k_2 = 1.3 \times 10^{-3}$ (25)[t]	HOAc	111
	(*cis*) (4.21)	$k_3 = 0.35$ (25)[u]	HOAc	120
		= 0.68 (50)	HOAc	120
X = *o*-Cl	(*trans*) (4.2)	$k_2 = 6.8 \times 10^{-5}$ (25)	HOAc	106
X = *p*-Cl	(*trans*) (4.2)	$k_2 = 6.0 \times 10^{-4}$ (25)	HOAc	106
PhCH=C(CO_2H)$_2$	(4.2)	$k_2 = 7.0 \times 10^{-6}$ (25)	HOAc	106
PhCH=CHCH_2Cl	(4.2)	$k_2 = 1.28$ (25)	HOAc	103
Camphene	(4.2)	$k_2 = 660$ (25)	HOAc	122
PhCH=CHCH=CHCO_2H	(4.2)	$k_2 = 3.3$ (25)	HOAc	106
PhCH=CHCH=C(CO_2H)$_2$	(4.2)	$k_2 = 7.0 \times 10^{-3}$ (25)	HOAc	106
n-1-Undecene	(4.2)	$k_2 = 1.07$ (−20)[v]	MeOH	127
		= 5.7 (0)	MeOH	127
		= 25 (20)	MeOH	127
trans-*n*-2-Undecene	(4.2)	$k_2 = 52.3$ (−20)[v]	MeOH	127
		= 232 (0)	MeOH	127
		= 737 (20)	MeOH	127
trans-*n*-3-Undecene	(4.2)	$k_2 = 82.6$ (−20)[v]	MeOH	127
		= 325 (0)	MeOH	127
		= 1108 (20)	MeOH	127

TABLE 5 (*Contd.*)

Olefin	*Kinetic form*	*k(°C)*	*Solvent*	*Ref.*
trans-n-4-Undecene	(4.2)	$k_2 = 53.1\ (-20)$[v]	MeOH	127
		$= 217\ (0)$	MeOH	127
		$= 697\ (20)$	MeOH	127
trans-n-5-Undecene	(4.2)	$k_2 = 61.5\ (-20)$[v]	MeOH	127
		$= 237\ (0)$	MeOH	127
		$= 797\ (20)$	MeOH	127
cis-n-2-Undecene	(4.2)	$k_2 = 97.8\ (-20)$[v]	MeOH	127
		$= 386\ (0)$	MeOH	127
		$= 1185\ (20)$	MeOH	127
cis-n-3-Undecene	(4.2)	$k_2 = 165\ (-20)$[v]	MeOH	127
		$= 610\ (0)$	MeOH	127
		$= 1866\ (20)$	MeOH	127
cis-n-4-Undecene	(4.2)	$k_2 = 118\ (-20)$[v]	MeOH	127
		$= 409\ (0)$	MeOH	127
		$= 1216\ (20)$	MeOH	127
cis-n-5-Undecene	(4.2)	$k_2 = 122\ (-20)$[v]	MeOH	127
		$= 414\ (0)$	MeOH	127
		$= 1353\ (20)$	MeOH	127
Allyl alcohol	(4.2)	$k_2 = 2.5\ (15)$[aa]	MeOH	128
Terminal olefins, $C_{6,7,10,\text{ and }12}$		$k_2 = 18.3\ (15)$[bb]	MeOH	128
Cyclohexene		$k_2 = 350\ (15)$[cc]	MeOH	128
Mesityl oxide		$k_2 = 12\ (15)$	MeOH	128
Hexen-2-ol-4-COOH-6-*t*-Bu ester		$k_2 = 30\ (15)$	MeOH	128
2- or 3-heptene		$k_2 = 760\ (15)$[dd]	MeOH	128
2,2,4-Me_3-4-pentene (diisobutylene)		$k_2 = 460\ (15)$[ee]	MeOH	128
2,6-Me_2-3-heptene		$k_2 = 160\ (15)$[ff]	MeOH	128
2,4-Me_2-2-hexen-4-ol-6-COOH *t*-Bu ester		$= 5000\ (15)$	MeOH	128
10-Undecenaldehyde	(4.2)	$k_2 = 18\ (15)$[w]	MeOH	128
Undecenoic acid	(4.2)	$k_2 = 20\ (25)$	HOAc	124
2-Vinylnaphthalene	(4.2)	$k_2 = 350\ (25)$	MeOH	125
Stilbene (PhCH=CHPh)	(4.21)	$k_2 = 1.3 \times 10^4\ (25)$	H_2O	116
	(4.2)	$k_2 = 0.30\ (25)$	HOAc	129
	(4.21)	$k_2 = 3.36 \times 10^{-2}\ (0)$	MeOH	130
		$k_3 = 3.33\ (0)$	MeOH	130
Phenanthrene and substituted phenanthrenes[y]				
X = H	(4.2)	$k_2 = 0.16\ (25)$	CCl_4	131
		$= 0.18\ (39)$	CCl_4	131
X = 2-Cl	(4.2)	$k_2 = 5 \times 10^{-4}\ (25)$	CCl_4	131
		$= 3 \times 10^{-4}\ (36)$	CCl_4	131
X = 3-Cl	(4.2)	$k_2 = 6 \times 10^{-3}\ (25)$	CCl_4	131
		$= 1.1 \times 10^{-2}\ (36)$	CCl_4	131
X = 2-Br	(4.2)	$k_2 = 6 \times 10^{-4}\ (25)$	CCl_4	131
		$= 6 \times 10^{-4}\ (36)$	CCl_4	131
X = 2-*t*-Bu	(4.2)	$k_2 = 5 \times 10^{-2}\ (25)$	CCl_4	131
		$= 7 \times 10^{-2}\ (36)$	CCl_4	131
X = 3-Br	(4.2)	$k_2 = 1.1 \times 10^{-2}\ (36)$	CCl_4	131
X = 2-*t*-Bu	(4.2)	$k_2 = 5 \times 10^{-2}\ (25)$	CCl_4	131
		$= 7 \times 10^{-2}\ (36)$	CCl_4	131
X = 3-*t*-Bu	(4.2)	$k_2 = 0.15\ (25)$	CCl_4	131
		$= 0.10\ (36)$	CCl_4	131
X = 2-CO_2Me	(4.2)	$k_2 = 1 \times 10^{-2}\ (25)$	CCl_4	131

References pp. 80–86

TABLE 5 (*Contd.*)

Olefin	*Kinetic form*	*k(°C)*	*Solvent*	*Ref.*
		$= 1.5 \times 10^{-2}$ (36)	CCl_4	131
X = 3-CO_2Me	(4.2)	$k_2 = 1 \times 10^{-2}$ (25)	CCl_4	131
		$= 2 \times 10^{-3}$ (36)	CCl_4	131
X = 2-CO_2Et	(4.2)	$k_2 = 6 \times 10^{-2}$ (25)	CCl_4	131
		$= 9 \times 10^{-2}$ (36)	CCl_4	131
X = 3-CO_2Et	(4.2)	$k_2 = 4 \times 10^{-2}$ (25)	CCl_4	131
		$= 6 \times 10^{-2}$ (36)	CCl_4	131
PhCH=CHCOPh and substituted chalcones				
X = H	(4.2)	$k_2 = 5.5 \times 10^{-3}$ (24)	HOAc	103
X = p′-Me	(4.2)	$k_2 = 5.5 \times 10^{-3}$ (24)	HOAc	103
X = p-Me	(4.2)	$k_2 = 0.13$ (24)	HOAc	103
Oleic acid	(4.2)	$k_2 = 500$ (25)	HOAc	124
Methyl oleate	(4.2)	$k_2 = 500$ (25)	HOAc	122
3-X,17-Y,-5-Androstenes				
X = OAc, Y = H	(4.2)	$k_2 = 2.15 \times 10^{-3}$ (25)	$CHCl_3$–HOAc	132
X = OAc, Y = β-OH	(4.2)	$k_2 = 2.40 \times 10^{-3}$ (25)	$CHCl_3$–HOAc	132
X = OAc, Y = β-6-metho-2-heptyl	(4.2)	$k_2 = 2.25 \times 10^{-3}$ (25)	$CHCl_3$–HOAc	132
X = OAc, Y = βOCOMe	(4.2)	$k_2 = 1.21 \times 10^{-3}$ (25)	$CHCl_3$–HOAc	132
X = OAc, Y = β-OCOPh	(4.2)	$k_2 = 1.09 \times 10^{-3}$ (25)	$CHCl_3$–HOAc	132
X = OAc, Y = α-OCOPh	(4.2)	$k_2 = 1.09 \times 10^{-3}$ (25)	$CHCl_3$–HOAc	132
X = OAc, Y = β-OTos[z]	(4.2)	$k_2 = 8.22 \times 10^{-4}$ (25)	$CHCl_3$–HOAc	132
X = OAc, Y = β-$OCOCCl_3$	(4.2)	$k_2 = 6.65 \times 10^{-4}$ (25)	$CHCl_3$–HOAc	132
X = β-OAc, Y = β-$OC_6H_4(p\text{-}NO_2)$	(4.2)	$k_2 = 9.3 \times 10^{-4}$ (25)	$CHCl_3$–HOAc	132
X = β-OAc, Y = β-$OC_6H_3(3,5\text{-}(NO_2)_2)$	(4.2)	$k_2 = 8.0 \times 10^{-4}$ (25)	$CHCl_3$–HOAc	132
X = β-OAc, Y = β-C_8H_{17}	(4.2)	$k_2 = 1.46 \times 10^{-3}$ (25)	$CHCl_3$–HOAc	132
X = β-OAc, Y = β-CO_2Me	(4.2)	$k_2 = 1.07 \times 10^{-3}$ (25)	$CHCl_3$–HOAc	132
X = β-OAc, Y = =O(17-ketone)	(4.2)	$k_2 = 9.8 \times 10^{-4}$ (25)	$CHCl_3$–HOAc	132
X = β-OBz[gg], Y = β-OBz[gg]	(4.2)	$k_2 = 1.06 \times 10^{-3}$ (25)	$CHCl_3$–HOAc	132
X = β-OBz[gg], Y = β-CO_2Me	(4.2)	$k_2 = 9.3 \times 10^{-4}$ (25)	$CHCl_3$–HOAc	132
X = β-OBz[gg], Y = =O (17-ketone)	(4.2)	$k_2 = 5.83 \times 10^{-4}$ (25)	$CHCl_3$–HOAc	132
X = β-OTos[z], Y = β-CO_2Me	(4.2)	$k_2 = 2.58 \times 10^{-4}$ (25)	$CHCl_3$–HOAc	132
X = β-OPNB[hh], Y = H	(4.2)	$k_2 = 9.55 \times 10^{-4}$ (25)	$CHCl_3$–HOAc	132
X = β-OPNB, Y = OPNB[hh]	(4.2)	$k_2 = 2.49 \times 10^{-4}$ (25)	$CHCl_3$–HOAc	132
X = β-OPNB[hh], Y = =O (17-ketone)	(4.2)	$k_2 = 2.81 \times 10^{-4}$ (25)	$CHCl_3$–HOAc	132
X = β-ODNB[jj], Y = OBz[gg]	(4.2)	$k_2 = 7.95 \times 10^{-4}$ (25)	$CHCl_3$–HOAc	132
X = β-ODNB[jj], Y = CO_2Me	(4.2)	$k_2 = 1.93 \times 10^{-4}$ (25)	$CHCl_3$–HOAc	132
X = β-ODNB[jj], Y = ODNB[jj]	(4.2)	$k_2 = 1.10 \times 10^{-4}$ (25)	$CHCl_3$–HOAc	132
X = β-ODNB[jj], Y = =O (17-ketone)	(4.2)	$k_2 = 1.06 \times 10^{-4}$ (25)	$CHCl_3$–HOAc	132
X = β-$OCOCCl_3$, Y = H	(4.2)	$k_2 = 3.22 \times 10^{-4}$ (25)	$CHCl_3$–HOAc	132
X = β-$OCOCCl_3$, Y = β-OBz[gg]	(4.2)	$k_2 = 1.40 \times 10^{-4}$ (25)	$CHCl_3$–HOAc	132
X = β-$OCOCCl_3$, Y = β-CO_2Me	(4.2)	$k_2 = 1.18 \times 10^{-4}$ (25)	$CHCl_3$–HOAc	132
X = β-$OCOCCl_3$, Y = β-$OCOCCl_3$	(4.2)	$k_2 = 7.3 \times 10^{-5}$ (25)	$CHCl_3$–HOAc	132
X = β-$OCOCCl_3$, Y = =O (17-ketone)	(4.2)	$k_2 = 8.67 \times 10^{-5}$ (25)	$CHCl_3$–HOAc	132
3-X-5-Cholestenes				
X = β-OH	(4.2)	$k_2 = 1.90 \times 10^{-2}$ (23)	$CHCl_3$–HOAc	132
X = β-OMe	(4.2)	$k_2 = 1.80 \times 10^{-2}$ (23)	$CHCl_3$–HOAc	132
X = β-O.tetrahydropyranyl	(4.2)	$k_2 = 8.33 \times 10^{-3}$ (23)	$CHCl_3$–HOAc	132
X = β-$OCOCMe_3$	(4.2)	$k_2 = 3.47 \times 10^{-3}$ (23)	$CHCl_3$–HOAc	132
X = β-OCOMe	(4.2)	$k_2 = 2.25 \times 10^{-3}$ (23)	$CHCl_3$–HOAc	132
X = β-$OCOCCl_3$	(4.2)	$k_2 = 2.43 \times 10^{-4}$ (23)	$CHCl_3$–HOAc	132
X = β-OBz[gg]	(4.2)	$k_2 = 1.97 \times 10^{-3}$ (23)	$CHCl_3$–HOAc	132
X = β-OTos[z]	(4.2)	$k_2 = 5.0 \times 10^{-3}$ (23)	$CHCl_3$–HOAc	132

TABLE 5 (*Contd.*)

Olefin	*Kinetic form*	*k(°C)*	*Solvent*	*Ref.*
X = β-Cl	(4.2)	$k_2 = 7.33 \times 10^{-4}$ (23)	$CHCl_3$–HOAc	132
X = β-OPNB[hh]	(4.2)	$k_2 = 8.77 \times 10^{-4}$ (23)	$CHCl_3$–HOAc	132
X = β-ODNB[jj]	(4.2)	$k_2 = 3.56 \times 10^{-4}$ (23)	$CHCl_3$–HOAc	132
X = β-$OCO(CH_2)_2CO_2H$	(4.2)	$k_2 = 2.13 \times 10^{-3}$ (23)	$CHCl_3$–HOAc	132
Recent work				
Allyl alcohol	(4.20)	4.7 (25.0)	MeOH–NaBr	148
	(4.20)	2.26×10^5 (25.0)	H_2O–NaBr	148
CH_2=$CH(CH_2)_2OH$	(4.20)	8.75 (25.0)	MeOH–NaBr	148
	(4.20)	1.03×10^6 (25.0)	H_2O–NaBr	148
CH_2=$CH(CH_2)_3OH$	(4.20)	44.0 (25.0)	MeOH–NaBr	148
	(4.20)	5.0×10^6 (25.0)	H_2O–NaBr	148
CH_2=$CH(CH_2)_4OH$	(4.20)	29.0 (25.0)	MeOH–NaBr	148
	(4.20)	6.1×10^6 (25.0)	H_2O–NaBr	148
Styrenes and substituted styrenes				
X = H	(4.20)	11.2 (25.0)	HOAc	149
X = *p*-F	(4.20)	14.5 (25.0)	HOAc	149
X = *p*-Cl	(4.20)	3.26 (25.0)	HOAc	149
X = *p*-Br	(4.20)	3.60 (25.0)	HOAc	149
X = *p*-NO_2	(4.20)	0.083 (25.0)	HOAc	149
X = *m*-Cl	(4.20)	0.67 (25.0)	HOAc	149
X = *m*-NO_2	(4.20)	0.140 (25.0)	HOAc	149
X = 3,4-dichloro	(4.20)	0.44 (25.0)	HOAc	149
X = α-Me	(4.20)	680 (25.0)	HOAc	149
X = *trans*-β-Me	(4.20)	12.3 (25.0)	HOAc	149
X = *cis*-β-Me	(4.20)	8.9 (25.0)	HOAc	149
X = β,β-Me_2	(4.20)	14.7 (25.0)	HOAc	149
X = αβ-Me_2 (*trans*)	(4.20)	300 (25.0)	HOAc	149
X = αβ-Me_2 (*cis*)	(4.20)	61.7 (25.0)	HOAc	149
X-αββ-Me_3	(4.20)	56 (25.0)	HOAc	149
3-Phenylindene	(4.20)	2.08×10^4 (25.0)	MeOH–NaBr	150
1-Substituted-phenyl-3,4-dihydronaphthalenes				
X = H	(4.20)	1.66×10^4 (25.0)	MeOH–NaBr	150
X = *p*-MeO	(4.20)	3.4×10^5 (25.0)	MeOH–NaBr	150
X = *p*-Me	(4.20)	6.45×10^4 (25.0)	MeOH–NaBr	150
X = *m*-Me	(4.20)	2.9×10^4 (25.0)	MeOH–NaBr	150
X = *p*-Br	(4.20)	2.2×10^3 (25.0)	MeOH–NaBr	150
X = *m*-Br	(4.20)	655 (25.0)	MeOH–NaBr	150
3-(Substituted-phenyl)-benzocyclohepta-1,3-dienes				
X = H	(4.20)	41.8 (25.0)	MeOH–NaBr	150
X = *p*-MeO	(4.20)	6.05×10^4 (25.0)	MeOH–NaBr	150
X = *p*-Me	(4.20)	470 (25.0)	MeOH–NaBr	150
X = *m*-Me	(4.20)	75.2 (25.0)	MeOH–NaBr	150
X = *m*-Cl	(4.20)	2.3 (25.0)	MeOH–NaBr	150
1-(*p*-MeO-phenyl)-3,4-dihydro-6-methoxynaphthalene	(4.20)	1.92×10^6 (25.0)	MeOH–NaBr	150
Iodine				
Ethylene	(4.24)[kk]	$k = 1.25 \times 10^{-5}$ (40)	CCl_4	133
		$= 2.48 \times 10^{-4}$ (75)	CCl_4	133
		$= 6.0 \times 10^{-3}$ (122)	CCl_4	133
		$= 9.5 \times 10^{-3}$ (129.8)	CCl_4	133
		$= 1.1 \times 10^{-2}$ (132)	CCl_4	133

References pp. 80–86

TABLE 5 (*Contd.*)

Olefin	*Kinetic form*	*k(°C)*	*Solvent*	*Ref.*
		$= 1.8 \times 10^{-2}$ (141)	CCl_4	133
		$= 1.9 \times 10^{-2}$ (142)	CCl_4	133
		$= 3.4 \times 10^{-2}$ (152)	CCl_4	133
	(4.20)	$k_2 = 0.001$ (25?)[ll]	HOAc	134
Propene	(4.20)	$k_2 = 1.5 \times 10^{-2}$ (25?)[ll]	HOAc	134
Allyl alcohol	(4.20)	$k_2 = 4.2 \times 10^{-2}$ (14.4)	$PhNO_2$	134
		$= 5.1 \times 10^{-2}$ (25)[mm]	$PhNO_2$	134
		$= 7.3 \times 10^{-2}$ (45)[mm]	$PhNO_2$	134
		$= 3.5 \times 10^{-3}$ (25)	HOAc	134
		$= 4.8 \times 10^{-3}$ (35)	HOAc	134
		$= 5.8 \times 10^{-3}$ (45)[nn]	HOAc	134
		$= 7.7 \times 10^{-3}$ (55)[nn]	HOAc	134
$CH_2{=}CHCH_2OH$	(4.20)	0.00018 (25.0)	H_2O–KI	148
$CH_2{=}CH(CH_2)_2OH$	(4.20)	0.00038 (25.0)	H_2O–KI	148
$CH_2{=}CH(CH_2)_3OH$	(4.20)	0.036 (25.0)	H_2O–KI	148
$CH_2{=}CH(CH_2)_4OH$	(4.20)	0.0063 (25.0)	H_2O–KI	148
Allyl acetate	(4.20)	$k_2 = 1.4 \times 10^{-3}$ (25)	HOAc	134
		$= 2.2 \times 10^{-3}$ (35)	HOAc	134
		$= 2.8 \times 10^{-3}$ (45)[pp]	HOAc	134
		$= 3.7 \times 10^{-3}$ (55)[pp]	HOAc	134
Allyl bromide	(4.20)	$k_2 = 4.6 \times 10^{-4}$ (25?)	HOAc	134
Allyl benzoate	(4.20)	$k_2 = 1.1 \times 10^{-3}$ (25?)	HOAc	134
Cyclohexene	(4.25)	$k_3 = 1.86$ (20)	DMF	135
		$k_2 = 0.063$ (20)	DMF	135
Cyclohexene	(4.20)	$k_2 = 1.95 \times 10^{-3}$ (30)	PhH	136
		$= 1.67 \times 10^{-3}$ (30)	PhMe	136
		$= 4.67 \times 10^{-3}$ (30)	PhCl	136
		$= 4.42 \times 10^{-3}$ (30)	$CHCl_3$	136
		$= 2.21 \times 10^{-3}$ (30)	CCl_4	136
		$= 1.94 \times 10^{-3}$ (30)	$(CH_2Br)_2$	136
		$= 2.94 \times 10^{-3}$ (30)	MeOH	136
		$= 2.4 \times 10^{-3}$ (30)	EtOH	136
		$= 9.06 \times 10^{-4}$ (30)	BuOH	136
Undecenoic acid	(4.20)	$k_2 = 5 \times 10^{-2}$ (25?)	HOAc	134
9-Allyl-fluorene-9-carboxylic acid 2,2-Diphenyl-4-pentenoic acid	Indefinite kinetic order			137
Iodine monochloride				
Acrylic acid	(4.12)	$k_3 = 8.0 \times 10^{-4}$ (25)	HOAc	88
Allyl acetate	(4.12)	$k_3 = 0.3$ (25)	HOAc	88
Undecenoic acid	(4.12)	$k_3 = 100$ (25)	HOAc	88
Stilbene			Various	138
Hypochlorous acid and hypobromous acid				
Allyl alcohol (HOCl)	(4.23)	$k_2^{I} = 2.78 \times 10^{-2}$ (25)	aq. (4.50)[qq]	139
		$= 2.37 \times 10^{-2}$ (25)	aq. (4.00)[qq]	139
		$= 2.15 \times 10^{-2}$ (25)	aq. (3.80)[qq]	139
		$k_2^{II} = 0.131$ (25)	aq. (4.50)[qq]	139
		$= 0.159$ (25)	aq. (4.00)[qq]	139
		$= 0.171$ (25)	aq. (3.80)	139
	(4.24)	$k_2^{III} = 0.242$ (25)	aq. (4.75)[qq]	140
	(4.24)	$k_3^{II} = 26$ (25)	aq. (4.75)[qq]	140
$CH_2{=}CBr(SO_3)^-$ (HOBr)	(4.25)	$k_2 = 0.39$ (0)	aq.	141

TABLE 5 (*Contd.*)

Olefin	*Kinetic form*	*k(°C)*	*Solvent*	*Ref.*
		= 0.68 (10)	aq.	141
		$\Delta H^{\ddagger}$ = 8.6 kcal.mole^{-1}		
MeCH=CHCHO (HOBr)	(4.13)	k = 0.159 (0)	aq.	142
		= 0.263 (10) $\Delta H^{\ddagger}$ = 7.7	aq.	142
		k' = 0.740 (0)[rr]	aq.	142
		= 1.38 (10)[rr] $\Delta H^{\ddagger}$ = 9.5	aq.	142
$MeCH{=}CHCO_2H$ (HOCl)	(4.25)	k_2^{I} = 0.075 (25)	aq. (4.73)[qq]	144
		= 0.116 (35) $\Delta H^{\ddagger}$ = 8.4	aq. (4.73)[qq]	145
		k_2^{II} = 0.067 (25)	aq. (4.73)[qq]	144
		= 0.115 (35) $\Delta H^{\ddagger}$ = 9.9	aq. (4.73)[qq]	145
		k_3^{II} = 35 (25)	aq. (4.73)[qq]	144
		= 67 (35) $\Delta H^{\ddagger}$ = 12.1	aq. (4.73)[qq]	145
	(4.16)	k_3 = 638 (0)	aq.	143
$MeCH{=}CHCO_2Et$(HOCl)	(4.26)	$k_2^{I} = 2.8 \times 10^{-4}$ (25)	aq. (4.74)[qq]	146
		k_3^{II} = 0.273 (25)	aq. (4.74)[qq]	146
		k_3^{III} = 0.114 (25)	aq. (4.74)[qq]	146
2-Butene-1,4-diol	(4.16)	k_3 = 15.8 (0)	aq.	147

[a] Containing 10%, 20%, or 40% water.
[b] Containing sodium acetate to prevent acid-catalysed reaction.
[c] LiBr-catalysed; $\Delta H^{\ddagger}$ = 8.6 kcal.mole^{-1}.
[d] LiCl-catalysed.
[e] E_1 = 20.4 kcal.mole^{-1}, $\log A_1$ = 14.80; E_2 = 9.8 kcal.mole^{-1}, $\log A_2$ = 7.86; E_3 = 17.0 kcal.mole^{-1}, $\log A_3$ = 10.85.
[f] Relative to ethylene (competition rates).
[g] E = 2.7 kcal.mole^{-1}.
[h] E = 1.6 kcal.mole^{-1}.
[j] E = 3.9 kcal.mole^{-1}.
[k] E = 3.6 kcal.mole^{-1}.
[l] Kinetic expression is in fact very complex.
[m] E = 9.2 kcal.mole^{-1}.
[n] E = 9.6 kcal.mole^{-1}.
[p] Mixture of 2,2,4-trimethyl-1- and 2-pentenes; apparently the former isomer was predominant in the mixture used.
[q] Reaction follows the kinetics of eqn. (4.3); k_2 refers to the second-order process. The data in ref. 103 contains results obtained *assuming* a pure second-order reaction.
[r] $\Delta H^{\ddagger}$ = 10.5 kcal.mole^{-1}; $\Delta S^{\ddagger}$ = −28 eu.
[s] $\Delta H^{\ddagger}$ = 11.3 kcal.mole^{-1}; $\Delta S^{\ddagger}$ = −27 eu.
[t] k_3 = 0.083 (eqn. 4.21).
[u] $\Delta H^{\ddagger}$ = 5.8 kcal.mole^{-1}.
[v] *n*-1-Undecene, ΔH = 11.0 kcal.mole^{-1}, ΔS = −14.6 eu
trans-*n*-2-Undecene, ΔH = 9.1 ΔS = −14.2
trans-*n*-3-Undecene, ΔH = 8.9 ΔS = −14.1
trans-*n*-4-Undecene, ΔH = 9.0 ΔS = −14.7
trans-*n*-5-Undecene, ΔH = 8.9 ΔS = −14.9
cis-*n*-2-Undecene, ΔH = 8.7 ΔS = −14.8
cis-*n*-3-Undecene, ΔH = 8.4 ΔS = −14.9
cis-*n*-4-Undecene, ΔH = 8.0 ΔS = −17.0
cis-*n*-5-Undecene, ΔH = 8.1 ΔS = −16.7
[w] $\Delta H^{\ddagger}$ = 10.5 kcal.mole^{-1}, $\Delta S^{\ddagger}$ = −16.0 eu.
[x] Calculated for a second-order reaction; eqn. (4.21) is probably more appropriate.

References pp. 80–86

5. Addition of other electrophiles

5.1 PERACIDS

The reaction of olefins with peracids involves firstly the formation of an epoxide, reaction (5.1); in acidic media the epoxide may suffer ring-opening to yield the mono-ester of a glycol, reaction (5.2)[151].

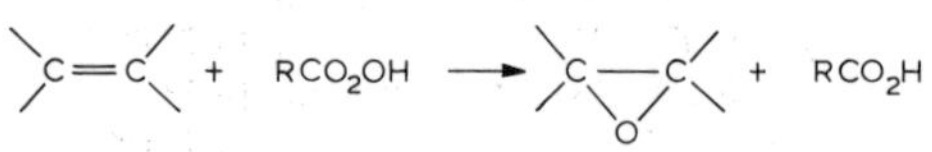

(5.1)

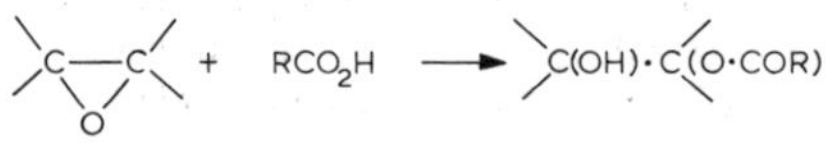

(5.2)

Boeseken *et al.*[152,153] first studied the kinetics of such reactions; the first report[152] dealt with the reaction of perbenzoic acid with derivatives of styrene in chloroform solution and despite the ease of decomposition of the reagent, styrene was proved to react less rapidly than β-methyl- or β-ethyl-styrene. In the second paper, the rates of reaction of a number of olefins with peracetic acid in acetic acid were shown to increase with substitution of the olefinic bond, *viz.*

$$CH_2{=}CH_2 < RCH{=}CH_2 < R_2C{=}CH_2 < R_2C{=}CHR$$

and the reactions were shown to be first-order with respect to each reactant, *viz.*

---

[y] From equilibrium measurements; the variation of k_2 with T may be due to disturbance of a pre-equilibrium forming ($ArHBr_2$).
[z] Tos = p-toluenesulphonyl (p-$MeC_6H_4SO_2$).
[aa] $\Delta H^{\ddagger} = 10.4$ kcal.mole^{-1}, $\Delta S^{\ddagger} = -20.6$.
[bb] $\Delta H^{\ddagger} = 10.5$ kcal.mole^{-1}, $\Delta S^{\ddagger} = -16.$ eu.
[cc] $\Delta H^{\ddagger} = 8.6$ kcal.mole^{-1}, $\Delta S = -16.4$ eu.
[dd] $\Delta H^{\ddagger} = 7.6$ kcal.mole^{-1}, $\Delta S = -18.7$ eu.
[ee] $\Delta H^{\ddagger} = 10.4$ kcal.mole^{-1}, $\Delta S^{\ddagger} = -14.7$ eu
[ff] $\Delta H^{\ddagger} = 7.9$ kcal.mole^{-1}, $\Delta S = -20.7$ eu.
[gg] Bz = COPh.
[hh] PNB = p-nitrobenzoyl.
[jj] DNB = 3,5-dinitrobenzoyl.
[kk] Rates derived from the equations published[133]. $\Delta H^{\ddagger} = 18.2$ kcal.mole^{-1}.
[ll] Rate coefficient with olefin in great excess.
[mm] $k_3 = 8k_2$. $\Delta H^{\ddagger} = 3.3$ kcal.mole^{-1}.
[nn] $k_3 = 8k_2$. $\Delta H^{\ddagger} = 4.8$ kcal.mole^{-1}.
[pp] $k_3 = 8k_2$. $\Delta H^{\ddagger} = 6.1$ kcal.mole^{-1}.
[qq] pH indicated in parenthesis.
[rr] m-Nitrobenzoate ion = X^-.

$$\text{Rate} = k[\text{olefin}][\text{MeCO}_2\text{OH}] \quad (5.3)$$

It was subsequently suggested that there was a parallel between the susceptibilities of olefins towards attack by bromine (a known electrophile) and that towards attack by peracids[154], and these suggestions were put on to a quantitative basis by the work of Lynch and Pausacker[155], who studied the effects of substituents upon both the peracid and the substrate (stilbene) in the reaction in benzene solution. Second-order kinetics were found in all cases, and electron-withdrawing groups increased the rate of reaction of the stilbene in the order

$$p\text{-OMe} < p\text{-Me} < \text{H} < p\text{-Cl} < p\text{-NO}_2$$

and decreased the reactivity of the substituted perbenzoic acid towards stilbene itself, in the order

$$p\text{-NO}_2 < p\text{-Cl} < \text{H} < p\text{-Me} < p\text{-OMe}$$

Such susceptibility towards electronic effects would indicate electrophilic attack by the peracid upon the olefin. However, the extent to which substituents affect the reaction rate is far less than in more characteristic electrophilic additions (*e.g.* bromine addition, acid-catalysed hydration) and the effect of solvent changes is not nearly as great. The simplicity of the reaction order might also be regarded as evidence against a polar mechanism, since the addition of hydrogen halides or of halogen to olefins in poorly solvating solvent like benzene is characterised by poorly reproducible kinetics involving high orders with respect to one of the reagents and showing great sensitivity to traces of impurities, whereas no added solvation appears necessary in the peracid oxidations other than that supplied by the two reacting species.

These observations suggested a cyclic intermediate (I) in which parts of the two reacting molecules supply the bulk of the stabilisation for the charged intermediates. The normal carboxylic acid formed in the oxidation was not

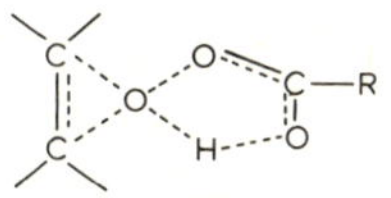

I

found to catalyse the reaction[155] and only interfered by subsequent fission of the epoxide ring; it seems not to act in any solvating capacity either, another point indicating the cyclic transition state. However, catalysis has been shown in the reaction of stilbene with perbenzoic acid in benzene containing trichloroacetic acid, when the simple rate law (5.3) becomes

$$\text{Rate} = -\text{d}[\text{PhCO}_3\text{H}]/\text{d}t = k[\text{olefin}][\text{ArCO}_3\text{H}][\text{CCl}_3\text{CO}_2\text{H}]^{1/2} \tag{5.4}$$

The order with respect to the catalysing acid was suggested[156] to reflect the dimerisation of this species in benzene solution, following the formation of an ion-pair (II) which is a reactive intermediate, *viz.*

$$(\text{CCl}_3\text{CO}_2\text{H})_2 + 2\ \text{PhCO}_2\text{OH} \rightleftharpoons 2\ [\text{CCl}_3\text{CO}_2^-\ .\ \text{PhCO}_3\text{H}_2^+]\ \text{II} \tag{5.5}$$

The product of the reaction with *cis*-stilbene was a trichloroacetate ester which yielded the *meso*-glycol upon reduction with lithium aluminium hydride; the *trans*-olefin yielded the racemic glycol after the same treatment[156]. Such reaction products are consistent with ring-opening of the epoxide by strong acid, and it was shown that (–)-*trans*-stilbene oxide reacted rapidly (relative to the rate of epoxidation) with trichloroacetic acid in benzene; the overall observations are indicative of acid-catalysed epoxidation of stilbene but may admit other interpretations, one of which involves thermodynamic intervention of trichloroacetic acid dimer in a solvating capacity rather than in a truly kinetic function.

An alternative suggestion, the intervention of OH^+ as a reactive intermediate, is less likely, since it has been shown that little exchange between peracids and the corresponding carboxylic acid anions (*i.e.* RCO_2OH and RCO_2^-) occurs in *aqueous* media under what were considered to be the most suitable conditions[157]. A sequence involving ionisation of the peracid

$$\text{RCOOOH} \rightarrow \text{OH}^+ + \text{RCOO}^- \tag{5.6}$$

therefore seems unlikely in the solvents of low dielectric constant in which epoxidations have been studied. The situation is by no means settled, and there has been evidence put forward for four- or six-centre transition states in some cases[158].

5.2 OZONE

The detailed mechanism of the reaction of ozone with olefins proposed by Criegee *et al.*[159] involves initial attack upon the olefin to form a "primary ozonide" which suffers oxygen–oxygen and then carbon–carbon bond fission to form a reactive zwitterionic intermediate (III) which provides the "normal ozonide", together with the products of rearrangement, polymerisation, or secondary decomposition, *viz.*

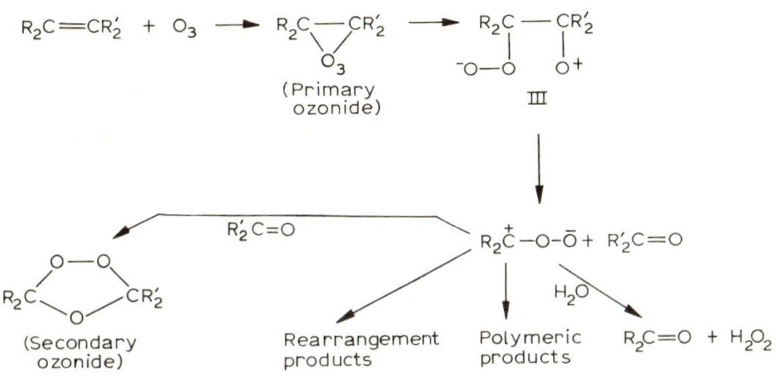

Such a sequence explains, for instance, the formation of the same secondary ozonide from *cis* or *trans*-stilbene[159], and the isolation of the secondary ozonide of isobutene from the ozonolysis of 2,3-dimethyl-2-butene in the presence of formaldehyde[159], *viz.*

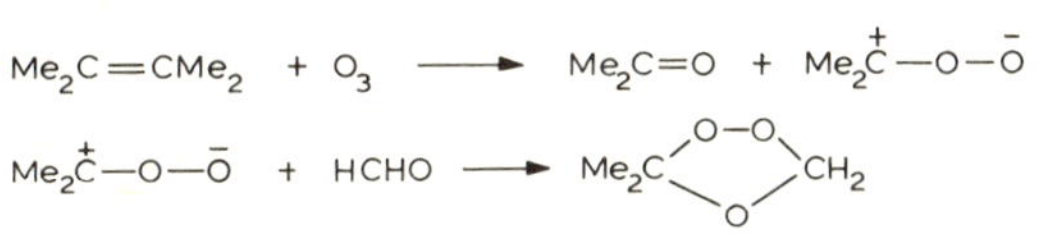

(5.7)

The relative extents to which the various reaction paths are controlled by structural changes in the olefin have been discussed by Bailey[160]; in some cases, the primary ozonide is sufficiently stable to be detected, as in the reaction of di-*t*-butylethylene. The ozonide formed at −70°C gives the glycol upon reduction with lithium aluminium hydride at this temperature; the "normal" ozonide is formed at −60°C and all reaction products are derived from this entity at higher temperatures[161]. The substituent effects are generally those expected from electrophilic attack upon the olefin, and much of the definitive kinetic results are due to Wibant and Sixma[162]. The rate-determining stage appears to be the reaction of one molecule of ozone with one molecule of olefin, although there is controversy over whether the immediate reaction product is the primary ozonide[160].

5.3 SULPHUR ELECTROPHILES

The reaction of elemental sulphur with olefins, a reaction of considerable industrial importance, was long thought to involve a homolytic mechanism[163]. At high temperatures such a process may well occur, but at relatively low temperatures there is evidence[164] of a heterolytic reaction which is characterised by electrophilic attack of the olefin by sulphur[165]. The reaction is a chain process involving polysulphides; although such processes are more usual in free-radical reactions, they are not excluded from heterolytic processes, and it has been shown[165] that the relative rates of reaction of different olefins with sulphur lie in the order RCH=CH$_2$ < RCH=CHR′ < RR′C=CHR″. Under these conditions, the rate of the reaction was unaffected by the presence of free-radical catalysts, by inhibitors, or by irradiation. Various polar components (thiols, carboxylic acids) accelerated the reaction, as did increase in the polar nature of the medium. Such observations are consistent with a heterolytic mechanism in which the olefin is undergoing electrophilic attack. The proposed mechanism[164] is complex and involves the formation and subsequent reaction of a number of sulphur electrophiles, including sulfenium ions and episulphenium ions[164].

By far the most evidence for electrophilic sulphur is found in the *sulphenyl halides* (RSCl, RSBr). Although these compounds may in theory react either as sources of $\overset{\delta+}{\text{RS}}$ or $\overset{\delta+}{\text{X}}$ (X = halogen), none of the observed reactions of the sulphenyl halides indicate the latter mode of heterolysis. Kharasch *et al.*[166] have presented good evidence for the existence of the 2,4-dinitrobenzenesulphenium ion ($\text{Ar}\overset{+}{\text{S}}$) in strongly acidic media; evidence has also been presented of a strong solvent effect upon the rate of reaction of 2,4-dinitrobenzenesulphenyl chloride and cyclohexene[167,168] and of a definite substituent effect in the reaction of this sulphenyl chloride with some substituted styrenes in acetic acid[167]. Such observations are entirely consistent with an electrophilic heterolytic addition mechanism involving attack by the sulphenyl chloride in the sense $\overset{\delta+}{\text{RS}}\text{–}\overset{\delta-}{\text{Cl}}$.

The kinetics of the reaction in acetic acid and in more polar solvents follows the simple law

$$-\text{d[RSCl]/d}t = k_2\text{[RSCl][olefin]} \tag{5.8}$$

but recently, Hogg and Campbell[170–172] have shown that quite complex kinetic forms can occur. Diverse, and often irreproducible, kinetic results are found when a polar reaction (*e.g.* halogen addition to olefins[169]) is studied in a solvent of low solvating power (*e.g.* benzene, carbon tetrachloride). This is usually attributed to the sensitivity of the reaction to traces of polar impurities (which may give rise to microscopically heterogenous conditions); an ionic transition state requires much stabilisation by solvation, and in the absence of a suitable medium the reagents, products, and impurities in the solvent may all be forced into this role. In carbon tetrachloride, the reaction between 2,4-dinitrobenzenesulphenyl bromide (shown to react by the same mechanism as the chloride in acetic acid[170]) and cyclohexene shows a marked dependence upon the concentration of organic bromide[171], *viz.*

$$\text{Rate} = k_2[\text{olefin}][\text{ArSBr}] + k_3[\text{olefin}][\text{ArSBr}]^2 + k_4[\text{olefin}][\text{ArSBr}]^3 \qquad (5.9)$$

Previously, a similar effect had been observed for the analogous reaction of the sulphenyl chloride, when, in the presence of excess of cyclohexene the rate is proportional to $[\text{ArSCl}]^2$, whereas in the presence of excess of sulphenyl chloride the kinetic order with respect to this reactant tended to unity[168].

In such a poorly solvating medium, other entities apart from the reactant could intervene. The kinetic equation then became more complex, and could be described by eqn. (5.10)

$$\text{Rate} = \Sigma k[\text{olefin}][\text{ArSBr}]^m[\text{X}]^n \qquad (5.10)$$

where X is the other solvating species, and m and n may have any integral value from 1 to 3 and from 0 to 2 respectively. The relative efficiencies of four such species, indicated in the term $k_6[\text{olefin}][\text{ArSBr}][\text{X}]^2$, was found to be 2,4-dinitroanisole > 2,4-dinitrophenyl 2-bromocyclohexyl sulphide (reaction product) > 2,4-dinitrobenzenesulphenyl chloride > 2,4-dinitrochlorobenzene[172]. This was interpreted as showing that solvation of the transition state was assisted by the *p*-nitro group since the order of efficiency of the catalysts fell in the sequence OMe > SR > SBr > Cl, the order of electron release by the substituents. The alternative hypothesis, that the reagent, product, or catalyst complexes with the olefin in a rapid pre-equilibrium sequence and that this complex is then more susceptible to attack by the reactant, *viz.*

$$\begin{aligned}&\text{ArSBr} + \text{olefin} \rightleftharpoons [\text{ArSBr.olefin}]\\&[\text{ArSBr.olefin}] + \text{ArSBr} \rightarrow \text{product} + \text{ArSBr}^*\end{aligned} \qquad (5.11)$$

does not seem to have been considered. An analogous suggestion has been made to explain high orders with respect to halogen in the addition of bromine to olefins[173].

5.4 ELECTROPHILIC NITROGEN

Electrophilic nitrogen, in the forms of NO_2^+, NO^+, or ArN_2^+, has been well authenticated in aromatic substitution. However, while there is evidence for electrophilic addition mechanisms involving nitrogen[174], very little kinetic work has been carried out on such systems. Of the two which have been studied, the addition of 2,4-*dinitrobenzenediazonium ion* to olefins seems to

*Other polynitrobenzenes may replace ArSBr in this scheme.

be well established. The reaction has been studied in acetonitrile, and involves addition of the diazonium ion across the olefinic bond to form (where distinction is possible) the more stable carbonium ion; this then undergoes hydrolytic cleavage to form a ketone and a 2,4-dinitrophenylhydrazone of another carbonyl fragment[175], *viz.*

$$ArN_2^+ + \,>C{=}C<\; \longrightarrow \;>C(N{=}NAr).C^+<$$

$$>C(N{=}NAr).C< \;\xrightarrow{H_2O}\; >C{=}N.NHAr + O{=}C<$$

The order of reactivity $Me_2C{=}CMe_2 > Me_2C{=}CHMe$ and the sequence $PhC(Me){=}CH_2 > PhCH{=}CHMe < PhCH{=}CH_2$ indicates that the rate-determining stage is assisted by electron-donating groups. The reaction is far more sensitive to steric effects around the reaction site than is hydration, oxidation with peracids, or chlorine addition; this may be rationalised by the greater size of the *ortho*-substituted diazonium ion, and is shown in the following table.

Olefin	$PhCH{=}CH_2$	$PhCH{=}CHPh$	$Ph_2C{=}CH_2$	$Me_2C{=}CHMe$
$k_{rel}MeCO_2OH$[a]	1.00	*cis* 1.0; *trans* 0.6	4.4	111
$k_{rel}ArN_2^+$[b]	1.00	*cis* 0.04; *trans* 0.2	20	140

[a] Peracetic acid rates measured in acetic acid at 25°C.
[b] 2,4-Dinitrophenyl diazonium ion rates as in ref. 176.

It is evident that substitution at the carbon atom undergoing attack is considerably diminished by a phenyl group, although this has the normal accelerating effect when steric hindrance is not serious. Solvent effects have not been studied in this system, nor have the effects of other diazonium ions been investigated kinetically[176].

The kinetics of addition of *nitrosyl chloride* to a number of olefins have been examined. In chloroform, methyl and phenyl substituents enhance the rate while electron-withdrawing groups (*e.g.* chlorine) retard it. Such behaviour would be consistent with electrophilic attack by the dipole $\overset{\delta+}{N}O{-}\overset{\delta-}{C}l$[177]. The effect of changes in the solvent was not simple, although more polar solvents (nitrobenzene, chloroform) speeded the reaction (compared with toluene or carbon tetrachloride). Methanol appeared to be a remarkably poor reaction medium, perhaps due to interaction with the reactant; the low reactivity of nitrosyl chloride in diethyl ether was attributed to strong bonding between the solvent and the incipient NO^+ cation.

While evidence has been presented of other electrophiles resulting from Group V elements[174], no kinetic studies appear to have been made.

5.5 ELECTROPHILIC BORON

The addition of boranes across unsaturated systems (olefins, acetylenes, carbonyl compounds, nitriles) proceeds in an "anti-Markowikow" sense, for the products of mild oxidation are not those of acid catalysed hydration of the olefin. Thus *p*-methoxystyrene gives 91% yields of 2-(*p*-methoxyphenyl)-ethanol from hydroboration, whereas acid-catalysed hydrolysis produces only 1-(*p*-methoxyphenyl)-ethanol[178]. Although such additions appear to involve orientation of boron to the more electron-rich carbon atom, it is not clear whether the rate-determining stage in the sequence is nucleophilic attack by hydride ion (as in the corresponding reaction of the aluminohydride ion) or electrophilic attack. In the one series of kinetic studies reported, Brown and Moerikofer[179], regarded the results as inconclusive. Certainly while there were some considerable variations in rate, the order of reactivity does not indicate an electrophilic reaction mechanism, and the mechanism of addition was suggested to be

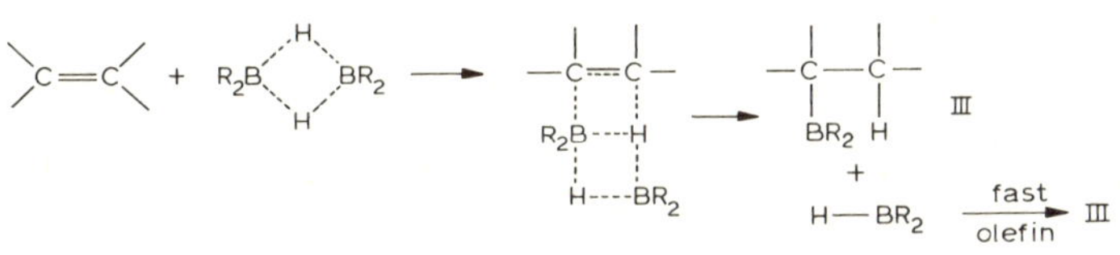

5.6 ELECTROPHILIC MERCURY

Electrophilic mercuration has been well established in aromatic substitution[180]. The corresponding mode of addition to olefins has been a source of continual controversy. While there is no doubt that a homolytic mechanism is involved in, for instance, the addition of mercuric acetate to cyclohexene in methanol, a reaction which is catalysed by peracetic acid[181], there seems no reason to discount heterolytic processes under more suitable conditions. The earlier literature has been critically reviewed and assessed on this point[182]. The first reaction product has been thought to involve a mercurinium ion (IV)[183].

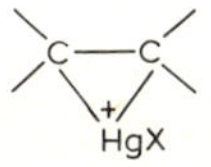

IV

Such a structure is analogous to those suggested for halogen addition, acid-catalysed hydration, and sulphenyl halide addition, but has suffered severe criticism. To follow the analogy, such an intermediate would be expected to encourage *trans* addition, but in the few cases where the stereochemistry of the addition has been identified (norbornene and related compounds) it is predominantly *cis*[184]. It may well be that two polar mechanisms can operate; in one, a four-centre transition state involving only the reagent and the olefin (V) would explain *cis* addition in sterically unfavourable environments, while the two-stage sequence involving a mercurinium ion or some kinetically equivalent species may be the more usual mode of reaction.

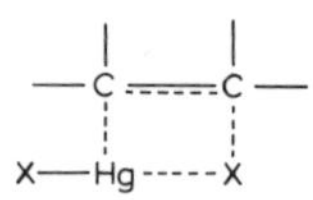

V

5.7 COINAGE METALS

While the existence of complexes between many heavy metals (Cu, Ag, Au, and the Pt series) and unsaturated hydrocarbons has been known for some years, and used to explain the increased aqueous solubility of olefins in the presence of these ions, relatively few measurements have been made on the position of the equilibrium (5.12) or the rate at which it is reached.

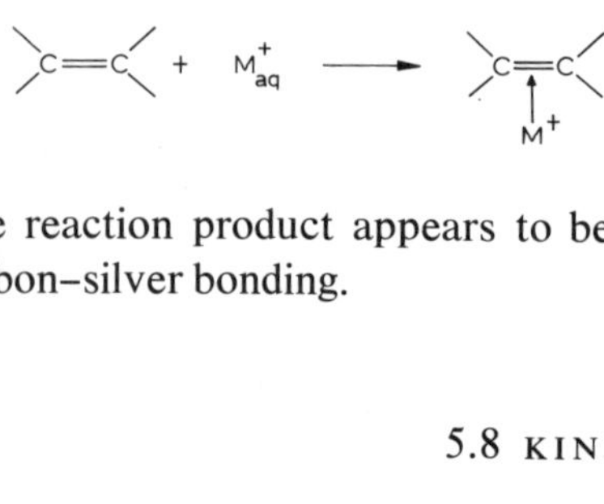

(5.12)

The reaction product appears to be a π-complex; there is no evidence[185] of carbon–silver bonding.

5.8 KINETIC DATA

The kinetic data for the electrophiles considered in this section are summarised in Table 6. Because of the diversity of electrophiles within this table, we have set out the electrophilic reagents in order of their position in the vertical rows of the Periodic Table, beginning at Group VI and proceeding back to Group I. Reactants having the same electrophilic atom are listed in order of increasing molecular weight (so that peracetic acid appears before perbenzoic acid). Within any one subdivision, the olefins are listed in order of increasing carbon content. It is assumed in the table that the rate coefficient (mole, l and sec units) refers to a second-order process, first-order with respect to olefin and to reagent, unless a specific designation (*e.g.* $k_1 = \cdots$) is used. Activation energies or enthalpies are listed in kcal.mole^{-1}; entropy units are cal. deg^{-1}.mole^{-1} (eu).

TABLE 6

ELECTROPHILIC ADDITION OF GROUPS I–VI ELEMENTS TO OLEFINS

Olefin	*k*	*Temp. (°C)*	*Solvent*		*Ref.*
Peracetic acid					
$H_2C{=}CH_2$	3.2×10^{-6}	25.8	HOAc		186
$MeCH{=}CH_2$[a]	$7{\cdot}0\times10^{-5}$	25.8	HOAc		186
$CH_2{=}CHCH_2OAc$	$5{\cdot}9\times10^{-7}$	24	EtOAc	$\Delta H^{\ddagger}=24.0$	188
	1.3×10^{-6}	30	EtOAc	$\Delta S^{\ddagger}=-6$	188
	9.1×10^{-6}	45	EtOAc		188
	2.7×10^{-6}	23	HOAc	$\Delta H^{\ddagger}=19.5$	188
	5.7×10^{-6}	30	HOAc	$\Delta S^{\ddagger}=-18$	188
	2.43×10^{-5}	45	HOAc		188
	3.4×10^{-6}	23	PhH	$\Delta H^{\ddagger}=18.8$	188
	3.1×10^{-5}	45	PhH	$\Delta S^{\ddagger}=-20$	188
$(CH_2{=}CHCH_2)_2O$	3.1×10^{-6}	20	EtOAc		188
	5.3×10^{-6}	25	EtOAc	$\Delta H^{\ddagger}=18.0$	
	8.6×10^{-6}	30	EtOAc	$\Delta S^{\ddagger}=-22$	
	1.4×10^{-5}	35	EtOAc		188
	5.5×10^{-5}	50	EtOAc		188
	3.8×10^{-5}	25	PhH	$\Delta H^{\ddagger}=15$	188
	8.9×10^{-5}	35	PhH	$\Delta S^{\ddagger}=-28$	188
$CH_2{=}CHCH_2O$-*n*-Pr	4.5×10^{-6}	30	EtOAc		188
	2.7×10^{-5}	50	EtOAc		188
Cyclobutene	3.3×10^{-4}	25.8	HOAc	$\Delta H^{\ddagger}=14.4$	189, 190
	1.0×10^{-3}	40.1	HOAc	$\Delta S^{\ddagger}=-26$	189, 190
2-Butene	1.55×10^{-3}	25.8	HOAc		186
2-Me-propene	1.63×10^{-3}	25.8	HOAc	$\Delta H^{\ddagger}=15.7$	186, 190
	5.00×10^{-3}	39.9	HOAc	$\Delta S^{\ddagger}=-19$	190
$MeCH{=}CHCO_2Et$	1.5×10^{-7}	25	EtOAc		188
	2.6×10^{-7}	30	EtOAc		188
	7.9×10^{-7}	40	EtOAc		188
	5.3×10^{-6}	60	EtOAc		188
Cyclopentene	3.1×10^{-3}	25.8	HOAc	$\Delta H^{\ddagger}=14.1$	186
	8.8×10^{-3}	39.6	HOAc	$\Delta S^{\ddagger}=-23$	190
2-Pentene	1.58×10^{-3}	25.8	HOAc	$\Delta H^{\ddagger}=15.8$	190
	5.15×10^{-3}	40.0	HOAc	$\Delta S^{\ddagger}=-19$	190
1-Pentene	7.2×10^{-5}	25.8	HOAc	$\Delta H^{\ddagger}=17.5$	186, 190
	2.95×10^{-4}	40.9	HOAc	$\Delta S^{\ddagger}=-19$	190
2-Me-2-butene	1.63×10^{-2}	25.8	HOAc	$\Delta H^{\ddagger}=13.8$	190
	5.00×10^{-2}	40.8	HOAc	$\Delta S^{\ddagger}=-21$	190
Cyclohexene	2.0×10^{-5}	0	EtOAc		188
	1.04×10^{-4}	15	EtOAc	$\Delta H^{\ddagger}=16.9$	188
	1.72×10^{-4}	20	EtOAc	$\Delta S^{\ddagger}=-16$	188
	2.83×10^{-4}	25	EtOAc		188
	8.2×10^{-4}	15	HOAc	$\Delta H^{\ddagger}=15.5$	188, 190
	2.15×10^{-3}	25.8	HOAc	$\Delta S^{\ddagger}=-19$	190
	6.74×10^{-3}	39.6	HOAc		190
1-Me-cyclopentene	3.70×10^{-2}	25.8	HOAc	$\Delta H^{\ddagger}=13.6$	190
	1.11×10^{-1}	40.9	HOAc	$\Delta S^{\ddagger}=-20$	190
3-(EtO_2C)-cyclohexene	1.23×10^{-4}	25	EtOAc	$\Delta H^{\ddagger}=16.9$	188
	3.13×10^{-4}	35	EtOAc	$\Delta S^{\ddagger}=-20$	188
	3.45×10^{-4}	15	HOAc		188
Cycloheptene	2.95×10^{-3}	25.8	HOAc	$\Delta H^{\ddagger}=16.8$	190
	1.02×10^{-2}	39.8	HOAc	$\Delta S^{\ddagger}=-14$	190

TABLE 6 (*Contd.*)

Olefin	*k*	*Temp.* (*°C*)	*Solvent*		*Ref.*
3-Heptene	1.84×10^{-3}	25.8	HOAc		191
1-Octene	8.2×10^{-5}	25.0	HOAc		192
Styrene	1.9×10^{-4}	25.8	HOAc	$\Delta H^{\ddagger}=13.8$	188, 190
	5.73×10^{-4}	40.8	HOAc	$\Delta S^{\ddagger}=-30$	190
	1.24×10^{-4}	20.5	HOAc		188
	2.2×10^{-5}	20	EtOAc		188
	3.4×10^{-5}	25	EtOAc	$\Delta H^{\ddagger}=16.2$	188
	5.5×10^{-5}	30	EtOAc	$\Delta S^{\ddagger}=-23$	188
	1.27×10^{-4}	40	EtOAc		188
	2.7×10^{-5}	0	PhH	$\Delta H^{\ddagger}=14.3$	188
	1.62×10^{-4}	20	PhH	$\Delta S^{\ddagger}=-27$	188
n-PrCH=C(Et)CO_2Et	1.9×10^{-5}	40	EtOAc	$\Delta H^{\ddagger}=18.8$	188
	1.18×10^{-4}	60	EtOAc	$\Delta S^{\ddagger}=-19$	188
PhCH=CHMe	7.7×10^{-4}	25.8	HOAc	$\Delta H^{\ddagger}=15.9$	190
	2.76×10^{-3}	40.8	HOAc	$\Delta S^{\ddagger}=-20$	190
$PhCH_2CH=CH_2$	3.2×10^{-5}	25.8	HOAc	$\Delta H^{\ddagger}=17.9$	190
	1.23×10^{-4}	39.6	HOAc	$\Delta S^{\ddagger}=-20$	190
Indene	7.8×10^{-4}	25.8	HOAc	$\Delta H^{\ddagger}=15.6$	190
	2.76×10^{-3}	41.0	HOAc	$\Delta S^{\ddagger}=-21$	190
4-Nonene	1.75×10^{-3}	25.8	HOAc		191
1-Decene	7.7×10^{-5}	25.0	HOAc		192
$H_2C=CH(CH_2)_8CO_2Me$	6.8×10^{-5}	25.0	HOAc		192
1,2-Dihydronaphthalene	1.93×10^{-3}	16.3	HOAc	$\Delta H^{\ddagger}=12.7$	190
	3.92×10^{-3}	25.8	HOAc	$\Delta S^{\ddagger}=-27$	190
	1.09×10^{-2}	40.9	HOAc		190
1,4-Dihydronaphthalene	6.2×10^{-4}	25.8	HOAc	$\Delta H^{\ddagger}=18.3$	190
	2.7×10^{-3}	40.9	HOAc	$\Delta S^{\ddagger}=-12$	190
Methylindene	1.0×10^{-2}	26.1	HOAc		190
$Ph_2C=CH_2$	8.0×10^{-4}	25.8	HOAc	$\Delta H^{\ddagger}=16.3$	190
	2.7×10^{-3}	39.8	HOAc	$\Delta S^{\ddagger}=-18$	190
PhCH=CHPh (*trans*)	8.5×10^{-5}	25.8	HOAc	$\Delta H^{\ddagger}=16.0$	190
	2.8×10^{-4}	39.6	HOAc	$\Delta S^{\ddagger}=-24$	190
PhCH=CHPh (*cis*)	1.85×10^{-4}	25.8	HOAc		190
Oleic acid[b]	6.0×10^{-5}	18	HOAc		193
Elaidic acid[b]	5.5×10^{-5}	18	HOAc		193
Ricinoleic acid[b]	4.3×10^{-5}	18	HOAc		193
Ricinelaidic acid[b]	2.7×10^{-5}	18	HOAc		193
$Ph_2C=CHPh$	9.6×10^{-5}	25.8	HOAc	$\Delta H^{\ddagger}=15.1$	190
	3.24×10^{-4}	41.2	HOAc	$\Delta S^{\ddagger}=-27$	190
3α,20β-Diacetoxy-17(20)-pregnen-11-one	6.4×10^{-4}	27	PhH		195
3β,20-Diacetoxy-5α,6β-dichloro-17(20)-pregnene	1.13×10^{-3}	27	PhH		195
3a,11,20-Triacetoxy-9(11), 17(20)-pregnadiene	7.0×10^{-4}	27	PhH		195
Perbenzoic acid					
$CH_2=CHCH_2Br$	1.73×10^{-5}	25	PhH		196
$CH_2=CHCH_2Cl$	1.71×10^{-5}	25	PhH		196
$CH_2=CHCH_2OH$	5.05×10^{-4}	25	PhH		196
$CH_2=CHCH_2CN$	4.4×10^{-6}	25	PhH		196
$CH_2=CHCH_2OPh$	7.94×10^{-5}	25	PhH		196

TABLE 6 (*Contd.*)

Olefin	*k*	*Temp. (°C)*	*Solvent*		*Ref.*
CH_2=$CHCH_2OPr$-*i*	1.65×10^{-4}	25	PhH		196
$BrCH_2CH$=$CHCH_2Br$	7.0×10^{-5}	25	PhH		196
$BrCH_2CH$=$CHCH_2OPh$	7.5×10^{-5}	25	PhH		196
$PhOCH_2CH$=$CHCH_2OPh$	1.27×10^{-4}	25	PhH		196
Sorbic acid and esters	3×10^{-5}	0	$CHCl_3$	$\Delta H^{\ddagger} = 12$	197
	2×10^{-4}	20	$CHCl_3$	$\Delta S^{\ddagger} = -43$	197
Sorbic acid chloride	2.3×10^{-4}	0	$CHCl_3$	$\Delta H^{\ddagger} = 15$	197
	1.5×10^{-3}	20	$CHCl_3$	$\Delta S^{\ddagger} = -32$	197
Cyclopentene	0.104	25	$CHCl_3$		198
Cyclohexene	4.73×10^{-2}	25	$CHCl_3$		198
	1.27×10^{-2}	20	PhH		155
	1.81×10^{-2}	25	PhH	$\Delta H^{\ddagger} = 12.7$	155
	2.63×10^{-2}	30	PhH	$\Delta S^{\ddagger} = -24$	155
	3.62×10^{-2}	35	PhH		155
	2.23×10^{-4}	15	Et_2O		155
	3.74×10^{-4}	20	Et_2O	$\Delta H^{\ddagger} = 15.3$	155
	5.84×10^{-4}	25	Et_2O	$\Delta S^{\ddagger} = -22$	155
	7.50×10^{-4}	28	Et_2O		155
1-Heptene	5.1×10^{-4}	25	PhH		196
$Ph(CH_2)_2CH$=CH_2	1.5×10^{-4}	15	$CHCl_3$		199
$PhCH_2CH$=$CHMe$	1.7×10^{-4}	15	$CHCl_3$		199
PhCH=CHEt	1.4×10^{-3}	15	$CHCl_3$		199
Eugenol	3.7×10^{-5}	0	$CHCl_3$		200
isoEugenol	2.1×10^{-3}	0	$CHCl_3$		200
Safrole	2.2×10^{-5}	0	$CHCl_3$		200
isoSafrole	2.47×10^{-3}	0	$CHCl_3$		200
Cinnamic acid and esters	2.1×10^{-6}	20	$CHCl_3$		201
Cinnamyl alcohol	3.38×10^{-3}	20	$CHCl_3$		201
Cinnamaldehyde	7.8×10^{-5}	20	$CHCl_3$		201
PhCH=CHCOMe	4.1×10^{-5}	20	$CHCl_3$		201
4-OH-norbornene	2.02×10^{-3}	5	PhH		202
4-OMe-norbornene	4.75×10^{-3}	5	PhH		202
4-OAc-norbornene	9.9×10^{-4}	5	PhH		202
4-CH_2OH-norbornene	9.5×10^{-2}	5	PhH		202
4-CH_2OMe-norbornene	2.88×10^{-3}	5	PhH		202
4-CH_2OAc-norbornene	2.11×10^{-3}	5	PhH		202
Styrene	6.16×10^{-4}	20	PhH		155
	9.65×10^{-4}	25	PhH	$\Delta H^{\ddagger} = 14.2$	155
	1.42×10^{-3}	30	PhH	$\Delta S^{\ddagger} = -25$	155
	2.13×10^{-3}	35	PhH		155
Pinene	1.8×10^{-2}	0	$CHCl_3$		200
Camphene	1.55×10^{-3}	0	$CHCl_3$		200
8-Menthene-1,2-diol	5.8×10^{-4}	0	$CHCl_3$		200
n-1-Undecene	2.55×10^{-4}	20	CCl_4	$\Delta H^{\ddagger} = 14.8$	203
	5.53×10^{-4}	30	CCl_4	$\Delta S^{\ddagger} = -25$	203
	8.81×10^{-4}	35	CCl_4		203
trans-*n*-2-Undecene	4.02×10^{-3}	20	CCl_4	$\Delta H^{\ddagger} = 14.6$	203
	8.48×10^{-3}	30	CCl_4	$\Delta S^{\ddagger} = -20$	203
	1.38×10^{-2}	35	CCl_4		203
trans-3-Undecene	5.10×10^{-3}	20	CCl_4	$\Delta H^{\ddagger} = 14.6$	203
	1.12×10^{-2}	30	CCl_4	$\Delta S^{\ddagger} = -19$	203
	1.82×10^{-2}	35	CCl_4		203

TABLE 6 (*Contd.*)

Olefin	*k*	*Temp. (°C)*	*Solvent*		*Ref.*
trans-4-Undecene	4.28×10^{-3}	20	CCl_4	$\Delta H^{\ddagger} = 14.6$, $\Delta S^{\ddagger} = -20$	203
	9.43×10^{-3}	30	CCl_4		203
	1.49×10^{-2}	35	CCl_4		203
trans-5-Undecene	4.31×10^{-3}	20	CCl_4	$\Delta H^{\ddagger} = 14.6$, $\Delta S^{\ddagger} = -19$	203
	9.33×10^{-3}	30	CCl_4		203
	1.54×10^{-2}	35	CCl_4		203
cis-*n*-2-Undecene	5.43×10^{-3}	20	CCl_4	$\Delta H^{\ddagger} = 14.6$, $\Delta S^{\ddagger} = -19$	203
	1.22×10^{-2}	30	CCl_4		203
	1.96×10^{-2}	35	CCl_4		203
cis-*n*-3-Undecene	6.31×10^{-3}	20	CCl_4	$\Delta H^{\ddagger} = 14.6$, $\Delta S^{\ddagger} = -19$	203
	1.38×10^{-2}	30	CCl_4		203
	2.30×10^{-2}	35	CCl_4		203
cis-*n*-4-Undecene	6.03×10^{-3}	20	CCl_4	$\Delta H^{\ddagger} = 14.6$, $\Delta S^{\ddagger} = -19$	203
	13.2×10^{-3}	30	CCl_4		203
	22.0×10^{-3}	35	CCl_4		203
cis-*n*-5-Undecene	6.22×10^{-3}	20	CCl_4	$\Delta H^{\ddagger} = 14.6$, $\Delta S^{\ddagger} = -19$	203
	1.33×10^{-2}	30	CCl_4		203
	2.23×10^{-2}	35	CCl_4		203
Undecylenic acid	2.67×10^{-4}	0	$CHCl_3$		204
Stilbene and substituted stilbenes X = H (*trans*)	2.70×10^{-4}	20	PhH	$\Delta H^{\ddagger} = 15.3$, $\Delta S^{\ddagger} = -23$	155
	4.27×10^{-4}	25[c]	PhH		155
	4.19×10^{-4}	25	PhH		205
	6.64×10^{-4}	30	PhH		155
	10.1×10^{-4}	35	PhH		155
	1.7×10^{-5}	15	Et_2O	$\Delta H^{\ddagger} = 17.9$, $\Delta S^{\ddagger} = -18$	155
	3.0×10^{-5}	20	Et_2O		155
	5.0×10^{-5}	25	Et_2O		155
	6.7×10^{-5}	28	Et_2O		155
X = H (*cis*)	5.75×10^{-4}	20	PhH	$\Delta H^{\ddagger} = 14.1$, $\Delta S^{\ddagger} = -25$	155
	8.8×10^{-4}	25	PhH		155
	1.34×10^{-3}	30	PhH		155
	1.97×10^{-3}	35	PhH		155
X = *p*,*p*′-(OMe)$_2$-(*trans*)	8.33×10^{-3}	25	PhH	$\Delta H^{\ddagger} = 14.0$, $\Delta S^{\ddagger} = -21$	155
	8.14×10^{-3}	25	PhH		205
	1.28×10^{-2}	30	PhH		155
	1.81×10^{-2}	35	PhH		155
	2.73×10^{-2}	40	PhH		155
X = *p*-OMe (*trans*)	1.31×10^{-3}	20	PhH	$\Delta H^{\ddagger} = 14.5$, $\Delta S^{\ddagger} = -22$	155
	2.02×10^{-3}	25	PhH		155
	1.95×10^{-3}	25	PhH		205
	3.13×10^{-3}	30	PhH		155
	4.80×10^{-3}	35	PhH		155
X = *p*-OMe (*cis*)	2.3×10^{-3}	25	PhH		205
X = *p*-Me (*trans*)	6.17×10^{-4}	20	PhH	$\Delta H^{\ddagger} = 15.0$, $\Delta S^{\ddagger} = -23$	155
	9.70×10^{-4}	25	PhH		155
	8.6×10^{-4}	25	PhH		205
	1.49×10^{-3}	30	PhH		155
	2.29×10^{-3}	35	PhH		155
X = *m*-Me (*trans*)	3.05×10^{-4}	20	PhH	$\Delta H^{\ddagger} = 15.2$, $\Delta S^{\ddagger} = -23$	155
	4.83×10^{-4}	25	PhH		155
	7.46×10^{-4}	30	PhH		155
	1.16×10^{-3}	35	PhH		155

TABLE 6 (*Contd.*)

Olefin	*k*	*Temp. (°C)*	*Solvent*		*Ref.*
X = *p*-Cl (*trans*)	1.71×10^{-4}	20	PhH		155
	2.69×10^{-4}	25	PhH	$\Delta H^{\ddagger} = 15.6$	155
	4.28×10^{-4}	30	PhH	$\Delta S^{\ddagger} = -23$	155
	6.54×10^{-4}	35	PhH		155
X = *m*-Cl (*trans*)	1.05×10^{-4}	20	PhH		155
	1.77×10^{-4}	25	PhH	$\Delta H^{\ddagger} = 16.0$	155
	2.76×10^{-4}	30	PhH	$\Delta S^{\ddagger} = -22$	155
	4.28×10^{-4}	35	PhH		155
X = *m*-NO_2 (*trans*)	7.3×10^{-5}	25	PhH		155
	1.14×10^{-4}	30	PhH	$\Delta H^{\ddagger} = 16.4$	155
	1.70×10^{-4}	35	PhH	$\Delta S^{\ddagger} = -23$	155
	2.65×10^{-4}	40	PhH		155
X = *p*-NO_2 (*trans*)	6.3×10^{-5}	25	PhH		155
	9.8×10^{-5}	30	PhH	$\Delta H^{\ddagger} = 16.7$	155
	1.56×10^{-4}	35	PhH	$\Delta S^{\ddagger} = -22$	155
	2.40×10^{-4}	40	PhH		155
$PhCH_2CH{=}CHCH_2Ph$	1.89×10^{-3}	25	PhH		196
Oleic acid	6.4×10^{-3}	0	$CHCl_3$		204
$Ph_2C{=}CHPh$	2.06×10^{-4}	25	PhH		155
	3.08×10^{-4}	30	PhH	$\Delta H^{\ddagger} = 15.7$	155
	5.01×10^{-4}	35	PhH	$\Delta S^{\ddagger} = -23$	155
	7.2×10^{-4}	40	PhH		155
$Ph_2C{=}CPh_2$	2.8×10^{-5}	30	PhH		155
	4.4×10^{-5}	35	PhH	$\Delta H^{\ddagger} = 17.2$ $\Delta S^{\ddagger} = -23$	155
	7.9×10^{-5}	40	PhH		155
3β-Acetoxycholest-2-ene	3.65×10^{-2}	25	$CHCl_3$		197
3β-Acetoxycholest-4-ene	5.34×10^{-2}	25	$CHCl_3$		197
3β-Acetoxycholest-5-ene	3.93×10^{-2}	25	$CHCl_3$		197
Cholest-5-ene	3.97×10^{-2}	25	$CHCl_3$		197
3β-Acetoxycholest-7-ene	1.275×10^{-1}	25	$CHCl_3$		197
3β-Acetoxyergost-8(14)-ene	1.359×10^{-1}	25	$CHCl_3$		197
3β-Acetoxyergost-14-ene	1.477×10^{-1}	25	$CHCl_3$		197
The original reports for the following were not available for more complete description					
α,α′-Me_2-stilbenes			PhH		206
Styrenes			PhH	$\rho = -1.3$	207
Substituted perbenzoic acids					
X = *p*-OMe					
Stilbene	8.1×10^{-5}	20	PhH		155
	1.33×10^{-4}	25	PhH	$\Delta H^{\ddagger} = 16.1$	155
	2.16×10^{-4}	30	PhH	$\Delta S^{\ddagger} = -23$	155
	3.45×10^{-4}	35	PhH		155
X = *p*-Me					
Stilbene	2.38×10^{-4}	25	PhH		155
	3.76×10^{-4}	30	PhH	$\Delta H^{\ddagger} = 15.6$	155
	5.65×10^{-4}	35	PhH	$\Delta S^{\ddagger} = -23$	155
	8.50×10^{-4}	40	PhH		155
X = *p*-Cl					
Stilbene	6.7×10^{-4}	20	PhH		155
	1.04×10^{-3}	25	PhH	$\Delta H^{\ddagger} = 14.9$	155
	1.56×10^{-3}	30	PhH	$\Delta S^{\ddagger} = -22$	155
	2.41×10^{-3}	35	PhH		155

TABLE 6 (*Contd.*)

Olefin	*k*	*Temp.* (°C)	*Solvent*	*Ref.*
X = *m*-Cl				
Ethyl crotonate	7.5×10^{-7}	30	5%Ac_2O/AcOH	208
	5.3×10^{-6}	30	10%H_2O/AcOH	208
	1.93×10^{-6}	25	EtOAc	208
	1.15×10^{-5}	25	CCl_4, CH_2Cl_2, or $(CH_2Cl)_2$	208
	1.73×10^{-5}	30	$(CH_2Cl)_2$	208
	1.73×10^{-5}	30	PhH	208
	$k_0 = 6.3 \times 10^{-7}$	25	MeCN[d]	208
	6.9×10^{-6}	50	MeCN[d]	208
Cyclohexene and substituted cyclohexenes				
X = H	1.88×10^{-3}	25	Et_2O	209
X = 4,5-Me_2 (*trans*)	1.49×10^{-3}	25	Et_2O	209
X = 4,5-Me_2 (*cis*)	8.7×10^{-4}	25	Et_2O	209
X = 3,3-Me_2	8.3×10^{-4}	25	Et_2O	209
X = 3-Me	1.77×10^{-3}	25	Et_2O	209
X = 4-Me	1.52×10^{-3}	25	Et_2O	209
X = 4-*t*-Bu	1.76×10^{-3}	25	Et_2O	209
X = 4,5-(CH_2OCH_2)-*trans*	5.5×10^{-4}	25	Et_2O	209
X = 4,5-(CH_2OCH_2)-*cis*	9.4×10^{-4}	25	Et_2O	209
cis-1,2,3,6-tetrahydrophthalic anhydride	5.0×10^{-3}	25	$CHCl_3$	210

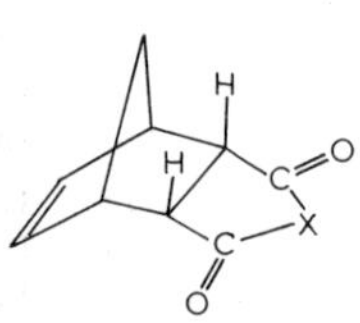

VI

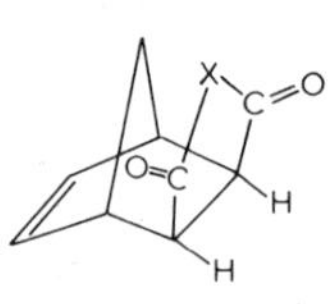

VII

Olefin	*k*	*Temp.* (°C)	*Solvent*	*Ref.*
Bicyclo[2,2,1]-hept-5-ene-*endo*-2,3,-dicarboxylic acid anhydride [(VI), X = O]	6.1×10^{-5}	25	$CHCl_5$	210
Bicyclo[2,2,1]-hept-5-ene-*exo*-2,3-dicarboxylic acid anhydride [(VII), X = O]	1.97×10^{-4}	25	$CHCl_3$	210
Bicyclo[2,2,1]-hept-5-ene-*endo*-2,3-dicarboxylic acid-*N*-(2-phenylethyl)-imide [(VI), X = NCH_2CH_2Ph]	4.6×10^{-4}	25	$CHCl_3$	210
Bicyclo[2,2,1]-hept-5-ene-*exo*-2,3-dicarboxylic acid *N*-(2-phenylethyl)-imide [(VII), X = NCH_2CH_2Ph]	1.13×10^{-3}	25	$CHCl_3$	210
PhCH=CHPh (*trans*)	1.46×10^{-3}	30	CCl_4	208
	3.60×10^{-3}	30	$CHCl_3 + 2\%\ CF_3COOH$	208
	3.61×10^{-3}	30	CH_2Cl_2	208
	3.27×10^{-3}	30	$(CH_2Cl)_2$	208
	4.2×10^{-5}	30	*t*-BuOH	208
	1.89×10^{-3}	30	PhH	208
	2.73×10^{-3}	30	PhCl	208
	3.63×10^{-3}	30	o-$C_6H_4Cl_2$	208

TABLE 6 (*Contd.*)

Olefin	*k*	*Temp. (°C)*	*Solvent*		*Ref.*
X = *m*-NO_2					
Stilbene	2.68×10^{-3}	20	PhH		155
	3.95×10^{-3}	25	PhH	$\Delta H^{\ddagger} = 13.9$	155
	5.90×10^{-3}	30	PhH	$\Delta S^{\ddagger} = -23$	155
	8.90×10^{-3}	35	PhH		155
X = *p*-NO_2					
Stilbene	2.93×10^{-3}	20	PhH		155
	4.42×10^{-3}	25	PhH	$\Delta H^{\ddagger} = 13.8$	155
	6.65×10^{-3}	30	PhH	$\Delta S^{\ddagger} = -23$	155
	9.78×10^{-3}	35	PhH		155
p-NO_2-stilbene	3.40×10^{-4}	20	PhH		155
	5.36×10^{-4}	25	PhH	$\Delta H^{\ddagger} = 15.1$	155
	8.27×10^{-4}	30	PhH	$\Delta S^{\ddagger} = -23$	155
	1.25×10^{-3}	35	PhH		155
Monoperphthalic acid					
1,4-Cyclohexadiene	4.1×10^{-5}	15	Et_2O		211
1,2-Dimethyl-1,4-cyclohexadiene					
	1.47×10^{-3}	0	Et_2O	$\Delta H^{\ddagger} = 15.1$	211
	4.10×10^{-3}	10	Et_2O	$\Delta S^{\ddagger} = -16$	211
	6.30×10^{-3}	15	Et_2O		211
4,7-Dihydroindane	2.76×10^{-3}	0	Et_2O		211
	6.70×10^{-3}	10	Et_2O	$\Delta H^{\ddagger} = 14.6$	211
	1.03×10^{-2}	15	Et_2O	$\Delta S^{\ddagger} = -17$	211
1,2,3,4,5,8-Hexahydronaphthalene					
	1.43×10^{-3}	0	Et_2O	$\Delta H^{\ddagger} = 15.1$	211
	3.85×10^{-3}	10	Et_2O	$\Delta S^{\ddagger} = -16$	211
	6.15×10^{-3}	15	Et_2O		211
4a,8a-Epoxy-1,2,3,4,5,8-hexahydronaphthalene					
	5.3×10^{-5}	15	Et_2O		211
MeCH=CH(C≡C)$_2$CH=CHMe (*cis* or *trans*)					
	3.6×10^{-4}	25	$CHCl_3$		212
MeCHCH(C≡C)$_2$CH=CHMe (*cis* or *trans*)					
O (epoxide)	9.1×10^{-5}	25	$CHCl_3$		212
Perlauric acid					
Vinyl laurate	2.0×10^{-4}	25	PhH		213
Vinyl oleate	4.50×10^{-3}	25	PhH		213
Methyl oleate	3.86×10^{-3}	25	PhH		213
Ozone					
PhH	3.23×10^{-4}	−40.0	$CHCl_3$		214
	4.08×10^{-4}	−38.0	$CHCl_3$		214
	5.62×10^{-4}	−35.2	$CHCl_3$		214
	8.34×10^{-4}	−31.0	$CHCl_3$	$\Delta H^{\ddagger} = 11.9$	214
	1.20×10^{-3}	−28.0	$CHCl_3$		214
	1.42×10^{-3}	−26.0	$CHCl_3$		214
	1.65×10^{-3}	−25.0	$CHCl_3$		214
	3.50×10^{-3}	−28.0	$MeNO_2$		214
PhMe	2.67×10^{-3}	−40.0	$CHCl_3$		215
	6.50×10^{-3}	−31.2	$CHCl_3$		215
	1.75×10^{-2}	−20.0	$CHCl_3$		215

References pp. 80–86

TABLE 6 (*Contd.*)

Olefin	k	*Temp.* (°C)	*Solvent*	*Ref.*
o-Xylene	1.07×10^{-2}	−40.0	$CHCl_3$	215
m-Xylene	1.15×10^{-2}	−40.0	$CHCl_3$	215
p-Xylene	1.75×10^{-2}	−40.0	$CHCl_3$	215
PhEt	9.00×10^{-3}	−31.2	$CHCl_3$	215
Ph.*n*-Pr	8.33×10^{-3}	−31.2	$CHCl_3$	215
Ph.*i*-Pr	2×10^{-2}	−31.2	$CHCl_3$	215
Ph.*n*-Bu	8.50×10^{-3}	−31.2	$CHCl_3$	215
Ph.*s*-Bu	8.84×10^{-3}	−31.2	$CHCl_3$	215
Ph.*t*-Bu	2.67×10^{-3}	−31.2	$CHCl_3$	215
$PhCH_2Cl$	8.50×10^{-4}	−31.2	$CHCl_3$	215
	2.50×10^{-3}	−20.0	$CHCl_3$	215
$PhCHCl_2$	1.83×10^{-4}	−31.2	$CHCl_3$	215
	5.8×10^{-4}	−20.0	$CHCl_3$	215
$PhCCl_3$	5.8×10^{-5}	−31.2	$CHCl_3$	215
	2.0×10^{-4}	−20.0	$CHCl_3$	215
PhF	2.67×10^{-4}	−31.2	$CHCl_3$	215
	8.17×10^{-4}	−20.0	$CHCl_3$	215
PhCl	1.67×10^{-4}	−31.2	$CHCl_3$	215
	6.0×10^{-4}	−20.0	$CHCl_3$	215
	1.05×10^{-3}	−28.0	$MeNO_2$	215
PhOMe	$>100\times10^{-3}$	−31.2	$CHCl_3$	215
$PhCO_2Et$	1.67×10^{-4}	−31.2	$CHCl_3$	215
$PhCH_2CO_2Et$	1.18×10^{-3}	−31.2	$CHCl_3$	215
Pyridine	2.17×10^{-4}	−31.2	$CHCl_3$	216
	8.17×10^{-4}	−20.0	$CHCl_3$	216
2-Me-pyridine	1.83×10^{-3}	−31.2	$CHCl_3$	216
3-Me-pyridine	2.08×10^{-3}	−31.2	$CHCl_3$	216
4-Me-pyridine	1.18×10^{-3}	−31.2	$CHCl_3$	216
4-Et-pyridine	1.30×10^{-3}	−31.2	$CHCl_3$	216
2,6-Me_2-pyridine	*ca.* 1.3×10^{-2}	−31.2	$CHCl_3$	216
2,4,6-Me_3-pyridine	*ca.* 1.0×10^{-2}	−31.2	$CHCl_3$	216
2,4-Dinitrobenzene sulphenyl chloride				
Vinyl acetate	2.94×10^{-4}	35	HOAc	217
Cyclohexene	4.17×10^{-3}	15.3	HOAc } $\Delta H^{\ddagger}=11.1$, $\Delta S^{\ddagger}=-33$	168
	4.89×10^{-3}	17.6	HOAc	168
	7.84×10^{-3}	25.0	HOAc	168
	1.48×10^{-2}	35.2	HOAc	168
	1.1×10^{-5}	35.2	CCl_4	168
	6.5×10^{-5}	35.2	$CHCl_3$	168
	1.46×10^{-2}	35.2	$(CH_2Cl)_2$	168
	3.02×10^{-2}	35.2	$PhNO_2$	168
	1.16×10^{-2}	30.1	HOAc	218
4-Me-cyclohexene	9.42×10^{-3}	30.1	HOAc	218
4,5-Me_2-cyclohexene	4.52×10^{-3}	30.1	HOAc	218
4-*t*-Bu-cyclohexene	10.4×10^{-3}	30.1	HOAc	218
4-OMe-cyclohexene	2.21×10^{-3}	30.1	HOAc	218
4-Br-cyclohexene	5.36×10^{-4}	30.1	HOAc	218
4-Cyano-cyclohexene	1.23×10^{-4}	30.1	HOAc	218
4-Carboxycyclohexene	1.13×10^{-3}	30.1	HOAc	218
4,5-Dicarboxycyclohexene (4-cyclohexenyl-1,2-dicarboxylic acid)	1.89×10^{-4}	30.1	HOAc	218

TABLE 6 (*Contd.*)

Olefin	*k*	*Temp. (°C)*	*Solvent*		*Ref.*
4-Cyclohexenyl-1,2-dicarboxylic acid anhydride	3.58×10^{-5}	30.1	HOAc		218
4-Carbomethoxycyclohexene	1.30×10^{-3}	30.1	HOAc		218
4,5-Dicarbomethoxycyclohexene	3.03×10^{-4}	30.1	HOAc		218
Norbornene	0.150	30.1	HOAc		218
exo-5-Me-norbornene	0.131	30.1	HOAc		218
endo-5-Me-norbornene	9.97×10^{-2}	30.1	HOAc		218
exo-5,6-$(CH_2)_3$-norbornene (*cis*)	0.208	30.1	HOAc		218
exo-5,6-$(CO_2H)_2$-norbornene (*cis*)	9.55×10^{-4}	30.1	HOAc		218
endo-5,6-$(CO_2H)_2$-norbornene (*cis*)	2.25×10^{-3}	30.1	HOAc		218
exo-5,6-(-CO.O.CO-)-norbornene (*cis*)	7.52×10^{-5}	30.1	HOAc		218
endo-5,6-(CO.O.CO)-norbornene (*cis*)	1.34×10^{-6}	30.1	HOAc		218
Styrene and substituted styrenes					
X = H	7.37×10^{-4}	25.0	HOAc	$\Delta H^{\ddagger} = 12.9$, $\Delta S^{\ddagger} = -32$	167
	1.48×10^{-3}	35.0	HOAc		167
	2.90×10^{-3}	45.0	HOAc		167
X = *p*-OMe	$\geqslant 0.5$	25.0	HOAc		167
X = *p*-Me	1.69×10^{-3}	20.0	HOAc	$\Delta H^{\ddagger} = 12.0$, $\Delta S^{\ddagger} = -32$	167
	2.43×10^{-3}	25.0	HOAc		167
	3.61×10^{-3}	30.6	HOAc		167
	4.67×10^{-3}	35.0	HOAc		167
	9.43×10^{-3}	44.9	HOAc		167
X = *p*-Cl	4.22×10^{-4}	34.6	HOAc	$\Delta H^{\ddagger} = 15.9$, $\Delta S^{\ddagger} = -24$	167
	4.32×10^{-4}	35.0	HOAc		167
	9.73×10^{-4}	45.0	HOAc		167
X = *p*-NO_2	2.67×10^{-5}	34.0	HOAc	$\Delta H^{\ddagger} = 16.9$, $\Delta S^{\ddagger} = -27$	167
	6.93×10^{-5}	45.0	HOAc		167
PhCH=CHPh (*trans*)	1.85×10^{-3}	45.0	HOAc	$\Delta H^{\ddagger} = 9.4$, $\Delta S^{\ddagger} = -42$	219
	3.00×10^{-3}	55.0	HOAc		219
	4.08×10^{-3}	65.0	HOAc		219
PhCH=CHPh (*cis*)	1.09×10^{-3}	45.0	HOAc	$\Delta H^{\ddagger} = 13.8$, $\Delta S^{\ddagger} = -29$	219
	2.11×10^{-3}	55.0	HOAc		219
	3.95×10^{-3}	65.0	HOAc		219
2,4-Dinitrobenzene sulphenyl bromide					
Cyclohexene	1.95×10^{-2}	25.0	HOAc	$\Delta H^{\ddagger} = 11.2$, $\Delta S^{\ddagger} = -31$	220
	3.49×10^{-2}	35.0	HOAc		220
	6.42×10^{-2}	45.0	HOAc		220
	$k_2 = 2.24 \times 10^{-4\text{f}}$	50.0	CCl_4		171
	$k_3 = 3.5 \times 10^{-4\text{f}}$	50.0	CCl_4		171
	$k_4 = 2.9 \times 10^{-3\text{f}}$	50.0	CCl_4		171
	$k_6 = 1.7 \times 10^{-2\text{f}}$	50.0	CCl_4		172
	$k_6 = 3.3 \times 10^{-2\text{f}}$	50.0	CCl_4		172
	$k_6 = 1.0 \times 10^{-3\text{f}}$	50.0	CCl_4		172
2-Nitro-4-X-benzene sulphenyl chloride					
X = OMe					
Cyclohexene	1.18×10^{-1}	25.0	HOAc		221

TABLE 6 (*Contd.*)

Olefin	*k*	*Temp. (°C)*	*Solvent*		*Ref.*
X = Me					
Cyclohexene	4.8×10^{-2}	25.0	HOAc		221
X = H					
Cyclohexene	2.6×10^{-2}	25.0	HOAc	$\Delta H^{\ddagger} = 11.7$	221
	8.93×10^{-2}	45.0	HOAc	$\Delta S^{\ddagger} = -29$	
X = Cl					
Cyclohexene	3.7×10^{-2}	25.0	HOAc		221
X = CO_2H					
Cyclohexene	8.37×10^{-3}	25.0	HOAc		221
X = CF_3					
Cyclohexene	1.24×10^{-2}	25.0	HOAc	$\Delta H^{\ddagger} = 11.0$	221
	4.00×10^{-2}	45.0	HOAc	$\Delta S^{\ddagger} = -32$	221
2-Nitro-4-X-benzene sulphenyl bromide					
X = Me					
Cyclohexene	9.9×10^{-2}	25.0	HOAc		222
X = Cl					
Cyclohexene	8.9×10^{-2}	25.0	HOAc		222
X = CF_3					
Cyclohexene	2.8×10^{-2}	25.0	HOAc		222
X = H					
Cyclohexene	6.9×10^{-2}	25.0	HOAc		222
2,4-Dinitrobenzene diazonium borofluoride					
$Me_2C{=}CHMe$	3.27×10^{-2}		MeCN		175
$Me_2C{=}CMe_2$	12.8		MeCN		175
$PhCH{=}CH_2$	2.33×10^{-4}		MeCN		175
1,2-Me_2-cyclopentene	11.6		MeCN		175
$Me_2C{=}CH.CH{=}CMe_2$	35.0		MeCN		175
$PhC(Me){=}CH_2$	1.42×10^{-2}		MeCN		175
PhCH=CHMe (*trans*)	3.26×10^{-3}		MeCN		175
PhCH=CHMe (*cis*)	6.3×10^{-5}		MeCN		175
Isopropylidene cyclopentene	65.3		MeCN		175
PhC(Me)=CHMe (*cis*)[g]	0.26		MeCN		175
PhC(Me)=CHMe (*trans*)[g]	3.3×10^{-3}		MeCN		175
$PhCH{=}CMe_2$	6.3×10^{-4}		MeCN		175
$Et_2C{=}CEt_2$	0.16		MeCN		175
$PhC(Me){=}CMe_2$	4.4×10^{-2}		MeCN		175
1-Ph-cyclohexene	3.0×10^{-2}		MeCN		175
$Ph_2C{=}CH_2$	4.7×10^{-3}		MeCN		175
PhCH=CHPh (*trans*)	4.7×10^{-5}		MeCN		175
PhCH=CHPh (*cis*)	9×10^{-6}		MeCN		175
$Ph_2C{=}CHMe$	1.4×10^{-2}		MeCN		175
PhC(Me)=CHPh	1.9×10^{-3}		MeCN		175
PhC(Me)=C(Me)Ph(*cis*)[g]	9.3×10^{-5}		MeCN		175
PhC(Me)=C(Me)Ph (*trans*)[g]	9.3×10^{-4}		MeCN		175
$Ph_2C{=}CMe_2$	7.0×10^{-4}	20	MeCN		175
$Ph_2C{=}CHPh$	1.4×10^{-4}	20	MeCN		175
$Ph_2C{=}CPh_2$	10^{-6}	20	MeCN		175
Nitrosyl chloride					
1-Butene	1.5×10^{-4}	0	$CHCl_3$		177
2-Butene	5.2×10^{-3}	0	$CHCl_3$		177

TABLE 6 (*Contd.*)

Olefin	*k*	*Temp. (°C)*	*Solvent*	*Ref.*
$Me_2C{=}CH_2$	3.0×10^{-3}	0	$CHCl_3$	177
Methallyl chloride	2.9×10^{-5}	0	$CHCl_3$	177
2-Pentene (*trans*)	1.08×10^{-2}	0	$CHCl_3$	177
2-Pentene (*cis*)	7.7×10^{-3}	0	$CHCl_3$	177
2-Methyl-1-butene	3.3×10^{-3}	0	$CHCl_3$	177
3-Methyl-1-butene	6.9×10^{-5}	0	$CHCl_3$	177
$Me_2C{=}CHMe$	0.1	0	$CHCl_3$	177
1-Hexene	1.6×10^{-4}	0	$CHCl_3$	177
3-Hexene	6.6×10^{-3}	0	$CHCl_3$	177
2-Methyl-1-pentene	2.3×10^{-3}	0	$CHCl_3$	177
2,3-Me_2-1-butene	2.0×10^{-3}	0	$CHCl_3$	177
3,3-Me_2-1-butene	3.2×10^{-5}	0	$CHCl_3$	177
3-Me-1-pentene	1.0×10^{-4}	0	$CHCl_3$	177
4-Me-1-pentene	1.2×10^{-4}	0	$CHCl_3$	177
$Me_2C{=}CMe_2$	0.1	0	$CHCl_3$	177
Cyclohexene	2.9×10^{-3}	0	$CHCl_3$	177
Cycloheptene	1.35×10^{-2}	0	$CHCl_3$	177
Norbornene	0.1	0	$CHCl_3$	177
4,4-Me_2-1-pentene	5.2×10^{-5}	0	$CHCl_3$	177
Styrene	6.5×10^{-4}	0	$CHCl_3$	177
Cyclooctene	0.1	0	$CHCl_3$	177
1-Octene	1.8×10^{-4}	0	$CHCl_3$	177
$PhC(Me){=}CH_2$	1.7×10^{-3}	0	$CHCl_3$	177
$PhCH{=}CHMe$	2.6×10^{-3}	0	$CHCl_3$	177
1-Dodecene	1.8×10^{-4}	0	$CHCl_3$	177
$Ph_2C{=}CH_2$	9.8×10^{-4}	0	$CHCl_3$	177
$PhCH{=}CHPh$ (*trans*)	2.9×10^{-4}	0	$CHCl_3$	177
$PhCH{=}CHPh$ (*cis*)	6.4×10^{-5}	0	$CHCl_3$	177
Mercuric salts				
Ethylene	4.38	20	1 *M* NO_3^- aq.	223
	5.26	25	1 *M* NO_3^- aq.	223
	5.58	27	1 *M* NO_3^- aq. } $\Delta H^{\ddagger} = 5.7$	223
	6.11	30	1 *M* NO_3^- aq.	223
	7.15	35	1 *M* NO_3^- aq.	223
	6.10	20	1 *M* ClO_4^- aq.	223
	7.26	25	1 *M* ClO_4^- aq. } $\Delta H^{\ddagger} = 5.8$	223
	8.50	30	1 *M* ClO_4^- aq.	223
	4.38	25	$HgCl^+$ electrophile	223
	4.30	25	$HgBr^+$ electrophile	223
	ca. 5.0	25	$Hg(NH_3)^{2+}$ electrophile	223
	2.2×10^{3} [j]	11	aq.[h]	224
	4.0×10^{3}	20	aq.[h]	224
	5.1×10^{3}	25	aq.[h] } $\Delta H^{\ddagger} = 9.2$, $\Delta S^{\ddagger} = -10$	224
	7.6×10^{3}	32	aq.[h]	224
	9.8×10^{3}	37	aq.[h]	224
Methyl acrylate	1.43×10^{-3}	35	$Hg(OAc)_2$–MeOH	225
	1.92×10^{-3}	40	$Hg(OAc)_2$–MeOH } $\Delta H^{\ddagger} = 9$	225
	2.25×10^{-3}	45	$Hg(OAc)_2$–MeOH	225
	3.05×10^{-3}	50	$Hg(OAc)_2$–MeOH	225
Propene	*ca.* 10^5	25	aq.[h]	224

References pp. 80–86

TABLE 6 (*Contd.*)

Olefin	*k*	*Temp.* (°C)	*Solvent*		*Ref.*
Allyl alcohol	2.4×10^2	4	aq.[h]		224
	4.3×10^2	11	aq.[h]	$\Delta H^\ddagger = 10.7$	224
	6.5×10^2	18	aq.[h]		224
	1.12×10^3	25	aq.[h]	$\Delta S^\ddagger = -8$	224
	1.8×10^3	32	aq.[h]		224
Allyl chloride	11	25	aq.[h]		224
Allyl cyanide	4.3	25	aq.[h]		224
Acrylonitrile	2.51×10^{-5}	45	$Hg(OAc)_2$–MeOH		226
Methyl methacrylate	1.58×10^{-4}	40	$Hg(OAc)_2$–MeOH		227
	2.15×10^{-4}	45	$Hg(OAc)_2$–MeOH	$\Delta H^\ddagger = 15$	227
	3.33×10^{-4}	50	$Hg(OAc)_2$–MeOH		227
Ethyl methacrylate	2.67×10^{-4}	45	$Hg(OAc)_2$–MeOH		227
1-Butene	8×10^4	25	aq.[h]		224
2-Me-propene	10^6	25	aq.[h]		224
cis-2-butene	3.9×10^3	18	aq.[h]		224
	5.8×10^3	25	aq.[h]	$\Delta H^\ddagger = 9.6$	224
	8.8×10^3	32	aq.[h]	$\Delta S^\ddagger = -9$	224
	1.2×10^4	37	aq.[h]		224
	ca. 0.75	25	$Hg(OAc)_2$–MeOH		228
trans-2-butene	9.0×10^2	18	aq.[h]		224
	1.7×10^3	25	aq.[h]	$\Delta H^\ddagger = 12.0$	224
	2.5×10^3	32	aq.[h]	$\Delta S^\ddagger = -3$	224
	3.3×10^3	37	aq.[h]		224
	ca. 0.075	25	$Hg(OAc)_2$–MeOH		228
1-Buten-3-ol	5.4×10^1	4	aq.[h]		224
	7.3×10^1	11	aq.[h]	$\Delta H^\ddagger = 13.5$	224
	1.3×10^2	18	aq.[h]	$\Delta S^\ddagger = -5$	224
	2.7×10^2	25	aq.[h]		224
	5.3×10^2	32	aq.[h]		224
1-Buten-4-ol	2.4×10^3	4	aq.[h]		224
	3.6×10^3	11	aq.[h]	$\Delta H^\ddagger = 8.7$	224
	5.6×10^3	18	aq.[h]	$\Delta S^\ddagger = -11$	224
	8.4×10^3	25	aq.[h]		224
	1.1×10^4	32	aq.[h]		224
2-Buten-1-ol (*trans*)	2×10^2	25	aq.[h]		224
2-Buten-1-ol (*cis*)	4.4×10^2	25	aq.[h]		224
2-Me-allyl alcohol	3×10^4	25	aq.[h]		224
2-Buten-1,4-diol (*cis*)	12	25	aq.[h]		224
1-Penten-3-ol	3.0×10^1	3	aq.[h]		224
	5.2×10^1	11	aq.[h]	$\Delta H^\ddagger = 11.7$	224
	9.6×10^1	18	aq.[h]	$\Delta S^\ddagger = -9$	224
	1.5×10^2	25	aq.[h]		224
1-Penten-4-ol	1.4×10^3	3	aq.[h]		224
	2.3×10^3	11	aq.[h]		224
	3.7×10^3	18	aq.[h]	$\Delta H^\ddagger = 10.7$	224
	6.1×10^3	25	aq.[h]	$\Delta S^\ddagger = -5$	224
	9.6×10^3	32	aq.[h]		224
1-Penten-5-ol	10^6	25	aq.[h]		224
Cyclohexene	6.3×10^{-3}	25	$Hg(OAc)_2$–MeOH		229
	5×10^3	25	aq.[h]		224

TABLE 6 (*Contd.*)

Olefin	*k*	*Temp. (°C)*	*Solvent*		*Ref.*
Cyclohexen-3-ol	8.6×10^{1}	3	aq.[h]		224
	1.6×10^{2}	11	aq.[h]	$\Delta H^{\ddagger}=11.6$	224
	2.7×10^{2}	18	aq.[h]	$\Delta S^{\ddagger}=-7$	224
	4.3×10^{2}	25	aq.[h]		224
1-Hexen-6-ol	10^{5}	25	aq.[h]		224
PhCH=CHCHO	5.0×10^{-4}	45	$Hg(OAc)_2$–MeOH		227
$PhCH=CHCO_2Me$ (*trans*)	*ca.* 2×10^{-4}	25	$Hg(OAc)_2$–MeOH		230
$PhCH=CHCO_2Me$ (*cis*)	*ca.* 6×10^{-4}	25	$Hg(OAc)_2$–MeOH		230
$PhCH=CHCO_2Et$	3.27×10^{-4}	35	$Hg(OAc)_2$–MeOH	$\Delta H^{\ddagger}=9$	227
	5.07×10^{-4}	45	$Hg(OAc)_2$–MeOH		227
2,6-Me_2-5-hepten-2-ol	1.62×10^{-2}	25	$Hg(OAc)_2$–$(CH_2OMe)_2$		231
Me 10-undecenoate	$k_1=1.11\times10^{-6k}$		$Hg(OAc)^2$–HOAc		232
n-1-Undecene	6.48×10^{-1}	0	$Hg(OAc)_2$–MeOH		233
	1.015	10	$Hg(OAc)_2$–MeOH	$\Delta H^{\ddagger}=8.1$	233
	1.78	20	$Hg(OAc)_2$–MeOH	$\Delta S^{\ddagger}=-30$	233
trans-*n*-2-Undecene	1.11×10^{-2}	0	$Hg(OAc)_2$–MeOH	$\Delta H^{\ddagger}=9.8$	233
	1.95×10^{-2}	10	$Hg(OAc)_2$–MeOH	$\Delta S^{\ddagger}=-32$	233
	3.82×10^{-2}	20	$Hg(OAc)_2$–MeOH		233
trans-*n*-3-Undecene	5.22×10^{-3}	0	$Hg(OAc)_2$–MeOH	$\Delta H^{\ddagger}=9.8$	233
	9.05×10^{-5}	10	$Hg(OAc)_2$–MeOH	$\Delta S^{\ddagger}=-33$	233
	1.78×10^{-2}	20	$Hg(OAc)_2$–MeOH		233
trans-*n*-4-Undecene	3.13×10^{-3}	0	$Hg(OAc)_2$–MeOH	$\Delta H^{\ddagger}=10.0$	233
	5.00×10^{-3}	10	$Hg(OAc)_2$–MeOH	$\Delta S^{\ddagger}=34$	233
	1.09×10^{-2}	20	$Hg(OAc)_2$–MeOH		233
trans-*n*-5-Undecene	2.80×10^{-3}	0	$Hg(OAc)_2$–MeOH	$\Delta H^{\ddagger}=9.8$	233
	4.92×10^{-3}	10	$Hg(OAc)_2$–MeOH	$\Delta S^{\ddagger}=-34$	233
	9.55×10^{-3}	20	$Hg(OAc)_2$–MeOH		233
cis-*n*-2-Undecene	5.25×10^{-2}	0	$Hg(OAc)_2$–MeOH	$\Delta H^{\ddagger}=8.6$	233
	8.90×10^{-2}	10	$Hg(OAc)_2$–MeOH	$\Delta S^{\ddagger}=-33$	233
	1.54×10^{-1}	20	$Hg(OAc)_2$–MeOH		233
cis-*n*-3-Undecene	3.52×10^{-2}	0	$Hg(OAc)_2$–MeOH	$\Delta H^{\ddagger}=8.4$	233
	5.87×10^{-2}	10	$Hg(OAc)_2$–MeOH	$\Delta S^{\ddagger}=-35$	233
	1.00×10^{-1}	20	$Hg(OAc)_2$–MeOH		233
cis-*n*-4-Undecene	2.40×10^{-2}	0	$Hg(OAc)_2$–MeOH	$\Delta H^{\ddagger}=8.5$	233
	4.17×10^{-2}	10	$Hg(OAc)_2$–MeOH	$\Delta S^{\ddagger}=-35$	233
	6.95×10^{-2}	20	$Hg(OAc)_2$–MeOH		233
cis-*n*-5-Undecene	2.17×10^{-2}	0	$Hg(OAc)_2$–MeOH	$\Delta H^{\ddagger}=8.6$	233
	3.77×10^{-2}	10	$Hg(OAc)_2$–MeOH	$\Delta S^{\ddagger}=-35$	233
	6.37×10^{-2}	20	$Hg(OAc)_2$–MeOH		233
PhCH=CHPh (*cis*)	*ca.* 1.6×10^{-4}	25	$Hg(OAc)_2$–MeOH		230
Thallic ion					
Ethylene	0.45	25?	0.25 *M* $HClO_4$ aq.		234
Propene	65	25?	0.10 *M* $HClO_4$ aq.		234
	75	25?	0.25 *M* $HClO_4$ aq.		234
	88	25?	0.50 *M* $HClO_4$,aq.		234
1-Butene	63	25?	0.10 *M* $HClO_4$,aq.		234
	73	25?	0.25 *M* $HClO_4$ aq.		234
	106	25?	0.50 *M* $HClO_4$ aq.		234
cis-2-Butene	23	25?	0.10 *M* $HClO_4$ aq.		234

TABLE 6 (*Contd.*)

Olefin	*k*	*Temp. (°C)*	*Solvent*	*Ref.*
	26	25?	0.25 *M* $HClO_4$ aq.	234
	31	25?	0.40 *M* $HClO_4$ aq.	234
trans-2-Butene	5.5	25?	0.10 *M* $HClO_4$ aq.	234
	6.1	25?	0.25 *M* $HClO_4$ aq.	234
	6.6	25?	0.50 *M* $HClO_4$ aq.	234
Isobutene (2-Me-propene)	*ca.* 10^5	25?	0.25 *M* $HClO_4$ aq.	234
Silver ion				
Ethylene	$\log_{10}k_1 = 14.88 - 20360/2.3RT$		aq.[m]	235

[a] k_2 reported recently as 8×10^{-5} l.mole^{-1}.sec^{-1} at 25°C[187].
[b] All unsaturated fatty acids are reported to give $k_2 =$ *ca.* 3×10^{-3} l.mole^{-1}.sec^{-1} at 30°C in HOAc[194].
[c] $(4.25–4.35) \times 10^{-4}$ l.mole^{-1}.sec^{-1} also reported[156] for k_2.
[d] Rate coefficients are in units of mole.sec. The reaction is of zero-order with respect to peracid; for a proposed mechanism, the original reference should be consulted.
[f] Rate $= k_2$[olefin][ArSBr] $+ k_3$[olefin][ArSBr]$^2 + k_4$[olefin][ArSBr]$^3 + k_5$[olefin][ArSBr][X] $+ k_6$-[olefin][ArSBr][X]^2; X = 2,4-dinitrophenyl 2-chlorocyclohexyl sulphide, 2,4-dinitroanisole, and 2,4-dinitrochlorobenzene respectively.
[g] With respect to the methyl groups.
[h] Solutions containing 0.01 *M* $HClO_4$; total ionic strength = 0.10 using $NaClO_4$.
[j] Suggested[224] that earlier rate measurements (*e.g.* ref. 223) refer to physically limited rather than chemically significant rates. Note also the acid-catalysis of this reaction
[k] Quoted as a first-order coefficient (sec^{-1}) but claimed to be a second-order process. The original paper was not available.
[m] 0.025*M* CF_3CO_2H, 0.875 *M* $NaCF_3CO_2$; see note *j* for a possible source of error.

6. Electrophilic attack on acetylenes

6.1 GENERAL DISCUSSION

The generalisation is often made that acetylenes undergo electrophilic attack less readily than olefins; this is usually attributed to the lower polarisability of the triple bond. In some cases, as in the acid-catalysed hydration of acetylenic ethers[236], the generalisation does not hold. However, the lower reactivity of acetylenic bonds in many electrophilic reactions explains the use of metal–ion catalysts, such as the well-known mercuric-ion-catalysed hydration of acetylene, and the greater tendency towards nucleophilic addition reactions.

Hydration of acetylenes will occur in acidic media. The products of the reaction are carbonyl compounds derived from rearrangement of the intermediate vinyl alcohol, *viz.*

$$-C{\equiv}C- + H^+ \rightarrow -CH{=}C^+- \xrightarrow[-H^+]{H_2O} -CH{=}C(OH)- \rightarrow -\underset{H}{\overset{H}{\underset{|}{\overset{|}{C}}}}-\underset{O}{\underset{\|}{C}}- \qquad (6.1)$$

the addition proceeding according to the Markownikow rule. Drenth *et al.*[237,238] have made a thorough investigation of the acid-catalysed hydration of a number of acetylenic ethers and thio-ethers. The reactions were subject to general acid-catalysis; the rate equation

$$\text{Rate} = k_1[\text{H}^+][\text{ether}] + k_2[\text{H}_2\text{PO}_4^-][\text{ether}] \tag{6.2}$$

showed that dihydrogen phosphate ion, present in the buffers used, also participated in the reaction. Vinyl thioethyne was used to show a linear dependence of $\log_{10} k$ upon Hammett's acidity function (H_0) rather than upon the stoichiometric hydrogen ion concentration; the slope of the graph was 1.07. Deuterium was not incorporated in the original olefin when the hydration was carried out in heavy water; the protonation stage was therefore irreversible. The mechanism was best described by the scheme

$$\text{HC}{\equiv}\text{CSR} \xrightarrow[\text{slow}]{\text{H}^+} [\text{H}_2\text{C}{=}\text{CSR}]^+ \xrightarrow[\text{fast}]{\text{X}} \text{Products} \tag{6.3}$$

where X may be the solvent or any other nucleophile. The acid-catalysed hydration of phenylacetylene and some phenylpropiolic acids can also be represented by this scheme, on similar evidence[239]. The use of heavy metals in such hydration reactions has been shown, in some cases, to be due to π-complexes between the acetylene and mercuric ion[240] or the aquo-Ru^{III} ion[241] but in neither case was any evidence of the detailed reaction mechanism given.

The addition of *halogens* to acetylenes has received rather more attention. In bromine addition, the principal kinetic form is

$$-\text{d}[\text{Br}_2]/\text{d}t = k_3[\text{acetylene}][\text{Br}_2]^2$$

while chlorine addition is predominantly second-order, *viz.*

$$-\text{d}[\text{Cl}_2]/\text{d}t = k_2[\text{acetylene}][\text{Cl}_2]$$

The order of susceptibility towards substituent effects is much less in the acetylene series than in the olefins, but its direction is similar. A similar statement may be made about the direction, rates, and kinetics of addition of *sulphenyl halides*. While the simple acetylenes, such as *sym*-diphenylacetylene and 3-hexyne, required catalytic quantities of aluminium chloride before undergoing addition by 2,4-dinitrobenzene sulphenyl chloride[242], the rates of reaction of acetylene itself with some *p*-substituted benzene sulphenyl chlorides alone have been measured. The effect of the *p*-substituents

$$\text{MeO} < \text{Me} < \text{H} < \text{Cl} < \text{NO}_2$$

is quantitatively similar to that found in the addition of *p*-substituted-2-nitrobenzene sulphenyl chlorides to cyclohexene in acetic acid[244]; subsequent work

CCK(9) – Pages 65–80 – 2nd Proof.

has failed to show a similar trend in the reaction of these *p*-substituted benzene sulphenyl chlorides with 3-hexyne[245].

It seems likely that, as the acetylenic bond is less readily polarised than the olefinic bond, substituent effects would be rather less important in determining the relative reactivities of a number of closely allied reagents such as the benzene sulphenyl chlorides; certainly the presence of substituents attached to the unsaturated system has a vastly different effect with different groups. For example, the rate of bromine addition to undecenolic acid is 10^{-5} times that to undecenoic acid, whereas that to phenylacetylene is 3×10^{-4} times that to styrene and diphenylacetylene reacts at 4×10^{-3} times the rate of stilbene[246]. Apparently the deactivating substituent (CO_2H) has a more pronounced effect upon the triple bond than upon the double bond, and this difference in intensity diminishes as more electron-donating (H, Ph) substituents are attached to the carbon atoms. The logical extension would be that, for a highly activated system, the acetylenic homologue would be more susceptible to attack than the olefinic compound; such a situation occurs with the ethers and thioethers, when the acid-catalysed hydration of vinyl thioethyne[236] or of *cis*-1-(1′-propenyloxy)-1-propyne[247] involves only attack at the acetylenic bond. The addition of trifluoroacetic acid is reported to occur at 0.2 times the rate with the corresponding olefin for 3-hexyne and some other acetylenes[248]. Perhaps the most general interpretation is to assume that the acetylene linkage is more susceptible to substituent effects and less susceptible to slight (substituent effect) changes in the electrophile than is the olefinic bond. The effects of changes of the solvent cannot be accurately assessed, although it would seem that acetylenes might be more susceptible than olefins.

The reaction of acetylenes with peracids has not received detailed investigation. The products of such oxidations are complex[249] and usually arise from an initial cleavage of the epoxide to give acids or esters, *e.g.*

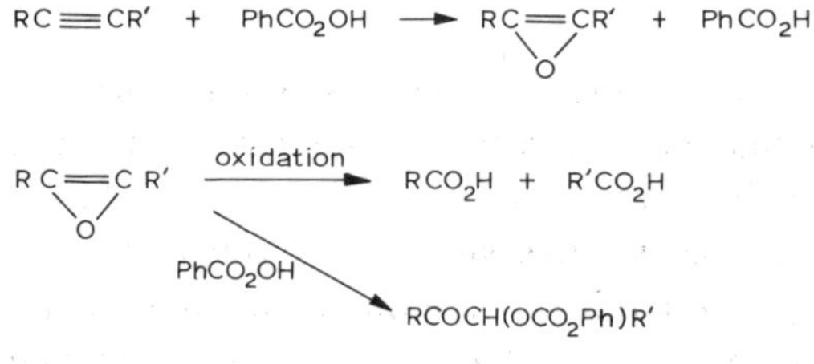

In a study of the effects of substituents upon reactivity, it was found that electron-withdrawing groups retarded the reaction while electron-releasing groups assisted it[250]. It is probable that the rate-determining stage, which involves one molecule of peracid and one molecule of acetylene[250], has similar requirements to that of the analogous reaction with olefins.

6.2 KINETIC DATA

The kinetic data on electrophilic addition to acetylenes are summarised in Table 7. As usual, the table is subdivided into sections depending upon the nature of the electrophile initiating the addition. Undesignated rate coefficients are moles, l and sec; activation enthalpies ($\Delta H^{\ddagger}$) are expressed in kcal. mole^{-1}, and entropies ($\Delta S^{\ddagger}$) in cal. deg^{-1}. mole^{-1} (eu), generally at 25°C.

TABLE 7

RATE DATA FOR ELECTROPHILIC ATTACK ON ACETYLENES

Acetylene	k_2	*Temp. (°C)*	*Solvent*	*Ref.*
Hydration				
HC≡CH	Complex		aq. H_2SO_4	251
HC≡COMe	77·7[a]	25	aq.	237
	0.0032[b]	25	aq.	237
HC≡CSCH=CH$_2$	0.0028	25	aq.	252
	$d(\log k_1)/dH_0 = 1.07$		aq. EtOH–HClO$_4$	253
HC≡COEt	329[a]	25	aq.	237
	0.0139[b]	25	aq.	237
	143	25	12.5% EtOH w/w	254
	70	25	42.5% EtOH w/w	254
HC≡CSEt[c]	0.21[a]	25	aq. $\Delta H^{\ddagger} = 16.5$	255
	2.0[a]	50	aq. $\Delta S^{\ddagger} = -6$	255
HC≡CO*i*-Pr	904[a]	25	aq.	256
	0.089[b]	25	aq.	256
HC≡CO*n*-Pr	296[a]	25	aq.	237
	0.0143[b]	25	aq.	237
MeC≡COEt	12.5	25	aq.	257
	$d(\log k_2)/dY_0 = 5.07$		aq. EtOH–HClO$_4$	258
HC≡CS*i*-Pr	0.24	25	aq.	252
MeC≡CSEt	0.0088	25	aq. $\Delta H^{\ddagger} = 19.1$	253
	0.118	50	aq. $\Delta S^{\ddagger} = -4$	253
MeC≡COCH=CHMe (*cis*)	$k_{H_3O^+}/k_{D_3O^+} = 1.7$[d]		aq.	258
HC≡CO*n*-Bu	116	25	42.5% EtOH w/w	256
HC≡CO*t*-Bu	5850[a]	25	aq.	256
	0.933[b]	25	aq.	256
	23.4	25	aq.	257
EtC≡COEt	0.30	25	aq. $\Delta H^{\ddagger} = 18.0$	255
HC≡CS*t*-Bu	3.4	50	aq. $\Delta S^{\ddagger} = -0.6$	255
EtC≡CSEt	0.0207	25	aq. $\Delta H^{\ddagger} = 17.9$	253
	0.235	50	aq. $\Delta S^{\ddagger} = -6$	253
i-PrC≡COEt	40	25	aq.	257
MeC≡CS*t*-Bu	0.0156	25	aq. $\Delta H^{\ddagger} = 18.8$	253
	0.196	50	aq. $\Delta S^{\ddagger} = -4$	253
	$d(\log k_2)/dY_0 = 5.2$	25	aq. EtOH–HCl	253
n-PrC≡CSEt	0.0202	25	aq.	252
i-PrC≡CSEt	0.0472	25	aq.	252
HC≡CPh and substituted phenylacetylenes				
X = H	$k_1 = 8.3 \times 10^{-4}$ sec^{-1}	25	44% H_2SO_4–5% EtOH[e]	259
	$= 1.32 \times 10^{-4}$	50.2	1.89 *M* H_2SO_4[f]	260

TABLE 7 (*Contd.*)

Acetylene	k_2	*Temp. (°C)*	*Solvent*	*Ref.*
	$= 2.03 \times 10^{-4}$	50.2	2.02 *M* H_2SO_4[f]	260
	$= 4.8 \times 10^{-4}$	50.2	2.37 *M* H_2SO_4[f]	260
	$= 1.41 \times 10^{-3}$	50.2	2.82 *M* H_2SO_4[f]	260
	$= 3.34 \times 10^{-3}$	50.2	3.20 *M* H_2SO_4[f]	260
X = *p*-MeO	$k_1 = 3.31 \times 10^{-3}$	50.2	0.42 *M* H_2SO_4[f]	260
X = *p*-Me	$k_1 = 1.82 \times 10^{-4}$	50.2	0.84 *M* H_2SO_4[f]	260
	$= 2.05 \times 10^{-3}$	50.2	1.89 *M* H_2SO_4[f]	260
	$= 3.16 \times 10^{-3}$	50.2	2.02 *M* H_2SO_4[f]	260
X = *m*-Me	$k_1 = 3.88 \times 10^{-4}$	50.2	2.02 *M* H_2SO_4[f]	260
	$= 9.23 \times 10^{-4}$	50.2	2.36 *M* H_2SO_4[f]	260
	$= 2.68 \times 10^{-3}$	50.2	2.82 *M* H_2SO_4[f]	260
X = *o*-Me	$k_1 = 1.01 \times 10^{-3}$	50.2	2.02 *M* H_2SO_4[f]	260
X = 2,4,6-Me_3-	$k_1 = 1.09 \times 10^{-3}$	50.2	0.42 *M* H_2SO_4[f]	260
	$= 3.66 \times 10^{-3}$	50.2	0.82 *M* H_2SO_4[f]	260
X = 2,3-Me_2-	$k_1 = 2.43 \times 10^{-3}$	50.2	2.02 *M* H_2SO_4[f]	260
X = *p*-*t*-Bu	$k_1 = 2.13 \times 10^{-3}$	50.2	2.02*M* H_2SO_4[f]	260
X = *p*-F	$k_1 = 2.96 \times 10^{-4}$	50.2	2.02 *M* H_2SO_4[f]	260
X = *o*-F	$k_1 = 1.98 \times 10^{-4}$	50.2	3.20 *M* H_2SO_4[f]	260
X = *p*-Cl	$k_1 = 4.0 \times 10^{-4}$	50.2	2.82 *M* H_2SO_4[f]	260
X = *m*-Cl	$k_1 = 2.06 \times 10^{-4}$	50.2	3.63 *M* H_2SO_4[f]	260
X = *o*-Cl	$k_1 = 1.26 \times 10^{-4}$	50.2	3.20 *M* H_2SO_4[f]	260
X = *p*-Br	$k_1 = 7.26 \times 10^{-4}$	50.2	3.20 *M* H_2SO_4[f]	260
X = *m*-Br	$k_1 = 1.83 \times 10^{-4}$	50.2	3.63 *M* H_2SO_4[f]	260
X = *o*-Br	$k_1 = 1.15 \times 10^{-4}$	50.2	3.20 *M* H_2SO_4[f]	260
X = *p*-I	$k_1 = 7.95 \times 10^{-4}$	50.2	3.20 *M* H_2SO_4[f]	260
X = *o*-I	$k_1 = 1.88 \times 10^{-4}$	50.2	3.20 *M* H_2SO_4[f]	260
X = *m*-MeO	$k_1 = 1.75 \times 10^{-3}$	50.2	3.20 *M* H_2SO_4[f]	260
X = *o*-MeO	$k_1 = 1.73 \times 10^{-3}$	50.2	3.20 *M* H_2SO_4[f]	260
DC≡CPh	$k_1 = 7.50 \times 10^{-4}$	25	44% H_2SO_4–5% EtOH[e]	259
MeC≡CPh	$k_1 = 2.82 \times 10^{-5}$	25	44% H_2SO_4–5% EtOH[e]	259
HC≡COPh	0.0092	25	42.7%–EtOH	254
n-BuC≡COEt	24.5	25	aq.	257
PhC≡CCO_2H and substituted phenylpropiolic acids				
X = H	$k_1 = 4.44 \times 10^{-6}$ sec^{-1}	25	50% H_2SO_4[g,h]	259
X = *p*-OMe	$k_1 = 2.82 \times 10^{-2}$	25	50% H_2SO_4[g,k]	259
X = *p*-Me	$k_1 = 1.35 \times 10^{-4}$	25	50% H_2SO_4[g,l]	259
X = *p*-Cl	$k_1 = 1.65 \times 10^{-6}$	25	50% H_2SO_4[g,m]	260
Iodine				
HC≡CH	1.27×10^{-4}	78.9	MeOH[n]	261
	4.0×10^{-4}	94.5	MeOH	261
	1.15×10^{-3}	109.2	MeOH	261
Bromine[p]				
$(O_2CC{\equiv}CCO_2)^{2-}$	2	25	aq.	262
EtC≡CCO_2Me	0.035	25	50% HOAc	262
HC≡CPh	2.3	25	HOAc	262
Undecenolic acid	0.13	25	HOAc	262
PhC≡CPh	0.042	25	HOAc	262
Stearolic acid	0.65	25	HOAc	262
Chlorine				
EtO_2C.C≡C.CO_2Et	6×10^{-3}	25	HOAc	262
PhCOC≡C.C_6H_4-*o*-Cl	0.079	25	HOAc–NaOAc	262

TABLE 7 (*Contd.*)

Acetylene	k_2	*Temp. (°C)*	*Solvent*	*Ref.*
Perbenzoic acid[q]				
HC≡CCH$_2$Br	10^{-8}	45	$CHCl_3$	250
HC≡CCH$_2$OH	8.3×10^{-7}	45	$CHCl_3$	250
MeC≡CMe	8.3×10^{-7}	45	$CHCl_3$	250
HOCH$_2$C≡CCH$_2$OH	4.7×10^{-6}	45	$CHCl_3$	250
HC≡CPh	1.75×10^{-7}	45	$CHCl_3$	250
AcOCH$_2$C≡CCH$_2$OAc	3.0×10^{-7}	45	$CHCl_3$	250
PrC≡CPr	6.7×10^{-6}	45	$CHCl_3$	250
MeC≡CCH(OH)*n*-C$_4$H$_9$	8.0×10^{-6}	45	$CHCl_3$	250
MeCH(OH)CH$_2$C≡CCH$_2$CH-(OH)Me	6.7×10^{-6}	45	$CHCl_3$	250
PrC≡CPh	6.0×10^{-6}	45	$CHCl_3$	250
(EtO)$_2$CH.C≡CCH(OEt)$_2$	10^{-8}	45	$CHCl_3$	250
PhC≡CPh	6.7×10^{-5}	45	$CHCl_3$	250
PhC≡CC≡CPh	3.5×10^{-6}[r]	45	$CHCl_3$	250
PhC≡CC≡CC≡CPh	1.47×10^{-7}[r]	45	$CHCl_3$	250
Benzenesulphenyl chlorides				
X = *p*-Me				
HC≡CH	3.5×10^{-5}	25	EtOAc	245
HC≡CEt	9.6×10^{-3}	25	EtOAc	245
HC≡C*n*-Bu	9.1×10^{-3}	25	EtOAc[s]	246
EtC≡CEt	0.73	25	EtOAc	245
X = 2,4-(NO$_2$)$_2$				
EtC≡CEt	1.29×10^{-3}	35	HOAc[t] } $\Delta H^\ddagger = 16$	263
	2.94×10^{-3}	45	HOAc } $\Delta S^\ddagger = -20$	263
HC≡CPh	2.67×10^{-5}	45	HOAc } $\Delta H^\ddagger = 24$	263
	8.61×10^{-5}	55	HOAc } $\Delta S^\ddagger = -3$	263
PhC≡CPh	1.6×10^{-5}	45	HOAc	263
Hydrogen chloride				
PhC≡CMe	$k_3 = 5\times10^{-7}$	25	HOAc[u]	264

[a] H_3O^+ as acid catalyst.
[b] $H_2PO_4^-$ as acid catalyst. In all other cases, total acidity is considered due to the proton alone.
[c] $k_{DC\equiv CSEt}/k_{HC\equiv CSEt} = 1.04, 1.02^{252}$.
[d] $k_1 = 3.36\times10^{-4}$ sec^{-1} at $[H_3O^+] = 0.00497$ *M* in H_2O at 25°C.
[e] $H_0 = -2.78$.
[f] Solvent = 2 vol. acetic acid to 1 vol. of aq H_2SO_4; molarities of acid quoted assume ideal mixing.
[g] $H_0 = -3.38; \rho = -4.79$.
[h] $d(\log k)/dH_0 = -0.96; k_{H_2SO_4}/k_{D_2SO_4} = 4.07$.
[k] $d(\log k)/dH_0 = -1.13; k_{H_2SO_4}/k_{D_2SO_4} = 3.93$
[l] $d(\log k)/dH_0 = -0.98; k_{H_2SO_4}/k_{D_2SO_4} = 3.70$
[m] $d(\log k)/dH_0 = -0.97; k_{H_2SO_4}/k_{D_2SO_4} = 3.75$.
[n] Rate = $k_2[C_2H_2][I_2]$ and is derived from consideration of the reverse reaction, *i.e.* the deiodination of 1,2-diiodoethylene.
[p] Regardless of the actual kinetic order, the coefficients are quoted as second-order.
[q] Derived from half-life values quoted in ref. 250.
[r] Overall k_2; more than one triple bond in substrate.
[s] $\Delta H^\ddagger = 7.4, \Delta S^\ddagger = -45$ (EtOAc); $\Delta H^\ddagger = 4.8, \Delta S^\ddagger = -47$ ($CHCl_3$).
[t] $\Delta H^\ddagger = 10.8, \Delta S^\ddagger = -40$ (EtOAc); $\Delta H^\ddagger = 10.2, \Delta S^\ddagger = -40$ ($CHCl_3$); as quoted in ref. 245.
[u] Rate = $k[HCl]^2$[PhC≡CMe].

7. Electrophilic additions in other systems

7.1 AROMATIC COMPOUNDS

Previous sections have dealt with electrophilic additions to olefins and to acetylenes. It is also true that addition can be a minor mode of reaction during electrophilic substitution. De la Mare *et al.*[265] have demonstrated that adducts, resulting from reactions such as (7.1) and (7.2), are isolated during

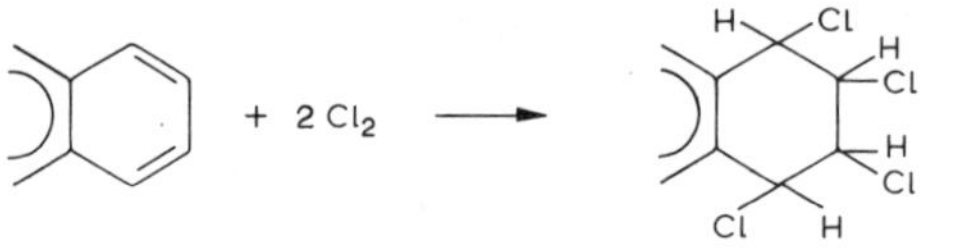

(7.1)

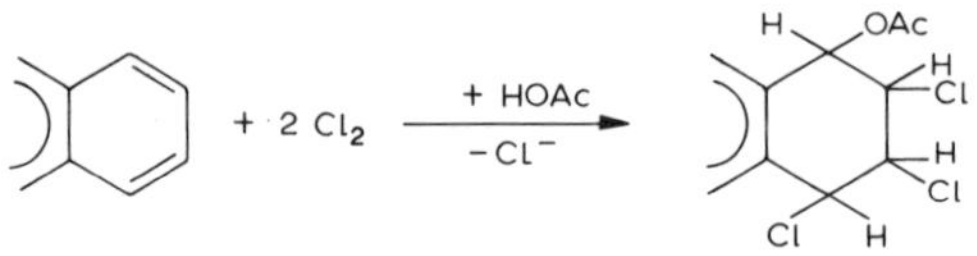

(7.2)

electrophilic substitution of chlorine in a number of aromatic species and there is considerable evidence for such additions occurring during nitration reactions, especially when acetic acid or acetic anhydride are solvents[266]. However, as the details of these additions have not yet been adequately determined, we shall not consider the reactions within the framework of the present article.

7.2 CARBONYL COMPOUNDS

Contrary to some reports, electrophilic addition reactions may occur in other multiple-bond systems. In many of the reactions of aldehydes and ketones the first stage involves the addition of some entity across the carbon–oxygen bond, *e.g.*, the formation of oximes, semicarbazones, hydrazones, hydrates (1,1-diols) and their ethers, and the "aldol" condensation. Most of these reactions entail a subsequent loss (elimination) of a small molecule (*e.g.* water, ammonia, ethanol) and, while one must be careful to determine whether the rate-determining stage involves attack on the carbonyl compound or elimination from the adduct[267], there are some systems in which it is evident that electrophilic attack is involved in the slow stage of the reaction sequence. Examples of such reactions are the acid-catalysed formation of oximes of aliphatic[268,269] and aromatic[270,271] carbonyl compounds, of furfural semicarbazone[272], and of 1,1-diols from aldehydes or ketones[273].

Similarly, the acid-catalysed hydrolysis of oximes[274] or of nitriles or amides[275] doubtless involve electrophilic addition to carbon–nitrogen multiple bonds.

Such reactions will not be dealt with (see Vols. 10 and 12). The addition of water to carbonyl compounds, however, deserves mention here. The reaction is reversible, *viz.*

$$RCHO + H_2O \rightleftharpoons RCH(OH)_2 \qquad (7.3)$$

and both forward and back reactions are catalysed by base and by acid. The acid-catalysis is general, since acetic acid, formic acid, and monohydrogen-phosphate ion are all active catalysts, although hydrogen ion is more effective[276]. The proton-catalysed reaction presumably involves prior attack upon carbonyl oxygen followed by a rate-determining attack by solvent upon the carbon atom bearing a partial positive charge, and the catalysis by other acids could involve a similar sequence, in which a complex (RCHO·HX) was attacked by solvent in the rate-determining stage[277]. The kinetics of such processes are complicated by such general acid catalysis, but the results may be meaningfully explained by a series of concurrent, catalysed, bimolecular processes*.

As well as such additions to aldehydes and ketones, there are a number of addition reactions involving other multiple bonds; some of these show definite electrophilic character[266].

7.3 ADDITION TO CARBON–NITROGEN DOUBLE BONDS

In addition to the acid-catalysed hydrolysis of oximes, a kinetic study has also been made of the addition of bromine to a number of substituted benzaldehyde phenylhydrazones[278]. The measurements were made in 70% acetic acid v/v, in the presence of 0.1 *M* potassium bromide. The attacking species was shown to be predominantly molecular halogen, and not the tribromide ion, since a graph of $k_2(1+K[Br^-])$ against $[Br^-]$ was linear, and of zero intercept, k_2 being defined by the equation

$$-d[Br_2]/dt = k_2[ArCH{=}NNHAr][Br_2]_{total} \qquad (7.4)$$

and K as the equilibrium constant for the reaction

$$Br_2 + Br^- \rightleftharpoons Br_3^- \qquad (7.5)$$

The observed values of k_2 also did not change with changing initial concen-

* The general form is Rate $= k$[RCHO], where $k = k_1[H^+] + k_2[HOAc] + k_3[OAc^-]$ in the case of acetic acid.

trations of the two reactants; thus, there is good evidence that tribromide ion is not involved in the reaction*.

The reaction rate was increased when electron-donating groups were present in the benzylidene system ($\rho = -0.62$) and decreased when electron-attracting species were present. Introducing similar substituents to the phenylhydrazine system had a much more apparent effect ($\rho = -2.17$) similar to that found in the styrene system[279]. The reaction evidently involves electrophilic attack; it would be interesting to know the mechanism of the process in more detail since it apparently involves substitution at unsaturated carbon, which may proceed through an addition–elimination sequence (assumed here) or a one-step substitution process.

Some preliminary kinetic work has been reported on the ozonolysis of nitrones. Ozone has been established as an electrophile towards carbon–carbon unsaturated systems; in its reaction with some *p*-substituted-*N*-phenylbenzaldoximes in chloroform at -60°C the order of reactivity *p*-OMe > H > *p*-Cl has been established[280]. Such a sequence could be accommodated by a process such as

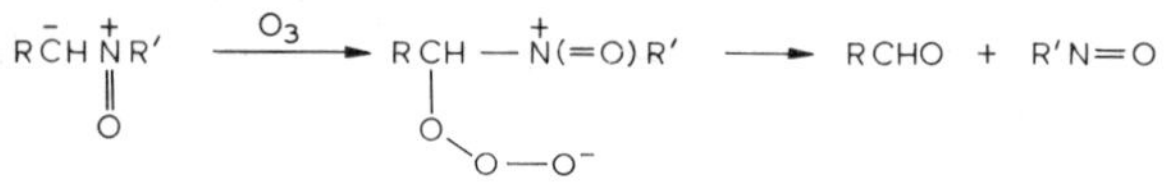

(7.6)

However, the similar reaction between aldehydes and ozone has been described as complex[281], and probably does not involve direct addition to the carbonyl group. A sequence such as (7.6) may be a gross simplification.

The most extensively studied addition to carbon–nitrogen multiple bonds in which breakdown of the molecule does not often occur (*cf.* hydrolysis of nitriles and oximes) is the hydration of heterocyclic substances such a pteridine and a number of polyazanaphthalenes. The uncharged entities, as well as their conjugate acids and bases, can add water across one or more of the multiple bonds. This process is usually reversible and does not generally involve any further disruption of the molecular skeleton. The addition, like hydration of carbon–carbon multiple bonds, is catalysed by acids, and the rates of hydration show a pronounced variation with the pH of the medium, being generally high at the acid and alkaline ends of the scale and having a minimum value slightly on the basic side of pH 7. The complexity of the system is reflected in the apparent first-order rate coefficients, *viz.*

$$\text{Rate} = k[\text{heterocycle}] \qquad (7.7)$$

* Such a method fails if the rate coefficient for the attack of tribromide ion upon the hydrazone is approximately unity. Since the detailed study was carried out using benzylidene *p*-nitrophenylhydrazone ($k_2 = 157$ l.mole^{-1}.sec^{-1}), and since Br_3^- would be expected to be less inherently reactive than the free halogen, it may be thought that attack by tribromide has not been completely disproved.

Here the coefficient k has three contributing factors, *viz.*

$$k = k_1[H^+] + k_2 + k_3/[H^+] \tag{7.8}$$

The detailed mechanisms suggested for these hydrations have been discussed in recent reviews[282].

7.4 ADDITION TO CARBON–PHOSPHORUS DOUBLE BONDS

The reaction of alkylidene phosphoranes with carbonyl compounds (Wittig reaction) has been widely used in preparative organic chemistry. Only one report of a kinetic study has been made, to our knowledge. In this, the rate of reaction of carbomethoxymethylene triphenylphosphorane ($Ph_3P{=}CHCO_2Me$) with a number of substituted benzaldehydes in benzene solution was studied. Such conditions are not the most favorable for heterolytic processes, but it was shown (*i*) that the rate of removal of aldehyde was the same as that of removal of phosphorane, (*ii*) that the rate-determining stage was first-order with respect to each reagent, and (*iii*) that substituents in the benzaldehyde had a considerable effect, electron-donating groups retarding and electron-withdrawing groups accelerating the reaction in the order[283]

$$p\text{-MeO} < p\text{-Me} < H < m\text{-Cl} < p\text{-NO}_2$$

On this basis an intermediate (I) was suggested, the result of electrophilic attack by carbon of the carbonyl group bonded to phosphorus, *viz.*

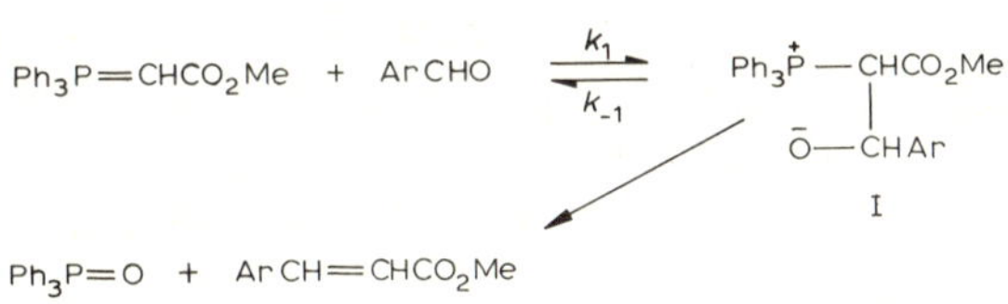

In keeping with such a mechanism, the rate was increased when more polar solvents were used, in the order

$$MeOH \gg CHCl_3 > PhH$$

which parallels the polarity of the three solvents[283]. The effects of substituents in the benzaldehyde group was marked; $\rho = 2.7$ for a number of p-substituted benzaldehydes reacting with the phosphorane in benzene. Such observations are entirely consistent with electrophilic attack by carbonyl carbon upon the carbon–phosphorus double bond.

References pp. 80–86

7.5 ADDITION TO NITROGEN–OXYGEN DOUBLE BONDS

The oxidation of nitroso groups to nitro is the last stage in a useful synthetic sequence by which aromatic amines may be oxidised to aromatic nitro-compounds. The oxidation is often accomplished by permonosulphuric acid (Caro's acid) or by pertrifluoroacetic acid, but recently an investigation has been made into the oxidation, by peracetic acid, of a number of nitrosobenzenes[284]. The reaction proceeded more rapidly in 16% aqueous ethanol than in 96% ethanol, following changes in the polarity of the mixed solvent, and the substituent effect of groups in the *para* position showed that electron-donating groups facilitated the reaction; *viz.*

$$p\text{-OMe} > p\text{-Me} > \text{H} > p\text{-Cl} \sim p\text{-Br}$$

Table 8 compares substituent effects in the epoxidation of stilbenes with those in the oxidation of nitrosobenzenes.

TABLE 8

LINEAR ENERGY RELATIONSHIPS FOR SOME PERACID ADDITION REACTIONS

		$\log_{10}(k_R/k_H)$	
Substituent	σ^+	*PhCO₂OH + stilbenes*[a]	*MeCO₂OH + PhN=O*[b]
p-MeO	−0.78	0.68	0.87
p-Me	−0.32	0.36	0.37
H	0.00	0.00	0.00
p-Cl	0.11	−0.20	−0.24

[a] Reaction of *trans*-stilbenes, *p*-X-C_6H_4CH=CHPh, with perbenzoic acid in benzene at 25°C[155].
[b] Reaction of peracetic acid with some nitrobenzenes, *p*-X-C_6H_4N=O, in 47% aq. ethanol at 50°C[284].

Insofar as the reaction conditions can be considered comparable, the two oxidations are similar; an electrophilic attack by the peracid seems to initiate the oxidation. It was also shown that peroxy-monochloroacetic acid ($ClCH_2$-CO_2OH) reacted more rapidly than peracetic acid itself[284] which is consistent with an electrophilic attack by the reagent upon the nitrosobenzene. A similar reaction mechanism could be proposed; *viz.*

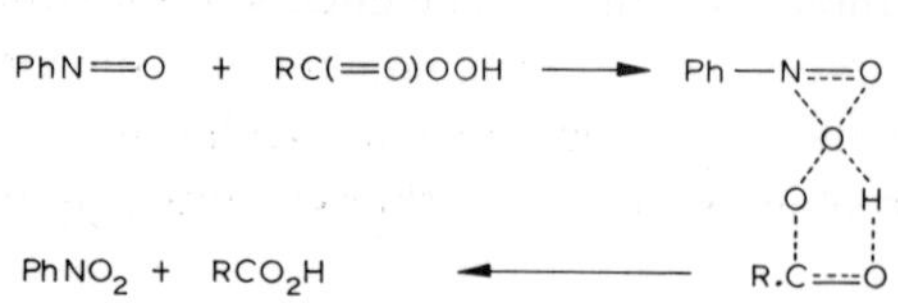

(7.9)

In this scheme electrophilic oxygen attacks the N=O bond in an analogous way to the epoxidation of olefins; the initially formed putative three-membered ring system then rearranges to the stable nitro-group.

Many of the Diels–Alder type reactions may be described in terms of heterolytic mechanisms, some of which involve electrophilic attack as the first stage. The evidence for such mechanisms is, as yet, incomplete and such additions are discussed in Chapter 2.

7.6 KINETIC DATA

Kinetic data on the additions discussed are summarised in Table 9. The sections are subdivided according to the electrophile undergoing addition to the multiple-bond system. Undesignated rate coefficients refer to second-order processes (units, $l.mole^{-1}.sec^{-1}$). Rate coefficients relevant to different orders are appropriately designated (k_1, etc.); the units are standard, *i.e.* mole, l and sec. In some cases, individual rate coefficients at differing acidities are quoted; in such cases, the pH of the solution is quoted in parenthesis after each – such pH values are expected to be accurate to two decimal places; where these are not quoted they are to be assumed (*e.g.* pH 3.00 appears in the table as 3).

Where the reaction does not proceed by a kinetically simple route, or where the reacting species are not immediately evident (*e.g.* in the acid-catalysed condensation of benzaldehyde with acetophenone), the kinetic equation is described in a footnote.

TABLE 9

ELECTROPHILIC ADDITIONS TO OTHER UNSATURATED SYSTEMS

Substrate	*k*	*Temp. (°C)*	*Solvent*	*Ref.*
Hydration				
MeCHO	$k_1 = 3.2 \times 10^{-4}$ sec^{-1}	25	92.5% aq. Me_2CO	273[a]
	$k_2 = 670$	25	aq.	276[a]
EtCHO	$k_2 = 490$	25	aq.	276[b]
n-PrCHO	$k_2 = 420$	25	aq.	276[b]
i-PrCHO	$k_2 = 370$	25	aq.	276[b]
CH_3COCO_2H	$\Delta H^{\ddagger} = 9.6$ kcal. mole^{-1}			285
Pteridine	$\alpha = 354$[c]	20	aq.	286
	$\beta = 2.40 \times 10^{-5}$	20		286
	$\gamma = 4.06 \times 10^{-13}$	20		286
2-Me-pteridine	$\alpha = 1079$[c]	20	aq.	286
	$\beta = 3.54 \times 10^{-5}$	20		286
	$\gamma = 1.90 \times 10^{-13}$	20		286
4-Me-pteridine	$\alpha = 191$[c]	20	aq.	286
	$\beta = 6.0 \times 10^{-6}$	20		286
	$\gamma = 8.1 \times 10^{-14}$	20		286

References pp. 80–86

TABLE 9 (*Contd.*)

Substrate	k	*Temp. (°C)*	*Solvent*	*Ref.*
7-Me-pteridine	$\alpha = 174^{c}$	20	aq.	286
	$\beta = 9.0 \times 10^{-6}$	20		286
	$\gamma = 1.13 \times 10^{-13}$	20		286
6-Cl-pteridine	$k_1 = 1.23 \times 10^{-4}$	20	aq., pH 6.89	287
7-Cl-pteridine	$k_1 = 1.55 \times 10^{-4}$	20	aq., pH 6.80	287
6,7-Cl_2-pteridine	$k_1 = 1.10 \times 10^{-4}$	20	aq., pH 7.00	287
2-OH-pteridine	$A = 6.3 \times 10^{4\,d}$	20	aq.	288
	$B = 1.1 \times 10^{-2}$	20	aq.	288
	$C = 1.0 \times 10^{-11}$	20	aq.	288
	$D = 3.0 \times 10^{-23}$	20	aq.	288
	$pK_a = 7.7$	20	aq.	288
6-Me-2-OH-pteridine	$A = 1.15 \times 10^{5}$	20	aq.[e]	288
	$B = 3.34 \times 10^{-3}$	20	aq.	288
	$C = 3.54 \times 10^{-12}$	20	aq.	288
	$D = 2.7 \times 10^{-23}$	20	aq.	288
	$pK_a = 7.95$	20	aq.	288
7-Me-2-OH-pteridine	$A = 6.49 \times 10^{4}$	20	aq.[e]	288
	$B = 2.66 \times 10^{-3}$	20	aq.	288
	$C = 1.85 \times 10^{-12}$	20	aq.	288
	$D = 1.1 \times 10^{-23}$	20	aq.	288
	$pK_a = 8.05$	20	aq.	288
6,7-Me_2-2-OH-pteridine	$A = 5.4 \times 10^{4}$	20	aq.[e]	288
	$B = 1.99 \times 10^{-3}$	20	aq.	288
	$C = 3.73 \times 10^{-12}$	20	aq.	288
	$D = 3.3 \times 10^{-23}$	20	aq.	288
	$pK_a = 7.95$	20	aq.	288
6,7-Et_2-OH-pteridine	$A = 6.73 \times 10^{4}$	20	aq.[e]	288
	$B = 2.55 \times 10^{-3}$	20	aq.	288
	$C = 2.95 \times 10^{-12}$	20	aq.	288
	$D = 2.28 \times 10^{-23}$	20	aq.	288
	$pK_a = 8.04$	20	aq.	288
6-OH-pteridine	$A = 1.81 \times 10^{3}$	20	aq.[e]	288
	$B = 9.81 \times 10^{-4}$	20	aq.	288
	$C = 1.13 \times 10^{-11}$	20	aq.	288
	$D = 1.03 \times 10^{-20}$	20	aq.	288
	$pK_a = 6.45$	20	aq.	288
2-Me-6-OH-pteridine	$A = 3.45 \times 10^{3}$	20	aq.[e]	288
	$B = 8.83 \times 10^{-4}$	20	aq.	288
	$C = 4.16 \times 10^{-11}$	20	aq.	288
	$D = 4.59 \times 10^{-21}$	20	aq.	288
	$pK_a = 6.53$	20	aq.	288
4-Me-6-OH-pteridine	$A = 1.96 \times 10^{3}$	20	aq.[e]	288
	$B = 1.19 \times 10^{-3}$	20	aq.	288
	$C = 6.49 \times 10^{-11}$	20	aq.	288
	$D = 5.79 \times 10^{-21}$	20	aq.	288
	$pK_a = 6.3$	20	aq.	288
7-Me-6-OH-pteridine	$A = 3.99 \times 10^{2}$	20	aq.[e]	288
	$B = 8.80 \times 10^{-5}$	20	aq.	288
	$C = 3.15 \times 10^{-12}$	20	aq.	288
	$D = 3.06 \times 10^{-22}$	20	aq.	288
	$pK_a = 7.09$	20	aq.	288
2-SH-pteridine	$A = 6.92 \times 10^{4}$	20	aq.[e]	288

TABLE 9 (*Contd.*)

Substrate	*k*	*Temp. (°C)*	*Solvent*	*Ref.*
	B = 3.49×10^{-2}	20	aq.	288
	C = 4.09×10^{-11}	20	aq.	288
	D = 2.41×10^{-21}	20	aq.	288
	pK_a = 6.52	20	aq.	288
2,6-$(OH)_2$-pteridine[f]	k_1 = 0.379(5.03), 0.274(5.26), 0.229(5.48), 0.140(5.78), 0.057(6.20), 0.056(6.30), 0.040(6.46), 0.022(6.64), 0.0152(6.84), 0.0086(7.06), 0.0071(7.28), 0.0039(7.58), 0.0025(7.84), 0.00174(8.09), 0.00114(8.43), 0.00107(8.67), 0.00104(8.95), 0.00125(9.25), 0.0012(9.46), 0.00123(9.90) at 20°C			295
2,6$(OH)_2$-4-Me-pteridine[f]	k_1 = 0.00363(5.26), 0.00240(5.48), 0.00129(5.78), 0.00066(6.20), 0.00036(6.46), 0.00022(6.84), 0.00016(7.06), 0.00169(10.19), 0.00207(10.50), 0.00451(11.00), 0.0092(11.24), 0.0157(11.43), 0.0230(11.63), 0.0352(11.79), 0.160(12.50) at 20°C			295
2-NH_2-quinazoline[g]	$t_{1/2}$ = 1.6(2), 13.7(3), 46.8(4), 72.6(5), 75.0(6), 72.8(7), 40.3(8), 8.1(9), 2.7(10), 0.4(11) at 20°C			296
2-NHMe-quinazoline[g]	$t_{1/2}$ = 1.0(2), 9.6(3), 42.5(6), 41.3(7), 29.3(8), 5.6(10) at 20°C			296
2-NMe_2-quinazoline[g]	$t_{1/2}$ = 2.1(2), 10.0(3) at 20°C			296
2-SMe-quinazoline[g]	$t_{1/2}$ = 0.5(3), 1.4(6), 8.7(7), 31.6(8), 1.8(9), 0.5(10) at 20°C			296
2-OMe-quinazoline[g]	$t_{1/2}$ = 0.6(3), 2.2(6), 7.5(7), 30.5(8), 2.0(9), 0.7(10) at 20°C			296
2-CH_2Cl-quinazoline[g]	$t_{1/2}$ = 1.2(3), 1.2(6), 12.5(7), 17.7(8), 11.8(9), 1.4(10) at 20°C			296
3-OH-1,4,6-Triazanaphthalene	A = 3.26×10^{3}	20	aq.[e]	288
	B = 6.17×10^{-5}	20	aq.	288
	C = 9.72×10^{-13}	20	aq.	288
	D = 1.49×10^{-23}	20	aq.	288
	pK_a = 7.32	20	aq.	288
2-OH-1,3,8-Triazanaphthalene	A = 8.59×10^{4}	20	aq.[e]	288
	B = 8.42×10^{-4}	20	aq.	288
	C = 8.29×10^{-13}	20	aq.	288
	D = 4.77×10^{-24}	20	aq.	288
	pK_a = 9.1	20	aq.	288
1,3,8-Triazanaphthalene	α = 248	20	aq.[c]	293
	$\beta = 4 \times 10^{-6}$	20	aq.	293
	$\gamma = 9.37 \times 10^{-15}$	20	aq.	293
8-Aza-purine (3H-1,2,3,4,6- -penta-azaindene[h]	k_1 = 14.10(0.60), 8.85(0.92), 6.35(1.2), 2.76(1.6) 1.89(2), 1.31(2.4), 1.07(2.8), 0.97(3.0), 1.09(3.4), 0.92(3.8), 0.82(4), 0.66(4.2), 0.478(4.73), 0.404(5), at 20°C			297
2-Me-8-azapurine[f]	k_1 = 15.8(0.80), 9.9(1.2), 4.93(1.6), 2.15(2), 0.91(2.4), 0.57(2.8) at 20°C			297
2-NH_2-8-azapurine[f]	k_1 = 4.09(0.4), 2.70(0.8), 1.74(1.3), 0.86(1.7), 0.403(2.2), 0.161(2.6), 0.063(3.0), 0.029(3.47), 0.0099(3.88), 0.0037(4.32), 0.0016(4.73), 0.00056(5.23) at 20°C			297
2-OH-8-azapurine[f]	k_1 = 3.67(3.78), 3.33(4.06), 1.98(4.35), 1.00(4.8), 0.86(5.1), 0.361(5.44), 0.178(5.84), 0.058(6.31), 0.020(6.81) at 20°C			297
Bromine				
X-C_6H_4CH=$NNHC_6H_4$-*p*(NO_2)				278
X = H	157	20.0	0.1 *M* KBr in 70% v/v HOAc	
X = *p*-OH	730	20.0	0.1 *M* KBr in 70% v/v HOAc	
X = *p*-NMe_2	565	20.0	0.1 *M* KBr in 70% v/v HOAc	
X = *p*-OMe	338	20.0	0.1 *M* KBr in 70% v/v HOAc	

TABLE 9 (*Contd.*)

Substrate	*k*	(°C)	*Solvent*	*Ref.*
X = *p*-Me	211	20.0	0.1 *M* KBr in 70% v/v HOAc	
X = *p*-$CHMe_2$	179	20.0	0.1 *M* KBr in 70% v/v HOAc	
X = *p*-F	148	20.0	0.1 *M* KBr in 70% v/v HOAc	
X = *p*-Cl	116	20.0	0.1 *M* KBr in 70% v/v HOAc	
X = *p*-Br	115	20.0	0.1 *M* KBr in 70% v/v HOAc	
X = *p*-CN	62.6	20.0	0.1 *M* KBr in 70% v/v HOAc	
X = *p*-NO_2	50.5	20.0	0.1 *M* KBr in 70% v/v HOAc	
X = *m*-Me	187	20.0	0.1 *M* KBr in 70% v/v HOAc	
X = *m*-Cl	97.5	20.0	0.1 *M* KBr in 70% v/v HOAc	
X = *m*-Br	97.0	20.0	0.1 *M* KBr in 70% v/v HOAc	
X = *m*-NO_2	65.5	20.0	0.1 *M* KBr in 70% v/v HOAc	
PhCH=$NNHC_6H_3$(*o*-X, *p*-Y)				
X = Y = NO_2	3.9[j]	20.0	0.1 *M* KBr in 70% v/v HOAc	278
X = Br, Y = NO_2	32[j]	20.0	0.1 *M* KBr in 70% v/v HOAc	278
X = H, Y = NO_2	158[j]	20.0	0.1 *M* KBr in 70% v/v HOAc	278
X = Y = Br	890[j]	20.0	0.1 *M* KBr in 70% v/v HOAc	278
X = Br, Y = Cl	980[j]	20.0	0.1 *M* KBr in 70% v/v HOAc	278
Peracetic acid				
PhNO	5.15×10^{-4}	50	96% EtOH	284
	1.84×10^{-3}	50	56% EtOH	284
	2.72×10^{-3}	50	47% EtOH	284
	4.02×10^{-3}	50	36% EtOH	284
	7.88×10^{-3}	50	16% EtOH	284
	1.21×10^{-3}	40	47% EtOH $\Delta H^\ddagger = 16.1$ kcal. mole^{-1}	284
	5.15×10^{-4}	30	47% EtOH $\Delta S^\ddagger = -22$ eu	284
p-$OMeC_6H_4NO$	2.00×10^{-2}	50[k]	47% EtOH	284
p-MeC_6H_4NO	6.34×10^{-3}	50	47% EtOH	284
p-ClC_6H_4NO	1.58×10^{-3}	50	47% EtOH	284
p-BrC_6H_4NO	1.47×10^{-3}	50	47% EtOH	284
Perbenzoic acid				
PhN=NPh (*trans*)	9.7×10^{-5}	15	PhH } $\Delta H^\ddagger = 14.8$ kcal. mole^{-1}	298
	1.5×10^{-4}	20	PhH	298
	2.32×10^{-4}	25	PhH	298
	3.48×10^{-4}	30	PhH	298
PhN=NPh (*cis*)	6.23×10^{-3}	15	PhH } $\Delta H^\ddagger = 12.3$ kcal. mole^{-1}	298
	8.78×10^{-3}	20	PhH	298
	1.25×10^{-2}	25	PhH	298
	1.81×10^{-2}	30	PhH	298
Monopersulphuric acid				
PhNO	3.5×10^{-3}	30	47% EtOH	299
Monoperoxychloroacetic acid				
PhNO	1.67×10^{-2}	30	47% EtOH	284
Ph_3P=$CHCO_2Me$				
XC_6H_4CHO				
X = *p*-OMe	7.1×10^{-5}	25	PhH } $\Delta H^\ddagger = 11.3$ kcal. mole^{-1}	
	2.36×10^{-4}	45	PhH } $\Delta S^\ddagger = -42$ eu	
	4.35×10^{-4}	25	$CHCl_3$	

TABLE 9 (*Contd.*)

Substrate	*k*	*Temp. (°C)*	*Solvent*	*Ref.*
	6.57×10^{-2}	25	$MeOH^3$	
X = *p*-Me	2.15×10^{-4}	25	PhH } $\Delta H^{\ddagger} = 10.7$ kcal. mole^{-1}	
	6.69×10^{-4}	45	PhH } $\Delta S^{\ddagger} = -42$ eu	
	1.24×10^{-3}	25	$CHCl_3$	
X = H	9.62×10^{-4}	25	PhH } $\Delta H^{\ddagger} = 10.2$ kcal. mole^{-1}	
	2.86×10^{-3}	45	PhH } $\Delta S^{\ddagger} = -40$ eu	
	4.36×10^{-3}	25	$CHCl_3$	
X = *p*-Cl	2.45×10^{-3}	25	PhH } $\Delta H^{\ddagger} = 9.5$ kcal. mole^{-1}	
	6.71×10^{-3}	45	PhH } $\Delta S^{\ddagger} = -41$ eu	
	1.24×10^{-2}	25	$CHCl_3$	
X = *p*-NO_2	1.06×10^{-1}	25	PhH	
X = *m*-Cl	1.20×10^{-2}	25	PhH	
Aldol condensations[m]				
(A) Acetaldehyde–RSH[n]				
Methyl mercaptoacetate	1.4	25	aq., 1 *M* ionic strength	300
$MeOCH_2CH_2SH$	8.8	25	aq., 1 *M* ionic strength	300
$HSCH_2CH_2SH$	32.6	25	aq., 1 *M* ionic strength	300
(B) Pivalaldehyde–RSH[n]				
$HOCH_2CH_2SH$	3.5	25	aq., 1 *M* ionic strength	300
$MeOCH_2CH_2SH$	3.0	25	aq., 1 *M* ionic strength	300
(C) Benzaldehydes–ketones[p]				
PhCHO–MeCOEt	$\log_{10} k_2 + H_0 = -6.05(25°)$		H_2SO_4–HOAc	301
p-$NO_2C_6H_4$CHO–MeCOEt	$\log_{10} k_2 + H_0 = -6.50(25°)$		H_2SO_4–HOAc	301
p-$OMeC_6H_4$CHO–MeCOEt	$\log_{10} k_2 + H_0 = -5.89(25°)$		H_2SO_4–HOAc	302
	$k_2 = 1.42 \times 10^{-4}(25.1°)$		H_2SO_4–HOAc	303
PhCHO–PhCOMe	$\log k_2 + H_0 = -4.85(28.0°)$		H_2SO_4–HOAc	304

[a] k_1 measured in the absence of acid: $k_2 = 50\ M^{-1}$, sec^{-1} for specific catalysis by H_3O^+.
[b] k_2 applies to specific acid-catalysis by H_3O^+; other catalysis constants for HOAc, HCOOH, and HPO_4^- are quoted in the original text.
[c] Rate = k_h [pteridine]$_{total}$; $k_h = \alpha \,.\, a_{H^+} + \beta + \gamma / a_{H^+}$.
[d] $k_h(a_{H^+} + K_a) = A(a_{H^+})^2 + Ba_{H^+} + C + D/a_{H^+}$; k_h has the same significance, and K_a refers to the acidity of the heterocycle under study. This treatment allows for changes in the nature of the substrate due to changes in the solvent acidity. Values for both the substrate and the 2-OH-pteridine anion, using the simple three parameter formula (footnote c) are quoted in the original paper[289].
[e] The symbols A–D have the same significance as in footnote d. These parameters only apply to the hydration stage, and not to the concomitant dehydration sequence. The original references (290–294) should also be consulted.
[f] Rate = k_1[heterocycle]$_{stoichiometric}$; pH of each determination is in parenthesis after each rate constant (sec^{-1}). The aqueous solutions were maintained at an ionic strength of 0.1.
[g] Half-lives of the apparent first-order reaction are quoted, with the pH of the aqueous medium in parenthesis.
[h] First-order rate coefficients (sec^{-1}) of the total reaction (hydration and dehydration); pH quoted in parenthesis.
[j] Hammett $\rho = -2.17$.
[k] Hammett $\rho = -1.58$.
[m] Some selected addition reactions to carbonyl groups are quoted here; the list is not exhaustive.
[n] Rate = $k_{H^+} a_{H^+}$[RSH][RCHO] + k_{RS^-}[RS$^-$][RCHO]; k_{H^+} is quoted here.
[p] Rate = *k*[protonated aldehyde][ketone enol]; k_2 (l. mole^{-1}. sec^{-1}) is given by the equations.

REFERENCES

1 M. S. KHARASCH AND F. R. MAYO, *J. Am. Chem. Soc.*, 55 (1933) 2468; 60 (1938) 3097; F. R. MAYO AND C. WALLING, *Chem. Rev.*, 27 (1940) 351.
2 C. K. INGOLD AND E. D. HUGHES, *J. Chem. Soc.*, (1935) 252.
3 H. C. BROWN AND F. R. JENSEN, *J. Am. Chem. Soc.*, 80 (1958) 2291.
4 S. WINSTEIN AND H. J. LUCAS, *J. Am. Chem. Soc.*, 59 (1937) 1461; D. PRESSMAN, L. BREWER AND H. J. LUCAS, *J. Am. Chem. Soc.*, 64 (1942) 1122.
5 R. W. TAFT, *O.N.R. Rep., 1960*, Cleveland Meeting, A.C.S., 1960, p. 6.
6 I. R. C. MCDONALD, R. M. MILBURN AND P. W. ROBERTSON, *J. Chem. Soc.*, (1950) 2836; P. B. D. DE LA MARE AND P. W. ROBERTSON, *J. Chem. Soc.*, (1950) 2838; L. D. SWINDALE, B. E. SWEDLUND AND P. W. ROBERTSON, *J. Chem. Soc.*, (1950) 812; P. W. ROBERTSON, J. K. HEYES AND B. E. SWEDLUND, *J. Chem. Soc.*, (1952) 1014.
7 P. B. D. DE LA MARE AND R. BOLTON, *Electrophilic Additions to Unsaturated Systems*, Elsevier, Amsterdam, 1966.
8 L. P. HAMMETT, *Physical Organic Chemistry*, McGraw-Hill, N.Y., 1940, pp. 292–3.
9 R. W. TAFT, *J. Am. Chem. Soc.*, 74 (1952) 5372.
10 L. P. HAMMETT AND A. J. DEYRUP, *J. Am. Chem. Soc.*, 54 (1932) 2721.
11 L. P. HAMMETT, *Chem. Rev.*, 16 (1935) 67.
12 See, however, M. A. PAUL AND F. A. LONG, *Chem. Rev.*, 57 (1957) 1, 935.
13 V. GOLD AND B. W. V. HAWES, *J. Chem. Soc.*, (1951) 2102; N. C. DENO, J. J. JARUZELSKI AND A. SCHREISHEIM, *J. Am. Chem. Soc.* 77 (1955) 3044.
14 B. C. CHALLIS, *Chem. Soc. (London), Ann. Rept.*, 63 (1966) 277.
15 E. M. ARNETT AND G. W. MACH, *J. Am. Chem. Soc.*, 88 (1966) 1177.
16 J. F. BUNNETT AND F. P. OLSEN, *Can. J. Chem.*, 44 (1966) 1899, 1917.
17 R. W. TAFT, *J. Am. Chem. Soc.*, 77 (1955) 1584.
18 D. PRESSMAN AND H. J. LUCAS, *J. Am. Chem. Soc.*, 64 (1942) 1953.
19 D. PRESSMAN AND H. J. LUCAS, *J. Am. Chem. Soc.*, 61 (1939) 2271.
20 E. L. PURLEE, R. W. TAFT AND C. A. DEFAZIO, *J. Am. Chem. Soc.*, 77 (1955) 837.
21 W. M. SCHUBERT, BO LAMM AND J. R. KEEFFE, *J. Am. Chem. Soc.*, 86 (1964) 4727.
22 N. C. DENO, F. A. KISH AND H. J. PETERSON, *J. Am. Chem. Soc.*, 87 (1965) 2157.
23 R. P. BELL, J. PRESTON AND R. B. WHITNEY, *J. Chem. Soc.* (1962) 1166.
24 L. TUCKER AND L. P. HAMMETT, *J. Am. Chem. Soc.*, 61 (1939) 2791.
25 A. J. KRESGE AND Y. CHIANG, *Proc. Chem. Soc.*, (1961) 81; A. J. KRESGE, G. W. BARRY, K. R. CHARLES AND Y. CHIANG, *J. Am. Chem. Soc.*, 84 (1962) 4343.
26 S. EHERNSON, S. SELTZER AND R. DIFFERBACH, *J. Am. Chem. Soc.* 87 (1965) 563.
27 R. H. HALL AND E. S. STERN, *J. Chem. Soc.*, (1950) 490.
28 G. R. LUCAS AND L. P. HAMMETT, *J. Am. Chem. Soc.*, 64 (1942) 1938.
29 H. J. LUCAS AND W. F. EBERZ, *J. Am. Chem. Soc.*, 56 (1934) 460.
30 J. B. LEVY, R. W. TAFT, JR., D. AARON AND L. P. HAMMETT, *J. Am. Chem. Soc.*, 73 (1951) 3792.
31 V. GOLD AND M. A. KESSICK, *J. Chem. Soc.*, (1965) 6718.
32 S. WINSTEIN AND H. J. LUCAS, *J. Am. Chem. Soc.*, 59 (1937) 1461.
33 R. P. BELL, J. PRESTON AND R. B. WHITNEY, *J. Chem. Soc.*, (1962) 1166.
34 D. PRESSMAN AND H. J. LUCAS, *J. Am. Chem. Soc.*, 61 (1939) 2271.
35 P. RIESZ, R. W. TAFT, JR. AND R. H. BOYD, *J. Am. Chem. Soc.*, 79 (1957) 3724.
36 R. W. TAFT, *J. Am. Chem. Soc.*, 77 (1955) 1584.
37 J. B. LEVY, R. W. TAFT, JR., D. AARON AND L. P. HAMMETT, *J. Am. Chem. Soc.*, 75 (1953) 3955.
38 YUN-PU LIU AND TIEN-CHI WEI, *J. Chinese Chem. Soc.*, 4 (1936) 297.
39 H. J. LUCAS AND YUN-PU LIU, *J. Am. Chem. Soc.*, 56 (1934) 2138.
40 E. L. PURLEE AND R. W. TAFT, JR., *J. Am. Chem. Soc.*, 78 (1956) 5807.
41 R. W. TAFT, JR., J. B. LEVY, D. AARON AND L. P. HAMMETT, *J. Am. Chem. Soc.*, 73 (1951) 3792.
42 D. PRESSMAN, L. BREWER AND H. J. LUCAS, *J. Am. Chem. Soc.*, 64 (1942) 1122.
43 W. SCHUBERT AND BO LAMM, *J. Am. Chem. Soc.*, 88 (1966) 120.

44 J.-P. Durand, M. Davidson, M. Hellin and F. Coussemant, *Bull. Soc. Chim. France*, (1966) 43.
45 H. J. Peterson, *Ph.D. Thesis*, Pennsylvania State University, 1960, quoted in ref. 47.
46 N. C. Deno, F. A. Kish and H. J. Peterson, *J. Am. Chem. Soc.*, 87 (1965) 2157.
47 J.-P. Durand, M. Davidson, M. Hellin and F. Coussemant, *Bull. Soc. Chim. France*, (1966) 52.
48 S. Cabani, *Bull. Soc. Fac. Chim. Ind. Bologna*, 20 No. 3–4 (1962) 85.
49 S. Cabani and N. Ceccanti, *J. Chem. Soc. (B)*, (1966) 77.
50 H. Kwart and L. B. Weisfeld, *J. Am. Chem. Soc.*, 80 (1958) 4670.
51 D. S. Noyce and R. A. Heller, *J. Am. Chem. Soc.*, 87 (1965) 4325.
52 L. R. Fedor and J. McLaughlin, *J. Am. Chem. Soc.*, 91(13), (1969) 3594.
53 Y. Pocker and M. J. Hill, *J. Am. Chem. Soc.*, 91(12), (1969) 3243.
54 P. B. D. de la Mare and R. Bolton, *Electrophilic Additions to Unsaturated Systems*, Elsevier, Amsterdam, 1966.
55 P. E. Peterson and G. Allen, *J. Am. Chem. Soc.*, 85 (1963) 3608.
56 A. G. Evans and J. Halpern, *Trans. Faraday Soc.*, 48 (1952) 1034.
57 F. Asinger, B. Fell, G. Steffan and G. Hadik, *Chem. Ber.*, 97 (1964) 1562.
58 F. R. Mayo and M. G. Savoy, *J. Am. Chem. Soc.*, 69 (1947) 1348.
59 *cf.* G. S. Hammond and T. D. Nevitt, *J. Am. Chem. Soc.*, 76 (1954) 4121.
60 *cf.* Y. Pocker, *J. Chem. Soc.* (1960) 1292.
61 O. Maass and C. H. Wright, *J. Am. Chem. Soc.*, 46 (1924) 2664; O. Maass and C. Swertz, *J. Am. Chem. Soc.*, 47 (1925) 2883.
62 F. R. Mayo and J. J. Katz, *J. Am. Chem. Soc.*, 69 (1947) 1339.
63 G. Williams and H. Bardsley, *J. Chem. Soc.*, (1952) 1707.
64 G. F. Hennion and C. F. Irwin, *J. Am. Chem. Soc.*, 63 (1941) 860.
65 D. H. R. Barton, A. da S. Campos-Neves and R. C. Cookson, *J. Chem. Soc.*, (1956) 3500.
66 Y. Pocker, K. D. Stevens and J. J. Champoux, *J. Am. Chem. Soc.*, 91 (1969) 4199.
67 Y. Pocker and K. D. Stevens, *J. Am. Chem. Soc.*, 91 (1969) 4205.
68 R. C. Fahey and C. A. McPherson, *J. Am. Chem. Soc.*, 91 (1969) 3865.
69 E. A. Jeffery and D. P. N. Satchell, *J. Chem. Soc.*, (1963) 2889.
70 I. V. Smirnov-Zamkov and G. A. Piskovitina, *Ukr. Khim. Zh.*, 28 (1962) 531.
71 M. J. S. Dewar and R. C. Fahey, *J. Am. Chem. Soc.*, 85 (1963) 3645.
72 F. Asinger, B. Fell, G. Steffan and G. Hadik, *Chem. Ber*, 97(6) (1964) 1562.
73 H. Cohn, E. D. Hughes, M. H. Jones and M. G. Peeling, *Nature*, 169 (1952) 291; E. A. Shilov and D. F. Mironva, *Proc. Acad. Sci. USSR, Phys. Chem. Sect. (Eng. Transl.)*, 115 (1957) 531.
74 P. E. Peterson, C. Casey, E. V. P. Tao, A. Agtarap and G. Thompson, *J. Am. Chem. Soc.*, 87 (1965) 5163.
75 P. E. Peterson and E. V. P. Tao, *J. Am. Chem. Soc.*, 86 (1964) 4503.
76 P. E. Peterson, *J. Am. Chem. Soc.*, 82 (1960) 5834.
77 P. E. Peterson and G. Allen, *J. Am. Chem. Soc.*, 85 (1963) 3608.
78 S. J. Cristol, T. C. Morrill and R. A. Sanchez, *J. Org. Chem.*, 31 (1966) 2719.
79 M. Mollard, B. Torck, M. Hellin and F. Coussemant, *Bull. Soc. Chim. France*, (1966) 83.
80 G. R. Buell, W. E. McEwen and J. Kleinberg, *J. Am. Chem. Soc.*, 84 (1962) 40.
81 A. G. Evans and J. Halpern, *Trans. Faraday Soc.*, 48 (1952) 1034.
82 H. G. Harris, Jr. and D. M. Himmelblau, *J. Chem. Eng. Data*, 9 (1964) 61; *J. Phys. Chem.*, 67 (1963) 802.
83 G. V. Korovina, S. G. Entelis and N. M. Chirkov, *J. Appl. Chem.*, 31 (1958) 597.
84 D. Lumbroso, M. Hellin and F. Coussemant, *Compt. Rend. Congr. Intern. Chim. 31e Liege, 1958*, 1 (1959) 624.
85 Dieman (1795) quoted in *Thorpe's Dictionary of Applied Chemistry*, Longman, Green, London, 1941.
86 A. W. Francis, *J. Am. Chem. Soc.*, 47 (1925) 2340.
87 H. D. C. Waters, A. R. Caverhill and P. W. Robertson, *J. Chem. Soc.* (1947) 1168.

88 E. D. White and P. W. Robertson, *J. Chem. Soc.*, (1939) 1509.
89 B. E. Swedlund and P. W. Robertson, *J. Chem. Soc.*, (1945) 131; (1947) 630.
90 J. Groh and J. Szelestey, *Z. Anorg. Allgem. Chem.*, 162 (1927) 333; J. Groh, *ibid.* p. 287; J. Groh and E. Takacs, *Z. Physik. Chem.*, 149A (1930) 195.
91 N. J. Bythell and P. W. Robertson, *J. Chem. Soc.*, (1938) 179.
92 J. C. Ghosh, S. K. Battacharyya, M. M. Dutt and M. J. Rao, *J. Indian Chem. Soc.*, 18 (1941), 171, 245, 253.
93 P. W. Robertson, J. B. Butchers, R. A. Durham, W. B. Healy, J. K. Heyes, J. K. Johanneson and D. A. Tait, *J. Chem. Soc.*, (1950) 2191.
94 K. Nozaki and R. A. Ogg, *J. Am. Chem. Soc.*, 64 (1942) 697, 704, 709.
95 I. R. C. McDonald, R. M. Milburn and P. W. Robertson, *J. Chem. Soc.*, (1950) 2837.
96 R. E. Buckles and J. P. Yuk, *J. Am. Chem. Soc.*, 75 (1953) 5048.
97 A. I. Popov and J. J. Mannion, *J. Am. Chem. Soc.*, 74 (1952) 222.
98 N. P. Kanyaev, *Zh., Fiz. Khim.*, 24 (1950) 154; *Sb. Statei Obshch. Khim.*, 2 (1953) 1172.
99 E. A. Shilov, N. P. Kanyaev and A. P. Otmennikova, *J. Gen. Chem. USSR (Eng. Transl.)*, 8 (1936) 909; E. A. Shilov and N. P. Kanyaev, *Compt. Rend. Acad. Sci. USSR*, 24 (1936) 890; E. A. Shilov, S. N. Solodushenkov and A. N. Kurakin, *J. Gen. Chem. USSR (Eng. Transl.)*, 10 (1937) 759; E. A. Shilov, N. P. Kanyaev, M. A. Domina and P. P. Ionina, *J. Gen. Chem. USSR (Eng. Transl.)*, 13 (1939) 1242; G. C. Israel, J. K. Martin and F. G. Soper, *J. Chem. Soc.*, (1950) 1283; D. A. Craw and G. C. Israel, *J. Chem. Soc.*, (1952) 550; K. D. Reeve and G. C. Israel, *J. Chem. Soc.*, (1952) 2327.
100 P. B. D. de la Mare and R. Bolton, ref. 7, p. 84.
101 B. E. Swedlund and P. W. Robertson, *J. Chem. Soc.*, (1947) 630.
102 J. R. Shelton and Liung-Huang Lee, *J. Org. Chem.*, 25 (1960) 428.
103 P. B. D. de la Mare, *Quart. Rev.*, 3 (1949) 126.
104 I. D. Morton and P. W. Robertson, *J. Chem. Soc.*, (1945) 129.
105 P. B. D. de la Mare, *J. Chem. Soc.*, (1960) 3823.
106 D. A. Evans, T. R. Watson and P. W. Robertson, *J. Chem. Soc.*, (1950) 1624.
107 P. B. D. de la Mare and P. W. Robertson, *J. Chem. Soc.*, (1945) 888.
108 D. A. Evans and P. W. Robertson, *J. Chem. Soc.*, (1950) 2834.
109 H. P. Rothbaum, I. Ting and P. W. Robertson, *J. Chem. Soc.*, (1948) 980.
110 I. Ting and P. W. Robertson, *J. Chem. Soc.*, (1947) 628.
111 I. Hartman and P. W. Robertson, *J. Chem. Soc.*, (1945) 891.
112 J. E. Dubois and G. Mouvier, *Tetrahedron Letters*, (1963) 1325.
113 S. V. Anantakrishnan and C. K. Ingold, *J. Chem. Soc.*, (1935) 984, 1396.
114 S. V. Anantakrishnan, *Proc. Indian Acad. Sci. Sect. A*, 27 (1948) 630.
115 H. S. Davis, *J. Am. Chem. Soc.*, 50 (1928) 2769.
116 J. R. Atkinson and R. P. Bell, *J. Chem. Soc.*, (1963) 3260.
117 B. E. Swedlund and P. W. Robertson, *J. Chem. Soc.*, (1947) 630.
118 K. Nozaki and R. A. Ogg, *J. Am. Chem. Soc.*, 64 (1942) 699, 704, 710.
119 J. G. Hanna and S. Siggia, *Anal. Chem.*, 37 (1965) 390.
120 P. W. Robertson, N. T. Clare, K. J. McNaught and G. W. Paul, *J. Chem. Soc.*, (1937) 335.
121 N. P. Kanyaev, *Izv. Vysshikh Uchebn. Zavedenii, Khim. i Khim. Tekhnol.*, 4 (1961) 781.
122 P. W. Robertson, B. E. Swedlund and J. K. Heyes, *J. Chem. Soc.*, (1952) 1014.
123 J. E. Dubois and G. Mouvier, *Tetrahedron Letters*, (1965) 1629.
124 P. W. Robertson, R. M. Milburn, W. E. Dasent and W. H. Oliver, *J. Chem. Soc.*, (1950) 1628.
125 J. E. Dubois and A. Schwarcz, *Tetrahedron Letters*, (1964) 2167.
126 K. Yates and W. V. Wright, *Tetrahedron Letters*, (1965) 1927.
127 F. Asinger, B. Fell, G. Hadik and G. Steffan, *Chem. Ber.*, 97 (1964) 1568.
128 J. E. Dubois, S. Molnarfi, F. Hennequin, M. H. Durand and R. Fellous, *Bull. Soc. Chim. France*, (1963) 786, 1491.
129 I. Ting and P. W. Robertson, *J. Chem. Soc.*, (1947) 628.
130 P. D. Bartlett and D. S. Tarbell, *J. Am. Chem. Soc.*, 58 (1936) 466.
131 C. C. Price, *J. Am. Chem. Soc.*, 58 (1936) 1834; C. C. Price and L. F. Fieser, *J. Am. Chem. Soc.*, 58 (1936) 1838.

132 V. Schwarz, S. Hermanek and J. Trojanek, *Chem. Ind. (London)*, (1960) 1212; V. Schwarz and S. Hermanek, *Tetrahedron Letters*, (1962) 809; *Collection Czech. Chem. Commun.*, 29 (1964) 2360.
133 M. J. Polissar, *J. Am. Chem. Soc.*, 52 (1930) 956.
134 N. J. Bythell and P. W. Robertson, *J. Chem. Soc.*, (1938) 179.
135 G. V. Chernyavskii, G. F. Dvorko, V. A. Shrubovich and O. M. Grishin, *Ukr. Khim. Zh.*, 32 (1966) 260.
136 Chen-Shen Chao, Ruey-Rong Lii, Lie-Chung Chao and Hei-Min Chen, *Hua Hsueh*, (1965) 95; *Chem. Abstr.*, 65 (1966) 3691c.
137 M. de Moura Campos, *J. Am. Chem. Soc.*, 76 (1954) 4480.
138 G. Heublein, *Z. Chem.*, 6 (1966) 186.
139 G. C. Israel, J. K. Martin and F. G. Soper, *J. Chem. Soc.*, (1950) 1282.
140 G. C. Israel, *J. Chem. Soc.*, (1950) 1286.
141 N. P. Kanyaev, *Izv. Vysshikh Uchebn. Zavendenii, Khim. i Khim. Tekhnol.*, 4 (1961) 225.
142 N. P. Kanyaev, *Izv. Vysshikh Uchebn. Zavendenii, Khim. i Khim. Tekhnol.*, 4 (1961) 78.
143 E. A. Shilov and N. P. Kanyaev, *J. Phys. Chem. USSR*, 5 (1934) 654.
144 D. A. Craw and G. C. Israel, *J. Chem. Soc.*, (1952) 550.
145 D. A. Craw, *J. Chem. Soc.*, (1954) 2510.
146 K. D. Reeve and G. C. Israel, *J. Chem. Soc.*, (1952) 2327.
147 E. A. Shilov and N. P. Kanyaev, *J. Phys. Chem. USSR*, 10 (1937) 123.
148 D. L. H. Williams, E. Brenvenue-Goetz and J. E. Dubois, *J. Chem. Soc. B*, (1969), 517.
149 J. H. Rolston and K. Yates, *J. Am. Chem. Soc.*, 91 (6) (1969) 1483.
150 J. E. Dubois and A. F. Hegarty, *J. Chem. Soc. B*, (1969) 639.
151 D. Swern, *Org. Reactions*, 7 (1953) 378.
152 J. Boeseken and J. S. P. Blumberger, *Rec. Trav. Chim.*, 44 (1925) 90.
153 J. Boeseken and J. Stuurman, *Rec. Trav. Chim.*, 56 (1937) 1034.
154 D. Swern, *J. Am. Chem. Soc.*, 69 (1947) 1692.
155 B. M. Lynch and K. H. Pausacker, *J. Chem. Soc.*, (1955) 1525; for similar observations on the methyl stilbenes, see Y. Ogata and I. Tabushi, *J. Am. Chem. Soc.*, 83 (1961) 3440.
156 G. Berti and F. Bottari, *J. Org. Chem.*, 25 (1960) 1286.
157 G. Levey, D. R. Campbell, J. O. Edwards and J. Maclachlan, *J. Am. Chem. Soc.*, 79 (1957) 1797.
158 W. E. Rosen, L. Dorfman and M. P. Linfield, *J. Org. Chem.*, 29 (1964) 1723.
159 R. Criegee, A. Kerchow and H. Zincke, *Chem. Ber.*, 88 (1955) 1878; R. Criegee, G. Blust and H. Zincke, *Chem. Ber.*, 87 (1954) 766.
160 P. S. Bailey, *Chem. Revs.*, 58, (1958), 925.
161 R. Criegee and G. Schröder, *Chem. Ber.*, 93 (1960) 689.
162 J. P. Wibaut and F. L. J. Sixma, *Rec. Trav. Chim.*, 71 (1952) 761 and earlier papers in this series.
163 E. H. Farmer and F. W. Shipley, *J. Chem. Soc.*, (1947) 1519.
164 L. Bateman and C. G. Moore, *Organic Sulfur Compounds*, N. Kharasch (Ed.), Pergamon, 1961, p. 210 ff.
165 L. Bateman, C. G. Moore, *et al.*, *J. Chem. Soc.*, (1958) 2838, 2866; G. W. Ross, *J. Chem. Soc.*, (1958) 2856.
166 C. M. Buess and N. Kharasch, *J. Am. Chem. Soc.*, 72 (1950) 3529; N. Kharasch and S. J. Assony, *J. Am. Chem. Soc.*, 75 (1953) 1081; N. Kharasch, C. M. Buess and W. King, *J. Am. Chem. Soc.*, 75 (1953) 6035.
167 W. L. Orr and N. Kharasch, *J. Am. Chem. Soc.*, 75 (1953) 6030; *J. Am. Chem. Soc.*, 78 (1956) 1201.
168 D. R. Hogg and N. Kharasch, *J. Am. Chem. Soc.*, 78 (1956) 2728.
169 P. B. D. de la Mare, *Quart, Rev.*, 3 (1949) 126.
170 D. S. Campbell and D. R. Hogg, *J. Chem. Soc.*, (1965) 5887.
171 D. R. Hogg and D. S. Campbell, *J. Chem. Soc. B*, (1966) 109.
172 D. R. Hogg and D. S. Campbell, *J. Chem. Soc. B*, (1966) 294.
173 E. P. White and P. W. Robertson, *J. Chem. Soc.*, (1939) 1509.
174 P. B. D. de la Mare and R. Bolton, *Electrophilic Additions to Unsaturated Systems*, Elsevier, Amsterdam, 1966, p. 182 ff.

175 H. MARXMEIER AND E. PFEIL, *Chem. Ber.*, 97(3) (1964) 815.
176 H. MARXMEIER AND E. PFEIL, *Ann. Chem.*, 678 (1964) 28.
177 T. BEIER, H. G. HAUTHAL AND W. PRITZKOW, *J. Prakt. Chem.*, 26 (1964) 304.
178 H. C. BROWN, *Hydroboration*, Benjamin Inc., New York, 1962.
179 H. C. BROWN AND A. W. MOERIKOFER, *J. Am. Chem. Soc.*, 85 (1963), 2063.
180 H. C. BROWN AND C. W. MCGARY, *J. Am. Chem. Soc.*, 77 (1955) 2300 ff.
181 J. ROMEYN AND G. F. WRIGHT, *J. Am. Chem. Soc.*, 69 (1947) 697.
182 J. CHATT, *Chem. Rev.*, 48 (1951) 7.
183 H. J. LUCAS, F. R. HEPNER AND S. WINSTEIN, *J. Am. Chem. Soc.*, 61 (1939) 3102.
184 T. G. TRAYLOR AND A. W. BAKER, *Tetrahedron Letters*, (1959) (19) 14; *J. Am. Chem. Soc.*, 85 (1963) 2746.
185 P. BRUCK, D. THOMPSON AND S. WINSTEIN, *Chem. Ind. (London)*, (1960) 405.
186 J. STUURMAN, *Proc. Acad. Sci. Amsterdam*, 38 (1935) 450.
187 YOSHIRO OGATA AND KEIZO AOKI, *J. Org. Chem.*, 31 (1966) 1625.
188 F. C. FROSTICK, B. PHILLIPS AND P. S. STARCHER, *J. Am. Chem. Soc.*, 81 (1959) 3350.
189 J. STUURMAN, *Thesis*, Delft, 1936.
190 J. BOESEKEN AND J. STUURMAN, *Rec. Trav. Chim.*, 56 (1937) 1034.
191 J. BOESEKEN AND J. STUURMAN, *Proc. Acad. Sci. Amsterdam*, 39 (1936) 2; J. BOESEKEN AND F. HANEGRAAFF, *Rec. Trav. Chim.*, 61 (1942) 69.
192 D. SWERN, G. N. BILLEN AND J. T. SCANLAN, *J. Am. Chem. Soc.*, 68 (1946) 1504.
193 J. SMIT, *Rec. Trav. Chim.*, 49 (1930) 686.
194 YASUO SUHARA, *Tokyo Kogyo Shikensho Hokoku*, 57 (1962) 370.
195 F. A. CUTLER, J. F. FISHER AND J. M. CHEMERDA, *J. Org. Chem.*, 24 (1959) 1626.
196 Y. OGATA AND I. TABUSHI, *J. Am. Chem. Soc.*, 83 (1961) 3444.
197 P. HEINÄNEN, *Ann. Acad. Sci. Fennicae: Ser. A*, 59 (1943) 3.
198 R. CETINA, J. L. MATEOS AND E. TRABULSE, *Bol. Inst. Quim. Univ. Nacl. Auton. Mex.*, 17 (1965) 56.
199 J. BOESEKEN AND J. S. P. BLUMBERGER, *Rec. Trav. Chim.*, 44 (1925) 90.
200 H. MEERWEIN, A. OGAIT, W. PRANG AND A. SERINI, *J. Prakt. Chem.*, 113 (1926) 9.
201 K. BODENDORF, *Arch. Pharm.*, 268 (1930) 491.
202 H. B. HENBEST AND B. NICHOLLS, *J. Chem. Soc.*, (1959) 221.
203 F. ASINGER, B. FELL, G. HADIK AND G. STEFFAN, *Chem. Ber.*, 97 (1964) 1568.
204 I. M. KOLTHOFF AND T. S. LEE, *J. Polymer Sci.*, 2 (1947) 206.
205 D. R. CAMPBELL, J. O. EDWARDS, J. MCLACHLAN AND K. POLGER, *J. Am. Chem. Soc.*, 80 (1958) 5310.
206 YOICHIRO NAGAI, *Yuki Gosei Kagaku Kyokai Shi*, 19 (1961) 527; *Chem. Abstr.* 55 (1961) 23441b, d, e.
207 YOSHIO ISHII AND YOSHIHIKO INAMOTO, *Kogyo Kagaku Zasshi*, 63 (1960) 765; *Chem. Abstr.*, 58 (1963) 4393b.
208 N. N. SCHWARTZ AND J. H. BLUMBERGS, *J. Org. Chem.*, 29 (1964) 1976.
209 B. RICKBORN AND SHIOW-YUEH LWO, *J. Org. Chem.*, 30 (1965) 2212.
210 A. P. GRAY AND D. E. HEITMEIER, *J. Org. Chem.*, 30 (1965) 1226.
211 E. GIOVANNINI AND H. WEGMULLER, *Helv. Chim. Acta*, 42 (1959) 1142.
212 F. BOHLMANN AND H. SINN, *Chim. Ber.*, 88 (1955) 1869.
213 C. RICCIUTI, L. S. SILBERT AND W. S. PART, *J. Am. Oil. Chemists Soc.*, 34 (1957) 134.
214 F. L. J. SIXMA, H. BOER AND J. P. WIBAUT, *Rec. Trav. Chim.*, 70 (1951) 1005.
215 F. L. J. SIXMA AND J. P. WIBAUT, *Rec. Trav. Chim.*, 71 (1952) 761.
216 F. L. J. SIXMA, *Rec. Trav. Chim.*, 71 (1952) 1124.
217 N. KHARASCH AND H. H. VON DUESEL, *J. Org. Chem.*, 24 (1959) 1806.
218 H. KWART AND L. J. MILLER, *J. Am. Chem. Soc.*, 83 (1961) 4552.
219 N. R. SLOBODKIN AND N. KHARASCH, *J. Am. Chem. Soc.*, 82 (1960) 5837.
220 D. S. CAMPBELL AND D. R. HOGG, *J. Chem. Soc.*, (1965) 5887.
221 D. R. HOGG AND C. BROWN, *J. Chem. Soc. B*, (1968) 1262.
222 C. BROWN AND D. R. HOGG, *J. Chem. Soc. B*, (1969) 1054.
223 P. BRANDT AND O. PLUM, *Acta. Chem. Scand.*, 7 (1953) 97.
224 J. HALPERN AND H. B. TINKER, *J. Am. Chem. Soc.*, 89 (1967) 6427.

225 K. L. MALLIK AND M. N. DAS, *J. Am. Chem. Soc.*, 82 (1960) 4269.
226 A. K. CHAUDHURI AND M. N. DAS, *Tetrahedron*, 21 (1965) 457.
227 A. K. CHAUDHURI, K. L. MALLIK AND M. N. DAS, *Tetrahedron*, 19 (1963) 1981.
228 M. H. THOMAS AND F. E. W. WETMORE, *J. Am. Chem. Soc.*, 63 (1941) 136.
229 J. ROMEYN AND G. F. WRIGHT, *J. Am. Chem. Soc.*, 69 (1947) 697.
230 G. F. WRIGHT, *J. Am. Chem. Soc.*, 57 (1935) 1993. In this and some other earlier papers (*e.g. J. Am. Chem. Soc.*, 62 (1940) 1991) the kinetic results are presented as graphs which do not allow accurate measurement of the second-order rate coefficients.
231 A. G. BROOK, A. RODGMAN AND G. F. WRIGHT, *J. Org. Chem.*, 17 (1952) 991.
232 KYO TAKAOKA AND YUJI IZUMISAWA, *Kogyo Kagaku Zasshi*, 67 (1964) 1244; *Chem. Abstr.*, 61 (1964) 11887h.
233 F. ASINGER, B. FELL, G. HADIK AND G. STEFFAN, *Chem. Ber.*, 97 (1964) 1568.
234 P. M. HENRY, *J. Am. Chem. Soc.*, 87 (1965) 4423.
235 P. BRANDT, *Acta. Chem. Scand.*, 13 (1959) 1639.
236 H. HOGEVEEN AND W. DRENTH, *Rec. Trav. Chim.*, 82 (1963) 410.
237 W. DRENTH AND E. J. STAMHUIS, *Rec. Trav. Chim.*, 80 (1961) 797.
238 W. DRENTH, H. HOGEVEEN AND E. J. STAMHUIS, *Rec. Trav. Chim.*, 82 (1963) 395 ff.
239 R. W. BOTT, C. EABORN AND D. R. M. WALTON, *J. Chem. Soc.*, (1965) 384; D. S. NOYCE, M. A. MATESICH, M. D. SCHIAVELLI AND P. E. PETERSON, *J. Am. Chem. Soc.*, 87 (1965) 2295.
240 W. L. BUDDE AND R. E. DESSY, *J. Am. Chem. Soc.*, 85 (1963) 3964.
241 J. HALPERN, B. R. JAMES AND A. L. W. KEMP, *J. Am. Chem. Soc.*, 83 (1961) 4097.
242 N. KHARASCH AND S. J. ASSONY, *J. Am. Chem. Soc.*, 75 (1953) 1081.
243 F. MONTANARI, *Gazz. Chim. Ital.*, 86 (1956) 406.
244 D. R. HOGG AND C. BROWN, *J. Chem. Soc. B*, (1968) 1262.
245 A. DONDONI, G. MODENA AND G. SCORRANO, *Ric. Sci. Rend., Sez. A*, 6 (1964) 665.
246 P. W. ROBERTSON, W. E. DASENT, R. M. MILBURN AND W. H. OLIVER, *J. Chem. Soc.*, (1950) 1628.
247 E. J. STAMHUIS AND W. DRENTH, *Rec. Trav. Chim.*, 82 (1963) 385.
248 P. E. PETERSON, *J. Am. Chem. Soc.*, 85 (1963) 2865.
249 H. H. SCHLUBACH AND V. FRANZEN, *Ann. Chem.*, 578 (1952) 220.
250 H. H. SCHLUBACH AND V. FRANZEN, *Ann. Chem.*, 577 (1952) 60.
251 R. H. FRIEMAN, E. R. KENNEDY AND H. J. LUCAS, *J. Am. Chem. Soc.*, 59 (1937) 722.
252 H. HOGEVEEN AND W. DRENTH, *Rec. Trav. Chim.*, 82 (1963) 375.
253 H. HOGEVEEN AND W. DRENTH, *Rec. Trav. Chim.*, 82 (1963) 410.
254 T. L. JACOBS AND S. SEARLES, *J. Am. Chem. Soc.*, 66 (1944) 686.
255 H. HOGEVEEN AND W. DRENTH, *Rec. Trav. Chim.*, 79 (1960) 1002.
256 E. J. STAMHUIS AND W. DRENTH, *Rec. Trav. Chim.*, 82 (1963) 394.
257 G. L. HEKLEERT AND W. DRENTH, *Rec. Trav. Chim.*, 82 (1963) 385.
259 D. S. NOYCE, M. A. MATESICH, M. D. SCHIAVELLI AND P. E. PETERSON, *J. Am. Chem. Soc.*, 87 (1965) 2295.
260 R. W. BOTT, C. EABORN, AND D. R. M. WALTON, *J. Chem. Soc.*, (1965) 384.
261 S. I. MILLER AND R. M. NOYES, *J. Am. Chem. Soc.*, 74 (1952) 3403.
262 P. W. ROBERTSON, W. E. DASENT, R. M. MILBURN AND W. H. OLIVER, *J. Chem. Soc.*, (1950) 1628.
263 N. KHARASCH AND C. N. YIANNIOS, *J. Org. Chem.*, 29 (1964) 1190.
264 R. C. FAHEY AND DO JAE LEE, *J. Am. Chem. Soc.*, 88 (1966) 5555.
265 P. B. D. DE LA MARE, N. V. KLASSEN AND R. KOENIGSBERGER, *J. Chem. Soc.*, (1962) 5285; P. B. D. DE LA MARE, E. A. JOHNSON AND J. S. LOMAS, *J. Chem. Soc.*, (1964) 5317; (1963) 5973; G. H. BEAVEN, P. B. D. DE LA MARE, E. A. JOHNSON AND N. V. KLASSEN, *J. Chem. Soc.*, (1962) 988.
266 For a discussion see P. B. D. DE LA MARE AND R. BOLTON, *Electrophilic Additions to Unsaturated Systems*, Elsevier, Amsterdam, 1966, p. 240 ff.
267 W. P. JENCKS, in *Progress in Physical Organic Chemistry*, S. G. COHEN, A. STREITWIESER AND R. W. TAFT, (Eds.) Interscience, New York, 1964, p. 63.
268 A. ÖLANDER, *Z. Physik. Chem.*, 129 (1927) 1.

269 P. G. KLETZKE, *J. Org. Chem.*, 29 (1964) 1363 and refs. therein.
270 J. D. DICKINSON AND C. EABORN, *J. Chem. Soc.*, (1959) 3036, 3641.
271 B. M. ANDERSON AND W. P. JENCKS, *J. Am. Chem. Soc.*, 82 (1960) 1773.
272 J. B. CONANT AND P. D. BARTLETT, *J. Am. Chem. Soc.*, 54 (1932) 2881.
273 See, for example, R. P. BELL AND J. C. CLUNIE, *Trans. Faraday Soc.* 48 (1952) 439; R. P. BELL AND B. DE B. DARWENT, *Trans. Faraday Soc.*, 46 (1950) 34.
274 R. W. JOHNSON AND J. GOLENTERNEK, *J. Am. Chem. Soc.* 59 (1937) 365; see also ref. 268.
275 B. S. RABINOVITCH AND C. A. WINKLER, *Can. J. Res.*, 20B (1942) 221; V. K. KRIEBLE AND A. L. PEIKE, *J. Am. Chem. Soc.*, 55 (1933) 2326; and refs. cited in both papers.
276 L. C. GRUEN AND P. T. MCTIGUE, *J. Chem. Soc.*, (1963) 5224.
277 E. S. GOULD, *Mechanism and Structure in Organic Chemistry*, Holt, New York, 1959, p. 540 ff.
278 A. F. HEGARTY AND F. L. SCOTT, *Tetrahedron Letters*, 42 (1965) 3801; *J. Chem. Soc. B*, (1966) 672.
279 K. YATES AND W. V. WRIGHT, *Tetrahedron Letters*, (1965) 1927.
280 R. E. ERICKSON AND T. M. MYSZKIEWICZ, *J. Org. Chem.*, 30 (1965) 4326.
281 R. E. ERICKSON, D. BAKALIK, C. RICHARDS, M. SCANLON AND G. HUDDLESTON, *J. Org. Chem.*, 31 (1966) 416.
282 A. ALBERT AND W. L. F. ARMEAREGO, *Advan. Heterocyclic Chem.*, 4 (1965) 1; D. D. PERRIN, *Advan. Heterocyclic Chem.*, 4 (1965) 43.
283 A. J. SPEZIALE AND D. E. BISSING, *J. Am. Chem. Soc.*, 85 (1963), 1888, 3878.
284 K. M. IBNE-RASA, C. G. LAURO AND J. O. EDWARDS, *J. Am. Chem. Soc.*, 85 (1963) 1165.
285 A. STREHLOW, *Z. Elektrochem.*, 66 (1962) 392.
286 Y. INOUE AND D. D. PERRIN, *J. Chem. Soc.*, (1963) 2648.
287 A. ALBERT AND J. CLARK, *J. Chem. Soc.*, (1964) 1666.
288 D. D. PERRIN, *Advan. Heterocyclic Chem.*, 4 (1965) 43.
289 Y. INOUE AND D. D. PERRIN, *J. Chem. Phys.*, 66 (1962) 1689.
290 Y. INOUE AND D. D. PERRIN, *J. Chem. Soc.*, (1963) 3936.
291 Y. INOUE AND D. D. PERRIN, *J. Chem. Soc.*, (1963) 4803.
292 A. ALBERT, Y. INOUE AND D. D. PERRIN, *J. Chem. Soc.*, (1963) 5151.
293 Y. INOUE AND D. D. PERRIN, *J. Chem. Soc.*, (1963) 5166.
294 Y. INOUE, *Thesis*, Australian National University, 1963.
295 Formation of the 3,4-hydrate in the rate-determining stage; the 7,8-adduct is thermodynamically more stable (ref. 292).
296 W. L. F. ARMAREGO AND J. I. C. SMITH, *J. Chem. Soc. C*, (1966) 234.
297 J. W. BUNTING AND D. D. PERRIN, *J. Chem. Soc. B*, (1966) 433.
298 G. M. BADGER, R. G. BUTTERY AND G. E. LEWIS, *J. Chem. Soc.*, (1953) 2143.
299 E. KOUBEK AND J. O. EDWARDS, *J. Org. Chem.*, 28 (1963) 2157.
300 G. E. LIENHARD AND W. P. JENCKS, *J. Am. Chem. Soc.*, 88 (1966) 3982.
301 D. S. NOYCE AND L. R. SNYDER, *J. Am. Chem. Soc.*, 81 (1959) 620.
302 D. S. NOYCE AND L. R. SNYDER, *J. Am. Chem. Soc.*, 80 (1958) 4324.
303 D. S. NOYCE AND L. R. SNYDER, *J. Am. Chem. Soc.*, 80 (1958) 4033.
304 D. S. NOYCE AND W. A. PRYOR, *J. Am. Chem. Soc.*, 77 (1955) 1397.

Chapter 2

Addition of Unsaturated Compounds to Each Other

P. BELTRAME

1. Introduction

Organic compounds with multiple bonds add to each other in a variety of ways. The following will be considered in detail.

(*i*) cycloaddition of conjugated dienes to unsaturated compounds (Diels–Alder reaction), *e.g.*

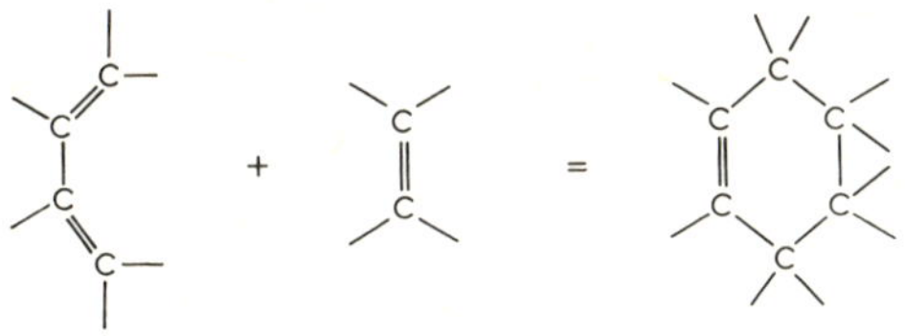

(*ii*) cycloaddition of "1,3-dipoles" to unsaturated compounds (1,3-cycloaddition), *e.g.*

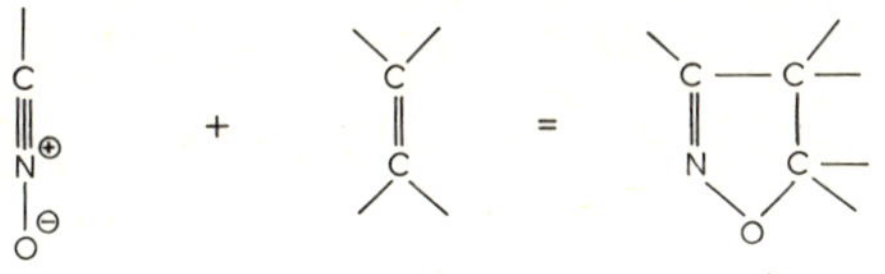

(*iii*) cycloaddition of simple unsaturated compounds to each other (1,2-cycloaddition), *e.g.*

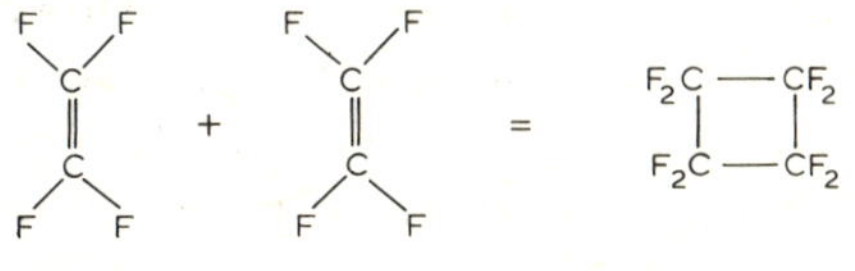

Additions involving hydrogen shifts, and isomerisations through intramolecular cycloaddition will not be covered.

It is intended to consider only reactions where the product corresponds to the sum of the reactants. A more ample definition of cycloaddition has been recently proposed[1], including reactions where elimination of a small molecule accompanies the addition.

Even the strict definition adopted here leaves some uncertainty, as sometimes the pure addition gives rise to an unstable product, whose fate involves further fast reaction, for instance fragmentation, so that the actually isolated products differ from the first adduct. In these cases the rate of formation of ultimate products coincides with the rate of addition only if the addition step is irreversible, *i.e.*

$$\mathrm{A+B} \xrightarrow{\text{slow}} \mathrm{A{-}B} \xrightarrow{\text{fast}} \text{stable products} \tag{1}$$

Some systems behave in a way compatible with (1) and have been included in this chapter. In contrast, the Wittig reaction of "stable" alkyliden-phosphoranes with carbonylic compounds has been shown to conform to a more complex mechanism[2–4], of the following type

$$\mathrm{A+B} \underset{\text{fast}\,(k_{-1})}{\overset{\text{slow}\,(k_1)}{\rightleftarrows}} \mathrm{A{-}B} \xrightarrow[(k_2)]{\text{fast}} \text{stable products} \tag{2}$$

so that a steady-state treatment yields

$$k_{\text{obs}} = \frac{k_1 k_2}{k_{-1} + k_2} \tag{3}$$

The observed rate coefficient k_{obs} gives insufficient information as to the kinetics of the addition step; therefore such a reaction has been omitted.

Less-common addition reactions involving two molecules, and the additions of three or more molecules, are not considered, simply because of the paucity or lack of kinetic data.

Nomenclature of some addition reactions is still a matter of taste. A rigorous classification of cycloadditions, based on the number of atoms entering the ring, puts Diels–Alder reactions in the (4+2) class, *i.e.* 4 from diene and 2 from dienophile, 1,3-cycloadditions in the (3+2) class, etc.[5,6], and has gained general acceptance for complex cases (see for instance ref. 7). However, current names are widely used for the simplest additions, and will be used here.

Kinetic data on addition reactions, and their implications as to the reaction mechanism, have been reviewed in the recent past; this is particularly true for the Diels–Alder reaction[5,8,9] and its reverse reaction[10], and for 1,3-cycloadditions[5,11]. As well, kinetics and mechanism have been reviewed for particular cases, like Diels–Alder reactions of cyanogen[12] and nitroso derivatives[13] as dienophiles, and of cyclopentadienones as dienes[14]. Other discussions of mechanism, of the Diels–Alder reaction[15] and of 1,2-cycloadditions[5,16], are also worth noting. Another particular case is that of 1,3-cycloadditions of nitrile oxides[274].

2. Analytical procedures

2.1 DETERMINATION OF RATE COEFFICIENTS

The unsaturated compounds whose additions have been the object of kinetic study are so numerous and diversified that several analytical methods have been used to follow the reactions.

For processes in the gas phase, static systems (constant volume and temperature) have been mostly used, and the manometric method has been widely employed for Diels–Alder reactions[17–21] and for 1,2-cycloadditions[22–24]; pressure measurements were associated with gravimetric[21], infrared[22] or gas-chromatographic[24] analyses whenever required by the presence of more than one product. For Diels–Alder reactions only nonmanometric methods were used, either weighing the product[25] or chemical analysis of it[26], or gas chromatography of reactants and product[27]. 1,2-Cyclo self-addition of gaseous benzyne has been kinetically followed by UV spectroscopy[28,29].

Reactions in the pure liquid phase, typically the cyclopentadiene dimerisation, have been sometimes studied by refractometry[19,30,31] or dilatometry[32]. In other cases, either determination of product and/or reactants has been preferred[33,34] or the same methods as for reactions in solution were employed. It is worth mentioning that kinetics of a few Diels–Alder reactions were determined by gas-chromatographic analysis of diene, the dienophile being the stationary phase inside the column[35].

Most kinetic studies have been carried in solution, and the main analytical methods are listed in Table 1, with a choice of references. Electronic absorption spectroscopy is the most common method, as in other fields of chemical kinetics. Frequently, unless it has been shown by preparative techniques that more than one product is present, analyses by visible–ultraviolet, as well as by infrared spectroscopy, are effected with wavelengths at which absorption by one of the reactants prevails. Gas-volumetric methods are advantageous for analysis of products in special systems where adducts quickly decompose producing small molecules (see also Section 1.). Gas chromatography could, in principle, be used to give analysis of both reactants and product(s); however it has been mainly used to determine one of them, as the other methods.

TABLE 1

ANALYTICAL METHODS USED FOR DETERMINATION OF RATE COEFFICIENTS OF ADDITIONS IN SOLUTION

Method	*Components analysed*[a]	*Type of reaction*[b]	*Ref.*
Chemical or mixed			
(a) Direct titration			
by iodine bromide	Cyclohexadiene	(i)	36
by iodide	Quinones	(i)	37
by amines	Phenyl isocyanate	(iii)	38, 39
(b) Chemical reaction followed by titration (reaction with benzoic acid)	Diazomethane	(ii)	40
(c) Chemical reaction followed by physical measurement			
gas volume	Aryl azides	(ii)	41
colorimetry–spectrophotometry	Cyclopentadiene	(i)	42, 43
Physical methods			
(d) Colorimetry	Quinone, azoesters	(i)	44, 45
(e) Visible–UV spectroscopy	Anthracene and derivatives	(i)	43, 46–48
	Cyclopentadiene (or various dienophiles)	(i)	49
	Aromatic nitroso compounds	(i)	50
	Furan derivatives	(i)	51, 52
	Tetracyanoethylene complexes	(i)	53
	Pyridinic products	(i)	54
	Triazolinic adducts	(ii)	55
	Diazoalkanes	(ii)	56
	Diphenylketene	(iii)	57
	Azoester	(i)	58
(f) Infrared spectroscopy	Aryl azides	(ii)	59, 60
	Aryl nitrile oxides	(ii)	61
	Aryl nitrile oxide and arylacetylenes	(ii)	62
(g) Gas evolution (N_2, CO_2, CO) from unstable adducts	Adducts from azides	(ii)	60
	Adducts from sydnones	(ii)	63
	Adducts from cyclopentadienones	(i)	64, 65
(h) Dilatometric methods	(Cyclopentadiene dimerisation)	(i)	66
(i) Gravimetric method	Volatile dienes	(i)	67
(j) Vapour pressure	Butadiene	(i)	68
(k) Gas chromatography	Various reagents or adducts	(i)	69, 70
		(iii)	71, 72
(l) Polarimetry	Levopimaric acid	(i)	73

[a] Taking into account, when required, the minor analytical response of other components (*e.g.* in spectrophotometry).

[b] See p. 87.

2.2 DETERMINATION OF RELATIVE RATES

Comparison of the reactivity of two reactants towards a common partner is often effected by the competition method, with the third reactant in the presence of excess quantities of the competing reactants (A and B). If the initial concentrations are $C_A{}^o$ and $C_B{}^o$, and after reaction the final concentrations of products PA and PB are C^e_{PA} and C^e_{PB}, respectively, it is easily shown that the ratio of the rate coefficients for second-order competing reactions is given by

$$\frac{k_A}{k_B} = \frac{\log C_A{}^o - \log (C_A{}^o - C^e_{PA})}{\log C_B{}^o - \log (C_B{}^o - C^e_{PB})} = \frac{\log (1 - C^e_{PA}/C_A{}^o)}{\log (1 - C^e_{PB}/C_B{}^o)} \tag{4}$$

When $C_A{}^o \gg C^e_{PA}$ and $C_B{}^o \gg C^e_{PB}$, the equation above simplifies to

$$\frac{k_A}{k_B} = \frac{C^e_{PA} \cdot C_B{}^o}{C^e_{PB} \cdot C_A{}^o} \tag{5}$$

Such a method requires separate analyses of the reaction products; for addition process this has usually been done by infrared spectroscopy[74–76], gas chromatography[71,77,78], isotopic dilution method[79] or NMR spectroscopy[80]. Analysis of the unreacted fraction of the competing reactants may be used as a substitute for analysis of products, when the latter (or one of them) are slowly decomposing.

Sometimes the aim is to compare different directions of addition in the reaction of two unsymmetrical reagents, that can give rise to mixtures of isomers, *e.g.*

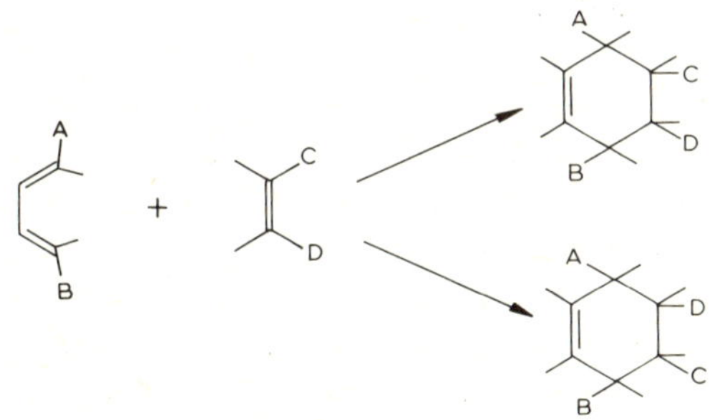

For some reactions of this kind, the products have been subjected to gas-chromatographic analysis[82–84] or to chemical treatment followed by infrared analysis[78], having checked their stability towards isomerisation during the treatment. Also ^{19}F-NMR[85], ^{1}H-NMR[76,86] and IR[86] direct spectroscopic analyses have been used.

Isomeric adducts can also derive from concurrent reactions at different positions of a reactant; a common case is the occurrence of both Diels–Alder and 1,2-cycloaddition products in photochemical and sometimes in thermal reactions between simple olefins and dienes. The analytical problem is usually solved by gas chromatography[87–91].

Often there are stereochemical alternatives, for instance in several Diels–Alder reactions both *endo* and *exo* (or *syn* and *anti*) adducts can be expected. Quantitative differentiation of such configurational isomers has been made possible by various methods: specific chemical treatment (iodolactonisation, to which only *endo*-acids respond) and gravimetry[92], IR or NMR spectroscopic analysis[92,93], gas chromatography[94–96]. Kinetic control was checked in these cases, *i.e.* it has been experimentally verified that under the reaction conditions there is no isomerisation of the products, as could happen if conditions were severe enough to allow reversion to reactants or intramolecular rearrangement.

A similar problem is the determination of the amounts of *cis* and *trans* (or *syn* and *anti*) di- or polysubstituted cyclobutanes from 1,2-cycloaddition of olefins: gas chromatography is frequently used in such cases[87,88,97,98] but also UV[99] and ^{19}F-NMR[98] spectroscopy has been employed.

3. Reproducibility of kinetic results

Several additions have been kinetically studied by more than one research group, but most of them were studied in solution and different workers used different solvents, so that the published data cannot be compared for reproducibility. Such a difficulty does not arise for reactions in the gas and in the pure liquid phases. Two popular Diels–Alder reactions, the dimerisation of cyclopentadiene to endo-dicyclopentadiene and that of butadiene to 4-vinylcyclohexene are suitable for a comparison of experimental results. Rate coefficients from 8 different sources for the former reaction in the pure liquid phase at various temperatures (in the range where comparison is possible) are given in Table 2. The coefficients are all extrapolated to zero time, as both secondary reactions and variation of the environment during the reaction cause a drift in the observed values of k. Rate coefficients for butadiene dimerisation in the gas phase, from different sources, are collected in Table 3. In this case also the temperature range has been limited to that where comparison is possible. In both Tables 2 and 3 activation parameters, as given by the authors (unless otherwise indicated) are also listed.

TABLE 2

EXPERIMENTAL RATE COEFFICIENTS (EXTRAPOLATED TO $t = 0$) ($10^8\,k$ IN L.MOLE^{-1}SEC^{-1}) FOR CYCLOPENTADIENE DIMERISATION IN THE PURE LIQUID PHASE IN THE RANGE $-2°$ TO $+50°$C, TAKEN FROM DIFFERENT SOURCES

Temperature (°C)	*Ref. 100*[a]	*Ref. 101*[a]	*Ref. 19*	*Ref. 102*	*Ref. 30*	*Ref. 103*	*Ref. 31*	*Ref. 32*
−2				6.73	(5.4)[b]			
0			4.50		6.75			
12		25			(24)[b]			
20	50			40.8	50.3			
25					(83)[b]	55	63.1	83.2
30					134.5			
35				192	(213)[b]			
40			282		333		273	
45			350		(512)[b]			
50					(775)[b]		806	
E(kcal.mole⁻¹)			17.1	16.2	17.0		18.3	
log A			6.3	5.8	6.4		7.2	

[a] k values given by ref. 102.
[b] Values in parentheses were calculated from experimental values at the nearest temperatures, by means of the Arrhenius equation, to facilitate comparison.

Two types of comparison are possible, between coefficients and between activation parameters. Rate coefficients at the same temperature agree in the worst case within a factor of 1.7, more often within a factor of 1.5. Taking into account the difficulty, that more or less exists in the whole field of chemical kinetics, of reproducing data by different experimenters in various laboratories, the agreement between rate coefficients is satisfactory. The same cannot be said of the activation energies and the frequency factors. Values of E in Table 2 vary from 16.2 to 18.3 kcal.mole^{-1}. In Table 3, E values between 23.7 and 26.8 kcal.mole^{-1} are reported; for the same reaction a value of 28.0 kcal.mole^{-1} has also been obtained from experiments in the range 400–700°C[104].

Similar variations are found for the dimerisation of gaseous cyclopentadiene. Two sources[19,20] give rate coefficients having a ratio of 1.7 (at 150°C), but activation energies as different as 14.9 and 16.7 kcal.mole^{-1}, and $\log A$ equal to 4.93 and 6.1, respectively.

For the butadiene dimerisation it has been suggested that the Arrhenius plot is non linear[21,25], higher E values being found in the high-temperature range. While such an explanation may be justified in the case of butadiene, where kinetic studies have explored a range of nearly 500°, usually differences arise from errors in evaluating E from coefficients obtained in much narrower temperature ranges. In most cases only a moderate weight should be given,

TABLE 3

EXPERIMENTAL RATE COEFFICIENTS (10^2k IN L.MOLE^{-1}.SEC^{-1}) FOR BUTADIENE DIMERISATION IN GAS PHASE IN THE RANGE 282–418°C, TAKEN FROM DIFFERENT SOURCES

Temperature (°C)	*Ref. 17*	*Ref. 18*	*Ref. 21*	*Ref. 25*
282		0.246	0.419	
291		0.337	(0.59)[a]	
305			0.983	
326	1.56		(2.0)[a]	
333			2.51	
342	2.65		(3.4)[a]	
349			4.27	
367			8.41	
369			8.44	
370	7.21		(8.9)[a]	
384			13.9	
386			15.0	
388	12.0		(15.7)[a]	
418			(34)[a]	41.3
E(kcal.mole^{-1})	26.3[b]	23.7[b]	23.69	26.8
log A	7.78[b]	6.7[b]	6.96	8.14

[a] Calculated values (see footnote b of Table 2)

[b] Activation parameters calculated by the present author (by the least-squares procedure in the case where four values were available).

when examining sets of activation parameters of addition reactions, to differences within the range of 2 kcal.mole^{-1} for E and 1.5 units for log A (that is 7 cal. deg^{-1}.mole^{-1} for activation entropy) in cases when such differences tend to compensate each other.

4. The Diels–Alder reaction

4.1 THERMAL REACTIONS

The thermal reaction of conjugated dienes and unsaturated compounds (dienophiles), with which the names of Diels and Alder are associated, was understood in a general way by these authors about forty years ago[105]. Much preparative, stereochemical and kinetic work on this reaction was done in the thirties, but interest in its mechanism has continued, and a large amount of quantitative data about the Diels–Alder reaction has been collected during recent years. In the following section, only a small fraction of all the data can be reported; the stress will be mainly on recent results.

The reactions are first-order with respect to both diene and dienophile, and overall second-order, except when catalysis introduces a further term in the rate law (see Section 4.1.6).

4.1.1 Stereochemistry and orientation

It is substantially unquestioned that thermal Diels–Alder reactions are strictly *cis* additions, configurations of both diene and dienophile being retained in the adduct[15].

Dienes cannot react with an unsaturated bond unless they are in a *cisoid* conformation; the latter is fixed in some cyclic dienes such as cyclopentadiene and cyclohexadiene, but is equilibrated with other conformations in open-chain dienes, the *transoid* being usually more favoured, *viz.*

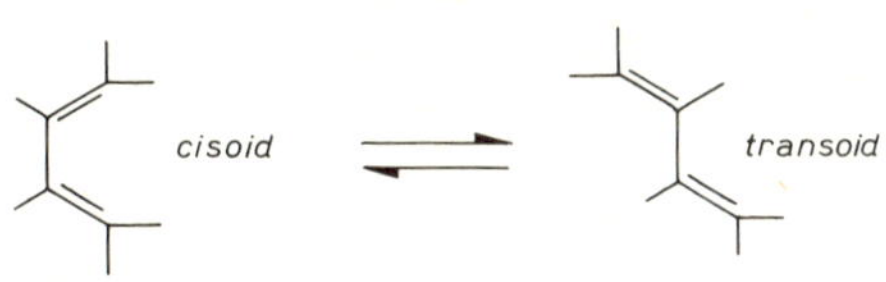

This is certainly one of the reasons why cyclic *cisoid* dienes react faster than open-chain dienes (see Section 4.1.3.).

The stereochemical course of the Diels–Alder reaction, with olefins as dienophiles, follows trends that can be better specified separating different kinds of additions as follows.

(*a*) Monocyclic dienes + cyclic dienophiles; in this case there is exclusive or strongly preferential formation of *endo* adducts as kinetic products[5,9,15], in agreement with the reaction scheme of Alder and Stein[106], commonly known as the "rule of maximum accumulation of unsaturation" in the transition state, *viz.*

(*b*) Monocyclic dienes + open-chain dienophiles; the latter can be (*i*) *cis*-1,2-disubstituted olefins; in this case there is preferential formation of *endo* adducts[15], that can be enhanced by Lewis acid catalysis (Table 4); (*ii*) mono-substituted olefins; a preference for *endo* adducts is common[15], and can be increased by Lewis acids and by lowering the reaction temperature (Table 4); (*iii*) 1,1- or *trans*-1,2- unsymmetrically disubstituted olefins; the stereochemical course depends, by a delicate balance, on the nature of the two substituents[15] and can be reversed by Lewis acids (Table 4).

TABLE 4

STEREOCHEMICAL COURSE OF SOME DIELS–ALDER REACTIONS

Type of addition (See text)	*Reaction*	*Solvent (Temperature °C)*	*Preferred product (% on the sum)*	*Ref.*
(b, i)	Cyclopentadiene +			
	maleic dinitrile	Dioxane (20)	*endo* (76 ± 3)	49
	dimethyl maleate	Benzene (30)	*endo* (73)	107
	dimethyl maleate + $AlCl_3$	Benzene (30)	*endo* (94)	107
(b, ii)	Cyclopentadiene +			
	methyl acrylate	(Various)(3 – 66)	*endo* (72 – 90)	94
	(–) menthyl acrylate	CH_2Cl_2 (0)	*endo* (84)[a]	95
	(–)menthyl acrylate + $AlCl_3.OEt_2$	CH_2Cl_2 (0)	*endo* (93)[b]	95
(b, iii)	Cyclopentadiene +			
	methyl methacrylate	(Various)(3 – 66)	*exo*-ester (59 – 77)	94
	methyl *trans*-crotonate	(Various)(3 – 66)	mostly *endo*-ester (46 – 70)	94
	trans-cinnamic acid	Acetone or toluene (55)	*exo*-acid (57 ± 3)	92
	p-NO_2-*trans*-cinnamic acid	Acetone or toluene (55)	*exo*-acid (70 ± 1)	92
	methyl methacrylate	Benzene (30)	*exo*-ester (69)	107
	methyl methacrylate + $AlCl_3$	Benzene (30)	*endo*-ester (60)	107
(c)	2-Nitro-anthracene + M.A.[c]	Benzene (reflux)	*anti* (61)	93
	2-(Me_2N)Anthracene + M.A.	Benzene(reflux)	*syn* (55)	93
	Naphthacene + acrylonitrile	Acrylonitrile (160 – 164)	*syn* to the benzene moiety (54 ± 1)	96

[a] With weak asymmetric induction: *endo*(+) adduct, 9.1% optical purity at the new asymmetric center[108].
[b] With strong asymmetric induction: *endo*(+) adduct, 49% optical purity, and *exo*(–) adduct, 36% optical purity at the new asymmetric center[108]. Preference for *endo* adduct and (+) optical purity of the latter are both increased by lowering the temperature to – 70°C[95,108].
[c] M.A. = maleic anhydride.

(*c*) Di- or poly-cyclic dienes* + monosubstituted or *cis*-disubstituted olefins as dienophiles; in this case both stereoisomeric adducts have *endo* character, but may present the substituent(s) on the dienophile as *syn* to different moieties, and there can be a detectable preference for one adduct (Table 4).

* Also monocyclic dienes having exocyclic double bonds, like fulvenes, fall in this class[106]: in their adducts the "diene" portion presents π-electron systems on both sides of the plane containing the four atoms involved in the addition, as for naphthalene (attacked in 1,4-positions), anthracene (attacked in positions 9, 10), etc.

(*d*) Open-chain dienes + cyclic dienophiles: the *endo* adduct commonly predominates in kinetic product[15] (a substituent on the diene is obviously required in order to have two possible products).

(*e*) Open-chain dienes + open-chain dienophiles (containing substituents with unsaturated bonds, *e.g.* carboxylic group): preference for *endo* addition (detectable when a substituent is present on the diene) is more evident the lower the temperature[5,9].

Data in Table 4 are results with kinetically controlled reactions, and indicate, therefore, the presence of different free energy barriers along the reaction paths through *endo* and *exo* transition states. Such difference in free energy is rather small; calculating it by the formula

$$\Delta G^{\ddagger}_{exo} - \Delta G^{\ddagger}_{endo} = 4.575T \log(endo/exo) \qquad (6)$$

one obtains from the figures in Table 4 [(*b*, *i*) and (*b*, *ii*)] values ranging from 0.5 to 1.7 kcal.mole^{-1}. Even if "exclusive" formation of *endo* adducts reported in other cases means 99.9% *endo* and 0.1% *exo* product, the free energy difference is only 4.1 kcal.mole^{-1} ($T = 300$°K).

Such *endo* preference is usually explained by attractive interactions (between π-electron systems or because of dipole induction forces) between the diene and the dienophile at non-bonding positions in the transition state (see Section 7); this seems to justify both its regularity and the small values of the relative stabilisation energy. Other interpretations point to the direction that steric repulsion in some *exo* transition states (for instance when the diene is cyclopentadiene, with its methylene hydrogen atoms) must give to the course of the addition[15].

The reactivity difference along *endo* and *exo* reaction paths has been shown by Berson *et al.*[94] to depend clearly, although slightly, on the solvent polarity in the case of cyclopentadiene addition to some acrylic esters; on increasing the polarity of the solvent, as defined by the Z parameter[109], the logarithm of the *endo*/*exo* ratio in the adduct increases in a roughly linear manner (Fig. 1).

An other correlation has been found between log (*endo*/*exo*) in the cyclopentadiene addition to ring-substituted *trans*-cinnamic acids[92] and Hammett-type substituent constants; the low absolute value of the slope (−0.30)[110] indicates very similar substituent effects on the two reaction paths. For the reaction of 2-substituted anthracenes with maleic anhydride[93] the influence of substituents on log (*syn*/*anti*) is even weaker[110].

Interesting cases of catalysis by Lewis acids are reported in Table 4 (see also Section 4.1.6).

Limited quantitative data on orientation of unsymmetrical dienes and dienophiles in kinetically controlled Diels–Alder reactions are available. Most of the work in this field is preparative, and a discussion based on relative yields

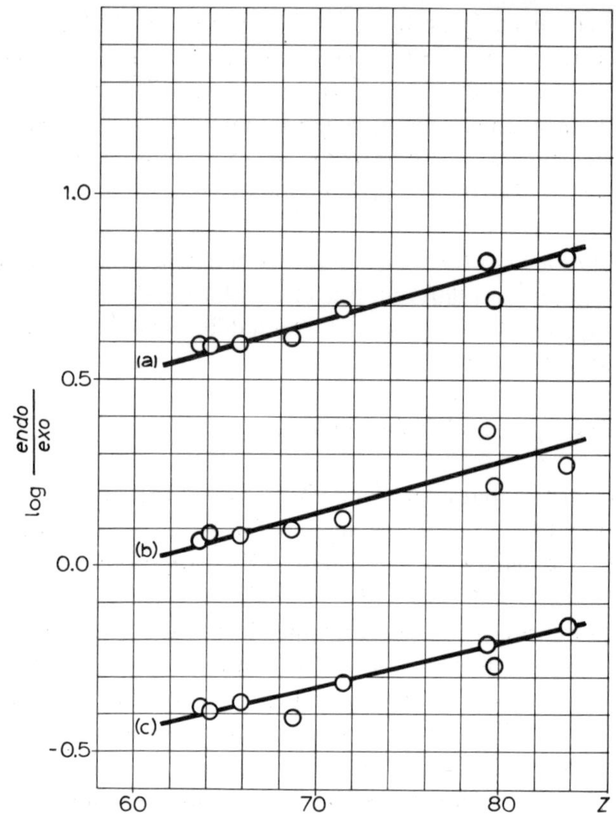

Fig. 1. The Diels–Alder addition of cyclopentadiene to (a) methyl acrylate, (b) methyl *trans*-crotonate, and (c) methyl methacrylate at 30°. Dependence of the stereochemistry on the solvent polarity[94]. Solvents are $C_2H_4Cl_2$, pyridine, acetone, DMF, MeCN, AcOH, EtOH and MeOH.

(often of derivatives of the primary adducts) is a bad substitute for one of relative rates. Taking mono-substituted butadiene derivatives as dienes, the following observations can be made.

(*a*) With open-chain monosubstituted olefins (and acetylenes) as dienophiles, there is preference for adducts with vicinal ("*ortho*") substituents from 1-substituted butadienes, *viz.*

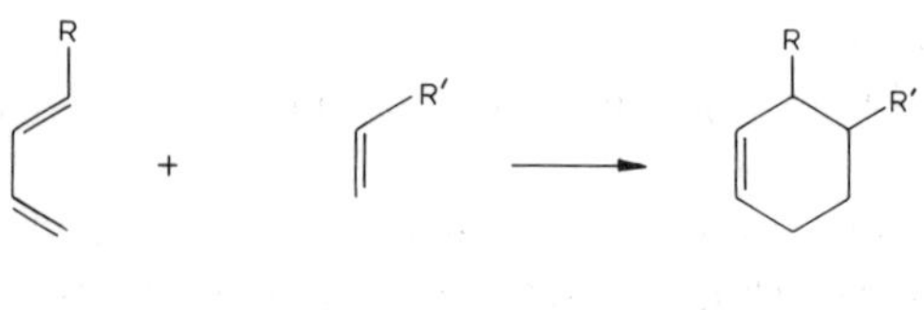

and for "*para*" disubstituted adducts from 2-substituted butadienes[5,8,9]. This preference is stronger when Lewis acid catalysts are used, as is shown by the following examples.

(*i*) *trans*-piperylene + Me-acrylate[111]
benzene (25°C) no catalyst; "*ortho*"/"*meta*" = 90:10
benzene (10–20°C) $AlCl_3$; "*ortho*"/"*meta*" = 98:2
(*ii*) isoprene + Me-acrylate[78]
(25°C) no catalyst; "*para*"/"*meta*" = 70:30
(10–20°C) $AlCl_3$; "*para*"/"*meta*" = 95:5
(*iii*) isoprene + Me-vinyl-ketone[83]
toluene (120°C) no catalyst; "*para*"/"*meta*" = 71:29
benzene (< 25°C) $SnCl_4.5H_2O$; "*para*"/"*meta*" = 93:7.

For systems like (*ii*) it has been proved that the ratio of products in uncatalysed reactions is only slightly dependent[82] or even independent[112] of temperature over a wide range.

(*b*) With nitroso-aromatics as dienophiles, asymmetry is extended from the field of substituents to the very centers of reaction in the polarised (δ+)N=O(δ−) bond; in this case 1-R-butadienes give on many occasions only adducts with R "*ortho*" to oxygen[113,114] (R = Ph, *p*-$MeOC_6H_4$, *p*-$NO_2C_6H_4$, COOMe, OCOMe, NMe_2, *t*-Bu),

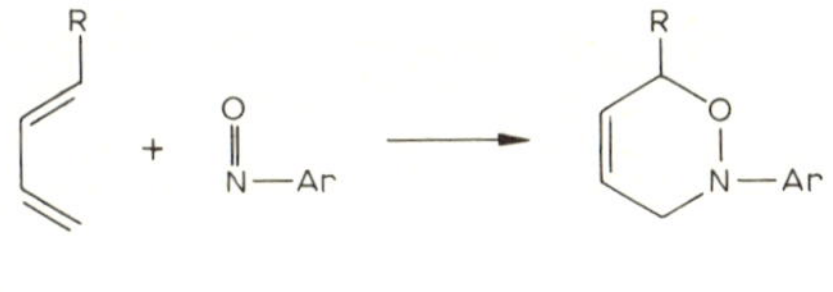

presenting what has been defined as *regiospecificity*[115], a notable exception being when R = Me, in which only a low regioselectivity in the same direction has been observed (58% "*ortho* to O" product)[114]. 2-R-butadienes usually give regiospecific adducts with R "*para*" to oxygen (R = Ph, *p*-$MeOC_6H_4$, *p*-ClC_6H_4, CN, Me, Cl, CF_3) but two isomeric products were obtained for R = OMe[116].

(*c*) With *N*-sulphinyl amines, the site of attack is the polarised (δ−)N=S(δ+) bond; preferred products from addition of 1-substituted and 2-substituted butadienes under kinetic control seem to be "*ortho* to N" and "*para* to N", respectively, substituents being aryl groups, methyl, chlorine[117].

(*d*) For carbon–nitrogen multiple bonds there are indications that a similar preference may exist. Acetonitrile adds to isoprene in the gas phase at 400°C at short contact times giving 2,4-dimethylpyridine[118]; from isoprene and CF_3CN two isomers were obtained, but the ratio of 4-Me-2-(CF_3) pyridine to the 5-Me isomer[12] was 3; the 1-substituted butadiene, piperylene, with CF_3CN

gave 84% 6-Me-2-(CF_3) pyridine[12]. If such reactions are accepted to be irreversible Diels–Alder additions followed by fast dehydrogenation[12], the prefered adducts seem to be again those with the diene substituent in the positions "*para*" or "*ortho*" to the electronegative pole of the dienophile. In the last two cases, however, only methyl has been used as substituent on the diene.

4.1.2 Medium effects

A large amount of experimental evidence has been cited to show that the rates of several Diels–Alder additions are little dependent on the medium. Comparison of rates in the gas and liquid phases can only occasionally be made. The dimerisation of cyclopentadiene has the following coefficients (l.mole^{-1}.sec^{-1}) at 20°C:

gas phase: $10^7 k = 5.3$ (calculated from the experimental value at the nearest temperature, 79°C, by means of Arrhenius equation[20])
pure liquid phase: $10^7 k = 4.1 - 5.0$ (Table 2)
in solution (8 different solvents)[5]: $10^7 k = 6.6 - 19.0$.

For the dimerisation of butadiene, some kinetic results for the reaction in the pure liquid phase are available. Rate coefficients can be calculated from experimental conversion data at low total conversion[33] to a good approximation*, and so compared with coefficients for the gas-phase reaction, calculated from kinetic results at higher temperatures[21], by using the appropriate Arrhenius equation; the k values (l.mole^{-1}.sec^{-1}) are:

gas phase: $10^9 k = 2.6$ (60°C); 24 (82°C)
pure liquid phase: $10^9 k = 3.4$ (60°C); 21 (82°C).

Thus the rate of both cyclopentadiene and butadiene dimerisations are remarkably little affected by a change of phase.

The reactions in solution proceed at slightly different rates in different solvents, but the variation, beside being small, is not apparently related to the polarity of the solvent. In Table 5 some recent results are presented: the reactions represent extreme cases of strongly electrophilic dienophile (a) and electrophilic diene (c), besides an addition of identical molecules (b) and a case of a heteroatomic dienophile (d). Hydroxylic solvents are separated from the others, which are ordered according to decreasing values of dielectric constant. No regular trend is apparent: the variation of rate coefficients along a column is always within one order of magnitude, and particularly small in case (d). In Table 5 k values are higher with hydroxylic than with aprotic solvents; however differences are small and furthermore the trend is not general for Diels–Alder reactions[8,9]. When measured for a given reaction at various temperatures, the effect of solvent proved slight and irregular in the whole range examined[119a].

* From the formula given in ref. 25 in a footnote on p. 202; density of butadiene taken from ref. 119.

TABLE 5

EFFECT OF SOLVENT ON THE RATE COEFFICIENTS ($L.MOLE^{-1}SEC^{-1}$) OF SOME DIELS–ALDER REACTIONS

(a) Anthracene + tetracyanoethylene[53]; (b) dimerisation of 3,4-dimethyl-*o*-benzoquinone[37]; (c) hexachlorocyclopentadiene + styrene[9], (d) 1,3-cyclohexadiene + nitrosobenzene[120].

Solvent (dielectric constant)	*(a) k (30°C)*	*(b) 10^4k (30°C)*	*(c) 10^5k (85°C)*	*(d) 10^2k (25°C)*
Methanol (ϵ_{25} = 32.6)		62		
Ethanol (ϵ_{25} = 24.3)		35		1.16
Dimethylformamide (ϵ_{20} = 37.7)		16	11.1	
Acetonitrile (ϵ_{20} = 37.5)	[a]	17		
Nitrobenzene (ϵ_{25} = 34.8)			10.2	
Cyclohexanone (ϵ_{20} = 18.3)	0.260			
Dichloromethane (ϵ_{20} = 9.08)	[a]	6		1.07
Chloroform (ϵ_{20} = 4.81)	[a]	8		1.11
Toluene (ϵ_{25} = 2.38)	0.184[b]	15	6.6	
Benzene (ϵ_{25} = 2.27)	0.497[b]	18		1.00
Carbon tetrachloride (ϵ_{20} = 2.24)				1.04
Dioxane (ϵ_{25} = 2.21)		27	6.8	

[a] Rate coefficients given for the reaction in some solvents are in disagreement with those obtainable from activation parameters given by the same authors[53]; such values are not accepted here.

[b] Calculated by means of Arrhenius equation from the experimental value at the nearest temperature.

4.1.3 *Influence of the structure of reactants*

Most of the classical dienes are hydrocarbons, like cyclopentadiene, butadiene, anthracene, 9,10-dimethylanthracene, isoprene, 2,3-dimethylbutadiene. For dienes of this kind it is generally true that electrophilic dienophiles are the most reactive, and actually acrylic derivatives, maleic anhydride, *p*-benzoquinone and similar compounds have found a large use as dienophiles both for preparative purposes and for kinetic studies. The latter demonstrated quantitatively the importance of electronegative groups on the dienophiles, and conversely of electron-releasing substituents on the diene, in order to accelerate such type of Diels–Alder reaction. It was also realised later that Diels–Alder additions with "inverse electron demand", that is between electrophilic dienes and nucleophilic dienophiles, do occur[121].

Some kinetic data, showing how a change of substituents at the diene unsaturated bonds can affect the rate of Diels–Alder reactions of the first kind, are collected in Table 6. Dienophiles are of the >C=C<, –N=N– and –N=O types; their reactions give the expected products, *e.g.*

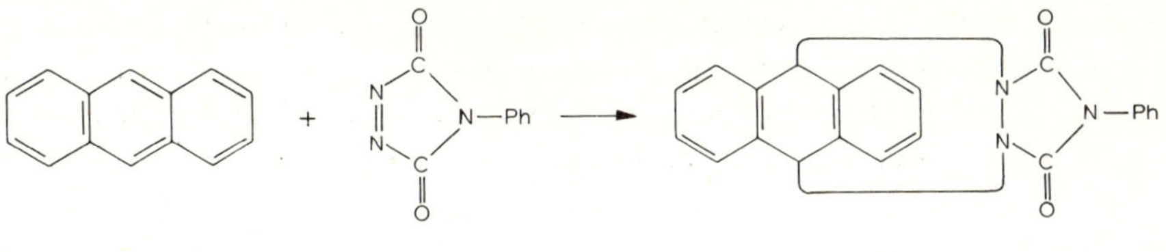

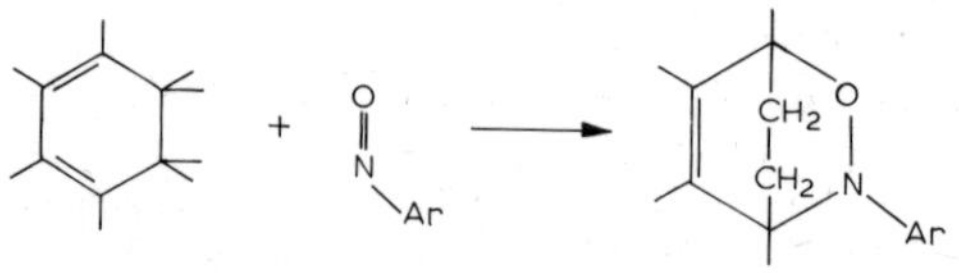

TABLE 6

REACTIVITY OF DIFFERENT DIENES WITH SOME ELECTROPHILIC DIENOPHILES

Electrophilic dienophiles: (i) maleic anhydride (benzene, 25°C)[67], unless otherwise indicated; (ii) tetracyanoethylene (dioxane, 20°C)[9]; (iii) 4-phenyl-1,2,4-triazolin-3,5-dione (dioxane, 20°C)[122]; (iv) *p*-chloro-nitrosobenzene (solvents as indicated, 10°C).

Diene	*(i)* 10^5k (*l.mole*$^{-1}$*sec*$^{-1}$)	*(ii)* 10^5k (*l.mole*$^{-1}$*sec*$^{-1}$)	*(iii)* 10^5k (*l.mole*$^{-1}$*sec*$^{-1}$)	*(iv)* 10^5k (*l.mole*$^{-1}$*sec*$^{-1}$)
Butadiene	5.37	519		
2-Chlorobutadiene	0.52	1.02[c]	2.36×10^4	
Isoprene	15.7	1130		
trans-Piperylene	25.6	2060		
2,3-Dimethylbutadiene	55.6	2.43×10^4		164[d,h]
1-Methoxybutadiene	165	5.98×10^5		
trans-1-Phenylbutadiene	4.28[a]			101[e,h]
trans,trans-1,4-Diphenylbutadiene	0.296[a]	840[c]	1.15×10^5	10.4[f,h]
2-Phenylbutadiene	*ca.* 60[a]			4080[e,h]
Anthracene	0.62[b]	2.76×10^4[c]	7100	
9,10-Dimethylanthracene	1600[a]	*ca.* 1.3×10^9		
Cyclopentadiene	9210[a]	*ca.* 4.3×10^7		
Hexachlorocyclopentadiene	1.14×10^{-3}[a]	*ca.* 0	27.6	
Cyclohexadiene	15.2	7290		1200[g]

[a] Dioxane, 30°C[9,123].
[b] Dioxane, 25°C[48].
[c] Ref. 122.
[d] Dichloromethane[124].
[e] Benzene[116,125].
[f] Chlorobenzene[125].
[g] 96% ethanol[125].
[h] Calculated by means of the Arrhenius equation from the experimental value at the nearest temperature.

The different kinetic behaviour of dienophiles, according to their substituents at the unsaturated bond, is illustrated in Table 7, by considering only carbon dienophiles (either of C=C or of C≡C type). Two out of four dienes chosen are of the electron-rich type, and two of the electrophilic type; Diels–Alder reactions of both kinds are therefore presented.

Substitution directly at the atoms involved in the cycloaddition can have, of course, both electronic and steric consequences. Direct steric effects are absent when substituents are varied at a *meta* or *para* position in an aryl group. The latter type of substituent effect is generally apt to be correlated by the Hammett equation: ρ values for some reactions are given in Table 8.

TABLE 7

REACTIVITY OF VARIOUS DIENOPHILES WITH SOME DIENES

Dienes: (i) butadiene (benzene, 25°C); (ii) cyclopentadiene (dioxane, 20°C)[49]; (iii) hexachlorocyclopentadiene (dioxane, 130°C)[126]; (iv) dimethyl-1,2,4,5-tetrazine-3,6-dicarboxylate (dioxane, 30°C)[9,127]

Dienophile	*(i)* 10^8k *(l.mole^{-1}sec^{-1})*	*(ii)* 10^8k *(l.mole^{-1}sec^{-1})*	*(iii)* 10^5k *(l.mole^{-1}sec^{-1})*	*(iv)* 10^8k *(l.mole^{-1}sec^{-1})*
Tetracyanoethylene		*ca.* 4.3×10^{10}		
1,1-Dicyanoethylene		*ca.* 4.6×10^{7}		
Maleic anhydride	{5370[a] 3500[b]	5.56×10^{6}	2.91	
p-Benzoquinone	306[c]	8.97×10^{5}		
Maleic dinitrile		9.10×10^{4}		
Dimethyl fumarate		7.42×10^{4}		
Dimethyl maleate		628		
Dimethyl acetylene dicarboxylate		3.13×10^{4}		
Acrolein	0.55[d,e]	7360[h,e]		
Acrylonitrile		1040		930
Methyl acrylate	1.68[f,e]	1180		1.17×10^{5}
Styrene			75	6.55×10^{6}
Ethylene	2.2×10^{-5}[g,e]			3.63×10^{7}
1-Octadecene			9.47[i,e]	
Cyclopentadiene		94	1520	
Ethyl vinyl ether				2.16×10^{7}

[a] Ref. 67.
[b] Ref. 128.
[c] Ref. 68.
[d] In the gas phase[18].
[e] Calculated by means of the Arrhenius equation from the experimental value at the nearest temperature.
[f] Ref. 70.
[g] In the gas phase[25].
[h] Benzene, 20°C[42].
[i] No solvent, 130°C[129].

Tables 7 and 8 contain, among ordinary Diels–Alder additions, some reactions of the type already discussed (Section 1) that give isolatable products different from the first adduct. This is true for the tetrazine diene (Table 7, iv), *viz.*

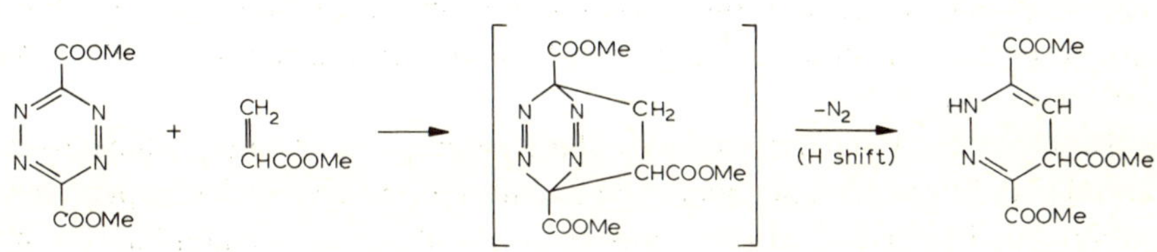

and for tetraphenyl-cyclopentadienone (Table 8,d), *viz.*

TABLE 8

ELECTRONIC EFFECTS OF SUBSTITUENTS TRANSMITTED THROUGH PHENYL RINGS

Reaction (solvent)	*Substituents employed*	*Temp. (°C)*	ρ	*Ref.*
(a) 1-Aryl-butadiene + maleic anhydride (dioxane)	*p*-OMe, *p*-Me, H, *p*-Cl, *p*-NO_2	25	−0.685	79
		35	−0.621	
		45	−0.606	
(b) 1-Aryl-butadiene + *p*-chloronitroso-benzene (benzene)	*p*-OMe, *p*-Me, H, *p*-Cl, *p*-NO_2	15	−0.25	125
		25	−0.24	
		35	−0.22	
(c) 2-Aryl-butadiene + *p*-chloronitroso-benzene (benzene)	*p*-OMe, *p*-Me, H, *p*-Cl, *m*-Cl	14	−0.43	116
		24	−0.40	
		34	−0.36	
		44	−0.33	
(d) Tetraphenyl-cyclopentadienone + methyl arylpropiolate (phenylcyclohexane)	*p*-OMe, *p*-Me, *m*-Me, H, *m*-OMe, *p*-Cl, *m*-Cl, *m*-NO_2, *p*-NO_2	175.6	+1.10[a]	64
(e) Cyclohexadiene + substituted nitrosobenzene (96% ethanol)	*p*-OMe, *p*-Me, H, *p*-Cl, *m*-NO_2, *p*-NO_2	10	+2.57	125
(f) 2,3-Dimethylbutadiene + substituted nitrosobenzene (dichloromethane)	*p*-Me, H, *p*-Cl, *p*-Br, *p*-I, *m*-Cl, *m*-Br	2	+2.53	124
		25	+2.50	
(g) Substituted cinnamaldehyde[b] + octyl vinyl ether	*p*-OMe, *p*-Me, H, *p*-Cl, *m*-Cl, *m*-NO_2	150	+0.71	130
(h) Cinnamaldehyde[b] + substituted styrene	*p*-Me, H	170	(−0.67)	131
(i) Dimethyl 1,2,4,5-tetrazine-3,6-dicarboxylate + substituted styrene (dioxane)	*p*-OMe, H, *p*-NO_2	30	*ca.* −1.5	9, 127

[a] Using a σ scale determined from ionisation of arylpropiolic acids in 50% ethanol. From the Hammett σ values, a ρ-value of +0.76 results.

[b] The reacting diene is the >C=C–C=O system, the product a dihydropyran derivative.

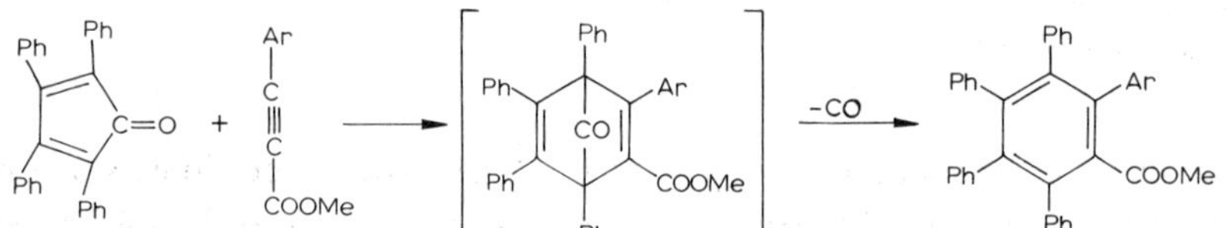

However, the actual addition step being slow and substantially irreversible, the measured rate of these reactions can be considered as a rate of Diels–Alder cycloadditions[9,64,127].

Both Tables 6 and 7 frequently allow comparison of rate coefficients measured in different solvents, and even in different phases. Such comparisons are acceptable because of general considerations about the small influence of the medium (Section 4.1.2), taking into account that only the order of magnitude of k is often required.

It is apparent from Table 6 that methyl and methoxy substituents enhance the reactivity of butadiene and anthracene, while chlorine atoms decrease that of butadiene and cyclopentadiene, towards electrophilic dienophiles. Phenyl substituents have very different effects in positions 1 and 2 [columns (i) and (iv)]: here the substituent is large enough to exert a marked steric influence on the *cisoid* ⇄ *transoid* equilibrium of the diene, so that the *cisoid* conformation of 2-phenyl-butadiene is favoured, and therefore the overall reactivity of 2-phenyl-butadiene is enhanced, while *trans*-1-phenyl-butadiene is little affected. Cyclic unsubstituted dienes are more reactive than butadiene; however, the difference between cyclopentadiene and cyclohexadiene is enormous: the high reactivity of the former must be in part attributed to factors other than the *cisoid* conformation.

A reactivity order of general validity, even if only for Diels–Alder reactions of the first kind, cannot be written. It will suffice to consider the sequence 2-chlorobutadiene, 1,4-diphenylbutadiene and anthracene, for which Table 6 gives the following relative rates.

(*i*) 1 : *ca.* 0.57 : *ca.* 1.2

(*ii*) $1:824:2.7 \times 10^4$

(*iii*) 1 : 4.87 : 0.30

It is apparent that specific factors interfere; for instance, anthracene is able to give charge-transfer complexes with some dienophiles, notably maleic anhydride[46] and tetracyanoethylene[53].

With this limitation in mind, and taking into account the differences in reaction temperature, one can nevertheless read Table 6 along the rows, to have an idea of the "strength" of different dienophiles in Diels–Alder reactions of the first kind: tetracyanoethylene (ii) is in any case more reactive than maleic anhydride (i), and the phenyl-triazolindione (iii) is, in three cases out of four, the most efficient of the dienophiles listed; *p*-chloro-nitrosobenzene is intermediate between (i) and (ii).

In Table 7 a large number of dienophiles are listed in order of roughly decreasing electrophilicity. Apart from some anomalies, rate coefficients decrease in the same order in cases (i) and (ii), while they increase in cases (iii) and (iv). This confirms the difference in electronic requirements of the two kinds of Diels–Alder reactions. Unfortunately few values are available in columns (iii) and (iv), as the chemical literature contains much more kinetic data about the classical type of Diels–Alder addition than about the reactions of electrophilic dienes.

From Table 7 (column ii) a notable difference of reactivity between dimethyl fumarate and dimethyl maleate is seen. This is not fortuitous, since, in most cases, *trans*-1,2-disubstituted ethylenes were found to be better or much better dienophiles than their *cis* isomers, in classical Diels–Alder reactions: k_{trans}/k_{cis} ratios up to 740 are reported[132]. Steric compression in adducts obtained from *cis*-isomers (already present, in part, in the corresponding transition states) should explain this fact[5].

Comparing the acetylene dicarboxylic ester with the fumaric ester, a slightly higher reactivity of the C=C over the C≡C dienophile is noted. The efficiency of C=C and C≡N bonds as dienophiles was compared by studying the reaction of butadiene with acrylonitrile[133]: mainly cyanocyclohexene and no vinylpyridine was found in the product, proving that the C=C bond is far more reactive than the C≡N bond, at least in the homogeneous gas-phase reaction (heterogeneous catalysts lower the reactivity ratio to *ca.* 10)[133].

Electronic effects transmitted through aromatic systems are in fair agreement with previous observations. Table 8 shows that, for the number of Diels–Alder reactions (a–f), substituents on the diene affect the reaction to give negative ρ's, while substitution on the dienophile corresponds to positive ρ values. On the contrary, the second group of reactions (g–i) shows a positive ρ in the case where substituents are on the diene and negative ρ's when they are on the dienophile; (although the last ρ values are obtained from only two or three points, their sign is well defined). The first group of reactions includes electron-rich dienes (only for tetraphenyl-cyclopentadienone may this classification be questionable because of the presence of a carbonyl group); electrophilic dienes are involved on the other hand in reactions of the second group.

In general, the sensitivity of Diels–Alder reactions to electronic effects transmitted through phenyl rings, as expressed by ρ values, is low but significant: it demonstrates that in the transition state there is some charge unbalance, but it proves also, at least in most of the cases, that the addition is not comparable to a nucleophilic (or electrophilic) attack of one neutral reagent on another to give a dipolar intermediate. Some doubt may arise about the reactions of nitroso-aromatic dienophiles, that show the highest ρ's, around 2.5 (see (e) and (f) in Table 8).

Sometimes correlations of kinetic results with Hammett σ constants present difficulties. Reactions (a), (d) and (i) of Table 8 give curved plots, all with the concavity upwards, independently of the signs of the average slopes. For reaction (d), however, it has been shown that a linear plot can be obtained using a σ^- scale, based on ionisation of phenols in various solvents: adopting the new abscissa, the ρ value is lowered[134] to something between 0.2 and 0.3.

On the other hand, the reaction of aryl-phenylfulvenes with maleic anhydride in benzene at 20.6°C seems to exhibit a maximum, although not well pronounced, in the Hammett plot; the unsubstituted diphenylfulvene has the highest rate coefficient, while *p*-Me, *p*-OMe and *p*-Cl substituents all decrease the rate[135]. However, in the reaction with methyl acetylene dicarboxylate in heptane at 60–80°C, (*p*-chlorophenyl)-phenylfulvene was found to be slightly *more* reactive than diphenylfulvene[136]. It is, in any case, a matter of small differences.

Some kinetic data on Diels–Alder reactions, substitution being mainly directly at the unsaturated bonds of the reactants, were correlated using an extended Hammett equation[110], *viz.*

$$\log k_X = \alpha\sigma_{I,X} + \beta\sigma_{R,X} + h \tag{7}$$

where $\sigma_{I,X}$ and $\sigma_{R,X}$ are constants characterising localised and delocalised effects, respectively, of a substituent X; additivity of effects was assumed in case of disubstitution. It has been found that parameters α and β are often little changed when going from a monosubstituted to a disubstituted (in symmetrical positions) diene or dienophile. This supports the hypothesis of a rate-determining step for addition in which two positions in each reactant are involved, as otherwise the sensitivity of the reaction to substituents would significantly change with the site(s) of substitution[110]. However, by an analogous line of reasoning, it has been shown that the two reacting positions in each reactant are not necessarily involved to the same extent[119a].

4.1.4 Activation parameters

The author has tried to collect in Table 9 a representative sample of Diels–Alder reactions with different types of reactants, for which the usual activation parameters could be determined with care (in most cases from kinetic runs at four or more different temperatures, in a range of at least 20°C). Other data (E and $\log A$ values) have been given in Tables 2 and 3 (pp. 93, 94); the relevant values of the activation entropy can be calculated from $\log A$ by the conversion formulae[137]

For bimolecular gas-phase reactions at constant volume:
$$\frac{\Delta S^{\ddagger}}{R} = \ln A - \ln \frac{kT}{h} - 2 \tag{8}$$

For liquid-phase reactions*:
$$\frac{\Delta S^{\ddagger}}{R} = \ln A - \ln \frac{kT}{h} - 1 \tag{9}$$

with unit concentration (1 mole.l^{-1}) as the standard state. Values of $\Delta S^{\ddagger}$ between −28 and −34 eu are obtained for pure liquid cyclopentadiene dimerisation (Table 2), and values between −27 and −33 eu for gas-phase butadiene dimerisation (Table 3).

Gas-phase, pure liquid-phase and solution reactions, involving either electron-rich or electrophilic dienes, and dienophiles with C=C, C≡C or N=O

* This formula holds also for gas-phase reactions at constant pressure and for unimolecular reactions.

TABLE 9

ACTIVATION PARAMETERS OF DIELS–ALDER REACTIONS[a]

Reaction (solvent)	*E (kcal.mole⁻¹)*	*log A*	$\Delta S^{\ddagger}$ *(eu)*	*Ref.*
(a) Butadiene + acrolein (gas phase)	19.7	6.2	−35	18
(b) Butadiene + ethylene (gas phase)	27.5	7.5	−30	25
(c) Butadiene + methyl acrylate (benzene)	18.0	5.4	−36.2	70
(d) Anthracene + tetracyanoethylene				
(cyclohexanone)	10.3	6.9	−29	53
(benzene)	9.8	6.7	−30	
(e) 9,10-Dimethylanthracene + methyl acrylate (dioxane)	15.5	5.25	−36	49
(f) Cyclopentadiene + dimethylfumarate (dioxane)	11.7	5.62	−35	49
(g) Cyclopentadiene + dimethyl maleate (dioxane)	14.6	5.72	−34	49
(h) Tetraphenyl-cyclopentadienone + methyl phenylpropiolate (phenylcyclohexane)	20.0	6.8	−30	134
(i) 2,3-Dimethylbutadiene + X–C_6H_4NO:				
(dichloromethane) X = *m*-Cl	11.0[b]	5.9[b]	−33[b]	124
(dichloromethane) X = *p*-Me	12.6[b]	6.0[b]	−33[b]	
(j) 1-Phenylbutadiene + *p*-Cl–C_6H_4NO (benzene)	12.2	6.4	−31	125
(k) 2-Phenylbutadiene + *p*-Cl–C_6H_4NO (benzene)	7.8	4.6	−39	116
(l) Hexachlorocyclopentadiene + maleic anhydride (dioxane)	19.1	5.80	−35	126
(m) Hexachlorocyclopentadiene + cyclopentadiene (dioxane)	14.0	5.77	−35	126

[a] Values of log A and $\Delta S^{\ddagger}$ were reported to as many figures as were given by the relevant authors. When a parameter was not directly available, it was calculated from the corresponding one (*e.g.* $\Delta S^{\ddagger}$ from log A) by the appropriate conversion formula at a temperature in the range of the experiments. A is in l.mole^{-1}.sec^{-1}.

[b] Activation parameters recalculated by the least-squares procedure from rate coefficients given in Ref. 124.

unsaturated bonds can now be compared. They are characterised by very different reactivities; at room temperature, rate coefficients (l.mole^{-1}.sec^{-1}) vary from of the order of 10^{-13} [(b) in Table 9] to that of 10^{-1}, (k). This is reflected in the large variation of activation energy, from 7.8 gradually up to 27.5 kcal.mole^{-1}, whereas the activation entropy is hardly affected. The great majority of all values of $\Delta S^{\ddagger}$ available in the literature lie between −30 and −40 cal. deg^{-1}.mole^{-1}. This is one of the typical features of Diels–Alder and other cycloadditions. It is specially interesting to check in Table 9 that differences of reactivity between isomeric compounds (fumarate compared to maleate as dienophiles, 1-phenyl-butadiene compared to 2-phenyl-butadiene as dienes) are mainly due to differences in E values.

The almost constant and low values of the entropy of activation, and the fact that they are of the same order for the gas phase and for solution, point to a mechanism dependent on the approach of the reactants to form a transition state with rather well-defined geometry and heavy loss of rotational entropy, but little dependent on solvation phenomena. These requirements are met by a cyclic transition state and a concerted mechanism of cycloaddition. Furthermore, reverse Diels–Alder reactions, which are all unimolecular processes[10], have high frequency factors ($A \approx 10^{11}$–10^{15} sec^{-1}), corresponding to $\Delta S^{\ddagger}$ values around zero; some examples are given in Table 10. The typical transition state of a Diels–Alder addition seems, therefore, to have a geometric form close to that of the corresponding adduct, although it differs very much from the adduct with respect to the bond orders, as is shown by the high E values of the reverse reaction (Table 10).

The cyclic transition state and the one-step mechanism are also consistent with the *cis*-stereospecificity of the reaction (Section 4.1.1), the small medium effects (Section 4.1.2) and the observations concerning substituent effects (Section 4.1.3).

However a typical Diels–Alder reaction, the butadiene dimerisation (Table 3, p. 94), and its reversal (case (g) in Table 10) have been recently considered by Benson[142] as examples of two-step processes. The initial argument is that 1,5-cyclooctadiene, when pyrolysed in the gas phase, gives 4-vinylcyclohexene as the main product but also a small amount of butadiene, which cannot be produced from 4-vinylcyclohexene under the conditions of the experiments. A kinetic and thermochemical analysis shows that the *cis-cis*-1,7-octadiene-3,6-diyl biradical (singlet) could be the common precursor of 4-vinylcyclohexene

TABLE 10

ENERGY AND ENTROPY OF ACTIVATION OF REVERSE DIELS–ALDER REACTIONS

Reaction	*Medium (temperature range, °C)*	*E (kcal.mole⁻¹)*	$\Delta S^{\ddagger}$ [a] *(eu)*	*Ref.*
(a) Decomposition of cyclopentadiene–acrolein adduct	Gas phase (192–242)	33.6	−5	18
(b) Decomposition of cyclopentadiene–benzoquinone adduct	Benzene (55–79)	29	−3	138
(c) *endo*-Dicyclopentadiene = 2-cyclopentadiene	Paraffin (145–175)	34.2	−2	138
(d) As in (c)	Gas phase (152–211)	34.0	−2	139
(e) Cyclohexene = butadiene + ethylene	Gas phase (541–629)	66.2	+7	140
(f) As in (e)	Gas phase (630–880)	66.7	+6	141
(g) 4-Vinylcyclohexene = 2 butadiene	Gas phase (630–780)	62.0	+7	141

[a] Calculated, whenever necessary, from log A by eqn. (9) at a temperature in the range of the experiments.

and butadiene, and the intermediate in butadiene dimerisation and, of course, in 4-vinylcyclohexene decomposition[142]. A similar scheme was proposed for chloroprene thermal dimerisation, which gives both Diels–Alder and 1,2-cycloadducts, the competing reactions having very similar activation parameters and solvent effects[142a].

Evidence in favour of a two-step mechanism has also been given in the case of the cycloaddition of perfluorocyclopentadiene to cyclopentadiene[143]. Two products are formed, the most abundant from the addition having cyclopentadiene as the diene and perfluorocyclopentadiene as the dienophile, the other from the opposite. The ratio of the two products maintains practically constant (81–84% to 19–16%) on changing the temperature from 20° to 120°C and the solvent from *n*-hexane to nitrobenzene, although they are kinetically stable products. A common intermediate whose formation is rate-determining and which is able to collapse quickly to either product, seems required to explain these facts. It could be represented as

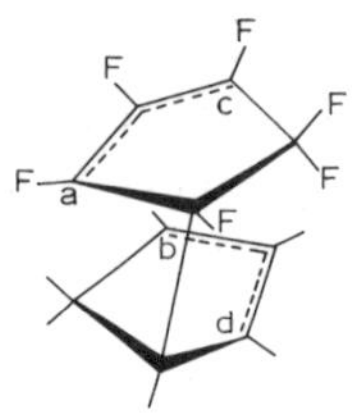

(interactions between the two allyl systems being omitted), and can give one product by bonding a with b, the other product by bonding c with d[143].

A less-common activation parameter, the volume of activation ($\Delta V^{\ddagger}$), has been determined for a few Diels–Alder reactions carried out under pressure in liquid phase. The processes are cyclopentadiene dimerisation[30,69], isoprene dimerisation[34], addition of 2,3-dimethylbutadiene to butyl acrylate[69], addition of cyclopentadiene to dimethyl acetylene dicarboxylate[143a], and addition of maleic anhydride to 1,3-cyclohexadiene, *trans*-1-methoxybutadiene and isoprene[143a]. Activation volumes are negative, i.e. the reacting systems contract on passing from the initial to the transition state. In some cases the transition state appears to be even smaller than the adduct, independently of the solvent[143a]. Some of these experimental results gave rise to controversial interpretations[69,144]; however, the most recent ones favour a concerted four-center mechanism for the reaction.

4.1.5 *Isotope effects*

Secondary kinetic isotope effects corresponding to change from H to D at the carbon atom site of attack (α effects) have been measured for a number of additions to olefins. The measured effects are approximately justified by an increase in H–C bending frequency involving the α proton, as the attacked carbon atom modifies its hybridisation[145] from sp^2 to sp^3; accordingly, the activation process is accompanied by a gain in zero-point energy and therefore $k_D > k_H$. Values between 1.03 and 1.09 were obtained for $(k_D/k_H)_{25°C}$ per D atom, for additions of CH_3 and CF_3 radicals to ethylene, propene and styrene[146,147]. Similar effects were noted for addition of other reactants[145].

For every Diels–Alder reaction the H/D α kinetic isotope effect of which has been measured, the qualitative findings are the same, i.e. (k_D/k_H) is larger than unity. The magnitude of the effect is also of the expected order, both when values per D atom are calculated assuming one-step addition and when attack at a single site is assumed to be rate-controlling (for symmetrical deuteration, the former assumption attributes the observed effect to twice as many atoms as the latter). A few typical examples are listed below (isotope effects per D atom, as calculated for one-step addition, are given).

Reaction	k_D/k_H *at 25°C*
d_8-isoprene + maleic anhydride (benzene)[67]:	1.06
1,1,4,4-d_4-butadiene + maleic anhydride (dichloromethane)[148]:	1.07
cyclopentadiene + d_2-maleic anhydride (dichloromethane)[148]:	1.03
9,10-d_2-anthracene + tetracyanoethylene (chlorobenzene)[149]:	1.06

Within experimental uncertainty, the effect is substantially independent of solvent[149].

A reverse Diels–Alder reaction, the decomposition of 2-methylfuran–maleic anhydride *exo*-adduct (in iso-octane) was the object of an extensive study of isotope effects[150]. Here rehybridisation is from sp^3 to sp^2 and therefore a ratio $k_H/k_D > 1$ is to be anticipated. This has, in fact, been found: $(k_H/k_D)_{50°C} = 1.08$ per D atom, both for all possible deuterations on the maleic anhydride moiety (see below), either at X or at Y or at X and at Y, and for deuteration of the furan moiety at Z.

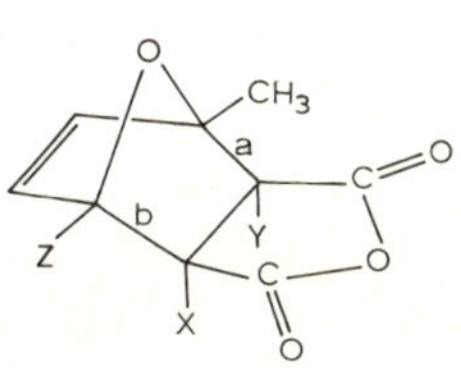

It was concluded that the breaking of bonds a and b takes place in a single step. Trideuteration of the methyl group also allowed the measurement of a small β secondary isotope effect ($k_{CH_3}/k_{CD_3} = 1.03 \pm 0.01$)[150].

An elegant study was the measurement of the intramolecular isotope effect in the addition of hexachlorocyclopentadiene to 1,1-d_2-allene through NMR analysis of the kinetic product[151]. It was found that $(k_D/k_H) = 1.08$ per D atom at 25°C. A two-step mechanism, including rate-determining attack at the central allene carbon atom with formation of an allyl system, and subsequent fast ring closure, can therefore be discarded; in such a case no α secondary effect would be involved in the first step, and during the second step, which would require rotation of a methylene, CD_2 would rotate more slowly than CH_2[151,152].

The one-step mechanism does not exclude some unbalance in the advancement of bond formation at the transition state. For the reaction of 9-*d*-anthracene and acrylonitrile the addition of the unsubstituted end of the dienophile to the deuterated position of the diene was indeed found to be favoured, the relative rates being 1.07[149].

4.1.6 *Action of catalysts*

Peroxides and other substances able to influence the rates of radical processes were proved to be without influence on Diels–Alder reactions[8]. Catalytic power was mainly recognized for normal acids, phenols and Lewis acids.

A general scheme for the interpretation of the catalysis is

$$\mathrm{P+C} \underset{k_{-1}}{\overset{k_1}{\rightleftharpoons}} \mathrm{P.C} \tag{10}$$

$$\mathrm{D+P.C} \xrightarrow{k_2} \mathrm{Adduct+C} \tag{11}$$

where P is the dienophile, D the diene and C the catalyst, and it has been assumed that the catalyst acts on the dienophile, either by proton transfer or hydrogen bonding or by some other chemical interaction, giving a complex P.C which is active in the catalytic reaction. A similar scheme could be given for bonding between catalyst and diene. In principle, the uncatalysed reaction (rate coefficient k_U) must be considered together with the catalytic process.

Various cases may occur.

(*a*) Equilibrium (10) is displaced to the left; applying the stationary-state approximation to the catalytic reaction one obtains

$$\text{rate} = k_U[\mathrm{D}][\mathrm{P}] + \frac{k_1 k_2[\mathrm{C}][\mathrm{D}][\mathrm{P}]}{k_{-1} + k_2[\mathrm{D}]} \tag{12}$$

Now, if step (11) is rate-controlling ($k_{-1} \gg k_2$[D])

$$\text{rate} = k_U[\text{D}][\text{P}] + \frac{k_1 k_2}{k_{-1}}[\text{C}][\text{D}][\text{P}]$$

and in condensed form ($k_C = k_1 k_2 / k_{-1}$)

$$\text{rate} = k_U[\text{D}][\text{P}] + k_C[\text{C}][\text{D}][\text{P}] \quad (13)$$

(*b*) Similar to case (*a*), but the forward reaction of step (10) is rate-controlling ($k_{-1} \ll k_2$[D]); then

$$\text{rate} = k_U[\text{D}][\text{P}] + k_1[\text{C}][\text{P}] \quad (14)$$

(*c*) Equilibrium (10) is displaced to the right, and substantially all of the catalyst (being present in much smaller concentration than the other reagents) is present in the form of a soluble complex; the rate of the catalytic reaction is therefore insensitive to the excess concentration of free dienophile and the rate-law becomes:

$$\text{rate} = k_U[\text{D}][\text{P}] + k_2[\text{C}][\text{D}] \quad (15)$$

When eq. (13) holds, the efficiency of the catalyst can be measured by the ratio k_C/k_U, which gives the relative rate of catalysed and uncatalysed reaction for a (hypothetical) catalyst concentration of unity. When eqn. (14) is valid, the ratio k_1/k_U compares the reactivity of the dienophile towards the catalyst with that towards the diene. Finally, when eqn. (15) is valid, the ratio k_2/k_U measures the relative reactivity of the dienophile–catalyst complex and free dienophile with the same diene. Often the uncatalysed reaction is negligible in the presence of the catalyst, but k_U is, anyway, known from separate experiments.

Several acids and phenols were used as catalysts in reactions of cyclopentadiene with itself and with quinones[66,153]. Cases (*a*) and (*b*) have been met with; examples are given below:

(*a*) cyclopentadiene + *p*-benzoquinone (benzene, 25°C)

$10^3 k_U = 8.7$ l.mole^{-1}.sec^{-1} (ref. 42)

Catalyst: PhOH[66,153]; $10^3 k_C = 130$ l^2.mole^{-2}.sec^{-1}; $k_C/k_U = 15$ l.mole^{-1}

$CH_2ClCOOH$[153]; $10^3 k_C = 200$ l^2.mole^{-2}.sec^{-1}; $k_C/k_U = 23$ l.mole^{-1}

CCl_3COOH[66]; $10^3k_C = 3170$ l^2.mole^{-2}.sec^{-1};
$k_C/k_U = 360$ l.mole^{-1}

(*b*) dimerisation of cyclopentadiene (CCl_4, 25°C)

$10^6k_U = 1.37$ l.mole^{-1}.sec^{-1} (ref. 102)
Catalyst: HCl[66]; $10^6k_1 = 117$ l.mole^{-1}.sec^{-1}; $k_1/k_U = 85$

It has recently been confirmed by differential thermal analysis that dimerisation of pure liquid cyclopentadiene is accelerated by hydrochloric acid[154].

Case (*b*) was also encountered in the cyclopentadiene + dialkylazodicarboxylate reaction, where solvents containing active hydrogen atoms act themselves as catalysts, and kinetics tend toward overall first-order (the uncatalysed reaction being negligible)[45].

Later, Lewis acids were suggested as catalysts for Diels–Alder additions[155–157]; in the case of the reaction of anthracene with maleic anhydride the acceleration due to $AlCl_3$ was estimated[156] to be of the order of 10^5. As has been mentioned (Section 4.1.1), in the presence of Lewis acids *endo* adducts are more favoured than in uncatalysed reactions, and there is more selectivity in orientation when unsymmetrical dienes and dienophiles add to each other; furthermore, with an optically active reactant, asymmetric induction can be stronger in the catalysed reaction (Table 4, footnote b). It should be appreciated that substrates sensitive to $AlCl_3$ and similar catalysts are always polar molecules, usually containing carboxyl or carbonyl groups, to which Lewis acids can become bound.

A kinetic study has been carried out on the methyl acrylate addition to butadiene in benzene, with and without $AlCl_3$ as catalyst[70]. The rate-law has the form of eqn. (15), case (*c*). The ratio k_2/k_U amounts to 10^5 at 20°C. The activation parameters are

from k_U: $E = 18.0$ kcal.mole^{-1}; $\log A$ (l.mole^{-1}.sec^{-1}) = 5.4;
$\Delta S^{\ddagger} = -36.2$ cal.deg^{-1}.mole^{-1}
from k_2: $E = 10.4$ kcal.mole^{-1}; $\log A$ (l.mole^{-1}.sec^{-1}) = 4.8;
$\Delta S^{\ddagger} = -37.7$ cal.deg^{-1}.mole^{-1}

showing that the catalytic action is due mainly to a reduction in the *E* value, the activation entropy being substantially unaltered. It has been proposed[70] that the acrylate.$AlCl_3$ complex may actually carry out a true electrophilic attack on the diene, but the dipolar intermediate so obtained is held in a fixed conformation by electrostatic and other electronic interactions, as suggested by Woodward and Katz[158] in their mechanistic model for Diels–Alder reactions.

The substituent effect in the $AlCl_3$-catalysed reaction of butadiene and methyl acrylate was investigated by competition experiments[78,111]. While for uncatalysed additions (neat) at room temperature $k_{butadiene}:k_{piperylene}:k_{isoprene} = 1:1.9:2.16$, for catalysed reactions (in benzene) at about 10°C it was found that $k_{butadiene}:k_{piperylene}:k_{isoprene}:k_{2,3\text{-}Me_2\text{-}butadiene} = 1:6.5:12.1:36.2$. The sensitivity to methyl-substitution on the diene has been increased by the presence of the catalyst: a Hammett-type correlation was found to give quantitative expression to this fact[158a]. For purpose of comparison, one may consider the reaction of butadiene with maleic anhydride, for which (in benzene, 25°C; case (*a*) in Table 6) $k_{butadiene}:k_{piperylene}:k_{isoprene}:k_{2,3\text{-}Me_2\text{-}butadiene} = 1:4.8:2.9:10.4$.

The catalyst clearly makes methyl acrylate more electrophilic, however not much more electrophilic than maleic anhydride. This fact, together with the $\Delta S^\ddagger$ value in the range typical of common Diels–Alder reactions, and with the stereochemical and orientational features mentioned, suggests that the catalysed reaction is similar in mechanism to the uncatalysed one[158b]. The electronic interactions that govern the course of the addition seem to be only strengthened. The stereochemistry of the $AlCl_3$ catalysed reaction of butadiene with 2-phenyl-2-cyclohexen-1-one[158c] favours, however, the existence of a dipolar intermediate.

Also the addition of the dicyanoacetylene.$AlBr_3$ complex to aromatics (exceptionally behaving as dienes) is very sensitive to methyl substituents on the diene: $k_{p\text{-}xylene}/k_{benzene}$ has been found to be 400 at 25°C in competition experiments[159].

Metal complexes can occasionally give good results as catalysts in Diels–Alder reactions, as for butadiene and isoprene dimerisations with $Fe(CO)_2(NO)_2$[160] and butadiene addition to 2-butyne in the presence of Fe (cyclooctatetraene)$_2$[160a]. However, these catalysts usually promote formation of several products in non-specific manner, through a two-step mechanism[161,162], not dissimilar to most photochemical processes (see following section).

4.2 PHOTOCHEMICAL REACTIONS

Additions of dienes to unsaturated compounds under the influence of direct irradiation or by photosensitised reaction are usually quite different from thermal Diels–Alder reactions. A most important difference is in the structure of the products: most photochemical additions give a mixture of products, among which the 1,4-cycloadduct, if present at all, is often not the major component, 1,2-cycloaddition being preferred. These reactions will be dealt with in Section 6, together with a few thermal reactions having the same characteristics*.

* For a review of photochemical cycloadditions of dienes see Dilling[162a].

However, it may be anticipated that the portion of photochemical product, which structurally corresponds to the Diels–Alder adduct, does not do so stereochemically. In the photosensitised dimerisation of cyclopentadiene, *exo*-dicyclopentadiene is obtained in about the same amount as the *endo*-isomer, together with an equal amount of *trans*-1,2-cycloadduct[87], while the kinetic product of the thermal reaction is exclusively *endo*-dicyclopentadiene. 1,3-Cyclohexadiene dimerisation is similar, the kinetic thermal product being 80% *endo* and 20% *exo* Diels–Alder dimer, while the photosensitised dimerisation gives, together with major amounts of cyclobutane derivatives (4 parts), the *exo* Diels–Alder dimer (1 part), with only traces of the *endo* isomer[88]. *Exo* Diels–Alder adducts were obtained also from photosensitised addition of cyclic dienes to maleic anhydride[163].

A third difference from thermal reactions is the lack of stereospecificity in photochemical 1,4-cycloadditions. This can be shown by reaction of *o*-quinones with olefins, a process that, with careful choice of the olefin, proceeds mainly as addition to give 1,4-dioxenes (although 1,2-cycloaddition to give oxetane derivatives is also possible)[164], *viz.*

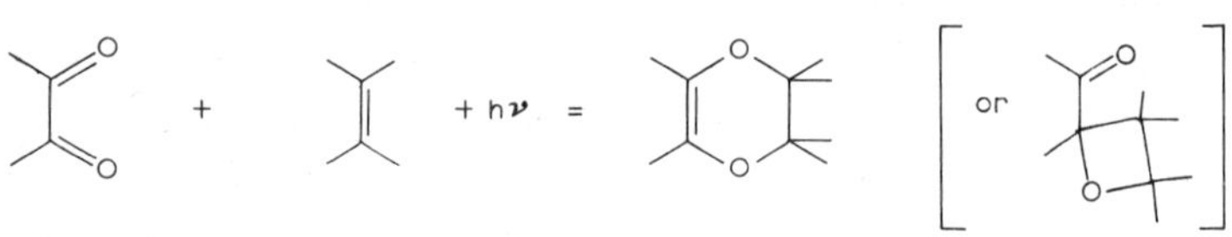

As quinones are typical photosensitisers, and irradiation is with light of long wavelength absorbed only by the quinone, these reactions proceed through triplet intermediates.

Phenanthroquinone was induced to react with *cis*- and *trans*-stilbene, separately, under irradiation ($\lambda > 390$ nm), and the stereochemistry of the 1,4-dioxenes produced was examined by NMR spectroscopy; after a small correction, because of some isomerisation of the stilbenes induced by quinone, the product distribution is[165]

addition to *cis*-stilbene: *cis* product from 72% (15°C) to 60% (70°C)
addition to *trans*-stilbene: *trans* product from 81% (15°C) to 65% (70°C).

In this case stereoequilibration is not attained, even at the highest temperature; the intermediate biradical, *viz.*

"*cis*" biradical "*trans*" biradical

produced by addition of an olefin to the triplet quinone, retains in part the conformation of the olefin, as the process of spin inversion and ring closure is apparently competitive with internal rotation. From data at four different temperatures, the difference in activation energy between internal rotation and adduct formation has been evaluated as 3.8 $kcal.mole^{-1}$ for the *cis*-biradical, and 4.8 $kcal.mole^{-1}$ for the *trans* biradical[165].

An even larger stereoselectivity has been found in the reaction of tetrachloro-*o*-benzoquinone with *trans*- and *cis*-stilbene under irradiation at 15°C ($\lambda > 400$ nm)[166]: from *trans*-stilbene in benzene, 88% *trans* product, and from *cis*-stilbene in acetone or acetonitrile 83% *cis* product are reported. A large measure of the steric control is thought to derive from the dipolar character of the intermediate[166].

When phenanthroquinone is added (under irradiation) to *cis*- and *trans*-2-butene, stereospecificity is completely absent: a mixture of 57% *cis* and 43% *trans* product is obtained in both cases[167], showing that it is easier for this biradical to attain conformational equilibrium.

Rates of the cycloaddition of photoexcited phenanthroquinone to some methyl-substituted ethylenes were measured[168]. Although the reaction gives other products, 1,4-dioxenes are the main ones; their rate of formation increases with the number of methyl substituents on the double bond.

An exception, with respect to the photosensitised reactions mentioned so far, is represented by the Diels–Alder addition of maleic anhydride to anthracene in dioxane, carried out under irradiation ($\lambda = 365$ nm) without sensitisers, at 26–45°C[48]. The rate of removal of anthracene was found to be linearly dependent on the total light absorption. As maleic anhydride quenches the fluorescence (transition from excited singlet to ground state) of anthracene, and in agreement with the kinetic evidence, a mechanism was suggested by which singlet anthracene is responsible for cycloaddition rather than triplet anthracene. According to this mechanism, the reaction of the excited diene with maleic anhydride has a rate coefficient of about 3×10^7 $l.mole^{-1}.sec^{-1}$, *i.e.* of the same order as the reported frequency factor of the thermal reaction[48,123].

5. 1,3-Cycloaddition

Several additions of unsaturated compounds to molecules with a resonant dipole, leading to five-membered rings according to

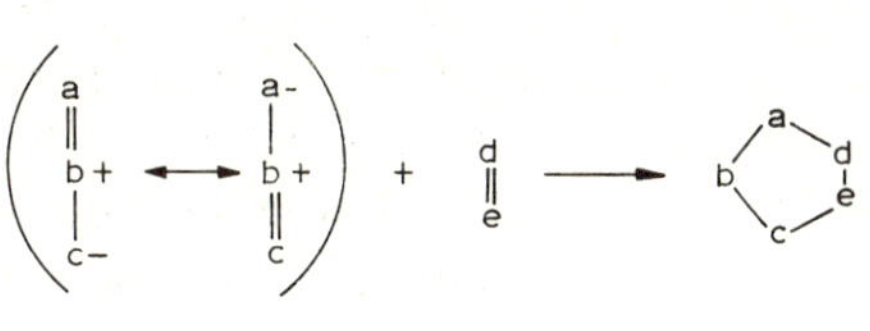

have been known for a long time[169]. The idea of sextet formulae, already occasionally considered as additional contributions to the resonance hybrid in order to justify, for instance, the 1,3-cycloadditions of diazomethane[170], *viz.*

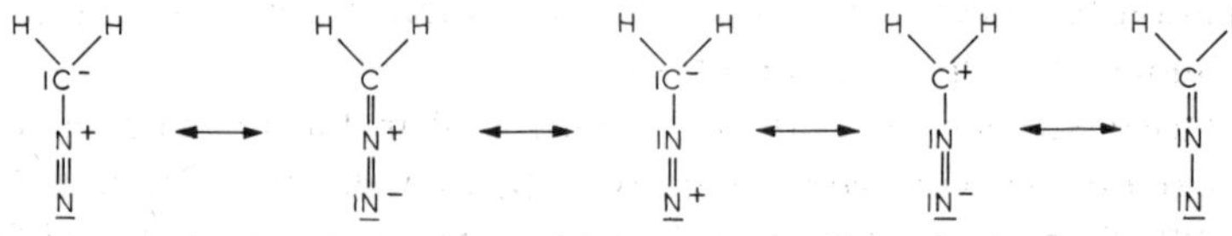

was developed into the generalized concept of "1,3-dipoles", and an extensive investigation into the whole field of 1,3-cycloadditions initiated by Huisgen *et al.*[171] in 1957–58.

Most experimental work in this field is recent; this is particularly true for kinetic measurements, reported only after 1960. A limited but significant part of these measurements will be considered here.

The following 1,3-dipoles will be considered: (*a*) aryl azides; (*b*) diazoalkanes; (*c*) aryl nitrile oxides; (*d*) nitrile imines; (*e*) azomethine imines; (*f*) azomethine oxides; (*g*) azomethine ylides. (*a*) to (*d*) represent 1,3-dipoles "with a double bond" in their sextet structure, while the last three, from (*e*) to (*g*), are "without a double bond"[171]. All of them have nitrogen as the central atom of the 1,3-dipole. They will be formulated as allyl-like systems, having their negative charge distributed (according to an unspecified balance) at the two sides of the positive nitrogen, *e.g.*

diazomethane azomethine oxide (nitrone)

The π-bond indicated as an ordinary double bond in 1,3-dipoles (*a*) to (*d*) is, of course, in a plane perpendicular to that of the allyl-like three-centered π-bond. The extra π bond must be lost if the 1,3-dipole has to bend in the activation process of a 1,3-cycloaddition[172,173].

1,3-Cycloadditions were found to be first-order reactions with respect to both 1,3-dipole and dipolarophile, in all cases so far investigated. There are some limits to kinetic studies of these reactions, as many 1,3-dipoles are very reactive substances. While aryl azides, diazoalkanes, some classes of azomethine imines (for instance sydnones), and some classes of azomethine oxides (nitrones) are stable and isolable, azomethine ylides are usually unstable, an exception being represented by a mesoionic oxazolone that has been used for kinetic investigations[174]; benzonitrile oxide has a very limited stability, although some substituted derivatives are stable for long periods; nitrile imines are not commonly isolable because of their strong tendency to dimerise. 1,3-Dipoles of the last type, and also some that belong to other classes, can be tested for their reactivity towards dipolarophiles only by preparing them in the actual presence of dipolarophiles, and are therefore useful at most for measurements of relative rates by competition experiments (see Section 2.2).

5.1 STEREOCHEMISTRY AND ORIENTATION

The course and orientation of 1,3-cycloadditions will be illustrated by listing examples of actual reactions. Most dipolarophiles have carbon–carbon unsaturated bonds, and give derivatives of 1,2,3-triazole, pyrazole, isoxazole and pyrrole according to the following reactions (unless otherwise indicated, reference is made to monosubstituted dipolarophiles; olefins and acetylenes usually, but not always, behave similarly):

(a)

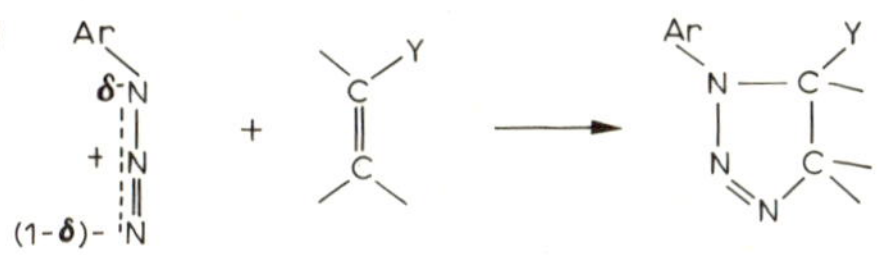

Y = NR_2, OR, Ph*
(ref. 59,175)

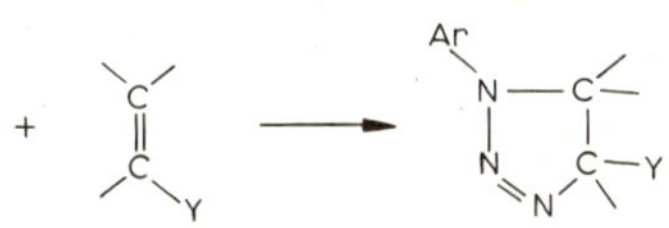

Y = COOR, CN, Ph*
(ref. 59)

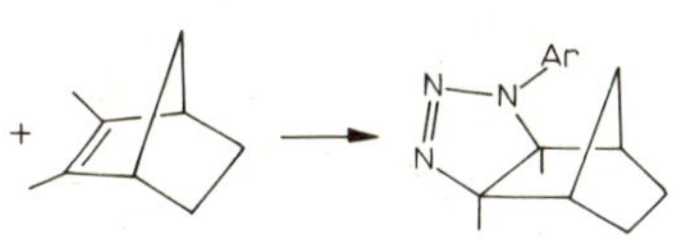

(ref. 41)

(b)

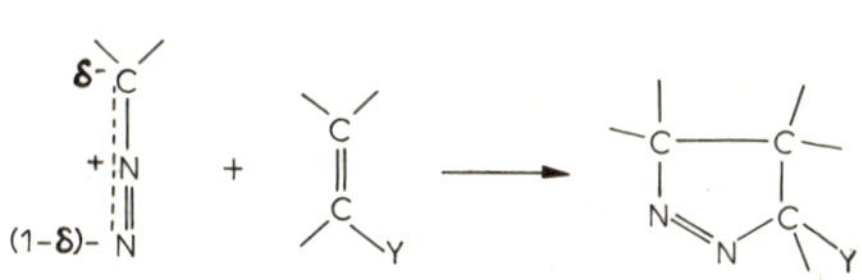

Y = COOR, CN, Ph, CH=CH_2
(ref. 5,171)

(c)

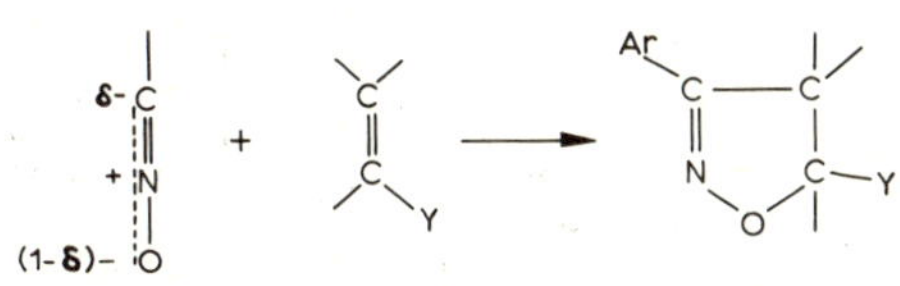

Y = COOR**, Ph, CH=CH_2,
Alkyl, OR
(ref. 176, 176a)

* The possibility of mixed orientation also exists for phenylacetylene and may extend to other monosubstituted alkynes[171].

** According to a recent report[177], high yields of 4-COOMe-substituted product can be obtained from methyl propiolate addition.

(d)

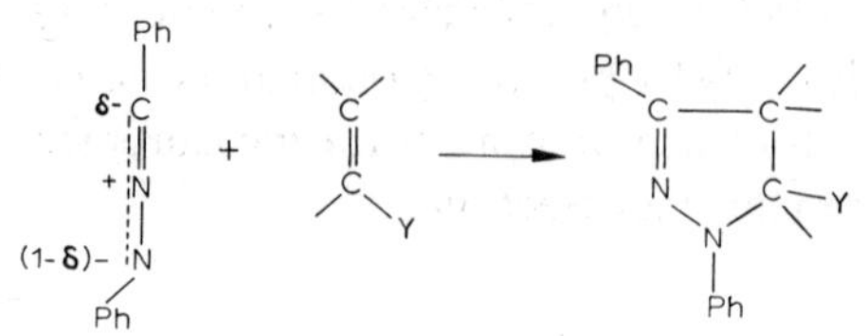

Y = COOR, CN, Ph, CH=CH$_2$, Alkyl (ref. 171)

(e_1)

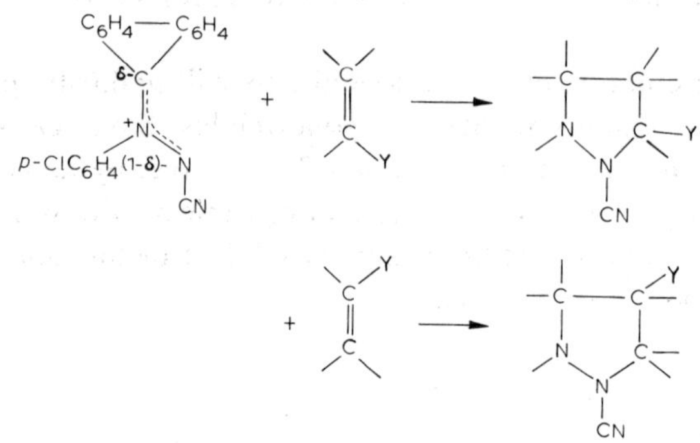

Y = Ph (ref. 171)

Y = COOR (ref. 5)

(e_2)

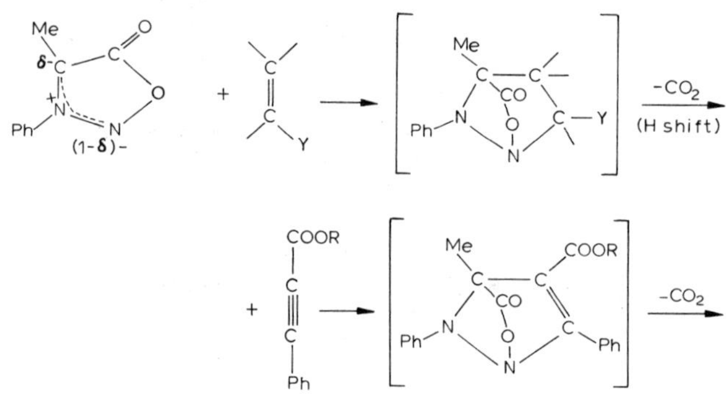

Y = COOR, Ph *

stable product

stable product (ref. 63)

(f)

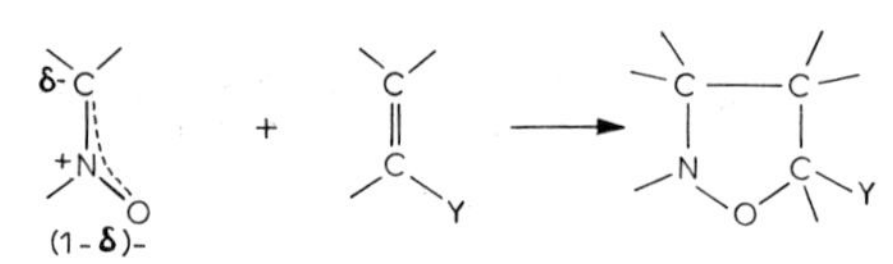

Y = COOR**, Ph, CH$_2$OH, Alkyl (ref. 171)

(g)

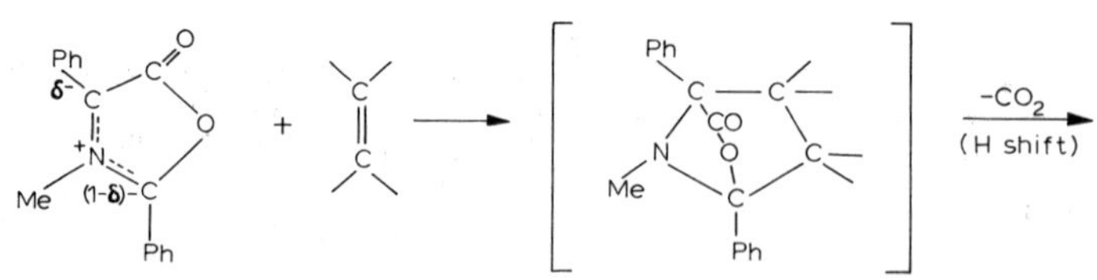

stable product (ref. 174)

* This orientation is indicated for olefin additions[171]; it is accompanied by the alternative orientation, although to a minor extent, in acetylene additions[174].

** The indicated orientation is that of the kinetic product; the reaction can be reversible and give a thermodynamic product of opposed orientation[178].

Less common is the use of dipolarophiles with heteroatoms. The following reactions may be mentioned, as they have been studied also from a kinetic point of view.

(H-b)

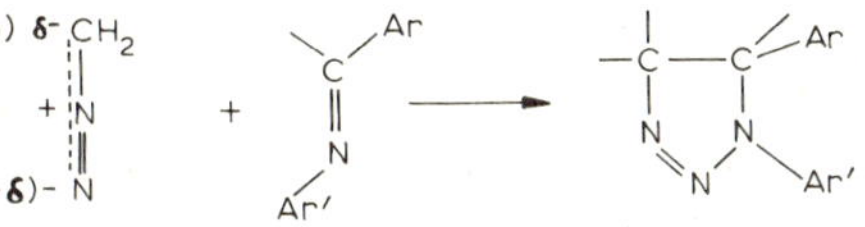

(ref. 179)

(H-c)

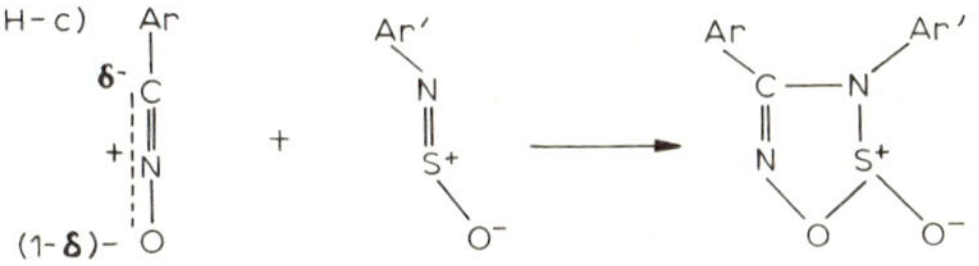

(ref. 180,181)

1,3-Cycloadditions are stereospecific *cis* additions, as has been shown in several cases using geometric isomers as dipolarophiles[11]. In addition to the rigid structure of norbornene, as well-defined approach is preferred, namely the one that gives an *exo* adduct, as is shown in (*a*) above for azides, but also occurs with *C*-phenyl-*N*-methyl-nitrone[182] and for diphenylnitrilimine[183]. The alternative approach of the reactants is sterically hindered in the case of norbornene; the steric course of reactions of norbornadiene can be different, as was found using phenyl azide as the 1,3-dipole[183a].

Examples given above show that several degrees of regio-selectivity[115] are possible in reactions with monosubstituted carbon dipolarophiles, *viz.*

(*i*) specificity or high selectivity in orientation, independent of the substituent's nature; (reactions (*b*), (*c*), (*d*), and (*f*):

(*ii*) specific orientation, but differing for different substituents; (e_1) and most of reactions like (*a*):

(*iii*) both orientations followed in the addition of the same dipolarophile, as for styrene and phenylacetylene in (*a*), for methyl propiolate in (*c*), and for phenylacetylene and methyl propiolate in (e_2) (see footnotes).

Regiospecificity has been found in several 1,3-cycloadditions to hetero-dipolarophiles, besides (***H**-b*) and (***H**-c*). This has been explained by a "principle of maximum gain in σ-bond energy"[11]: those five-membered rings should close that have the new σ-bonds of highest bond energy. More correctly, the orientation corresponding to the highest "partial bond energy" in the transition state should be favoured, and this more probably coincides with the highest bond energy in the product if the two possible products are strongly different in bond energy. This is not the case for reaction (***H**-b*), which gives a 1,2,3-triazoline, while benzalanilines and diphenylnitrilimine are reported to give 1,2,4-triazolines[11]; in one case a C–C and an N–N bond, in the other case two C–N

References pp. 156–162

bonds are closed, and in both cases regiospecificity is reported. Contradictory behaviour is also observed in the cycloaddition of benzonitrile oxide to C≡N bonds; when the reactant is benzonitrile, the product is 3,5-diphenyl-1,2,4-oxadiazole[184], but in the dimerisation of neutral benzonitrile oxide the product is 3,4-diphenyl-1,2,5-oxadiazole-5-oxide (diphenyl furoxan). The last orientational scheme is also found in reaction (*H-c*).

If one uses dipolarophiles with carbon–carbon unsaturated bonds, when the difference in energy of the two possible products is less marked, the orientation often is specifically determined by the type of 1,3-dipole and does not depend on the polarity of the dipolarophile (*i*). This proves that we are not dealing simply with polar reactions. Sensitivity of orientation to the nature of the substituent Y appears in case (*ii*) where the intrinsic difference between atoms 1 and 3 of the 1,3-dipole is attenuated, as for azides and apparently also for azomethine imines: both classes of 1,3-dipoles show some basically similar responsiveness of atoms 1 and 3, by presenting cases of double orientation (*iii*). Such cases are relatively rare for 1,3-cycloadditions with monosubstituted dipolarophiles; however, the situation is more complex in the case of disubstituted olefins[177].

The case of azides is the one where the difference of orientation according to the polar character of Y is best defined: These 1,3-dipoles behave as if they had a stable polarisation in the sense Ar–N̄–N̄=N̄, the nitrogen atom on the left receiving electrophilic attack by acrylic esters and acrylonitrile, the terminal atom nucleophilic attack by enamines and vinyl ethers. While it has been argued that the main factor controlling orientation in 1,3-cycloadditions to olefins and acetylenes is a steric one[11], phenyl azide can add in the usual orientation even to a bulky enamine[185]; *viz.*

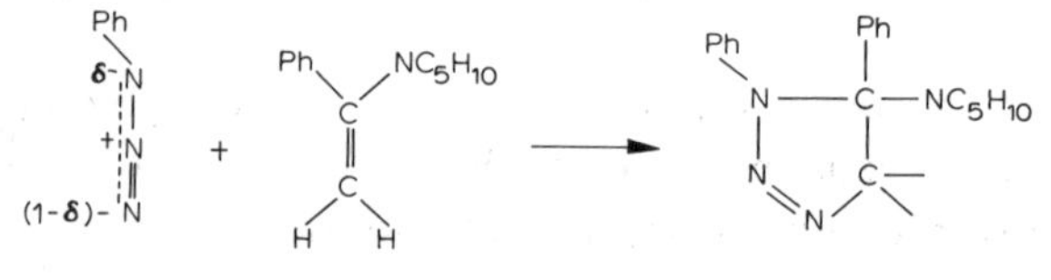

showing that at least in this case electronic control predominates.

Reactions classified under (e_2) and (*g*) are of the type that does not allow isolation of the first adduct (see Section 1). They are included here because there are good reasons to believe that the addition is a slow irreversible step, as in scheme (1), p. 88[63,174].

5.2 MEDIUM EFFECTS

1,3-Cycloadditions are exclusively reactions in solution, and no comparison of kinetics in the liquid and gas phase can be made here.

TABLE 11

EFFECT OF SOLVENT ON THE RATE COEFFICIENTS ($l.mole^{-1}.sec^{-1}$) OF SOME 1,3-CYCLOADDITIONS

(a) Picryl azide + dicyclopentadiene[60]; (b) phenyl azide + 1-pyrrolidino-cyclopentene[59]; (c) diazomethane + methyl methacrylate[56,186]; (d) *C*-methyl-*N*-phenylsydnone + ethyl phenylpropiolate[63]; (e) *C*-phenyl-*N*-methylnitrone + ethyl acrylate[186a].

Solvent (dielectric constant)	*(a)* 10^3k *(24–27°C)*	*(b)* 10^3k *(25°C)*	*(c)* 10^2k *(25°C)*	*(d)* 10^4k *(140°C)*	*(e)* 10^4k *(85°C)*
Ethanol ($\epsilon_{25} = 24.3$)	2.56				0.86
Dimethylformamide ($\epsilon_{20} = 37.7$)		32.9			1.64
Acetonitrile ($\epsilon_{20} = 37.5$)			2.70		1.63
Nitrobenzene ($\epsilon_{25} = 34.8$)		19.2		4.1	
Benzonitrile ($\epsilon_{25} = 25.2$)				4.7	
Acetone ($\epsilon_{25} = 20.7$)			1.70		1.88
o-Dichlorobenzene ($\epsilon_{25} = 9.93$)				8.4	
Dichloromethane ($\epsilon_{20} = 9.08$)	8.45				
Ethyl acetate ($\epsilon_{25} = 6.02$)	1.50	8.15			
Chlorobenzene ($\epsilon_{25} = 5.62$)	9.64				
Chloroform ($\epsilon_{20} = 4.81$)	13.0				
Toluene ($\epsilon_{25} = 2.38$)			2.00		4.80
Benzene ($\epsilon_{25} = 2.27$)	5.59	11.5			4.24
Dioxane ($\epsilon_{25} = 2.21$)	2.66				2.77
Tetrahydrofuran ($\epsilon_{20} = 2.20$)			1.60		
Decalin ($\epsilon_{20} = 2.2$)				22	
Cyclohexane ($\epsilon_{20} = 2.02$)		3.42			
n-Hexane ($\epsilon_{20} = 1.89$)			0.92		

Changes of solvent have usually a limited effect on reaction rate, as is shown by examples in Table 11. This table lists solvents ordered in the same way as in Table 5 (p. 101) and reactions of 1,3-dipoles "with and without double bond" [(a)–(c) and (d)–(e), respectively], and of both electrophilic and nucleophilic dipolarophiles. The observed solvent effect is always within a factor of 10, and only in some cases can a fairly regular trend be recognised, rate coefficients slightly increasing (d) or decreasing (b, c) along the columns. For types (b) and (c), apparently the transition state is slightly more polar than the initial state and reaction is favoured by polar solvents. Although (b) is the reaction of phenyl azide with an enamine, so sensitive to polar factors as far as orientation is concerned, the rate enhancement due to polar solvents is remarkably poor. For type (d) the transition state appears less polar than the initial state, a fact attributable to the high value of the sydnone dipole moment ($\mu = 6.57$ D)[11]. A similar conclusion can be reached for case (e), when solvent polarity is expressed by a parameter other than dielectric constant[186a].

In general, solvent polar effects seem to play a secondary role. Even in case (b), where a fair amount of charge separation in the transition state can be reasonably accepted, most of the charge is probably compensated in the form of an intimate "internal" ion pair.

Type (a) reactions are notable because of an unexplained effectiveness of halogenated solvents in increasing the rate. Little is known about kinetics of 1,3-cycloadditions in hydroxylic solvents; for type (a) and (e) reactions ethanol seems as good a medium as other solvents.

Reactions in which the dipolarophile acts through a hetero-unsaturated bond, like C=S[187] or N=S[181] also have rate coefficients little affected by the solvent polarity.

5.3 INFLUENCE OF THE STRUCTURE OF REACTANTS

Under the unifying term of 1,3-dipoles we are considering substances differing in their degree of unsaturation and in the atoms involved in the cycloadditions (N ··· N, C ··· N, C ··· O, C ··· C). Keeping this in mind, a general order of reactivity of 1,3-dipoles and of dipolarophiles is hardly to be expected. However, choosing simple electrophilic olefins and acetylenes, that are among the most common dipolarophiles, it has been possible to collect in Table 12 enough data for a rough comparison of several 1,3-dipoles. An unavoidable approximation has been that of listing rate coefficients measured with different solvents trusting that solvent effects are small (see previous

TABLE 12

REACTIVITY OF DIFFERENT 1,3-DIPOLES WITH SOME DIPOLAROPHILES

Dipolarophiles: (i) styrene (50°C); (ii) ethyl acrylate (25°C); (iii) dimethyl acetylene dicarboxylate (50°C); (iv) ethyl phenylpropiolate (50°C).

1,3-Dipole	*(i)* 10^5k ($l.mole^{-1}.sec^{-1}$)	*(ii)* 10^5k ($l.mole^{-1}.sec^{-1}$)	*(iii)* 10^5k ($l.mole^{-1}.sec^{-1}$)	*(iv)* 10^5k ($l.mole^{-1}.sec^{-1}$)
Phenyl azide		0.0985[e]	2·62[a,e]	
C-methyl-*N*-phenyl-sydnone			13.1[a,h]	0.15[a,h]
C-biphenylene-*N*(α)-*p*-chlorophenyl-*N*(β)cyanoazomethine imine	3.1[a,b]	*ca.* 14[f]	3500[a,i]	
Diazomethane		2.9×10^4[g]		
Diphenyldiazomethane	1.40[c]	707[c]	9680[j,c]	3.33[c]
Anhydro-5-hydroxy-2,4-diphenyl-3-methyloxazolium hydroxide	970[d]		1.42×10^6[d]	340[k,d]

[a] Calculated by means of the Arrhenius equation from experimental value at the nearest temperature. [b] In chlorobenzene[11]. [c] At 40°C, in dimethylformamide[188]. [d] In benzonitrile[174]. [e] In carbon tetrachloride[59]. [f] In chlorobenzene[189] calculated from activation parameters by means of Eyring equation. [g] *Methyl* acrylate, in toluene[56]. [h] In *p*-cymene[63]. [i] In chlorobenzene[174]. [j] *Diethyl* acetylene dicarboxylate. [k] *Methyl* phenylpropiolate.

section). Furthermore there are temperature differences, that must be mentally compensated for in an approximate way when comparing the figures.

Phenyl azide appears as a rather unreactive 1,3-dipole. Of the azomethine imines that follow, sydnone is more stable than the open cyano-derivative. Diazomethane is a very reactive compound, but double phenyl substitution reduces its reactivity. The highest rate coefficients are those of reactions of the mesoionic oxazolone. Reactivity varies by five or six orders of magnitude from azide to mesoionic oxazolone.

Looking at Table 12 along the rows, it is possible to appreciate the activating action of carboxylic groups in dipolarophiles, and the comparatively low reactivity of dipolarophiles having a phenyl ring adjacent to the multiple bond. These observations are substantiated by considering Table 13, where several dipolarophiles, differently substituted at their carbon–carbon unsaturated bonds, are listed according to the same criteria used for dienophiles in Table 7 (p. 103). Available rate coefficients are given for reactions with five 1,3-dipoles, in order of roughly increasing reactivity from left to right (taking into account differences in reaction temperature). The last 1,3-dipole is benzonitrile oxide which is often even more reactive than the mesoionic oxazolone.

Along the columns of Table 13 no regular trend is immediately recognizable and the conclusion that dienophilic and dipolarophilic power are not synonymous must be drawn. A more accurate inspection shows that both electronegative and electron-rich substituents on the multiple bonds can accelerate some of these reactions, at variance with what had been found for Diels–Alder additions (Table 7). The lowest rate coefficients in (i) (iii) and (v) correspond to cyclohexene as dipolarophile. Styrene and phenylacetylene are one or two orders of magnitude more reactive. A similar or larger effect is observed with alkyl vinyl ethers. For a further increase in rate more polar substituents are required, and the best dipolarophiles appear to be acrylic, propiolic, acetylene dicarboxylic and similar esters, as well as 1-pyrrolidino-cyclohexene.

In type (i), however, we are not dealing with a single reaction, but rather with two different reactions, giving rise to 3-aryl-4-substituted or to 3-aryl-5-substituted triazolines, according to the polar nature of the substituent on the dipolarophile (see Section 5.1), and it is not surprising that one of them is faster the more electron-rich is the substituent, while the other one is assisted by electronegative substituents. What is really surprising is that such behaviour seems common also to benzonitrile oxide, reaction type (v), although in cycloadditions to monosubstituted acetylenic or olefinic dipolarophiles only 3-phenyl-5-substituted isoxazoles or isoxazolines are known to be produced, whichever the substituent, apart from a recent report to the contrary[177].

Cyclic olefins are rather interesting in their dipolarophilic behaviour. Table 13 reveals that in (i), (iii) and (v) the reactivity order is invariably cyclohexene $<$ cyclopentene $\ll$ norbornene. Release of angular strain during the

TABLE 13

REACTIVITY OF VARIOUS ALKENES AND ALKYNES WITH SOME 1,3-DIPOLES

1,3-Dipoles: (i) phenyl azide (carbon tetrachloride, 25°C)[41,59]; (ii) *C*-methyl-*N*-phenyl-sydnone (*p*-cymene, 140°C)[63]; (iii) picryl azide (chloroform, 25°C)[60]; (iv) anhydro-5-hydroxy-2,4-diphenyl-3-methyloxazolium hydroxide, *viz.* *C*,*C*-diphenyl-*N*-methyl-münchnone, (benzonitrile, 50°C)[174]; (v) benzonitrile oxide (carbon tetrachloride, 25°C)[5,11,190a].

Dipolarophile	*(i)* 10^8k *(l.mole⁻¹.sec⁻¹)*	*(ii)* 10^4k *(l.mole⁻¹.sec⁻¹)*	*(iii)* 10^8k *(l.mole⁻¹.sec⁻¹)*	*(iv)* 10^7k *(l.mole⁻¹.sec⁻¹)*	*(v)* 10^8k *(l.mole⁻¹.sec⁻¹)*
Maleic anhydride	72.0[a]				
Dialkyl fumarate	83.6[b]	4.0[e]			*ca.* 1.7×10^7[c,g]
Dialkyl maleate	3.4[c]	2.6[e]			*ca.* 2.8×10^5[c,g]
Dimethyl acetylene dicarboxylate	254	258		1.42×10^8	
Alkyl acrylate	98.5[b]	28.3[e]		2.9×10^7[c]	*ca.* 1.2×10^7[b,g]
Alkyl propiolate	104[c]	82.3[c]		2.93×10^6[c]	*ca.* 2.0×10^6[f,g]
Alkyl phenylpropiolate	2.1[b]	9.9[b]		3.4×10^4[c]	
Styrene	4.0	2.5	1050	9.7×10^4	1.10×10^6
Phenylacetylene	2.9	1.8		1.03×10^4	1.78×10^5
Cyclopentene	18.6		1.08×10^4		*ca.* 1.8×10^5[g]
Cyclohexene	0.33		255		*ca.* 1.0×10^4[g]
Norbornene	1880		2.04×10^6		*ca.* 1.7×10^7[g]
Alkyl vinyl ether	4.0[d]		2.92×10^5[b]		*ca.* 2.6×10^6[d,g]
1-Pyrrolidino-cyclohexene	9.93×10^4[a]		$>10^6$		

[a] In benzene. [b] Alkyl = Et. [c] Alkyl = Me. [d] Alkyl = n-Bu. [e] Alkyl = hexyl. [f] Alkyl unspecified. [g] From relative rates in ether at 20°C[5,11].

cycloaddition explains the rate variations[59,75]; a similar observation has been made about the effectiveness of the three olefins as dienophiles in Diels–Alder reactions[191]. However it has been found that 1,3-dipoles "without a double bond" react with norbornene at about the same rate as with cyclopentene[11].

Another possible comparison is that between fumaric and maleic esters, as an example of a more general observation that higher rate coefficients are usually found for *trans*-1,2-disubstituted ethylenes than for their *cis* isomers (for Diels–Alder reactions see Section 4.1.3). Rate ratios up to 110 were reported and the phenomenon was proved to have a steric origin[192]. Rate ratios are particularly low in the case of 1,3-dipoles "without a double bond", as for instance in type (ii) reactions.

Table 13 offers also the opportunity of a comparison of olefins and acetylenes as dipolarophiles. In some cases the C=C dipolarophile is more effective, for instance styrene appears more reactive than phenylacetylene; in other cases the contrary is true, for instance acetylene dicarboxylic ester reacts faster than fumaric ester. When C≡C and C≡N dipolarophiles were compared, the former were found to be more reactive; for instance with diphenylnitrilimine (in benzene, 80°C) phenylacetylene reacts 18 times as rapidly as benzonitrile[75]. Nitriles also are less efficient than acetylenes as dienophiles (Section 4.1.3).

Disubstituted acetylenes and 1,2-disubstituted olefins are comparatively bad dipolarophiles. If the introduction of the second substituent was expected to have a strong accelerating effect, this is lacking (fumaric ester is sometimes less reactive than acrylic ester) or is small, c.f. acetylene dicarboxylate with propiolates. When the second substituent is a phenyl group, it deactivates; phenylpropiolate is less reactive than propiolate in reaction types (i) (ii) and (iv), Table 13. The introduction of methyl groups into the α or β position of acrylic ester, has a similar effect on the rate of several different 1,3-cycloadditions[11]. Steric hindrance appears to intervene in these cases.

It is possible to compare the reactivity of some 1,3-dipoles inserted in cyclic structures with that of the same dipoles in open chains. This comparison, recalling that between cyclic and open-chain dienes (Section 4.1.3), is limited to two classes of 1,3-dipoles "without a double bond", namely azomethine imines and oxides. For the former, there are data in Table 12 for the reaction with dimethyl acetylene dicarboxylate (iii): the cyclic azomethine imine (sydnone) is about 300 times less reactive than the open-chain cyanoazomethine imine. In the case of nitrones, a typical comparison is, for reaction with ethyl crotonate in toluene at 100°C[5,186a]

TABLE 14

ELECTRONIC EFFECTS OF SUBSTITUENTS TRANSMITTED THROUGH PHENYL RINGS

Reaction (solvent)	*Substituents employed*	*Temp. (°C)*	ρ	*Ref.*
(a) Aryl azide + norbornene (ethyl acetate)	*p*-OMe, *p*-Me, *m*-Me, H, *m*-OMe, *p*-Br, *m*-Br, *m*-NO_2	20 25 36	+0.79 +0.84[a] +0.77	55
(b) Aryl azide + 1-pyrrolidino-cyclohexene (benzene)	*p*-OMe, H, *p*-Cl, *m*-Cl, *p*-NO_2	25	+2.54	59
(c) Aryl azide + maleic anhydride (benzene)	*p*-OMe, *p*-Me, H, *p*-Cl, *m*-Cl, *p*-NO_2	25	−1.1	59
(d) Aryl nitrile oxide + phenylacetylene (carbon tetrachloride)	*p*-OMe, *p*-Me, H, *p*-Cl, *m*-Cl	25	*ca.* +0.6	190
(e) *N*-Aryl-sydnone + ethyl phenylpropiolate (*p*-cymene)	*p*-OMe, *p*-Me, H, *p*-Cl	140	+0.8	63
(f) *C*-Phenyl-*N*-methylnitrone + substituted styrene (toluene)	*p*-OMe, *p*-Me, H, *p*-Cl, *p*-NO_2	120	+0.8	11, 186a
(g) Diphenylnitrilimine + substituted styrene (benzene)	*p*-OMe, *p*-Me, H, *p*-Cl, *p*-NO_2	80	+0.8	75
(h) 3,5-Dichloro-2,4,6-trimethylbenzonitrile oxide + arylacetylene (carbon tetrachloride)	*p*-OMe, *p*-Me, H H, *p*-Cl, *p*-Br, *p*-NO_2	25 25	−0.3 +0.6	193, 62
(i) Diazomethane + benzal-(substituted aniline) (dimethylformamide)	*p*-Me, H, *p*-Cl, *m*-Cl	25	*ca.* +1.8	179
(j) 3,5-Dichloro-2,4,6-trimethylbenzonitrile oxide + substituted *N*-sulfinyl aniline (carbon tetrachloride)	*p*-OMe, *p*-OEt, *p*-Me, H, *p*-Cl, *p*-Br, *m*-Cl	25	*ca.* +0.9	b

[a] For the same reaction in benzene[59] (substituents: *p*-OMe, H, *p*-Cl, *m*-Cl, *p*-NO_2): $\rho = +0.88$ at 25°C; for the reaction in chloroform[60] (substituents: H, *m*-NO_2, *p*-NO_2) kinetic data at 25°C give $\rho = ca. +1.1$.

[b] From data in ref. 194; this ρ value is only indicative, since the correlation coefficient is quite low ($r = 0.83$).

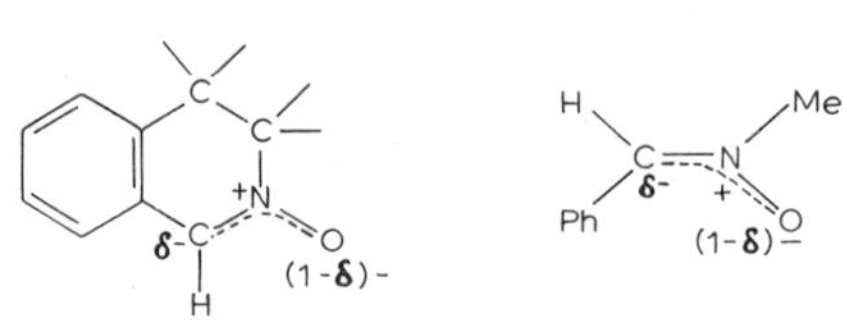

$10^4 k$ (l.mole^{-1}.sec^{-1}) = 240   1.07

The stability of sydnones, due to their aromatic character, and that of *trans*-aldonitrones with respect to their *cis* isomers should be important factors in keeping the reactivity of these compounds low in 1,3-cycloadditions.

Electronic effects transmitted through aromatic systems to the reacting centers in 1,3-dipoles and in dipolarophiles can often be evaluated by means of the Hammett relation. Table 14 shows some typical results. The same 1,3-dipole (aryl azide) can be rather strongly activated by electron-withdrawing substituents ($\rho = +2.54$) when reacting with an electron-rich dipolarophile (b), and moderately activated by electron-releasing substituents ($\rho = -1.1$) when the dipolarophile is electrophilic (c); for norbornene as dipolarophile, the ρ value is intermediate (a). Cycloadditions of aryl azides are, however, exceptional in some respects, and the role played in these reactions by polar factors has been already stressed (Section 5.1). Other reactions, provided only carbon–carbon dipolarophiles are involved, (d)–(h), have lower absolute ρ values, showing that there is a very limited charge separation in their transition states.

Reaction type (h) is at the limits of applicability of the $\rho\sigma$ correlation: both electron-releasing and electron-withdrawing substituents are able to slightly accelerate 1,3-cycloaddition of aryl-acetylenes to a nitrile oxide, although the addition has been proved to maintain always the same orientation[62]. This is not an isolated case of a Hammett plot having a minimum: unsubstituted benzonitrile oxide adds to aryl acetylenes with similar substituent effects[190]; *C*-biphenylene-*N*(α)-*p*-chlorophenyl-*N*(β)cyanoazomethine imine adds to both *p*-methoxy and *p*-nitrostyrene roughly twice as fast as to styrene[11], and a similar effect seems to be present even in the addition of phenyl azide to substituted styrenes[59]. As these variable ρ's have low absolute values, it appears that the corresponding cycloadditions are mainly due to non-polar interactions, and can flexibly accommodate polar influences, irrespective of their sign (at least within the explored range).

The assumption of a singlet diradical as intermediate in 1,3-cycloadditions[195] would account for the feeble polar character of several such reactions, as well as for other of their features. However, the generality of this two-step mechanism is questionable[196], mainly on stereochemical grounds. It seems favoured in particular cases[196a].

When the reacting centers on the dipolarophile are atoms of different elements, having different electronegativities, one might expect an increased polar character of the addition. Experimental results are scarce, cases (i) and (j) refer to C=N and N=S bonds, and do not support this view. High positive ρ values had been reported for (j), from experiments with few electron-releasing substituents[181], but extension of the kinetic study to a wider range of substituents has shown that on the whole the Hammett relationship is poorly obeyed; roughly, electron-withdrawing substituents favour the reaction, but not in a very impressive way[194].

A two-step 1,3-cycloaddition may have been found in the case of *p*-anisyl-diazomethane reacting with *p*-methoxy-styrene[197]. While with other substituents only *trans*-3,5-diaryl-pyrazolines are produced, *p*-methoxy-substituents on both 1,3-dipole and dipolarophile give rise to a mixture of *cis* and *trans* isomers in roughly equal amounts. Overberger *et al.*[197] suggest a dipolar species, produced in a nucleophilic attack of the olefin on the diazocompound, as a precursor of the two products.

5.4 ACTIVATION PARAMETERS

In the field of 1,3-cycloaddition a wealth of data about activation parameters comparable with that for the Diels–Alder reaction is not available. Published values were obtained from experiments rarely carried out at more than three, sometimes at only two, different temperatures, and this may be a source of error.

Several types of 1,3-cycloaddition are represented in Table 15. Reaction velocities at room temperature are expressed by rate coefficients ($l.mole^{-1}.sec^{-1}$) ranging from the order of 10^{-6}, cases (f) and (i) to about 10^{-2}, (c) and (e). This variation is due (Table 15) to values of the activation energy ranging from 8.6 to 18.0 $kcal.mole^{-1}$, while entropies of activation span from −40 to −24 eu, a range similar to that typical of the Diels–Alder reaction. Very few published $\Delta S^{\ddagger}$ values of 1,3-cycloadditions are outside the range indicated.

TABLE 15

ACTIVATION PARAMETERS OF 1,3-CYCLOADDITIONS[a]

Reaction (solvent)	*E (kcal.mole⁻¹)*	*log A*	*ΔS‡ (eu)*	*Ref.*
(a) $X–C_6H_4N_3$ + norbornene (ethyl acetate):				55
X = H[b]	15.3	6.4	−31.0	
X = *p*-NO_2	13.1	5.6	−34.9	
(b) Phenyl azide + 1-morpholino-cyclopentene (benzene)	12.3	5.4	−36	59
(c) Picryl azide + norbornene (chloroform)	10.4	5.9	−33.4	60
(d) Picryl azide + cyclopentene (chloroform)	12.9	5.5	−35.5	60
(e) Diazomethane + methyl methacrylate (tetrahydrofuran)	8.6	4.5	−40	56
(f) Phenyl azide + dimethyl acetylene dicarboxylate (carbon tetrachloride)	18.0	7.6	−26	59
(g) 3,5-Dichloro-2,4,6-trimethylbenzonitrile oxide + *p*-nitrophenylacetylene (carbon tetrachloride)	15.1	8.0	−24	62
(h) 3,5-Dichloro-2,4,6-trimethylbenzonitrile oxide + *p*-OEt–C_6H_4N=SO (carbon tetrachloride)	12.3	5.8	−34	194
(i) *C*-Phenyl-*N*-methylnitrone + methyl methacrylate (toluene)	16.4	6.3	−32	11
(j) *C*-Biphenylene-*N*(α)-*p*-chlorophenyl-*N*(β)-cyanoazomethine imine + dimethyl acetylene dicarboxylate (chlorobenzene)	14.1	8.1	−24	174
(k) *C*-methyl-*N*-phenyl-sydnone + dimethyl acetylene dicarboxylate (*p*-cymene)	15.5	6.6	−31	63

[a] See footnote a of Table 9.
[b] Other measurements: in $CHCl_3$[60] $E = 14.5$, $\log A = 6.0$, $\Delta S^{\ddagger} = -33.1$; in CCl_4[59] $E = 15.8$, $\log A = 6.9$, $\Delta S^{\ddagger} = -29$.

Remarkably, substituent effects and reactivity differences due to angle strain in the dipolarophile are mainly reflected in E values. Reactions that could be suspected, because of their high ρ value, of being kinetically controlled by a polar attack, like reaction (b), have $\Delta S^{\ddagger}$ values within the range of the others.

In contrast with Section 4.1.4, no discussion can be given here of any reverse reaction, for lack of experimental data, although 1,3-cycloadditions should also in principle be reversible, and in fact are so in some cases[178,198].

No determination of volume of activation of 1,3-cycloadditions has been found in the literature.

5.5 ACTION OF CATALYSTS

Only qualitative information is available about the action of catalysts on 1,3-cycloadditions. Reactions of diazomethane with benzal-anilines were carried out in dioxane containing a few percent of water or methanol, but the efficacy of such "catalysts" was not measured[40,179]. Cycloadditions of nitrile oxides to acetylenic dipolarophiles were found to give better results, on the preparative scale, in the presence of traces of alkalies[176]. However it has been discovered that in some instances the direct 1,3-cycloaddition is accompanied by a mechanism of addition with hydrogen shift, giving rise, from acetylenes of the type R–C≡C–H, to acetylenic oximes, which in alkaline medium easily rearrange to isoxazoles[198a]. This may explain, at least in part, the better yields of isoxazolic adducts obtained in the presence of bases.

The most interesting series of observations in this field have been made on the catalytic action of Lewis acids, which are also effective catalysts of Diels–Alder reactions (Section 4.1.6). In the presence of BF_3 or its derivatives it has been possible to obtain 1,3-cycloadditions that do not proceed under uncatalysed conditions, *e.g.* those of benzonitrile oxide to aliphatic nitriles and carbonylic compounds[199] and to oximes[200]. The method has been applied also to the cycloaddition of diazomethane to CN multiple bonds, by using alkyl aluminium halides[201]. Lewis acid catalysts probably act by making the dipolarophiles more polar, through complexing.

6. 1,2-Cycloaddition

6.1 THERMAL REACTIONS

A few classes of unsaturated compounds are able to give thermal 1,2-cycloadditions, yielding four-membered rings. The latter can be C_4 rings (from fluoro- and cyano-ethylenes, ketenes, allenes, benzyne, etc. dimerising or adding to olefins) as well as β-lactones (from ketenes and carbonylic compounds), β-lactams (from ketenes and Schiff bases or from isocyanates and olefins), 1,2-thiazetidine-1-oxides (from *N*-sulfinyl-compounds and olefins) or other cyclic compounds.

Apart from a steady interest in fluoroethylenes, dating from the fifties, most of the work on the mechanism of these reactions has been done after 1960. A notable portion is devoted to the direction, orientation and stereochemistry of the reactions, while kinetic studies are just beginning, and in some cases not yet initiated.

When the second reactant is a diene, it is interesting to investigate the possible competition between 1,4- and 1,2-cycloadditions. Tetrafluoroethylene adds to cyclopentadiene giving about 60% 1,2-adduct and 40% 1,4-(Diels–Alder)-adduct[202]; 1,1-dichloro-2,2-difluoroethylene gives with the same diene about 16% 1,2-adduct and 84% 1,4-adduct[203]. Tetracyanoethylene and 4-methyl-1,3-pentadiene give predominantly 1,2-addition but also some 1,4-addition, in proportions markedly dependent on solvent[204]. 1-Cyanovinylacetate gives with butadiene 80–90% 1,4-adduct and 10–20% 1,2-adduct, depending on the temperature and solvent[205], but the same olefin can cycloadd to *trans,trans*-2,4-hexadiene in a pure stereospecific 1,4 fashion[205a]. Since hexadiene is more reactive than butadiene in Diels–Alder addition to maleic anhydride, while the contrary is true for 1,2-cycloaddition to $Cl_2C{=}CF_2$, the different behaviour of the two dienes towards 1-cyanovinylacetate is indicative of the concomitant operation of different reaction mechanisms for the two kinds of cycloaddition by butadiene[205a]. In the reaction of cyclopentadiene with *cis*- and *trans*-1,2-dichloro-1,2-difluoroethylene, a 1,2-cycloaddition with large loss of configuration competes with a stereospecific 1,4-cycloaddition[205b].

Butadiene gives a few percent of *trans*-1,2-divinyl-cyclobutane when dimerised without added catalysts at 150°C and 100 atm[206], while the dimerisation of 2-chlorobutadiene at 35–40°C yields in nearly equal amounts a cyclohexene and a cyclobutane derivative[207].

2-Alkylbutadienes react with excess 1,1-dichloro-2,2-difluoroethylene at 80°C to give mainly 1,2-cycloaddition (in variable proportions at the two double bonds), but also 1,4-cycloadducts, in the following yields[90]; alkyl = Me, 1.6%; Et, 5%; iso-Pr, 16%; *t*-Bu, 45%; *neo*-Pe, 10%; Ph, 12%. Butadiene at the same temperature gives 1.0% 1,4-adduct[91], and the entire sequence is (apart from the neopentyl case) in rough agreement with the rates of Diels–Alder addition of the dienes to maleic anhydride (see Table 6, p. 102 and ref. 67). On increasing the temperature of the butadiene reaction gradually to 176°C, the fraction of 1,4-cycloaddition increases up to 2.3%, maintaining a constant ratio with the fraction of butadiene which is thought to be present at equilibrium in *cisoid* conformation (from 6 to 13% in the range investigated)[91].

This is regarded as evidence that, of two different biradicals formed from *cisoid* and *transoid* butadiene reacting with $Cl_2C{=}CF_2$, the latter gives only 1,2-adduct and the former collapses to 1,2- and 1,4-adducts in a ratio not dependent on temperature[91]. However, more attention should probably be paid to the skewed conformations of dienes[207a,207b]. As (1,2-adduct/1,4-adduct) ratios for addition of $Cl_2C{=}CF_2$ to dienes fixed in non-*transoid* conformation vary from less than 0.2 for cyclopentadiene to more than 99 for 1,2-dimethylenecyclobutane, Bartlett *et al.*[207a] have suggested that the distance between the ends of the diene system is a significant variable affecting such ratio, since a shorter distance obviously favours 1,4-cycloaddition.

On the other hand diphenylketene does not give Diels–Alder products, but only 1,2-cycloadducts, on reaction with cyclopentadiene, piperylenes, isoprene, and other dienes[208,209].

The orientation of 1,2-cycloadditions is usually dictated by the condition of lowest energy for biradical or polar intermediates, often with regioselectivity rather than complete specificity. In the case of fluorohalogenoethylenes, the more fluorinated carbon atom is commonly bound in the adduct to the less substituted atom from the other reactant (or to the more fluorinated atom of the second molecule if the reaction is a dimerisation)[24,210], *e.g.*

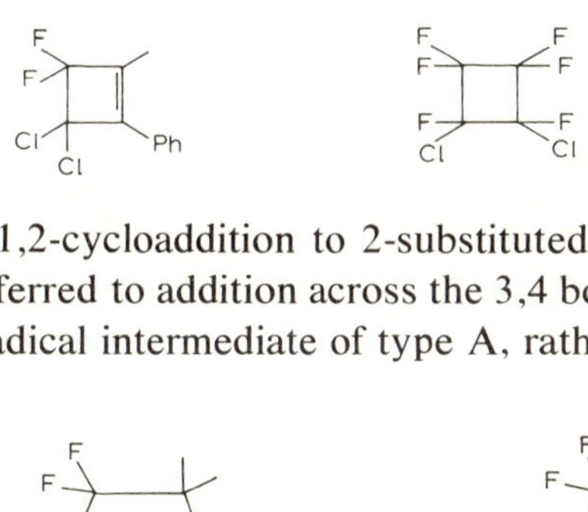

In 1,2-cycloaddition to 2-substituted dienes, addition across the 1,2 bond is preferred to addition across the 3,4 bond, consistent with the stabilisation of a diradical intermediate of type A, rather than of type B,

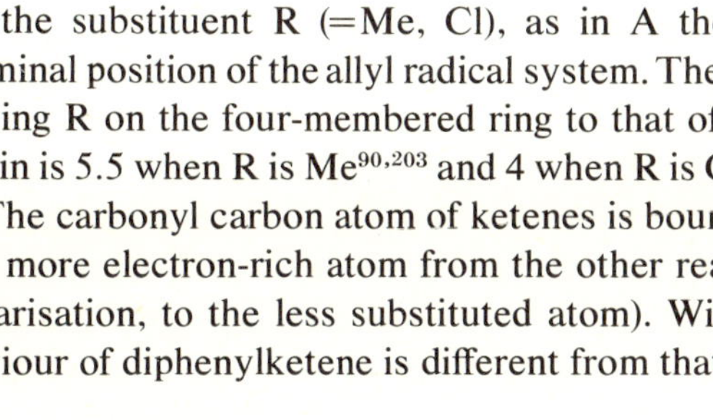

by the substituent R (=Me, Cl), as in A the substituent is bonded at the terminal position of the allyl radical system. The ratio of the yield of the product having R on the four-membered ring to that of the isomer with R on the side-chain is 5.5 when R is Me[90,203] and 4 when R is Cl[203].

The carbonyl carbon atom of ketenes is bound, in their 1,2-cycloadducts, to the more electron-rich atom from the other reactant (in the absence of strong polarisation, to the less substituted atom). With 2-substituted dienes, the behaviour of diphenylketene is different from that of fluorinated olefins: isoprene

References pp. 156–162

contributes to the C_4 ring mainly (60–70%) with its 1,2-double bond[208,209], but 2-chlorobutadiene preferentially (80%) uses its 3,4-double bond[208], indicating a polar substituent effect, *viz.*

The 1,2-adduct obtained from 1-cyanovinylacetate and butadiene (1-acetoxy-1-cyano-2-vinylcyclobutane) has again the orientation compatible with the maximum stability of a possible reaction intermediate[205].

Allenes appear to react in such a way that their central carbon atom is bound, in the most abundant product, to the central carbon atom of a second allene molecule, as in dimerisations[16,72], to the terminal carbon of a vinyl derivative, or to the most fluorinated atom of a fluoro-halogenoethylene, as has been found in the reactions of propadiene with $F_2C{=}CFCl$ and $F_2C{=}CCl_2$[85], where 85% and 95% of the product, respectively, conformed to this orientation.

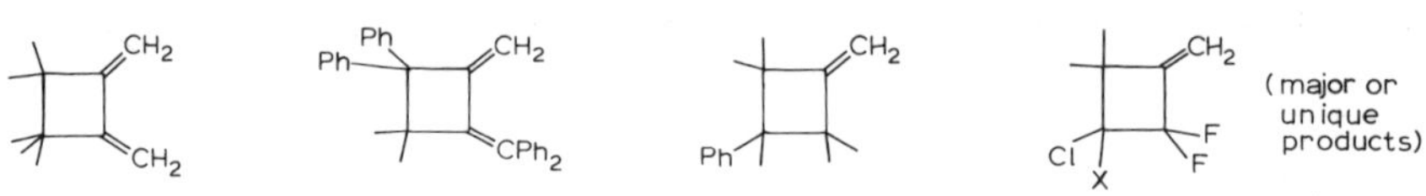

When 1,2-disubstituted or polysubstituted ethylenes are employed in 1,2-cycloadditions, retention of the olefin configuration is not as general and complete as in Diels–Alder and 1,3-cycloadditions. Typical examples are some reactions of *cis*- and *trans*-1,2-dicyano-1,2-(trifluoromethyl)ethylene with olefins[98]: stereospecificity was observed when the second reactant was ethyl vinyl ether or *t*-butyl vinyl sulfide or *trans*-propenyl *n*-propyl ether; however, the *cis* isomer of the last olefin gave non-stereospecific cycloaddition, a considerable change being registered in the configuration of the cyano-ethylene moiety, although the ether moiety retained its own configuration. Using ethyl acetate as solvent, from *trans*-$CN(CF_3)C{=}C(CF_3)CN$ the product was 46% of *trans* and 54% of *cis*(dicyano) configuration; from *cis*-$CN(CF_3)C{=}C(CF_3)CN$ the product was 60% of *cis* and 40% of *trans*(dicyano) configuration. Additions are more stereoselective in more polar solvents. A dipolar intermediate

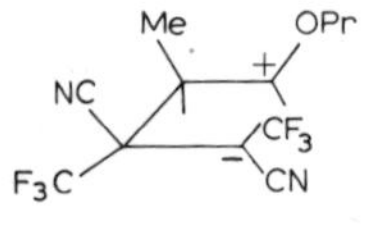

which would lose its stereochemistry only at the carbanion side because of steric crowding (CF_3 facing OPr in the example above) has been suggested[98]. A polar reaction mechanism is supported by a strong solvent effect on the kinetics of these reactions, expressed by $k_{MeOH}/k_{CCl_4} \approx 10^3$ in one of the above reactions[211], and by $k_{HCOOH}/k_{PhMe} \approx 10^5$ in a similar case, involving tetracyanoethylene and *p*-methoxystyrene[212].

In the 1,2-cycloaddition of the *trans-trans*, *trans-cis*, and *cis-cis* isomers of 2,4-hexadiene to 1,1-dichloro-2,2-difluoroethylene, under conditions of stereochemical stability of reactants and products, it has been ascertained that there is *cis–trans* isomerisation at the site of addition, and retention of configuration only on the side chain[213]. An equilibrium composition of the product is approached but not attained, proving that in the open-chain (diradical) intermediate there is competition between ring closure and internal rotation around the C–C bond at the reaction site, the latter process being about 10 times faster than the former, at 80°C[213].

Benzyne, when added to the geometric isomers of propenyl methyl ether (in dichloromethane at 40°C) gives mixtures of 1,2-adducts, approaching within 65% of the equilibrium composition[214]. Non-stereospecific 1,2-cycloadditions of benzyne to *cis* and *trans* isomers of 1,2-dichloroethylene and propenyl ethyl ether have also been reported; in both cases the products are far from completely equilibrated, indicating a stepwise addition with a relatively fast ring closure[214a].

Stereospecific 1,2-cycloadditions are given by *cis* and *trans* propenyl propyl ether with diphenylketene and dimethylketene[215,216], and furthermore by diphenylketene with *cis*- and *trans*-2-butene[216] and by dichloroketene with *cis*- and *trans*-cyclooctene[217]. "Stereospecific" means that contamination by the other isomer was not found; the available methods would have detected a few percent of it, and in the most favourable case[216] even 1%.

Monosubstituted ketenes (aldoketenes) were found to add to benzalaniline yielding β-lactams of fixed (*trans*) stereochemistry[218].

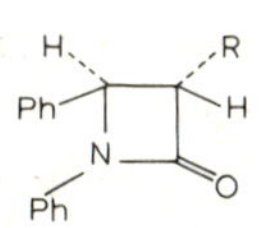

R = Me , Et , *iso*-Pr, *t*-Bu , Ph

The situation is more involved with unsymmetrically disubstituted ketenes, like alkyl-halogenoketenes[219].

Sulfonyl isocyanates seem able to undergo stereospecific or highly stereoselective reactions. The cycloadditions of chlorosulphonyl isocyanate to *cis* and *trans* isomers of β-methylstyrene and of 3-hexene were carried with retention of the olefin configuration[219a]. The initial products of the following reaction (R = p-$MeC_6H_4SO_2$–)

were analysed by NMR spectroscopy and showed no *trans*-adduct from *cis*-ether, and only 10% *cis*-adduct from *trans*-ether. However, during the progress of the reaction, the composition of the adduct was gradually modified to reach an equilibrium value (27% *cis*- and 73% *trans*-adduct)[220].

In the case above, the loss of the original stereochemistry (with opening of the β-lactam ring at a to give a dipolar species) appears to follow instead of accompanying the cycloaddition[220a]. However for the addition of diphenylketene to benzalaniline, which also gives a β-lactam, a dipolar intermediate is invoked, to explain side reactions when the addition is effected in the presence of water or of methanol as scavengers[221]. Trapping of a dipolar intermediate by water addition was also reported for the cycloaddition of diphenylketene to a carbodiimide[221a].

Another case, in which the capture of a dipolar intermediate is reported, is the cycloaddition of an arene-sulfonyl isocyanate to a dialkyl-carbodiimide. One of the ultimate products is a 2:1 adduct, and it is suggested that an open dipolar species, the product of the first addition, adds a second molecule of isocyanate in a subsequent step[222]. Formation of 2:1 adducts is regarded as strong evidence of cycloadditions with polar intermediates[222a].

Cis- and *trans*-1-butenyl ethyl ether react with *N*-sulfinyl-(*p*-chlorophenyl)-sulfonamide in $CDCl_3$ in a stereospecific manner[220]. 1,1-Dimethylallene was tested in cycloaddition to dimethyl fumarate and maleate; in the former case stereospecificity (to 99%) was found and in the latter more than 90% stereoselectivity[223].

Summing up about stereochemistry, the most consistent group of stereospecific 1,2-cycloadditions has so far been found in the case of ketenes, although other cumulenes and heterocumulenes also seem to share this behaviour.

Measurements of rate coefficients and activation parameters were carried in some cases and the results are given in Table 16. Other kinetic studies were made at one temperature, for purpose of evaluation of medium and/or structural effects; some of the relevant rate coefficients are gathered in Table 17. For the reaction between $F_2C{=}CCl_2$ and butadiene derivatives, partial rate coefficients attributable to every double bond were also calculated, taking into account the product analyses[90,203], and are given in parentheses. All reactions were found to be first-order with respect to both reactants.

In Table 16 it is remarkable how small is the change in dimerisation rate of tetrafluoroethylene when one chlorine is substituted for a fluorine atom: according to some results[22] the difference appears to be due only to an entropy

TABLE 16

RATE COEFFICIENTS AND ACTIVATION PARAMETERS OF 1,2-CYCLOADDITIONS[a]

Reaction (medium)	Temp. (°C)	k ($l.mole^{-1}.sec^{-1}$)	E ($kcal.mole^{-1}$)	log A	$\Delta S^{\ddagger}$ (eu)	Ref.
(i) $CF_2{=}CF_2 + CF_2{=}CF_2$ (gas phase)	400	0.46[b]	26.3	8.22	−26.5	22
		0.61	25.4	8.01	−27.5	224, 23
(ii) $CF_2{=}CF_2 + CF_2{=}CFCl$ (gas phase)	400	0.29[b]	26.3	7.93	−27.8	22
(iii) $CF_2{=}CFCl + CF_2{=}CFCl$ (gas phase)	400	0.092[b]	26.3	7.55	−29.6	22
		0.090[b]	26.6	7.63	−29.2	24
(iv) Benzyne dimerisation (gas phase)	90–200	4.6×10^{9}	0	9.7	−24.2[d]	29
(v) $Me_2C{=}C{=}O + Me_2C{=}C{=}O$ (PhCN)	35	3.49×10^{-4}	10.7[c]	4.1	−42	225, 225a
(vi) $Ph_2C{=}C{=}O + CH_2{=}CHOBu$ (PhCN)	35	3.45×10^{-3}[b]	9.9	4.5	−40	57
(vii) $Ph_2C{=}C{=}CH_2 + Ph_2C{=}C{=}CH_2$ (EtOH)	35	8.5×10^{-7}[b]	18.4	6.96	−29.1	72
(viii) p-$MeC_6H_4SO_2N{=}C{=}O$ + $HC{=}CH$ (O–CH_2–O) (CH_2Cl_2)	35	2.05×10^{-4}[b]	10.5	3.74	−43.5	225b

[a] See footnote a of Table 9.
[b] Calculated by means of Arrhenius equation from experimental values at the nearest temperatures.
[c] The correct value of activation enthalpy is 10.1 $kcal.mole^{-1}$ (ref. 225a).
[d] The value −10.5 eu given by the authors[29] for a standard state of one $mole.ml^{-1}$ has been converted to a standard state of one $mole.l^{-1}$ by subtracting $R \ln 10^3$.

TABLE 17

STRUCTURAL AND SOLVENT EFFECTS ON SOME 1,2-CYCLOADDITIONS

Reactants (solvent)(temp., °C)	*10^5k (l.mole^{-1}sec^{-1})*		*Ref.*
(a) 1,1-Dichloro-2,2-difluoroethylene (in excess) +			71
(81.9) butadiene	0.906	($k_{12} = k_{34} = 0.453$)	
(82.0) isoprene	1.23	($k_{12} = 1.04$; $k_{34} = 0.19$)	
(81.9) 2-chlorobutadiene	0.678	($k_{12} = 0.542$; $k_{34} = 0.136$)	
(79.5) isoprene	0.776	($k_{12} = 0.655$; $k_{34} = 0.121$)	
(79.5) *cis*-piperylene	0.664	(k_{12} = very small; k_{34} = *ca.* 0.664)	
(79.5) *trans*-piperylene	0.475	(k_{12} = very small; k_{34} = *ca.* 0.475)	
(80) cyclopentadiene	0.11	($k_{12} = k_{34} = 0.055$)	
(b) Diphenylketene (tetrahydrofuran + *n*-heptane)(30) +			208
butadiene	0.117		
cyclopentadiene	28.0		
(c) Diphenylketene (benzonitrile)(40)+			57
$CH_2{=}C(R_1)(R_2)$	2.3	(R_1 = H; R_2 = Ph)	
	452	(R_1 = H; R_2 = OEt)	
	0.38	(R_1 = Me; R_2 = Ph)	
	2270	(R_1 = Me; R_2 = OEt)	
	116	(R_1 = Ph; R_2 = OMe)	
$Me_2C{=}CHR$	2.9	(R = OEt)	
	1690	(R = morpholino)	
	ca. 2.4×10^6	(R = pyrrolidino)	
cyclopentene	0.125		
norbornene	1.57		
2,3-dihydrofuran	1.04×10^4		

(d) Diphenylketene (bromobenzene) (120)+		226a
(p-$CF_3C_6H_4)_2C=CH_2$	0.167	
(p-$MeC_6H_4)_2C=CH_2$	2.12	
p-$CF_3C_6H_4CH=CH_2$	28.5	
p-$MeC_6H_4CH=CH_2$	94.3	
(e) Diphenylketene + styrene (120)		226a
(bromobenzene; $\epsilon_{25}=5.40$)	70.0	
(o-dichlorobenzene; $\epsilon_{25}=9.93$)	66.7	
(di(β-chloroethyl)ether; $\epsilon_{20}=21.2$)	65.0	
(dimethylformamide; $\epsilon_{20}=37.7$)	60.3	
(f) Dimethylketene dimerisation (35)		225
(carbon tetrachloride; $\epsilon_{20}=2.24$)	2.31	
(chlorobenzene; $\epsilon_{25}=5.62$)	6.5	
(dichloromethane; $\epsilon_{20}=9.08$)	26	
(acetonitrile; $\epsilon_{20}=37.5$)	67	
(g) Sulphonyl isocyanates $XSO_2N{=}CO$ + 1,3-dioxole(dichloromethane) (20)		225b
X = p-MeC_6H_4	9	
MeO	63	
2,4,6-$Cl_3C_6H_2O$	1×10^3	
Cl	1.2×10^5	
(h) Chlorosulphonyl isocyanate + 2-ethylhexene (25)		225c
(n-hexane; $\epsilon_{20}=1.89$)	3–5	
(carbon tetrachloride)	10–15	
(chlorobenzene)	$(1.2–1.5)\times10^3$	
(dichloromethane)	$(6–7.5)\times10^3$	
(nitromethane; $\epsilon_{20}=38.6$)	$(5–10)\times10^4$	

factor, while other measurements[23,24] attribute it partly to entropy and partly to energetic factors. Also, some reverse reactions have been studied, *i.e.* the decomposition of octafluorocyclobutane[23] and that of *cis*- and *trans*-1,2-dichloro-hexafluorocyclobutane[24]. Such reactions have very high E values (74.1 kcal.mole^{-1} in the first case, 65.3 kcal.mole^{-1} in the other two cases), and positive values of the activation entropy (10.6, 8.7, and 7.7 eu, respectively). An open-chain diradical intermediate has been suggested for these additions[23,24].

The kinetics of benzyne dimerisation to biphenylene have been the object of two studies. In the first a limit was determined for the rate coefficient ($k \geqslant 7 \times 10^8$ l.mole^{-1}.sec^{-1})[28], while the authors of the second arrived at a more accurate value ($k = (4.6 \pm 1.2) \times 10^9$ l.mole^{-1}.sec^{-1}) and checked its invariancy over the range 90–200°C[29] (benzyne was obtained *in situ* by decomposition of aromatic derivatives).

Data about reactions in solution, from Tables 16 and 17, concern mainly one fluorinated ethylene and two ketenes. With dichloro-difluoroethylene the rate coefficient is only slightly affected, Table 17(a), by substituent changes on the butadiene partner; 2-Me and 2-Cl both mildly accelerate the addition to the 1,2-double bond and depress the rate at the 3,4-positions, while in Diels–Alder reactions (see Table 6, p. 102) isoprene and 2-chlorobutadiene are well separated in reactivity. In both piperylenes the methyl substituent has a strongly negative effect on reactivity at the 1,2 double bond, but slightly accelerates the addition to the other position. The close reactivity of the two isomers is another point of difference from the Diels–Alder addition, where the *cis*-isomer reacts much more slowly than the *trans*[5], having steric difficulties in reaching a planar *cisoid* conformation. Evidently a *cisoid* conformation is not required for these 1,2-cycloadditions and the *transoid* is perhaps favoured, as the low rate of addition of cyclopentadiene (Table 17) confirms. When 2,3-dimethylbutadiene was a partner in the same reaction, it was found[226], by competition experiments, $k(2,3\text{-Me}_2)/k(\text{H, H}) = 0.90$, that the depressing and accelerating effects of the two methyl groups almost compensate.

The situation is very different with ketenes: diphenylketene reacts with cyclopentadiene 240 times faster than with butadiene, Table 17(b), the *transoid* conformation of the latter being probably the origin of steric hindrance[208]. Effect of substituents directly bound to the reacting center on olefinic reactants, Table 17(c), is large, be it steric or electronic. It is surely polar in the case of $-OR$ and $-NR_2$ groups; the activating ability of these groups affects rate coefficients by several orders of magnitude, *c.f.* some cases of 1,3-cycloaddition (see Table 13, p. 126). Norbornene is also here more reactive than cyclopentene, although only by one power of ten. Increased levels of substitution on both positions of the ethylenic reactant seem to depress the rate when the additional substituents are of small polarity (Me, Ph), and particularly when they are bulky (Ph), revealing a steric origin of the effect. Electron-withdrawing substituents have such a depressing effect on rates that no adduct of this type has so far been obtained[57].

Electronic effects transmitted through conjugated systems are scarce, Table 17(d), corresponding to ρ-values around -0.7 for reactions of 1,1-diarylethylenes and of *p*-substituted styrenes with diphenylketene[226a]. In the 1,2-cycloaddition of diphenylketene to benzalaniline, a *p*-methoxy substituent on either phenyl group of benzalaniline increases[221] the rate by a factor of 2.3, corresponding to $\rho = -1.4$.

Rate coefficients for two reactions of diphenylketene have been found to increase with the polarity of the solvent[57]:

$$Ph_2C{=}C{=}O + H_2C{=}CHOBu\text{-}n, \frac{k_{\text{acetonitrile}}}{k_{\text{cyclohexane}}} = 160$$

$$Ph_2C{=}C{=}O + \text{2,3-dihydropyran}, \frac{k_{\text{acetonitrile}}}{k_{\text{cyclohexane}}} = 48^*$$

Also the dimerisation of dimethylketene is quite sensitive to solvents, Table 17(f), in a measure unknown with Diels–Alder reactions and 1,3-cycloadditions. However, for the reaction of diphenylketene with styrene the rate is slightly decreased with increasing dielectric constant of the medium, Table 17(e).

Activation entropies of the ketenes reactions, Table 16(e) and (f), are around -40 eu.

Comparing *cis* and *trans* isomers of propenyl propyl ether in their reactivity towards diphenylketene, a ratio k_{cis}/k_{trans} of about 180 has been found (in PhCN, 40°C)[57], confirmed by the value of 172 ± 20 from experiments without solvent[216]; with dimethylketene the ratio is 60 ± 10[216]. The opposite phenomenon has been observed in Diels–Alder reactions of fumarate/maleate and other 1,2-disubstituted ethylenes (Section 4.1.3). The fact that *cis* isomers are thermodynamically less stable than *trans* isomers seems to prevail in determining relative reactivities of these 1,2-cycloadditions.

A measure of the isotope effect was obtained for the reaction of diphenylketene with 1-*d*-cyclohexene, at 100°C without solvent. The two possible products are A and B; a preference for the former was found.

* A different result is given by other authors[227]. However, they studied the reaction in unusually concentrated (*ca.* 40%) solutions in mixed solvents; also their data on the activation parameters of the reaction should be viewed with caution, as they do not correspond to the rate coefficients published.

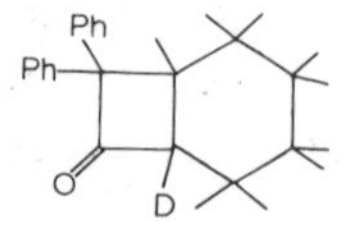

(A)

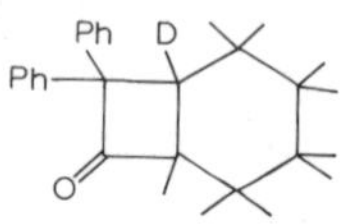

(B)

$(k_A/k_B) = 1.13 \pm 0.03$[228]. The results should be interpreted as evidence (see Section 4.1.5) that bond formation with the carbonyl carbon is more advanced in the transition state than formation of the other bond. Measurement of isotope effects for the addition of diphenylketene to ordinary, α-deuteriated and β-dideuteriated styrene led to similar conclusions[228a].

Stereochemistry, substituent effects and activation parameters of most ketene reactions are consistent with a one-step cycloaddition; polar effects of substituents and solvents, as well as the isotope effect, often require, however, that a fair amount of charge separation (that is, unequal bond formation) characterises the transition state. It has been kinetically proved that cycloadditions of enamines to ketenes can also proceed through a dipolar intermediate: this is so for the reaction between dimethylketene and *N*-isobutenylpyrrolidine[228b]. In the latter case, the rate coefficient for the formation of the intermediate strongly depends on solvent polarity[228b]: $k_{\text{acetonitrile}}/k_{\text{cyclohexane}} = 560$. Use of the same criteria used for ketenes (as far as experimental data allow it) in the case of the 1,2-cycloadditions of fluorinated olefins results, instead, in the conclusion that a two-step biradical mechanism is operating. Results for 1,2-cycloaddition of sulfonyl isocyanates to olefins, cases (g) and (h) in Table 17, give indications of dipolar intermediates during the course of these reactions.

Few results are available for allene derivatives. Dimerisation of 1,1-diphenylallene is remarkably slower than the analogous reaction of dimethylketene, see (vii) and (v) in Table 16; besides, the geometries of the two adducts do not fully correspond. The change from a C=O to a $C{=}CH_2$ double bond has obvious polar consequences. These are also apparent in the slight dependence on solvent of the rate of 1,1-dimethylallene cycloaddition to dimethyl fumarate and maleate[223]: $k_{\text{dimethylsulphoxide}}/k_{\text{cyclohexane}} = ca.\ 2$. Competition experiments showed fumarate to react faster than maleate: $k_{trans}/k_{cis} = 9.3$ at 170°C[223], the direction of the effect being in this case the same as for Diels–Alder reactions.

An intramolecular secondary isotope effect has been measured by analysing the product of 1,2-cycloaddition of 1,1-d_2-allene to acrylonitrile in benzene at 210–225°C[151]. The result is $(k_H/k_D)_{25°C} \geqslant 1.18$ per D atom; it has the opposite direction of that found for a Diels–Alder reaction of 1,1-d_2-allene, and is consistent with a two-step mechanism (see Section 4.1.5). A diradical intermediate seems probable. A similar result was obtained from the study of isotope effects for the dimerisation of allene[228c]. Since the reaction of *R*(-)-

1,3-dimethylallene with acrylonitrile gives optically active isomeric ethylidene-cyclobutanes, a dissymmetric intermediate was suggested[228d].

Some data are available about catalysis in 1,2-cycloadditions. Tributyl phosphine catalyses dimerisation of phenyl isocyanate to uretidinedione in toluene[38]. The reaction is kinetically of first order with respect to catalyst and overall third order; the reverse process is first order with respect to catalyst and overall second order. The mechanism is complex, as revealed by the value of the apparent activation energy of the forward reaction ($E = 1.1 \pm 0.7$ kcal.mole^{-1})[38], which presumably results from the combined temperature dependence of two or more steps, including formation of an isocyanate–phosphine complex (see eqn. (13), p. 113).

Butadiene dimerisation is very sensitive to catalysis by zero-valent nickel complexes, which can direct the reaction towards 1,2-cycloaddition or (4+4) cycloaddition, through a bis-π-allyl intermediate, with small amounts of Diels–Alder product[162]. Larger quantities of the latter were obtained in other cases[229]. This and other examples of activation of olefins by transition metal complexes have been associated with excitation of the coordinated π-system[230].

6.2 PHOTOCHEMICAL REACTIONS

Photochemical cycloadditions are carried out at temperatures where the thermal reaction has a negligible rate. Under these conditions, reaction may take place only if one of the reactants is promoted to an excited state.

When UV–visible radiation is absorbed by a reactant, it usually excites it from its singlet ground state to a singlet (commonly the first) excited state; the latter may be the reactive species, or it may convert, by intersystem crossing, to the lower triplet state, which is capable of reaction. The concentration of excited species in solution derives from an equilibrium between rates of excitation, spontaneous deactivation, intersystem crossing, possible energy transfer to other species, and chemical reactions with ground-state molecules, which may give rise to (singlet) excited adducts or to (triplet) biradical intermediates, able to convert into adducts by spin inversion and ring closure.

The simplest reaction scheme, assuming that the triplet state of reactant A is the reactive species, is

$$\mathrm{A} \underset{k_{d,1}}{\overset{h\nu}{\rightleftarrows}} {}^1\mathrm{A}^* \qquad (16)$$

$${}^1\mathrm{A}^* \xrightarrow{k_{ic}} {}^3\mathrm{A}^* \qquad (17)$$

$${}^3\mathrm{A}^* \xrightarrow{k_{d,3}} \mathrm{A} \qquad (18)$$

$${}^3\mathrm{A}^* + \mathrm{B} \xrightarrow{k_r} \uparrow\cdot\mathrm{A{-}B}\cdot\uparrow \qquad (19)$$

$$\uparrow\cdot\mathrm{A-B}\cdot\uparrow \;\rightarrow\; \uparrow\cdot\mathrm{A{-}B}\cdot\downarrow \;\rightarrow\; \text{adduct} \qquad (20)$$

If steps (20) are assumed to be fast compared with step (19), this may be formally rewritten as

$$^3A^* + B \xrightarrow{k_r} \text{adduct} \tag{21}$$

A stationary-state treatment gives the following expression for the product quantum yield Φ (molecules of product formed per quantum of light absorbed, that is moles of product per einstein of light absorbed), as a function of the concentration of B

$$\frac{1}{\Phi} = \left(\frac{k_{ic} + k_{d,1}}{k_{ic}}\right)\left(1 + \frac{k_{d,3}}{k_r[B]}\right) = \frac{1}{\Phi_{ic}}\left(1 + \frac{k_{d,3}}{k_r[B]}\right) \tag{22}$$

having defined Φ_{ic} as the quantum yield for intersystem crossing. A plot of Φ^{-1} *versus* $[B]^{-1}$ is therefore linear.

In case triplet quenchers, which can remove energy from triplet states $^3A^*$, are present, the previous scheme must include the step

$$^3A^* + Q \xrightarrow{k_q} A + {}^3Q^* \tag{23}$$

and a stationary-state treatment gives

$$\frac{\Phi_0}{\Phi} = 1 + \frac{k_q[Q]}{k_{d,3} + k_r[B]} \tag{24}$$

where Q is the quencher and Φ_0 is the quantum yield in the absence of quencher, as defined by eqn. (22). A plot of (Φ_0/Φ) *versus* $[Q]$, other quantities being constant, is accordingly linear (Stern–Volmer plot of the quenching efficiency). A substance can act as triplet quencher by reaction (23) if the excitation energy of its triplet state is lower than the excitation energy of $^3A^*$, vertical excitation being usually considered. A possible way of deviating an excited state, when the excitation energy of the quencher is too high for a process like (23), is through complexing of $^3A^*$ with Q.

Finally, in several cases the radiation has a frequency lower than required for the direct excitation of reactants. In this case excitation is still possible if a sensitiser is present; this can absorb the radiation, be converted from a singlet to a triplet excited state and transfer energy to a reactant, say A (the respective triplet excitation energies being, of course, in the correct relation). A simple scheme in this case (S is the sensitiser) is

$$S \underset{k_{ds,1}}{\overset{h\nu}{\rightleftarrows}} {}^1S^* \tag{25}$$

$$^1S^* \xrightarrow{k_{ics}} {}^3S^* \tag{26}$$

$$^3S^* \xrightarrow{k_{ds,3}} S \tag{27}$$

$$^3S^* + A \xrightarrow{k_t} {}^3A^* + S \tag{28}$$

$$^3A^* + B \xrightarrow{k_r} \text{adduct} \tag{29}$$

$$^3A^* \xrightarrow{k_{d,3}} A \tag{30}$$

Employing the stationary-state approximation, and assuming a highly efficient energy transfer in reaction (28) ($k_{ds,3} \ll k_t[A]$) one obtains

$$\frac{1}{\Phi} = \frac{1}{\Phi_{ics}}\left(1 + \frac{k_{d,3}}{k_r[B]}\right) \tag{31}$$

where Φ_{ics} is the quantum yield for intersystem crossing of the sensitiser. Equation (31) is completely analogous to eqn. (22).

Investigation of the mechanism of photocycloadditions can be carried out with the help of the above or similar schemes, including analysis of the effect of varying reactant concentrations, and of adding quenchers and sensitisers, on the quantum yield of the processes at different wavelength of the radiation employed. A few cases will be mentioned.

When a diene is involved, there is often competition between ring closure to 1,2- and 1,4-cycloadduct (see also Section 4.2). For instance, the photosensitised reactions of 1,1-dichloro-2,2-difluoroethylene, 1,1-dichloroethylene and tetrachloroethylene with cyclopentadiene at 0°C give the following products[89]

(*a*)	(X = F)	2	:	1 (for thermal reaction, see Section 6.1)
(*b*)	(X = H)	3	:	2
(*c*)	(X = Cl)	3	:	2

On the basis of the triplet energy of the sensitisers employed, it was deduced that the diene, and not the halogenoethylene, is excited to its triplet state. Under direct excitation, a greater number of products were obtained in very low yields. Apparently the triplet state is not attained efficiently and the singlet state is both inefficient in cycloaddition (because of its short lifetime) and unselective[89].

Photosensitised butadiene dimerisation at 27–29°C gives *trans*- and *cis*-1,2-divinylcyclobutane, and 4-vinylcyclohexene, in proportions remarkably dependent on triplet excitation energy, E_T, of the sensitizer[231] as shown in Fig. 2 (the *trans/cis* ratio of the cyclobutane derivatives maintains constant). This result, which was not observed in the case of cyclic dienes[88], has been rationalised as follows[231]. For $E_T > 60$ kcal.mole^{-1} energy is transferred from $^3S^*$ to both *transoid*- and *cisoid*-butadiene at diffusion-controlled rates, so that most butadiene triplets are *trans* and ring closure to four-membered rings is preferred in a way largely independent of E_T; when $50 < E_T < 60$ kcal.mole^{-1} the difference in triplet excitation energy of *transoid*-butadiene (59.6 kcal.mole^{-1}) and *cisoid*-butadiene (*ca.* 53 kcal.mole^{-1})[232] becomes significant, *cisoid*-butadiene is preferentially excited and the corresponding *cis* triplet gives a high percentage of 1,4-adduct. As for the region where $E_T < 50$ kcal.mole^{-1}, a non-vertical excitation was suggested, and actually reaction rates become very low in this region[231].

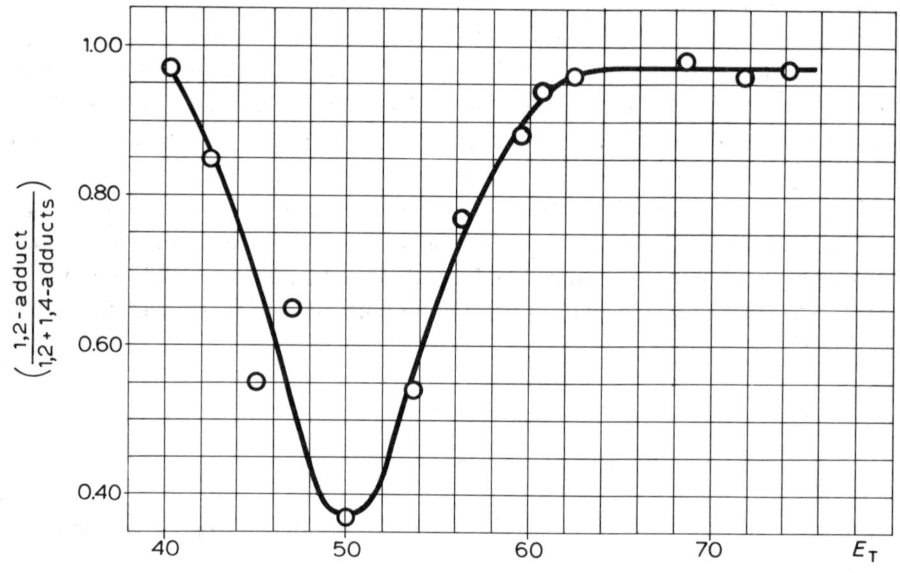

Fig. 2. 1,2-Cycloaddition *versus* 1,4-cycloaddition in the photosensitised butadiene dimerisation, using sensitisers of different triplet excitation energy[231].

Photosensitised cycloaddition of butadiene to cyclopentadiene at −15°C gives primary products from 1,2 addition (55%), 1,4 addition in which cyclopentadiene acts as diene (42%), and 1,4 addition where butadiene acts as diene (3%)[233]. As the sensitiser was benzophenone ($E_T = 68.5$ kcal.mole^{-1}), the result is in agreement with the previous discussion.

Butadiene and 1-cyanovinylacetate give, from the photosensitised cycloaddition at about 30°C[234], the same products as in the thermal reaction[205], but in different ratios and with a dependence of the proportions on the excitation energy of the sensitiser similar to that shown in Fig. 2, although with a less marked minimum[232]. This dependence can be explained as for butadiene dimerisation[231], assuming that butadiene, and not 1-cyanovinylacetate, gives a reactive triplet[232]. Analogous considerations apply to the photosensitised reaction of 1-cyanovinylacetate with isoprene, but not with cyclopentadiene[234a], in agreement with the fixed *cisoid* conformation of the latter.

Examples of 1,2-photocycloaddition *followed* by 1,4-cycloaddition are known, the reaction of maleic anhydride with aromatics being the most studied. In the case of benzene the product is

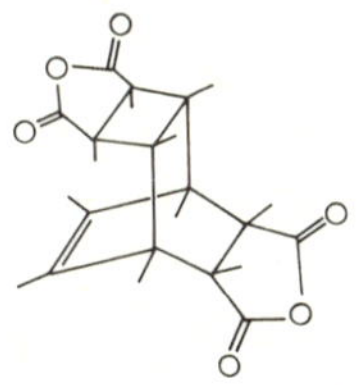

Reactants are able to give 1 : 1 molecular complexes of *exo*-stereochemistry[235], and there is evidence that the reaction involves excitation of these charge-transfer complexes, rate-controlling formation of a 1,2-monoadduct and subsequent fast thermal Diels–Alder reaction[236,237]. Substituted benzenes are less reactive than benzene, as is shown by the following relative rates at 30°C for unsensitised and sensitised reactions[236]

	Unsensitised	*Sensitised*
Benzene	0.38	(≡ 1.0)
Toluene	0.12	0.5
o-Xylene	0	0.15
p-Xylene	0	0.08

The aromatics listed above each give two structurally isomeric products; in benzophenone-sensitised reactions at 20°C, 1,2-cycloaddition is exclusively or predominantly at unsubstituted positions on the ring, but at 100°C the proportions of products are changed and *p*-xylene is preferentially (10 : 1) attacked at the 1,2-positions[238]. Steric effects and effects of temperature on equilibria involving charge-transfer complexes are suggested[236,238].

An increase of quantum yield with the concentration of maleic anhydride has been observed in its sensitised reaction with benzene[237]. This can be explained assuming that the Diels–Alder reaction of the first adduct (M) is in competition with a decomposition process, *viz.*

$$M + M.A. \xrightarrow{k'} 2:1 \text{ adduct}$$

$$M \xrightarrow{k''} \text{benzene} + M.A.$$

where M.A. = maleic anhydride. The unsensitised reaction is quenched by oxygen (a triplet quencher) and should therefore be, at least in part, a triplet process too[237].

While the first adduct of the maleic anhydride–benzene reaction has not been isolated, an analogous monoadduct has been obtained, from benzene and acrylonitrile by irradiation at 0°C; it is relatively stable and can give a thermal Diels–Alder reaction with acrylonitrile, yielding a 2:1 adduct[239]. Photochemical 1,2-cycloaddition in this case is not quenched by air, a fact that supports the hypothesis of a singlet reactive species[239].

Maleic anhydride and its methyl derivatives can dimerise to cyclobutane rings under direct irradiation or in the presence of benzophenone as sensitiser, in benzene or dioxane as solvents. The rate of dimer production is little affected by methyl substitution (a slight decrease is reported) in the sensitised reaction, but more strongly in the unsensitised process[240]. In analogous sensitised 1,2-cycloadditions of dimethylmaleic anhydride to several olefins, effects of substituents are also limited, as appears from the few examples below[241].

Olefin	*Rate (mole × $10^{-3}h^{-1}$)*
(a) Cyclohexene	0.94
(b) Tetramethylethylene	0.56
(c) *cis*-1,2-Dichloroethylene	0.26
(d) *trans*-1,2-Dichloroethylene	0.20
(e) Trichloroethylene	0.77

The *cis* or *trans* relation of the chlorine atoms in reactions (c) and (d) is reported to be retained in the corresponding cyclobutane adducts by Schenck *et al.*[241]. These authors favour a mechanism involving a biradical complex of sensitiser and anhydride as the actual entity attacking the olefin.

The stereochemistry of other photocycloadditions can be examined by considering, for instance, the reaction of *cis* and *trans* 1,2-dichloroethylene with indene[242], and the addition of dimethyl maleate and fumarate to acetone[243]. In the former process (sensitised by benzophenone) two products (*syn* and *anti*) are possible for each stereochemical relation of the two chlorine atoms. Taking together the two *cis* and the two *trans* products, the behaviour of the systems can be described as follows[242]

from *cis* ClHC=CHCl: 51.1% *cis* products at 60°C
73.4% *cis* products at −60°C

from *trans* ClHC=CHCl: 76.4% *trans* products at 50°C
85.1% *trans* products at −40°C

Biradical intermediates have a rate of internal rotation comparable, particularly at low temperatures, with the rate of spin inversion and ring closure.

In the other cycloaddition (yielding oxetanes) under kinetic control fumarate gives only a *trans* product, while maleate gives *cis* + *trans* products (1:6.35). Initial reaction rates (r_0) were measured under two conditions, *viz.* (*a*) with radiation filtered through quartz (λ > 200 nm) and (*b*) with radiation filtered through borosilicate glass (λ > 280 nm), and are[243]

	r_0 (*cis-product*) (*mole.l*$^{-1}$*.h*$^{-1}$)	r_0 (*trans-product*) (*mole.l*$^{-1}$*.h*$^{-1}$)
From fumarate (*a*)	0	0.0447
(*b*)	0	0.00166
From maleate (*a*)	0.00095	0.00604
(*b*)	0	0

For (*b*) most of the radiation is absorbed by acetone, little by fumarate and even less by maleate. Analysis of the data indicates that they are compatible with excitation of the esters and attack of excited esters on ground-state acetone[243]. The multiplicity of the reactive excited state is not yet specified.

α-β-Unsaturated ketones are suitable reactants for 1,2-photocycloadditions to olefins, which usually yield cyclobutane rings. Cyclopentenone dimerises to two *anti* derivatives, a head-to-head (h–h), and a head-to-tail (h–t) dimer. Using a triplet quencher (*cis*-piperylene) a linear Stern–Volmer plot was obtained and the rate was substantially insensitive to addition of a sensitiser (xanthone)[84]. These results strongly favour a reaction mechanism involving an excited triplet state.

On the other hand, the ratio (h–t)/(h–h) in the product composition is a function of reactant concentration, at least in some solvents: in cyclohexane, (h–t)/(h–h) = 2.0 at 6.2 *M*, 4.3 at 1.2 *M* and 6.4 at 0.12 *M*[244]; in benzene, (h–t)/(h–h) = 1.9 at 5 *M*, 3.3 at 1.1 *M* and 5.7 at 0.093 *M*[84]. As processes *via* singlet excited states are certainly unimportant at low concentration (because of strong competition by the deactivation process), this effect could mean that a singlet excited state contributes to the reaction at high concentration[244]. It may, however, be due to modification of the solvent when the solute is present in high concentration, as the ratio (h–t)/(h–h) is solvent-dependent[84].

The similar dimerisation of cyclohexenone under irradiation (λ = 366 nm) gives mostly a (h–t) and a (h–h) *anti* dimer, in a ratio dependent on solvent (in benzene it is about 2.3)[245]. Determination of quantum yield as a function of the ketone and quencher concentrations suggested a triplet mechanism, although there are some unexplained anomalies[245].

Other cases of solvent dependence of the product composition have been reported. In the photocycloaddition

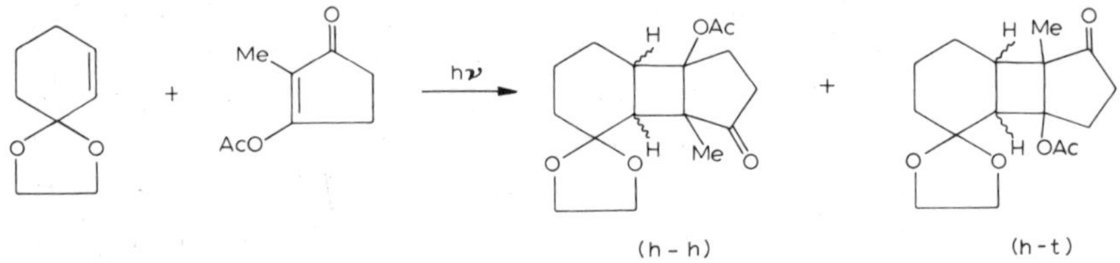

the ratio (h–h)/(h–t) can be as high as 49 (in iso-octane), but is only 0.82 in methanol (values extrapolated to zero concentration); its logarithm is a linear function of the solvent polarity, as expressed by $(\epsilon-1)/(2\epsilon+1)$, with a negative slope[246]. Dipole interactions appear to be important in determining orientation in such a reaction. An investigation has been initiated on the dimerisation of dimethylthymine[247,248]; this gives 4 dimers, one of which is reproduced (others being easily deduced), *viz.*

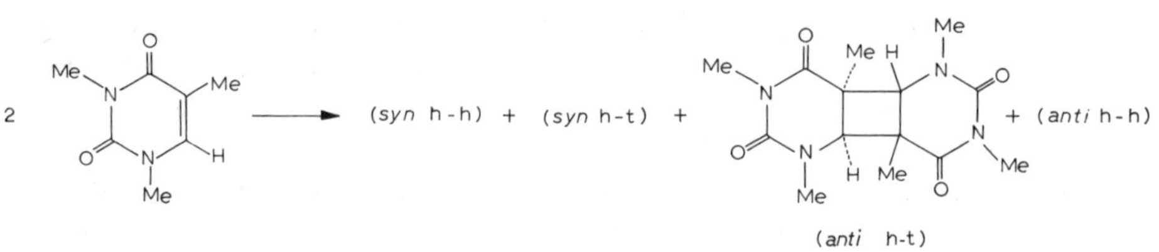

The rate is relatively high in water (quantum yield around 5.0×10^{-2}) and one order of magnitude lower in other solvents tested, but scarcely dependent on their apolar, protic or aprotic nature. Large variations in the product composition are found and using a triplet quencher on the one hand, and benzophenone as sensitiser on the other hand, Morrison *et al.* could isolate two different processes, through singlet and triplet excited states, respectively. The composition of the product of both processes is solvent-dependent, *e.g.*[248] (in percentages)

		syn (h–h)	*syn (h–t)*	*anti (h–t)*	*anti (h–h)*
In benzene	singlet reaction (%)	11	63	25	0
	triplet reaction (%)	30	51	13	6
In acetonitrile	singlet reaction (%)	44	40	12	3
	triplet reaction (%)	74	16	4	6

Both singlet and triplet reactions give higher yields of (h–h) products in the more polar acetonitrile than in benzene; the (h–h) orientation should actually correspond to the highest dipole moment in the dimer.

p-Naphthoquinone can react with olefins under irradiation in benzene at

20°C to give both cyclobutane and oxetane derivatives, and an analogous behaviour is found with acetylenes[249]. Reaction at the C=C bond accounts for 9–28% of product (NMR spectroscopy), depending on the substituents of the acetylene: $k(C{=}C)/k(C{=}O)$ = from 0.1 to 0.4. Methyl substitution at the site of reaction on the quinone favours attack at C=C, which is the only addition observed in the case of dimethyl naphthoquinone[249].

Various studies have been carried out on the photodimerisation of acenaphthylene, which yields a *syn* and an *anti* adduct[99,250–253], *viz.*

Two different reactions occur together in solution. One of them is quenched by oxygen and by cyclooctatetraene, and becomes the only significant process when a sensitiser is used. Its products are both *syn* and *anti* dimers, in a ratio dependent on solvent: log (*anti*/*syn*) is a linear function of $[(\epsilon-1)/(2\epsilon+1)]\rho$ (where ρ is the molar density of solvent), with a negative slope[253]. The rate of this reaction is particularly high in heavy-atom solvents, like *n*-PrBr, MeI and EtI, that are known to catalyse intersystem crossing and increase the concentration of triplet states; in one case yields are directly related to the square of the spin-orbit coupling factor of bromine and iodine atoms[252].

The other reaction can be isolated by using air-saturated solutions or by adding cyclooctatetraene, without sensitisers; it gives only *syn*-dimer. In the molten phase, this is the only reaction that has been observed[250]. As to the mechanism of the two competing processes, it seems convenient to attribute the former to an excited triplet state of acenaphthylene, and the latter to the excited singlet state of a π-complex, that gives only *syn* adduct because of its initial geometry.

1,2-Photocycloaddition of ketenimines to ketones yields oxetane derivatives. The kinetics and mechanism have been studied for alkyl- and phenyl-substituted ketenimines reacting with benzophenone[254] and fluorenone[255,256]. Some relative rates (from quantum yields, after correction for light absorption by ketenimines) are given below.

$R_1R_2C{=}C{=}N{-}R_3$			+ *Benzophenone* (*in benzene at room temperature*)	+ *Fluorenone* (*in benzene at room temperature*)
R_1	R_2	R_3	Φ/Φ_0	Φ/Φ_0
Me	Me	cyclohexyl	(1.00)	(1.00)
Me	Me	Ph	0.46	0.27
Ph	Ph	*sec*-Bu	0.097	0.80
Ph	Ph	Ph	0.0089	0.32

For some ketenimines, determination of the dependence of quantum yields on ketenimine and quencher concentration (see eqns. 22 and 24) has been carried; linear plots are found, apart from minor anomalies. Triplet states of ketones seem responsible for these reactions, but while triplet fluorenone is rather unselective towards different ketenimines, triplet benzophenone is quite sensitive to aryl-substitution on the ketenimine. The n–π^* triplet state of benzophenone is quenched by some ketenimines, and therefore it is suggested that photocycloaddition (or a complex formation step preceding it) is in competition with an energy-transfer step, which is favoured by phenyl substituents[254]. Accepting this reaction scheme, the detailed analysis of the quantum yield plots for dimethyl-*N*-(cyclohexyl)ketenimine gives approximate numerical values for the rate coefficients of the addition step (19): $k_r = 4 \times 10^8$ l.mole^{-1}.sec^{-1} (benzophenone triplet)[254] and 2.7×10^8 l.mole^{-1}.sec^{-1} (fluorenone triplet)[256].

7. Theoretical considerations

As is well known, Woodward and Hoffmann[257] have rationalised the stereochemistry of electrocyclic reactions (intramolecular cycloadditions) by considering the symmetry of molecular orbitals of the open-chain reactant. The more rigorous correlation of molecular orbitals of reactant and product, on the basis of their symmetry about the elements of symmetry that are preserved during the reaction[258], strengthened their argument. When applying this point of view to intermolecular cycloadditions[259], it was assumed that symmetric reactants, like butadiene and ethylene, approach in parallel planes, directly above each other (*cis–cis* or *supra,supra* addition), and it was shown that thermal and photochemical processes can easily take place under certain circumstances (symmetry-*allowed* reactions) and are highly unfavourable in other cases (symmetry-*forbidden* reactions). For cycloadditions involving two reactants, that participate with m and n π-electrons, respectively, the following *selection rules* were given (q is an integer, including zero):

$m+n = 4q+2$	thermally allowed, photochemically forbidden
$m+n = 4q$	thermally forbidden, photochemically allowed

This classification, valid for concerted *cis–cis* (or *supra,supra*) processes, confirmed that thermal Diels–Alder reactions ($m+n = 6$) can be (but not, must be) one-step reactions, while predicting that photochemical 1,4-cycloadditions should be multistep reactions; 1,3-cycloadditions should behave analogously, since 1,3-dipoles are four π-electron systems. Common 1,2-

cycloadditions ($m+n=4$) should be in the opposite situation, *i.e.* a thermal reaction should occur more easily by a multi-step pathway, but if one reactant is in its first excited state a concerted addition becomes possible. Furthermore, the preference of Diels–Alder reactions for *endo*-approach of reactants has been justified by symmetry-allowed secondary orbital interactions among unsaturated centers[260].

Also *cis–trans* (or *supra,antara*) cycloadditions of some π-electron systems are sterically feasible (*i.e.* the top face of one component reacts with the bottom face of the other, at one side, but the two bottom faces react with each other at the other side of the cycloaddition). In such cases the above selection rules are exactly reversed[261].

These predictions involve some assumptions and approximations when applied in a generalised form. Thus symmetry is lacking in most reactants in cycloadditions, either because of different substitution at identical atoms (for instance vinyl derivatives instead of ethylene) or because different atoms are present as reactions centers (as in most 1,3-dipoles). In the former case the substance of the previous considerations should be unaltered[261], but in the latter case the selection rules (*e.g.* derived for the allyl anion taken as model of 1,3-dipole) may be less stringent when applied only on the basis of analogy.

The largest body of data on cycloadditions concerns thermal processes *allowed* by the selection rules; differences of opinion about concerted or two-step pathways for Diels–Alder reactions[158,262] and for 1,3-cycloadditions[195,196] cannot be settled by virtue of the rules.

A more interesting field is that of the thermal 1,2-cycloadditions: here concerted processes are *forbidden*, and experimental results can be compared with this prediction. A discrepancy arises in the case of ketenes, as there is considerable evidence in favour of a concerted mechanism for their thermal 1,2-cycloadditions (Section 6.1). However, it is possible to envisage[263] the intervention of the perpendicular π-system of the C=O bond of ketene, in such a way as to surmount the steric difficulties of an orthogonal approach of the reactants, required[263] by a *cis–trans* (or *supra,antara*) cycloaddition: the latter is symmetry-allowed as a thermal process when it is $m+n=4$.

Theoretical quantitative treatments of cycloadditions mainly concern the Diels–Alder reaction. A possible approach is that of calculating the *para*-localisation energy, that is the variation of π-electron energy of the conjugated diene system, when two π-electrons are localised upon the atoms, in 1,4-relation to each other, which must form the new σ-bonds. This has been done by Brown[264], using the molecular orbital method: some successful predictions of reactivity and of the positions of addition for polycyclic aromatic hydrocarbons and polyenes were made. This method has also been used by other authors[119a,264a].

A different approach regards a π-complex of diene and dienophile as a precursor of the transition state sufficiently similar to it from the electronic point of view to be chosen as a model[265]. Values of the delocalisation energy of the complex (ΔE^{π}) have been calculated by the Hückel MO method for cyclopentadiene addition to different dienophiles; such values are linearly related to the activation energy of the reactions. When there is a choice between *endo* and *exo* addition, ΔE^{π}(*endo*) is larger than ΔE^{π}(*exo*), in agreement with the preferred *endo* stereochemistry[265].

A common theoretical approach during the last few years has been the application of perturbation theory to the HMO method (PMO method), as suggested by Dewar[266], and specifically applied to Diels–Alder reactivity[267,268]. A brief descriptive account will be given here. The separate (both conjugated) molecules are regarded as the initial system; their Hückel (or extended-Hückel) molecular orbitals and energies are known. Interaction of the two compounds, that is the activation stage of their addition, brings about a perturbation of the system, and there is a computable change in its π-electron energy. Here, the following symbols are chosen.

	1st compound	*2nd compound*
Indexes of atoms interacting	r ------	r'
	s ------	s'
Indexes of M.O.'s	$\ldots i \ldots$	$\ldots i' \ldots$
Energy of M.O.'s	$\ldots E_i \ldots$	$\ldots E_{i'} \ldots$
Coefficients of atomic orbitals in the LCAO M.O.'s	$\ldots c_{ir}, c_{is} \ldots$	$\ldots c_{i'r'}, c_{i's'} \ldots$

Interaction between two atomic orbitals of different compounds is expressed by an exchange integral $\gamma_{rr'}$ or $\gamma_{ss'}$. Considering only the interaction at the two reaction sites of a cycloaddition, the second-order perturbation theory gives the interaction energy as

$$\Delta E = 2\left(\sum_{i}^{\text{occ}} \sum_{i'}^{\text{unocc}} - \sum_{i}^{\text{unocc}} \sum_{i'}^{\text{occ}}\right) \frac{(c_{ir}c_{i'r'}\gamma_{rr'} + c_{is}c_{i's'}\gamma_{ss'})^2}{E_i - E_{i'}} \tag{32}$$

Summations are taken over occupied (occ) or unoccupied (unocc) molecular orbitals of the two compounds, as indicated. A similar formula gives the interaction energy when all possible interactions of each atom of the first compound with any atom of the second compound have to be considered, as is the case when *endo* and *exo* approaches are compared.

Exchange integrals can be evaluated by approximate methods, taking into account the geometry of the interacting molecules. Finally it is assumed that differences in reactivity along a series of compounds derive mainly from differences in ΔE terms.

Using the PMO method, in the approximation that considers only interactions at the reaction centers (eqn. 32), with $\gamma_{rr'} = \gamma_{ss'}$, calculations were made on the Diels–Alder reaction of some 1- or 2-substituted butadienes with vinyl derivatives as dienophiles[269], and in all cases it has been correctly found that "*ortho*" and "*para*" orientations are preferred to "*meta*" orientation in the adduct (see Section 4.1.1).

In the previous, and almost in every calculation on Diels–Alder reactions, it has been assumed that diene and dienophile lie on top of each other in parallel (or roughly parallel) planes. Under this condition, the *endo* approach is theoretically better than the *exo* approach only because of secondary interactions[268]. However, it has been shown that, for cyclopentadiene dimerisation, if the two molecules are allowed to approach in a non-parallel way, the *endo* preference can be mainly attributed to a more favourable primary interaction, due to an approach at an angle of *ca.* 60°, which could be allowed only to the *endo*-oriented dienophile because of steric reasons[270]. The PMO method has also been applied to a simplified treatment of some competing 1,4 and 1,2 thermal cycloadditions involving diradical intermediates[270a].

The concept of intermolecular orbitals, built out of the LCAO molecular orbitals of two interacting conjugated systems by means of perturbation theory, has been recently introduced[271]. Analytical expressions for the interaction energy involve a function of the overlap integral between pairs of atoms ($S_{rr'}$), of the charge density q_r and $q_{r'}$ at such atoms, and of an interaction integral $\eta_{rr'}$, taking into account coefficients c_{ir}, $c_{i'r'}$ . . . and energies E_i, $E_{i'}$ . . . as previously defined. For instance, molecules in their ground states have to a good approximation the following interaction energy.

$$E_{\text{int}} = -\sum_{rr'} (q_r + q_{r'})\eta_{rr'}S_{rr'} + 2\left(\sum_{i}^{\text{occ}} \sum_{i'}^{\text{unocc}} - \sum_{i}^{\text{unocc}} \sum_{i'}^{\text{occ}}\right) \frac{\left(\sum_{rr'} c_{ir}c_{i'r'}\eta_{rr'}\right)^2}{E_i - E_{i'}} \qquad (33)$$

As the interaction integrals $\eta_{rr'}$ are negative, the first term is a positive repulsive term, while the second term will usually be negative and attractive (its correspondence with eqn. (32) is obvious). Thus attraction between the two molecules derives, as in previous treatments, from mixing of the occupied orbitals of one reactant with the vacant orbitals of the other, and is stronger if the corresponding energy levels are relatively close. The approximation of considering only the upper occupied and the lowest unoccupied molecular orbital of reactants, as in Fukui's "frontier electron theory"[272] has therefore some justification[269]. A different origin is predicted for attractive interaction between an excited and a ground-state molecule[271].

Application of intermolecular orbital theory to some thermal and photochemical 1,2- and 1,4-cycloadditions[273] made possible the construction of reaction paths along directions of minimum energy; in this way two-step and concerted processes, and also intermediate cases of "quasi-concerted"

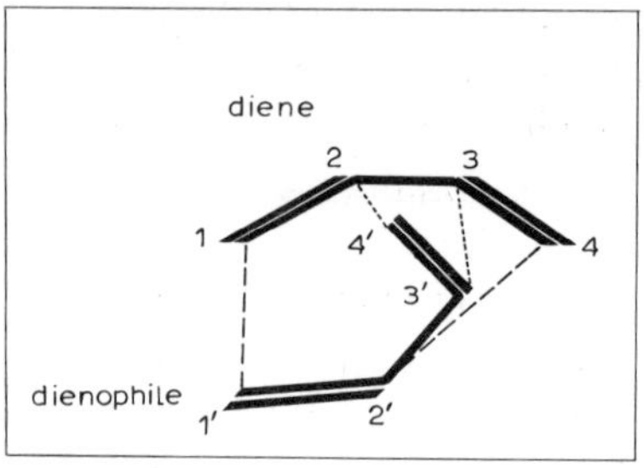

Fig.3. Scheme of primary (11', 2'4) and secondary (33', 24') interactions in butadiene dimerisation[273]. The molecules lie on top of each other in parallel planes.

reactions like acrolein dimerisation were predicted. Extending the $\Sigma_{rr'}$ summations in eqn. (33) to secondary interactions, the *endo* mechanism of Diels–Alder additions has been fully justified, importance being given to the interactions with the diene of the atoms of dienophile which are neighbour and next-nearest neighbour to the double bond involved in the primary interaction (Fig. 3).

The conclusions are in agreement with the Hoffmann–Woodward rules, at least for cycloadditions between alternant hydrocarbons; exceptional deviations are shown to be possible in particular photocycloadditions[273].

REFERENCES

1 J. E. Baldwin, *J. Org. Chem.*, 32 (1967) 2438.
2 A. J. Speziale and D. E. Bissing, *J. Am. Chem. Soc.*, 85 (1963) 1888.
3 A. J. Speziale and D. E. Bissing, *J. Am. Chem. Soc.*, 85 (1963) 3878.
4 C. Rüchardt, P. Panse and S. Eichler, *Chem. Ber.*, 100 (1967) 1144.
5 R. Huisgen, R. Grashey and J. Sauer, in *The Chemistry of Alkenes*, S. Patai (Ed.), Interscience, London, 1964, p. 739.
6 R. Huisgen, *Angew. Chem.*, 80 (1968) 329; *Angew. Chem. Intern. Ed. Engl.*, 7 (1968) 321.
7 G. B. Gill, *Quart. Rev.*, 22 (1968) 338.
8 A. Wassermann, *Diels–Alder Reactions*, Elsevier, Amsterdam, 1965.
9 J. Sauer, *Angew. Chem.*, 79 (1967) 76; *Angew. Chem. Intern. Ed. Engl.*, 6 (1967) 16.
10 H. Kwart and K. King, *Chem. Rev.*, 68 (1968) 415.
11 R. Huisgen, *Angew. Chem.*, 75 (1963) 742; *Angew. Chem. Intern. Ed. Engl.*, 2 (1963) 633.
12 G. J. Janz, in *1,4-Cycloaddition Reactions*, J. Hamer (Ed.), Academic Press, New York, 1967, p. 97.
13 J. Hamer and M. Ahmad, in *1,4-Cycloaddition Reactions*, J. Hamer (Ed.), Academic Press, New York, 1967, p. 419.
14 M. A. Ogliaruso, M. G. Romanelli and E. I. Becker, *Chem. Rev.*, 65 (1965) 261.
15 J. G. Martin and R. K. Hill, *Chem. Rev.*, 61 (1961) 537.
16 J. D. Roberts and C. M. Sharts, *Org. Reactions*, 12 (1962) 1.
17 W. E. Vaughan, *J. Am. Chem. Soc.*, 54 (1932) 3863.
18 G. B. Kistiakowsky and J. R. Lacher, *J. Am. Chem. Soc.*, 58 (1936) 123.
19 J. B. Harkness, G. B. Kistiakowsky and W. H. Mears, *J. Chem. Phys.*, 5 (1937) 682.
20 G. A. Benford and A. Wassermann, *J. Chem. Soc.*, (1939) 362.
21 G. B. Kistiakowsky and W. W. Ransom, *J. Chem. Phys.*, 7 (1939) 725.
22 J. R. Lacher, G. W. Tompkin and J. D. Park, *J. Am. Chem. Soc.*, 74 (1952) 1693.
23 B. Atkinson and A. B. Trenwith, *J. Chem. Soc.*, (1953) 2082.

24 B. Atkinson and M. Stedman, *J. Chem. Soc.*, (1962) 512.
25 D. Rowley and H. Steiner, *Discussions Faraday Soc.*, 10 (1951) 198.
26 P. J. Hawkins and G. J. Janz, *J. Am. Chem. Soc.*, 74 (1952) 1790.
27 J. L. Skinner and C. M. Sliepcevich, *Ind. Eng. Chem. Fundamentals*, 2 (1963) 168.
28 M. E. Schafer and R. S. Berry, *J. Am. Chem. Soc.*, 87 (1965) 4497.
29 G. Porter and J. I. Steinfeld, *J. Chem. Soc. A*, (1968) 877.
30 B. Raistrick, R. H. Shapiro and D. M. Newitt, *J. Chem. Soc.*, (1939) 1761.
31 N. M. Seidov, R. A. Kadyrov and A. A. Bakhshi-Zade, *Azerb. Khim. Zh.*, (1964) 81; *Chem. Abstr.*, 63 (1965) 6879.
32 A. G. Turnbull and H. S. Hull, *Australian J. Chem.*, 21 (1968) 1789.
33 R. F. Robey, H. K. Wiese and C. E. Morrell, *Ind. Eng. Chem.*, 36 (1944) 3.
34 C. Walling and J. Peisach, *J. Am. Chem. Soc.*, 80 (1958) 5819.
35 E. Gil-Av and Y. Herzberg-Minzly, *Proc. Chem. Soc.*, (1961) 316.
36 Y. Yukawa and A. Isohisa, *Mem. Inst. Sci. Ind. Res. Osaka Univ.*, 10 (1953) 191; *Chem. Abstr.*, 48 (1954) 7598.
37 L. Horner and T. Burger, *Ann. Chem.*, 710 (1967) 102.
38 R. E. Buckles and L. A. McGrew, *J. Am. Chem. Soc.*, 88 (1966) 3582.
39 T. Yamabe, A. Nagasawa, H. Kitano and K. Fukui, *Kogyo Kagaku Zasshi*, 66 (1963) 821; *Chem. Abstr.*, 60 (1964) 8032.
40 P. K. Kadaba and J. O. Edwards, *J. Org. Chem.*, 26 (1961) 2331.
41 R. Huisgen, L. Möbius, G. Müller, H. Stangl, G. Szeimies and J. M. Vernon, *Chem. Ber.*, 98 (1965) 3992.
42 A. Wassermann, *J. Chem. Soc.*, (1936) 1028.
43 J. A. Berson and W. A. Mueller, *J. Am. Chem. Soc.*, 83 (1961) 4940.
44 A. Wassermann, *J. Chem. Soc.*, (1935) 828.
45 A. Rodgman and G. F. Wright, *J. Org. Chem.*, 18 (1953) 465.
46 L. J. Andrews and R. M. Keefer, *J. Am. Chem. Soc.*, 77 (1955) 6284.
47 W. R. Vaughan and K. S. Andersen, *J. Org. Chem.*, 21 (1956) 673.
48 J. P. Simons, *Trans. Faraday Soc.*, 56 (1960) 391.
49 J. Sauer, H. Wiest and A. Mielert, *Chem. Ber.*, 97 (1964) 3183.
50 J. Hamer, M. Ahmad and R. E. Holliday, *J. Org. Chem.*, 28 (1963) 3034.
51 J. Gillois-Doucet and P. Rumpf, *Bull. Soc. Chim. France*, (1959) 1823.
52 D. C. Ayres and J. R. Smith, *Chem. Commun.*, (1967) 886.
53 P. Brown and R. C. Cookson, *Tetrahedron*, 21 (1965) 1977.
54 J. M. S. Jarvie and G. J. Janz, *J. Phys. Chem.*, 60 (1956) 1430.
55 P. Scheiner, J. H. Schomaker, S. Deming, W. J. Libbey and G. P. Nowack, *J. Am. Chem. Soc.*, 87 (1965) 306.
56 A. Ledwith and D. Parry, *J. Chem. Soc. C*, (1966) 1408.
57 R. Huisgen, L. A. Feiler and P. Otto, *Tetrahedron Letters*, (1968) 4485; *Chem. Ber.*, 102 (1969) 3444.
58 B. Franzus, *J. Org. Chem.*, 28 (1963) 2954.
59 R. Huisgen, G. Szeimies and L. Möbius, *Chem. Ber.*, 100 (1967) 2494.
60 A. S. Bailey and J. E. White, *J. Chem. Soc. B*, (1966) 819.
61 A. Dondoni, A. Mangini and S. Ghersetti, *Tetrahedron Letters*, (1966) 4789.
62 P. Beltrame, C. Veglio and M. Simonetta, *J. Chem. Soc. B*, (1967) 867.
63 R. Huisgen and H. Gotthardt, *Chem. Ber.*, 101 (1968) 1059.
64 I. Benghiat and E. I. Becker, *J. Org. Chem.*, 23 (1958) 885.
65 T. Jaworski, *Roczniki Chem.*, 35 (1961) 1309; *Chem. Abstr.*, 57 (1962) 5888.
66 W. Rubin, H. Steiner and A. Wassermann, *J. Chem. Soc.*, (1949) 3046.
67 D. Craig, J. J. Shipman and R. B. Fowler, *J. Am. Chem. Soc.*, 83 (1961) 2885.
68 B. Eisler and A. Wassermann, *J. Chem. Soc.*, (1953) 979.
69 C. Walling and H. J. Schugar, *J. Am. Chem. Soc.*, 85 (1963) 607.
70 T. Inukai and T. Kojima, *J. Org. Chem.*, 32 (1967) 872.
71 P. D. Bartlett and L. K. Montgomery, *J. Am. Chem. Soc.*, 86 (1964) 628.
72 P. Beltrame, D. Pitea, A. Marzo and M. Simonetta, *J. Chem. Soc. B*, (1967) 71.
73 W. D. Lloyd and G. W. Hedrick, *J. Org. Chem.*, 26 (1961) 2029.

74 C. A. Stewart, Jr., *J. Org. Chem.*, 28 (1963) 3320.
75 A. Eckell, R. Huisgen, R. Sustmann, G. Wallbillich, D. Grashey and E. Spindler, *Chem. Ber.*, 100 (1967) 2192.
76 J. S. Clovis, A. Eckell, R. Huisgen and R. Sustmann, *Chem. Ber.*, 100 (1967) 60.
77 A. R. Monahan and G. J. Janz, *J. Phys. Chem.*, 69 (1965) 1070.
78 T. Inukai and T. Kojima, *J. Org. Chem.*, 31 (1966) 1121.
79 E. J. De Witt, C. T. Lester and G. A. Ropp, *J. Am. Chem. Soc.*, 78 (1956) 2101.
80 D. C. F. Law and S. W. Tobey, *J. Am. Chem. Soc.*, 90 (1968) 2376.
81 K. R. Henery-Logan and R. A. Clark, *Tetrahedron Letters*, (1968) 801.
82 V. F. Kucherov, A. S. Onishchenko, B. A. Rudenko and E. A. El'Perina, *Dokl. Akad. Nauk SSSR*, 158 (1964) 397; *Chem. Abstr.*, 62 (1965) 7630.
83 E. F. Lutz and G. M. Bailey, *J. Am. Chem. Soc.*, 86 (1964) 3899.
84 P. E. Eaton and W. S. Hurt, *J. Am. Chem. Soc.*, 88 (1966) 5038.
85 D. R. Taylor and M. R. Warburton, *Tetrahedron Letters*, (1967) 3277.
86 J. S. Clovis, A. Eckell, R. Huisgen, R. Sustmann, G. Wallbillich and V. Weberndörfer, *Chem. Ber.*, 100 (1967) 1593.
87 N. J. Turro and G. S. Hammond, *J. Am. Chem. Soc.*, 84 (1962) 2841.
88 D. Valentine, N. J. Turro and G. S. Hammond, *J. Am. Chem. Soc.*, 86 (1964) 5202.
89 N. J. Turro and P. D. Bartlett, *J. Org. Chem.*, 30 (1965) 1849.
90 P. D. Bartlett, G. E. H. Wallbillich, A. S. Wingrove, J. S. Swenton, L. K. Montgomery and B. D. Kramer, *J. Am. Chem. Soc.*, 90 (1968) 2049.
91 J. S. Swenton and P. D. Bartlett, *J. Am. Chem. Soc.*, 90 (1968) 2056.
92 C. S. Rondestvedt, Jr. and C. D. Ver Nooy, *J. Am. Chem. Soc.*, 77 (1955) 4878.
93 F. Kaplan and H. Conroy, *J. Org. Chem.*, 28 (1963) 1593.
94 J. A. Berson, Z. Hamlet and W. A. Mueller, *J. Am. Chem. Soc.*, 84 (1962) 297.
95 J. Sauer and J. Kredel, *Tetrahedron Letters*, (1966) 731.
96 J. S. Meek, F. M. Dewey and M. W. Hanna, *J. Org. Chem.*, 32 (1967) 69.
97 R. Robson, P. W. Grubb and J. A. Barltrop, *J. Chem. Soc.*, (1964) 2153.
98 S. Proskow, H. E. Simmons and T. L. Cairns, *J. Am. Chem. Soc.*, 88 (1966) 5254.
99 R. Livingstone and K. S. Wei, *J. Phys. Chem.*, 71 (1967) 541.
100 H. Stobbe and F. Reuss, *Ann. Chem.*, 391 (1912) 151.
101 E. G. V. Barrett and L. J. Burrage, *J. Phys. Chem.*, 37 (1933) 1029.
102 G. A. Benford, H. Kaufmann, B. S. Khambata and A. Wassermann, *J. Chem. Soc.*, (1939) 381.
103 H. Schmid, F. Kubassa and R. Herdy, *Monatsh. Chem.*, 79 (1948) 430; *Chem. Abstr.*, 43 (1949) 4934.
104 V. G. Moor, N. V. Strigaleva and L. V. Shilyaeva, *J. Gen. Chem. USSR*, 5 (1935) 818; *Chem. Abstr.*, 30 (1936) 365.
105 O. Diels and K. Alder, *Ann. Chem.*, 460 (1928) 98.
106 K. Alder and G. Stein, *Angew. Chem.*, 50 (1937) 510.
107 T. Inukai and T. Kojima, *J. Org. Chem.*, 31 (1966) 2032.
108 J. Sauer and J. Kredel, *Tetrahedron Letters*, (1966) 6359.
109 E. M. Kosower, *J. Am. Chem. Soc.*, 80 (1958) 3253; 82 (1960) 2188.
110 M. Charton, *J. Org. Chem.*, 31 (1966) 3745.
111 T. Inukai and T. Kojima, *J. Org. Chem.*, 32 (1967) 869.
112 H. E. Hennis, *J. Org. Chem.*, 28 (1963) 2570.
113 G. Kresze and J. Firl, *Tetrahedron*, 19 (1963) 1329.
114 G. Kresze and J. Firl, *Tetrahedron Letters*, (1965) 1163.
115 A. Hassner, *J. Org. Chem.*, 33 (1968) 2684 and literature cited therein.
116 G. Kresze and O. Korpiun, *Tetrahedron*, 22 (1966) 2493.
117 G. Kresze, in *1,4-Cycloaddition Reactions*, J. Hamer (Ed.). Academic Press, New York, 1967, p. 453.
118 G. J. Janz and S. C. Wait, Jr., *J. Am. Chem. Soc.*, 76 (1954) 6377.
119 M. R. Dean and T. W. Legatski, *Ind. Eng. Chem. Anal. Ed.*, 16 (1944) 8.
119a M. J. S. Dewar and R. S. Pyron, *J. Am. Chem. Soc.*, 92 (1970) 3098.
120 M. Ahmad and J. Hamer, *J. Org. Chem.*, 31 (1966) 2831.

121 W. E. Bachmann and N. C. Deno, *J. Am. Chem. Soc.*, 71 (1949) 3062.
122 J. Sauer and B. Schröder, *Chem. Ber.*, 100 (1967) 678.
123 J. Sauer, D. Lang and A. Mielert, *Angew. Chem.*, 74 (1962) 352; *Angew. Chem. Intern. Ed. Engl.*, 1 (1962) 268.
124 M. Ahmad and J. Hamer, *J. Org. Chem.*, 31 (1966) 2829.
125 G. Kresze, J. Firl, H. Zimmer and U. Wollnik, *Tetrahedron*, 20 (1964) 1605.
126 J. Sauer and H. Wiest, *Angew. Chem.*, 74 (1962) 353; *Angew. Chem. Intern. Ed. Engl.*, 1 (1962) 269.
127 J. Sauer and G. Heinrichs, *Tetrahedron Letters*, (1966) 4979.
128 B. Eisler and A. Wassermann, *J. Chem. Soc.*, (1953) 1943.
129 E. K. Fields, *J. Am. Chem. Soc.*, 76 (1954) 2709.
130 M. Mizuta and Y. Ishii, *Kogyo Kagaku Zasshi*, 66 (1963) 1442; *Chem. Abstr.*, 60 (1964) 11880.
131 M. Mizuta, T. Kato and Y. Ishii, *Kogyo Kagaku Zasshi*, 67 (1964) 1382; *Chem. Abstr.*, 62 (1965) 8963.
132 J. Sauer, D. Lang and H. Wiest, *Chem. Ber.*, 97 (1964) 3208.
133 G. J. Janz and N. E. Duncan, *J. Am. Chem. Soc.*, 75 (1953) 5389.
134 D. N. Matthews and E. I. Becker, *J. Org. Chem.*, 31 (1966) 1135.
135 A. I. Konovalov, E. G. Yarkova, I. Sh. Salikhov and R. G. Izmailova, *Zh. Org. Khim.*, 3 (1967) 1319; *Chem. Abstr.*, 67 (1967) 99387.
136 G. Kresze, S. Rau, G. Sabelus and H. Goetz, *Ann. Chem.*, 648 (1961) 57.
137 S. Glasstone, K. J. Laidler and H. Eyring, *The Theory of Rate Processes*, McGraw-Hill, New York, 1941, p. 199.
138 B. S. Khambata and A. Wassermann, *J. Chem. Soc.*, (1939) 375.
139 W. C. Herndon, C. R. Grayson and J. M. Manion, *J. Org. Chem.*, 32 (1967) 526.
140 M. Uchiyama, T. Tomioka and A. Amano, *J. Phys. Chem.*, 68 (1964) 1878.
141 W. Tsang, *J. Chem. Phys.*, 42 (1965) 1805.
142 S. W. Benson, *J. Chem. Phys.*, 46 (1967) 4920.
142a N. C. Billingham, J. R. Ebdon, R. S. Lehrle, J. L. Markham and J. C. Robb, *Trans. Faraday Soc.*, 65 (1969) 470.
143 R. E. Banks, A. C. Harrison and R. N. Haszeldine, *Chem. Commun.*, (1966) 338; R. E. Banks, A. C. Harrison, R. N. Haszeldine and K. G. Orrell, *J. Chem. Soc. C.*, (1967) 1608.
143a R. A. Grieger and C. A. Eckert, *J. Am. Chem. Soc.*, 92 (1970) 2918, 7149; *Trans. Faraday Soc.*, 66 (1970) 2579.
144 S. W. Benson and J. A. Berson, *J. Am. Chem. Soc.*, 84 (1962) 152; E. Whalley, *Advan. Phys. Org. Chem.*, 2(1964) 93.
145 E. A. Halevi, *Progr. Phys. Org. Chem.*, 1 (1963) 109.
146 M. Matsuoka and M. Szwarc, *J. Am. Chem. Soc.*, 83 (1961) 1260.
147 M. Feld, A. P. Stefani and M. Szwarc, *J. Am. Chem. Soc.*, 84 (1962) 4451.
148 D. E. Van Sickle and J. O. Rodin, *J. Am. Chem. Soc.*, 86 (1964) 3091.
149 P. Brown and R. C. Cookson, *Tetrahedron*, 21 (1965) 1993.
150 S. Seltzer, *J. Am. Chem. Soc.*, 87 (1965) 1534.
151 W. R. Dolbier, Jr. and S.-H. Dai, *J. Am. Chem. Soc.*, 90 (1968) 5028.
152 R. J. Crawford and D. M. Cameron, *J. Am. Chem. Soc.*, 88 (1966) 2589.
153 A. Wassermann, *J. Chem. Soc.*, (1942) 618.
154 E. J. Barrett, H. W. Hoyer and A. V. Santoro, *Tetrahedron Letters*, (1968) 603.
155 A. N. Nesmeyanov, M. I. Rybinskaya and N. K. Kochetkov, *Izvest. Akad. Nauk SSSR, Otdel. Khim. Nauk*, (1955) 817; *Chem. Abstr.*, 50 (1956) 9360, 11303.
156 P. Yates and P. Eaton, *J. Am. Chem. Soc.*, 82 (1960) 4436.
157 G. I. Fray and R. Robinson, *J. Am. Chem. Soc.*, 83 (1961) 249.
158 R. B. Woodward and T. J. Katz, *Tetrahedron*, 5 (1959) 70.
158a T. Inukai and T. Kojima, *Chem. Commun.*, (1969) 1334.
158b T. Kojima and T. Inukai, *J. Org. Chem.*, 35 (1970) 1342.
158c H. W. Thompson and D. G. Melillo, *J. Am. Chem. Soc.*, 92 (1970) 3218.
159 E. Ciganek, *Tetrahedron Letters*, (1967) 3321.

160 J. P. Candlin and W. H. Janes, *J. Chem. Soc. C*, (1968) 1856.
160a A. Carbonaro, A. Greco and G. Dall'Asta, *J. Org. Chem.*, 33 (1968) 3948.
161 G. Wilke, *Angew. Chem. Intern. Ed. Engl.*, 2 (1963) 105.
162 P. Heimbach and W. Brenner, *Angew. Chem.*, 79 (1967) 813; *Angew. Chem. Intern. Ed. Engl.*, 6 (1967) 800.
162a W. L. Dilling, *Chem. Rev.*, 69 (1969) 845.
163 G. O. Schenck, J. Kuhls and C. H. Krauch, *Ann. Chem.*, 693 (1966) 20.
164 G. Pfundt and G. O. Schenck, in *1,4-Cycloaddition Reactions* J. Hamer (Ed.), Academic Press, New York, 1967, p. 345.
165 S. Farid, *Chem. Commun.*, (1967) 1268.
166 D. Bryce-Smith and A. Gilbert, *Chem. Commun.*, (1968) 1701, 1702.
167 Y. L. Chow and T. C. Joseph, *Chem. Commun.*, (1968) 604.
168 S. Farid and K.-H. Scholz, *Chem. Commun.*, (1968) 412.
169 L. I. Smith, *Chem. Rev.*, 23 (1938) 193.
170 F. Bergmann, H. E. Eschinazi and M. Neeman, *J. Org. Chem.*, 8 (1943) 179.
171 See R. Huisgen, *Angew. Chem.*, 75 (1963) 604; *Angew. Chem. Intern. Ed. Engl.*, 2 (1963) 565.
172 R. Huisgen, *Proc. Chem. Soc.*, (1961) 357.
173 J. D. Roberts, *Chem. Ber.*, 94 (1961) 273.
174 R. Huisgen, in *Aromaticity*, The Chemical Society, London, 1967, p. 51.
175 R. Fusco, G. Bianchetti and D. Pocar, *Gazz. Chim. Ital.*, 91 (1961) 849.
176 A. Quilico, in *Five- and Six-membered Compounds with Nitrogen and Oxygen* R. Wiley (Ed.), Interscience, New York, 1962, p. 1.
176a. R. Paul and S. Tchelitcheff, *Bull. Soc. Chim. France*, (1962) 2215.
177 M. Christl and R. Huisgen, *Tetrahedron Letters*, (1968) 5209.
178 G. R. Delpierre and M. Lamchen, *J. Chem. Soc.*, (1963) 4693.
179 P. K. Kadaba, *Tetrahedron*, 22 (1966) 2453.
180 P. Rajagopalan and B. G. Advani, *J. Org. Chem.*, 30 (1965) 3369.
181 P. Beltrame, A. Comotti and C. Veglio, *Chem. Commun.*, (1967) 996; see also ref. 194.
182 R. R. Fraser and Y. S. Lin, *Can. J. Chem.*, 46 (1968) 801.
183 R. Huisgen, H. Knupfer, R. Sustmann, G. Wallbillich and V. Weberndörfer, *Chem. Ber.*, 100 (1967) 1580.
183a S. McLean and D. M. Findlay, *Tetrahedron Letters*, (1969) 2219.
184 G. Leandri and M. Pallotti, *Ann. Chim. (Rome)*, 47 (1957) 376.
185 M. E. Munk and Y. Ki Kim, *J. Am. Chem. Soc.*, 86 (1964) 2213.
186 A. Ledwith and Y. Shin-Lin, *J. Chem. Soc. B*, (1967) 83.
186a R. Huisgen, H. Seidl and I. Brüning, *Chem. Ber.*, 102 (1969) 1102.
187 D. Martin and W. Mucke, *Z. Chem.*, 3 (1963) 347; *Chem. Abstr.*, 59 (1963) 12792.
188 R. Huisgen, H. Stangl, H. J. Sturm and H. Wagenhofer, *Angew. Chem.*, 73 (1961) 170.
189 R. Huisgen, *Dechema Monograph.*, 49, 1964, p. 201.
190 A. Dondoni, *Tetrahedron Letters*, (1967) 2397.
190a A. Battaglia and A. Dondoni, *Ric. Sci.*, 38 (1968) 201.
191 F. Bergmann and A. Weizmann, *J. Org. Chem.*, 9 (1944) 352.
192 R. Huisgen, H. J. Sturm and H. Wagenhofer, *Z. Naturforsch.*, 17b (1962) 202.
193 P. Beltrame, C. Veglio and M. Simonetta, *Chem. Commun.*, (1966) 433.
194 P. Beltrame and C. Vintani, *J. Chem. Soc. B*, (1970) 873.
195 R. A. Firestone, *J. Org. Chem.*, 33 (1968) 2285.
196 R. Huisgen, *J. Org. Chem.*, 33 (1968) 2291.
196a P. Beltrame, P. Sartirana and C. Vintani, *J. Chem. Soc. B*, (1971) 814.
197 C. G. Overberger, N. Weinshenker and J.-P. Anselme, *J. Am. Chem. Soc.*, 87 (1965) 4119.
198 R. Huisgen, H. Hauck, R. Grashey and H. Seidl, *Chem. Ber.*, 101 (1968) 2568.
198a S. Morrocchi, A. Ricca, A. Zanarotti, G. Bianchi, R. Gandolfi and P. Grünanger, *Tetrahedron Letters*, 1969, 3329.

199 S. Morrocchi, A. Ricca and L. Velo, *Tetrahedron Letters*, (1967) 331.
200 S. Morrocchi and A. Ricca, *Chim. Ind. (Milan)*, 49 (1967) 629.
201 H. Hoberg, *Ann. Chem.*, 707 (1967) 147.
202 J. J. Drysdale, W. W. Gilbert, H. K. Sinclair and W. H. Sharkey, *J. Am. Chem. Soc.*, 80 (1958) 3672.
203 P. D. Bartlett, L. K. Montgomery and B. Seidel, *J. Am. Chem. Soc.*, 86 (1964) 616.
204 C. A. Stewart, Jr., *J. Am. Chem. Soc.*, 84 (1962) 117.
205 J. C. Little, *J. Am. Chem. Soc.*, 87 (1965) 4020.
205a P. D. Bartlett and K. E. Schueller, *J. Am. Chem. Soc.*, 90 (1968) 6071, 6077.
205b R. Wheland and P. D. Bartlett, *J. Am. Chem. Soc.*, 92 (1970) 3822.
206 H. W. B. Reed, *J. Chem. Soc.*, (1951) 685.
207 N. C. Billingham, P. A. Leeming, R. S. Lehrle and J. C. Robb, *Nature*, 213 (1967) 494.
207a P. D. Bartlett, A. S. Wingrove and R. Owyang, *J. Am. Chem. Soc.*, 90 (1968) 6067.
207b J. Dale, *Angew. Chem.*, 78 (1966) 1070; *Angew. Chem. Intern. Ed. Engl.*, 5 (1966) 1000; J. Dale and P. O. Kristiansen, *Chem. Commun.*, (1968) 1293.
208 W. T. Brady and H. R. O'Neal, *J. Org. Chem.*, 32 (1967) 2704.
209 R. Huisgen and P. Otto, *Tetrahedron Letters*, (1968) 4491.
210 J. D. Park, H. V. Holler and J. R. Lacher, *J. Org. Chem.*, 25 (1960) 990.
211 S. Proskow, H. E. Simmons and T. L. Cairns, *J. Am. Chem. Soc.*, 85 (1963) 2341.
212 J. K. Williams, D. W. Wiley and B. C. McKusick, *J. Am. Chem. Soc.*, 84 (1962) 2210.
213 L. K. Montgomery, K. Schueller and P. D. Bartlett, *J. Am. Chem. Soc.*, 86 (1964) 622.
214 I. Tabushi, R. Oda and K. Okazaki, *Tetrahedron Letters*, (1968) 3743.
214a M. Jones, Jr. and R. H. Levin, *Tetrahedron Letters*, (1968) 5593; *J. Am. Chem. Soc.*, 91 (1969) 6411; H. H. Wasserman, A. J. Solodar and L. S. Keller, *Tetrahedron Letters*, (1968) 5597.
215 R. Huisgen, L. Feiler and G. Binsch, *Angew. Chem.*, 76 (1964) 892; *Angew. Chem. Intern. Ed. Engl.*, 3 (1964) 753.
216 G. Binsch, L. A. Feiler and R. Huisgen, *Tetrahedron Letters*, (1968) 4497; R. Huisgen, L. A. Feiler and G. Binsch, *Chem. Ber.*, 102 (1969) 3460.
217 R. Montaigne and L. Ghosez, *Angew. Chem.*, 80 (1968) 194; *Angew. Chem. Intern. Ed. Engl.*, 7 (1968) 221.
218 J.-L. Luche and H. B. Kagan, *Bull. Soc. Chim. France*, (1968) 2450.
219 W. T. Brady and B. M. Holifield, *Tetrahedron*, 23 (1967) 4251.
219a E. J. Moriconi and J. F. Kelly, *Tetrahedron Letters*, (1968) 1435.
220 F. Effenberger and G. Kiefer, *Angew. Chem.*, 79 (1967) 936; *Angew. Chem. Intern. Ed. Engl.*, 6 (1967) 951.
220a F. Effenberger, *Angew. Chem.*, 81 (1969) 374; *Angew. Chem. Intern. Ed. Engl.*, 8 (1969) 295.
221 H. B. Kagan and J. L. Luche, *Tetrahedron Letters*, (1968) 3093.
221a W. T. Brady and E. D. Dorsey, *Chem. Commun.*, (1968) 1638; *J. Org. Chem.*, 35 (1970) 2732.
222 H. Ulrich, B. Tucker and A. A. R. Sayigh, *J. Am. Chem. Soc.* 90 (1968) 528.
222a R. Gompper, *Angew. Chem.*, 81 (1969) 348; *Angew. Chem. Intern. Ed. Engl.*, 8 (1969) 312.
223 E. F. Kiefer and M. Y. Okamura; *J. Am. Chem. Soc.*, 90 (1968) 4187.
224 B. Atkinson and A. B. Trenwith, *J. Chem. Phys.*, 20 (1952) 754.
225 R. Huisgen and P. Otto, *J. Am. Chem. Soc.*, 90 (1968) 5342.
225a R. Huisgen, private communication.
225b R. Lattrell, *Ann. Chem.*, 722 (1969) 132.
225c K. Clauss, *Ann. Chem.*, 722 (1969) 110.
226 P. D. Bartlett, G. E. H. Wallbillich and L. K. Montgomery, *J. Org. Chem.*, 32 (1967) 1290.
226a J. E. Baldwin and J. A. Kapecki, *J. Am. Chem. Soc.*, 92 (1970) 4868.
227 W. T. Brady and H. R. O'Neal, *J. Org. Chem.*, 32 (1967) 612.
228 T. J. Katz and R. Dessau, *J. Am. Chem. Soc.*, 85 (1963) 2172.
228a J. E. Baldwin and J. A. Kapecki, *J. Am. Chem. Soc.*, 91 (1969) 3106; 92 (1970) 4874.

228b P. OTTO, L. A. FEILER AND R. HUISGEN, *Angew. Chem.*, 80 (1968) 759; *Angew. Chem. Intern. Ed. Engl.*, 7 (1968) 737; R. HUISGEN AND P. OTTO, *J. Am. Chem. Soc.*, 91 (1969) 5922.
228c W. R. DOLBIER, JR. AND S.-H. DAI, *J. Am. Chem. Soc.*, 92 (1970) 1774.
228d J. E. BALDWIN AND U. V. ROY, *Chem. Commun.*, (1969) 1225.
229 H. W. B. REED, *J. Chem. Soc.*, (1954) 1931.
230 S. CARRA' AND R. UGO, *Inorg. Chim. Acta Rev.*, 1 (1967) 49.
231 R. S. H. LIU, N. J TURRO AND G. S. HAMMOND, *J. Am. Chem. Soc.*, 87 (1965) 3406.
232 W. L DILLING, *J. Am. Chem. Soc.*, 89 (1967) 2742.
233 G. SARTORI, V. TURBA, A. VALVASSORI AND M. RIVA, *Tetrahedron Letters*, (1966) 4777.
234 W. L. DILLING AND J. C. LITTLE, *J. Am. Chem. Soc.*, 89 (1967) 2741.
234a W. L. DILLING AND R. D. KROENING, *Tetrahedron Letters*, (1968) 5101, 5601.
235 D. BRYCE-SMITH, B. E. CONNETT AND A. GILBERT, *J. Chem. Soc. B*, (1968) 812, 816.
236 D. BRYCE-SMITH AND A. GILBERT, *J. Chem. Soc.*, (1965) 918.
237 W. M. HARDHAM AND G. S. HAMMOND, *J. Am. Chem. Soc.*, 89 (1967) 3200.
238 D. BRYCE-SMITH AND A. GILBERT, *Chem. Commun.*, (1968) 19.
239 B. E. JOB AND J. D. LITTLEHAILES, *J. Chem. Soc. C*, (1968) 886.
240 G. O. SCHENCK, W. HARTMANN, S.-P. MANNSFELD, W. METZNER AND C. H. KRAUCH, *Chem. Ber.*, 95 (1962) 1642.
241 G. O. SCHENCK, W. HARTMANN AND R. STEINMETZ, *Chem. Ber.*, 96 (1963) 498.
242 W. METZNER, *Tetrahedron Letters*, (1968) 1321.
243 E. S. ALBONE, *J. Am. Chem. Soc.*, 90 (1968) 4663.
244 J. L. RUHLEN AND P. A. LEERMAKERS, *J. Am. Chem. Soc.*, 88 (1966) 5671.
245 E. Y. Y. LAM, D. VALENTINE AND G. S. HAMMOND, *J. Am. Chem. Soc.*, 89 (1967) 3482.
246 B. D. CHALLAND AND P. DE MAYO, *Chem. Commun.*, (1968) 982.
247 H. MORRISON, A. FEELEY AND R. KLEOPFER, *Chem. Commun.*, (1968) 358.
248 H. MORRISON AND R. KLEOPFER, *J. Am. Chem. Soc.*, 90 (1968) 5037.
249 S. FARID, W. KOTHE AND G. PFUNDT, *Tetrahedron Letters*, (1968) 4147.
250 K. S. WEI AND R. LIVINGSTONE, *J. Phys. Chem.*, 71 (1967) 548.
251 D. O. COWAN AND R. L. DRISKO, *Tetrahedron Letters*, (1967) 1255.
252 D. O. COWAN AND R. L. DRISKO, *J. Am. Chem. Soc.*, 89 (1967) 3068.
253 I.-M. HARTMANN, W. HARTMANN AND G. O. SCHENCK, *Chem. Ber.*, 100 (1967) 3146.
254 L. A. SINGER AND G. A. DAVIS, *J. Am. Chem. Soc.*, 89 (1967) 598.
255 L. A. SINGER AND G. A. DAVIS, *J. Am. Chem. Soc.*, 89 (1967) 941.
256 L. A. SINGER AND G. A. DAVIS, *J. Am. Chem. Soc.*, 89 (1967) 158.
257 R. B. WOODWARD AND R. HOFFMANN, *J. Am. Chem. Soc.*, 87 (1965) 395.
258 H. C. LONGUET-HIGGINS AND E. W. ABRAHAMSON, *J. Am. Chem. Soc.*, 87 (1965) 2045.
259 R. HOFFMANN AND R. B. WOODWARD, *J. Am. Chem. Soc.*, 87 (1965) 2046.
260 R. HOFFMANN AND R. B. WOODWARD, *J. Am. Chem. Soc.*, 87 (1965) 4388.
261 R. HOFFMANN AND R. B. WOODWARD, *Accounts Chem. Res.*, 1 (1968) 17.
262 J. A. BERSON AND A. REMANICK, *J. Am. Chem. Soc.*, 83 (1961) 4947.
263 R. B. WOODWARD AND R. HOFFMANN, *Angew. Chem.*, 81 (1969) 797; *Angew. Chem. Intern. Ed. Engl.*, 8 (1969) 781.
264 R. D. BROWN, *J. Chem. Soc.*, (1950) 691, 2730.
264a E. RAGHAVAN AND P. T. NARASIMHAN, *Tetrahedron*, 25 (1969) 4643.
265 N. TYUTYULKOV AND P. MARKOV, *Monatsh. Chem.*, 96 (1965) 2030.
266 M. J. S. DEWAR, *J. Am. Chem. Soc.*, 74 (1952) 3341.
267 K. FUKUI, in *Molecular Orbitals in Chemistry, Physics and Biology*, P. O. LÖWDIN AND B. PULLMANN (Eds.), Academic Press, New York, 1964, p. 525.
268 W. C. HERNDON AND L. H. HALL, *Theoret. Chim. Acta*, 7 (1967) 4.
269 J. FEUER, W. C. HERNDON AND L. H. HALL, *Tetrahedron*, 24 (1968) 2575.
270 W. C. HERNDON AND L. H. HALL, *Tetrahedron Letters*, (1967) 3095.
270a W. C. HERNDON AND J. FEUER, *J. Org. Chem.*, 33 (1968) 417.
271 L. SALEM, *J. Am. Chem. Soc.*, 90 (1968) 543.
272 K. FUKUI, T. YONEZAWA AND H. SHINGU, *J. Chem. Phys.*, 20 (1952) 722.
273 L. SALEM, *J. Am. Chem. Soc.*, 90 (1968) 553.
274 CH. GRUNDMANN AND P. GRÜNANGER, *The Nitrile Oxides*, Springer, Berlin, 1971, p. 85.

Chapter 3

Elimination Reactions

A. F. COCKERILL

1. Introduction

Although it is often convenient to classify the reactions of organic compounds according to their "functional groups", a division into "reaction types" can be equally informative[1]. On the basis of mechanistic principles, most reactions in organic chemistry can be reduced to a relatively small number of classes. This chapter is concerned with the discussion of the mechanisms of one of the most common "reaction types", namely "elimination reactions".

An organic molecule undergoes an elimination reaction when it loses two atoms or groups of atoms which are subsequently not replaced. Such a definition embraces the reactions of a wide variety of organic molecules, but these can be combined into a few principal types of elimination reaction.

Most frequently the fragments are removed from adjacent atoms and the processes are called beta-eliminations (or 1,2-eliminations). Many well-known synthetic organic reactions fall into this category: *e.g.* (*1*) Hofmann elimination, (*2*) Cope elimination and (*3*) Chugaev reaction.

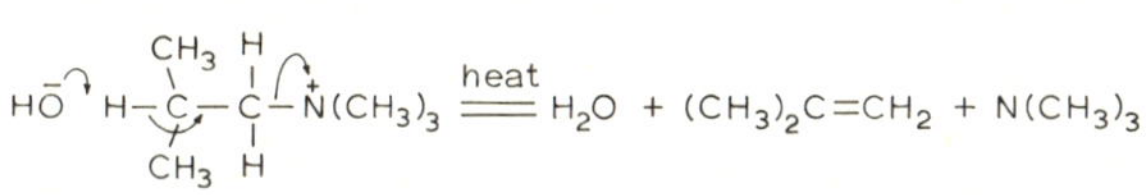

(1)

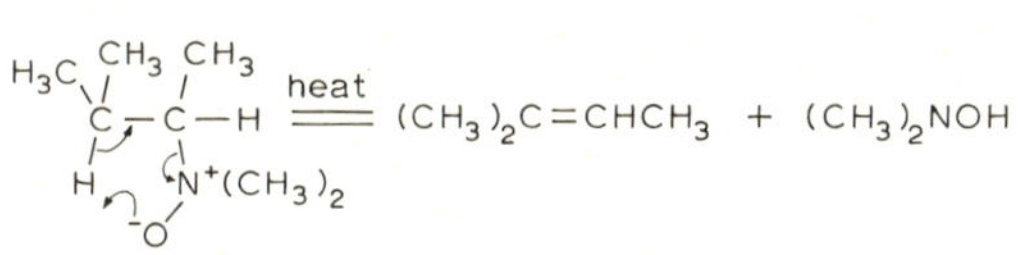

(2)

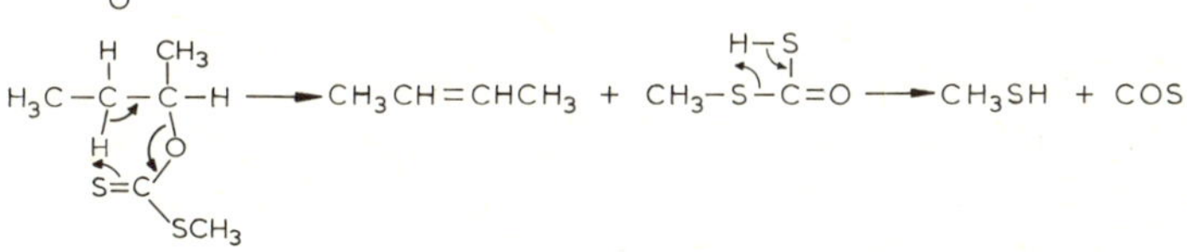

(3)

All these examples are olefin-forming processes and many organic chemists associate the term "elimination reactions" predominantly with 1,2-elimination. This is not surprising as the investigation of the Hofmann elimination represented the first kinetic study of elimination reactions[2].

References pp. 350–364

Usually one or two steps are involved in beta-eliminations. The number of steps, the stereochemistry of the reaction and the relative importance of the bond-breaking and bond-making processes in the transition state are greatly affected by both the choice of the substrate and the reaction conditions. For simplicity or discussion, beta-elimination has been divided into categories determined primarily by the reaction products, (*e.g.* olefins (C=C), acetylenes (C≡C), imides (C=N–R), nitriles (C≡N) and carbonyls (C=O), and secondly by the reaction conditions.

When the fragments are removed from the same atom, the process is termed an alpha-elimination (or 1,1-elimination), *viz.*

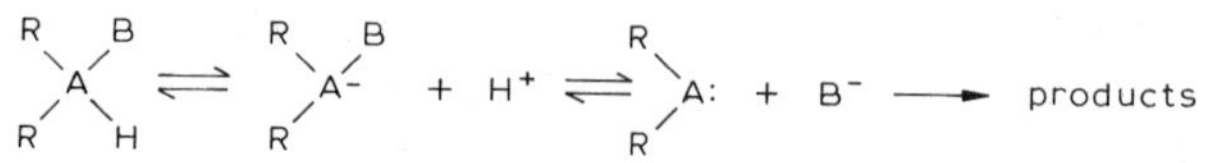

(4)

The initial product of elimination is a highly electron-deficient species, usually a carbene or a nitrene, which subsequently reacts rapidly to give products. The base-catalysed hydrolysis of chloroform is an alpha-elimination[3]. The properties of the electron-deficient species represent a rapidly expanding field of organic chemistry and this subject is discussed in the following chapter. In this chapter, only apparent alpha-eliminations, (*e.g.* reaction (5)), are included. The nature of the products indicates the intermediacy of electron-deficient species, but kinetic studies show the processes occur via reactions of the precursor carbanions, *viz.*

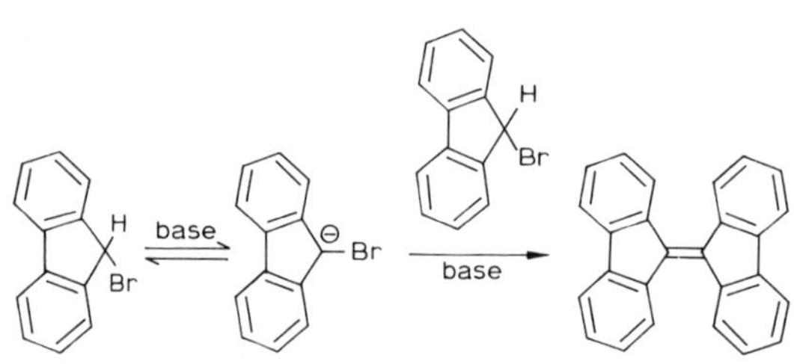

(5)

Occasionally the fragments are removed from more distant atoms and the processes are termed gamma-(or 1,3-) eliminations (reaction (6)), the products being three-membered ring systems, and delta-(or 1,4-) eliminations which give rise to four-membered rings. The latter type of elimination often involves a bond migration (reaction (7)).

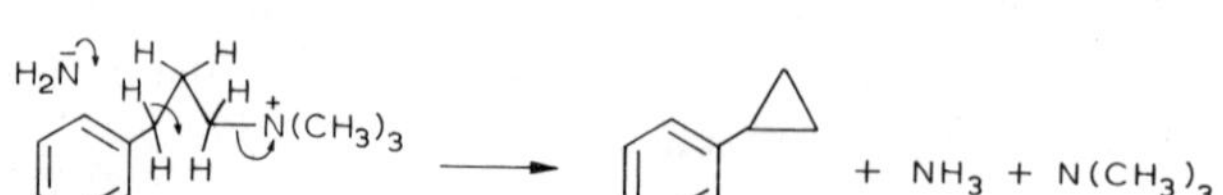

(6)

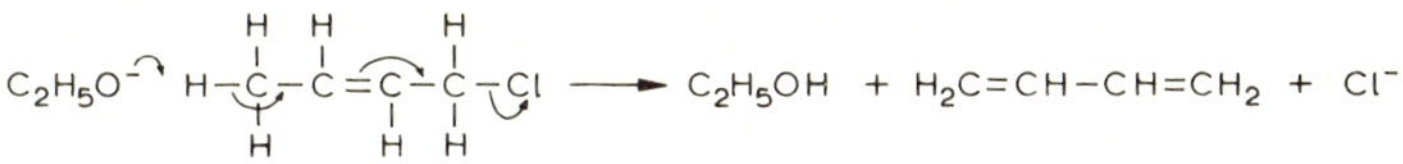

(7)

Discussion of "fragmentation reactions" (8), which have been considered apart from the more usual description of elimination processes[4], is also included, but "extrusion reactions" (9)[5], on which investigations are at present mainly restricted to preparative studies, have been omitted.

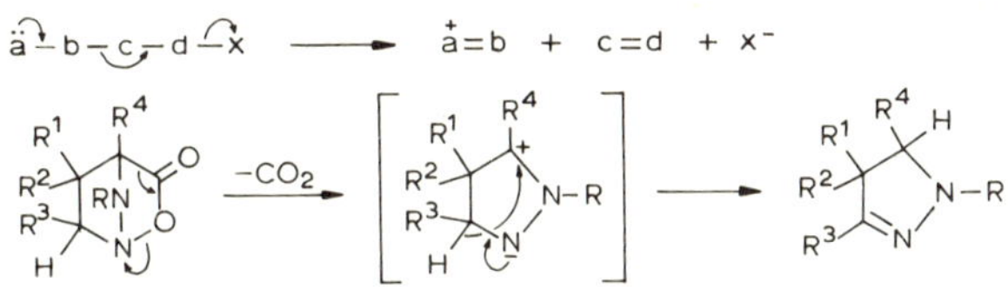

(8)

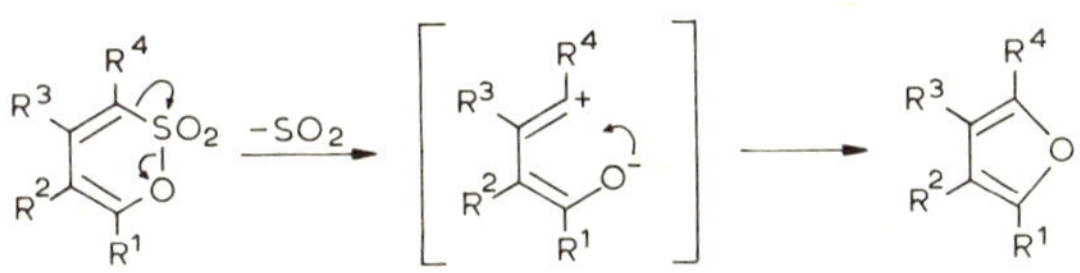

(9)

2. Olefin-forming beta-eliminations

The mechanisms of reactions leading to olefinic products which occur by beta-elimination processes are conveniently discussed under sections determined by reactions conditions.

2.1 SOLVOLYTIC AND BASE-CATALYSED ELIMINATION REACTIONS

The majority of olefin-forming reactions of synthetic importance are beta-eliminations, which usually involve the removal of the elements HX, from adjacent carbon atoms. The processes are often base-catalysed but will also occur in the absence of bases under solvolytic conditions. The mechanisms of these types of reaction have aroused great interest and the subject has been reviewed on numerous occasions[6–14]. The terminology was developed by Hughes and Ingold and their collaborators who developed a great deal of the fundamental knowledge. Although a number of mechanisms have some current credibility, by far the larger number of elimination reactions appear to follow either of two fundamental processes namely the E1 and E2 reactions. Initially these mechanisms are defined and then discussed subsequently in greater detail, whereas the less commonly encountered E1cB and ylid mechanisms are considered more fully in this section.

2.1.1 *The bimolecular mechanism*

The first mechanism to be classified, the bimolecular elimination (E2), was suggested to account for the Hofmann degradation (1), p. 163, of quaternary ammonium hydroxides to olefins and tertiary amines[2]. A single-step reaction between the substrate and a base is involved and the C_β–H and C_α–X bonds break and the C_α–C_β double bond forms in a concerted fashion, *viz.*

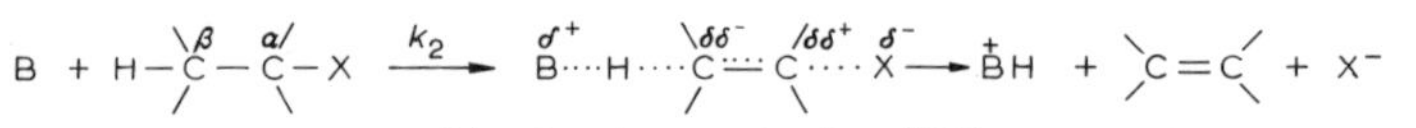 (10)

bimolecular mechanism (E2)

The reaction thus follows second-order kinetics, unless, as often is the case, the base is present in a large excess, when apparent-first-order kinetic behaviour is observed, *viz.*

$$\begin{aligned} \text{rate} &= k_2\,[\mathrm{B}][\text{substrate}] \\ &= k_1'\,[\text{substrate}] \text{ for } [\mathrm{B}] \geqslant 10[\text{substrate}]. \end{aligned} \quad (11)$$

The E2 mechanism is widely applicable to reactions of organic molecules. The bases usually employed are hydroxide or alkoxide ions, but a much wider range have been used successfully, *viz.*

Base B = $\bar{\mathrm{O}}$H, $\bar{\mathrm{O}}$R, NR_3, H_2O, $\bar{\mathrm{O}}$Ac, $\bar{\mathrm{O}}$Ar, $\bar{\mathrm{S}}$Ar, $\bar{\mathrm{N}}H_2$, CO_3^{2-}, I^-, $\bar{\mathrm{C}}$N, and organic bases (12)

When the C_β–H bond is only weakly acidic, stronger bases are used. A wide range of substrates utilise the E2 mechanism, these including both charged (13) and uncharged molecules (14).

$\overset{+}{N}R_3$ (refs. 2, 15), $\overset{+}{P}R_3$ (refs. 16–18) and $\overset{+}{S}R_2$ (ref. 19) represent charged groups X (13)

SO_2R (ref. 20), halogen (refs. 21–24), sulphonate ester (refs. 25, 26) and sterically hindered carboxylate (ref. 27) represent neutral groups X (14)

The phosphonium hydroxides usually give a saturated hydrocarbon and a phosphine oxide, rather than an olefin and a phosphine, *viz.*

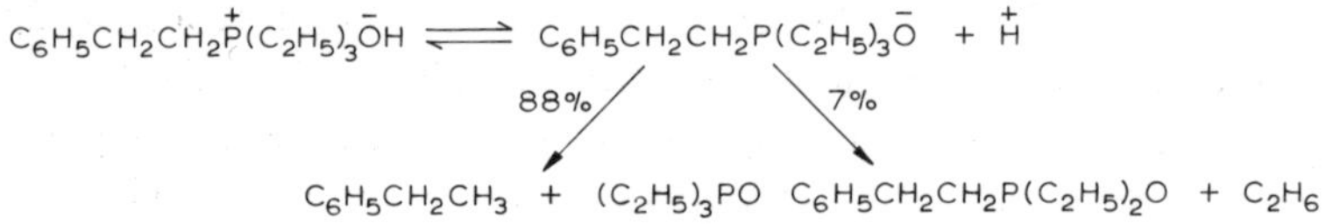

2.1.2 *The unimolecular elimination*

In 1935, a second mechanism for elimination was recognised and was called the unimolecular (E1) mechanism[28]. The presence of a base is not required and the reaction proceeds through a two-step process (16). Initially, dissociation of the C_α–X bond, aided by an ionising solvent, occurs and the carbonium ion formed undergoes rapid unimolecular decomposition to the olefin, *viz*.

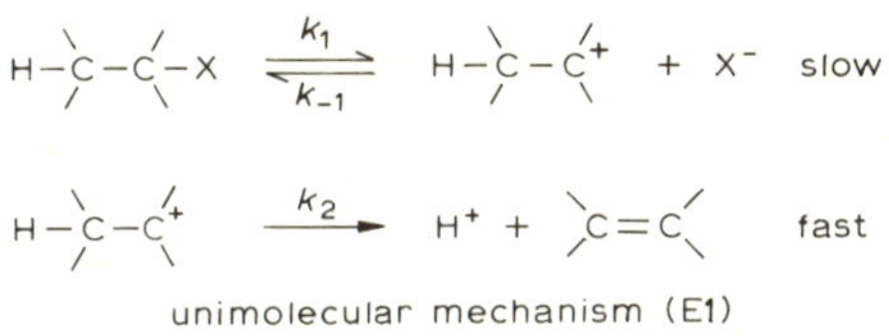

(16)

The reaction is first-order in the substrate. The steady-state approximation for the carbonium-ion concentration gives

$$\text{rate}\ (v) = \frac{k_1 k_2[\text{substrate}]}{k_{-1}[\text{X}^-] + k_2} \tag{17}$$

$$\text{if } k_2 \gg k_{-1}[\text{X}^-], \quad v = k_1[\text{substrate}]$$

$$\text{if } k_2 \ll k_{-1}[\text{X}^-], \quad v = \frac{k_1 k_2[\text{substrate}]}{k_{-1}[\text{X}^-]}$$

Evidence for the mechanism often depends on the observation that the elimination is not accelerated by the addition of base, other than the effect of the latter on the reaction medium. Addition of the common ion (*e.g.* X^- = halide) also slows the reaction by a mass action effect on the first step.

Although base plays no part in the E1 process, the yield of olefin can be improved by its presence, without causing the mechanism to change. During the unimolecular reaction the medium gradually becomes more acidic and as acids easily add to olefinic double bonds, an equilibrium between reactant and product is established. Weak bases, such as pyridines, absorb the acid and prevent the equilibrium, thus simplifying kinetics and improving synthetic yields. The solvolysis reactions of some alkyl halides[29], alkyl sulphonates[30], trialkylsulphonium salts[31] and tetraalkylammonium salts[32] follow the unimolecular mechanism, although the substrates must be secondary or tertiary to afford stability to the intermediate carbonium ion.

2.1.3 *The carbanion mechanism*

Instead of the reaction being initiated by C_α–X bond rupture, a mechanism in which C_β–H bond breaking occurred first would seem reasonable in the presence of a highly basic reagent. Such a mechanism, called the E1cB (or carbanion) mechanism (18) was proposed before the unimolecular process[33] and has recently been reviewed[34].

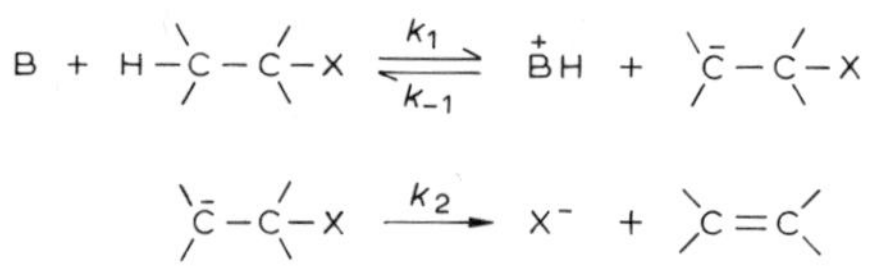

(18)

carbanion mechanism (E1cB)

From a steady-state approximation for the carbanion concentration, expression (19) is obtained.

$$\text{rate}\ (v) = \frac{k_1 k_2[\text{substrate}][\text{B}]}{k_{-1}[\overset{+}{\text{BH}}] + k_2} \tag{19}$$

Two limiting conditions can be applied, the first in which a rapid equilibrium producing the carbanion is followed by a slow unimolecular decomposition. This is the circumstance for which the mechanism E1cB was originally defined, and the kinetics are first-order in each reactant, provided that if the base B is not the lyate ion of the solvent, its conjugate acid, $\overset{+}{\text{B}}\text{H}$, is present in excess[34], *viz.*

$$\text{if}\ k_{-1}[\overset{+}{\text{B}}\text{H}] \gg k_2, \qquad v = \frac{k_1 k_2[\text{substrate}][\text{B}]}{k_{-1}[\overset{+}{\text{B}}\text{H}]} \tag{20}$$

A second condition is achieved when the carbanion decomposes to the olefin more rapidly than it protonates, *viz.*

$$\text{if}\ k_2 \gg k_{-1}[\overset{+}{\text{B}}\text{H}] \qquad v = k_1[\text{substrate}][\text{base}] \tag{21}$$

Under this circumstance the E1cB mechanism degenerates into the E2 mechanism and this part of the topic is considered more fully under discussion of the concerted processes (Section 2.2).

Despite the fact that both the carbanion and bimolecular mechanisms have a similar kinetic form, it is possible to distinguish them. Two different approaches have been used, namely isotopic exchange and the type of base catalysis.

Although the carbanion mechanism offers an attractive alternative to the bimolecular process, it has proved extremely elusive and has been demonstrated for only a few organic reactions. The presence of an electron pair on the *beta* carbon appears to be too powerful for the C_α–X bond to remain intact and concerted processes seem to predominate. The conditions under which the carbanion mechanism might occur rather than the concerted elimination have been summarised[7] and are (*i*) the leaving group X is not easily displaced with its bonding electrons, (*ii*) the intermediate carbanion is reasonably stable relative to the reactants and products and (*iii*) the unsaturated product is comparatively unstable relative to the reactants. Various combinations of these factors can be recognised as important in the few elimination reactions that follow the carbanion mechanism.

When the carbanion is formed in a rapid pre-equilibrium, isotopic exchange between the substrate and the solvent should be a faster process than formation of the olefin. The elimination reaction of 2-(*p*-nitrophenyl)ethyltrimethylammonium ion in aqueous solution is accelerated by hydroxide ion but retarded by acid[15]. This observation led to the postulation of the carbanion mechanism[33], acid depressing the rate by reversing the equilibrium step (22).

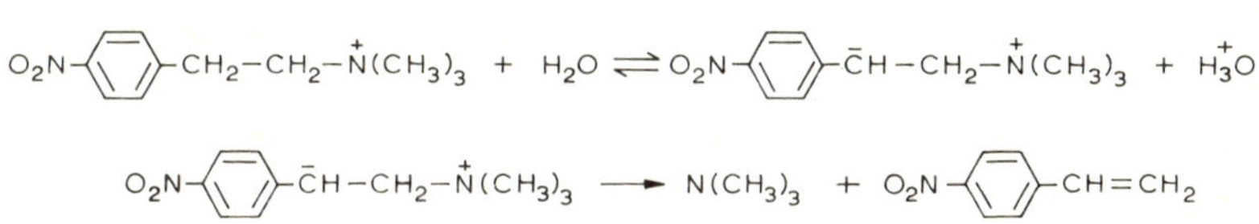

(22)

However, in neutral solution (pH = 7.0), the formation of *p*-nitrostyrene from 2(*p*-nitrophenyl)2-tritio-ethyltrimethylammonium ion is not accompanied by tritium exchange of the reactant[35]. The mechanism in acidic solution remains unsolved but reversible carbanion formation is excluded in neutral solution. Increasing acid concentration retards eliminations occurring by the E2 mechanism by reducing the concentration of hydroxide ion from autoprotolysis of the solvent. The ethoxide-catalysed elimination of 2-phenylethyl bromide[36] or 2-phenylethyltrimethylammonium ion[37] does not follow the carbanion mechanism as uptake of deuterium from the solvent, deuteroethanol, by the substrates is not observed. The exchange criterion also illustrates the absence of the carbanion mechanism even when activating β-phenyl groups[38] or highly basic media[39] (*e.g.* $NaNH_2$ in ammonia) are employed in the alkyl bromide series. Tracer studies also favour the concerted mechanism for elimination from some sulphonium[40] and ammonium ions[41].

Hydrogen exchange alpha to a carbonyl group has been observed in the pyrolysis of 9-methyl-3,9-diazabicyclo-[4,2,1]nonan-4-one methiodide in the presence of hydroxide ion at 150°C[42]. This is not surprising in view of the ketonic function, but as to whether the carbanion is involved in elimination (23) remains unsolved.

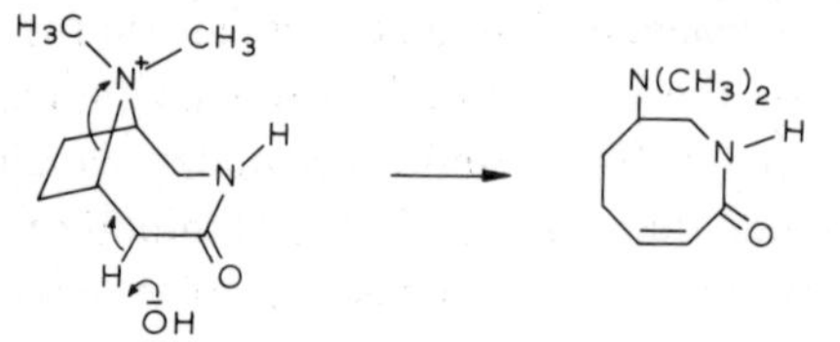 (23)

p-Nitrostyrene formation by elimination from 2-(*p*-nitrophenylethyl)thioacetate ion is accompanied by hydrogen exchange[43], but relative rates of exchange and elimination are not reported and further details of this unusual reaction are required before the carbanion mechanism (24) can be confidently assigned.

$$\bar{O}H + O_2NC_6H_4CH_2CH_2SCH_2COO^- \underset{k_{-1}}{\overset{k_1}{\rightleftharpoons}} O_2NC_6H_4\bar{C}HCH_2SCH_2COO^- + H_2O \quad (24)$$

$$O_2NC_6H_4\bar{C}HCH_2SCH_2COO^- \xrightarrow[\text{slow}]{k_2} O_2NC_6H_4CH{=}CH_2 + \bar{S}CH_2COO^-$$

Hydrogen exchange as conclusive proof of the carbanion mechanism has been questioned[44]. It is argued that if the transition state for anion formation has a lower energy than that for either elimination (Fig. 1), the substrate will be reversibly equilibrated with the anion and exchange could be an irrelevant side reaction to a concerted elimination. From a consideration of enthalpies, the E2 mechanism should be favoured as its transition state should have a lower energy, having less negative charge on the beta-carbon atom and hence more resembling the stable hydrocarbon than its less stable anion. However, as the reactions are second-order, the base energetics have also to be considered. In attaining the transition state for proton abstraction the base has to become

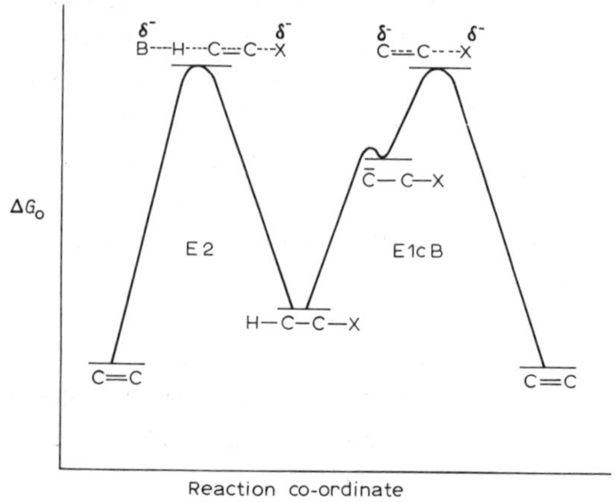

Fig. 1. Free energy profile for E2 and E1cB reactions. (Reproduced with permission from Breslow[44].)

desolvated[45]. The base is only partially neutralised in the transition state for the E2 mechanism but has become a solvent molecule before the transition state in the carbanion mechanism. Although the difference in the energies of the partly neutralised base and solvent molecules may be small, it may be sufficient in cases in which the carbanion electron pair is extensively delocalised to tilt the energetics of the reaction to favour the carbanion route. In any case, entropy considerations favour the carbanion rather than the bimolecular mechanism as the latter tends to restrict the base.

Earlier, in anticipation of the above argument, it was suggested that if carbanions are formed, they are most likely to be intermediates in the elimination[46], as transfer of their free electron pairs should be more effective in promoting formation of the olefin than should transfer of an incipient electron pair in the transition state for bimolecular elimination. In the highly acidic fluorene series, where extensive ionisation is possible under strongly basic conditions, the carbanions possessing aromatic stability, it should be possible to test this postulation. In the region of extensive ionisation, if carbanion formation is a side reaction, an increase in base concentration by increasing the carbanion concentration should retard or have little effect on the rate of the bimolecular elimination as the available concentration of unionised substrate is decreased. For the ElcB reaction, however, a rise in rate should be observed with increasing concentration of the carbanion until the substrate becomes almost fully ionised when further increase in basicity of the medium should not affect the rate.

The few cases of the carbanion mechanism that have been detected by the isotopic exchange approach involve fluorine as the leaving group. Of the halogens, fluorine bonds most strongly to carbon and the small fluoride ion requires the most solvation. These two factors undoubtedly contribute to the fact that fluorides exhibit the carbanion mechanism. 9-Trifluoromethylfluorene anion slowly decomposes to the substituted fulvene by ejection of fluoride ion[47], the carbanion being stabilised by the aromaticity of the five-membered ring and the electron-withdrawing nature of the trifluoromethyl group. The olefin, however, is not isolated but reacts rapidly with the medium and 9-trimethoxymethylfluorene is obtained, *viz.*

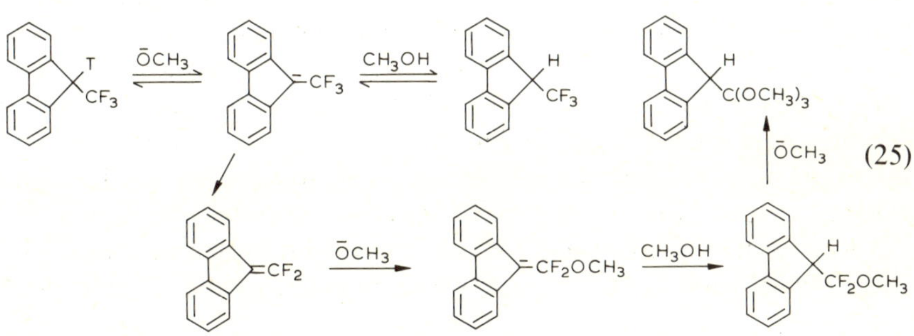

(25)

Dehydrofluorination of 1,1,1-trifluoro-2,2-dihalogenoethanes in alkaline methanol[46] is a much slower process than hydrogen exchange[48], *viz.*

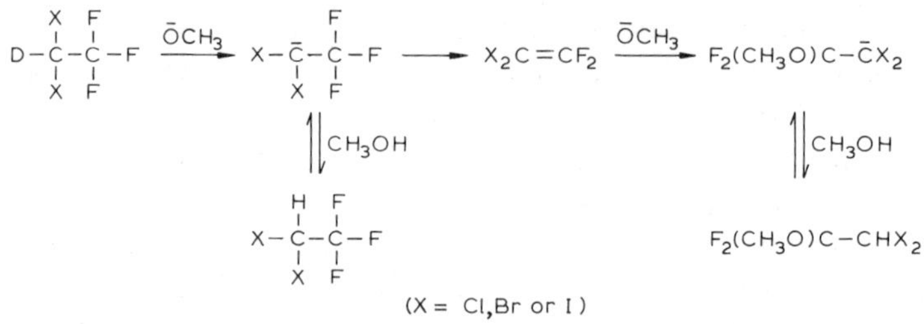

(26)

The activating effects of the halogen atoms on the beta-hydrogen coupled with the reluctance of fluorine to depart as an anion from saturated carbon, especially in the presence of other alpha-fluorine atoms[49,50], make the intermediacy of carbanions in these eliminations highly probable. Unfortunately the olefinic product adds alcohol too rapidly to be isolated[51] and this necessitated careful considerations of alternative mechanisms. A minor fraction of the decomposition may follow an alpha-elimination, *viz.*

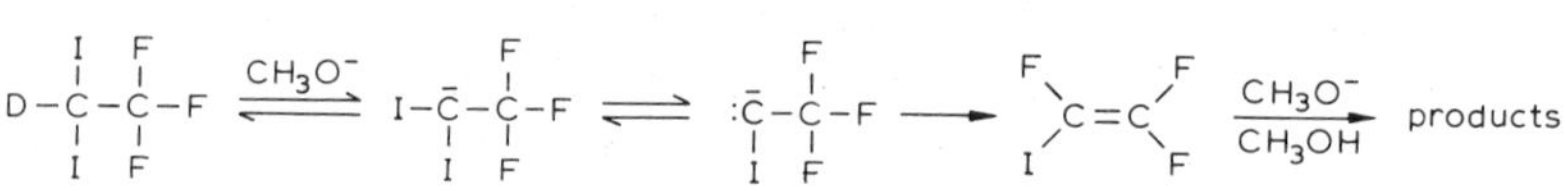

(27)

as, although fluoride is the predominant halide ion eliminated in the slow step, 11% iodide and 4% of the other halides was detected. A rate-determining nucleophilic displacement of fluoride by methoxide can be excluded, as rates of fluoride liberation were 100 to 1000 times more rapid than reactions of typical primary alkyl fluorides with ethoxide in ethanol[24]. Ethoxide in ethanol is twice as reactive as methoxide in methanol towards ethyl bromide[49] and moreover, the presence of the alpha-fluorine atoms will retard the S_N2 reactivity of the carbon–fluorine bond[49,50,52]. It appears that the presence of more than one alpha-fluorine atom is required for hydrogen exchange to be a faster process than elimination. Neither 2-pentyl or 2-phenylethyl fluorides exhibit hydrogen exchange during elimination with sodium ethoxide in deuteroethanol[53]. If the alpha carbon is substituted by three fluorine atoms, then as the electron pair forms on the beta carbon during removal of a proton, it will be delocalised into the carbon–fluorine bonds. The charge will distribute evenly among the carbon–fluorine bonds such that insufficient electron density is present in any single bond to cause instant fragmentation. With only a single alpha fluorine, delocalisation into this bond has to be more extensive, to achieve a similar carbanion stability and this extra electron density causes a concerted elimination.

The activation parameters of the substrates following the carbanion mechanism are vastly different from those which eliminate by the E2 process (Table

TABLE 1

ACTIVATION PARAMETERS FOR SOME DEHYDROHALOGENATION REACTIONS

No.[a]	*Substrate*	*Mechanism*	$\Delta H^{\ddagger}_{ex}$ *(kcal.mole⁻¹)*	$\Delta S^{\ddagger}_{ex}$ *(cal.mole⁻¹.deg⁻¹)*	$\Delta H^{\ddagger}_{elim}$ *(kcal.mole⁻¹)*	$\Delta S^{\ddagger}_{elim}$ *(cal.mole⁻¹.deg⁻¹)*
1	CF_3CDCl_2	E1cB	21.4	5.2	37.2	28
2	$CF_3CDBrCl$	E1cB	20.0	2.1	34.5	21
3	CF_3CDBr_2	E1cB	22.3	12.3	32.9	17
4	CF_3CDI_2	E1cB	23.6	14.8	32.0	19
5	$C_6H_5CH_2CH_2F$	E2			25.3	−5.4
6	$C_6H_5CH_2CH_2Cl$	E2			23.2	−5.6
7	$C_6H_5CH_2CH_2Br$	E2			20.4	−6.8

[a] Compounds 1–4, solvent methanol, base methoxide; exchange parameters from rate studies at 0° and 20°C, error in $\Delta H^{\ddagger}_{ex} \pm 1$ kcal.mole⁻¹, in $\Delta S^{\ddagger}_{ex} \pm 3$ cal.mole⁻¹.deg⁻¹ (ref. 48); elimination parameters from rate studies at 55° and 70.5°C, error in $\Delta H^{\ddagger}_{elim} \pm 3$ kcal.mole⁻¹, $\Delta S^{\ddagger}_{elim} \pm 9$ cal.mole⁻¹.deg⁻¹ (ref. 46); Compounds 5–7, solvent ethanol, base ethoxide; elimination parameters from rate studies at 60° and 80°C (5 and 6) and 30° and 50°C (7), (ref. 23).

1)[23]. As predicted, the entropies of activation are much greater for the two-step process. The activation enthalpies for exchange are very similar and attributable to a combination of polar and steric factors. However, the activation enthalpies for elimination in the dihalo-trifluoro alkanes increase with the electronegativity of the beta-halogen atoms. (Due to large errors, this agreement could be fortuitous and it would be desirable to have more precise data.) This would be expected for elimination from the carbanion, as the electron density in the trifluoromethyl group and hence the ease of departure of fluoride ion, will be reduced if the inductive effect of the beta halogens is increased. The greater values of enthalpy and entropy of activation for the elimination than for the exchange reaction are in agreement with more solvation of the carbanion than of its transition state leading to the olefin.

Elimination reactions are not truly base-catalysed as the base is consumed during the reaction. However, the kinetics do adhere to rate laws which are consistent with the recognised classes of base catalysis[54–56]. Under first-order conditions ($B, \overset{+}{B}H > 10$[substrate]) the observed rate coefficient (k_{obs}), is dependent on the ratio of $[B]/[\overset{+}{B}H]$ for the carbanion mechanism, (28). Consequently, in a $[B]/[\overset{+}{B}H]$ buffer system, changes in the base concentration at constant buffer ratio should not affect the observed rate coefficient and specific base catalysis should be observed.

$$SH + B \underset{k_{-1}}{\overset{k_1}{\rightleftharpoons}} S^- + \overset{+}{B}H$$

$$S^- \xrightarrow{k_2} \text{products}$$

$$k_{obs} = \frac{v}{[SH]} = \frac{k_1k_2[B]}{k_{-1}[\overset{+}{B}H] + k_2} = \frac{k_1k_2[B]}{k_{-1}[\overset{+}{B}H]} \quad \text{if } k_2 \ll k_{-1}[\overset{+}{B}H] \tag{28}$$

For the bimolecular mechanism, general base catalysis is observed and the observed rate coefficient (k'_{obs}) should vary in a linear fashion with the base concentration at constant buffer ratio, *viz.*

$$SH + B \xrightarrow[k'_2]{\text{slow}} [S\ldots H\ldots B] \xrightarrow{\text{fast}} \text{products}$$

$$k'_{obs} = \frac{v}{[SH]} = k'_2[B] \qquad (29)$$

for a number of bases; $k'_{obs} = \sum_i k'_i [B_i]$.

When the carbanion decomposes more readily than it reprotonates, kinetic behaviour intermediate between that of the carbanion and bimolecular mechanism is predicted. For only a small extent of substrate ionisation in low conjugate acid concentration ($k_2 \gg k_{-1}[\overset{+}{B}H]$), general base catalysis is observed. At constant buffer ratio, an increase in base concentration causes a linear increase in observed rate coefficient until $k_{-1}[\overset{+}{B}H]$ approaches k_2. Under this condition the rate coefficient attains a maximum with further increase in base concentration, the kinetics parallel the carbanion mechanism and specific base catalysis is observed[34].

The three different second-order processes thus exhibit widely different kinetic behaviour towards the varying base concentration at constant buffer ratio. In theory this dependence should provide a means of assigning the mechanism. An advantage over the isotopic exchange approach is that it should be possible to detect carbanion intermediates that eliminate more rapidly than they protonate. Unfortunately, the kinetics are not always clear-cut. The E2 mechanism can, under certain conditions, follow specific base catalysis, especially if one base is of much greater "catalytic" efficiency than the other bases present (*e.g.* the E2 reaction of 1,1,1-trichloro-2,2-di-*p*-chlorophenylethane with sodium thiophenoxide in methanol)[57]. Alternatively, the base may be sufficiently powerful to produce a kinetically significant concentration of lyate ions (*e.g.* the E2 reaction of alkyl bromides with phenoxides in ethanol)[57,58].

In agreement with earlier studies on the chalcone dibromides[59], the dehydrochlorinations of 4,4′-dichlorochalcone dichlorides are not stereospecific[60]. The *threo* isomer yields the *trans* olefin chiefly (see (30) below), as expected for elimination of HX from an "*anti*" conformation (see Section 2.3), but the *erythro* isomer gives a mixture of olefins in which the *trans* form predominates over the *cis* by a factor of 2:1 (31).

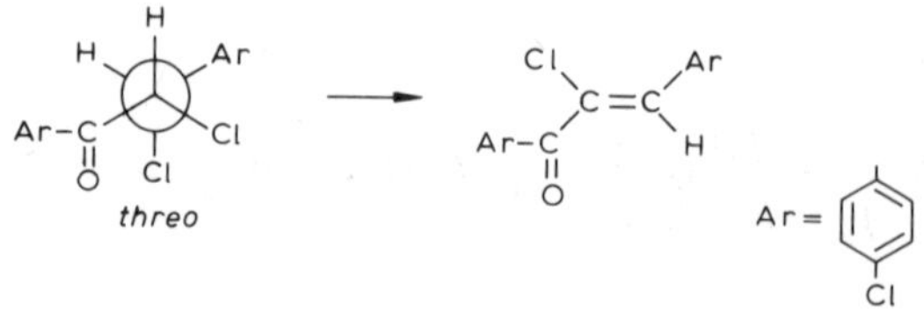

(30)

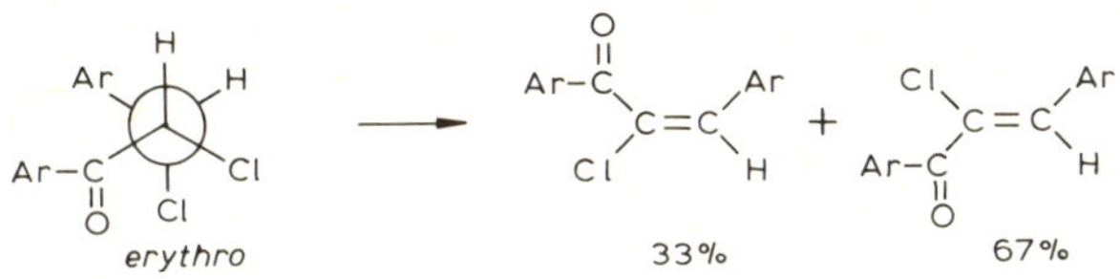

(31)

The *erythro* isomer cannot eliminate entirely from an *anti* conformation because only the *cis* olefin would be produced. Of the two hydrogen atoms which could be abstracted by a base, the more acidic atom is removed and the elimination probably proceeds through an enolate anion (32). These anions however, have the same configuration as their precursors as isomerisation of the *erythro* to *threo* substrate before elimination is not observed, presumably due to steric interactions which cause rotation about the 2–3 carbon–carbon bond to be a slower process than proton capture from the solvent. To minimise steric interactions in the *erythro* dichloride, the bulky aryl residues will separate to a maximum by occupying the "*anti* positions" (32).

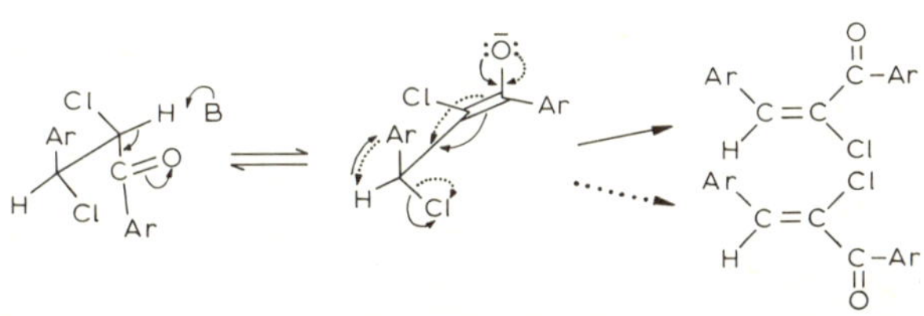

(32)

The resulting enolate ion then eliminates to give a mixture of products in which the more stable *trans* olefin predominates.

Stereochemical evidence alone does not provide sufficient evidence for the carbanion mechanism but the kinetics of the reaction, in buffered solutions in ethanol, of the *erythro* isomer are consistent with a refined form of eqn. (19), *viz.*

$$k_{obs} = \frac{k_2 K_{SHX} \Sigma k_1 [B]}{k_2 K_{SHX} + [\overset{+}{H}] \Sigma k_1 [B]} \qquad (33)$$

where K_{SHX} is the dissociation constant of the substrate. The experimental data clearly demonstrate general base catalysis modified by a pH-dependent term in the denominator, consistent with a "carbanion" mechanism in which a balance between carbanion decomposition to olefin and reprotonation to substrate exists (Fig. 2). In acidic solution, the denominator is dominated by the second term, rate is proportional to pH, the reaction is self-decelerating in the absence of buffer and the elimination of X^- (18, p. 168) is rate-controlling (I). In the central region (II), the second term in the denominator becomes small and the ionisation gradually becomes rate-controlling, the base being ethanol in presence of acetate buffers. The dominant base in region (III) is ethoxide, and deuterium exchange would no longer be anticipated as $k_2 \gg k_{-1}[\overset{+}{B}H]$.

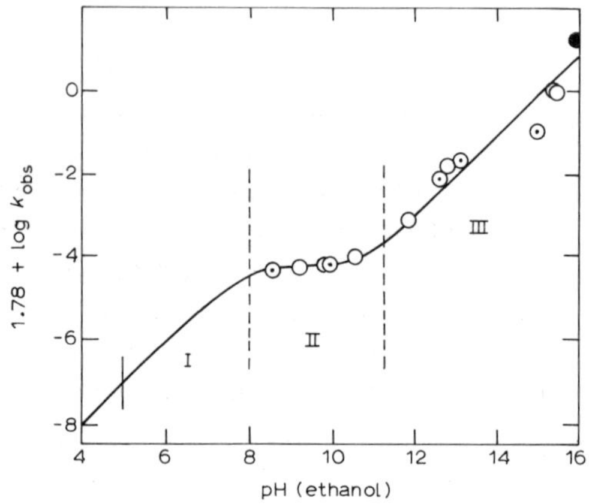

Fig. 2. The dehydrochlorination of *erythro*-4,4′-dichlorochalcone dichloride in ethanol at 25°C. Relationship of rate coefficient to pH (ethanol). k_{obs} defined in eqn. (33) and ref. 60. (Reproduced with permission from Crowell *et al.*[60].)

Similar observations of base catalysis have been used to invoke the E1cB mechanism for elimination from 4,4-dicyano-3-*p*-nitrophenyl-1-phenylbutan-1-one in neutral and acidic methanol (34)[61], and 1,1,1,3-tetranitro-2-phenylpropane in methanol in the presence of hydrochloric acid and pyridine–pyridine hydrochloride buffers (36)[62]. In the former reaction, an example of a reverse Michael addition, the carbanion intermediate with the electron pair alpha to the carbonyl rather than in the gamma position is favoured, as the methyl isomer (35) eliminates more rapidly than the parent compound.

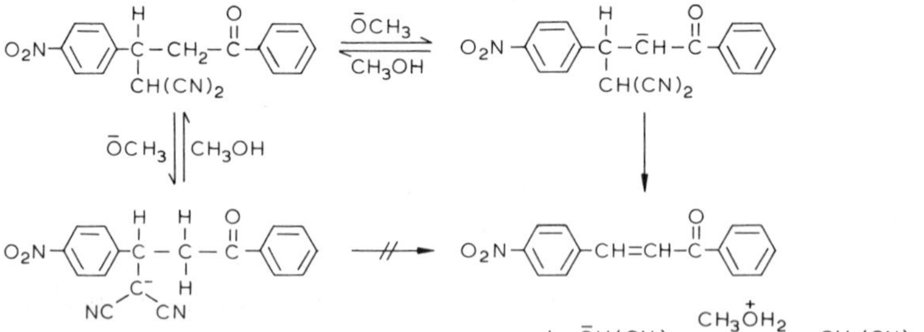

(34)

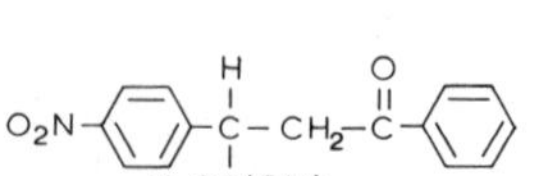

(35)

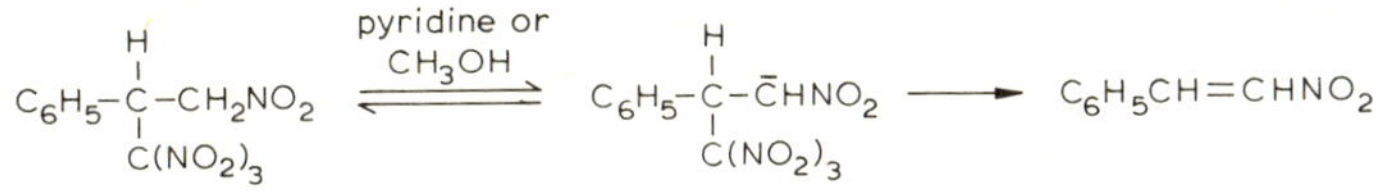 (36)

The inconvenience of having to utilise constant buffer ratios can be avoided by relating the observed rate coefficient to the acidity function, H_-[63–65], which measures the tendency of the medium to accept a proton from a neutral acid (37). The observed rate coefficients for the carbanion and bimolecular eliminations are related to H_- by the final equations in (38) and (39) respectively. Both are examples of linear free energy relationships (see also Chapter 4, Volume 2).

$$\mathrm{AH} + \mathrm{B}^- \rightleftharpoons \mathrm{BH} + \mathrm{A}^-$$

$$\frac{[\bar{\mathrm{B}}]}{[\mathrm{BH}]} = \frac{R}{K_\mathrm{a}} \frac{f_{\mathrm{BH}} f_{\mathrm{A}^-}}{f_{\mathrm{B}^-} f_{\mathrm{AH}}} \quad R = \frac{[\mathrm{A}^-]}{[\mathrm{AH}]}; f \text{ are activity coefficients} \qquad (37)$$

$$H_- = -\log \frac{a_{\mathrm{H}^+} f_{\mathrm{B}^-}}{f_{\mathrm{BH}}}$$

$$= \mathrm{p}K_\mathrm{a} + \log R$$

$$\mathrm{B}^- + \mathrm{RHX} \overset{K_c}{\rightleftharpoons} \mathrm{BH} + \mathrm{RX}^- \qquad \textit{carbanion mechanism}$$

$$\mathrm{RX}^- \overset{K^\ddagger}{\rightleftharpoons} [\text{activated complex}] \xrightarrow{k^\ddagger} \text{olefin}$$

$$v_\mathrm{c} = k^\ddagger[\text{activated complex}]$$

$$= k^\ddagger K^\ddagger [\mathrm{RX}^-] f_{\mathrm{RX}^-} / f^\ddagger \qquad (38)$$

but

$$K_\mathrm{c} = \frac{[\mathrm{BH}][\mathrm{RX}^-]}{[B^-][\mathrm{RHX}]} \frac{f_{\mathrm{BH}} f_{\mathrm{RX}^-}}{f_{\mathrm{RHX}} f_{\mathrm{B}^-}}$$

$$\therefore v_\mathrm{c} = \frac{k^\ddagger K^\ddagger K_\mathrm{c} [\mathrm{B}^-][\mathrm{RHX}]}{[\mathrm{BH}]} \frac{f_{\mathrm{B}^-} f_{\mathrm{RHX}}}{f_{\mathrm{BH}} f^\ddagger}$$

$$\therefore k^\mathrm{c}_{\mathrm{obs}} = \frac{v_\mathrm{c}}{[\mathrm{RHX}]} = k^\ddagger K^\ddagger K_\mathrm{c} \frac{R}{K_\mathrm{a}} \frac{f_{\mathrm{RHX}} f_{\mathrm{A}^-}}{f^\ddagger f_{\mathrm{AH}}}$$

$$\text{or } \log k^\mathrm{c}_{\mathrm{obs}} = H_- + \log k^\ddagger K^\ddagger K_\mathrm{c} + \log \frac{f_{\mathrm{RHX}} f_{\mathrm{A}^-}}{f^\ddagger f_{\mathrm{AH}}}$$

$$\mathrm{B^- + RHX \overset{K^\ddagger}{\rightleftharpoons} [B\ldots H\ldots R\ldots X]^- \xrightarrow{k^\ddagger} BH + R + X^-}$$

bimolecular mechanism

$$v_b = k^\ddagger[\mathrm{B\ldots H\ldots R\ldots X}]^- = k^\ddagger K^\ddagger[\mathrm{B^-}][\mathrm{RHX}]\frac{f_{\mathrm{RHX}}f_{\mathrm{B^-}}}{f^\ddagger} \tag{39}$$

$$\therefore k^{\mathrm{b}}_{\mathrm{obs}} = \frac{v_b}{[\mathrm{RHX}]} = k^\ddagger K^\ddagger \frac{R}{K_a}[\mathrm{BH}]\frac{f_{\mathrm{RHX}}f_{\mathrm{BH}}f_{\mathrm{A^-}}}{f^\ddagger f_{\mathrm{AH}}}$$

or

$$\log k^{\mathrm{b}}_{\mathrm{obs}} = H_- + \log k^\ddagger K^\ddagger + \log\frac{f_{\mathrm{RHX}}f_{\mathrm{BH}}f_{\mathrm{A^-}}[\mathrm{BH}]}{f^\ddagger f_{\mathrm{AH}}}$$

For the carbanion mechanism (38), $\log k^{\mathrm{c}}_{\mathrm{obs}}$ is a linear function of H_-, of unit slope, if $f_{\mathrm{RHX}}/f^\ddagger$ varies in the same way as $f_{\mathrm{AH}}/f_{\mathrm{A^-}}$ with changes in the reaction medium. Both activity coefficient ratios contain a neutral and a similarly charged term and the required condition should thus be more or less fulfilled. It is unlikely that the activity coefficient ratio, $f^\ddagger/f_{\mathrm{RHX}}$ responds to changes in the medium in the same way as $f_{\mathrm{BH}}f_{\mathrm{A^-}}[\mathrm{BH}]/f_{\mathrm{AH}}$ and a graph of $\log k^{\mathrm{b}}_{\mathrm{obs}}$ against H_- should not have unit slope for the E2 mechanism (39). More appropriately, $\log k^{\mathrm{b}}_{\mathrm{obs}}$ should be related to $H_- + \log(\mathrm{BH})$ and this has been demonstrated for the E2 reaction of *DL*-serine phosphate with hydroxide ion in water (40)[66].

$$\log k^{\mathrm{b}}_{\mathrm{obs}} = H_- + \log(\mathrm{H_2O}) + \log k^\ddagger K^\ddagger + \log\frac{f_{\mathrm{RHX}}f_{\mathrm{A^-}}}{f^\ddagger f_{\mathrm{AH}}} \tag{40}$$

$$(\mathrm{H_2O}) = f_{\mathrm{H_2O}}[\mathrm{H_2O}]$$

An identical slope (0.98) was observed for a similar correlation for the E2 reaction of 2-phenylethyldimethylsulphonium ion in aqueous dimethyl sulphoxide containing dilute sodium hydroxide[67]. In the former case, the water activity, $(\mathrm{H_2O})$, was estimated as the free water concentration, *viz.*

$$(\mathrm{H_2O}) = d - 0.001\ (M + 18.0n)[\bar{\mathrm{O}}\mathrm{H}] \tag{41}$$

(where $M = 40$ for NaOH and 56 for KOH, d is the solution density and $n(=3)$ the number of water molecules solvating the $\bar{\mathrm{O}}\mathrm{H}$ ion) and in the latter case as the total water concentration. The generality of the expression was demonstrated by the range of rate coefficient measured, 10^2 in the former case and 10^8 in the latter.

Despite the obvious differences in behaviour of the rates of carbanion and bimolecular mechanisms towards acidity functions, the latter have seldom been used to illustrate E1cB reactions in probable cases. One drawback is that acidity functions have only been established for a limited number of basic solutions and many of them are founded on the ionisation of nitrogen acids such as substituted anilines and diphenylamines[68,69], whereas carbanion mechanisms are initiated by carbon acid ionisation. However, unlike the acidity functions established in strong acid solution (*e.g.* H_0 (ref. 70), H_R (ref. 71), H_A (ref. 72), H_0''' (ref. 73) and H_I (ref. 74)) which all behave differently due to large variations in activity coefficient behaviour of the different types of base and their conjugate acids), acidity functions established from ionisation of carbon and nitrogen acids in strongly basic media appear to differ very little[75,76]. It is thus reasonable to correlate rates of carbanion mechanisms with acidity functions based on ionisation of nitrogen acids.

In concentrated methanolic solutions of sodium methoxide, due to a shortage of base-solvating solvent molecules, the activity of methoxide as measured by H_- rises more rapidly than methoxide concentration[77]. The rate of methanolysis of chloroform ($\log k^c_{obs}$), an E1cB-alpha-elimination, is proportional to the H_- value of the medium, which is not surprising as the reaction is under thermodynamic control. Observed rate coefficients ($\log k^b_{obs}$) for the E2 reactions of 2-phenylethyl chloride and 1-chloro-3,3-dimethylbutane, which are under kinetic control, show a much slower response to medium changes and in fact increase linearly with methoxide concentration[78].

The remaining examples of the E1cB reaction which lead to olefin products are supported by less sound evidence. To explain the "abnormal stereochemistry" of a beta-elimination, the E1cB mechanism has often been invoked rather than the E2 mechanism, the classic example being the base induced dehydrochlorination of β-benzene hexachloride in which all the adjacent chlorine atoms are situated *trans* to each other[79]. Stereochemical evidence alone is insufficient proof of the carbanion mechanism; this aspect is discussed more fully elsewhere (see Section 2.3).

In line with the reasoning that stabilisation of the carbanion will favour the E1cB mechanism, the two-step process has been suggested for substrates possessing sulphone substituents on the beta carbon. Exchange of deuterium is up to 2000 times faster than inversion at the asymmetric carbon when a carbanion-stabilising alpha sulphone group is present[80]. Powerful electron-withdrawing groups such as phenylsulphone favour carbanion formation by loosening the beta hydrogen and stabilising the liberated electron pair on the beta carbon atom by "*d*-orbital resonance"[81]. However, elimination from $PhSO_2CHCH_3CHCH_3I$ shows complete "*anti*" stereospecificity and a concerted E2 mechanism is favoured[82]. The carbanion-stabilising ability of the phenylsulphone group in this case may be nullified by the ease of ionisation of the carbon–iodine bond. However, a two-step elimination is favoured in the elimination–addition reaction (42), which is not a direct substitution as the thioether (43) does not react[81].

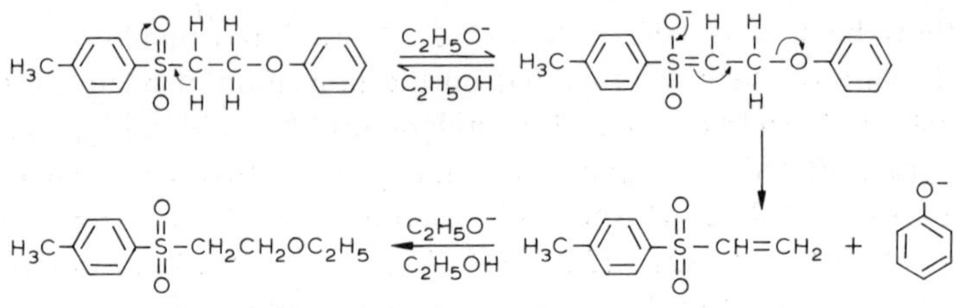

(42)

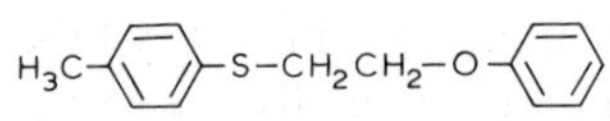

(43)

A carbanion mechanism is also favoured for elimination from the sulphonium salt (44)[83], as it undergoes facile elimination compared to (45) and (46) which possess alpha substituents lacking a heteroatom with low-lying *d*-orbitals[84].

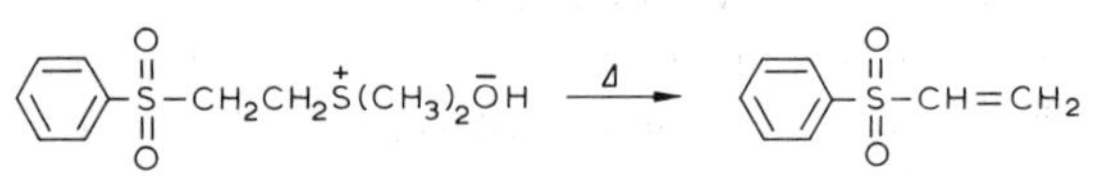

(44)

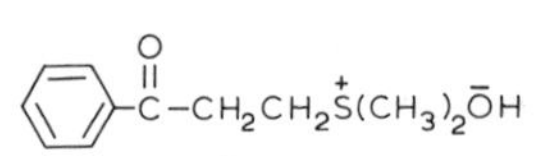

(45)

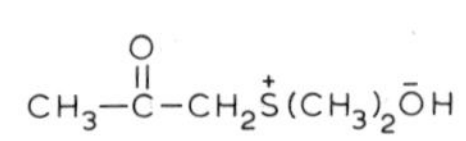

(46)

These examples of the carbanion mechanism based on substrates possessing alpha sulphone groups would be more certain if founded on kinetic evidence. In all the cases quoted above, the enhanced reactivity of the sulphones relative to the other model compounds may be explicable in terms of beta hydrogen activation and a normal concerted elimination reaction.

2.1.4 *The ylid mechanism*

Unlike the three fundamental mechanisms of olefin formation already outlined, the α'–β (or ylid) mechanism is only applicable to elimination from "onium salts". The base abstracts an alpha hydrogen from the leaving group to form an ylid which subsequently acts as an internal base and abstracts the beta hydrogen atom, *viz.*

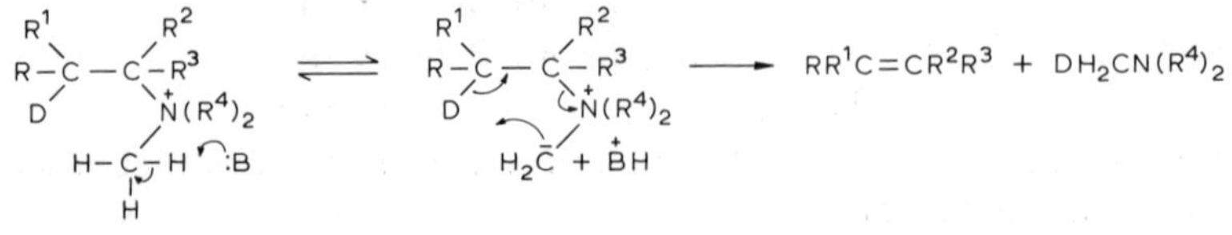

(47)

The ylid mechanism was first postulated to account for the presence of iodobenzene among the reaction products when elimination from isopropyldimethyl(iodomethyl)ammonium ion was induced by phenyl lithium[85], *viz.*

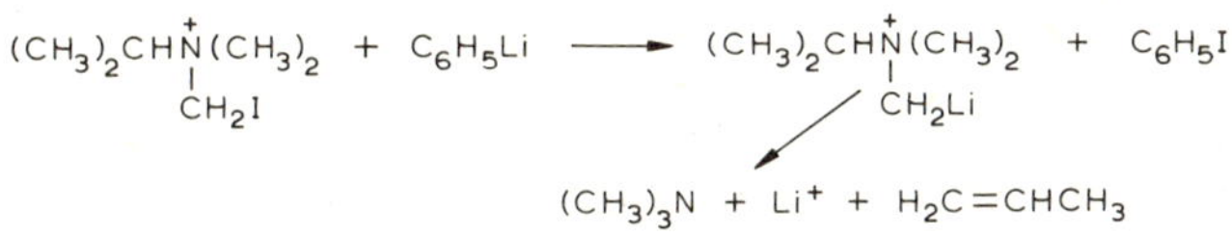

(48)

An E2 reaction would have produced benzene. Under the same conditions it was suggested that eliminations from isopropyltrimethylammonium ion followed the same route.

Subsequent work revealed that the ylid mechanism was induced by metal alkyls and aryls but not by metal alkoxides and amides[86–88]. Stereochemical evidence was cited, metal alkyls giving more *cis*-cyclooctene and metal amides and alkoxides more *trans*-cyclooctene for elimination from cyclooctyltrimethylammonium ion (49). Elimination from cyclooctyldimethyl(bromomethyl)ammonium ion with methyl lithium gave the same predominance of the *cis*-cyclooctene.

	Base	*Cyclooctene* % *trans*	% *cis*	*Mechanism*	*Ref.*
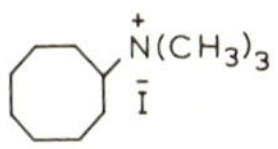 (49)	C_6H_5Li/ether	20	80	ylid	87
	KNH_2/NH_3	85	15	E2	87
	Hofmann degradation	60	40	E2	252
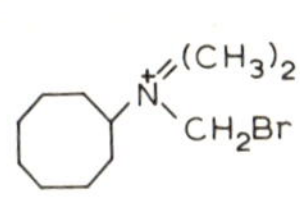	$LiCH_3$	10	90	ylid	85, 87, 88
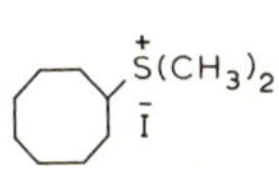	Various bases in dipolar aprotic solvents	0	100	ylid	98, 99

Five- and six-membered alicyclic ammonium ions, (50) below, undergo facile elimination under the normal conditions of the Hofmann reaction

(50)

(pyrolysis with hydroxide or alkoxide ion), to give acyclic dimethylamino-olefins. However, treatment with *n*-butyl lithium yields stable ylids which regenerate the starting material on hydrolysis and only eliminate at elevated temperatures, producing distinctly different product ratios from those obtained under normal Hofmann reaction conditions.

These results imply that the α'–β mechanism does not operate under the usual conditions for the Hofmann reaction and is only encountered in dipolar aprotic and not in protic solvents. Presumably in dipolar aprotic solvents the ylid formed is stabilised by ion pairing for sufficient time to allow the planar arrangement of the bonds (C_β–H, C_β–C_α, C_α–$\overset{+}{N}$ and $\overset{+}{N}$–$\bar{C}H_2$) necessary for elimination via the ylid, to be attained.

A more elegant and convincing demonstration of the ylid mechanism is accomplished by tracer studies. In the ylid mechanism, the trimethylamine produced contains one deuterium atom per molecule if the beta hydrogen atom is labelled in the substrate (47), whereas the E2 mechanism yields unlabelled material. Analysis by mass spectroscopy or nuclear magnetic resonance affords a convenient diagnosis of mechanism. Tracer studies confirm the absence of the ylid mechanism for the Hofmann reactions of a series of quaternary ammonium salts (51)[89], (52)[90].

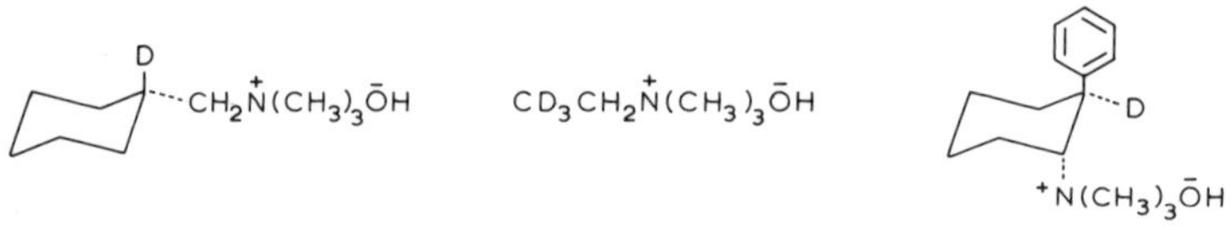

(51) (51) (52)

Despite the rigid coplanarity in the substrate of all the requisite bonds except $\overset{+}{N}$–$\bar{C}H_2$, the Hofmann elimination of the bicyclic ammonium salt (53) occurs predominantly (94%) by a "*syn*" E2 mechanism rather than the ylid alternative[91], *viz.*

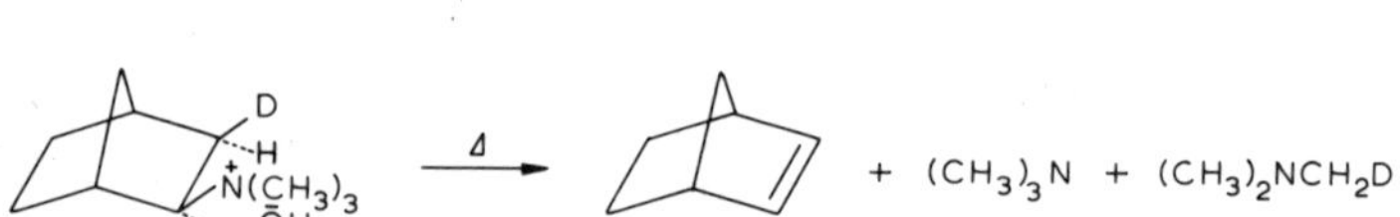

(53)

Entropy considerations may favour the concerted process as the rotation of the trimethylammonium group is restricted during the ylid mechanism.

Elimination from *N*,*N*,*N*-trimethyl-2-*t*-butyl-3,3-dimethylbutylammonium-2d_1 hydroxide[92] produces trimethylamine, 75% of which is monodeuterated, *viz.*

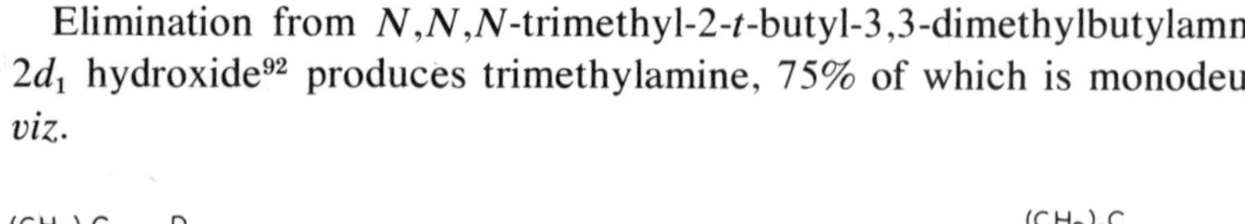

(54)

In this case, the bulky *t*-butyl groups inhibit the approach of the base to the C_β–H bond. Consequently the E2 mechanism is greatly retarded and decomposition by the ylid mechanism is predominantly followed. A small amount of dideuterated material is also isolated. The percentage yield of this species increases with the total reaction time, indicating exchange of trimethylamine-d_1 with the medium, which gradually becomes enriched with deuterium during the course of the reaction. Hence, despite the simplicity of the tracer technique, care has to be taken when interpreting the results, especially in deciding on the exact percentage of reaction following the ylid mechanism. Complications arising from exchange of the trimethylamine and the reaction medium can be minimised by isolating the required samples during the early stages of elimination or corrected for by taking samples at various stages of reaction and extrapolating the percentage of trimethylamine, which is monodeuterated, to zero reaction time.

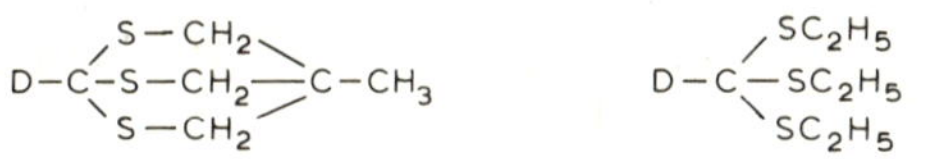

(55)

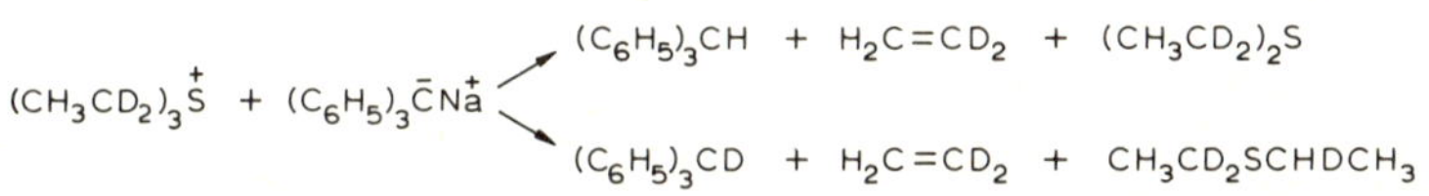

(56)

Deuterium exchange with the solvent occurs readily when the thioethers (55) but not their oxygen analogues are reacted with either ethanol or *t*-butyl alcohol containing their respective lyate ions[93]. During its E2 decomposition with hydroxide ion to styrene, 2-phenylethyl-1,1-d_2-dimethylsulphonium ion exhibits considerable deuterium exchange as a side reaction[94], while dideuterated styrene is obtained from elimination of the corresponding trimethylammonium ion[95]. These observations reflect the ability of sulphur to stabilise carbanions by $d\pi$–$p\pi$ orbital overlap, a process not available to the first-row elements, which lack 3*d*-orbitals of low energy. Thus dimethylsulphonium compounds offer more attractive possibilities for the ylid mechanism than do the corresponding trimethylammonium ions.

However, unlabelled dimethyl sulphide is obtained from the reaction of 2-phenylethyl-2d_2-dimethylsulphonium ion with hydroxide in water[96]. Elimination from tri(ethyl-1,1-d_2)-sulphonium ion with triphenylmethylsodium in ether gives triphenylmethane, 75% of which contains deuterium, indicating that the predominant route for the decomposition is via an ylid mechanism. The analyses of the other products are also in agreement with this conclusion (56)[97,98].

A useful guide to the ylid mechanism should be provided by the activation entropy. In a reaction following the E2 mechanism, in attaining the transition state in polar solvents a reduction in solvation of the ammonium ion occurs, resulting in a large positive activation entropy. However, the ylid mechanism, having a cyclic transition state should exhibit a high negative activation entropy.

An apparent beta-elimination to give an olefin can occur by an alpha-elimination followed by hydrogen migration, *viz.*

$$\begin{array}{l} RCH_2CH_2X + B \rightleftharpoons RCH_2\bar{C}HX + \overset{+}{B}H \\ RCH_2\bar{C}HX \xrightarrow{\text{slow}} RCH_2\ddot{C}H + \bar{X} \\ RCH_2\ddot{C}H \longrightarrow RCH{=}CH_2 \end{array} \qquad (57)$$

Although, in the absence of beta-hydrogen atoms, there is ample evidence of carbene intermediates[3], substrates possessing beta-hydrogen atoms appear to disfavour the carbene mechanism, as evidenced by the use of deuterium tracers[99,100]. Carbene intermediates would seem most probable for substrates possessing weakly acidic beta-hydrogen atoms and reactions involving highly reactive bases. The reaction of phenyl sodium with deuterated butyl chloride, *viz.*

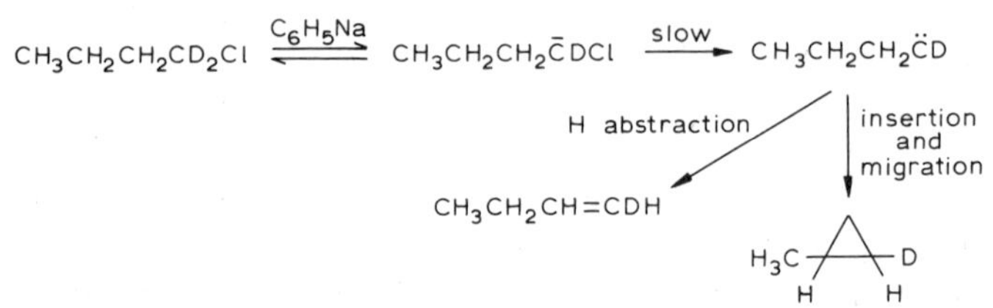

(58)

and the action of sodium or potassium on isobutyl chloride appear to fall into this category[101,102]. Cyclopropanes, typical carbene insertion products, are isolated. The base-catalysed decomposition of isobutyldiphenylsulphonium fluoroborate with triphenylmethylsodium, yielding isobutylene and methylcyclopropane, is again indicative of an alpha-elimination[103]. Present available results indicate that, like the ylid mechanism, the alpha-elimination followed by hydrogen migration to give an olefin seems important only under aprotic conditions and with organometallic bases.

2.2 TRANSITION STATES OF BIMOLECULAR ELIMINATIONS

In the previous section, three general mechanisms of olefin formation by beta-elimination of the "elements" HX from adjacent carbon atoms, were outlined. Whereas the E1 and E1cB mechanisms involve preliminary breaking of one bond, the E2 process is concerted, both the C_β–H and C_α–X bonds being partially broken at the transition state. When initially designating the latter mechanism[2], care was taken to indicate that the reaction could be concerted without being fully synchronous. However, for many years the E2 mechanism was considered to imply equal breaking of the two relevant bonds at the transition state and it was not until 1956 that the concept of a variable transition state for E2 reactions with changing reaction conditions was invoked to explain the experimental observations[104]. The modern view is to regard the E1 and E1cB mechanisms as the extremes of a continuous spectrum of transition states of the E2 mechanism. Thus, like substitution reactions, elimination reactions can have transition states with widely varying degrees of bond making and bond breaking. Since 1956, many interesting kinetic approaches have been used to confirm this concept. The subject has also been reviewed on several occasions[10,14,105–107].

The complexity of a beta-elimination makes a precise assignment of the transition state for an E2 reaction difficult and most kinetic approaches give information on the relative position rather than the exact nature. Four major types of transition state, each being concerted, but differing in degrees of bond making and bond breaking can be envisaged[106], these being illustrated by the use of Newman projections (59).

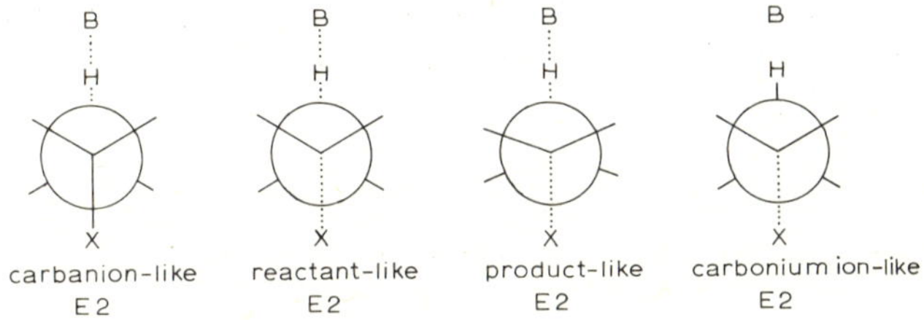

(59)*

Negative charge accumulates on C_β in both the carbanion-like E2 and reactant-like E2 transition states and positive charge develops on C_α in the carbanion ion-like E2 and reactant-like E2 transition states. Bond breaking can be small or extensive in three of the transition states but is well-developed always in the product-like E2 transition state, which alone has considerable double-bond character. For the remaining transition states, double-bond character approximates to the least stretched of the bonds undergoing fragmentation.

The choice of transition state utilised by a particular reaction is greatly dependent on the environmental factors, these being substrate structure, base and solvent. Usually it is safer to assign a reaction to a particular range of the spectrum rather than to a specific type. This approach was followed by

* Dotted lines represent partial bonds; full lines represent complete bonds; in each projection an *anti*-elimination of the groups H and X is assumed in line with the usual stereochemistry of E2 reactions (see Section 2.3).

Bunnett[105] in 1962, when the relevant factors affecting the choice of transition state were summarised. Since that date, kinetic techniques have supported the ideas which are listed below.

(*i*) A shift towards the carbonium ion-like E2 transition state is favoured by the introduction of alpha aryl or alpha alkyl substituents which stabilise the developing carbonium ion. A similar trend is also predicted by a change to a more-ionising leaving group and a change to a more-ionising solvent when the leaving group is uncharged. All these factors aid C–X bond rupture to attain a more advanced position in the transition state.

(*ii*) Carbanion-like E2 transition states are encouraged by electron-attracting beta substituents, stronger bases in poorer ionising media and leaving groups which are more electron-attracting.

(*iii*) A reactant-like E2 transition state is encouraged by a compromise of the factors mentioned previously and a product-like E2 transition state is promoted by substituents at the alpha or beta positions which can conjugate with the developing double bond. Steric interaction between substituents at the alpha and beta positions will tend to oppose the development of extensive double-bond character if the substituents are linked *cis* across the developing double bond.

Support for predicted effects has come from two major types of kinetic approach, namely the use of kinetic isotope effects and the application of linear free energy relationships.

2.2.1 *Kinetic isotope effects* (see Chapter 4, Volume 2)

Although the substitution of an atom of a bond by another isotope does not qualitatively alter the chemical reactivity of that bond, a quantitative change can be induced if the force constant of the bond undergoes a change in the rate-determining step of the reaction. The rate is reduced by substitution of a heavier isotope if the force constant of the bond is decreased in attainment of the transition state. As a number of bonds exhibit force-constant changes when a substrate undergoes a beta-elimination, the measurement of kinetic isotope effects by successively labelling each atom affords an attractive method of assessing E2 transition state structure. Various applications of kinetic isotope effects to elimination reactions have been reviewed in recent years[10,14,34,105–109].

(*a*) *Theoretical interpretation*

Primary kinetic isotope effects are predominantly due to differences in the masses of the isotopes and the resulting differences in the zero-point energies of the bonds[110–116]. The larger the relative difference between the masses of the two isotopes, then the more pronounced are the isotope effects. Consequently, isotope effects involving hydrogen are most easily measured and have been used most frequently in studies of elimination reactions.

According to absolute rate theory[117], the kinetic isotope effect is equal to the quotient of the equilibrium constants for the formation of the transition states in the labelled and unlabelled substrates, respectively, if symmetry factors are neglected and the transmission coefficients are independent of isotopic substitution (60). Substitution of equilibrium constants reduces the calculation to solving the complete partition functions (61).

$$\begin{aligned} A_1 &\overset{K_1^{\ddagger}}{\rightleftharpoons} X_1 \xrightarrow{k_1} \text{products} \\ A_2 &\overset{K_2^{\ddagger}}{\rightleftharpoons} X_2 \xrightarrow{k_2} \text{products} \end{aligned} \tag{60}$$

A_1 and A_2 are the isotopic substrates and X_1 and X_2 their corresponding activated complexes.

$$k_1 = K_1\left(\frac{kT}{h}\right)K_1^{\ddagger}$$

but K, the transmission coefficient, is assumed to be independent of isotopic substitution. Therefore

$$\frac{k_1}{k_2} = \frac{K_1^{\ddagger}}{K_2^{\ddagger}} = \frac{Q_1^{\ddagger}}{Q_2^{\ddagger}}\frac{Q_{A_2}}{Q_{A_1}}$$

where $Q^{\ddagger}$ and Q_A are the partition functions for the activated complex and the substrate, respectively.

$$Q = Q_{trans}Q_{\text{rot}}Q_{\text{vib}}Q_{\text{elec}} \tag{61}$$

$$\left[\frac{Q_1^{\ddagger}}{Q_2^{\ddagger}}\right]_{trans}\left[\frac{Q_{A_2}}{Q_{A_1}}\right]_{trans} = \left[\frac{M_2}{M_1}\right]^{3/2}\left[\frac{M_1^{\ddagger}}{M_2^{\ddagger}}\right]^{3/2}$$

where M_1 and M_2 are the molecular masses.

$$\left[\frac{Q_1^{\ddagger}}{Q_2^{\ddagger}}\right]_{\text{rot}}\left[\frac{Q_{A_2}}{Q_{A_1}}\right]_{\text{rot}} = \left[\frac{I_{X_2}I_{Y_2}I_{Z_2}}{I_{X_1}I_{Y_1}I_{Z_1}}\right]^{1/2}\left[\frac{I^{\ddagger}_{X_1}I^{\ddagger}_{Y_1}I^{\ddagger}_{Z_1}}{I^{\ddagger}_{X_2}I^{\ddagger}_{Y_2}I^{\ddagger}_{Z_2}}\right]^{1/2}$$

where I is the moment of inertia about the X, Y and Z axes.

$$\left[\frac{Q_1^{\ddagger}}{Q_2^{\ddagger}}\right]_{\text{vib}}\left[\frac{Q_{A_2}}{Q_{A_1}}\right]_{\text{vib}} = \prod_{i}^{3n-7} e^{-\Delta\mu_i^{\ddagger}/2}\left[\frac{1-e^{-\mu_i^{\ddagger}(2)}}{1-e^{-\mu_i^{\ddagger}(1)}}\right] \times \prod_{i}^{3n-6} e^{\Delta\mu_i/2}\left[\frac{1-e^{-\mu_i(1)}}{1-e^{-\mu_i(2)}}\right]$$

References pp. 350–364

where $\mu_i = h\nu_i/kT$ in which ν_i is the vibrational frequency and $\Delta\mu_i = h(\nu_i(1) - \nu_i(2))/kT$.

Q_{elec} is isotopically independent and can therefore be neglected in the isotope ratio equation. Fundamental vibrational frequencies of the reactant can be estimated for simple molecules from analysis of the infrared and Raman spectra. Transition state vibrations must be calculated from an assumed model. Only in recent years with the advent of high-speed computers have complete calculations for any but the simplest molecules become feasible. More often, the equation is simplified by assuming particular models.

For most organic molecules, the substitution of hydrogen for its heavier isotopes materially has little influence on the molecular mass or moment of inertia. The isotopic rate ratio thus reduces, essentially, to a dependence on the vibrational partition functions, *viz.*

$$\frac{k_H}{k_D} = \left[\frac{Q_{H^\ddagger}}{Q_{D^\ddagger}}\right]_{vib}\left[\frac{Q_D}{Q_H}\right]_{vib} = \prod_i^{3n-7} e^{\frac{-\Delta_i^\ddagger}{2}}\left[\frac{1-e^{-\mu_i^\ddagger(D)}}{1-e^{-\mu_i^\ddagger(H)}}\right]\prod_i^{3n-6} e^{\frac{\Delta\mu_i}{2}}\left[\frac{1-e^{-\mu_i(H)}}{1-e^{-\mu_i(D)}}\right] \qquad (62)$$

At room temperature most oscillators are in their zero-point energy levels, vibrational mode spacing being greater than kT and a further simplification results, *viz.*

$$e^{-\mu} = e^{-h\nu/kT} \ll 1$$
$$\therefore \frac{k_H}{k_D} = \prod_i^{3n-7} e^{\frac{-\Delta\mu_i^\ddagger}{2}}\prod_i^{3n-6} e^{\frac{\Delta\mu_i}{2}} \qquad (63)$$

Usually it is assumed that the bending frequencies of the isotopic reactants are conserved in their activated complexes and only the changes in the stretching vibrations are considered to contribute to the isotope effect. Thus

$$\frac{k_H}{k_D} = e^{\frac{h}{2kT}(\nu_H{}^s-\nu_D{}^s)-(\nu_H{}^{s\ddagger}-\nu_D{}^{s\ddagger})}$$
$$= e^{\frac{h}{2kT}(\nu_H{}^s-\nu_D{}^s)} \qquad (\text{if } \nu_H{}^{s\ddagger} = \nu_D{}^{s\ddagger}) \qquad (64)$$
$$= 7.0 \text{ at } 25°\text{C}$$

where ν^S are stretching frequencies of the isotopically labelled bond.

If complete bond dissociation occurs in attaining the transition state, then isotopic substitution does not affect the stretching frequency in the transition state and the maximum isotope effect is observed, whereas partial bond dissocation leaves an energy difference between the isotopic activated complexes and gives rise to a smaller isotope effect (Figure 3). This approach was used to predict a maximum isotope effect (k_H/k_D) of 8.3 at 0°C, 6.9 at 25° and 4.7 at 100° for bonds to carbon[118], from the assumption that the

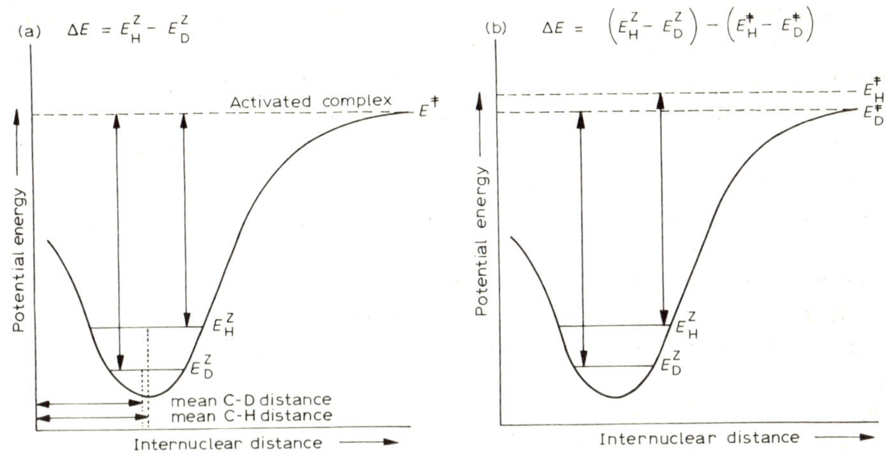

Fig. 3. Potential energy profiles for C–H and C–D bond dissociations. (a) Complete dissociation; (b) partial dissociation in the transition state.

C–H(D) bonds behave as simple harmonic oscillators and hence $\nu_Z{}^H/\nu_Z{}^D = 1.414$. Experimental observations lend support to this approximation but give a lower maximum isotope effect[119,120] (Table 2). Consideration of zero-point energies associated with the bending frequencies in the reactants imposes a considerable change on the calculated maximum isotope effect.

TABLE 2

CALCULATION OF THE MAXIMUM ISOTOPE EFFECT (k_H/k_D) FROM INFRARED SPECTRA OF REACTANTS

Solvent	ν_{C-H} (cm^{-1})	ν_{C-D} (cm^{-1})	(k_H/k_D)	*Temp.* (°C)	ν_{C-H}/ν_{C-D}
1-bromo-2-phenylpropane[120]					
CCl_4	2904(s)	2160(s)	6.79[a]	5	1.346
			6.02[a]	25	
			5.33[a]	45	
	1299(b)	934(b)	2.42[b]	25	1.391
			14.55[c]	25	
9-bromo-9,9′-bifluorenyl[121]					
CS_2	2896(s)	2166(s)	5.64[a]	30	1.336
	1291(b)	962(b)	2.18[b]	30	1.342
			12.9[c]	30	

[a] Calculated from the stretching frequencies.

[b] Calculated from the bending frequencies.

[c] Calculated from one stretching and one bending frequency, assuming complete loss of the isotopic zero-point energy difference of the reactant in attaining the transition state.

The mere observation of an isotope effect ($k_H/k_D > 1$) serves to demonstrate that a base-catalysed elimination reaction is following the E2 mechanism rather than the carbanion mechanism and much of the early data was obtained for this purpose. However, the magnitude of the effect can give useful information concerning the extent of proton transfer in the transition state, but interpretation is not simple and unambiguous. Before 1961, the interpretation of proton-transfer reactions was rather naïve, a greater isotope effect being associated with greater carbon–hydrogen bond breaking in the transition state. This approach is reasonable for bond dissociations but when a proton is abstracted by a base, a base-hydrogen bond is formed as the C–H bond breaks in the transition state and any isotopic zero-point energy differences associated with this forming bond have to be considered in the calculation of the isotope effect[112,121].

Neglecting bending vibrations, a proton transfer can be represented as passing through a linear three-centre transition state, *viz.*

$$A-H+B \rightleftharpoons [A\ldots H\ldots B] \rightarrow A+HB$$

$$\underset{\text{symmetric stretch}}{\overleftarrow{A}\ldots\ldots H\ldots\ldots\overrightarrow{B}} \quad \underset{\text{antisymmetric stretch}}{\overleftarrow{A}\ldots\ldots\overrightarrow{H}\ldots\ldots\overrightarrow{B}} \tag{65}$$

If A . . . H . . . B were a linear molecule, rather than an activated complex, it would have two stretching vibrations. Neither of these corresponds to the stretching vibration of A–H or B–H. Transition-state theory requires that one of these vibrations becomes translation but the symmetric stretching vibration is retained. If the H is equally bonded to both A and B then it will not move during the symmetric stretch and hence replacement of hydrogen by deuterium will not affect the zero-point energy of this vibration. Consequently, the isotope effect will be essentially that predicted for complete loss of the stretching vibrations of the A–H and A–D bonds in the substrates. When H is more strongly bonded to A or B it will tend to move during the symmetric stretch and consequently a zero-point energy difference will be associated with the isotopic transition states, which partly cancels the difference for the isotopic reactants and a lower isotope effect results. A large isotope effect thus reflects approximately half transfer of the proton to the base in the transition state, but a smaller effect can be associated with more or less extensive proton transfer. This interpretation was supported on mathematical grounds[116,121] and has since been confirmed by experimental studies[122–124]. A plot of isotope effect (k_H/k_D) against the difference in pK_a between the base and substrate in the ionisation of a series of alkanes containing ketone, nitro, or ester functions, passes through a maximum in the region of $\Delta pK_a = 0$, gradually decreasing values being observed as ΔpK_a becomes more negative or positive[123]. When $\Delta pK_a = 0$, the force constants of the C–H and B–H bonds should be of a similar magnitude in the transition state. Unfortunately, the profile of the curve is rather flat at the maximum, and such an insensitivity poses a serious threat to the use of kinetic isotope effects to estimate the position of the proton in the transition state with certainty. Mathematical calculations also indicate a flat maximum[125], and the only elimination reaction in which a variable isotope effect has been observed by simply modifying the solvent supports this possibility. The maximum isotope effect is observed whilst the solvent composition is changed from 30 to 60 mole % dimethyl sulphoxide in water, despite a large continual increase in the rate of the elimination reaction of 2-phenylethyldimethylsulphonium bromide with hydroxide ion[67] (Fig. 4). Accepting that the proton is more than half transferred to the base in aqueous solution, then the observed maximum must arise by a gradual reduction in the extent of proton transfer and a change to a more reactant-like transition state for elimination.

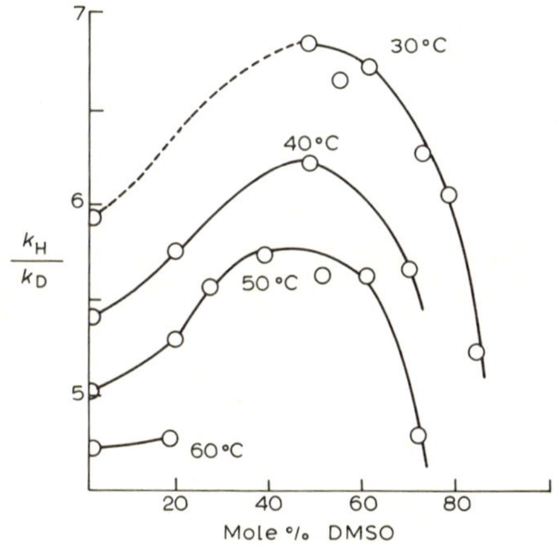

Fig. 4. Variation of the primary kinetic isotope effect in the elimination reaction of 2-phenylethyldimethylsulphonium bromide with hydroxide ion as the solvent composition is changed from water to 84% dimethyl sulphoxide. (A similar profile has recently been obtained for the elimination reaction of 2-phenylethyltrimethylammonium ion under the above reaction conditions[198b]). (Reproduced with permission from Cockerill[67].)

This would seem highly reasonable as the addition of dimethyl sulphoxide to aqueous alkali produces an increase in medium basicity[64].

The diversity in isotope effects for elimination reactions (Table 3) is indicative of widely varying transition states, but these effects alone provide insufficient evidence of the extent of proton transfer and have to be considered in conjunction with alternative kinetic evidence.

TABLE 3

PRIMARY KINETIC HYDROGEN ISOTOPE EFFECTS FOR A NUMBER OF E2 REACTIONS

No.	Substrate	Base	Temp. (°C)	k_H/k_D[a]	Ref.
Solvent: ethanol					
1	$C_6H_5CH_2^*CH_2\overset{+}{N}(CH_3)_3$	$\bar{O}C_2H_5$	50	2.98 ± 0.08	40
2	$(p\text{-}Cl\cdot C_6H_4)_2CH^*CCl_3$	$\bar{O}C_2H_5$	25	3.8	57
	(DDT)	$\bar{O}C_2H_5$	45	3.4	57
		$\bar{S}C_6H_5$	45	3.1	57
3	$C_6H_5CH_2^*CH_2\overset{+}{S}(CH_3)_2$	$\bar{O}C_2H_5$	30	5.07 ± 0.22	40
4	$C_6H_5CH_2^*CH_2OTs$	$\bar{O}C_2H_5$	30	5.66 ± 0.52	40
5	$C_6H_5CH_2^*CH_2Br$	$\bar{O}C_2H_5$	30	7.11 ± 0.17	40
6	$C_6H_5C(CH_3)H^*CH_2Br$	$\bar{O}C_2H_5$	25	7.82	119, 126
			35	6.81	119, 126
Solvent: methyl cellosolve					
7	$(C_6H_5)_2CH^*CH_2OSO_2\cdot C_6H_4\cdot$ $p\text{-}OCH_3$	$\bar{O}CH_3$	50	5.27	133
8	$(C_6H_5)_2CH^*CH_2OSO_2\cdot C_6H_5$	$\bar{O}CH_3$	50	5.42	133
9	$(C_6H_5)_2CH^*CH_2OSO_2\cdot C_6H_4\cdot$ $p\text{-}NO_2$	$\bar{O}CH_3$	50	6.70	133
Solvent: ethylene glycol					
10	$CH_3^*CH_2\overset{+}{N}(CH_3)_3$	$\bar{O}H$	137	3.90 ± 0.5	41
Solvent: methanol					
11	$C_6H_5CH_2^*C(CH_3)_2Cl$	$\bar{O}CH_3$	76	2.6	134
		$\bar{S}C_2H_5$	76	2.4	134
Solvent: 50% aqueous ethanol					
12	$C_6H_5CH_2^*CH_2\overset{+}{N}(CH_3)_3$	$\bar{O}H/\bar{O}C_2H_5$	50	3.02 ± 0.09	40
Solvent: water					
13	$C_6H_5CH_2^*CH_2\overset{+}{S}(CH_3)_2$	$\bar{O}H$	30	5.93	40
Solvent: t-butyl alcohol					
14	$C_6H_5CH_2CH^*(CH_3)CF_3$	$KOBu^t$	124	1.2	135
15	$C_6H_5CH_2^*CH_2\overset{+}{N}(CH_3)_3$	$KOBu^t$	30	6.96 ± 0.13	136
16	$C_6H_5CH_2^*CH_2OTs$	$KOBu^t$	30	8.01 ± 0.32	40
17	$C_6H_5CH_2^*CH_2Br$	$KOBu^t$	30	7.89 ± 0.46	40
				8.16	45
		$KOBu^t$ with 0–2.32 M DMSO	30	8.19 ± 0.05	45
18	9-bromo-9,9′-bifluorenyl	$\overset{+}{K}$, $\overset{+}{Na}$ or $C_6H_5CH_2\overset{+}{N}\text{-}(CH_3)_3OBu^t$	30	8.0	120
		$KOBu^t$	30	8.0	120
		$C_6H_5CH_2\overset{+}{N}(CH_3)_3$ $\bar{O}H/H_2O$	30	7.0	120
		$C_6H_5CH_2N(CH_3)_3$ $\bar{O}CH_3/CH_3OH$	30	6.2	120
		Piperidine	30	6.0	120
19	$p\text{-}NO_2\cdot C_6H_4\cdot CHTCH_2\overset{+}{N}(CH_3)_3$				
		$CH_3OH/\bar{O}CH_3$	22	13[b]	131, 132
			22	2.20 ± 1.6	131, 132
			60	8 ± 1.6[b]	131, 132
		H_2O of pH 7	60	12 ± 4[b]	35
		H_2O of pH 6	98	1.76 ± 0.28[b]	128
				2.12 ± 0.31	128

TABLE 3 (*Contd.*)

No.	Substrate	Base	Temp (°C)	k_H/k_D[a]	Ref.
		H_2O of pH 6.6	98	1.99±0.31[b]	128
				2.47±0.12	128
		C_2H_5OH/6% formamide	98	2.28±0.22	128
				2.35±0.25[b]	128
		C_2H_5OH/5% dimethyl formamide	98	2.20±0.30	128
				2.27±0.31[b]	128
20	$TCH_2CH_2\overset{+}{N}(CH_3)_3$	Under vacuo/$\bar{O}H$	60	1.60±0.05	131, 132
				2.96±0.02[b]	131, 132
21	$CH_3CHTCH_2\overset{+}{N}(CH_3)_3$	Under vacuo/$\bar{O}H$	60	2.00±0.02	131, 132
				2.86±0.05[b]	131, 132

[a] k_H/k_D ratios are derived from comparison of the rate coefficients for elimination of the unlabelled substrate and the substrate labelled fully with deuterium in the position which is starred.
[b] k_H/k_T ratios are intermolecular isotope effects except for the values thus indicated (see 66–68).

(*b*) *Measurements of primary kinetic hydrogen isotope effects*

Primary kinetic hydrogen–deuterium isotope effects are usually calculated from the ratio of the rates of elimination of isotopically labelled substrates. An uncertainty in rate coefficients of ±2% arises from a titrimetric determination[23,26,40] whereas a spectrophotometric approach, usually the monitoring of the olefin concentration under conditions of first-order kinetics, gives an improved precision (±1%)[45,67,120]. During attainment of the transition state, a change from sp^3 towards sp^2 hybridisation occurs at the beta carbon and a variation in the force constant of all the bonds to the beta carbon occurs. Consequently, comparison of the rates of elimination of the substrates labelled fully at the beta carbon with hydrogen and deuterium, respectively, only represents the primary kinetic isotope effect for tertiary structures. For substrates in which the beta carbon is of a secondary or primary nature, the changes in the force constants of the "non-reacting" beta-carbon–hydrogen bonds in attaining the transition state differ from the corresponding changes in beta-carbon–deuterium bonds. The observed rate ratio thus represents a combined primary and secondary kinetic isotope effect. Secondary isotope effects are usually between one and two, approximating near to unity most often[115]. However, they may account for the observed isotope effects in the 2-phenylethyl series exceeding the value calculated from the stretching frequency of the carbon–hydrogen and carbon–deuterium bonds in the substrates. Secondary isotope effects vary only slightly with temperature and their effect on the observed rate ratio is thus more critical at higher temperatures. Obviously primary substrates are affected more seriously than secondary substrates. These points have to be considered carefully when comparing results for different types of substrate.

The use of tritium produces less accurate isotope effects as its high specific activity curtails its use unless in a diluted form. Rate coefficients, k_T, have to be evaluated in competition with hydrogen or deuterium and the technique involves isolation, at various reaction times, of either the products or reactants in which the tritium concentration is estimated by measuring the molar activity (the number of radioactive disintegrations per mole per unit time). Although radioactivity estimates can be as good as $\pm 1\%$, uncertainty is increased by the sampling procedure. Preferentially, deuterium should be used as the diluent[126], as the smaller isotope ratio, k_D/k_T enables a more precise value of k_T to be obtained. Tritium isotope effects can be converted to deuterium isotope effects by use of the Swain equation (65) which is based on the zero-point energy differences arising from the isotopic masses[127].

$$\frac{k_H}{k_D} = \left[\frac{k_D}{k_T}\right]^{2.26}, \quad \frac{k_H}{k_T} = \left[\frac{k_D}{k_T}\right]^{3.26} \quad \text{and} \quad \left[\frac{k_H}{k_D}\right]^{1.44} = \frac{k_H}{k_T} \tag{65}$$

Despite practical difficulties, tritium tracers can be used to dissect the various isotope effects which are incorporated in the direct rate comparisons for substrates having secondary or primary beta carbon. The requisite rate coefficients in reaction (66) can be calculated from a knowledge of the mole fraction reacted (f), the molar activity of the reactant (M_1) and the product (M_2) at that time, and the initial molar activity (M_0) of the reactant (67, 68)[128,129].

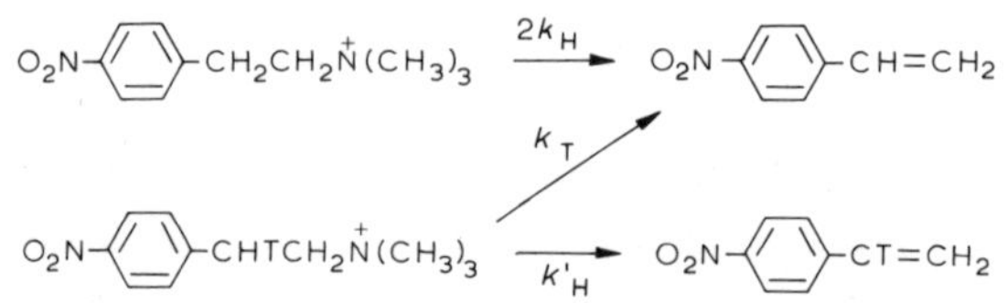

(66)

$$\frac{k_T + k'_H}{2k_H} = 1 - \log\left[\frac{M_0/M_1}{1-f}\right]$$
$$\frac{k'_H}{2k_H} = \frac{M_2}{M_0} \quad \text{for } f < 5\% \tag{67}$$

$$\frac{k_H}{k_T} = \text{intermolecular (kinetic) isotope effect}, \quad \frac{k'_H}{k_T} = \text{intramolecular isotope effect},$$
$$\frac{k_H}{k'_H} = \text{secondary isotope effect} \tag{68}$$

The intramolecular isotope effect reflects zero-point energy differences in the isotopic transition states[130] and also contains a secondary isotope effect contribution as the "non-reacting" atom at the beta carbon is tritium for k'_H and hydrogen for k_T. The intermolecular isotope effect is independent of the secondary phenomenon in this case.

(c) *Factors complicating hydrogen isotope effects*

There is an alarming discrepancy between the values of the intramolecular isotope effect for the elimination from 2-*p*-nitrophenylethyltrimethylammonium ion (Table 3) which have been reported[35,109,128,131,132]. The disagreement is too large to be attributable to a solvent effect and it has been suggested that it may have arisen from polymerisation of *p*-nitrostyrene during the work-up procedure[109], although verification of this point is really required. A similar discrepancy exists for elimination from ethyltrimethylammonium ion. The deuterium isotope effect (k_H/k_D) from the 2,2,2-trideutero substrate is larger than the tritium isotope effect (k_H/k_T) from 2,2,2-tritritioethyltrimethylammonium ion, even though the latter value was measured at a much lower temperature. However, in the former case, the isotope ratio predominantly reflects the secondary isotope effect.

The magnitude of the kinetic isotope effect is highly dependent on the solvent. A change from ethanol to *t*-butyl alcohol causes an increase in the observed isotope effect for substrates in the 2-phenylethyl series, and could represent a movement towards a more symmetrical transition state for proton transfer. However, the majority of the isotope effects in *t*-butyl alcohol are greater than the normal predicted maximum, even with the substrate 9-bromo-9,9′-bifluorenyl, which has a tertiary beta carbon. Moreover, the isotope effect rises with increasing steric size of the base. It is possible that these observations arise, in part, from contributions to the isotope effect from the zero-point energy differences due to bending vibrations. Another explanation involves the tunnelling phenomenon (see Chapter 4, Volume 2).

As the proton is a small particle it must be treated quantum mechanically rather than classically. The possibility of the proton tunnelling through the potential energy barrier has to be considered[137]. Tunnelling is more important for hydrogen than deuterium and is least important for tritium due to the successive increase in mass[119]. Steric hindrance in either the substrate or the base appears to favour tunnelling presumably because such effects promote high potential energy barriers with narrow cross-sections. The exact magnitude of the tunnel factor appears to vary markedly. Values of k_H/k_D as great as 24 have been reported for proton abstraction from nitroalkanes by sterically hindered pyridine bases[123,138]. On theoretical grounds, the tunnel factor has been shown to broaden the range of proton transfer for which the maximum isotope effect is observed[125]. Consequently, tunnelling not only introduces uncertainty in the magnitude of the isotope effect but also reduces its sensitivity to changes in the force constants of the B–H and C–H bonds in the transition state. Such an insensitivity is reflected in eliminations induced in *t*-butyl alcohol containing its lyate ion in the presence of varying amounts of dipolar aprotic solvents[45,121]. The gradual increase in the concentration of the dipolar aprotic solvent not only increases the medium basicity[139] but reduces the activation energy markedly and causes a sharp rise in the rate of elimination[45]. The isotope effect, however, remains unchanged[45].

References pp. 350–364

Transition-state theory predicts limits of 1 to 2 for the ratio of the pre-exponential factors (A_D^*/A_H^*), with values close to unity most probable[137]. Experimental values greater than 2 provide strong evidence of an appreciable degree of tunnelling[137]. In addition, the activation energy differences between reactions of isotopically labelled substrates should differ from the values calculated from the stretching frequencies of the isotopic bonds to carbon in the substrates. These differences will be greatest for isotope measurements concerning hydrogen and least for those involving deuterium and tritium which, having higher masses, will tend to tunnel to a much smaller extent. On the strength of these predictions, Shiner *et al.*[119,126] have shown that the high isotope effect observed in the elimination reaction of 1-bromo-2-phenylpropane is due, in part, to the presence of proton tunnelling. The ratio of the pre-exponential factors indicates tunnelling is extensive for hydrogen, less for deuterium and least important for tritium (Table 4). Theoretical calculations were used to illustrate that the zero-point energy term and the tunnel factor contributed approximately equally to the kinetic isotope effect[126]. On the other hand, deuterium and tritium isotope effects were used to exclude quantum mechanical tunnelling in the elimination from 2,2-diphenylethyl benzenesulphonates as all the pre-exponential factor ratios approximated to unity[140]. Moreover, the activation energy differences are close to those calculated from the spectral data of the tertiary 2-carbon–hydrogen and 2-carbon–deuterium

TABLE 4

DEUTERIUM AND TRITIUM ISOTOPE EFFECTS FOR SOME E2 REACTIONS

Substrate	$(C_6H_5)_2CH^*CH_2OSO_2C_6H_5$		$C_6H_5CH^*(CH_3)CH_2Br$	
Base	$\bar{O}CH_3$		$\bar{O}C_2H_5$	
Solvent	Methyl cellosolve		Ethanol	
Temp. (°C)	40	50	25	45
k_H/k_D	5.61	5.42	7.82	6.43
k_H/k_T	13.4	12.2	20.7	15.9
$\frac{\log (k_H/k_T)}{\log (k_H/k_D)}$	1.50	1.48	1.47	1.49
A_D^*/A_H^*	0.905		2.53	
A_T^*/A_H^*	1.03		3.01	
A_T^*/A_D^*	1.14		1.19	
$(E_D-E_H)(cal.mole^{-1})$	1016 ± 111		1766 ± 115 (1071)[a]	
$(E_T-E_H)(cal.mole^{-1})$	1617 ± 214		2445 (1548)[a]	
$(E_T-E_D)(cal.mole^{-1})$	601		679 ± 25 (477)[a]	
Ref.	140		126	

The position of the isotope labels is indicated by*. *A* is the Arrhenius pre-exponential factor in the rate equation; *E* is the experimental activation energy; *k* is the rate coefficient for elimination from the isotopically labelled substrate.
[a] Values calculated from the infrared stretching frequencies of the C–H and C–D bonds in the substrates and derived for C–T by assuming harmonicity and using the reduced mass relationship.

bonds in the 1-bromo-2-phenylpropanes (which should be reasonable models) and the ratios of the rate coefficients for the three isotopically labelled substrates were found to fit the Swain equation (65) closely. The latter equation, however, may not be a sensitive measure of tunnelling as an equally good approximation is shown by the elimination from the isotopic 1-bromo-2-phenylpropanes.

Another restriction which may inhibit the use of isotope effects in predicting the variable nature of the transition state is the profile of the isotope effect curve with the extent of proton transfer to a series of bases. Recent theoretical calculations, which considered both tunnelling and zero-point energy contributions for both bending and stretching vibrations, have shown that such curves deviate markedly from a symmetrical form and that the magnitude of the maximum effect and its position with respect to the order of the reacting C–H bond in the transition state varies quite markedly with changes in the base (Fig. 5)[141]. Variations in the solvent have not yet been considered.

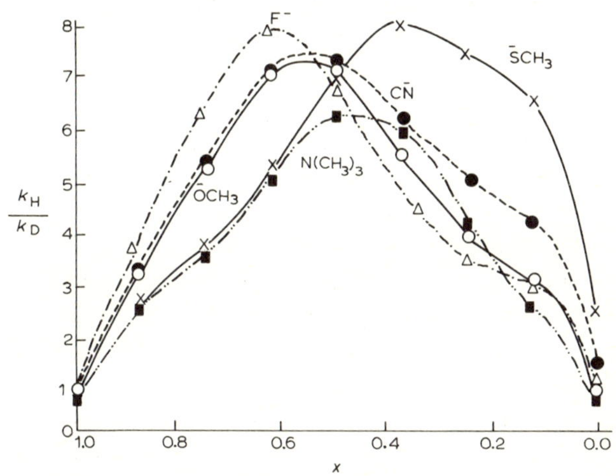

Fig. 5. Model calculations for hydrogen isotope effects for proton transfer reactions from carbon to various bases. x is the order of the reacting hydrogen bond (*e.g.* if C–H $= x$, H–B $= 1-x$).

In view of the complications imposed on interpretation of kinetic isotope effects by quantum mechanical tunnelling and a variable profile of isotope effect with proton transfer to different bases, a more certain prediction would seem most probable if comparisons are restricted to reactions of a series of similar substrates within a given reaction medium. Within this framework it is possible to make reasonable predictions of the effect of substrate structure on the nature of the transition state for elimination using only primary kinetic hydrogen isotope effects.

The primary kinetic isotope effects (Table 3, p. 192) have been listed in order of predicted decreasing extent of proton transfer within a given solvent. Proton transfer to the base should be favoured by an increase in acidity of the beta hydrogen and should be decreased by a change to a more-ionising leaving group. Under the latter circumstance, the breaking of the C_α–X bond should be more sensitive to small changes in electron density at the beta carbon and a shift to a carbonium-ion-like E2 transition state is encouraged. In ethanol, a gradual decrease in the electronegativity of the leaving group is accompanied by an increase in the observed isotope effect in the 2-phenylethyl series, most probably indicating less extensive proton transfer along the series $\overset{+}{N}(CH_3)_3 >$ $\overset{+}{S}(CH_3)_2 > OSO_2C_7H_7(OTs) > Br$, the latter representing approximately half transference of the proton to the base in the transition state. The beta methyl group in 1-bromo-2-phenylpropane should reduce acidity of the beta hydrogen and the two electron-withdrawing *p*-chlorophenyl substituents and the three alpha chlorine atoms should favour a transition state with a high degree of carbanion character in the elimination from DDT. In the latter reaction, a change to a weaker base ($\bar{O}C_2H_5$ to $\bar{S}C_6H_5$), gives a lower isotope effect, indicative of more proton transfer in the transition state.

The elimination studies of Willi[133] on the 2,2-diphenylethyl benzenesulphonates are of interest. The introduction of electron-withdrawing substituents into the leaving group causes a slight increase in rate and makes the C_α–X bond more ionisable. However, the electronegativity of the leaving group is also increased and the observed isotope effects can be interpreted as representing either more or less extensive proton transfer to the base in the transition state. It would be highly informative in this case to measure the oxygen isotope effect at the leaving group and the Hammett reaction constant for substituents in the beta phenyl groups (Section 2.2.2(a)) with each benzenesulphonate already examined by Willi.

Elimination from 2-benzyl-1,1,1-trifluoropropane with potassium *t*-butoxide undoubtedly occurs through a carbanion-like E2 transition state. The structure of this substrate (14, Table 3, p. 192) is closely related to the pentahaloethanes known to react by the carbanion mechanism[46]. The lesser tendency of the beta methyl and beta benzyl groups to withdraw electrons than the beta halogens accounts for a reduction in carbanion stability and a change to a concerted elimination.

The Hofmann eliminations (19–21) give very low isotope effects, excepting the controversial intramolecular isotope effects for the 2-*p*-nitro-phenylethyltrimethylammonium ion. The conditions vary widely and comparisons are thus difficult. The enhanced acidity of this substrate, however, should promote extensive proton transfer and the small tritium isotope effects would seem most plausibly interpreted as indicative of highly carbanionic transition states for these eliminations. The alkyl ammonium salts, possessing much more weakly acidic beta hydrogen, probably react via less carbanion-like E2 transition states[132].

(*d*) *Carbon isotope effects*

After hydrogen, isotope effects involving carbon have been most frequently measured. Direct rate measurements are inadequate to estimate carbon isotope effects as the relatively small difference in the molecular masses of the isotopic substrates results in much smaller isotope effects (< 15%) than are observed for the hydrogen isotopes. The intramolecular isotope effect is used and with ^{14}C, isotope effects as high as 15% have been reported, ^{13}C giving half this magnitude in cases where both ratios (k_{12}/k_{14}, k_{12}/k_{13}) have been measured[142]. The greater magnitude of the ^{14}C effects is counterbalanced by the increased precision of the mass spectrometric approach (±0.1% and better) used to estimate ^{13}C as against the reduced precision of radio tracer techniques (±1%).

Complex organic molecules give mass spectra in which the parent peaks are surrounded by many "isotopic satellites" (due to traces of ^{2}H, ^{13}C, ^{15}N, ^{18}O etc.), complicating determination of isotope ratios. Consequently the product or reactant is degraded and the labelled atom is converted into carbon dioxide. The isolated parent peaks, 44 ($^{12}CO_2$) and 45 ($^{13}CO_2$) are surrounded only by predictable ones from traces of ^{17}O and ^{18}O in the oxygen and the isotope ratio can be measured with a high degree of certainty, especially if a simultaneous estimation using a dual collector is employed (±0.01–0.02%)[108]. More often, a repetitive scanning technique is used, 20 to 30 scans of the region 44–45 being measured, during which time the sample pressure decreases inside the spectrometer. This change can give rise to varying isotope ratios and a drift in peak height, but this latter complication on isotope ratio variation can be minimised by comparing each parent peak with the parent plus one peaks on either side and *vice versa*[143]. In any case, samples should be compared only under similar experimental conditions of pressure and total scanning time. The isotope effect is calculated from equation (69) which simplifies to (70) if one sample is isolated in the early stages of the reaction[144,145].

$$\frac{k_{12}}{k_{13}} = \frac{\log(1-f)}{\log(1-rf)} \qquad r = \left[\frac{M_{44}}{M_{45}}\right]_{100\%} \Big/ \left[\frac{M_{44}}{M_{45}}\right]_{x\%} \tag{69}$$

$$\frac{k_{12}}{k_{13}} \simeq \frac{[M_{44}/M_{45}]_{x\%}}{[M_{44}/M_{45}]_{100\%}} \qquad \text{(if } x\% \text{ less than 5\% reaction)} \tag{70}$$

where f = mole fraction reacted and M represents the isotope abundance of ^{12}C and ^{13}C in carbon dioxide samples isolated at various reaction times.

Carbon isotope effects (k_{12}/k_{14}) have been measured for both the beta and alpha carbon atoms in some Hofmann degradations of quaternary ammonium compounds (Table 5). Three methods were used to determine the isotope effects[146]; (*a*) comparison of molar specific activity of reactant remaining after a measured extent of reaction, (*b*) comparison of molar specific activity of the accumulated reaction product up to a measured extent of reaction with the specific activity of the reactant and (*c*) comparison of molar specific activity of a small increment of product with molar specific activity of the reactant. When the results were obtained by more than one approach, a close agreement was observed[132].

TABLE 5

CARBON-14 ISOTOPE EFFECTS FOR SOME HOFMANN DEGRADATIONS[109,132]

No.	*Substrate*[a]	*Conditions*	*Temp. (°C)*	k_{12}/k_{14}[b]
1	$CH_3C^*H_2\overset{+}{N}(CH_3)_3$	Diethylene glycolate	139	1.033 (1.05)
2	$CH_3C^*H_2\overset{+}{N}(CH_3)_3$	$\bar{O}H$ under vacuo	40	1.065 (1.06)
3	$CH_3CH_2C^*H_2\overset{+}{N}(CH_3)_3$	$\bar{O}H$ under vacuo	50	1.068 (1.073)
4	p-$NO_2.C_6H_4.CH_2C^*H_2\overset{+}{N}(CH_3)_3$	Phosphate buffer, p*H* 7	100	1.026 (1.034)
5	$(CH_3)_3\overset{+}{N}C^*(CH_3)_3$	Triethylene glycolate	91	1.052 (1.064)
6	$CH_3C^*H_2CH_2\overset{+}{N}(CH_3)_3$	$\bar{O}H$ under vacuo	51	1.036 (1.038)

[a] The labelled carbon atoms are starred. The mean standard deviations in the isotope effects vary from 0.003 to 0.006.

[b] The figures in parentheses are the isotope effects after conversion to a common reaction temperature of 40°C.

Initially it was hoped that carbon isotope effects would provide an easy method of distinguishing between S_N1 and S_N2 reactions, *viz.*

$$\text{(a)}\ S_N1\quad (CH_3)_3C^*Cl \rightleftharpoons (CH_3)_3\overset{*}{C}{}^+ + \bar{C}l \rightarrow (CH_3)_3\overset{*}{C}Y \qquad (71)$$

$$\text{(b)}\ S_N2\quad \overset{*}{C}H_3I \xrightarrow{Y^-} Y^{\delta-}\cdots\underset{H}{\overset{H\quad H}{\overset{*}{C}}}\cdots I^{\delta-} \longrightarrow \overset{*}{C}H_3Y + I^-$$

The unimolecular reaction (S_N1) was predicted to give a large isotope effect as extensive bond breaking occurs in the transition state leading to the carbonium ion. However, the S_N2 reactions of methyl iodide[147] gave ^{14}C isotope effects varying from 9 to 14% whilst the S_N1 hydrolysis of *t*-butyl chloride[148] gave an effect of only 3%. Ionisation studies of triphenylmethyl chloride in liquid sulphur dioxide furnished the answers to these somewhat surprising observations[149]. Replacement of the central ^{12}C by ^{13}C is accompanied by a 1.7% increase in the dissociation constant. This means that the carbon–phenyl bonds are stronger in the carbonium ion than in the halide, which is not surprising as a change from sp^3 to sp^2 hybridisation should increase the force constants at the central carbon, especially when there is partial double-bond character to the carbon–phenyl bond due to delocalisation of the charge into the benzene rings. Thus an inverse isotope effect on the ionisation equilibrium ($k_{12}/k_{14} < 1$)

partially cancels the slower breaking of the bond to the heavier isotope and a small isotope effect is observed in S_N1 reactions. This situation is analogous to proton-transfer reactions but is somewhat more complicated as the central atom (carbon) is no longer of negligible mass compared to the bonding fragments in the transition state. Obviously, several combinations of changing force constants could give the same carbon isotope effect.

The above conclusion must be remembered when considering carbon isotope effects for elimination reactions. Certainly the alpha- and beta-carbon isotope effects for elimination from propyl trimethylammonium ion indicate a different extent of rehybridisation at the two carbon atoms in the transition state, but a more definite conclusion requires additional kinetic evidence.

(*e*) *"Leaving group" isotope effects*

"Leaving group" isotope effects have been measured for elimination from dimethylsulphonium[143,145,150,151] and trimethylammonium salts[95,128,152,153] making use of the natural abundance of the elements. The eliminated dimethyl sulphide is analysed directly by mass spectrometry while the eliminated trimethylamine is usually converted into nitrogen by the Kjeldahl method before analysis[95,152,153].

The maximum isotope effect for rupture of a trivalent sulphur bond to carbon[145] is calculated to be 1.0186 (k^S_{32}/k^S_{34}) and the experimental value of 1.0180 for the S_N1 hydrolysis of *t*-butyldimethylsulphonium ion in water at 60°C is in close agreement with this prediction[145]. A maximum nitrogen kinetic isotope effect (k^N_{14}/k^N_{15}) of 1.032 has been proposed on theoretical grounds[111]. Increasing isotope effect is associated with increasing bond breaking, and the results in Table 6 illustrate only partial bond breaking of the C_α–"leaving group" bond in the transition state leading to elimination. The present treatment of kinetic isotope effects does not cater for the effect of solvation and this would seem important in consideration of positively charged substrates. Certainly the activation parameters vary greatly with changes in the solvent and the observed second-order rate coefficient increases markedly with decreasing polarity of the reaction medium[40,136]. With the present understanding of isotope theory, it is wisest to restrict comparison to a common solvent.

The nitrogen isotope effects indicate that C_α–N bond breaking is much more developed in the transition state for elimination from ethyltrimethylammonium ion than in the 2-phenylethyl salt. In the more acidic 2-phenylethyl compound, the proton is more extensively transferred to the base in the transition state and the greater electron density at C_β more effectively aids the breaking of the C_α–N bond, causing the latter to fragment with a smaller change in force constant. Substituents in the beta phenyl ring appear to have little influence on the nitrogen isotope effect. In aqueous solution the large difference between the reported isotope effects for the parent compound and 2-*p*-nitrophenylethyl-

TABLE 6

LEAVING GROUP ISOTOPE EFFECTS FOR SOME E2 REACTIONS

No.	*Substrate*	*Base*	*Temp.* (°C)	k_{14}/k_{15} *or* k_{34}/k_{35}[a]	*Ref.*
(a) Nitrogen isotope effects (k_{14}/k_{15})					
1	$C_6H_5CH_2CH_2\overset{+}{N}(CH_3)_3$	$\bar{O}C_2H_5$	40	1.015±0.001	95
2	$C_6H_5CH_2CH_2\overset{+}{N}(CH_3)_3$	OC_2H_5	40	1.0117±0.0007	153
3	$C_6H_5CH_2CH_2\overset{+}{N}(CH_3)_3$	$\bar{O}C_2H_5$	60	1.0094±0.0004	153
4	$p\text{-}OCH_3\cdot C_6H_4\cdot CH_2CH_2\overset{+}{N}(CH_3)_3$	$\bar{O}C_2H_5$	40	1.014±0.001	95
5	$p\text{-}Cl\cdot C_6H_4\cdot CH_2CH_2\overset{+}{N}(CH_3)_3$	$\bar{O}C_2H_5$	40	1.011±0.001	95
6	$C_6H_5CH_2CH_2\overset{+}{N}(CH_3)_3$	$\bar{O}H$	99	1.009±0.001	95
7	$p\text{-}NO_2.C_6H_4.CH_2CH_2\overset{+}{N}(CH_3)_3$	$\bar{O}H$	98	1.024±0.004	128
8	$CH_3CH_2\overset{+}{N}(CH_3)_3$	$\bar{O}C_2H_5$	60	1.0173±0.0001	153
			95	1.0154±0.0004	153
9	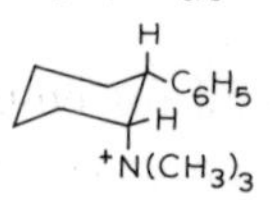	OH/OC_2H_5	60	1.0120	152
10	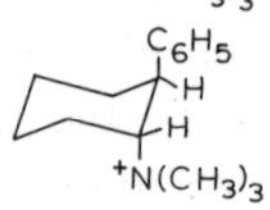	$\bar{O}H/\bar{O}C_2H_5$	60	1.0020	152
(b) Sulphur isotope effects (k_{32}/k_{34})					
11	$C_6H_5CH_2CH_2\overset{+}{S}(CH_3)_2$	$\bar{O}H$	59	1.0064±0.00012	151
12	$C_6H_5CH_2CH_2\overset{+}{S}(CH_3)_2$	$\bar{O}H$	40	1.0074	143
13	$C_6H_5CH_2CH_2\overset{+}{S}(CH_3)_2$	$\bar{O}H$	40	1.0065±0.00025	143
14	$C_6H_5CH_2CH_2\overset{+}{S}(CH_3)_2$	$\bar{O}H$	40	1.0064±0.00023	143
15	$C_6H_5CH_2CH_2\overset{+}{S}(CH_3)_2$	$\bar{O}H$	40	1.0038±0.00021	143
16	$C_6H_5CH_2CH_2\overset{+}{S}(CH_3)_2$	$\bar{O}H$	40	1.0011±0.00056	143
17	$(CH_3)_3C\overset{+}{S}(CH_3)_2$	$\bar{O}H/\bar{O}C_2H_5$	24	1.0072±0.0004	150

[a] With mean standard deviation.

Solvent: ethanol (1–5, 8); water (6, 7, 11, 12); 95% aqueous ethanol (9, 10); 97% aqueous ethanol (17); water containing DMSO: 1.4 *M*(13), 2.8 *M*(14), 4.2 *M*(15), 7.0 *M*(16).

trimethylammonium ion could be due to the dubious procedure used in the estimation of isotope effect for the latter reaction. Unreacted 2-*p*-nitrophenylethyltrimethylammonium ion was isolated and degraded, the isolated nitrogen arising from two sources within the molecule. This approach seems less sound than that based on analysis of the nitrogen from the liberated trimethylamine.

The results for the eliminations giving 1-phenylcyclohexene are very interesting. In the *cis*-2-phenylcyclohexyltrimethylammonium ion, the extent of C_α–$\overset{+}{N}$ bond breaking is slightly greater than that for the 2-phenylethyl series. However, in the *trans* substrate (10, Table 6), in which the C_β–H and C_α–$\overset{+}{N}$ bonds are unable to attain an *anti*-conformation, little bond breaking at the C_α–$\overset{+}{N}$ bond occurs in the transition state. Consequently, if the acidity of the beta hydrogen is the same in both cases, the latter elimination has a transition state with considerably more carbanion character.

The increased strength and electron-withdrawing power of the $C_\alpha-\overset{+}{N}$ bond relative to the $C_\alpha-\overset{+}{S}$ bond should cause reactions of the former to proceed via a more carbanionic transition state. Under the same reaction conditions, the sulphur isotope effect for 2-phenylethyldimethylsulphonium ion (11, Table 6) indicates more $C_\alpha-\overset{+}{X}$ bond breaking than the nitrogen isotope effect in elimination from the corresponding ammonium salt.

The similarity in the sulphur isotope effect for 2-phenylethyl- and *t*-butyl-dimethylsulphonium ions may be misleading due to different reaction conditions. The drastic decrease in the sulphur isotope effect for the elimination from 2-phenylethyldimethyl-sulphonium ion in aqueous alkali upon addition of increasing amounts of dimethyl sulphoxide indicates a gradual change towards a more reactant-like transition state. This interpretation accords well with the activation parameters and the primary kinetic hydrogen isotope effects which were measured over a more extensive variation in solvent composition, for this reaction (Fig. 4, p. 191). With increasing concentration of dimethyl sulphoxide, a large decrease in both the enthalpy and entropy of activation is observed[67]. In part, these changes can be attributed to replacement, at the sulphonium ion, of water by dimethyl sulphoxide solvating molecules. The latter solvent can associate effectively with the large polarisable sulphonium ion and also solvate the charge-dispersed transition state. However, in aqueous solution, the solvating water molecules have to be discarded as the sulphonium group becomes less charged and degenerates into neutral dimethyl sulphide.

(*f*) *Secondary isotope effects*

If a substrate contains more than one beta hydrogen or an alpha hydrogen, then secondary hydrogen isotope effects can be used as a probe for assessing transition-state structure of E2 reactions. The change in hybridisation at C_α and C_β causes a variation in the frequencies of all the bonds to these reaction centres. By analogy with stable molecules such as acetaldehyde, the out-of-plane vibration of a carbon–hydrogen bond to an sp^2 carbon is of lower frequency than the bending motions of a carbon–hydrogen bond to an sp^3 carbon. Replacement of hydrogen by deuterium on the carbon undergoing hybridisation change thus gives rise to a so-called alpha-deuterium secondary isotope effect. Being concerned with changes in the lower-energy bending frequencies rather than the higher-energy stretching frequencies, secondary hydrogen isotope effects are considerably smaller than the corresponding primary isotope effects. Streitwieser *et al.*[154], has predicted a maximum alpha-deuterium secondary isotope effect of 1.35 which, from the harmonicity assumption, gives a tritium effect of 1.47 (ref. 132). The magnitude of these values is sufficient for assessment from measurement of rate coefficients by conventional techniques. The tritium effects[132] are measured by the intramolecular isotope approach (68, p. 194).

Hybridisation appears to have progressed further towards sp^2 at the alpha carbon in the transition state in the elimination from 2-phenylethyl bromide than in the corresponding ammonium salt and least in the sulphonium salt (compounds 1, 2, 3, Table 7). The isotope effect of unity for the sulphonium salt is surprising when considered in conjunction with the greater extent of carbon–sulphur bond breaking compared to carbon–nitrogen bond breaking in the transition state, as indicated from the leaving-group isotope effects. However, the value for the sulphonium salt is uncertain since the substrate exchanged 75% of its deuterium with the solvent before the elimination commenced[94]. A reduction in the acidity of the beta hydrogen is usually accompanied by an increase in C_α–X bond breaking in the transition state and in agreement with the nitrogen isotope effects (Section 2.2.1 (e)), a greater hybridisation change occurs in the elimination from propyltrimethylammonium ion, an alkyl onium salt, than from 2-phenylethyltrimethylammonium ion (compounds 3 and 4, Table 7).

TABLE 7

SECONDARY HYDROGEN ISOTOPE EFFECTS FOR SOME E2 REACTIONS

No.	*Labelled substrate*	*Conditions*	*Temp. (°C)*	k_H/k_D [a]	*Ref.*
1	$C_6H_5CH_2CD_2Br$	$\bar{O}C_2H_5/C_2H_5OH$	60	1.09	94
2	$C_6H_5CH_2CD_2\overset{+}{S}(CH_3)_2$	$\bar{O}H/H_2O$	80	1.00	94
3	$C_6H_5CH_2CD_2\overset{+}{N}(CH_3)_3$	$\bar{O}H/H_2O$	80	1.01–1.02	155
				k_H/k_T	
4	$CH_3CH_2CHT\overset{+}{N}(CH_3)_3$	$\bar{O}H$ in vacuo	50	1.10	132
5	$CH_3CHTCH_2\overset{+}{N}(CH_3)_3$	$\bar{O}H$ in vacuo	60	1.33	132
6	$TCH_2CH_2\overset{+}{N}(CH_3)_3$	$\bar{O}H$ in vacuo	60	1.26	132
7	p-$NO_2.C_6H_4.CHTCH_2\overset{+}{N}(CH_3)_3$	$\bar{O}CH_3/CH_3OH$	22	1.18	132
8	$TCH_2CH_2CH_2\overset{+}{N}(CH_3)_3$	$\bar{O}H$ in vacuo	40	1.12	132
			60	1.11	132
			80	1.13	132

[a] k_H/k_D evaluated per deuterium atom.

For the Hofmann degradation, the secondary isotope effect at the beta carbon decreases along the series, n-propyl > ethyl > 2-(p-nitro-phenyl)ethyl (compounds 5–7). This is the order of predicted increasing acidity of the beta hydrogen and could reflect a series of transition states of gradually increasing carbanion character. Counterbalancing the increase in the extent of proton transfer is possibly a decrease in the extent of carbon–nitrogen bond breaking. Approximating the extent of double-bond formation to the least-stretched bond in the transition state, then hybridisation change will have progressed most towards sp^2 for the first and least for the last substrate in the series.

Explanations other than hybridisation change have been suggested to account for secondary isotope effects[115]. The anharmonicity of the potential energy *versus* bond length profiles of the C–H and C–D bonds (Fig. 3, p. 189) imparts a shorter mean bond length on the bond to deuterium. Consequently, the carbon–deuterium bond has a smaller spatial requirement than the carbon–hydrogen bond, and differing steric requirements in a substrate and its activated complex have been invoked to explain observed secondary isotope effects[156,157]. Such steric variations give rise to only small isotope effects[158], and more often electronic factors have been assumed responsible. In a shorter bond, the average electron density is greater and consequently inductive electron release from a carbon–deuterium bond is greater than that from the corresponding bond to hydrogen[159]. However, in the shorter bond the electrons are more confined and less polarisable. Hyperconjugative electron release is thus more facile from the carbon–hydrogen bond[160–162].

The beta-tritium secondary isotope effect for elimination from the propyltrimethylammonium ion (8, Table 7) seems most probably explained in terms of reduced acidity of the beta hydrogen caused by the greater inductive effect of the bond to the heavier isotope. A hyperconjugative effect also fits the observed data as in the transition state the developing double bond would be more effectively stabilised by the methyl than the monotritiomethyl group. However, the latter explanation seems less likely as the elimination from the propyl compound is slightly slower than that from the ethyl derivative[163].

(*g*) *Solvent isotope studies*

A change in solvent from water to deuterium oxide can alter the rate of a reaction and provide useful information concerning acid- and base-catalysed processes[118,164]. As $\bar{O}D$ is a stronger base than $\bar{O}H$ by a factor of 1.88 at 80.45°C[165], an effect of this magnitude would be expected for complete proton transfer to the base in the transition state. For a transition state in which the proton is half transferred between the substrate and the base, an isotope effect of 1.37, *i.e.* $(1.88)^{1/2}$, is predicted.

The ratios of rate coefficients of elimination (Table 8) induced by $\bar{O}D$ in D_2O to those for $\bar{O}H$ in H_2O represent a secondary isotope effect due to base change and also a solvent isotope effect. The latter, however, should have little influence in the present case as the solvent isotope effect, $[(k_{H_2O}/k_{D_2O})-1]$ 100, is only about 5% for the solvolysis of *t*-butyldimethylsulphonium ion[166] and a smaller effect would seem reasonable in an E2 reaction in which C–X bond breaking is less developed at the transition state.

The magnitude of the effect for the elimination reactions indicates a high degree of bond formation between the base and the proton in the transition state. Proton transfer is extensive in the ammonium compound and the proton is just more than half transferred to the base for the sulphonium salt. These

TABLE 8

COMPARISON OF THE RATE COEFFICIENTS FOR ELIMINATION FROM SOME 2-PHENYLETHYL DERIVATIVES IN H_2O TO D_2O CONTAINING THEIR RESPECTIVE LYATE IONS AT 80.45°C[165]

No.	*Substrate*	$k_{\bar{O}H}$ [a] (*l.mole^{-1}.sec^{-1}*)	$k_{\bar{O}D}/k_{\bar{O}H}$ [a]
1	$C_6H_5CH_2CH_2\overset{+}{S}(CH_3)_2\bar{Br}$	130 ± 2	1.57 ± 0.01
2	$C_6H_5CH_2CH_2\overset{+}{N}(CH_3)_2C_6H_5\bar{Br}$	28.5 ± 0.1	1.62 ± 0.04
3	*p*-Cl.$C_6H_4CH_2CH_2\overset{+}{N}(CH_3)_3\bar{Br}$	21.9 ± 0.5	1.73 ± 0.05
4	$C_6H_5CH_2CH_2\overset{+}{N}(CH_3)_3\bar{Br}$	5.63 ± 0.12	1.79 ± 0.04

[a] With standard deviation.

observations support the predictions drawn from the primary kinetic hydrogen isotope effects (Section 2.2.1 (c)). A change of the leaving group from $\overset{+}{N}Me_3$ to $\overset{+}{N}Me_2C_6H_5$ produces an increase in rate of elimination but a decrease in the extent of proton transfer. Presumably the C–$\overset{+}{N}$ bond becomes more labile due to the electron-withdrawing properties or increased steric size of the phenyl group relative to the methyl group. The measurement of the nitrogen isotope effect for the latter reaction would be useful. A change from a phenyl to a *p*-chlorophenyl substituent appears to produce a reduction in the extent of proton transfer, but further substituent studies are necessary before more certain conclusions can be reached.

Solvent isotope studies offer an attractive means of confirming primary kinetic hydrogen isotope data with aqueous solutions. However, the expense of deuterated solvents could seriously restrict their use with other media.

Solvent isotope studies have been used recently by Stirling and Crosby[411] as support that the conversion of a series of substrates, XCH_2CH_2OPh, into $XCH{=}CH_2$, is an E1cB mechanism. For cases in which X = $\overset{+}{S}(CH_3)_2\bar{I}$ and CH_3SO, the isotope effect, $(k_{XCH_2CH_2OPh})_{H_2O/\bar{O}H}/(k_{XCD_2CH_2OPh})_{D_2O/\bar{O}D}$, at 25°C was 0.66 and 0.78, respectively. These values are close to the ratio predicted from the difference in the basicity of the hydroxide and deuteroxide ions[165]. Certainly, they indicate a very low value for the primary kinetic isotope effect. Furthermore, for a series of substrates of widely differing reactivity (X = Ac, $C_6H_5SO_2$, $COOC_2H_5$,) the ratios of the rate coefficients obtained with *t*-butyl thiolate and ethoxide remained almost constant (see Section 2.4), and the addition of *t*-butyl thiol to solutions of *t*-butyl thiolate was found to depress the rate coefficient (see Section 2.1.3). In addition, the influence of substituents in the leaving group, was found to have a similar influence for substrates possessing various groups X. This factor suggests that in each case, the substituent is influencing essentially the same step. All these facts accord better with a carbanion rather than a concerted mechanism for elimination. To confirm this view even more substantially, a knowledge of the activation parameters would be invaluable.

2.2.2 *Application of linear free energy relationships to elucidate E2 transition state structure*

Kinetic isotope effects provide useful information concerning the transition state structure of E2 reactions but they suffer from some serious shortcomings. Not all elements exist in more than one stable isotopic form and isotope effects for heavier atoms are extremely small and cannot be measured using presently available experimental techniques. Moreover, even carbon and hydrogen isotope effects can give ambiguous results. An alternative approach makes use of the linear free energy relationship concept, namely the relation of rate or equilibrium constant of an unknown reaction to those of known standard reactions.

(*a*) *The Hammett equation*

Rates of elimination of substrates possessing suitably substituted alpha or beta phenyl groups can be correlated by the Hammett equation[167,168], *viz.*

$$\log \frac{k}{k_0} = \rho\sigma \qquad (72)$$

where k and k_0 are the rate constants for the reaction of the substituted and unsubstituted compounds, respectively, σ is the substituent constant, defined from a standard reaction (*e.g.* ionisation of benzoic[168] or phenylacetic acids[169]) and dependent on the substituent and its orientation only and ρ is the reaction constant. Substituents are restricted to the meta and para positions of the benzene ring to enable variation of electronic factors without steric complications. The reaction constant (ρ) which is a measure of the sensitivity of the reaction to changes in electron density at the reaction site, is positive for reactions favoured by electron-withdrawing substituents (σ positive) and is negative for reactions favoured by electron-releasing substituents (σ negative). Thus electron-withdrawing substituents should favour reactions in which there is an increase in electron density at the reaction site in the transition state and hinder reactions for which the converse is true.

Comparisons of reaction constants are more certain if common, and preferably more than three, substituents are used[168]. The correlation coefficient, r, reflects a measure of the success of the relationship, having the value of unity for a perfect correlation. The value of ρ has been determined as a function of the leaving group in base-catalysed elimination reactions of the 2-phenylethyl series, most comprehensive studies being reported for ethanol as solvent. (compounds 1–13, Table 9). The sign and magnitude of ρ reflect predominantly the difference in electron density of the beta carbon in the transition state to that in the ground state. Increasingly positive values of

ρ suggest that the poorer the leaving group, the more C–H bond breaking runs ahead of C–X bond breaking and the greater the carbanion character of the transition state. The rate coefficients for elimination from 2-*p*-nitro- and 2-*p*-acetyl-phenylethylbromides in ethanol correlate with the Hammett equation only if exalted substituent constants[170], (σ^-) are employed for these substituents. Such parameters are used for reactions in which direct mesomeric interaction between the reaction site and an electron-withdrawing substituent in either the reactant or transition state, but not both, is present. This observation suggests extensive carbanion character in the transition state even for a reaction exhibiting one of the lower ρ values.

In all solvents, the ρ value for the ammonium salts is greater than for the sulphonium salts, a result in agreement with the isotope studies (Section 2.2.1) which show that elimination from the former proceeds through a more carbanion-like E2 transition state.

Comparison of reaction constants obtained under differing reaction conditions is complicated as ρ is dependent on both solvent and temperature. In general for many reactions ρ varies inversely with both temperature and medium dielectric constant[168]. The slight decrease in ρ in changing from ethanol to water could reflect predominantly a solvent effect on ρ although an alternative explanation is required for the transition from ethanol to *t*-butyl alcohol (*e.g.* 13, 17, 23, Table 9). The decrease in ρ for the halide and ammonium ion elimination suggests a decrease in carbanion character. The increased base strength of *t*-butoxide relative to ethoxide[64] should produce a more reactant-like transition state[171]. A reduction in solvation of the ammonium ion, rendering the C–$\overset{+}{N}$ bond more labile could be responsible for the observed shift in transition state. For the halides, ion pairing of the developing halide ion[172] with the cation in *t*-butyl alcohol could impart a reduction in carbanion character of the transition state. In general, *t*-butyl alcohol will tend to depress charge localisation relative to ethanol and promote more double-bond formation in the transition state.

The tosylate elimination has a much higher ρ in *t*-butyl alcohol (4, 16, Table 9). This is a surprising result and presumably reflects a decrease in the extent of C_α–X bond breaking in the transition state. Ion pairing seems less likely in this case than for the halides as any developing negative charge can be delocalised into the *p*-tolyl ring. The results for the cyclic tosylates are very interesting (25, 26 and 27, Table 9). In both cases the presence of alpha and beta alkyl substituents reduces the carbanion character of the transition state. The reaction constant, however, is much smaller for the substrate which undergoes an *anti*-elimination (Section 2.3) suggesting that in this conformation the C_α–X bond is more susceptible to changes in the electron density of the beta carbon. In fact, in ethanol the *anti*-elimination appears to have considerable carbonium ion character, exhibiting only a small positive

TABLE 9

HAMMETT REACTION CONSTANTS FOR ELIMINATION FROM 2-PHENYLETHYL AND RELATED DERIVATIVES

No.	X	ρ[a]	r[b]	Y	Temp. (°C)	Ref.
Solvent, ethanol; base, ethoxide						
1	I	2.07 ± 0.09[c]	0.996	*p*-Cl, *m*-Br, *p*-OCH_3, H	30	22
2	Br	2.14 ± 0.15	0.999	*p*-Cl, *m*-Br, *p*-OCH_3, H	30	22
3	Br	2.15 ± 0.24	0.987	*m*-Cl, *p*-OCH_3, *p*-CH_3, H	30	179
4	OTs	2.27 ± 0.08	0.999	*p*-OCH_3, *p*-Cl, H	30	26
5	Cl	2.58 ± 0.33	0.991	*p*-Cl, *m*-Br, H	60	23
6	Cl	2.61		*p*-Cl, *m*-Br, H	30[d]	23
7	$\overset{+}{S}(CH_3)_2$	2.64 ± 0.16	0.996	*p*-OCH_3, *p*-CH_3, *m*-Cl, H	30	179
8	$\overset{+}{S}(CH_3)_2$	2.75 ± 0.21	0.989	*p*-Cl, *p*-CH_3, H	30	22
9	F	3.10 ± 0.07	0.999	*p*-Cl, *m*-Br, H	60	23
10	F	3.12		*p*-Cl, *m*-Br, H	30[d]	23
11	$\overset{+}{N}(CH_3)_3$	3.77 ± 0.21	0.997	*p*-OCH_3, *p*-Cl, *p*-CH_3, H	30	136
12	$\overset{+}{N}(CH_3)_3$	3.94 ± 0.55	0.990	*p*-OCH_3, *p*-CH_3, H	30	136
13	$\overset{+}{N}(CH_3)_3$	3.09 ± 0.15	0.998	*p*-OCH_3, *p*-CH_3, *p*-Cl, H	60	136
Solvent, t-butyl alcohol; base, KOBu^t						
14	I	1.88 ± 0.06	0.999	*p*-OCH_3, *p*-Cl, *m*-Br, H	30	26
15	Br	2.08 ± 0.02	0.999	*p*-OCH_3, *p*-Cl, *m*-Br, H	30	26
16	OTs	3.39 ± 0.29	0.995	*p*-Cl, *p*-OCH_3, *m*-Br, H	30	26
17	$\overset{+}{N}(CH_3)_3$	3.04 ± 0.03	0.998	*p*-OCH_3, *p*-CH_3, H	30	136
Solvent, water; base, hydroxide						
18	$\overset{+}{S}(CH_3)_2$	2.21 ± 0.08		*p*-CH_3, *p*-OCH_3, *p*-Cl, H	30	179
19	$\overset{+}{S}(CH_3)_2$	2.11		*p*-CH_3, *p*-OCH_3, *p*-Cl, H	50	67
20	$\overset{+}{S}(CH_3)_2$	2.53 ± 0.05	0.998	*p*-CH_3, *p*-OCH_3, *p*-Cl, H	50[c]	67
21	$\overset{+}{S}(CH_3)_2$	2.55 ± 0.04	0.998	*p*-CH_3, *p*-OCH_3, *p*-Cl, H	50[c]	67
22	$\overset{+}{S}(CH_3)_2$	2.59 ± 0.05	0.997	*p*-CH_3, *p*-OCH_3, *p*-Cl, H	50[c]	67
23	$\overset{+}{N}(CH_3)_3$	2.6		H, *p*-Cl	80	165

No.	Substrate	ρ	r	Solvent/base	Temp. (°C)	Ref.
24	Y-$C_6H_4CH_2C(CH_3)_2Cl$	0.999 ± 0.09	0.965	CH_3OH / $\bar{O}CH_3$	66.5	177
25	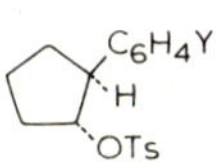	2.77 ± 0.03		*t*-BuOH / $\bar{O}Bu^t$	50	178
26	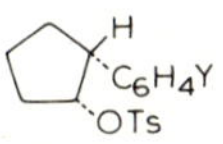	1.48 ± 0.09		*t*-BuOH / $\bar{O}Bu^t$	50	178
27	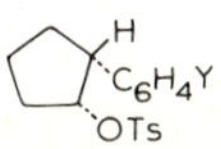	0.99 ± 0.06		C_2H_5OH / $\bar{O}C_2H_5$	50	178

[a] ρ (with standard deviation) obtained from the slope of a graph of log k_Y/log k_H against the substituent constant, σ.

[b] Correlation coefficient (see appendix of ref. 168).

[c] Examples 20–22, the solvent was a mixture of water and DMSO; the molar concentration of DMSO was 20, 19.1; 21, 38.0, 22, 60.3.

[d] Interpolated from rate coefficients measured at higher temperatures.

reaction constant (27, Table 9). A similar interpretation is reasonable for elimination from substituted benzyldimethylcarbinyl chlorides (24, Table 9) to give a substituted styrene, *viz.*

$$Y{-}C_6H_4CH_2C(CH_3)_2Cl \xrightarrow[CH_3OH]{\bar{O}CH_3} Y{-}C_6H_4CH{=}C(CH_3)_2 + Y{-}C_6H_4CH_2C(CH_3){=}CH_2 \quad (73)$$

The tertiary alpha carbon should promote extensive C_α–X bond breaking in the transition state. Rates of elimination into the methyl branch do not correlate with the Hammett equation[177] ($\rho = -0.103$, $r = 0.311$) and the predominant effect of electron-withdrawing substituents is to inhibit the reaction by opposing ionisation of the carbon–chlorine bond. The insulating CH_2 group prevents conjugation with any developing carbanion or developing double bond.

With aqueous dimethyl sulphoxide (19–22, Table 9) the sudden increase in ρ, when considered in conjunction with the decrease in the sulphur isotope effect (Table 7, p. 204) suggests an increase in carbanion character. Further addition of dimethyl sulphoxide has little influence and the constancy of ρ may reflect a delicate balance between the solvent effect on ρ and a gradual decrease in the extent of proton transfer as the strength of the base increases (Fig. 4, p. 191)[69,76].

In the absence of the knowledge of the reaction constant for a completely concerted reaction, an uncertainty arises in the interpretation of ρ. Though under such conditions no change in charge occurs at the beta carbon, a change in electron affinity accompanies the variation in hybridisation[173]. Likewise the magnitude of ρ for a fully developed carbanion is unknown. A limitation of 5, the ρ value for anionic polymerisation initiated by styrene monomers in tetrahydrofuran[174], has been suggested, but a model more closely related to elimination processes is desirable. Reaction constants ranging from 3 to 7 have been reported for kinetic[175] and equilibrium ionisations[76,176] of carbon acids in the substituted fluorene series. A predicted maximum ρ is really needed for each solvent.

(*b*) *The Brönsted equation*

The catalytic constants (k) for a series of related bases reacting by general base catalysis with a given substrate can be related to their base strengths (K) by the Brönsted equation (74)[180].

$$\log k = \beta \log K + \text{constant} \quad (74)$$

The proportionality constant, β, normally has values between zero and one and is taken as a measure of the degree of proton transfer from the substrate to the base in the transition state[181]. For a carbanion mechanism, the Brönsted component, β, should have the value of one, as rate coefficients of specific base-catalysed reactions are directly proportional to the basicity constants of the attacking bases[181]. Gradually decreasing values of β are predicted as the extent of proton transfer, from the substrate, decreases in the transition state. The Brönsted equation thus offers an attractive means of assessing the extent of proton transfer in E2 reactions and unlike primary kinetic hydrogen isotope effects, β is a linear function of the transition state[182]. The most serious limitation to the use of the Brönsted equation is the need to use bases of closely related structure in order to limit steric variations and adhere to the restrictions of an isoentropic relationship. Consequently, the only valid reported uses of the Brönsted equation in elimination reactions are concerned with base catalysis by substituted phenoxides and thiophenoxides. To test the Brönsted equation rigorously, both rate coefficients and basicity constants should be measured with the same solvent and at the same temperature.

In elimination reactions, the value of β increases with the increasing electron-withdrawing nature of the substituents at both the beta and alpha carbon atoms and decreases with increasing tendency of the C_α–X bond to undergo ionisation. These observations suggest a gradual shift from a carbonium ion-like E2 towards a carbanion-like E2 transition state (Table 10, 1–8, 9–12) as the beta hydrogen becomes more acidic and the leaving group less prone to ionise.

TABLE 10

BRÖNSTED COMPONENTS, β, FOR SOME E2 REACTIONS

No.	*Substrate*	β	*Temp. (°C)*	*Ref.*
1	*t*-Butyl chloride	0.17	45	184
2	Cyclohexyl tosylate	0.27	35	185
3	Cyclohexyl bromide	0.36	55	185
4	Cyclohexyl chloride	0.39	55	185
5	*t*-Butyl $\overset{+}{S}(CH_3)_2$	0.46	55	184
6	1,1-Dibromocyclohexane	0.51	55	185
7	1,1-Dichlorocyclohexane	0.58	55	185
8	$(p\text{-Cl}\cdot C_6H_4)_2CHCCl_3$	0.77	75	57
9	$(n\text{-Pr})_2CHBr$	0.39[a]	75	58
10	$C_6H_5CH_2CH_2Br$	0.54[a]	75	58
11	$p\text{-NO}_2\cdot C_6H_4\cdot CH_2CH_2Br$	0.67[a]	75	58
12	$(p\text{-Cl}\cdot C_6H_4)_2CHCCl_3$	0.88[a]	75	57

Solvent, ethanol; base, substituted thiophenoxides (1–8), substituted phenoxides (9–12).

[a] pK_a values of the substituted phenols in 95% aqueous ethanol[183] used in the Brönsted correlation.

(c) The ratio k_{OTs}/k_{Br}

Although certain toluenesulphonate esters undergo faster nucleophilic substitution (k_{OTs}) than their corresponding bromides (k_{Br}), the reverse order of reactivity is observed for elimination reactions. This observation led DePuy and Bishop[26] to postulate the ratio k_{OTs}/k_{Br} as a measure of the extent of bond breaking in the transition state, a hypothesis which has since been put on a more quantitative basis by Hoffmann[186,187].

The ratio, k_{OTs}/k_{Br}, is large for S_N1 reactions which involve extensive bond breaking in the transition state, smaller for S_N2 processes in protic solvents and less than unity for substitution reactions in dipolar aprotic media and in concerted elimination reactions (Table 11). Solvation[188] and steric[189] factors have been proposed to explain the rate enhancement of the tosylates but these explanations do not account for most of the low ratios observed and have been shown to be unimportant in most cases[186]. For extensive bond breaking, the developing charge is more effectively delocalised in the transition state for the tosylate. For less extensive bond breaking, the beneficial delocalisation of the tosylate is outweighed by the greater polarisability of the carbon–halogen bond[190]. The ratio, k_{OTs}/k_{Br}, thus increases with increasing bond breaking in the transition state. In the Finkelstein reactions of *n*-butyl and 1-phenylethyl bromides with bromide ion, a symmetrical transition state would seem reasonable in view of the principle of microscopic reversibility. The k_{OTs}/k_{Br} ratios are 0.5 and 4.0, respectively. For the bromide, the electron is half transferred in the

TABLE 11

THE RATIO k_{OTs}/k_{Br} FOR SOME SUBSTITUTION AND ELIMINATION REACTIONS[186,191]

No.	*Substrate*	*Solvent*	*Nucleophile or base*	*Temp. (°C)*	k_{OTs}/k_{Br}	*Reaction type*
1	1-Phenylethyl-X	Ethanol	Ethanol	0	2270	S_N2
2	1-Phenylethyl-X	Ethanol	Ethanol	50	845	S_N2
3	*t*-Butyl-X	CH_3CN		0	5126	E1
4	*t*-Butyl-X	CH_3CN		50	1670	E1
5	*i*-Butyl-X	Acetone	Iodide ion	25	0.2	S_N2
6	1-Phenylethyl-X	Acetone	Bromide ion	25	4	S_N2

No.	*Y in $Y\cdot C_6H_4CH_2CH_2X$*	*Solvent*	*Nucleophile or base*	*Temp. (°C)*	k_{OTs}/k_{Br}	*Reaction type*
7	*p*-OCH_3	Ethanol	Ethoxide	30	0.065	E2
					> 40	S_N2
8	H	Ethanol	Ethoxide	30	0.10	E2
					> 10	S_N2
9	*p*-OCH_3	*t*-BuOH	*t*-Butoxide	30	0.15	E2
10	H	*t*-BuOH	*t*-Butoxide	30	0.22	E2
11	*p*-Cl	*t*-BuOH	*t*-Butoxide	30	0.44	E2
12	*m*-Br	*t*-BuOH	*t*-Butoxide	30	1.19	E2
13	*p*-NO_2	*t*-BuOH	*t*-Butoxide	30	1.57	E2

transition state. Charge development has progressed further in the reaction of 1-phenylethyl tosylate and less for the *n*-butyl tosylate. Strictly speaking, k_{OTs}/k_{Br} indicates the extent of electron transfer intermediate between those for the transition states of the bromide and the tosylate[186].

The Brönsted coefficient, β, (Table 10) suggests, as expected, that electron-withdrawing substituents in the beta phenyl ring promote increased proton transfer to the base in the transition state for elimination in the 2-phenylethyl series. A concomitant increase in k_{OTs}/k_{Br} is also observed[191], and an interpretation has been forwarded that increased beta hydrogen bond breaking is accompanied by increased C_α–X bond breaking. Consequently the E1cB mechanism is extremely rare and has been considered as a near-paradox[191,192]. Two peculiar features of these results warrant further comment.

Electron-withdrawing groups surprisingly promote ionisation of the C_α–X bond. Furthermore, as k_{OTs}/k_{Br} is larger with *t*-butyl alcohol than with ethanol as solvent, C_α–X bond breaking is more advanced in the poorer solvating media. It could be argued that the charge density of the beta carbon is reduced by delocalisation into the C_α–X bond giving more charge dispersal in the transition state in *t*-butyl alcohol. This suggestion however, is in conflict with the Hammett reaction constants (Table 9) which suggest much greater carbanion character in *t*-butyl alcohol than ethanol for elimination from the tosylate. As all the primary kinetic hydrogen isotope effects (Table 3, p. 192) suggest transition states with the proton approximately half transferred to the base, the increase in carbanion character must arise predominantly from a decrease in C_α–X bond breaking in the less polar medium. An alternative explanation of the ratio k_{OTs}/k_{Br} is possible which avoids the discrepancy in the conclusions from different experimental approaches, and this is now considered.

For a complete understanding of the k_{OTs}/k_{Br} ratio it is necessary to know what magnitude this ratio has when no C–X bond breaking occurs in the transition state[193]. Under this condition, the reactivity of the alpha carbon is not determined by polarisability or electron delocalising properties of the leaving group. The critical factor is the imposition of electrophilic nature on the carbon by the electron-withdrawing properties of the leaving group. The tosylate group possessing four electronegative atoms to the single atom of the halogen should exert a stronger electron-withdrawing inductive effect. In agreement with this prediction, the substituent constant σ^* for the tosylate group is more positive than that for the bromine atom (1.31 and 1.00, respectively)[327], and σ_I for the ethanesulphonate group is more positive than that for the bromine atom (0.63 and 0.45, respectively)[194]. Substitution of a *p*-tolyl group for ethyl should not materially change the substituent constant (σ_I) of the sulphonate group, and for no bond breaking a k_{OTs}/k_{Br} ratio of greater than unity is predicted. A gradual increase in bond breaking will cause a downward trend

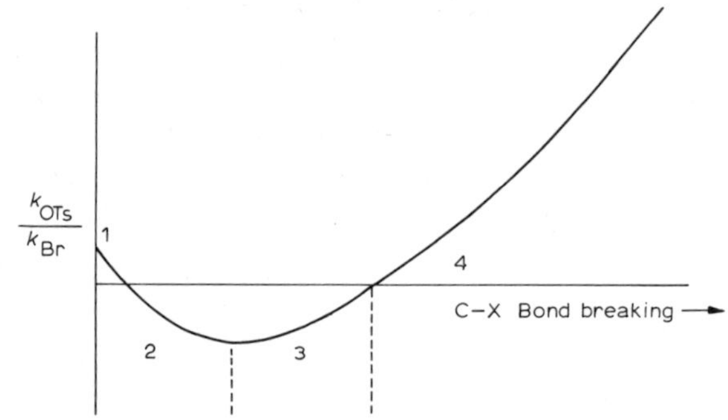

Fig. 6. The predicted variation of the ratio of the rate coefficients, k_{OTs} (tosylate) to k_{Br} (bromide), with the extent of C–X bond breaking in the transition state for S_N1, S_N2 and E2 reactions. In the above regions (1–4) the k_{OTs}/k_{Br} ratio is dependent mainly on: (1) the inductive effect of X; (2) the inductive effect of X and the polarisability of the C–X bond; (3) the polarisability of the C–X bond and the delocalising ability of $\bar{X}$; (4) the delocalising ability of $\bar{X}$.

in k_{OTs}/k_{Br} and a minimum is observed when the ratio becomes dominated by polarisability considerations (Fig. 6). Further increases in bond breaking are accompanied by an upward trend in k_{OTs}/k_{Br}.

Using this alternative interpretation of k_{OTs}/k_{Br}, the low value for the *p*-methoxy compound in ethanol is presumably near the minimum in Fig. 6, and the remaining results for the 2-phenylethyl series represent transition states (region 2, Fig. 6) with less C_α–X bond breaking.

The ambiguity in interpretation of ratios small and less than unity of k_{OTs}/k_{Br} warrants further investigation. It would be useful to correlate k_{OTs}/k_{Br} as a function of the oxygen isotope effect on the C_α–O bond of the sulphonate ester. Alternatively, using the analogous k_{OTs}/k_{Cl} ratio, the chlorine isotope effect[195] could be used as a measure of C–X bond breaking.

Since completion of this section, the secondary isotope effects (k_H/k_D) for the elimination from a series of substituted 2-phenylethyl-1,1-d_2-bromides have been measured[686]. The isotope effect increases with the electron-releasing power of the substituent. Thus, hybridisation at C_α and hence C_α–X bond breaking increases with the electron-releasing power of the aromatic substituent. This result supports the alternative hypothesis presented above for the interpretation of the ratio k_{OTs}/k_{Br}.

(*d*) *The ratio* $k_{trans\text{-}}/k_{cis\text{-}olefin}$

Clear evidence that the geometry of the transition state is not constant has been provided by studies of eclipsing effects[104]. Assuming anti stereospecificity, the *threo* and *erythro* isomers of the 1,2-diphenyl-1-propyl-X systems undergo bimolecular elimination to give the *trans*- and *cis*-α-methyl stilbenes respectively (75).

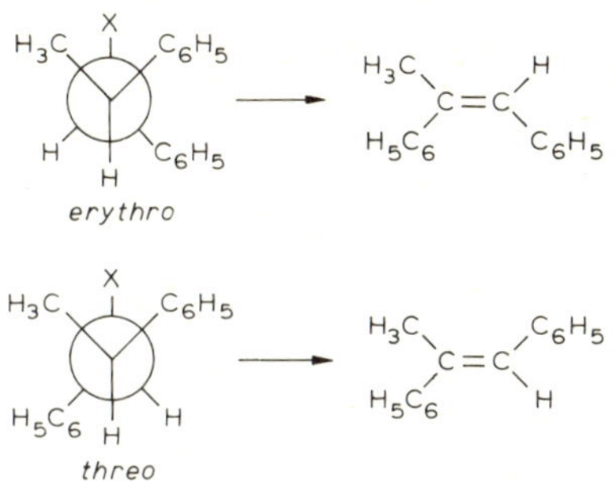

(75)

Although the *trans*-olefin is at least fifty times more stable than the *cis*-olefin, the reactants differ little in stability from each other. Consequently, for a reactant-like transition state, the two isomers should eliminate at similar rates, whereas with increasing double bond character, $k_{threo}/k_{erythro}$ should approach a value of about 50. Ratios of $k_{threo}/k_{erythro}$ of about unity were observed for the elimination from the chloride and bromide with ethoxide in ethanol, but a value of 57 was noted for the trimethylammonium salt. Increased eclipsing effects* were favoured in the transition states for the onium compounds and less double-bond character, presumably due to a carbonium ion-like transition state, in the halide eliminations. A change of solvent to *t*-butyl alcohol containing *t*-butoxide ion increased the ratio to 5.4 for the bromide and 15 for the chloride. This increased eclipsing most probably arises from an increase in the extent of double-bond character in the transition state, brought about by a change to a less ionising but more strongly basic medium. This induces an increase in C_β–H bond breaking but a reduction in the attenuation of C_α–X bond breaking.

Most often, eclipsing effects are estimated by studying the product distribution of geometrical isomeric olefins arising from elimination under kinetic control. The k_{trans}/k_{cis} ratios so determined offer an attractive means of assessing the double-bond character of an elimination but care must be taken when interpreting the results. Often the two geometrical isomers arise from transition states of differing stereochemistry[196,197] and the ratios can exceed the maximum differences predicted from thermodynamic stabilities of the products. Of course this may be due to eclipsing influences exerting greater distortion on the partly formed than on the fully formed double bonds. More difficult to interpret are cases in which the less thermodynamically stable *cis*-olefin is the major product. These aspects are discussed in more detail later (Section 2.5).

---

* The term eclipsing effects incorporates non-bonded interactions of substituents at C_α and C_β, steric inhibition of resonance and differences in degrees of solvation.

References pp. 350–364

2.2.3 Summary

There is little doubt that the transition-state geometry of E2 reactions varies widely with changing reaction conditions. On a more quantitative basis, the extent of proton transfer is possibly most safely predicted using the Brönsted equation or the isotopic base approach. Unfortunately, these methods are of only limited applicability and a resort to isotope studies is often required. As well as the factors which complicate studies of simple proton abstraction reactions (*e.g.* tunnelling, importance of bending and stretching frequency contributions, unsymmetrical isotope effect against proton transfer curves), recent theoretical investigations imply that an even more complex situation surrounds primary kinetic hydrogen isotope effects in elimination reactions[198a]. Using the model

$$(\mathrm{H{-}O}\underset{F_{11}}{\cdots}\mathrm{H}\underset{F_{22}}{-}\mathrm{C}\underset{F_{33}}{\cdots}\mathrm{C}\underset{F_{44}}{-}\overset{+}{\mathrm{S}}\langle),$$

the magnitude of the maximum primary kinetic hydrogen isotope effect was found to be markedly depressed as the concerted nature of the transition state was increased (by modifying the proportionality constants, α in particular and to a smaller extent β and γ, which relate the off-diagonal force constants to the diagonal force constants of the F matrix in the Bigeleisen equation, *e.g.* $F_{12} = \alpha(F_{11}F_{22})^{1/2}$, $F_{23} = \beta(F_{22}F_{33})^{1/2}$, $F_{34} = \gamma(F_{33}F_{44})^{1/2}$: for a proton transfer, $\alpha = 1$, $\beta = 0$, $\gamma = 0$; for concerted reactions, $\alpha = 0.95$–0.50, $\beta = \gamma = 0.40$–0.20). Of course the variations in the concerted nature of the transition state may be unrealistic, but the calculations do suggest an additional complicating factor which has hitherto been ignored. Extended calculations show that the sulphur isotope effect is not a linear function of C_α–S bond breaking. For negligible bond breaking a small inverse isotope effect is predicted, which increases very gradually until the C_α–S bond is more than 60% broken. With further bond breaking a sharp rise in the isotope effect is to be expected. These findings do not invalidate the conclusions deduced from the presently available experimental data, but they reflect the danger in interpreting small leaving group isotope effects as indicative of little bond breaking in the transition state of beta-eliminations.

In cases where the more established kinetic approaches become insensitive, mixed solvent studies on rate of elimination may prove enlightening. Although ρ and k_H/k_D remain unchanged in the elimination of 2-arylethyl bromides with *t*-butoxide in *t*-butyl alcohol on the addition of dimethyl sulphoxide (0–2.23 *M*), a rate increase of 120 times is noted[45]. Smaller increases in rate are observed for similar variations in the concentration of added dipolar aprotic solvent in elimination from 1-phenylethyl bromide[199] and 9-bromo-9,9′-bifluorenyl[120], but a steeper rise in rate is observed for 2,2-diphenylethyl tosylate[199]. A correlation between anticipated increasing carbanion character and increasing solvent influence on rate is apparent, but further studies are necessary to consolidate this criterion. At present, with the exception of leaving groups attached by elements of low atomic weight, kinetic techniques of assessing C_α–X bond breaking are rather uncertain. A similar conclusion applies to estimating the degree of double bond character in the transition state.

2.3 STEREOCHEMISTRY

A knowledge of the stereochemistry of an elimination reaction is of major importance not only to a more complete understanding of the mechanism but also in connection with the use of beta-eliminations in synthetic processes. A variation in the dihedral angle between the bonds undergoing fragmentation not only influences the rate but can also modify the yield of olefin and influence the direction of elimination in secondary and tertiary structures. The terms* "*anti*", "*syn*", "*anti-clinal*" and "*syn-clinal*" are used to describe the stereochemistry of the transition state when the dihedral angle between the parts being eliminated is 180°, 0°, 120° and 60°, respectively (76). The stereochemistry of beta-eliminations, the subject of numerous papers, has been extensively reviewed[6,7,10,14,34,107].

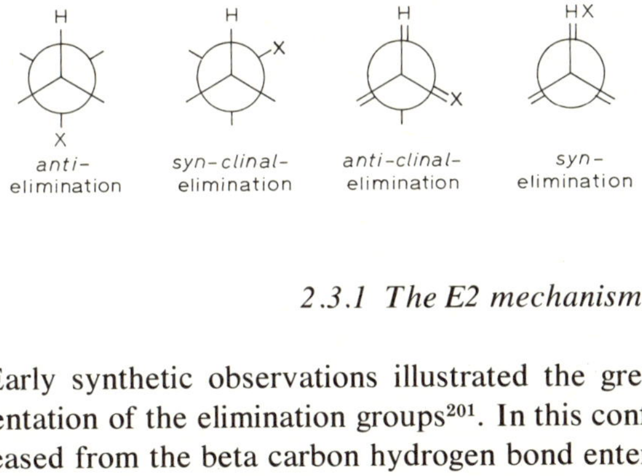

(76)

2.3.1 *The E2 mechanism*

Early synthetic observations illustrated the great preference for an *anti*-orientation of the elimination groups[201]. In this conformation the electron pair released from the beta carbon hydrogen bond enters the C_α-octet on the side remote from the leaving group. Repulsion energy between the electron pairs is thus minimised in the transition state and by analogy to the known stereochemistry of S_N2 reactions, the C_β–H electrons are most favourably disposed

* The more familiar terms "*trans*" and "*cis*" have been replaced by "*anti*" and "*syn*", respectively in this text, following the example of several recent authors[34,196,200]. This approach avoids extensive use of the former terms which are used only in connection with geometrical isomerism in the reactants or products. "*Syn*" is used in cases in which a distinction between *syn* and *syn-clinal* is not possible or has not been made.

to assist the displacement of the leaving group. If *anti*-elimination is not possible, *syn*-elimination is often encountered. Many workers have regarded this change in stereochemistry as indicative of a change in mechanism from E2 to E1cB but kinetic studies do not usually support this bold assumption. For both *anti*- and *syn*-eliminations, the partial bonds to the eliminating portions are coplanar and *p* orbital overlap to form the developing π bond is at a maximum. Recently, the predominance of *anti*-elimination has been justified by consideration of the magnitude of the resonance integral between the fragments whose departure effects double-bond formation[202]. Earlier workers reached a similar conclusion on the basis of a conceptual mechanistic dissection[107] and from molecular orbital calculations[203,204]. With a simple ethyl halide model, by applying the principle of least motion, Hine[205] has shown that the energy change required when the non-reacting bonds at C_β and C_α are rehybridised is least for an *anti*-elimination, and increases markedly in the order; *syn* < *anti-clinal* < *syn-clinal*.

(a) The use of diastereoisomers

The preference for *anti*-stereospecificity is demonstrated experimentally in the acyclic series by use of suitable diastereoisomeric pairs. The *meso*-stilbene dibromide eliminates to give the *cis*-bromostilbene and the (±)-dibromide yields the *trans*-olefin (77)[206].

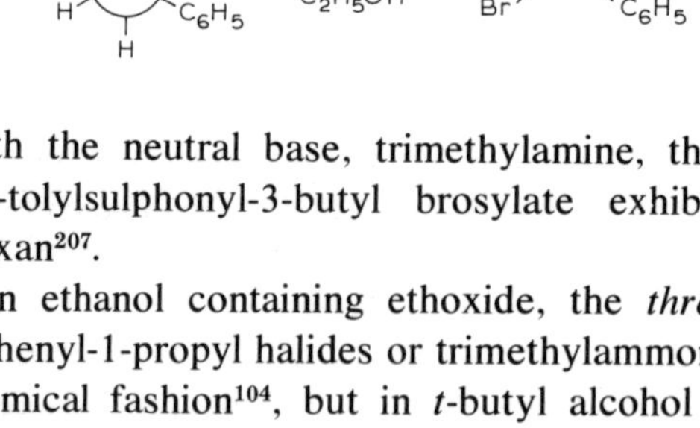

(77)

With the neutral base, trimethylamine, the *threo* and *erythro* isomers of 2-*p*-tolylsulphonyl-3-butyl brosylate exhibit *anti*-elimination in aqueous dioxan[207].

In ethanol containing ethoxide, the *threo* and *erythro* isomers of 1,2-diphenyl-1-propyl halides or trimethylammonium ion react in a similar stereochemical fashion[104], but in *t*-butyl alcohol containing *t*-butoxide, both the ammonium salts give the *trans*-olefin under kinetically controlled conditions[104]. As both isomers have similar thermodynamic stabilities and react at similar rates, the anticipated comparable energies of the transition states have been attributed to the possible incursion of the E1cB mechanism[34]. This seems unlikely as in the analogous elimination from the 2-phenylethyltrimethylammonium ion the primary kinetic hydrogen isotope effect and the Hammett reaction constant suggest less carbanion character of the transition state in *t*-butyl alcohol than in ethanol[136]. Moreover, the imposition of an alpha phenyl and beta methyl group should cause a shift to a more central type of E2 transition state. The apparent *syn*-elimination from the *erythro*-isomer might be attributable to the incursion of an ylid mechanism in view of the tertiary beta carbon and sterically hindered base. The similarity in rate coefficients could be fortuitous as no allowance was made for the known salt effect on rates of elimination from ammonium ions[136] and the activation parameters were not reported[104]. Alternatively, the *erythro* compound may undergo elimination with *syn-clinal* stereochemistry, thus reducing steric interactions involving the leaving group in the reactant (78).

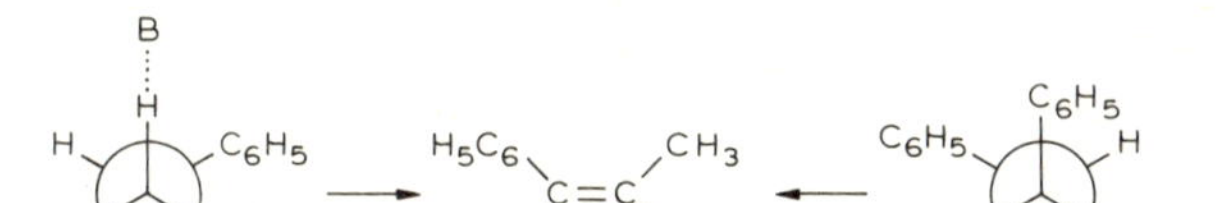

(78)

which might balance the advantages of coplanarity of the eliminating fragments in the transition state. *Syn-clinal*-eliminations have been invoked to account for the increase in the percentage yield of the *cis*-olefin from 5-nonyltrimethylammonium and sulphonium salts[208,209] with increasing solvating power of the medium (79)

X	$\overset{+}{N}(CH_3)_3$	$\overset{+}{N}(CH_3)_3$	$\overset{+}{N}(CH_3)_3$	$\overset{+}{S}(CH_3)_2$	$\overset{+}{S}(CH_3)_2$
Base	$\bar{O}Bu^t$	$\bar{O}C_2H_5$	$\bar{O}CH_3$	$\bar{O}Bu^t$	$\bar{O}C_2H_5$
Solvent	*t*-BuOH	C_2H_5OH	CH_3OH	*t*-BuOH	C_2H_5OH
% cis-olefin	26	74	81	9	64

(79)

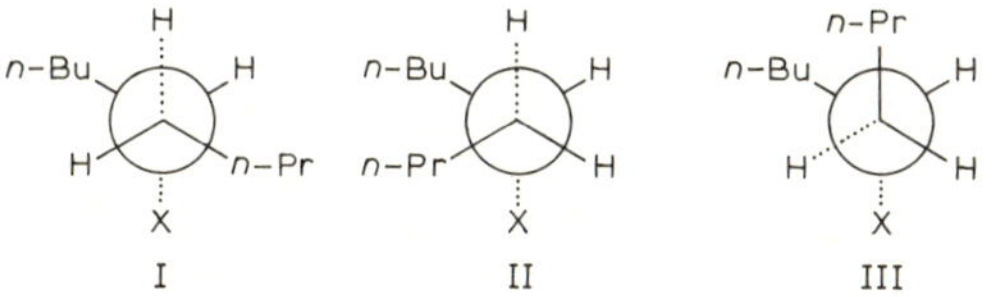

The poor solvating property of *t*-butyl alcohol coupled with the strong basic properties of *t*-butoxide encourage a transition state with well-developed double-bond character and transition state I and to a lesser extent II are favoured. In the more solvating media, C_α–X bond breaking is depressed and a shift towards a carbanion-like E2 transition state is encouraged. Conformation III, in which non-bonding interactions are reduced as the bulky "onium" group is flanked by hydrogen atoms, becomes dominant and advantages accruing from *anti* coplanarity are no longer decisive. However, it is not clear why the hydrogen leading to *cis*-olefin is abstracted when the alternative hydrogen abstraction, which would seem less sterically hindered, would give the more stable *trans*-olefin. Possibly the *cis*-olefin is formed by an *anti*-elimination but the *trans*-olefin arises from the *syn-clinal*-conformation (see Section 2.3.1(e)).

In the acyclic series the usual predominance of *anti-* over *syn*-elimination can be attributed to the difference in the energetics of the reactant conformations. The *anti*-elimination proceeds through a low-energy staggered conformation whereas the transition state for *syn*-elimination is completely eclipsed, thus maximising non-bonded interactions. If the 3 kcal.mole^{-1} potential barrier to rotation of ethane is considered as the minimum difference in energy between the *anti-* and *syn*-transition states, then *anti*-elimination is favoured by a factor of at least 160 (ref. 34). Of course eclipsing greater than that in ethane could lead to a much greater selectivity. It should be noted that reactions with transition states near the E1cB or E1 extremes may show lowered stereochemical preferences, double-bond character being so slight that the energetic advantage of a coplanar transition state giving maximum orbital overlap is small and alternative activation processes may dominate.

The *threo* and *erythro* chlorides both yield the *cis*-stilbene on dehydrochlorination, *viz.*

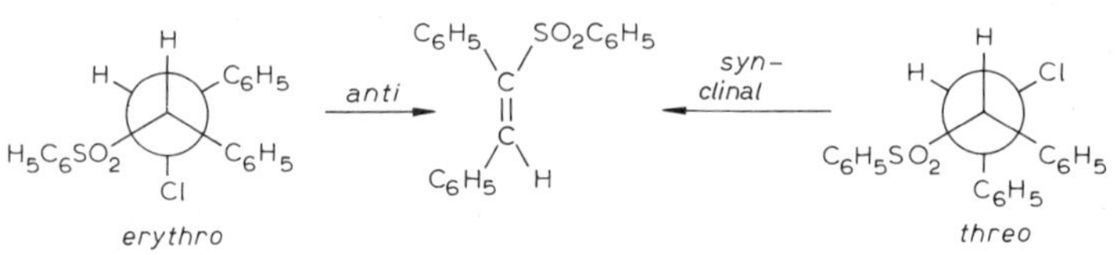

(80)

with *anti*-elimination from the *erythro* chloride being much faster than *syn-clinal*-elimination from the *threo* compound. Cristol and Pappas[210] claim a carbanion "*syn*"-elimination and E2 *anti*-elimination but more extensive kinetic data is needed to justify the former mechanism.

Electrostatic effects can influence the stereochemistry of the elimination (81).

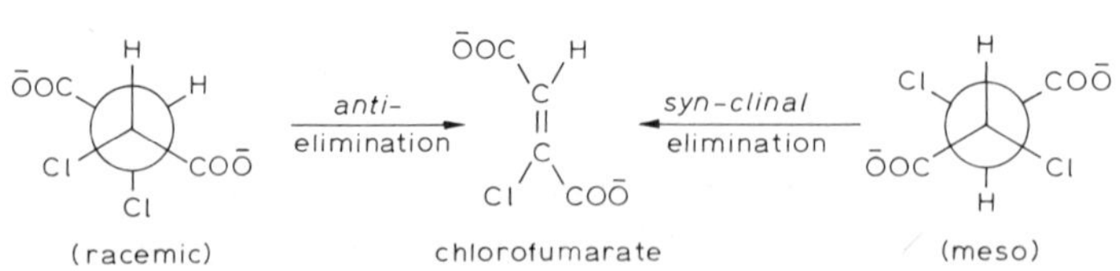

(81)

In aqueous alkaline solution both the racemic- and *meso*-1,2-dichloro-succinnic acids yield chlorofumarate ion although *anti*-elimination from the *meso* compound should give chloromaleate[211]. In the absence of beta hydrogen exchange, the "*syn*"-elimination was attributed to a carbanion-like E2 transition state in which the carboxylate groups acted as reservoirs for the electron pair partially released by the stretching of the C–H bond, by virtue of the potential prototropy they impart into the system. Electrostatic repulsion between the two carboxylate groups causes them to separate to a maximum thus favouring a transition state leading to the chloromaleate. This factor is depressed in neutral solution and the E2 reaction of the *meso* substrate yields comparable amounts of chlorofumarate and chloromaleate.

(*b*) *Cyclohexyl systems*

The cyclohexyl systems have proved extremely instructive in stereochemical investigations of bimolecular beta-eliminations. By the use of stereospecific synthesis, the hydrogen and leaving group can be introduced into known fixed positions due to the rigidity of the ring systems, and the mechanism can be framed to fit the structure rather than as in the acyclic series where the conformation of a molecule can be changed by free rotation in order to fit the mechanism. The surest models are multifused ring systems, joined by *trans* ring-junctions which are practically inflexible and assignment of the conformation of a group can be made with a high degree of certainty. In the *cis*-decalyl and cyclohexyl series, easier inversion from one chair form to another casts doubt on the eliminating conformation of the molecule. Experimental results indicate a great preference for an axial conformation for the leaving group in beta-eliminations. The 3-α-isomer (82) undergoes Hofmann degradation to the olefin in excellent yield, whereas the 3-β-epimer, in which the trimethylammonium group is equatorial, is converted into the substitution product, the dimethylamine[212]. In the cyclohexyl series,

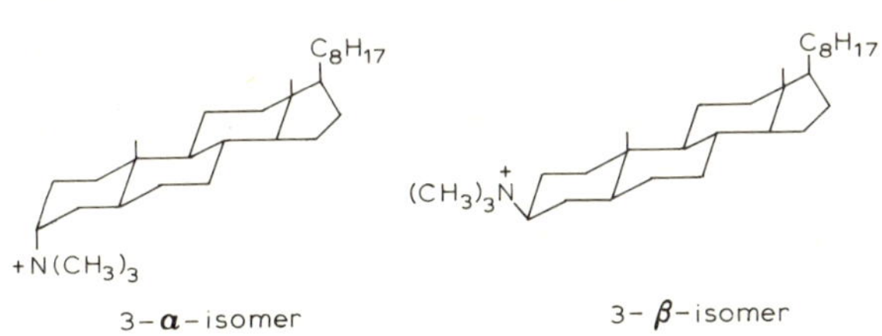

(82)

due to large non-bonded interactions in the axial conformation, a bulky substituent prefers to exist in an equatorial conformation. For the *t*-butyl and trimethylammonium groups an energy difference of 5–6 kcal.mole^{-1} between the two conformations has been estimated[213], although slightly lower values have also been suggested[214,215]. Consequently, such bulky substituents tend to exist predominantly in the equatorial position, thereby essentially freezing the conformation of the cyclohexane ring. Product analysis (83)[216] and rate studies

(84)[217] of eliminations from appropriately substituted *t*-butylcyclohexanes again illustrate the preference for an axial conformation of the leaving group.

	Base	*% Elimination*	*% Substitution*
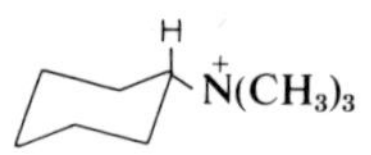	KOBu^t/*t*-BuOH	7	93
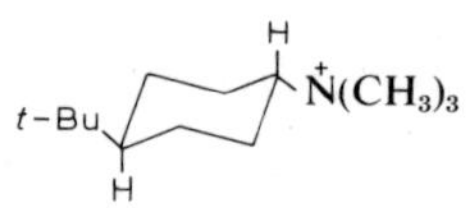	KOBu^t/*t*-BuOH	—	100
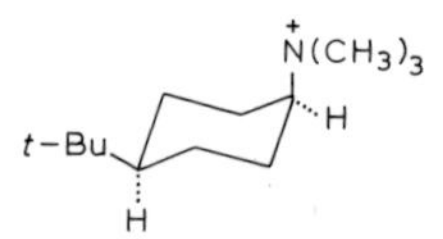	KOBu^t/*t*—BuOH	92	8

(83)

(84) Base, piperidine

	10^{3}k(elimination)	*($k_{axial}/k_{equatorial}$)**
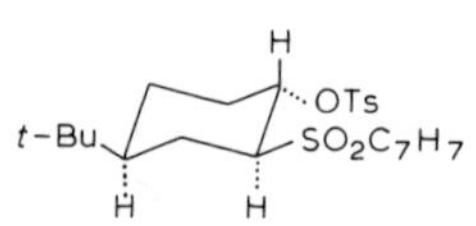	1.3	5
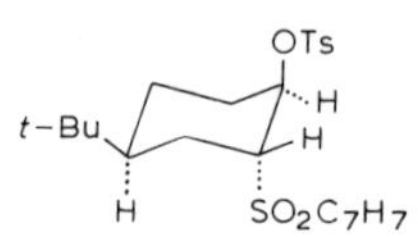	6.5	
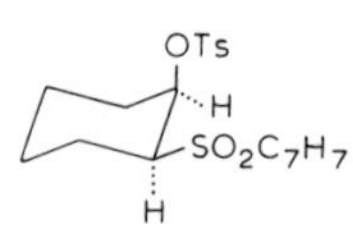	79	32.5
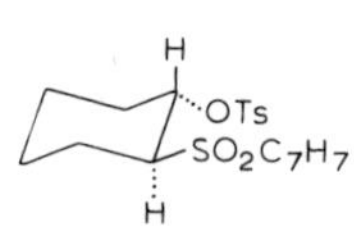	2.43	

* Ratio of rates of corresponding isomers in which the leaving group, OTs, is axial and equatorial.

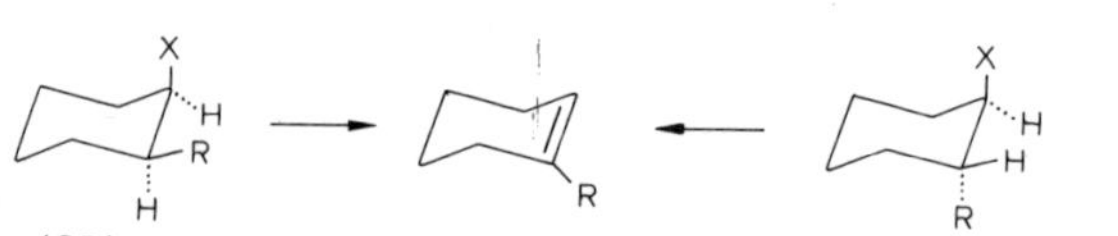

(85) *anti*-elimination from *cis*-isomer

(86) *syn-clinal*-elimination from *trans*-isomer

(R not H but alkyl, aryl etc.)

Restricting considerations to cases in which the leaving group has an axial conformation, *anti*-elimination from the *cis* isomer (85) is diaxial and coplanar whereas elimination from the *trans* isomer (86) is *syn-clinal* and is axial–equatorial. In the latter case it is necessary for the two larger groups to occupy the axial positions in the reactive conformation, which will consequently be sparsely populated. Neglecting elimination from conformations with an equatorial leaving group and assuming that elimination occurs at the same rate from both *syn-clinal* and *anti* orientations, McLennan[34] has shown by assuming a free energy difference between conformations (85) and (86) of 3.5 kcal.mole^{-1}, that elimination from the *cis* isomer is about 370 times greater than that from the *trans* isomer. Considering the additional advantage of coplanarity, it is not surprising that *anti*-elimination is predominant in this series. Transition states for *syn*-elimination are only available when the less stable boat conformations are attainable and the greater energy of these appears to accentuate the preference for *anti*-elimination[218].

The kinetics of dehydrohalogenation of five of the possible eight isomeric benzene hexachlorides have been measured[79,219–221]. The beta isomer, which alone does not possess a pair of hydrogen and chlorine atoms situated *trans* on adjacent carbon atoms [see (87) below], reacts 10^4 times more slowly than the other substrates. In deuteroethanol containing ethoxide, the unreacted beta isomer recovered after 50% reaction was found to contain 0.08% deuterium. This observation led to the postulation of the carbanion mechanism[222]. Both the beta and delta isomers eliminate to give similar yields of the corresponding products, and it was suggested that initial ionisation to a carbanion was followed by subsequent inversion and reprotonation to give the delta isomer, which subsequently underwent normal *anti*-elimination, *viz.*

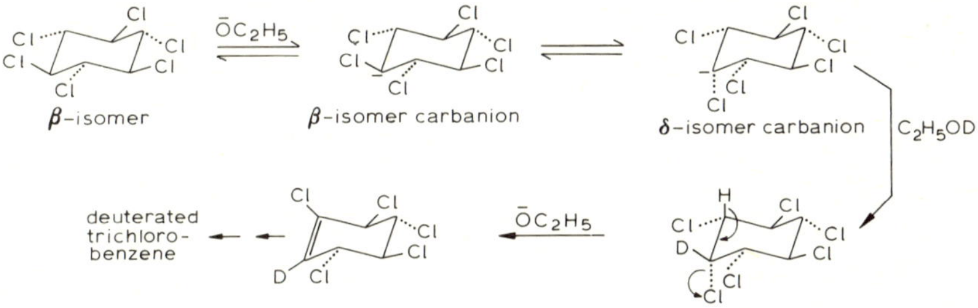

(87)

Such a mechanism requires that the major product, 1,2,4-trichlorobenzene should contain deuterium when the reaction is performed in a deuterated medium[223], but the amount of label found[224] was no greater than predicted from the rate of exchange of the product under the reaction conditions[225]. Presumably fairly extensive double-bond character is exhibited in the transition state for bimolecular elimination from the benzene hexachlorides and the abnormally high activation energy for the beta isomer[220] is most plausibly attributed to the lack of coplanarity in the *syn-clinal*-elimination or the high energy and low population of the boat conformer required for *syn*-elimination. Alternatively, a sparsely populated conformation for a concerted *syn-clinal*-

elimination with a favourable axial chlorine is likely due to the necessity of having the six larger substituents in the axial conformation.

Neomenthyl derivatives eliminate to give a mixture of 2- and 3-menthenes but menthyl derivatives only yield 2-menthene under E2 conditions as *anti* stereospecificity is only possible in this case (88)[31,32,226].

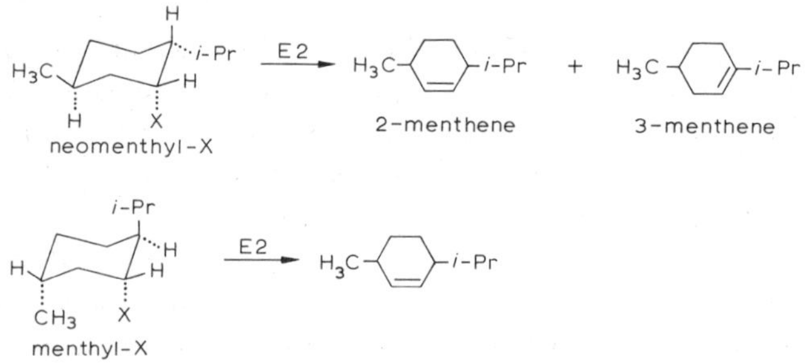

(88)

The accuracy of the analytical method to detect small quantities of the minor product determines the stringency with which stereospecificity is established and as a few per cent of 3-menthene might have been detected, a ratio of 50 : 1 as the lower limit for *anti* relative to that of *syn-clinal* stereospecificity can be estimated[14].

In the presence of beta acid strengthening substituents, the stereochemical preference for *anti*-elimination in cyclohexyl systems becomes less marked. Thermal degradation of *trans*-2-phenylcyclohexyltrimethylammonium hydroxide yields 1-phenylcyclohexene. Originally initial formation of 3-phenylcyclohexene followed by isomerisation was proposed[227], but the unconjugated olefin is stable under the reaction conditions[228]. Kinetic studies show that the *cis*-2-phenyl isomer (*anti*-elimination) is only 133 times more reactive than the *trans* isomer[229] (*syn-clinal*-elimination); but the activated *syn-clinal*-elimination is more favourable than an unactivated *anti*-elimination. Deuterium tracer experiments show the absence of the ylid or alpha-elimination[152] and in view of the much smaller nitrogen isotope effect for elimination from the *trans* isomer than the *cis* isomer (see Table 6, p. 202) the apparent *syn-clinal*-elimination may be either an E1cB or an E2 reaction close to the carbanion extreme. It may be informative to study the Hammett reaction constant and the primary kinetic hydrogen isotope effect for this reaction, as the nitrogen isotope effect shows that C_α–X bond breaking is less advanced in the transition state for the *syn-clinal*-elimination than for any other known E2 reaction.

Activated *syn-clinal*-elimination dominates over unactivated *anti*-elimination in the 2-*p*-tolylsulphonylcyclohexyl tosylate series[230], but although both isomers give the same product the activated *anti*-elimination occurs 435 times faster (89)[231].

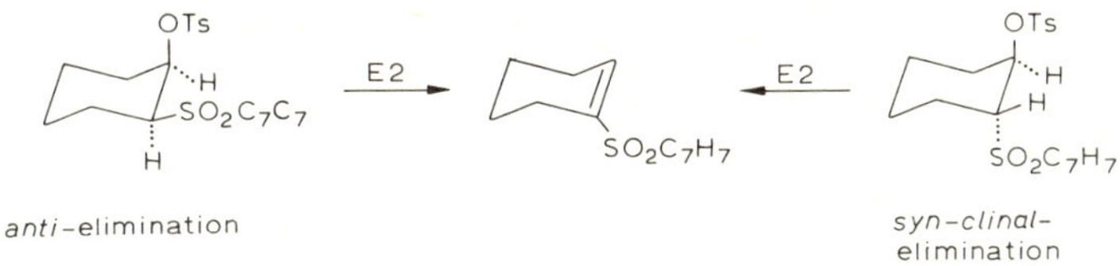

(89)

0.324	$k_{N(C_2H_5)_3}/k_{N(CH_3)_3}$	0.068
5220	$k_{\bar{O}H}/k_{N(CH_3)_3}$	308

Neither elimination follows the carbanion mechanism as general base catalysis is observed and base-catalysed deuterium exchange from the cyclohexyl *p*-tolylsulphones is 10^{-3}–10^{-5} times slower than elimination[232], indicating assistance in the elimination by C_α–X bond breaking and hence an E2 reaction. The catalytic efficiencies of the bases employed afford interesting results (see above). Despite the predictions from the Brönsted catalysis law, triethylamine reacts more slowly than trimethylamine due to steric hindrance to proton abstraction. This is more serious in the *syn-clinal* elimination due to the closer proximity of the bulky leaving group. Electrostatic arguments were proposed to explain the observed change in selectivity of hydroxide relative to trimethylamine in the two eliminations. The electronegative oxygens of the tosylate group exert an unfavourable repulsive force on the hydroxide ion but a favourable attractive force on the partially neutralised amine in the transition state for *syn-clinal*-elimination[231].

In one case (90) *syn-clinal*-elimination occurs faster than *anti*-elimination

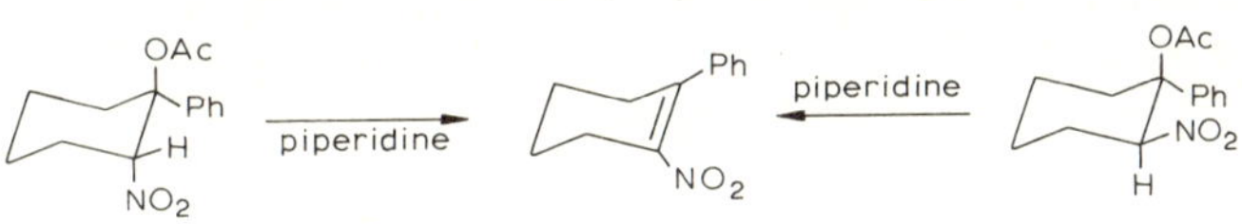

(90)

by a factor of 3.5 times[233]. A change of leaving group to bromoacetate increases the rate difference to 4.3 and propionate causes a decrease to 2.8. Both acetate eliminations have the same activation energy, the primary kinetic isotope effect is 4.9 and the Hammett reaction constant at the alpha carbon is 1.45. The base was piperidine in chloroform–ethanol mixtures and the rate of elimination was found to increase markedly with increasing polarity of the solvent or on the addition of salts. Bordwell *et al.*[233] invoked the E1cB mechanism, suggesting steric assistance to carbanion formation in the *syn-clinal*-elimination and noting that the strongly electron-withdrawing nitro-group would facilitate this mechanism. However, in view of the high isotope effect and positive ρ value, proton abstraction is undoubtedly rate-determining and elimination from the conjugate base is certainly a fast process. The possible involvement of a concerted mechanism involving an *aci*-nitro group may be worth consideration (91).

References pp. 350–364

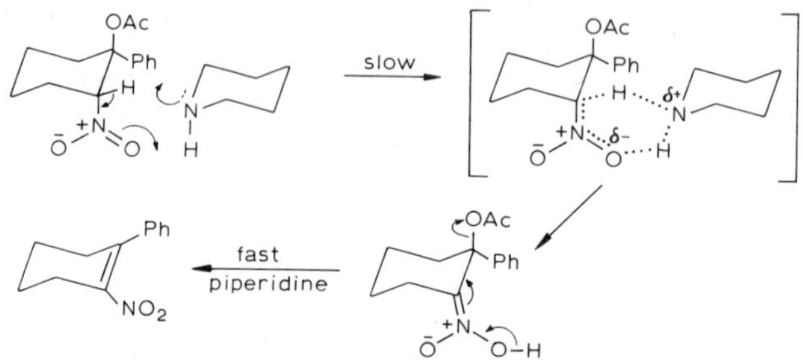

(91)

Such an intermediate could arise in a rate-determining internal proton transfer. The greater rate of apparent *syn-clinal*-elimination could be explained in terms of easier proton abstraction from an equatorial than an axial conformation. A similar cyclic transition state is not possible with *N*-methylpiperidine as the base and this suggests an obvious experiment.

(*c*) *Cyclopentyl systems*

Five-membered alicyclic ring substrates eliminate more readily than the six-membered analogs (92, below)[234]. This may be due to the greater thermodynamic stability of cyclopentene (heat of hydrogenation 1.7 kcal/mole less than cyclohexene[235]) or the reduction in steric interactions on elimination in the cyclopentene series due to a decrease in the eclipsing interactions between substituents on adjacent carbons[236]. Alternatively, a lower population of the favourable conformation for elimination from the six-membered systems, due to the preference of large axial groups to occupy the equatorial position, accounts for the observed differences in reactivity.

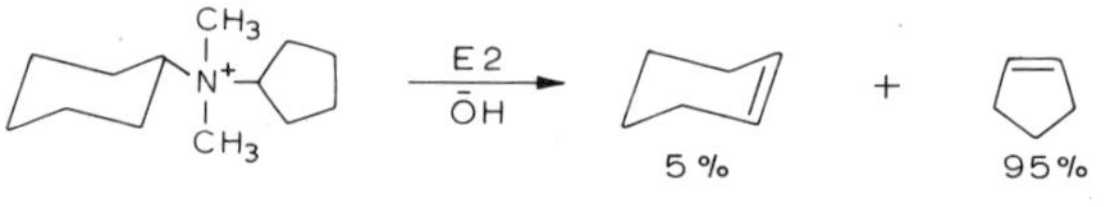

(92)

In the cyclopentyl series, *syn*-elimination occurs from a planar transition state, but *anti*-elimination is slightly distorted from a dihedral angle of 180°. Consequently, the preference for *anti*-elimination is less marked in the five- than in the six-membered ring systems (Table 12). With the small neutral base trimethylamine, electrostatic forces of attraction between the sulphonate ester and the partially neutralised base balance the normal preference for *anti*-elimination and the reaction is almost non-stereospecific. None of the reactions follows the carbanion mechanism, as general base catalysis is observed[231] for the 2-*p*-tolylsulphonylcyclopentyl tosylate elimination and the studies on 2-phenylcyclopentyl tosylates revealed large isotope effects (k_H/k_D) and ρ values smaller than in the 2-phenylethyl series (see Table 9, p. 209)[178].

TABLE 12

COMPARISON OF THE RATIOS OF THE RATE COEFFICIENTS OF ELIMINATION FROM *cis* AND *trans* ISOMERS IN THE CYCLOHEXENE AND CYCLOPENTANE SYSTEMS

Substrate	*Base*	$(k_{anti}/k_{syn})^{a}$	$(k_{anti}/k_{syn\text{-}clinal})^{b}$	*Ref.*
2-*p*-Tolylsulphonyl-cycloalkyl tosylate	$N(CH_3)_3$	1.2	22	231
	$N(C_2H_5)_3$	6.5	115	231
	$\bar{O}H$	20	435	231
2-Phenylcycloalkyl tosylate	$\bar{O}Bu^t$	9.1	10^4	178

[a] Cyclopentane system.
[b] Cyclohexane system.

(*d*) *Bridged ring systems*

In bridged ring systems, the conformation of a molecule can be rigidly fixed. However, the 1,2-dichloroacenaphthenes (93a) are apparently flexible enough to permit an almost planar *anti*-elimination, which proceeds 750 times faster than *syn*-elimination from the *trans*-dichloro substrate[237]. In the dihydrophenanthrene series (93b) *anti*-elimination again predominates, despite the lack of coplanarity of the benzene rings in the transition state, and *syn*-elimination from the *trans*-dichloroisomer is slower than dehydrohalogenation from *β*-benzene hexachloride[238]. Alkaline dehydrochlorination of *trans*-11,12-dichloro-9,10-ethanoanthracene (93c) proceeds eight times faster than for the

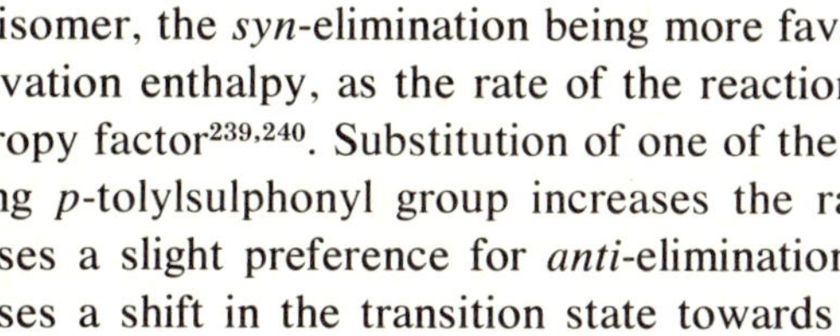

(93)

cis isomer, the *syn*-elimination being more favourable despite having a higher activation enthalpy, as the rate of the reaction is controlled by a favourable entropy factor[239,240]. Substitution of one of the chlorines by the acid strengthening *p*-tolylsulphonyl group increases the rate of elimination by 10^6, and causes a slight preference for *anti*-elimination[249]. The activating substituent causes a shift in the transition state towards the carbanion extreme, where bond changes are relatively uncoupled and the stereochemical consequences of geometric rigidity are overcome. An E1cB mechanism has also been suggested.

In the rigid 2,3-dihalogenonorbornane system, the only possible eliminating conformations in the substrate are planar *syn* and *anti-clinal*. Rate studies on the dibromo and dichloro substrates[241,242], the use of deuterium tracers[241–243] and the relative ease of elimination of bromine to chlorine suggest that the preferred conformational requirements for elimination are *exo-syn* > *anti-clinal* (with *exo*-hydrogen) > *endo-syn* > *anti-clinal* (with *endo*-hydrogen) (Table 13). Elimination is more facile with an *exo* hydrogen as steric factors inhibit the abstraction of an *endo* proton and despite the unfavourable isotope effect and lack of coplanarity of the eliminating atoms, *exo*-deuterium is abstracted six times faster than the *endo*-hydrogen from *exo*-2,3-dideutero-2-chloronorbornane[244]. The magnitude of the primary kinetic isotope effects (k_H/k_D) for *exo*-, *syn*- and *anti-clinal*-dehydrobrominations are very similar and other than illustrating a rate-determining proton transfer is involved, they afford little information concerning the mechanism[245]. An E2 rather than an E1cB reaction for the elimination from these dihalogenosubstrates seems more likely in view of the preference for coplanarity of the eliminating groups. Elimination from the hydroxide of *exo*-2-trimethylammonium-3-deuteronor-

TABLE 13

RELATIVE RATES OF ELIMINATION FROM 2,3-DIHALOGENONORBORNANES[241,242]

No.	*Substrate*	*Relative rate*[a]	k_H/k_D[b]	*Temp. (°C)*[c]
1	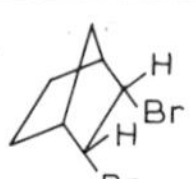	2.8	3.4±0.1	127
2	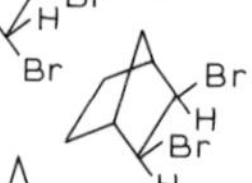	1.0	3.6±0.1	96
3	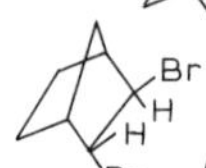	86		
4[d]		4.2		
5[e]	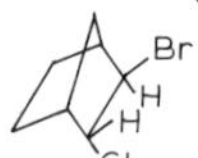	28		
6	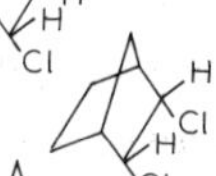	0.02		
7		1.6		

[a] Relative rate of elimination at 110°C in *n*-pentanol containing sodium pentoxide.
[b] Deuterated substrate labelled fully at both the 2 and 3 positions, with standard deviation quoted.
[c] Temperature of isotope effect.
[d] HCl mainly eliminated.
[e] HBr mainly eliminated.

bornane proceeds exclusively by *syn*-elimination[91,246,247] and the small isotope effect ($k_H/k_D = 1.86$ at 50°C) when considered in the light of the results in the 2-phenylethyl series (Section 2.2.1(c)) suggests a concerted process with a transition state in which proton transfer to the base is more extensive than in the eliminations from the dihalo derivatives. Despite the coplanarity, the ylid mechanism contributes only 6% of the total path for decomposition. Perhaps the ylid mechanism may be more competitive in elimination from the *endo* substrate in which proton abstraction from the beta carbon is more hindered, especially if *t*-butoxide is used as the base rather than hydroxide which promotes mainly substitution in this case[248]. Even in the bicyclooctane series (94) the ylid mechanism accounts for only 13% of the total *syn*-elimination[246]. In all of these reactions elimination involves a secondary

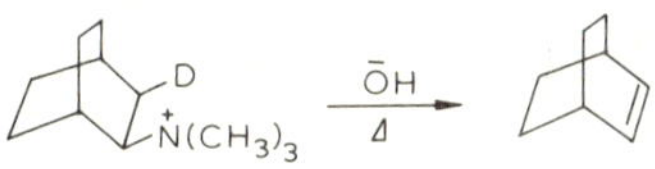

(94)

(95)

rather than a tertiary hydrogen, as the latter would result in double-bond formation at a bridgehead, in violation of Bredt's rule[249]. In small rigid cage-like structures, it is not possible to permit two C–C bonds linked at the 1-position to become co-planar with the rest of the olefin system. The inability to form a planar carbonium ion also depresses unimolecular elimination from (95), although some olefin, which may be rearranged, can be isolated in the presence of Lewis acids[250]. Bredt's rule is relaxed in larger ring systems (*e.g.* bicyclononene) in which any strain in the products is distributed over a larger number of bonds[251].

(*e*) *Competitive anti- and syn-eliminations*

Hofmann degradations of alicyclic trimethylammonium ions to give *cis*- and *trans*-cycloalkenes (96)

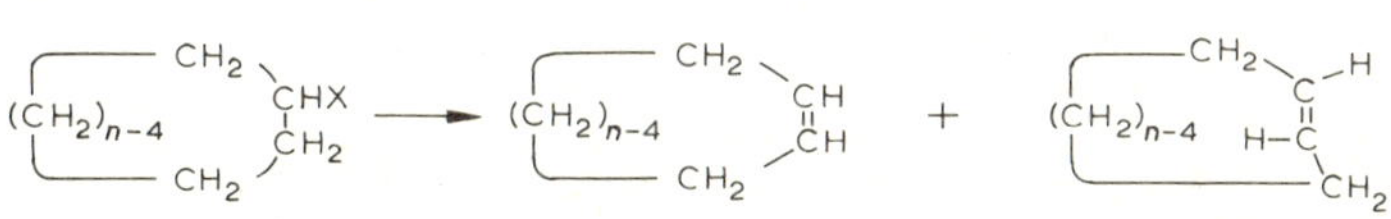

(96)

have for some time been considered as *anti*-eliminations[252]. Bonding-strain considerations permit only the formation of the *cis*-olefin for the small-membered rings but for $n > 8$ the *trans* isomer is predominantly formed[253–255]. This result is surprising as cyclooctene, cyclononene and cyclodecene are much more thermodynamically stable in the *cis* form and the *trans* arrangement is only slightly favoured for the larger ring systems[256]. The stereochemistry of the competing Hofmann eliminations offered an explanation, and as a mechanistic criterion the variation in rate with ring size was compared to a similar profile for a reaction of known stereochemistry[256–258]. Similar profiles (rate *versus* ring size) are predicted for different reactions which possess similar salient steric features. The well-characterised intramolecular *syn*-elimination of alicyclic amine oxides (Cope reaction)[259] was chosen as the reference reaction (2, p. 163). The rate profiles for the formation of *cis*- and *trans*-cycloalkenes from the Cope and Hofmann degradations[258,260,261] are illustrated in Fig. 7. The two curves for *trans*-olefin formation are strikingly

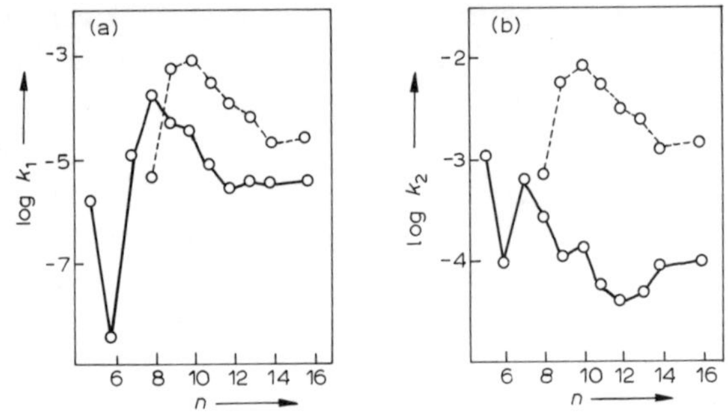

Fig. 7. The effect of ring size on the rates of formation of *cis*- and *trans*-cycloalkenes. (a) From cycloalkyldimethylamine oxides in *t*-butyl alcohol at 70°C; (b) from cycloalkyltrimethylammonium chlorides and potassium *t*-butoxide in *t*-butyl alcohol at 55°C. *n* is the number of carbon atoms in the alicyclic ring. ——— *cis*-cycloolefins; ----- *trans*-cycloolefins. (Reproduced with permission from Sicher *et al.*[258].)

similar but those for the *cis*-olefin differ considerably. In fact rates of formation of *cis*-olefins correspond closely to the known dependence of rate on ring size for S_N2 reactions of cycloalkyl bromides with potassium iodide in acetone[262]. These findings suggest that *trans*-cycloalkenes arise by a *syn*-elimination and *cis*-cycloalkenes by an *anti*-elimination in the Hofmann degradation. A similar approach was subsequently used to show that in elimination from cycloalkyl bromides both olefins arise by an *anti*-elimination when ethanol–ethoxide is the reaction mixture, but in *t*-butanol containing *t*-butoxide, the *trans*-cycloalkenes are formed by a *syn*-elimination[172]. In this latter case, an explanation involving an ion-paired intermediate was invoked, *viz.*

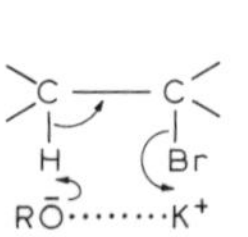

(97)

bearing in mind the poorer dissociating power of the medium. The same reasoning has recently been applied to explain the increase in *syn*-elimination relative to *anti*-elimination of *trans*-3-deuterobicyclo[2,2,2]oct-2-yl toluene *p*-sulphonate with potassium *t*-butoxide as a change to a less dissociating solvent was made[680].

To substantiate their stereochemical findings, Závada *et al.*[263] used a deuterium labelling technique. Stereospecifically labelled 1,1,4,4-tetramethyl-7-cyclodecyltrimethylammonium ions were selected as the most appropriate models, the *gem* dimethyl groups acting as positional markers in the olefin analysis. From a knowledge of the composition of the olefin fraction, the isotope effect can be calculated (98, below). The results in Table 14 undoubtedly illustrate that the same beta "hydrogen" is eliminated predominantly from a single substrate to give both olefins. Such a result can only be accommodated by competing *syn* and *anti* processes. The E2 reactions of the corresponding tosylates with potassium *t*-butoxide in dimethyl formamide showed a similar duality of mechanism.

TABLE 14

OLEFINS FORMED FROM THE HOFMANN DEGRADATION OF 1,1,4,4-TETRAMETHYLCYCLODECYLTRIMETHYLAMMONIUM IONS IN METHANOLIC POTASSIUM METHOXIDE AT 140°C[263]

		A 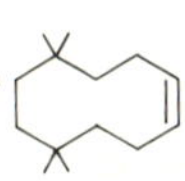				B 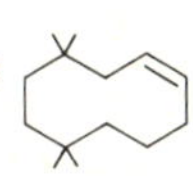	
		trans		*cis*		*trans*	*cis*
No.	*Substrate*	%	k_H/k_D[a]	%	k_H/k_D[a]	%	%
1	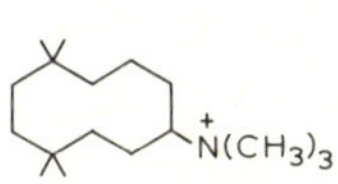	48.9		5.92		36.7	8.48
2	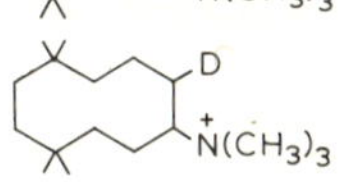	47.3	1.1	5.91	1.0	38.3	8.59
3	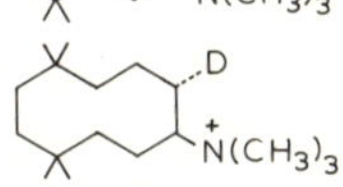	34.4	2.3	3.24	2.8	51.2	11.2

[a] Corrected for 12% non-deuterated material in 3, and 6% in 2. Dideuterated material between 0.5 and 1.0%. For calculation of the isotope effect, see eqn. (98).

For the formation of *trans* A from isomer 2 (see Table 14)

$$\frac{k_H}{k_D} = \frac{(\%\ \textit{trans}\text{-A from 1})/(\%\ \textit{trans}\text{-B from 1})}{(\%\ \textit{trans}\text{-A from 2})/(\%\ \textit{trans}\text{-B from 2})} \tag{98}$$

The preference for *syn*-elimination leading to the *trans*-olefins is not too difficult to understand in the alicyclic series. The use of models clearly indicates that the non-reacting alkyl residue exerts a severe limitation on the freedom of rotation of the larger leaving groups and hinders the approach of the attacking base to the beta proton in the conformation for *anti*-elimination leading to the *trans*-cycloalkene. Such limitations are greatly reduced in *syn*-elimination and in *anti*-elimination to yield the *cis*-olefins.

The duality of stereochemical mechanism persists in the Hofmann elimination from the beta deuterium-labelled derivatives of the *threo* and *erythro* isomers of the compound (99), under a variety of reaction conditions[264].

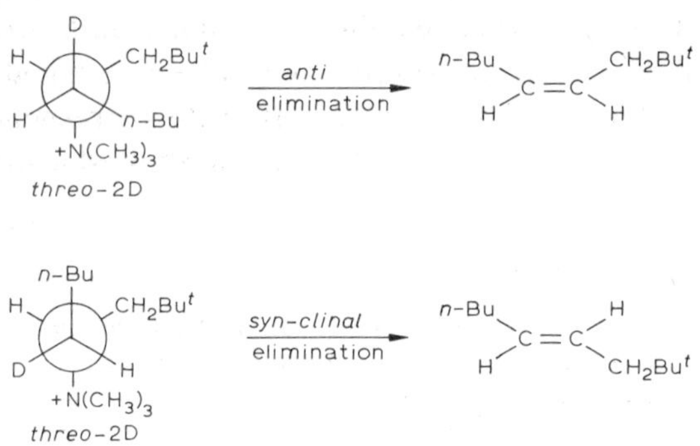

(99)

The *cis*- and *trans*-olefins are both formed with loss of deuterium from the *threo* isomer but with retention from the *erythro* isomer and the ylid mechanism is excluded by the absence of tracer in the eliminated trimethylamine. Elimination from *threo*- and *erythro*-3-deuterated-2-butyl-*p*-toluene sulphonates and trimethylammonium ions shows "normal" *anti*-stereospecificity[265]. However, elimination from *threo*- and *erythro*-3-deuterated-4-hexylammonium ions [(100) below] shows duality of mechanism[198c]. With methoxide/methanol and *n*-butoxide/*n*-butanol both isomeric 3-hexenes are formed by *anti*-eliminations, but with the larger bases, *t*-butoxide and *t*-pentoxide, the *trans*-olefin is formed by "*syn*"-elimination and the *cis*-olefin by *anti*-elimination. Secondary bases induce intermediate behaviour. Assuming elimination occurs from lower energy staggered rather than eclipsed conformations, the change in stereochemistry can be explained in terms of steric inhibition of solvation of the leaving group and steric hindrance to beta proton abstraction. *cis*-3-Hexene always arises from an *anti*-elimination of

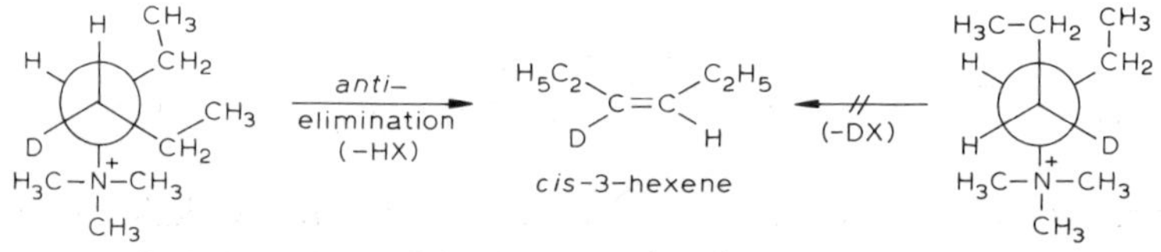

erythro-3-deutero-4-hexyltrimethylammonium ion

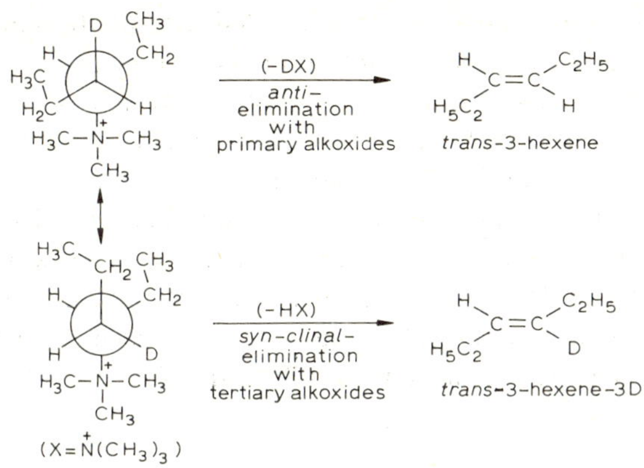

(100)

HX rather than a *syn-clinal*-elimination of DX from the *erythro* substrate (100) as in the former conformation the beta hydrogen is only flanked by one large substituent and the base approach is less impeded than in the alternative conformation. On the other hand, *trans*-3-hexene is formed by loss of DX in an *anti*-elimination with primary alkoxides, but with the larger base solvent systems steric inhibition to solvation and proton abstraction offsets the advantages arising from coplanarity and *syn-clinal*-elimination becomes dominant. The size of the reactants rather than the carbanion character of the transition state is the salient factor as *n*-butyl and *t*-butyl alcohol induce contrasting stereochemical behaviour. Recent results of Sicher *et al.*[681] on the Hofmann elimination leading to *cis*- and *trans*-5-decenes afford the same conclusions as the above results of Saunders and Bailey[198c]. These discoveries, possibly one of the major advances in the study of elimination reactions in recent years, underline the danger in generalising the most favourable stereochemistry of all alicyclic and acyclic compounds from studies on a few select models.

Coke *et al.*[266] have challenged the Sicher–Závada generalisation that *cis*-cycloalkenes are derived from *anti*-elimination and *trans*-cycloalkenes from *syn*-elimination. They agree with the latter deduction but regard *cis*-cycloalkene formation as more complex, arising from both *syn*- and *anti*-elimination. The *cis*-olefins were analysed from the Hofmann degradation of alicyclic stereospecifically deuterated trimethylammonium hydroxides, *viz.*

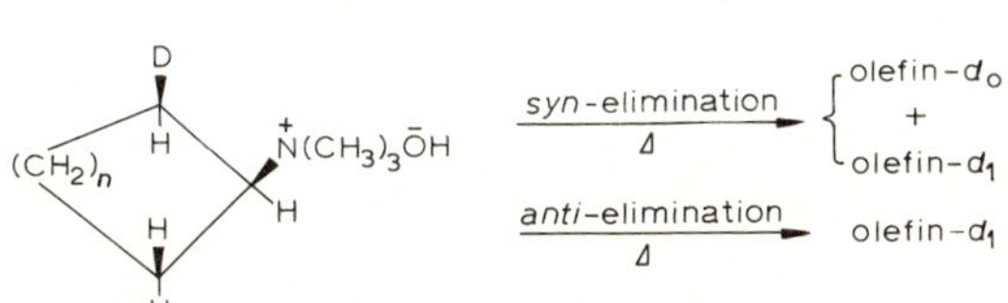

(101)

Since only one of the two *cis beta* positions was labelled with deuterium, the isotope effect had to be known in order to interpret the deuterium analysis of the olefins. A value of 1.84 was assumed, this being the average of the isotope effects (k_H/k_D) for *syn*-elimination from the *exo*-norbornyl system[246] (1.86), the 3,3-dimethylcyclopentyl system[267] (1.71) and the cyclooctyl system[266] (1.94). In view of the approximate nature of the calculation, secondary isotope effects and variation of the primary isotope effect with temperature were ignored. A change from 1.84 to 1.71 for the *syn* (k_H/k_D) does not materially affect the conclusions. The extent of reaction following the *syn*-elimination in non-deuterated amines was calculated using eqn. (102) and typical results are listed below this equation.

$$\frac{k_{syn}}{syn\ (k_H/k_D)} + k_{syn} + k_{anti} = 100$$

$$\text{and} \quad \frac{k_{syn}}{syn\ (k_H/k_D)} = \%\ \text{olefin-}d_0 \qquad (102)$$

Ring size	*% Olefin-d_1*	*% Olefin-d_0*	*% Syn-elimination in nondeuterated substrate*
4	69	31	90
5	86	14	46
6	99	1	4
7	89	11	37

The olefin distribution varied slightly with reaction conditions but the results show the competitive nature of *syn*- relative to *anti*-elimination in these systems. Only in the cyclohexyl system is *anti*-elimination clearly predominant. Isotope-labelling studies obviously afford a more accurate method of estimating stereochemistry of elimination reactions than do rate profile approaches. Subsequent work questioned the validity of using dimethyl-substituted compounds as models for the parent alicyclic substrates as the substituents can influence the stereochemical outcome of the reaction[267], *viz*.

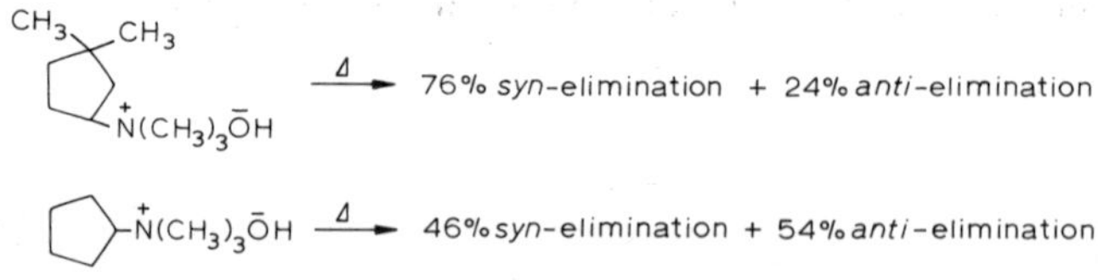

(103)

As a finale to this work, it would be worthwhile completely to exclude the admittedly unlikely possibility of the ylid mechanism[268] as the explanation of *syn*-elimination in the alicyclic series. In any event, excepting the cyclohexyl series it is no longer safe to assume a preference for *anti* stereospecificity for Hofmann degradations and elimination reactions in solvents of low dielectric constant.

2.3.2 The unimolecular mechanism

In principle, a planar carbonium ion should lose a beta proton from either side of the plane with equal facility and consequently there should be no stereoselectivity in E1 reactions. This simple interpretation can be distorted, however, in solvents which do not favour ionic dissociation, or if neighbouring group participation in the ionisation is afforded by a beta hydrogen or another group.

The ratio of substitution to elimination products in the solvolysis in 80% aqueous ethanol of a series of *t*-butyl and *t*-amyl substrates is little influenced by a change of leaving group from Cl to Br, I and $\overset{+}{S}(CH_3)_2$[29,30]. A similar insensitivity is demonstrated in the hydrolysis in water of *t*-butyl compounds[269]. These observations seemingly suggest that the origin of the carbonium ion is of little significance with respect to its mode of reaction. However, in less dissociating media, such as ethanol and acetic acid, the product distribution is highly sensitive to the leaving group and the counter ion obviously greatly influences the fate of the carbonium ion, elimination being favoured by the smaller negatively charged gegenions and substitution by the neutral leaving group, dimethyl sulphide. Using the dimethylsulphonium substrate as a model for the reaction of a dissociated carbonium ion, it was estimated that most of the reaction of *t*-butyl chloride proceeded through a carbonium ion-pair[269]. Solvolytic displacement of 2-octyl tosylate by acetate occurs with 100% inversion in glacial acetic acid, again confirming ion-paired intermediates[270].

A change to a less basic solvent can influence the stereochemistry of unimolecular elimination. In nitrobenzene or acetic acid, *erythro*-3-deutero-2-butyl tosylate yields deuterium-free *cis*-2-butene and monodeuterated *trans*-2-butene, suggestive of a *syn*-elimination (104). In more basic solvents, such as water, acetamide and 80% aqueous ethanol, the labelling of olefins is reversed and *anti*-elimination is followed. The change in stereochemistry is attributed to the relative basicities of the tosylate anion and solvent[271] and hence the tendencies of these species to accept a beta proton from the carbonium ion (105).

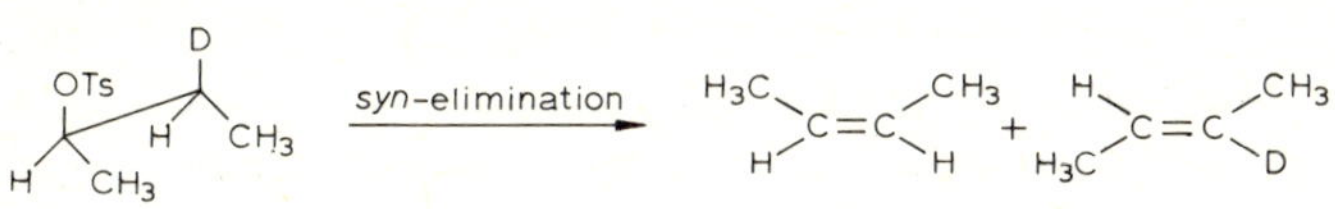

(104)

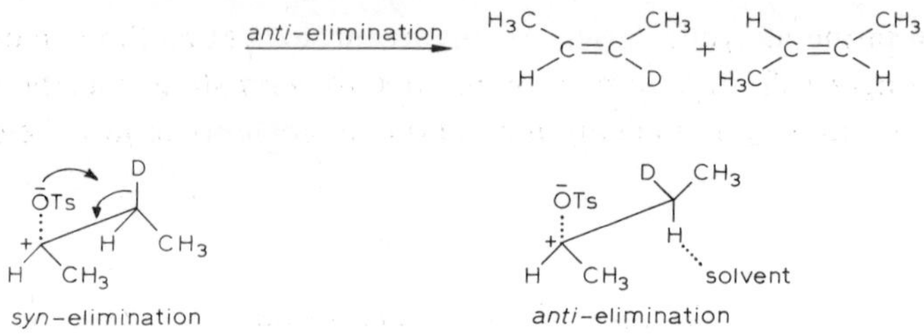

(105)

A similar explanation has been suggested to account for the variation in the product distribution from the decomposition of the 2-phenyl-2-butyl cation generated from a variety of substrates[272]. In solvolytic elimination reactions of 1,1,4,4-tetramethylcyclodecyl-6- or -7-tosylates *syn*-elimination appears to predominate in both acidic and basic solvents, an ion pair mechanism being suggested to account for the observed results[273]. It is obviously unsound to generalise when considering stereochemistry of E1 reactions.

If planar carbonium ions were the intermediates in E1 reactions in the cyclohexyl series, menthyl and neomenthyl compounds should give the same product ratios. However, the olefin distribution is quite different in the two E1 processes and the stereospecificity is less marked than in the E2 reactions of these substrates (Table 15). Whereas the concerted eliminations always show *anti* stereospecificity, the unimolecular eliminations only exhibit this preference when a tertiary beta hydrogen is *trans* to the ionising group (*e.g.* neomenthyl series). Possibly in this case the tertiary hydrogen aids ionisation by forming a type of non-classical bridged intermediate, *viz.*

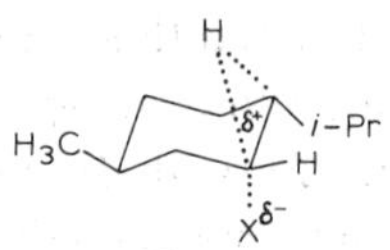

(106)

Participation of this kind in the rate-determining step should cause considerable weakening of the C–H bond and result in an isotope effect primarily due to variations in the zero-point energy associated with C–H bending vibrations[272]. In the acyclic series, the substitution of deuterium for hydrogen slows the ionisation of tosylates by a factor of 1.5–2.3 (refs. 274, 275) considerably larger than normal beta isotope effects in which participation is not possible. Hyperconjugative postulates also favour the reaction with the lighter isotope. For effective participation by either neighbouring group or hyperconjugative effects, coplanarity of the beta carbon–hydrogen and carbon–leaving-group bonds would seem preferable. Certainly a greater isotope effect is observed with the *anti* orientation of the axial beta hydrogen in the solvolysis in 50% aqueous ethanol of *cis*-4-*t*-butylcyclohexyl-*p*-bromobenzene sulphonate

TABLE 15

BIMOLECULAR AND UNIMOLECULAR ELIMINATION FROM MENTHYL AND NEOMENTHYL COMPOUNDS

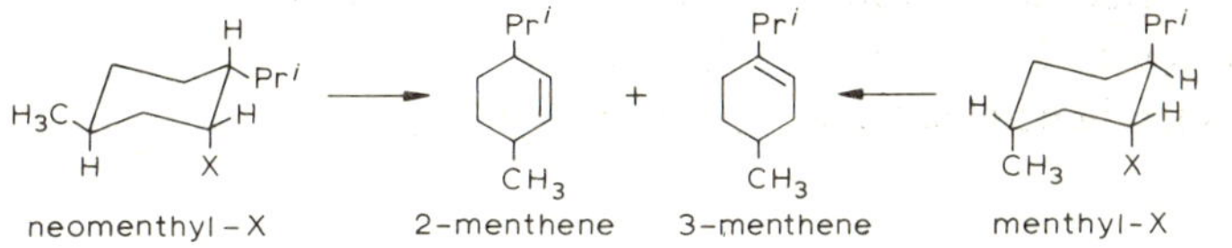

Reaction type	Reaction product	X					
		$\overset{+}{N}(CH_3)_3$[a]	*Cl*[b]	*OTs*[c]	$\overset{+}{N}(CH_3)_3$[a]	*Cl*[b]	*OTs*[c]
E1	%2-menthene	2	1	0	32	32	30
	%3-menthene	98	99	100	68	68	70
E2	%2-menthene	12	25		100	100	
	%3-menthene	88	75		0	0	
Ref.		32	226	31	32	226	31

Solvent: [a] water (containing $\bar{O}H$ for the E2 reaction); [b] 80% aqueous ethanol (E1), ethanol/ethoxide (E2); [c] ethanol (E1 reaction assumed).

($k_H/k_D = 1.44$) than in the *syn-clinal* orientation of the equatorial beta hydrogen ($k_H/k_D = 1.10$)[276]. A larger isotope effect is also associated with the *syn*-elimination in the cyclodecyl tosylate series ($k_H/k_D = 1.4 \rightarrow 1.9$)[273]. Because of the non-equivalence of beta protons in the presence of neighbouring group participation, Shiner and Jewett[276] have suggested the isotope effects arising from this cause should be non-accumulative, unlike the expected effect for hyperconjugation. Bearing this approach in mind, the isotope effects arising from axial beta hydrogen substitution in *cis*-4-*t*-butylcyclohexyl tosylate are attributed to a neighbouring group participation. In the solvolysis of *trans*-4-*t*-butylcyclohexyl tosylate, the reduced rate of ionisation of the equatorial tosylate group is accompanied by an almost accumulative isotope effect upon equatorial hydrogen substitution ($k_H/k_D = 1.34$) and a smaller non-accumulative isotope effect for the axial beta position ($k_H/k_D = 1.13$)[277]. The non-equivalence of the axial protons was interpreted as neighbouring group participation from a twist-boat conformation[278] in which the orientation of the axial beta carbon–hydrogen bond and carbon–leaving-group bonds approach coplanarity. In the more flexible system, cyclohexyl tosylate, a similar explanation was invoked to explain the similarity in the isotope effects of axial and equatorial beta protons upon rates of acetolysis, although ion-pair intermediates are more probable under these conditions[279].

Neomenthyl compounds ionise more rapidly than the corresponding menthyl derivatives[226] and although this observation is attributable to neighbouring group participation it may also be an artefact arising from a greater population of the axial conformation of the leaving group in the former series. The greater yield of 3-menthene in E1 than in E2 reactions can in part be attributed to the incursion of a tertiary carbonium ion intermediate (107) arising from a hydride shift (route C). However, experiments with optically active chlorides illustrates the minor role this route contributes to the overall reaction as the 3-menthene isolated is only partially racemised (107)[10,226].

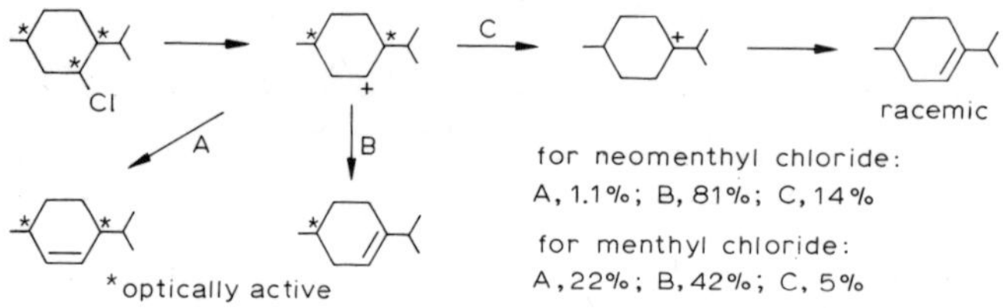

(107)

The uncertainty in the stereochemistry of E1 reactions and the tendency of carbonium ions to undergo rearrangement greatly reduces the synthetic utility of these processes and E2 reactions are generally more useful.

The lack of stereochemical evidence for E1cB reactions renders comments of dubious value. Presumably ion pairing and neighbouring group interactions will exert complicating factors in an analogous manner to those observed in E1 reactions.

2.4 COMPETING SUBSTITUTION AND ELIMINATION REACTIONS

Reagents with basic properties (affinity for protons) often also possess nucleophilic character (affinity to form bonds to carbon). Consequently, bimolecular elimination reactions occur frequently in competition with substitution reactions, *viz.*

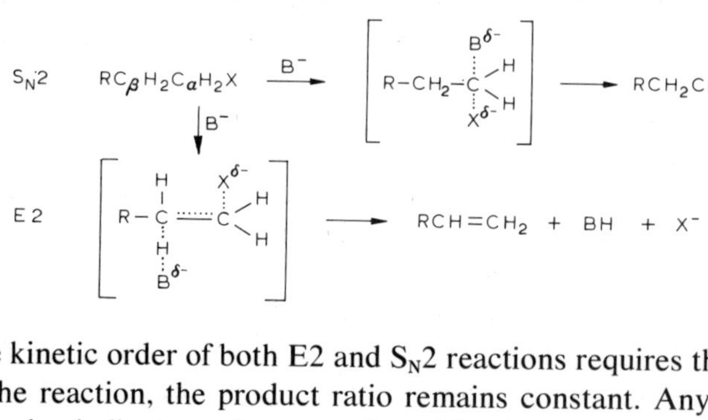

(108)

The kinetic order of both E2 and S_N2 reactions requires that during the course of the reaction, the product ratio remains constant. Any deviation from this behavior indicates a departure from kinetic control due to product instability under the reaction conditions or more frequently, the incursion of competing unimolecular processes in the later stages of reaction. The salient features affecting the elimination/substitution ratio have been studied systematically and summarised in earlier reviews[6,10,280]. As the substitution reactions of "onium" salts often differ from those of the halides and tosylates, the substitution occurring at the methyl substituents of the leaving group, *viz.*

$$RCH_2CH_2\overset{+}{S}(CH_3)_2 \xrightarrow[Y^-]{S_N2} RCH_2CH_2Y + S(CH_3)_2 \quad \text{and} \quad RCH_2CH_2SCH_3 + CH_3Y \tag{109}$$

comparison between the two types of substrate has to be made with great care. Although bimolecular and unimolecular processes often occur simultaneously, they are more conveniently discussed separately. Typical variations in substitution/elimination ratios are illustrated in Table 16.

Within the bimolecular category, the proportion of reaction leading to elimination increases with alkyl-substitution at the alpha carbon independently of the leaving group[281,282]. Such a variation increases the steric interaction in the transition state for the S_N2 reaction but accelerates the E2 process as a thermodynamically more stable olefin is produced. Similar factors explain the accelerating effect of substitution on the beta carbon in elimination from alkyl halides. However, beta alkyl substitution has an unfavourable effect on rates of elimination from "onium" salts and this observation has been attributed to

TABLE 16

THE INFLUENCE OF REACTION CONDITIONS ON THE YIELD OF OLEFINS ARISING FROM SOME ALKYL HALIDES AND SULPHONIUM SALTS

Substrate	*Reaction types*	*Temp. (°C)*	*Conditions*	*%E*	*Ref.*
i-PrBr	$E2/S_N2$	50	a	58	292
		80	a	61	292
		100	a	66.5	292
		55	b	79	281
C_2H_5Br	$E2/S_N2$	55	b	1	281
t-BuBr	$E2/S_N1$	55	b	100	284, 294
$C_2H_5\overset{+}{S}(CH_3)_2$	$E2/S_N2$	45	b	12	293
$(CH_3)_2CH\overset{+}{S}(CH_3)_2$	$E2/S_N2$	45	b	61	293
$(CH_3)_3C\overset{+}{S}(CH_3)_2$	$E2/S_N2$	45	b	97–100	293
t-BuCl	$E1/S_N1$	25	c	17	6
		50	c	24	6
		65	c	36	6
$(CH_3)_2C(C_2H_5)Cl$	$E1/S_N1$	50	c	40	6
$(CH_3)_2C(C_2H_5)\overset{+}{S}(CH_3)_2$	$E1/S_N1$	50	c	48	6

[a] 80% aqueous ethanol containing lyate ions.
[b] Ethanol containing ethoxide.
[c] 80% aqueous ethanol.

steric considerations or electronic factors which cause a shift in the nature of the transition state (Section 2.5). Aryl substituents which can conjugate with either a developing double bond or with developing *p*-orbitals on the alpha or beta carbons appear to exert a uniformly favourable effect on all E2 reactions. Obviously electronic rather than steric considerations are of primary importance in this case.

Higher yields of olefin are obtained in elimination from cyclohexyl tosylates and halides than from analogous secondary acyclic substrates. With an axial leaving group, substitution is more hindered by the axial hydrogen than is elimination as a closer approach of the nucleophile to the ring system is required in the transition state for substitution than elimination. The tendency of the bulky substituents to occupy an equatorial position can be highly detrimental for elimination and the failure to achieve the necessary coplanarity results in substitution at the methyl groups of dimethylsulphonium and trimethylammonium substituents becoming more competitive than in the acyclic series.

Many elimination reactions have been carried out in protic media containing their lyate ions as the base, a change in solvent thus being accompanied by a simultaneous change in base. Olefin yields usually increase along the series H_2O < ethanol < isopropyl alcohol < *t*-butyl alcohol, an order of increasing basicity, increasing size of the base and decreasing nucleophilicity, factors which certainly play a major role. The simultaneous increase in rate and olefin yield can also be attributed to a reduction in the polarity of the reaction medium. Charge dispersal is more extensive in the transition state leading to elimination than that leading to substitution as it occurs over more bonds and consequently elimination is favoured by a change to less solvating media[283]. This effect is most marked for reactions between oppositely charged reactants and the rates of base-catalysed eliminations of trimethylammonium ions are greatly enhanced by a change from ethanol/ethoxide to *t*-butyl alcohol/*t*-butoxide[136].

A dissection of the influence of solvent and base should be possible from studies in dipolar aprotic solvents containing various bases in which the two reactants can be varied independently. Although bases whose conjugate acids have pK_a values of less than 11 rarely induce elimination in protic media[10], halide ions have been used successfully in this connection in dipolar aprotic solvents[284,285]. Of course in the latter solvents, which lack exchangeable hydrogens but possess atoms with lone pairs capable of solvating cations, pK_a values may differ appreciably from those already derived in protic solvents[286]. There is obviously ample scope for future research in this field.

In general, as more bond changes occur in attaining the transition state for elimination than substitution, elimination reactions have a higher activation energy and consequently are favoured by a rise in reaction temperature.

Within the bounds of classical carbonium ions, secondary and tertiary substrates undergo competing S_N1 and E1 reactions, *viz.*

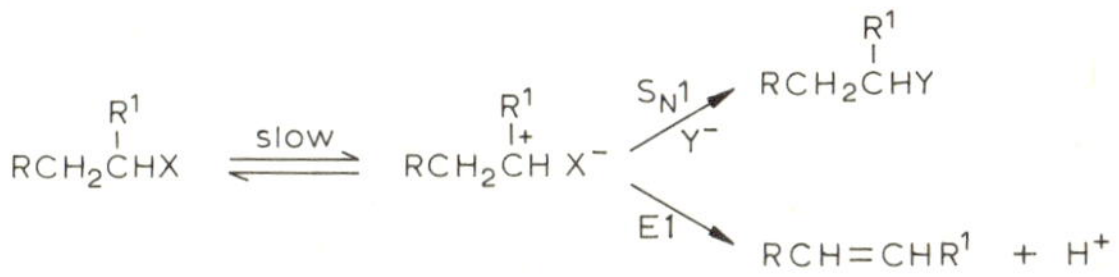

(110)

The rate- and product-determining steps are now different and there is no necessary relationship between rates and product proportions, the latter being dependent on the relative rates at which the intermediate carbonium ion loses a proton or combines with a solvent molecule.

Solvolytic reactions are rarely applied in synthetic studies as lower olefin yields result in unimolecular than in bimolecular reactions and often the intermediate carbonium ions can also rearrange. In polar media, the leaving group has little influence on the product distribution, but in less polar solvents ion pairing between the leaving group and the carbonium ion inhibits substitution, which in part simulates an S_N2 mechanism, and greater olefin yields result[269]. On the basis of increasing olefin yields with a rise in temperature, the activation energy for elimination from the carbonium ion has been estimated as 2–4 kcal.mole^{-1} greater than the unimolecular substitution[6,10]. Of course in solvents promoting ion-pair formation care has to be taken in assessing activation energy differences, as a change in temperature, independent of the relative energetics will cause a change in product composition by shifting the equilibria between the various types of ion pair[287]. Even in 80% aqueous ethanol, the *t*-amyl cation may be ion-paired to a small extent as a different product distribution results on solvolysis of the chloride than is obtained from the sulphonium salt[6].

In a variety of solvents, solvolysis of a series of alkyl halides gives increasing yields of olefin with increasing size and complexity of the alkyl group. A regular variation of rate of solvolysis with size of substrate is, however, not observed (Table 17). The interpretation of these results has been a matter of great controversy. Hughes and Ingold prefer polar control (*e.g.* Ingold[6]), explaining the variations in terms of inductive and hyperconjugative properties of the alkyl substituents and their subsequent effect on the rate-determining heterolysis and on the stability of the transition state leading to olefin[288]. Alternatively, Brown *et al.*[280,289–291] have invoked a steric explanation. Increasing ramification around the alpha carbon produces greater internal steric repulsions. These are less in the carbonium ion and olefin than in the substrate or substituted product as the bonding angle is 120° in the former two but only 109°28′ in the saturated structures. The steric theory adequately accounts for

TABLE 17

SOLVOLYTIC RATE COEFFICIENTS AND PRODUCT DISTRIBUTION IN UNIMOLECULAR REACTIONS OF SOME ALKYL BROMIDES[290] AND CHLORIDES[289] AT 25°C.

R in			
$RCH_2CBr(CH_3)_2$[a]	*$R(CH_3)_2CCl$*[b]	*k_1 (h^{-1})*	*Elimination (%)*
CH_3		0.387	27
C_2H_5		0.297	32.5
i-Pr		0.697	46
t-Bu		4.71	57
	CH_3	0.033	16
	C_2H_5	0.055	34
	i-Pr	0.029	62
	t-Bu	0.040	61
	$(CH_3)_3CCH_2$	0.74	65

[a] Solvent, 85% aqueous *n*-butyl cellosolve.
[b] Solvent, 80% aqueous ethanol.

the product variations, steric retardation of substitution being greater than of elimination which involves the more exposed beta hydrogens. The rate data are more difficult to reconcile, although the relatively small and somewhat irregular variation produced with large changes in substrate structure may in part be due to more difficult solvation of the leaving group in heterolysis of the hindered substrates.

Although a common intermediate is generally accepted for unimolecular substitution and elimination, the bimolecular mechanisms are usually regarded as distinct processes. In recent years, however, a number of authors have argued in favour of a "merged mechanism" of substitution and elimination[284]. Originally, the common intermediate was postulated to account for the similar bimolecular rates of elimination with halide ions in acetone from *cis*- and *trans*-4-*t*-butyl-cyclohexyl *p*-toluenesulphonates, *viz.*

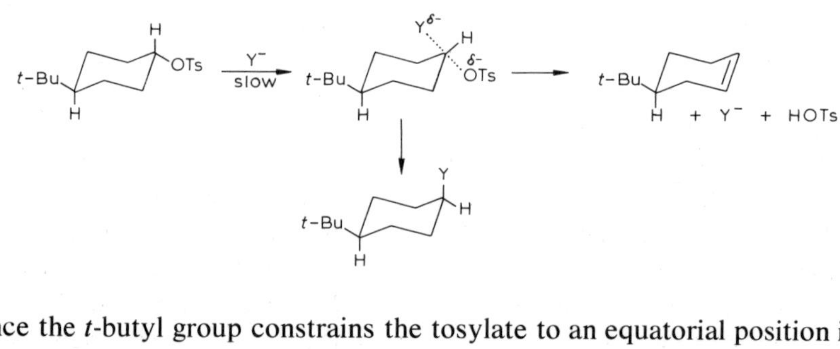

(111)

Since the *t*-butyl group constrains the tosylate to an equatorial position in the *trans* isomer, from stereochemical considerations an E2 reaction would be expected to proceed much more rapidly from the *cis* isomer. In addition, the rate of elimination did not correlate with the basicity of the anion and to accommodate these facts an intermediate similar to the carbonium ion but solvated by the nucleophile and tosylate anion was envisaged, which subsequently partitioned to products, beta hydrogen abstraction being by either the nucleophile or the tosylate anion. To substantiate these views isotope labelling of the beta hydrogens and the primary kinetic isotope effect should prove useful. Of course anion basicity in dipolar aprotic solvents differs quite considerably from that in protic media and even the order of reactivity of the halide ions is reversed. Experiments designed to measure thermodynamic basicity in these solvents are obviously required.

The merged mechanism was subsequently adopted by Eliel *et al.*[295,296] to explain observations with thiophenoxide in 87% aqueous ethanol. Both cyclohexyl tosylate and its 4,4′-dimethyl derivative react at nearly the same rate, the parent compound, however, gives 45% elimination whereas the substituted compound undergoes entirely substitution. The similarity in rates was attributed to a common rate-determining step and the lack of elimination to hindrance by the methyl group to beta proton abstraction. Merging was also suggested to account for the earlier observation that thiophenoxide[297], despite being less basic but definitely more nucleophilic than ethoxide, more effectively induces elimination in *t*-butyl chloride[296]. A merged mechanism has also been invoked to account for elimination reactions that occur when alpha haloketones are treated with amines or bromide ion in non-hydroxylic solvents[298], and to explain so-called distorted $k_{\mathrm{OTs}}/k_{\mathrm{Br}}$ ratios in substitution and elimination reactions[299]. One author has reviewed bimolecular substitution and elimination in general from a standpoint of the common intermediate[300].

Essentially for the merged mechanism, the rate of elimination is expected to relate to carbon nucleophilicity of the attacking reagent. Using the rate of substitution as a measure of nucleophilicity, with a variety of bases in a series of solvents, a good linear correlation between rates of elimination and substitution in cyclohexyl tosylate is observed (Fig. 8). To explain the correlations and deviations a spectrum of transition states was suggested[301], ranging from the normal E2 for the highly basic reagents to the various forms of the merged mechanism for the more nucleophilic reagents (112). Excluding the basic

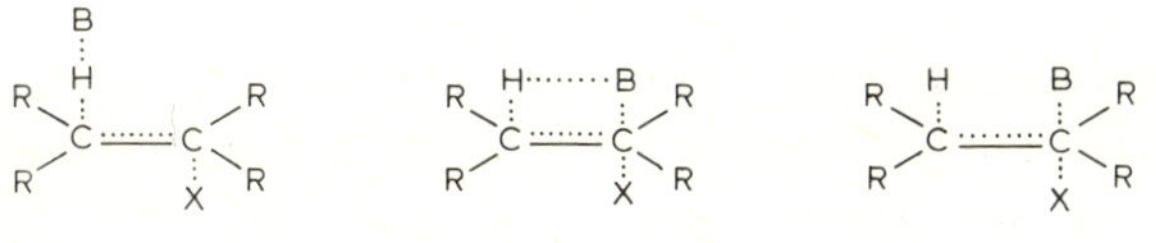

(112)

reagents, the unit slope of the figure suggests E2 and S_N2 transition states

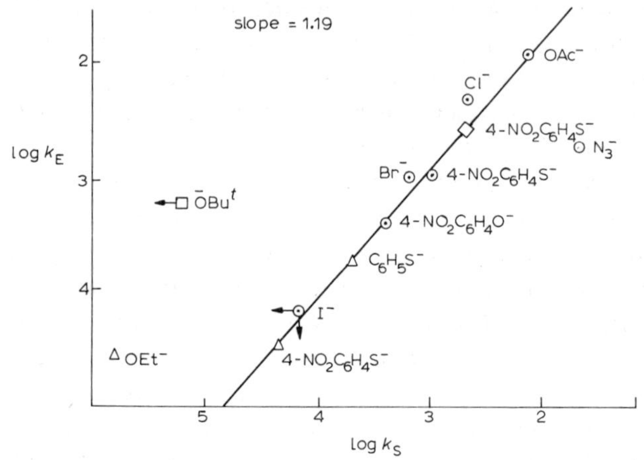

Fig. 8. The relationship between the rate coefficients for the bimolecular elimination (k_E) and substitution (k_S) of cyclohexyl *p*-toluenesulphonate with various nucleophiles in a number of solvents. Solvent: ⊙, acetone at 75°C; □, *t*-butyl alcohol at 75°C; △, ethanol at 35°C; ◊, methanol at 75°C. (Reproduced with permission from Bialé *et al.*[301].)

respond similarly to changes in carbon nucleophilicity. This could be fortuitous as the reacting conformations for E2 and S_N2 reactions would be highly similar in this case, an axial conformation of the tosylate group being favoured. It would be interesting to see if a similarly effective correlation exists between competitive substitution and elimination in the acyclic series in which substrate conformation need not be the same for both single processes. Measurement of the primary kinetic isotope effect will be equally informative for the elimination reactions from cyclohexyl tosylate with the variety of bases as the transition states for the merged mechanism will involve little weakening, but the normal E2 reaction more considerable weakening, of the beta hydrogen bonds. For reactions in acetone, the rate coefficient for elimination does not obey the Brönsted equation, *i.e.* log k_E is not proportional to pK_a (base), seemingly suggesting the lack of importance of basicity of the reagent. However, even for a simple proton abstraction, it is unlikely that these widely differing structural reagents would follow the Brönsted equation due to considerable variations in activation entropy. A relationship between activation enthalpy rather than rate of elimination with pK_a might be more successful and appropriate.

Although certain experimental observations are more easily interpreted by the merged mechanism than the more conventional transition states of bimolecular elimination and substitution, they do not require it. The unusual effectiveness of weak bases to promote elimination reactions is unconvincing, and Bunnett has stated that equilibrium and kinetic basicity need not parallel each other[105]. Equally important is the variation in the relative catalytic efficiencies of thiophenoxide and ethoxide to induce elimination with varying substrate structure. As the extent of proton transfer to the base is increased in the transition state, as measured by the Brönsted component, the relative effectiveness of thiophenoxide decreases (Fig. 9)[57,184,185]. This decrease parallels a change from a carbonium ion-like towards a carbanion-like E2 transition state as predicted for a gradual increase in electronegativity and tightness of binding of the leaving group. Along this series the enhanced polarisability of thiophenoxide becomes less effective and the greater basicity of ethoxide becomes the dominant factor. As pointed out by McLennan[185], the peculiar reactivity of thiophenoxide towards *t*-butyl chloride is unlikely to reside in the merged mechanism as *t*-butyldimethylsulphonium ion undergoes substitution at the methyl groups and not at the alpha carbon.

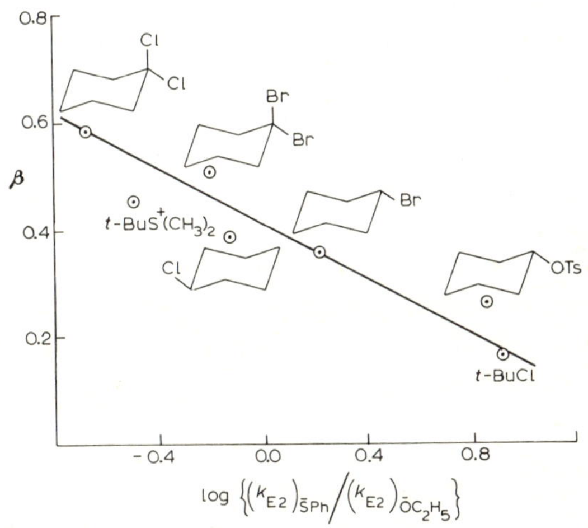

Fig. 9. The relationship between the Brönsted coefficient, β for E2 reactions (catalysed by substituted thiophenoxides) and the ratio of the rate coefficients for elimination by thiophenoxide to that by ethoxide.

It would be unreasonable to expect a nucleophilic species not to interact with both an electrophilic alpha carbon and beta hydrogen at some stage during the reaction. However, whether this interaction is representative of the rate-determining step is a matter of conjecture. Most probably this dual interaction occurs early in the reaction profile and is followed by partitioning to give the two different transition states commonly accepted for bimolecular elimination and substitution. The closer the dual interaction and the transition states are on the reaction profile, then the more closely the elimination rate will respond to carbon nucleophilicity.

2.5 ORIENTATION

In many organic substrates, the alpha carbon bearing the leaving group is flanked by two beta carbons bearing hydrogen atoms and elimination can consequently occur in either of two directions. In the previous century two rules relating to the orientation of elimination were formulated. The Hofmann rule[302] was followed if the least alkylated olefin was formed, this in the extreme case being ethylene, whereas elimination giving the more substituted olefin was said to follow the Saytzeff rule[303]. In more modern terminology the Saytzeff product is regarded as the more thermodynamically stable and the Hofmann product as the less stable olefin. Most authors agree on the factors governing the Saytzeff rule but great controversy has arisen over interpretation of the salient features causing the Hofmann product formation. The majority of the studies relate to alkyl and cycloalkyl substrates and discussion is limited predominantly to this field of interest.

The orientation of elimination is greatly influenced by the choice of the leaving group, the substituents at the alpha and beta carbons and the base and solvent. Investigators usually keep two of these variables constant whilst changing the third and a discussion of the relevant factors is most conveniently approached in this way. Often conclusions are reached from measurements of the olefin proportions only. Occasionally more certain deductions are possible when both product study and rate data at several temperatures are reported. The advent of vapour-phase chromatography in the last decade has enabled easier and more certain analysis of product mixtures, and with due respect to the earlier pioneers in this field, who had to use more primitive techniques, the most reliable results have been derived in more recent years.

In the acyclic series, E2 reactions of ammonium and sulphonium salts show a marked preference for Hofmann orientation[2,19,163,304] but secondary and tertiary alkyl halides generally follow the Saytzeff rule[6,305] and all unimolecular eliminations show an even greater tendency to give the more stable olefin (113)[6].

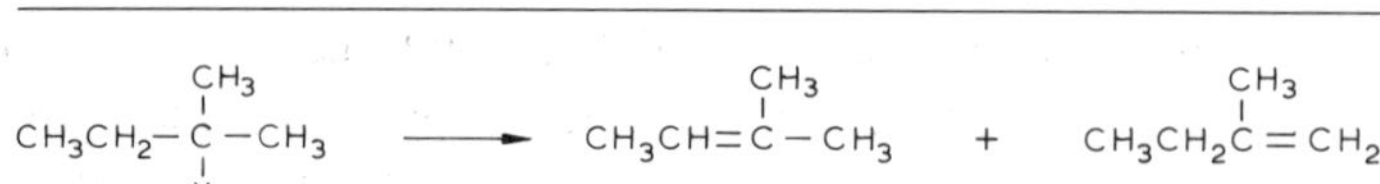

(113)

Substrate	*Saytzeff product*	*Hofmann product*
(X = $\overset{+}{S}(CH_3)_2$ ethanol E1	87%	13%
ethanol/ethoxide E2	14%	86%
(X = Br) ethanol E1	82%	18%
ethanol/ethoxide E2	70%	30%

The Saytzeff product is also favoured by both charged[32,306] and uncharged substrates[31,226] in the cyclohexyl series. It should, however, be noted that the Hofmann rule is not limited to charged acyclic substrates as illustrated by bimolecular eliminations of sulphones[20], nitro[307] and fluoro[308] compounds, *e.g.*

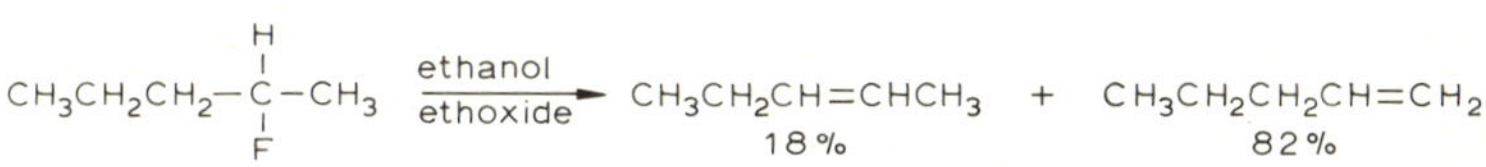
(114)

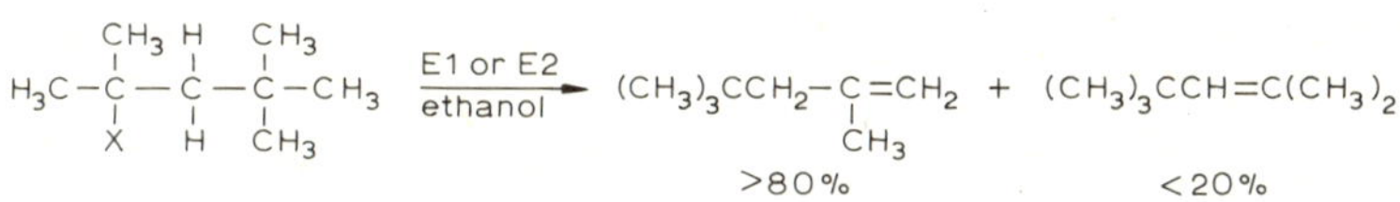
(115)

Modification of base and solvent can also induce a change from Saytzeff to Hofmann orientation even in alkyl bromide eliminations[309]. In one unimolecular elimination[3,4], the least substituted olefin is formed predominantly[310,311], but in this case it is the thermodynamically more stable product by about 1.8 kcal.mole^{-1}, due to eclipsing between the *t*-butyl and methyl substituents in the more substituted olefin[312], and consequently the Saytzeff pattern is not violated (115).

2.5.1 *Theories of orientation*

Under kinetic control, the influence of alkyl substituents on orientation in bimolecular elimination is determined by the rate of elimination into each branch. Without studying the competitive process, it should be possible using kinetic data from single olefin-forming reactions to predict the orientation in the more complex cases. In "onium" salts the rate of elimination is facilitated by the substitution of a methyl group for an alpha hydrogen but retarded by beta methyl substitution. The Hofmann rule should thus predict the behaviour of this type of compound. On the other hand alpha and beta methyl substitution enhances the rate of elimination from alkyl bromides and compliance with the Saytzeff rule is predicted. The inconsistent influence of a beta methyl group on the rate coefficient in the two series (Table 18) cannot obviously be explained by the operation of only a single factor.

From the earlier discussion on the nature of the transition state for E2 reactions, two salient factors affecting reactivity can be recognised, these being polar and steric effects. The polar effect can be divided into inductive and conjugative or electromeric components[314]. The influence of a substituent will depend principally on the nature of the transition state, which to a large extent is determined by the leaving group and the base and solvent. A reaction

TABLE 18

RATE COEFFICIENTS FOR BIMOLECULAR ELIMINATION FROM SOME ALKYL BROMIDES AND ALKYL ONIUM IONS

$$RCH_2CH_2X + \bar{O}C_2H_5 \rightarrow C_2H_5OH + \bar{X} + RCH{=}CH_2$$

	$10^5 k_{E2}$ (l.mole^{-1}.sec^{-1})		
R	$X = \overset{+}{N}(CH_3)_3$[a]	$X = \overset{+}{S}(CH_3)_2$[b]	$X = Br$[c]
H	71.3	79	1.6
CH_3	5.16	29	5.3
C_2H_5	2.82	21	4.3
$CH_3CH_2CH_2$			3.5
$(CH_3)_2$*	1.68	10	8.6
C_6H_5	179[d]	34,000	561

* R_2 in R_2CHCH_2X. Temp. (°C): [a] 104 [b] 64 [c] 55 [d] 50. References: [a] 313 [b] 6, 313 [c] 6, 314 [d] 40.

$$CH_3CHRX + \bar{O}C_2H_5 \rightarrow C_2H_5OH + X + RCH{=}CH_2$$

	$10^5 k_{E2}$ (l.mole^{-1}.sec^{-1})	
R	$X = \overset{+}{S}(CH_3)_2$[e]	$X = Br$[f]
H	5.0	0.025
CH_3	114	0.118
C_2H_5		0.065
$CH_3CH_2CH_2$		0.080
$(CH_3)_2$*	2930	1.00
C_6H_5	≃570**	0.79

* R_2 in CH_3CR_2X. Temp. (°C). [e] 45 [f] 25: Reference 314.

** Calculated using data from ref. 315 from $10^5 (k_{S_N2} + k_{E2}) = 961$ l.mole^{-1}.sec^{-1}, at 45°C and % olefin formation at 64°(59%) assumed appropriate for 45°; this rate coefficient may thus be slightly high.

possessing a carbanion-like transition state is favoured by an electron-withdrawing substituent and retarded by an electron-releasing substituent at the beta carbon. For a carbonium ion-like transition state, the same substituent at the alpha carbon will exhibit reverse effects. For a completely concerted reaction, substitution will tend to shift the transition state towards the carbanion or carbonium ion extremes to enable the benefit of any possible stabilisation[14]. Substituents which can exert conjugative influences should stabilise any developing double bond, an effect which can operate from both alpha and beta carbons[314]. This is illustrated by the influence of phenyl substitution on the rate coefficient (Table 18). Both alpha and beta phenyl groups cause acceleration but the beta phenyl group is more effective as in this case the inductive and conjugative influences enforce each other.

Alkyl substituents exert an electron-releasing inductive effect which increases along the series methyl, ethyl, isopropyl, *t*-butyl, values of the substituent constants σ_I and σ^* supporting this order[316]. In the absence of unusual non-bonding interactions, the more alkylated olefins are in general of greater thermodynamic stability. Although other explanations have been advanced[317], this increased stability is generally attributed to hyperconjugative electron release from the carbon–hydrogen bonds adjacent to the olefinic centre (116)[314]. Consequently, the importance of this

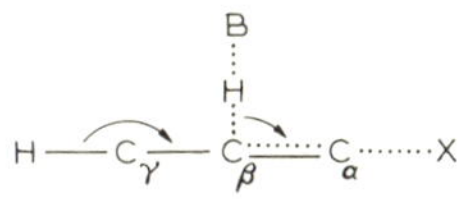

(116)

conjugative stabilisation decreases along the series methyl, ethyl, isopropyl, *t*-butyl[6]. The two polar effects act in the same direction on rate from the *alpha* position as both inductive electron release by aiding C_α–X heterolysis and hyperconjugative electron release by stabilising the developing double bond should enhance elimination. However, in the beta position, the hyperconjugative effect is beneficial but an electron-releasing inductive effect decreases carbon–hydrogen acidity and retards the elimination. Consequently, the influence of beta alkyl substitution depends largely on which of these factors is more dominating and the variation in rate along the alkyl series possibly reflects an increasing competitive nature of the inductive effect as the size of the alkyl substituent is increased (Table 18).

Additional to the polar factors, the steric requirements of the alkyl substituents may play a dominating role by hindering approach of the base to a beta proton or alternatively causing a divergence from the requisite coplanarity of reacting bonds in the transition state for elimination. Often, along a series of alkyl substituents, both steric and inductive effects vary in the same direction and the interpretation of the salient orientating features is a matter of personal preference.

Transition states possessing extensive double-bond character are presumably stabilised by the same factors which affect olefin stability and these consequently lead to Saytzeff product formation. Typical results indicate electromeric control along the alkyl series although elimination from the *t*-butyl isomer [see (118), p. 251] is predominantly controlled by steric influences. In the unimolecular eliminations, the highly electron deficient vacant *p* orbital on the alpha carbon may lead to considerable weakening of the beta carbon–hydrogen bond and more extensive double-bond character than is exhibited in bimolecular eliminations in which double-bond character is largely limited by the less stretched of the reacting bonds. This leads to a greater predominance of Saytzeff orientation in unimolecular reactions.

It is generally agreed that the dominating factor affecting Saytzeff orientation is the mesomeric or more particularly the hyperconjugative influence of alkyl substituents. Factors other than those stabilising double-bond character presumably influence the stability of transition states leading to the less stable olefin. To explain the change from Saytzeff to Hofmann orientation accompanying changes in substrate and reaction conditions the Hughes–Ingold school of thought[313,314] regard a balance between inductive and mesomeric effects as the governing factors, steric hindrance being influential only above determinable thresholds of molecular complexity. On the other hand Brown *et al.*[280,318–323] consider their arguments in terms of mesomeric and steric influences only.

The use of linear free energy relationships and isotope studies in the 2-phenylethyl series shows that elimination from the trimethylammonium ion passes through a transition state with considerably more carbanion character than that for the bromide in which C_α–X heterolysis is more advanced and double-bond character is more developed (Section 2.2). The dominating mesomeric influence of beta alkyl substituents consequently induces Saytzeff orientation in the alkyl bromide eliminations but in reactions of the ammonium salts the inductive effect is the more important and results in a reduction in the acidity of the beta hydrogen at the more alkylated carbon and preferential abstraction occurs at the less alkylated centre producing the Hofmann product. In more modern terminology, the ammonium salts eliminate through carbanion-like transition states and the stability of such intermediates is reduced by electron donating alkyl substituents.

For alpha alkyl substitution, the rate of elimination from ammonium and sulphonium salts increases sharply along the series ethyl < isopropyl < *t*-butyl when ethoxide in ethanol is the base[293,313,315,324]. Product studies reflect a similar accelerating effect as isopropyl and *t*-butyl groups are more easily eliminated from $R^1R^2(CH_3)_2\overset{+}{N}\,\overset{-}{O}H$ than ethyl[163]. These alpha substituents aid C_α–X heterolysis and shift the transition state away from the carbanion-like extreme towards one possessing more double-bond character.

The steric explanation is most easily illustrated with the use of Newman projections (117).

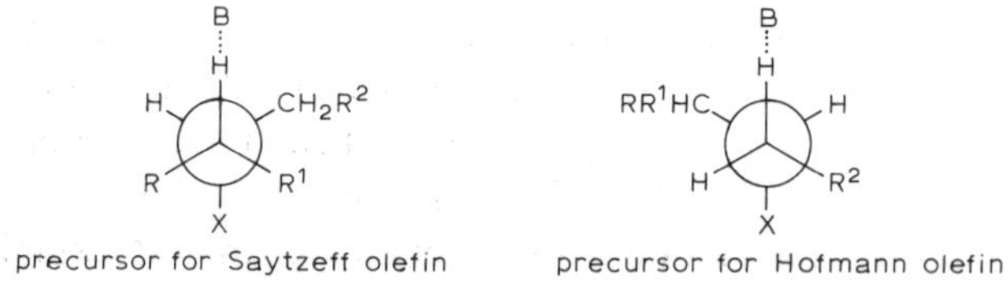

(117)

The theory has been outlined in a series of papers[320] and hinges on the fact that steric interactions are less in the conformation leading to the Hofmann product. The advantages of olefin stability have to be considered alongside disadvantages arising from non-bonding interactions between alkyl substituents becoming *cis* across a developing double bond, alkyl substituents and the leaving group, and alkyl substituents and the base. In all cases, unfavourable non-bonding interactions may lead to a lower population, and possibly higher energy level, of the conformer leading to the Saytzeff product. Of course serious retardations on the rate of elimination into the more alkylated branch may result from a failure to attain coplanarity of the reacting bonds, this latter factor being more critical for bimolecular than unimolecular eliminations. The idea of steric influences becoming dominant is not unreasonable as bulky substituents occupy equatorial rather than axial conformations in cyclohexane rings and steric inhibition of beta proton abstraction can cause elimination to follow an ylid rather than the bimolecular mechanism (Section 2.1.4)[92].

2.5.2 *The influence of alkyl substituents*

The direction of elimination in the reaction of a series of alkyl bromides (118)[311] changes from Saytzeff to Hofmann as the size of a beta alkyl substituent is increased, indicating a gradual increase in importance of steric interaction.

R in $RCH_2C(CH_3)_2Br$ with $\bar{O}C_2H_5/C_2H_5OH$	CH_3	C_2H_5	$(CH_3)_2CH$	$(CH_3)_3C$
alk-2-ene% per beta hydrogen	35		23	7
alk-1-ene% per beta hydrogen	5		9	14

(118)

Whether this is undisputed evidence for the steric theory is not clear as the inductive effect of the alkyl substituent increases simultaneously with steric size and a more complicating factor is the decreased thermodynamic stability of the more substituted olefin of the last member of the series due to the aforementioned unfavourable steric interactions. Ingold *et al.*[313] have argued that owing to a steep bank in the non-bonding energy curve, a steric effect once it starts should build up very rapidly with increasing material density about the reaction site. On the other hand, an electrostatic effect is subject to the attenuation of relay and is expected to show a gentler and more monotonous form of variation. Only in the last substrate of (118) do they regard steric influences as predominant. Rate studies on the more hindered trimethylammonium salts ($RCH_2CH_2\overset{+}{N}(CH_3)_3$) were quoted as supporting evidence for these assumptions[313].

Steric interactions have been invoked to explain orientation in some unimolecular eliminations[280]. In a series of tertiary bromides (Table 19) a gradual increase in the size of one of the branches causes an increase in the amount of

TABLE 19

RATE COEFFICIENTS AND OLEFIN YIELDS FOR E1 REACTIONS OF SOME ALKYL BROMIDES[290] AND *p*-BROMOBENZENESULPHONATE ESTERS[291]

Solvent, 85% aqueous *n*-butylcellosolve. Temp., 25°C.

Substrate	k_1[a] (h^{-1})	% *Olefin*[b]	% *1-Olefin*[c]	% *2-Olefin*[c]
$CH_3CH_2C(CH_3)_2Br$	0.387	27	21 (3.5)	79 (39)
$CH_3CH_2CH_2C(CH_3)_2Br$	0.297	32.5	29 (5)	71 (35)
$(CH_3)_2CHCH_2C(CH_3)_2Br$	0.697	46	41 (7)	59 (30)
$(CH_3)_3CCH_2C(CH_3)_2Br$	4.71	57	81 (13.5)	19 (9)

Solvent, acetic acid. Temp., 70°C.

Substrate	10^4k_1[a] (sec^{-1})[a]	% *1-Olefin*	*trans-2-Olefin* / *cis-2-Olefin*
$CH_3CH_2CH(CH_3)OBs$[d]	1.29	10.3[e]	1.08[e]
$CH_3CH_2CH_2CH(CH_3)OBs$[d]	1.23	15.7	1.39
$(CH_3)_2CHCH_2CH(CH_3)OBs$[d]	1.13	19.7	1.94
$(CH_3)_3CCH_2CH(CH_3)OBs$[d]	2.64	24.3	83

[a] Rate of unimolecular ionisation to a carbonium ion intermediate.
[b] Total yield of elimination products.
[c] Figures in parentheses are statistically corrected.
[d] OBs is an abbreviation for *p*-bromobenzenesulphonate.
[e] These figures refer to the reaction of the corresponding tosylate ester.

carbonium ion undergoing elimination and results in a gradual reduction in the yield of the more alkylated olefin. In all cases the more thermodynamically stable olefin is the major product. It is not clear whether the dominant factor is a changing steric requirement or a gradual reduction in the hyperconjugative influence with increasing alkyl substitution. Certainly the variation in the rate of ionisation of the first three members changes little and does not suggest that any reduction in steric compression accompanies ionisation.

An even smaller spread in the rate of ionisation is noted in the secondary series of brosylates despite the apparent greater steric size of the leaving group. Little change in the Hofmann–Saytzeff distribution is likewise noted in this series, the more alkylated olefin being more thermodynamically stable throughout. The only suggestive evidence for a steric controlling factor is the phenomenal change in the *trans*/*cis* ratio for the last member of the series. The remaining substrates can rotate into a conformation so as to minimise any non-bonding interaction between methyl and alkyl coming *cis* across a double bond.

2.5.3 *The influence of the leaving group*

The gradual trend from Saytzeff to Hofmann predominance in elimination from the 2-pentyl series has been attributed to the steric size of the leaving group (119)[319].

	Br	I	OTs	SO_2CH_3	$\overset{+}{S}(CH_3)_2$	$\overset{+}{N}(CH_3)_3$	
X in $CH_3CH_2CH_2CHXCH_3$ *with* OC_2H_5/C_2H_5OH*; % 1-pentene*	31	30	48	89	87	98	(119)

To explain the similar product ratio for the two halogeno compounds it was pointed out that covalent radius and halogen size increase simultaneously and consequently the larger halogens are further from the site of interference. In opposition to the steric theory, the eliminations involving a nitro leaving group follow Hofmann orientation predominantly despite this group being much smaller than the sulphone and "onium" substituents[307]. Moreover, steric size of the leaving group, its inductive effect and the ease of heterolysis of the bond to the leaving group run parallel in the 2-pentyl series of substrates, and the trend towards Hofmann orientation could equally well be explained by the gradual changes in the latter two properties.

Using the more accurate gas chromatographic technique rather than a combination of refractive index and infrared methods to analyse the olefin composition, Saunders *et al.*[308,325] showed that the initial halogeno elimination results were incorrect, (120) below. The results for the three heavier halogens are not in disagreement with a steric interpretation as a study of the conformational equilibrium in halocyclohexanes reveals a trend of apparent steric requirements, as judged by increasing axial isomer at equilibrium[326], which follows the order I < Br < Cl. Van der Waals' radius rather than actual geometrical size is the critical factor determining "steric size", as pointed out earlier by Brown *et al.*[319,322]. However, in any case, the smallest group is certainly fluorine and the increase in the relative yield of the 1-olefin along the series I < Br < Cl < F seems more easily interpreted in terms of ease of heterolytic cleavage of C_α–X and the inductive effect of X. These factors favour a gradual change to a transition state with more carbanion character, but less double-bond character, along the series I, Br, Cl to F. The concomitant increase in the *trans-/cis*-2-pentene ratios (120) and the Hammett reaction

X in $CH_3CH_2CH_2CHXCH_3$	F	Cl	Br	I	
with OC_2H_5/C_2H_5OH*; % 1-pentene*	82	35	25	20	
trans-/cis-2-pentene	2.6	3.5	3.7	4.1	(120)
σ^* *for X*	1.10	1.05	1.00	0.85	

constants for elimination from the analogous 2-phenylethyl substrates (Section 2.2.2(a)) support the prediction that carbanion character of the transition state plays a major role in determining orientation in the halogeno-eliminations[325]. An argument based on the acidity of the beta hydrogens as affected by the inductive effect of the leaving groups is less satisfactory in this case, as the polar effect of fluorine does not differ significantly enough from that of the other halogens (as measured by σ^*)[316,327] to account for the considerable change from Saytzeff to Hofmann orientation induced by a change in the leaving group from chlorine to fluorine[328].

The low conversion to olefin and the use of drastic reaction conditions caused Brown to query the reliability of the 2-pentyl fluoride elimination[322]. To accommodate the fluoro substituent into the steric theory of orientation, it was suggested that the necessary extra solvation required in the transition state for the smaller developing fluoride ion could impart a larger steric requirement to this leaving group than to the other halogens. If isomerisation of olefinic products under the reaction conditions can be prevented, then this latter point could be verified by studying the olefin distribution resulting from elimination in dipolar aprotic solvents in which anion solvation is lacking.

Bartsch and Bunnett's recent extensive investigations[328] on elimination from 2-hexyl halides with sodium methoxide in methanol support the proposals suggested by Saunders for the 2-pentyl halide results. The results, listed in Table 20, show that slightly lower yields of the Hofmann product are obtained from the 2-hexyl series than the 2-pentyl series, suggesting slightly more double-bond character and slightly less carbanion character in the transition states leading to the hexenes. This small change could result from the difference in reaction media or greater electron release by the inductive effect from

TABLE 20

KINETICS OF REACTION OF 2-HEXYL HALIDES WITH SODIUM METHOXIDE IN METHANOL[328]

X in 2-hexyl-X	*1-Hexene*			*trans-2-Hexene*			*cis-2-Hexene*		
	Yield (%)	$\Delta H^{\ddagger}$	$\Delta S^{\ddagger}$	*Yield (%)*	$\Delta H^{\ddagger}$	$\Delta S^{\ddagger}$	*Yield (%)*	$\Delta H^{\ddagger}$	$\Delta S^{\ddagger}$
F	69.9	30.2	−13.5	21.0	29.1	−16.6	9.1	29.0	−18.7
Cl	33.3	27.1	−9.0	49.5	25.1	−11.2	17.1	26.2	−10.6
Br	27.6	25.3	−6.2	54.5	23.7	−7.1	17.9	24.1	−8.5
I	19.3	24.7	−6.0	63.0	22.1	−7.8	17.6	23.1	−7.5

The data above refers to a temperature of 100°C and the results were obtained by extrapolation of kinetic results measured at a number of different temperatures. Activation enthalpies are quoted in kcal.mole^{-1} and activation entropies in cal.deg^{-1}.mole^{-1}. The activation entropies for 1-hexene formation have been statistically corrected. The average standard deviation in the activation enthalpy is ±0.5 and in the activation entropy is ±1.5. The information concerning the competing substitution reaction is not included.

a butyl than a propyl group. Surprisingly the *trans/cis* ratios are at variance with this interpretation, emphasising the danger in drawing conclusions from small differences in reactivities. For formation of all three hexenes, the activation enthalpy decreases along the series F > Cl > Br > I and the activation entropy becomes progressively less negative along the series I ≃ Br > Cl > F. In agreement with recent findings[329], the difference of 7–11 cal.deg^{-1}.mole^{-1} between the activation entropies for elimination from 2-hexyl fluoride and 2-hexyl iodide can be attributed to the incorporation of an extra methanol molecule in the transition state for elimination from the fluoride. As the authors point out, this extra molecule is most probably associated by hydrogen bonding to the developing fluoride ion and it is evident from Courtauld's models that this solvating methanol molecule does not contact the beta propyl group in any of the transition states leading to 2-hexene provided that the propyl and the methyl group of methanol are rotated into favourable conformations. Some rotational entropy loss may result from this restriction but solvation of the fluoride does not require increased steric requirements.

An interesting linear free energy relationship between the rate coefficients for the formation of 1-olefin and *cis*- and *trans*-2-hexenes along the series of halogens was noted. Such linear plots suggest the special increment of $\Delta G^{\ddagger}$ for *trans*-2-hexene formation is imposed halogen by halogen in linear proportion to the general increase in $\Delta G^{\ddagger}$ for the 1-hexene formation. The general increase in $\Delta G^{\ddagger}$ is not regarded as of steric origin, but is attributed to the energy required to break the C_{α}–X bond, a decrease in transition-state entropy owing to solvent incorporation, and other related electronic adjustments. As pointed out by the authors, it would be an incredible coincidence if steric and non-steric increments of $\Delta G^{\ddagger}$ were imposed in constant proportion through two series each of four reactions.

The present available evidence seems to exclude steric control as determined by the size of the halogen leaving groups, but this does not exonerate the more bulky substituents. It may be difficult to formulate precisely the regions dominated by steric control. For instance, in the elimination from the series of sulphonium salts (121)[313],

R in $RCH_2CH_2\overset{+}{S}(CH_3)_2$	H	CH_3	$(CH_3)_2CH$	$(CH_3)_3C$	
10^5k_2 *(l.mole*$^{-1}$*.sec*$^{-1}$*)* *with* $\bar{O}C_2H_5/C_2H_5OH$ *at 64°C*	79	29	16	0.43	(121)
R in $RCH_2CH(C_6H_5)\overset{+}{S}(CH_3)_2$	H	CH_3	$(CH_3)_2CH$	$(CH_3)_3C$	
10^5k_2 *(l.mole*$^{-1}$*.sec*$^{-1}$*)* *with* $\bar{O}C_2H_5/C_2H_5OH$ *at 35°C*	158	40	50	123	

the sharp decrease in the rate of elimination observed with a *t*-butyl group is offset by the incorporation of an alpha phenyl substituent, which causes a shift to a transition state possessing more double-bond character in which the interference between the *t*-butyl group and X is lessened by a flattening of the system and an increase in the attenuation of the C_α–X bond. Steric influences in this case may be masked as they tend to aid the formation of the transition state.

In an attempt to separate the electronic and steric effects of the leaving group, Colter and Johnson[330] investigated the orientation of elimination from substituted 2-pentyl sulphonate esters. Their results, using ethanol containing ethoxide as the base, were complicated by competing unimolecular decomposition for which a correction was carefully applied. Despite an eighty-fold increase in the bimolecular rate coefficient in changing the leaving group from the *p*-amino- to the *p*-nitro-benzenesulphonate, the ratio of Saytzeff to Hofmann product varied from only 1.37 to 1.46 and the *trans*-2-pentene/*cis*-2-pentene ratio from 1.98 to 2.14. These results, from which a Hammett reaction constant of 1.35 at 50°C was calculated, suggest a small role for the electronic properties of the leaving group in orientation of elimination. Presumably the transition state is in the carbonium-like region in agreement with the ease of unimolecular decomposition. Whether the rate enhancement is dependent on increased C–X heterolysis, or increased acidity and hence greater carbon–hydrogen bond breaking in the transition state with the more electron-withdrawing substituents, is not clear. Possibly the solvent dominates electronic variations on orientation in this case. A change to a less polar solvent (*t*-butyl alcohol, and *t*-butyl alcohol mixtures with dioxan and dimethyl sulphoxide) produced more significant variations in the olefin proportions arising from the bimolecular elimination of 2-methyl-3-pentyl arenesulphonates[331]. Competing unimolecular elimination was slightly more depressed in this case. Hammett reaction constants of 1.51 (50°C) and 2.40 (25°) in 50% *t*-butyl alcohol–dioxan and 25% *t*-butyl alcohol–dimethyl sulphoxide, respectively, were observed. In the same media, a change from *p*-methoxy- to *p*-nitro-benzenesulphonate was accompanied by changes from 48.9% to 59.3% and 47.9% to 61% of the Saytzeff product, respectively. These variations were interpreted in terms of increased C_α–X heterolysis and possibly some greater weakening of the C_β–H bond, producing greater double-bond character, with increasing electron-withdrawing power of the leaving group. In apparent support of this conclusion, in 50% *t*-butyl alcohol–dioxan, the *trans*/*cis* Hofmann olefin ratio changes from 1.48 for the *p*-methoxy ester to 3.20 for the *p*-nitro substrate. The significance of these results is not certain, as in 25% *t*-butyl alcohol–dimethyl sulphoxide *trans*/*cis* ratios of 20.7 (*p*-$N(CH_3)_2$ ester) and 34.8 (3,4-dichloro ester) were reported, these values being greater than the olefin distribution observed under equilibrium conditions in dimethyl sulphoxide containing potassium *t*-butoxide at 55°C (*trans*/*cis* = 6)[332]. Factors other than double-bond character obviously affect the *trans*/*cis*-olefin ratios. Moreover, although the Hofmann–Saytzeff olefin distribution differs little in *t*-butyl alcohol from that observed in *t*-butyl alcohol–dimethyl sulphoxide containing base, the *trans*/*cis* Hofmann olefin ratios are quite markedly different. The danger of comparing these two apparent mechanistic criteria in solvents of low dielectric constant is thus emphasised, and both parameters are obviously not a simple function of the double-bond character of the transition state.

2.5.4 *The influence of the base and solvent*

A change in the reaction medium can greatly influence the orientation of elimination. A gradual increasing tendency towards formation of the least-substituted olefin from tertiary pentyl bromide is observed with apparent increasing steric requirements of the base (122)[309]. However,

% l-olefin from t-pentyl bromide	30	72	78	89	(122)
Base	KOC_2H_5	$KOBu^t$	$KOAm^t$	$KOC(C_2H_5)_3$	
Solvent	C_2H_5OH	*t*-BuOH	*t*-AmOH	$(C_2H_5)_3COH$	

the polarity of the solvent decreases and the strength of the base increases along the same series. As further support for the steric interpretation, studies involving heterocyclic bases of differing steric environment but similar pK_a were quoted (123)[318].

% l-olefin from t-pentyl bromide	25	25	30	44.5	(123)
Base	4-picoline	pyridine	2-picoline	2,6-lutidine	
pK_a of base	6.02	5.17	5.97	6.75	

Unfortunately, the kinetics of elimination were not measured, and in view of the tertiary nature of the substrate, the incursion of a unimolecular decomposition is likely in the "pyridine" bases which were also employed as the solvents.

The variation in the product distribution could equally well have been influenced by the change in solvent which might have induced a change in the type of transition state. Further evidence, again interpreted in favour of the steric theory, concerns the increased yield of the Hofmann product resulting from the elimination of 2-butyl halides when the base is changed from *t*-heptoxide to the smaller *t*-butoxide (Table 21)[323]. However, despite drastic changes in the steric environment of the anion, only small differences are noted in the product distributions and in tetrahydrofuran and dimethyl sulphoxide these differences become very insignificant. The solvent certainly plays a major role in the determination of the direction of elimination, but

References pp. 350–364

TABLE 21

PRODUCTS OF ELIMINATION FROM 2-BUTYL HALIDES WITH *t*-BUTOXIDE AND *t*-HEPTOXIDE IN A VARIETY OF SOLVENTS[323]

2-Butyl-X		*X = chloride*		*X = iodide*	
Base	*Solvent*	*1-Butene (%)*	*trans-/cis-2-Butene*	*1-Butene (%)*	*trans-/cis-2-Butene*
$KOBu^t$	*t*-BuOH	67	1.28	34	2.04
$KOBu^t$	Toluene	66.8	1.03	36.1	1.70
$KOBu^t$	THF	48.6	2.92	20.4	3.57
$KOBu^t$	DMSO	41.2	4.07	19.6	3.47
$KOC(C_2H_5)_3$	Triethyl-carbitol	80.1	1.14	49.3	1.50
$KOC(C_2H_5)_3$	Toluene	77	1.29	46.8	1.75
$KOC(C_2H_5)_3$	THF	49.1	2.72	25.1	4.06
$KOC(C_2H_5)_3$	DMSO	43.6	4.22	21.4	3.79

The standard deviation in the yield is 1% and in the *trans-/cis*-2-butene ratio is 0.20%.

whether this is due to a modification in the extent of ion association and solvation of the base, and hence its reactivity, or due to a solvent effect on the transition state, is not clear. It should be noted that the comparisons are not all at the same temperature and knowledge of the activation parameters under the various reaction conditions would seem desirable. The results, however, do again show a lack of correlation (in the less polar solvents) between orientation and *trans-/cis*-olefin ratios.

In an attempt to estimate the relative importance of the solvent and the base, a systematic study, predominantly in the 2-butyl series, has been undertaken by Froemsdorf *et al.*[333–336]. Elimination from 2-butyl-*p*-toluenesulphonate gives an increased yield of 1-butene with an increase in the base strength (Table 22, 1–6 and 7–10 and 11–13). Base strength, as measured by the pK_a of the conjugate acid, is more important than base size as both 4-nitro- and the more sterically hindered 2-nitro-phenoxide give a similar distribution of olefins. A similar trend towards Hofmann orientation is induced by changing to a less polar solvent [Table 22, base $\bar{O}C_2H_5$ in C_2H_5OH (11), in *t*-BuOH (8)], but in this case the activity of the base is also changing. Similar trends are shown in the eliminations from the halides, but with the sulphonium salt, both the influence of base strength (26, 27) and solvent polarity (25, 26) are reversed. The latter difference indicates that steric factors cannot be dominant in all the cases and also excludes a theory dependent on substrate acidity as the same directional trend would always be predicted. However, solvent changes on solvation of cations and neutral substrates act in an opposite direction and the changes in orientation can be explained by the influence of

TABLE 22

OLEFIN PRODUCTS FROM BASE-INDUCED ELIMINATIONS OF 2-BUTYL, 2-PENTYL AND 3-PENTYL DERIVATIVES AT 55°C

$$CH_3CH_2CHXCH_3 \xrightarrow{\text{base}} \text{1-butene} + \text{2-butene } (cis \text{ and } trans)$$

No.	*X*	*Base* (pK_a)[a]	*Solvent*	*1-Olefin (%)*	*trans-/cis-2-Olefin*
1[b]	OTs	$KOBu^t$(19–20)	DMSO	61	2.53
2[b]	OTs	KOC_2H_5(16)	DMSO	54	2.34
3[c]	OTs	KOC_6H_5(9.95)	DMSO	31±2	1.96
4[c]	OTs	KOC_6H_4.*p*-OCH_3(10.2)	DMSO	33±3	1.97
5[c]	OTs	KOC_6H_4.*p*-NO_2(7.14)	DMSO	16±2	1.81
6[c]	OTs	KOC_6H_4.*o*-NO_2(7.23)	DMSO	16±2	1.85
7[c]	OTs	$KOBu^t$	*t*-BuOH	64	0.58
8[c]	OTs	KOC_2H_5	*t*-BuOH	54	0.80
9[c]	OTs	KOH(14)	*t*-BuOH	50±1	0.83
10[c]	OTs	KOC_6H_5	*t*-BuOH	34±2	1.24
11[b]	OTs	KOC_2H_5	C_2H_5OH	35	1.95
12[b]	OTs	KOC_2H_5	DMF	52	2.60
13[b]	OTs	KOC_2H_5	THF	59	1.65
14[d]	Br	KOC_2H_5	C_2H_5OH	19	3.35
15[d]	Br	KOC_2H_5	*t*-BuOH	38	2.19
16[d]	Br	KOC_2H_5	DMSO	27	3.57
17[d]	Br	$KOBu^t$	DMSO	31	3.65
18[d]	Br	$KOBu^t$	*t*-BuOH	53	1.64
19[e]	Cl	$KOBu^t$	*t*-BuOH	67	1.29
20[e]	I	$KOBu^t$	*t*-BuOH	33	2.19
21[e]	OTs[f]	$KOBu^t$	*t*-BuOH	74	0.44
22[e]	OTs[g]	$KOBu^t$	*t*-BuOH		0.54
23[e]	OTs[g]	$KOBu^t$	THF		2.22
24[e]	OTs[g]	$KOBu^t$	Diglyme		2.33
25[h]	$\overset{+}{S}(CH_3)_2$	KOC_2H_5	C_2H_5OH	86	0.95
26[h]	$\overset{+}{S}(CH_3)_2$	KOC_2H_5	*t*-BuOH	82	1.89
27[h]	$\overset{+}{S}(CH_3)_2$	$KOBu^t$	*t*-BuOH	79	2.81
28[h]	$\overset{+}{S}(CH_3)_2$	KOC_2H_5	DMSO	73	3.40
29[h]	$\overset{+}{S}(CH_3)_2$	$KOBu^t$	DMSO	73	3.40

[a] pK_a of conjugate acid.
References: [b], 333; [c], 336; [d], 334; [e], 338; [h], 337
[f] Substrate is a 2-pentyl compound.
[g] Substrate is a 3-pentyl compound.

basic strength and solvent polarity on the carbanion character of the transition state.

A particular noteworthy point is the random variation in the Saytzeff *trans-/cis*-olefin ratios with regular trends in the Saytzeff–Hofmann distribution. In particular, eliminations with *t*-butoxide in *t*-butyl alcohol on the toluenesulphonate esters give a greater yield of the thermodynamically less stable *cis* product. To account for this peculiarity Brown and Klimisch[338] offered an explanation which depends on the steric factors, (124) below. Interactions between the alkyl substituents are regarded as minor effects compared with the greater freedom of rotation of the base and the leaving group in the transition state leading to *cis*-olefin as against that giving the *trans*-olefin. It is surprising that this explanation does not apply to the studies in tetrahydrofuran and diglyme (Table 21, p. 258). Ease of solvation and the possibility of ion pairing have also to be considered.

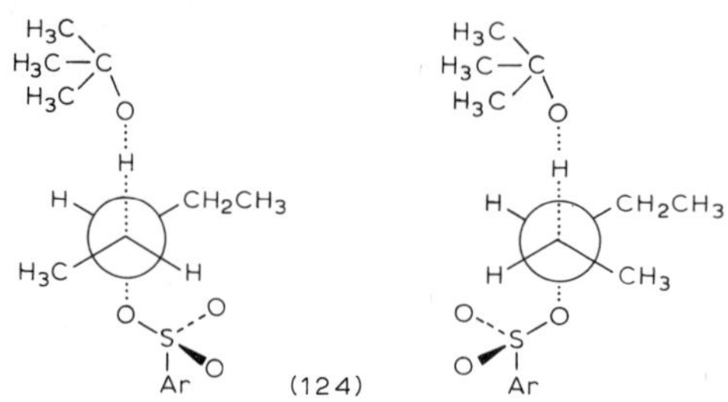

(124)

Ar = p-$CH_3 \cdot C_6H_4$–

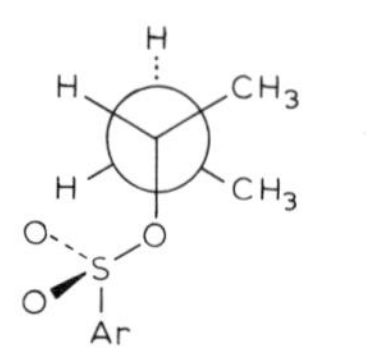

(125)

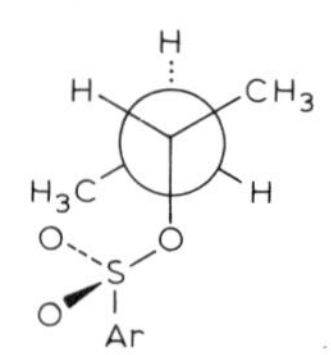

(126)

Other investigators prefer to consider the preference for *cis*-olefin formation arises as an artefact of the solvent, as the predominance of *cis*-olefin persists with bases other than *t*-butoxide as long as the solvent is *t*-butyl alcohol (Table 22, 7, 8, 9). The greater Hammett reaction constant indicates that C_α–X bond breaking is much less advanced in the transition state for bimolecular elimination from 2-phenylethyl tosylate in *t*-butyl alcohol than in ethanol (Section 2.2.2(a)). Consequently, in the former solvent, steric effects of the leaving group assume a major role in the transition state and as rotation about the S–O bond is more facile in (125) than (126), see above, *cis*-olefin formation occurs more easily than *trans*-olefin formation. Ethyl–tosylate interaction is greater than methyl–tosylate interaction, so the tendency for *cis*-olefin formation is greater for 2-pentyl than 2-butyl tosylate. A change to a weaker and more polarisable base or a more ionising solvent leads to increased C_α–X bond breaking in the transition state and a concomitant increase in double-bond character. Alkyl–alkyl interaction becomes more important than alkyl–tosylate interactions, resulting in preferential formation of the *trans* isomer.

It would be interesting to see how the *trans*-/*cis*-olefin ratio varies in *t*-butyl alcohol with changing electron-withdrawing properties of the sulphonate esters in the 2-butyl and 2-pentyl series. The Hofmann olefin *trans*/*cis* ratios reported for elimination from 3-methyl-2-butyl-*p*-*N*,*N*-dimethylaminobenzenesulphonate ($k_{trans}/k_{cis} = 3.3$ at 50°C) and 3-methyl-2-butyl-*p*-toluenesulphonate ($k_{trans}/k_{cis} = 2.0$ at 50°) are at apparent variance with the results for the 2-butyl ester. However, in this series methyl–isopropyl interaction may be more serious than methyl–tosylate interaction and consequently the more stable *trans*-olefin is formed predominantly.

With a gradual increase in the complexity of the alcohol/alkoxide medium, elimination from 2-methylpent-2-yltrimethylammonium ion yields progressively more of the less substituted olefin, (127).

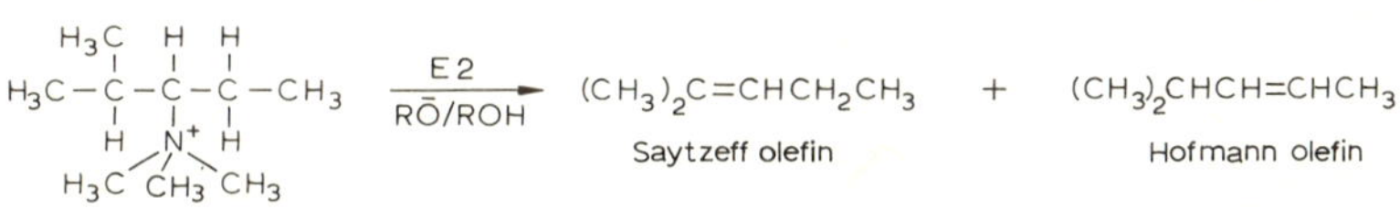

(127)

% Hofmann product	61 ± 1	62	58 ± 1	64 ± 2	72 ± 1	76 ± 1	84 ± 2
trans-/cis-Hofmann olefin	5.2	4.9	4.3	4.7		9.0	12.7
R in RŌ/ROH	C_2H_5	*n*-Pr	allyl	*n*-Bu	*i*-Pr	$C_2H_5CHCH_3$	*t*-Bu

The Hofmann product *trans-/cis*-olefin ratios are again quite large and the shift towards more extreme Hofmann orientation follows the same trend as that for elimination from the uncharged substrates. Bearing in mind the greater carbanion character associated with the transition state for elimination from 2-phenylethyltrimethylammonium ion in ethanol than in *t*-butyl alcohol (Section 2.2.2(a)), Saunders and Feit[339] prefer Brown's steric hypothesis rather than an explanation in terms of carbanion character to explain the observed results. Undoubtedly the difference in behaviour between these ammonium and the sulphonium salts (see Table 22) with changing reaction medium is associated with the much larger size of the leaving group in the former series.

Elimination from 2-pentyltrimethylammonium ion gives almost quantitative yields of 1-pentene with all bases but interestingly the recovered 2-pentene is mainly the *cis* isomer. Saunders and Feit[339], suggest that when double-bond character is little developed and steric interaction between alkyl substituents coming *cis* across a double bond is minimal, then the ease of solvation of the leaving group and approach of the base to the beta proton may become the controlling factors. Such factors will be more easily accomplished in the transition state leading to *cis* alkene as both alkyl substituents are on one side of the molecule.

There is little doubt that both the base and the solvent influence the transition state and hence the orientation of elimination. In a proton transfer reaction, a more reactant-like transition state is expected for a stronger base as the force constants in the C–H and H–B bonds equalise with a smaller change in the bonding strength of the C–H bond[341]. However, in an elimination reaction the situation is more complex. Polarisability complications arise as simultaneous weakening of the C_α–X bond can modify the extent of proton transfer. Leaving groups which are more sensitive to changes in electron density at the beta carbon will exert a greater distortion. The tendency of the C_α–X bond to dissociate into ions and the transmission of charge from the beta carbon are dependent on solvation of the transition state. Consequently any change in solvent must be carefully considered. An investigation of the extent of proton transfer to a series of bases, in a given solvent in an elimination reaction is a worthwhile topic for investigation.

2.5.5 Cyclic systems

In the more rigid cyclohexyl systems, steric strain imposed by the leaving group cannot be relieved by simple bond rotation as in the acyclic systems, and with the leaving group in an axial position, elimination occurs to give both Saytzeff and Hofmann olefins from the same substrate molecular conformation. The steric theory thus predicts little influence of the size of the leaving group on the orientation in cyclohexyl systems, and control by mesomeric factors[322]. These postulates are supported on experimental grounds, as unlike the acyclic trimethylammonium ions, the cyclohexyl analogues undergoes predominantly bimolecular elimination in accordance with the Saytzeff rule. In fact both neomenthyl chloride and neomenthyltrimethylammonium ion give a similar predominance of the Saytzeff product, 3-menthene[32,226,306].

Hughes and Wilby[32] prefer an alternative explanation to account for the difference in orientation pattern between the cyclohexyl and acyclic "onium" salts. They envisage steric compression between the bulky trimethylammonium ion and the cyclohexyl ring and this gives rise to an increased attenuation in the C_α–N bond in the transition state. Consequently, double-bond character is greater in the cyclohexyl eliminations and mesomeric rather than inductive control is imposed by the beta alkyl substituents. In support of this interpretation, the nitrogen isotope effect (Section 2.2.1(f)) is greater in the elimination from *cis*-2-phenylcyclohexyltrimethylammonium ion than from the 2-phenylethyl analogue (part of this difference may arise from the alpha and beta alkyl residues in the cyclohexyl substrate), and cyclic trimethylammonium ions more readily undergo unimolecular decompositions than the acyclic derivatives[32]. The bulky trimethylammonium ion may be an anomalous leaving group in this case and it would be worthwhile to study the elimination of other more "normal" Hofmann leaving substituents in the cyclohexyl series.

The Hughes and Wilby explanation has recently been challenged by Booth *et al.*[455]. The Hofmann elimination of a series of *cis*-2-alkylcyclohexyltrimethylammonium ions all gave similar yields (91–95%) of 1-alkylcyclohexene despite the alkyl substituent being varied from methyl to ethyl and isopropyl. These results are not explicable in terms of electromeric influences of the alkyl substituents, and elimination occurring from a twist-boat conformation, with a dihedral angle of 165° between the eliminating fragments in the transition state, was suggested.

Elimination from 1-methylcycloalkyltrimethylammonium ions has attracted interest, but in view of stereochemical complexities in the parent alicyclic series[266,267] the results warrant only brief mention. Despite the greater thermodynamic stability of the *endo*-olefin[340], the *exo*-olefin is the major product in all cases except elimination from the nine-membered ring (Table 23)[252]. The ease of attaining coplanarity in the transition state undoubtedly plays a decisive role, this factor being emphasised by the cyclohexyl results. The *endo*-olefin has to arise from an unfavourable axial conformation of the bulky trimethylammonium ion while the *exo* product can be formed with the leaving group in the equatorial conformation.

TABLE 23

HOFMANN ELIMINATION PRODUCTS OF 1-METHYLCYCLOALKYLTRIMETHYLAMMONIUM HYDROXIDES[252]

(Reproduced with permission from Cope and Trumbull[252].)

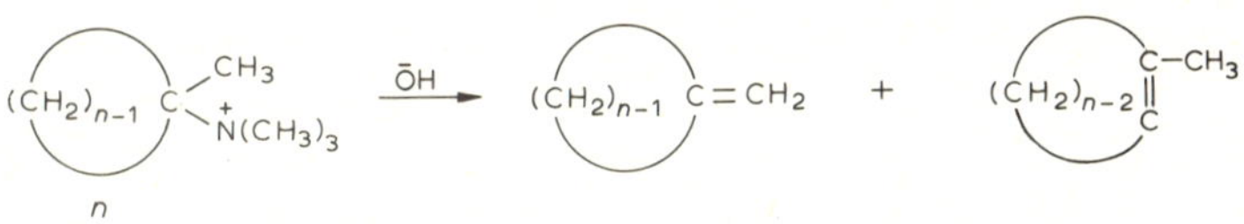

n	*exo-Olefin* (%)	*endo-Olefin* (%)
5	91	9
6	99	1
7	78	22
8	64	37
9	48	52
10	66	34

2.5.6 *Summary*

The acceptance of mesomeric control to account for the Saytzeff rule appears to be almost universal. In connection with the Hofmann rule, there is experimental evidence to support both steric and inductive influences of substituents but it is not possible to define rigid limits within which either of these influences can be assigned with certainty. It is not surprising that the direction of elimination cannot be predicted within the bounds of a simple theory in view of the relatively small activation energy differences (approximately 0.8 kcal.mole^{-1} at room temperature) required to induce a change from a 2:1 to a 1:2 product ratio[14].

Most arguments assume an *anti* stereospecificity for bimolecular elimination and this may be one of the major failings for the lack of understanding of the meaning of the *trans-/cis*-olefin ratios[682]. Bearing in mind the recent stereochemical discoveries (Section 2.3.1(e)), with the larger leaving groups in medium size and large acyclic molecules (*e.g.* $\geqslant$ 3-hexyl substrates), a change in orientation may reflect competition between *anti* Hofmann elimination and *anti* changing to *syn-clinal* Saytzeff elimination. Thus another theory may be appropriate to explain orientation in some of the more cumbersome substrates.

2.6 PYROLYTIC ELIMINATIONS: GENERAL TYPES AND REACTION CONDITIONS

A wide variety of organic substrates decompose to form almost quantitative yields of olefins at elevated temperatures in the absence of added reagents[342]. Under the feebly acidic reaction conditions the olefins are unlikely to undergo isomerisation which is often encountered in strongly acidic and basic solutions, and being free from solvent they are obtained in a high degree of purity. Although mechanisms of some pyrolytic processes have been compared to those which occur in solution, they are in general quite distinct as evidenced by the contrasting stereochemical and orientational behaviour. Most frequently preparative and kinetic investigations have been carried out on esters (128), xanthates (129), amine oxides (130), and alkyl halides (131), and discussion in the text is limited to these four classes.

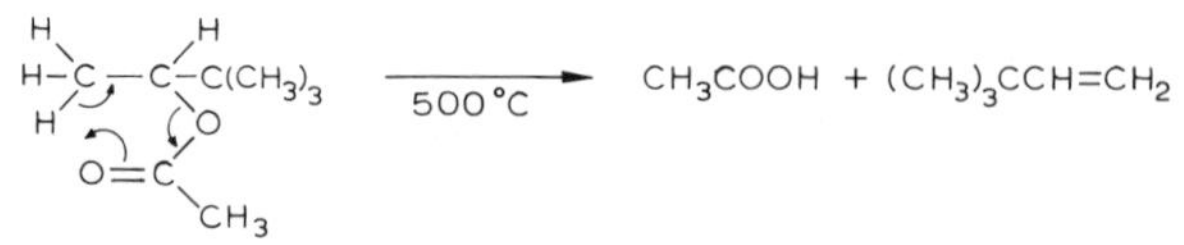

(128)

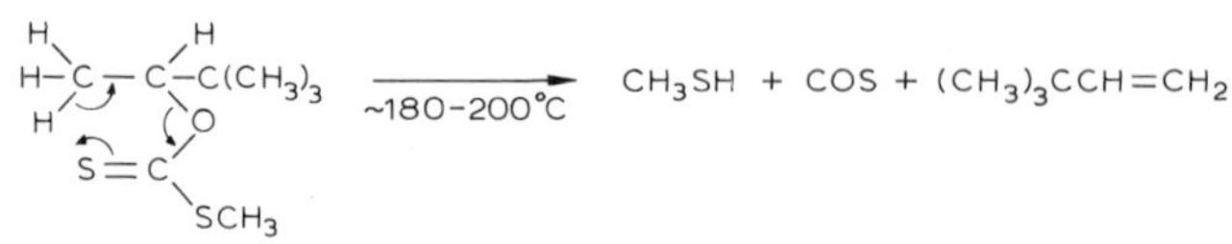

(129)

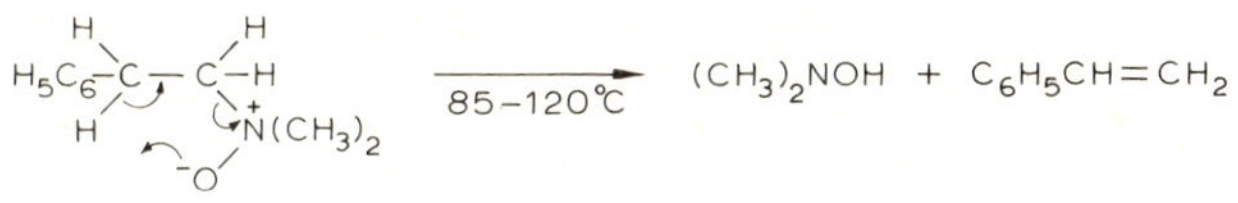

(130)

(131)

The first three types are important as synthetic methods for alkene preparations. Amine oxides (usually the dimethylamine oxide) generally prepared *in situ* are easily decomposed on concentration of the solution[252]. Xanthates (usually the *S*-methyl ester) decompose on distillation, which is often performed under reduced pressure to minimise sulphur contamination of the olefin[343]. Higher temperatures are required to decompose esters and most often the ester vapours are rapidly swept in a stream of nitrogen through a hot tube packed with glass wool or glass beads[344]. Decompositions of alkyl halides are also performed in the vapour phase, but generally the results are of more theoretical than practical interest due to the reversibility of the decomposition. At the high reaction temperatures the equilibrium favours the alkene but cooling to room temperature to isolate the product shifts the equilibrium in favour of the alkyl halide. From a preparative point of view it is thus necessary to "freeze" the high temperature equilibrium by performing the pyrolysis in the presence of a base such as ammonia. Alkyl iodides are omitted from the discussion as they react to give a mixture of alkenes and alkanes, *e.g.*

$$(CH_3)_2CHI \xrightarrow{290-357^\circ C} CH_3CH{=}CH_2 + HI$$
$$HI + (CH_3)_2CHI \longrightarrow CH_3CH_2CH_3 + I_2 \qquad (132)$$

the alkanes arising from a side reaction of the substrate with hydrogen iodide[345,346]. Although esters are often the most easily prepared, the preparation of thermally unstable olefins is more successfully accomplished by the lower temperature pyrolyses of the xanthates or amine oxides.

The intramolecular E_i mechanism is most commonly accepted to account for these pyrolytic processes. Considering the usual electronegativity of the atoms, the cyclic transition states as designated (128–131) simulate the last step of the base-induced ylid mechanism. Of course, for a completely concerted cyclic process a distinction between heterolytic and homolytic reactions is not possible and the direction of electron transport as either single or paired electrons is uncertain[347]. However, if considerable charge separation occurs in the transition state, due to a departure from a completely synchronous process, then specific experiments can be designed to determine the direction of electron transport. For example, the variation in the value of the isotope effect

(k_{RCOOH}/k_{RCOOD}) with the changing of the substituent in the benzene ring in the decarboxylation of substituted benzoylacetic acids[384] is in accord with a cyclic proton transfer (133) rather than a cyclic hydride transfer (134).

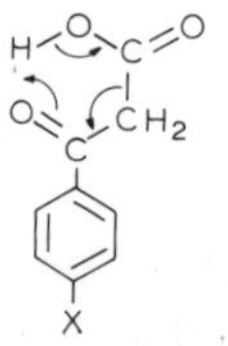

(133)

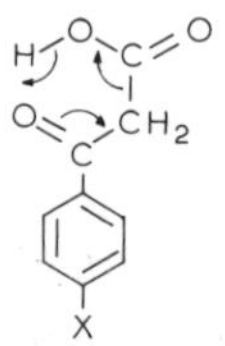

(134)

In a hydride transfer the two electrons contributed by the hydrogen cement the three nuclei together in the transition state in a strong short bond which is highly covalent and relatively non-polarisable. In the transition state for a proton transfer, a non-bonding or anti-bonding orbital is involved and this results in repulsion between the electron clouds of the two larger atoms giving rise to weaker, larger, more ionic and more polarisable bonding between the three nuclei which is consequently more susceptible to substituent effects[349].

2.7 STEREOCHEMISTRY

2.7.1 *Esters and xanthates*

The pyrolyses of esters and xanthates involve six-membered cyclic transition states and these are presumably almost strainless and "chair-shaped" rather than planar and eclipsed. Thus, they are more realistically regarded as occurring from *syn-clinal* rather than a *syn* conformation of the eliminating fragments in the substrate.

(a) Acyclic systems

Comparison of diastereoisomeric pairs has been used to demonstrate the "*syn*" nature of pyrolytic elimations. Both acetates of 2-deutero-1,2-diphenylethanol undergo pyrolysis to give *trans*-stilbene[27] but the product from the *erythro* substrate is labelled with deuterium (135) while that from the *threo* substrate only contains hydrogen (136), *viz.*

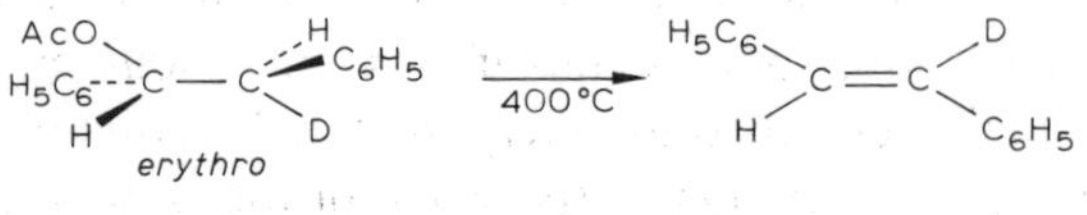

(135)

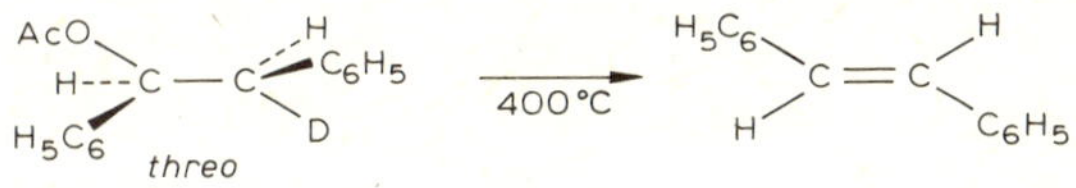

(136)

The same acetates give products labelled *vice versa* when *anti*-elimination is induced with potassium *t*-butoxide. The great preference for *trans*-stilbene formation is caused by the necessity to avoid eclipsing of the phenyl groups in the transition states. Similar "*syn*" stereospecificity is demonstrated for pyrolysis of 2-butyl-3-phenylxanthates[350], the *threo* isomer yielding *cis*-dimethylstyrene (137) while the *erythro* isomer gives the *trans*-olefin (138).

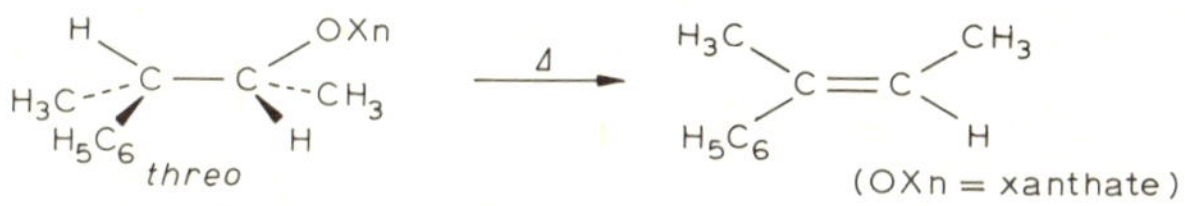

(137)

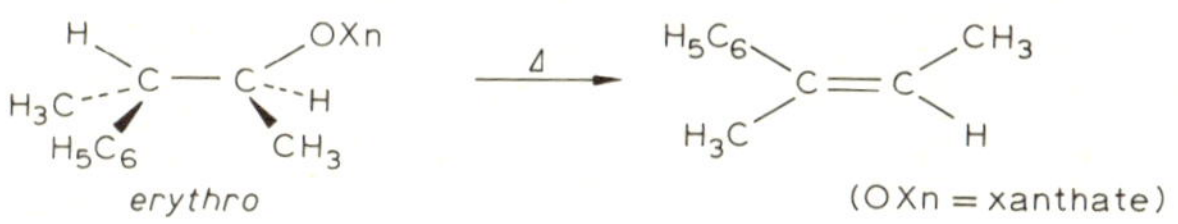

(138)

(*b*) *Alicyclic systems*

In the rigid indanyl system, *syn*-elimination from the *trans*-2-methyl-1-acetate occurs at a much lower temperature than the decomposition of the *cis* substrate[351,352], *viz.*

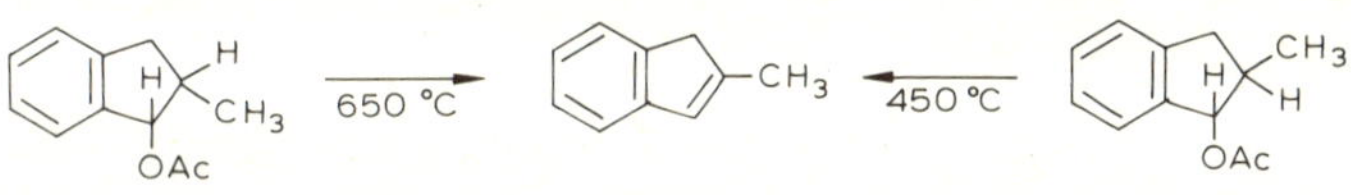

(139)

Realising the significance and necessity of "*syn*"-elimination in E_i processes, Barton[353] showed that this concept correctly predicted the direction of elimination of xanthates and esters of many natural products, based on 5- and 6-membered alicyclic rings and he subsequently used this method to predict unknown configurations of certain compounds.

In the cyclohexyl series, stereochemical studies are performed on molecules possessing a "marker" substituent adjacent to the carbon bearing the leaving group, and elimination can give rise to two olefins, 1-R- and 3-R-cyclohexene, *viz.*

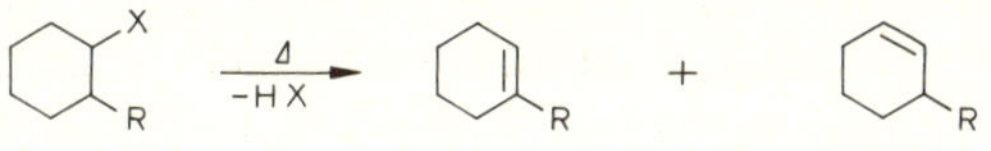

(140)

Assuming that the lower energy chair conformations rather than boat conformers are utilised, (it must be remembered, however, that at elevated temperatures the chair–boat and chair–chair equilibria will be considerably displaced from the ratios observed at room temperature), then the conformation of the *cis* and *trans* related substrates leading to the cyclohexenes can conveniently be listed as *syn-clinal-* or *anti*-eliminations (141).

Substrate conformation	*HX conformation leading to*		*Stereo-chemistry*
	1-R-cyclohexene	*3-R-cyclohexene*	
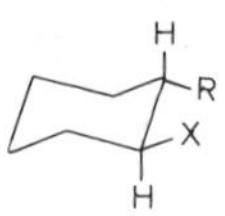 *trans*	ae	ae or ee	*syn-clinal*
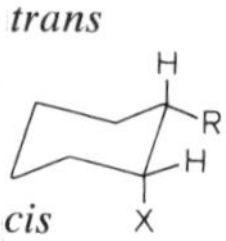 *cis*	aa	ae or aa	*anti, syn-clinal, anti*
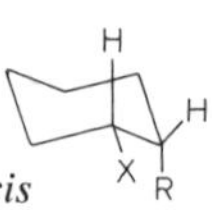 *cis*	ee	ee or ae	*syn-clinal*

(141)

If the leaving group X can attain an equatorial conformation, then the relative tendencies of the *cis* and *trans* isomers to give the 1-R-cyclohexene reflects the relative favourabilities of the axial–equatorial and diequatorial conformations of the eliminating fragments in the transition state.

Pyrolysis of *cis*-2-*t*-butylcyclohexyl xanthate gives exclusively 3-*t*-butylcyclohexene and serves to show the preponderence of *syn-clinal* over *anti* stereochemistry (Table 24). Molecular conformation is less rigidly fixed with an isopropyl group and a diequatorial *syn-clinal*-elimination leads to some 1-isopropylcyclohexene from *cis*-2-isopropylcyclohexyl xanthate. Although *trans*-2-methylcyclohexyl acetate eliminates to a mixture of alkenes[358], the *cis* substrate gives predominantly the 3-alkene[359]. Considering the similar size of methyl and acetate substituents and the elevated reaction temperature, a reasonable proportion of the *cis* substrate should exist with the acetate group in an equatorial position. The poor competitive nature of elimination into the more substituted branch presumably reflects the lower preference for *syn-clinal*-elimination from a diequatorial conformation than from an axial–equatorial disposition. It has been suggested that the restriction to diequatorial elimination may arise from buckling of the carbon skeleton which occurs as the 6-membered ring develops a certain amount of planarity in the transition state[10]. Pyrolysis of *cis*- and *trans*-1,2-dimethylcyclohexyl acetates affords similar stereochemical conclusions[360].

TABLE 24
PRODUCTS OF PYROLYSIS OF *cis*- AND *trans*-2-SUBSTITUTED CYCLOHEXYL ESTERS AND XANTHATES

2-substituent (R)	*X*	*1-R-cyclohexene (%)*	*3-R-cyclohexene (%)*	*Ref.*
cis-i-Pr	Xanthate	30	70	356
cis-*t*-Bu	Xanthate	0	100	356
cis-$SO_2C_7H_7$	Xanthate	90	10	355
cis-SC_7H_7	Xanthate	10	90	356
cis-C_6H_5	Xanthate	7	93	352
cis-$COOCH_3$	Xanthate	0	100	354
trans-$SO_2C_7H_7$	Xanthate	90	10	355
trans-SC_7H_7	Xanthate	90	10	356
trans-C_6H_5	Xanthate	88	12	352
trans-$COOCH_3$	Xanthate	97	3	354
cis-C_6H_5	Acetate	7	93	357
trans-C_6H_5	Acetate	87	13	357

Double-bond character in pyrolytic eliminations is much less developed than in E2 reactions as evidenced by the inability of strongly conjugative substituents such as phenyl and carbo-ethoxy ester to overcome the influence of unfavourable stereochemistry, Only with the strongly acid-strengthening *p*-tolylsulphonyl substituent do the *cis* substrates eliminate mainly to the 1-R-cyclohexene[361]. In this case the double-bond character of the transition state, being increased by the greater attenuation of C_β–H bond breaking, may become the dominating influence and elimination from a small steady-state concentration of the more active diequatorial isomer occurs, rather than the axial–equatorial elimination to give the 3-R-cyclohexene through a transition state possessing less double-bond character. It would be interesting to see if an increase in the 1-R-/3-R-cyclohexene ratio could be induced by the incorporation of electron-withdrawing substituents into the *para* position of a 2-phenyl substituent, which modifies the electronic properties but has little influence on the steric properties of the substrate.

Pyrolysis of the xanthates of 3-*p*-tolylthio-2-butanols shows "*syn*" stereospecificity, the *threo* and *erythro* substrates yielding the *cis*- and *trans*-2-*p*-tolylthio-2-butenes, respectively. However, both of the xanthates of the corresponding 3-*p*-tolylsulphonyl-2-butanols give the *cis*-2-butene as the major product. To explain this observation and the elimination from *cis*-2-*p*-tolylsulphonylcyclohexyl xanthate, Bordwell and Landis[356] suggested a two-step mechanism involving a dipolar intermediate, which in the case of the *erythro* isomer undergoes rotation about the C_α–C_β bond more rapidly than it eliminates (142). It is surprising that the opposite charges of the dipole should tend to separate to a maximum, although in this resulting conformation they are most favourably disposed for effective overlap of orbitals to give the double bond. Alternatively, the sulphone substituent may stabilise the dipolar intermediate which collapses to give the more stable olefin (143). Possibly more than one mechanism operates in this series.

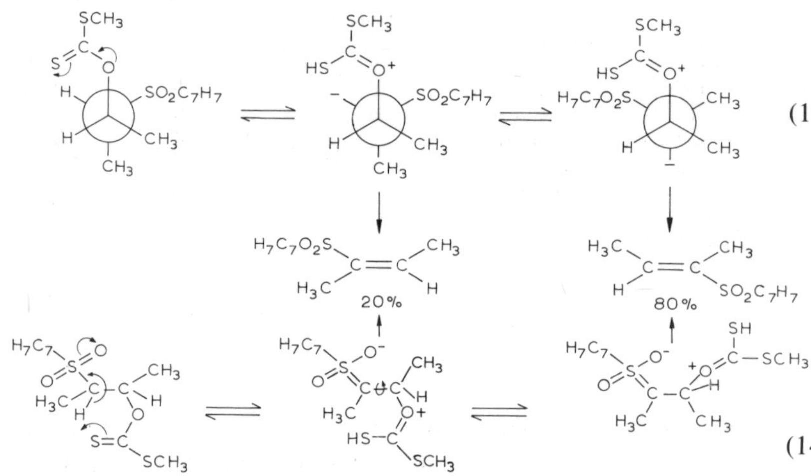

(142)

(143)

The cyclic esters (lactones) of five- and six-membered rings are stable to pyrolysis[362,363], but the larger analogues are more flexible and can attain the transition state for elimination, *viz.*

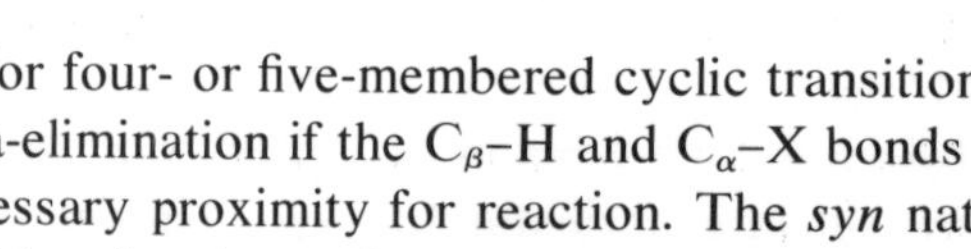

(144)

2.7.2 Amine oxides and alkyl halides

For four- or five-membered cyclic transition states, pyrolysis only leads to beta-elimination if the C_β–H and C_α–X bonds can eclipse so as to ensure the necessary proximity for reaction. The *syn* nature of the amine oxide decomposition has been demonstrated by elimination from the diastereoisomeric pairs of the 1,2-diphenylpropyl system[269]. In the cyclohexyl series, an eclipsed transition state is only achieved if elimination proceeds through the boat conformer and consequently the pyrolysis of 1-methylcyclohexylamine oxide gives an almost quantitative yield of the less stable *exo*-olefin. This orientation contrasts markedly with that of the six-membered cyclic transition states of the corresponding esters and xanthates which yield predominantly the *endo*-olefin, *syn-clinal* stereochemistry being preferred. For all the other ring systems, the relevant bonds are already eclipsed, or sufficient flexibility is present to enable coplanarity to be achieved without excessive expenditure of energy, that amine oxides decompose to yield the more stable *endo*-olefins. In the larger ring systems (*e.g.* nine- and ten-membered), in agreement with the lower non-bonded eclipsed interactions in the transition state, amine oxide pyrolyses give mainly the *trans*- rather than the *cis-endo*-olefin[364].

The stereochemistry of alkyl halide pyrolyses has been less thoroughly investigated. Considering the greater electronegativity of the halogens than oxygen and nitrogen, it would seem reasonable to expect the transition states for the E_i reactions of alkyl halides to possess more carbonium-ion character than the states utilised by the other substrates. There is considerable evidence to support this assumption (see later) but carbonium-ion character is certainly different and presumably less than is encountered in E1 reactions. Menthyl chloride decomposes to yield 2- and 3-menthenes in a similar ratio in both solvolytic and pyrolytic reactions[365] (145), but the corresponding decompositions of neomenthyl chloride give vastly different product ratios[366], *viz.*

(145)

Reaction	*Menthyl chloride*		*Neomenthyl chloride*	
	% 3-Menthene	*% 2-Menthene*	*% 3-Menthene*	*% 2-Menthene*
Pyrolysis	75	25	15	85
E1 solvolysis (see Table 15, p. 237)	68	32	99	1

These results support the "*syn*" nature of the pyrolyses, the menthyl substrate eliminating through a *syn*- (boat conformer) or *syn-clinal* conformation of HX (chair conformers) and orientation in favour of the Saytzeff rule indicating development of some carbonium-ion character in the transition state. The 3-menthene most likely arises from a *syn-clinal* diequatorial conformation of HX in the neomenthyl series. Despite the smaller size of the ring in the transition state in halide than in amine oxide pyrolyses, the necessity for eclipsing is reduced by a shift from a concerted to a more carbonium-ion-like transition state. It is worth noting that even though H and X are eclipsed in bornyl and isobornyl chloride, the products are indicative of Wagner–Meerwein rearrangements and bornylene is only a minor product[367], *viz.*

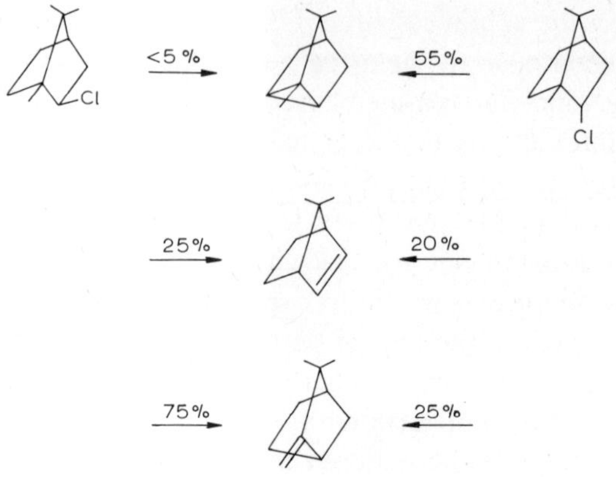

(146)

2.8 ORIENTATION

2.8.1 Acetates and xanthates

The majority of orientational studies refer to the pyrolysis of acetates although in cases in which corresponding xanthates have been decomposed the product distribution is found to be almost identical. Of course, bearing in mind the considerable difference in reaction temperature, the two pyrolyses may not always lead to the same product proportions and the more reactant-like nature of the xanthate reactions may cause them to be less sensitive to thermodynamic effects than the pyrolyses of the acetates.

The principal factors affecting orientation in acetate decompositions have been adequately summarised by DePuy and King[344]. Essentially three influences were recognised, these being termed statistical, steric and thermodynamic effects. Statistical control is observed in pyrolysis of simple aliphatic esters which under the elevated reaction temperatures experience little resistance to conformational rotation and the number of beta hydrogen atoms in each branch determines the direction of elimination (147)[368–370]. A distortion in statistical control is imposed by the steric influence of a *t*-butyl substituent (148), and is also illustrated by the predominance of *trans*- over *cis*-olefin formation (148, 149) due to eclipsing effects[371]. The latter example, however, may also arise from thermodynamic influences which are more certainly demonstrated by preferential elimination towards a phenyl rather than an alkyl substituent (150)[372]. The influence of substituents on olefin stability rather than beta hydrogen acidity seems more critical as elimination occurs more often towards a *p*-methoxyphenyl rather than a phenyl substituent (151)[373].

$$CH_3CH_2CH(OAc)CH_3 \xrightarrow{\Delta} \underset{43\%}{CH_3CH{=}CHCH_3} + \underset{57\%}{CH_3CH_2CH{=}CH_2} \quad (147)$$

$$CH_3CH_2C(CH_3)_2OAc \xrightarrow{\Delta} \underset{76\%}{CH_3CH_2C(CH_3){=}CH_2} + \underset{24\%}{CH_3CH{=}C(CH_3)_2}$$

$$CH_3CH_2CH(OAc)CH_2C(CH_3)_3 \xrightarrow{\Delta} \underset{70\%(5\%\ cis,\ 65\%\ trans)}{(CH_3)_3CCH{=}CHCH_2CH_3} + \underset{30\%(9\%\ cis,\ 21\%\ trans)}{(CH_3)_3CCH_2CH{=}CHCH_3} \quad (148)$$

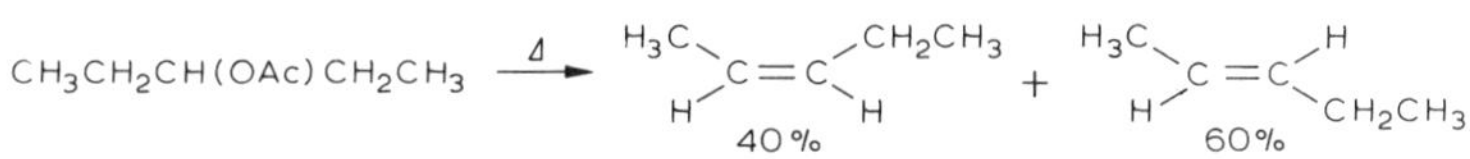
(149)

$$C_6H_5CH_2CH(OAc)CH_3 \xrightarrow{\Delta} \underset{75\%}{C_6H_5CH{=}CHCH_3} + \underset{25\%}{C_6H_5CH_2CH{=}CH_2} \quad (150)$$

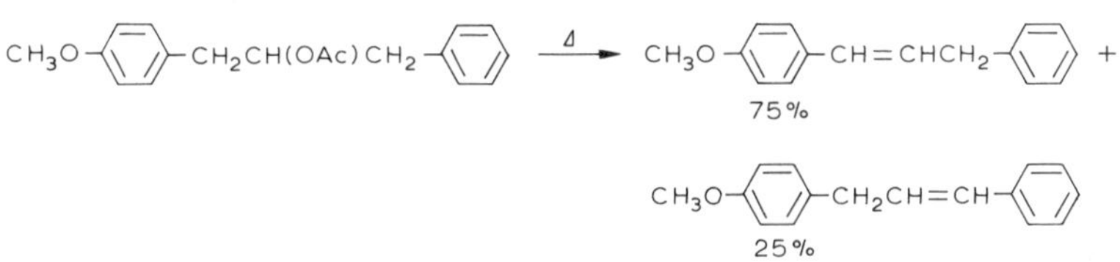
(151)

Unfavourable stereochemistry and restricted conformational rotation can impose complicating factors on orientation in the alicyclic series. However, for substrates in which elimination into either branch can occur with *syn-clinal* stereochemistry, Saytzeff orientation can be recognised as illustrated by pyrolysis of menthyl acetate (or xanthate) to give 2-menthene (35%) and 3-menthene (65%). Elimination from *trans*-2-carbomethoxycyclohexyl acetate (Table 24, p. 269) gives mainly the 1-R-cyclohexene[354]. The ester group causes more than a one hundred-fold increase in rate and the ratio of conjugated to non-conjugated olefins (1-R-/3-R-cyclohexene) is greater than predicted from the thermodynamic stabilities of the olefins alone. Consequently, additional to its influence on double-bond stability the ester function also enhances the acidity of the beta hydrogen atom.

2.8.2 *Amine oxides and alkyl halides*

The orientational behaviour of amine oxide pyrolyses has been adequately summarised by Cope and Trumbull[252]. As for the acetate decompositions, orientation in the simple alkyl systems is controlled primarily by statistical factors but departure from this influence is noted with the bulky *t*-butyl substituent and the acid strengthening beta phenyl substituents[163]. Eclipsing effects are greater in the planar five-membered transition states than in the puckered six systems and this is borne out by the greater preference for *trans*-olefin formation from amine oxides than esters and xanthates (152, *cf.* 149).

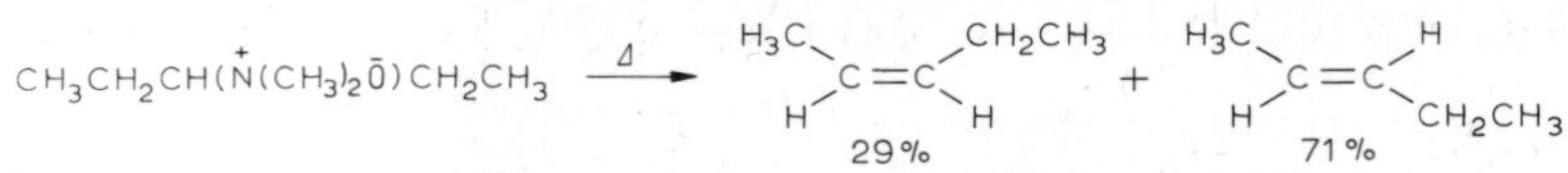

(152)

With increasing size of substituents, a greater tendency for *trans*- rather than *cis*-olefin formation should be exhibited and this is found to be the case as dimethylamine oxides of the larger alicyclic rings give mainly the *trans* isomer.

Menthyl dimethylamine oxide eliminates to give 2-menthene (64%) and 3-menthene (36%)[410], markedly contrasting behaviour to the corresponding ester, xanthate and halide pyrolyses which all exhibit Saytzeff orientation. Consideration of the stereochemistry of the transition state affords the explanation of the change to Hofmann control. Unfavourable non-bonding interactions between either hydrogen and isopropyl or dimethylamine oxide and methyl substituents are present in the boat conformers (153) leading to 3-menthene as compared to pseudo-axial methyl-hydrogen interaction in the conformation (153a) leading to 2-menthene. Neomenthyl dimethylamine oxide eliminates quantitatively to 2-menthene, a failure to achieve *syn* coplanarity of the tertiary hydrogen and amine oxide functions preventing 3-menthene formation.

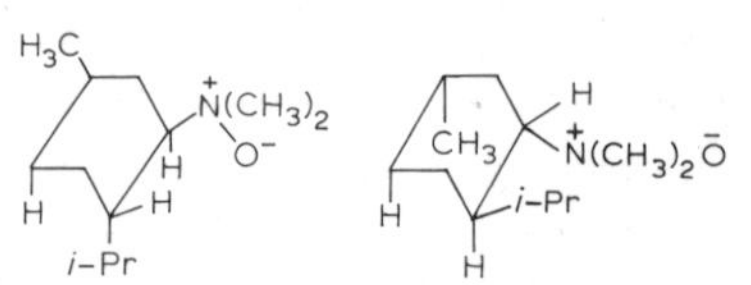

(153)

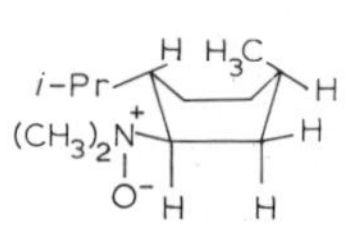

(153a)

Orientational studies of alkyl halide pyrolyses are complicated by practical difficulties. To prevent competing radical chain processes, the decompositions can be performed in "aged" vessels in the absence of oxygen and in the presence of inhibitors. Even so, the alkenes formed by a unimolecular homogeneous reaction are susceptible to isomerisation by the liberated hydrogen halides or iodine. In the presence of ammonia, isomerisation by hydrogen chloride can be prevented but for pyrolysis of alkyl bromides and iodides an extrapolation technique to zero reaction time has to be used to assess the isomer distribution of the olefins. In agreement with the greater influence of electronic factors in alkyl halide pyrolyses than ester pyrolyses, alkyl halides usually eliminate to give greater yields of the Saytzeff olefins. As in the case of E2 reactions, the weaker the carbon–halogen bond, the greater the tendency to follow the Saytzeff rule (154)[374].

X in $CH_3CH_2CHXCH_3$	acetate	chloride	bromide	iodide
% 2-butenes (Saytzeff product)	43	60	66	83

(154)

2.9 THE TRANSITION STATE OF E_i REACTIONS

The cyclic nature of transition states of E_i reactions imposes a greater influence of molecular stereochemistry on reactivity and orientation than is experienced in E2 reactions with changes in the leaving group. A relationship between ring size of the transition state and orientation is apparent for esters and xanthates and also amine oxides but halide pyrolyses do not fit the anticipated pattern. The influence of *alpha*-alkyl substituents on the rate of elimination varies markedly with the nature of the leaving group (Table 25). The effect of alpha substitution increases markedly along the series; $\overset{+}{N}Me_2\bar{O}$, acetate, chloride, bromide, indicating a corresponding increase in double bond character or C_α–X bond breaking in the transition state. The latter is probably mainly responsible as beta methyl substitution has little influence on the rate even when a statistical correction is applied, although the greater rate enhancement for the alkyl halide reactions suggest greater double-bond character in these eliminations. Extensive kinetic data on alkyl halides and alkyl acetate reactions have been compiled and these serve to illustrate the nature of E_i processes. Information concerning xanthate and amine oxide reactions is less soundly based and pyrolyses could involve mixed mechanisms in some cases.

TABLE 25

RELATIVE RATES OF PYROLYTIC ELIMINATION OF SOME ALKYL SUBSTRATES, RX

	X				
R	$\overset{+}{N}(CH_3)_2\bar{O}$	CH_3CO_2	CH_3CO_2	*Br*	*Cl*
alpha-series					
C_2H_5	1	1	1	1	1
$(CH_3)_2CH$	2.64	10^2	26 (24)	170	130
$(CH_3)_3C$	6.06	10^5	1660 (2000)	32000	11800
Temperature (°C)	100	100	400	400	400
Ref.	163	375	375 (376)	377–381	382–385
beta-series					
C_2H_5	1	1		1	1
$CH_3CH_2CH_2$	0.60	0.90		3	5
$(CH_3)_2CHCH_2$	0.48	0.45		5	5
Temperature (°C)	100	489		400	400
Ref.	163	376		379, 386	387, 388

2.9.1 *Alkyl halides*

From a kinetic standpoint, this field has been extensively studied and reviewed by Maccoll[383,389,390]. Heterogeneous reactions, which give irreproducible kinetics, can be minimised by the use of seasoned vessels and the rigid exclusion of oxygen. Homogeneous reactions can be assumed to occur if the rate of elimination is not affected by a variation in the surface material or a change in the volume: surface area ratio. Under homolytic conditions, three mechanisms have been recognised to account for thermolysis of alkyl halides (*i*) the E_i mechanism with synchronous elimination of HX; or an initial homolytic fission of the C–X bond followed by (*ii*) a radical chain-reaction or (*iii*) a radical non-chain reaction, *viz.*

$$R_2CHCH_2X \xrightarrow{\text{slow}} R_2CH\dot{C}H_2 + \dot{X}$$

$$\dot{X} + R_2CHCH_2X \longrightarrow HX + R_2\dot{C}CH_2X \qquad (155)$$

radical chain

$$R_2\dot{C}CH_2X \rightarrow R_2C{=}CH_2 + \dot{X}$$

radical non-chain

$$R_2\dot{C}CH_2X + R_2CH\dot{C}H_2 \rightarrow R_2C{=}CH_2 + R_2CHCH_2X$$

A more detailed analysis of the radical mechanisms has been presented[380]. Generally, all three processes show first-order kinetics but E_i reactions do not exhibit an induction period and are unaffected by radical inhibitors such as nitric oxide, propene, cyclohexene or toluene. For the non-chain mechanism, the activation energy should be equivalent to the homolytic bond dissociation energy of the C–X bond and within experimental error this requirement is satisfied for the thermolysis of allyl bromide[391,392]. For the chain mechanism, a lower activation energy is postulated, hence its more frequent occurrence, as the observed rate coefficient is now a function of the rate coefficients for the individual steps. Most alkyl halides react by a mixture of chain and E_i mechanisms, but the former can be suppressed by increasing the addition of an inhibitor until a constant rate is observed. Under these conditions a mass of reliable reproducible data has been compiled for E_i processes. Necessary conditions for this unimolecular mechanism are (*a*) first-order kinetics at high pressures, (*b*) Lindemann fall-off at low pressures, (*c*) the absence of induction periods and the lack of effect of inhibitors and (*d*) the absence of stimulation of the reaction in the presence of atoms or radicals.

For a whole range of substituted alkyl chlorides and bromides, the pre-exponential factor, *A*, of the Arrhenius equation, has a value of the order of $10^{13}\ \text{sec}^{-1}$. Variations in rate with changing substrate structure thus reflect changes in activation energy rather than activation entropy. A linear correlation between the activation energy for elimination of HX from a series of alkyl halides (R = C_2H_5, i-Pr, *t*-Bu; X = Cl, Br, I) and the heterolytic bond dissociation energy, $D(\overset{+}{R}\overset{-}{X})$, but not the homolytic bond dissociation energy, $D(\dot{R}\dot{X})$, was noted (bond dissociation energies can be measured by mass spectrometry[393]) and this observation led Maccoll[393] to postulate a "quasi-heterolytic" transition state for the E_i reaction, which involved considerable charge separation (156) rather than a unimolecular homolytic process (157). Moreover, the beta carbon–hydrogen bond was regarded as exerting a similar role in stabilising the forming X^- as the solvent in the corresponding solvolytic reaction.

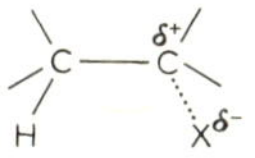

(156)

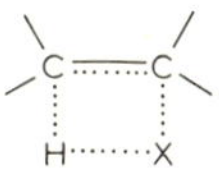

(157)

For a "quasi-heterolytic" transition state, substituent effects in pyrolytic reactions should simulate those observed in E1 reactions, whereas for a homolytic transition state a resemblance to effects noted in known homolytic processes is expected. The experimental facts strongly support the former suggestion.

Although alpha methyl substitution markedly influences the rates of pyrolysis of alkyl halides, it has virtually no effect on the homolytic decomposition of cyclobutane (*cf.* 158 and Table 25)[394,395].

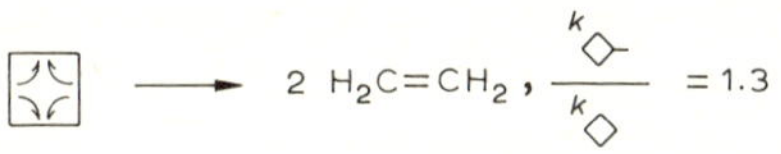

(158)

As for E1 reactions, alpha methyl substituents, but to a much less extent beta methyl substituents increase the rate of elimination. Substituents possessing atoms with lone pairs, such as methoxyl or halogen, can influence the rate coefficient for pyrolysis of alkyl halides. From the alpha position, such substituents can stabilise the developing carbonium ion by the mesomeric effect, and large increases in the rate coefficient are observed. However, from the beta position the substituent effect is limited to the inductive or direct field effect and a modest variation in the rate coefficient is noted. Thus, beta hydrogen acidity is of minor importance. Likewise, little double-bond character is developed in the transition state as, although α-phenylethyl bromide eliminates at a rate similar to that of *t*-butyl bromide, β-phenylethyl bromide reacts at a similar rate to that of ethyl bromide. Often rearrangements characteristic of carbonium ions in a polar solvent are observed in gas-phase eliminations, a particular example being the pyrolysis of neopentyl chloride, which does not possess a beta hydrogen atom[396], *viz.*

$$(CH_3)_3CCH_2Cl \xrightarrow[-HCl]{75\%} (CH_3)_2C{=}CHCH_3 + (CH_3)_2CHCH{=}CH_2 + H_2C{=}C(CH_3)CH_2CH_3 \quad (a)$$

$$\xrightarrow{25\%} \begin{cases} \longrightarrow (CH_3)_2C{=}CH_2 + CH_3Cl \\ \xrightarrow{-CH_4} (CH_3)_2C{=}CHCl + H_2C{=}C(CH_3)CH_2Cl \end{cases} \quad (b) \qquad (159)$$

A number of kinetic isotope studies have been reported for gas-phase eliminations. Isopropyl bromide-d_6 decomposes more slowly than isopropyl bromide[397] and the intramolecular isotope effects ($k_{(C_2D_4HX \to C_2D_4)}/k_{(C_2D_4HX \to C_2D_3H)}$) have been recorded for pyrolyses of ethyl acetate[398] (2.0), chloride[399] (2.20), and bromide[400] (2.10) at 500°C. At the elevated reaction temperatures, these values correspond to the maximum predicted for complete loss of the C–H stretching vibration and they have been interpreted as indicating considerable weakening of the C–H bond in the transition state. Whether this is a homolytic or heterolytic bond fission, it is remarkably insensitive to beta substituent effects on rate. The intramolecular isotope effects in these cases could reflect predominantly a secondary isotope effect rather than the intended primary effect and dissection into an intermolecular and secondary isotope effect would prove more fruitful. (Section 2.2.1.)

Ingold[401] has taken the heterolytic view of gas reactions to an extreme in suggesting an ion-pair formation. Although the energy required to cause ion separation in the gas phase may well be 100–120 kcal.mole^{-1}, ion-pair formation has been estimated to require only about 50–60 kcal.mole^{-1}, in probably furtuitous agreement with the experimental activation energies. Ion-pair formation does not seem unreasonable for tertiary and secondary structures but appears less likely for primary substrates in view of the low stability of primary carbonium ions. It seems highly probable that a spectrum of transition states similar to that observed in E2 reactions exists for gas phase heterolysis. The extremes are probably represented by the homolytic and heterolytic bond cleavages of the C_α–X bond which from the present experimental evidence appears to break ahead of the C_β–H bond. Possibly useful supporting evidence for this idea of a continuum of transition states may be gleaned from a combination of primary and secondary hydrogen isotope studies on a series of substrates of widely differing structure.

2.9.2 Esters

Some primary esters and those lacking beta hydrogen atoms eliminate via radical intermediates[402], which may intrude into many of the product studies for which cyclic mechanisms are usually proposed. In fact, mixed mechanistic behaviour may account for the disagreements in isomer distribution in some cases[368,403]. Using inhibitors, a quantity of reliable kinetic data has been produced and as many similar substituent effects have been employed in both ester and alkyl halide pyrolyses, it is possible to compare directly the transition states of these two E_i reactions. The cyclic nature of the E_i reaction for the esters accords well with the observed stereochemistry and orientational studies suggest little double-bond character except in the presence of strongly conjugating or "beta hydrogen-activating" substituents.

The influence of an alpha methyl substituent is much less pronounced on the rate of gas-phase thermolysis of esters than of halides but it is still quite marked, suggesting a certain amount of C_α–O bond heterolysis occurs in the transition state. On the other hand, beta methyl substitution has a slight retarding effect in the primary series but a slightly rate enhancing influence for secondary and tertiary acetates[376,404]. Possibly double-bond character is more developed in the transition states in the secondary and tertiary series and the mesomeric influence of beta methyl substituents stabilises the transition state. In the primary series, double-bond character is less developed and a balance between mesomeric stabilisation of this more incipient double bond and unfavourable deactivation of beta hydrogen acidity by the inductive effect of the alkyl substituent may explain its observed influence on the rate coefficient.

Slightly greater variations in the activation entropy are noted in ester than the halide thermolyses, but the A values again approximate to 10^{13} and indicate the cyclic unimolecular nature of the transition states. With changing ester, the rate of pyrolysis and strength of the liberated acid qualitatively increase in the same direction[405]. The dichloroacetate, chloroacetate and acetate of t-butyl alcohol exemplify this relationship and at 250°C their rates of pyrolysis follow the order 18.6; 4.4; 1 (ref. 406). Nucleophilicity of the carbonyl function cannot be the dominating influence as a reverse order of reactivity would be predicted[375].

Smith *et al.*[407–409] have studied the influence of alpha and beta phenyl substituents on the rate of pyrolysis of acetates and using substituted derivatives have measured the Hammett reaction constants for variations at both alpha and beta carbon and also in the ester function by studying pyrolysis of substituted benzoates. Their careful dissection of substituent effects thus affords a means of assessing the relative amounts of bond making and breaking in the transition state. The ratios of the reaction rates for ethyl acetate, α-phenylethyl acetate and α,β-diphenylethyl acetate (1:45:130) suggests that C_α–O bond breaking is of more importance than C_β–H bond breaking and double-bond character is little developed in the transition state. This picture of the transition state is supported by the Hammett reaction constants. The heterolytic nature of C_α–O bond breaking is emphasised by the rates of substituted α-phenylethyl acetates and substituted α,β-diphenylethyl acetates which correlate with the Hammett substituent constant σ^+, rather than the "normal" substituent constant, σ. However, carbonium ion character is only developed to a small extent at C_α as the reaction constants at 600°K are small, being only −0.66 and −0.62, respectively. The rate of elimination of the substituted beta phenylethyl acetates correlate with the normal Hammett equation; ρ is positive, indicating a favourable electron-withdrawing effect on beta hydrogen acidity, but its magnitude is much smaller than the alpha phenyl reaction constant, suggesting that less charge is developed at the beta carbon than at the alpha carbon in the transition state. Rates of elimination of substituted ethyl benzoates give a reaction constant of 0.20 and the sign of this parameter supports the idea that anion stability is more important than carbonyl nucleophilicity.

Although the reaction constants are all small, it was possible to measure all the reactions under the same experimental conditions, and first-order kinetics were observed for up to 95% reaction in most cases. Charge separation is obviously developed to only a small extent, and a transition state along the following lines can be visualised (160).

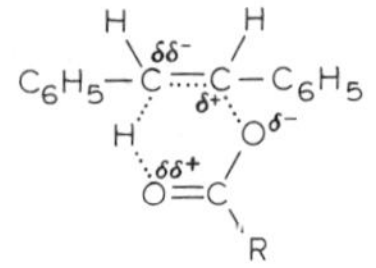

(160)

That C_α–X heterolysis is less developed in the transition state for thermolysis of esters than alkyl halides is supported by recent studies on substituted 1-phenylethyl chlorides[481]. At 608°K a Hammett reaction constant of −1.36 was observed, this being much smaller than that of −4.95 at 318°K observed for the solvolysis of the same substrates in 80% aqueous acetone. Although part of the difference between these latter two values is attributable to the difference in reaction temperature, the results suggest that C_α–X bond breaking is less developed in pyrolyses than in solvolytic reactions.

Isotope studies have been reported for acetate pyrolyses but the results are complicated as they represent a combination of primary and secondary hydrogen isotope effects. There is undoubtedly a need to separate these effects and measurements of oxygen and carbon isotope effects might prove instructive. However, it may be easier to measure isotope effects for the xanthate reactions as the lower reaction temperature will afford greater numerical ratios.

2.9.3 Xanthates and amine oxides

Few kinetic investigations of xanthate pyrolyses have been reported. The mechanism was in doubt for some time, some workers[350,357,412] favouring proton abstraction by the thion sulphur (161) and others by the thiol sulphur which gave the isolated products in a single step (162)[31,413].

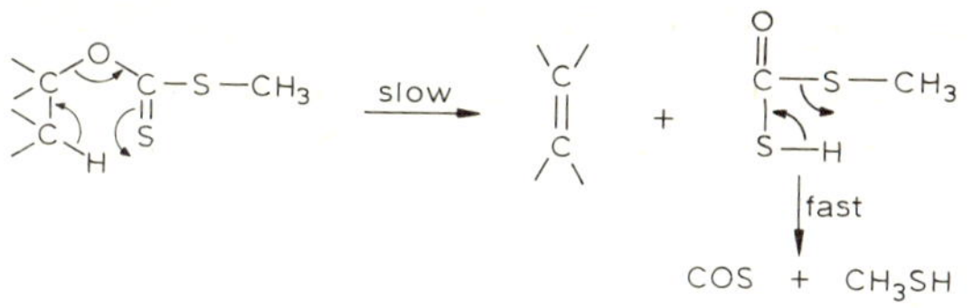

(161)

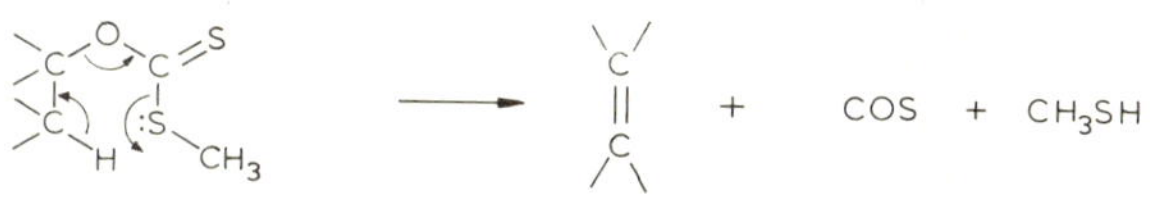

(162)

As dithiocarbonates are more stable than the corresponding xanthates[414], a route involving initial isomerisation of the xanthate to its dithiocarbonate can be discounted[415]. A careful investigation of the carbon and sulphur isotope effects showed the thion sulphur mechanism to be correct (163)[416].

	Thiol sulphur (k_{32}/k_{34})	*Thion sulphur* (k_{32}/k_{34})	*Carbonyl carbon* (k_{12}/k_{13})
Experimental results at 78°C (with standard deviation)	1.0021 ± 0.0007	1.0086 ± 0.0016	1.0004 ± 0.0006
Calculated isotope effects (mechanism 161)	Unity	1.007 − 1.010	Unity
Calculated isotope effects (mechanism 162)	1.012	Unity	1.03 − 1.04

(163)

The small carbonyl carbon isotope effect suggests C=O formation and C=S → C–S have progressed to a similar extent in the transition state. As xanthates decompose at a much lower temperature than acetates, their activation enthalpies are presumably smaller. Activation entropies, however, should be similar, as both decompositions give similar product distributions. Kinetic studies in the cholesteryl series support these predictions, the difference in activation entropies being negligible but the activation enthalpies being 8–9 $kcal.mole^{-1}$ lower for the xanthates[343]. The extra driving force in the xanthate decompositions arises from the O–C=S → O=C–S conversion for which summation of the bond energies suggests an exothermic reaction[344] to the extent of 20 $kcal.mole^{-1}$. This is only partly realised in the transition state which is undoubtedly much more reactant-like than that for the acetates. The analogy between xanthate and acetate pyrolyses is further demonstrated by the observations that the rates of elimination of *S*-alkyl-substituted cholesteryl xanthates parallel the pK_a values of the corresponding acetic acids[417].

Normally, menthyl xanthate decomposes to menthenes on distillation, but, after washing with ferrous salts which removes peroxide traces, a higher reaction temperature for elimination is required, although a similar product distribution is observed. The actual function of the peroxide impurities remains unsolved[418].

For the even more exothermic amine oxide pyrolyses, no quantitative data is available. From product studies[163], other than their effect on the statistical factor, changes in the alpha and beta alkyl substituents have little influence on the rate of elimination into a particular branch. These observations, as might be expected for the most exothermic of the pyrolytic decompositions, are consistent with little development of double-bond character or C_β–H and C_α–N bond breaking in the transition state. A beta phenyl substituent promotes either an increase in C_β–H bond breaking or double-bond formation as $C_6H_5CH_2CH(\overset{+}{N}(CH_3)_2\bar{O})CH_2CH_3$ eliminates to give a substituted styrene 70 times more rapidly than it gives a substituted ethylene. Such a rate enhancement is, however, much smaller than is observed in E2 reactions of alkyl halides or alkyl "onium" salts and serves to emphasise the more reactant-like nature of the amine oxide reactions. In agreement with the well-known ability of dipolar aprotic solvents to enhance medium basicity, the amine oxides are found to decompose at 25°C in dimethyl sulphoxide[419].

2.10 ELIMINATION OVER SOLID CATALYSTS

Olefins can be prepared by the dehydrogenation of paraffins, dehydration of alcohols, or decomposition of ethers and halides, if vapours of these substrates are passed over metals or metal oxides at elevated temperatures (300–600°C). Dehydration reactions have been most widely studied and by careful selection of the catalyst and the reaction conditions the direction and stereochemistry of elimination can be controlled. However, dehydration often has to compete with dehydrogenation, and isomerisation of olefinic products by the acidic sites on the catalyst can reduce the synthetic utility of these reactions. Most frequently alumina has been used as the catalyst and the advantages and complexities of the method are amply illustrated by the dehydration of alcohols. Surface-catalysed eliminations have been the subject of several reviews[420–424].

2.10.1 Nature of alumina catalysts

Many of the earlier investigations are difficult to interpret as the importance of the chemical nature of the alumina was not appreciated and analytical techniques lacked the accuracy of present available methods. The numerous reports of the nature of alumina catalysts have been summarised by Winfield[421] and Pines and Manassen[424]. Ample evidence is presented to show that pure alumina, prepared from aluminium isopropoxide or aluminium nitrate and ammonia and calcined at 600–800°C, contains intrinsic acidic sites. Acidity has been estimated by titration against basic solvents[425,426], chemisorption of gaseous bases[426–428] and by indicators which form coloured complexes with typical Lewis acids[427,428]. Lewis acidity of the dehydrated surface is best attributed to the presence of aluminium atoms which are not fully coordinated. Increasing the concentration of adsorbed sodium or potassium ions or moisture deactivates the catalyst as the Lewis acid sites are converted into less active Brönsted acid sites. The catalytic sites for dehydration and isomerisation appear to be the same, as an approximately parallel decrease in these processes is noted with increasing alkali metal ion incorporation[427–429]. The active sites for dehydrogenation and dehydration are thought to be different. For the reactions of isopropanol, dehydrogenation is favoured by catalysts which form the most stable bonds to phenol whilst dehydration catalysts adsorb pyridine more strongly[430]. Phenol presumably associates with the basic sites and the basic pyridine with the Brönsted or Lewis acid sites. Alumina is generally assumed to exist as alternate layers of aluminium and oxygen atoms[431,432]. Consequently, it would seem reasonable to expect dehydrogenation to be a surface process but to regard dehydration as occurring within the pores of the catalyst. In agreement with this, the dehydrating activity of alumina is decreased when it is heated to high temperatures as healing of the irregularities in the crystal lattice occurs[433].

A study of the ease of the isomerisation of cyclohexene and 3,3-dimethylbutene has provided information concerning the acidity of alumina[427]. Strong-acid sites are believed to be required to induce rearrangement of olefins which proceed through a primary carbonium ion but strong and weak-acid sites can induce rearrangement through secondary and tertiary carbonium ions. The number of acidic sites is thus demonstrated by the *extent* of isomerisation of 3,3-dimethylbut-1-ene and the intrinsic acid strength can be estimated from the *depth* of the isomerisation (164)[427].

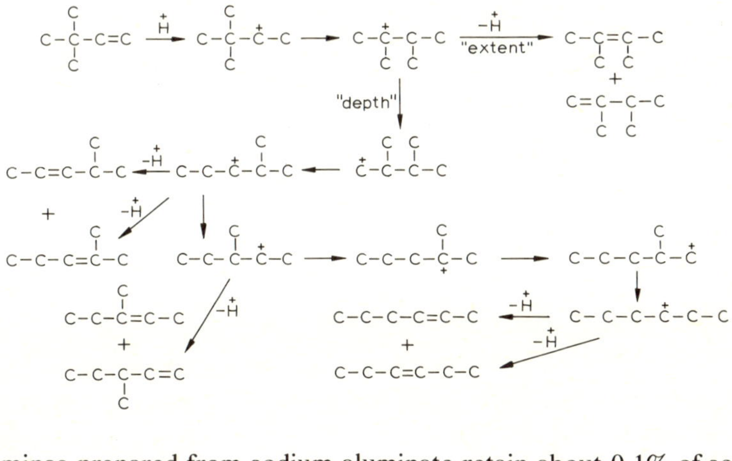

(164)

Aluminas prepared from sodium aluminate retain about 0.1% of sodium ions, have a large number of weakly acidic sites and consequently are good dehydration catalysts as, owing to the absence of strong-acid sites which are neutralised by the alkali metal ions, they do not isomerise olefins such as cyclohexene[427].

2.10.2 Separation of primary and secondary processes

In order to be able to evaluate the mechanism of dehydration it is necessary to be able to distinguish between the initially formed primary products and the secondary products which subsequently arise by isomerisation. Although cyclohexanol yields only cyclohexene when dehydrated over alumina containing 0.4% of alkali metal ions[434], high-purity alumina from aluminium isopropoxide gives large amounts of methylcyclopentenes. However, the methylcyclopentenes are not formed directly in a parallel reaction with cyclohexene from cyclohexanol as the product composition varies markedly with the contact time. Moreover, extrapolation of the product composition to zero reaction time indicates that a quantitative yield of cyclohexene is formed from cyclohexanol and this product then undergoes rearrangement. The lower acidic activity of the catalyst required to dehydrate an alcohol than to isomerise an olefin is understandable in terms of the greater basicity of the hydroxyl group than the olefinic double bond. To minimise the isomerisation, the strongly acidic sites of the highly-purified alumina can be neutralised by the passage of ammonia or organic bases over the catalyst surface during the dehydration.

2.10.3 Stereochemistry

The majority of stereochemical investigations refer to cyclic and rigid substrates. Unlike the corresponding simple pyrolytic eliminations (Section 2.7), the dehydration of menthol and neomenthol enhibits mainly *anti* rather than *syn* or *syn-clinal* stereospecificity, *viz.*[435]

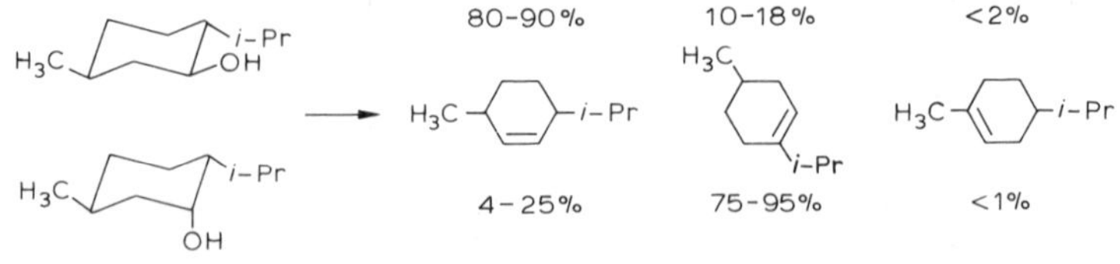

(165)

Traces of 1-menthene are found among the primary products and are thought to arise by abstraction of the *gamma* hydrogen and migration of the *beta* hydrogen, *viz.*

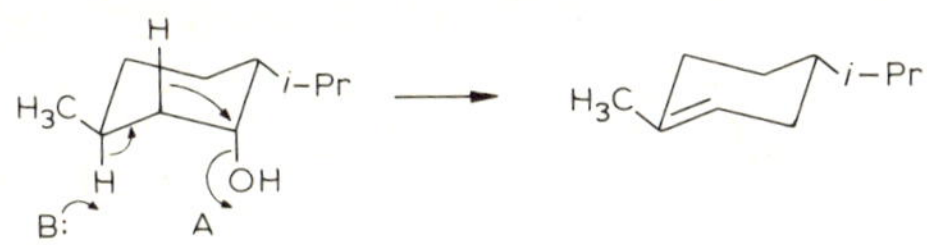

(166)

The importance of coplanarity of the eliminating fragments is emphasised by the large difference in the Arrhenius activation energies for elimination from *cis*-4-*t*-butylcyclohexanol ($E_A(cis)$ = 27 kcal.mole^{-1}) and *trans*-4-*t*-butylcyclohexanol ($E_A(trans)$ = 38 kcal.mole^{-1})[436]. In the *trans* isomer, elimination must occur from a diequatorial *syn-clinal* conformation, or a diaxial conformation with the bulky *t*-butyl substituent in an unfavourable axial conformation, or an initial epimerisation to the *cis* isomer, or via a boat conformation, all possibilities which involve much higher energy intermediates than the diaxial elimination from the *cis* isomer. The products of elimination from the four stereoisomers of 1-decalol are also consistent with *anti* stereospecificity[424].

To explain the predominance of *anti* stereospecificity in dehydration over alumina, participation by both acidic and basic sites of the catalyst is invoked. Furthermore, as the alumina must surround the substrate, it has been regarded as a "solvating" agent and an analogy to solvolytic elimination has been drawn[424].

The dehydration of 1,4-cyclohexane diols emphasises the critical importance of stereochemical arrangement of the hydroxy functions in a substrate[437,438]. The *cis*-diol eliminates to give the cyclohexenol as the major product but the *trans* isomer yields mainly 1,4-epoxycyclohexane. As elimination from *trans*-1,4-cyclohexanediol occurs more rapidly than from *t*-butyl alcohol, anchimeric assistance was suggested, the intramolecular concerted ring closure occurring through a boat conformation, *viz*.

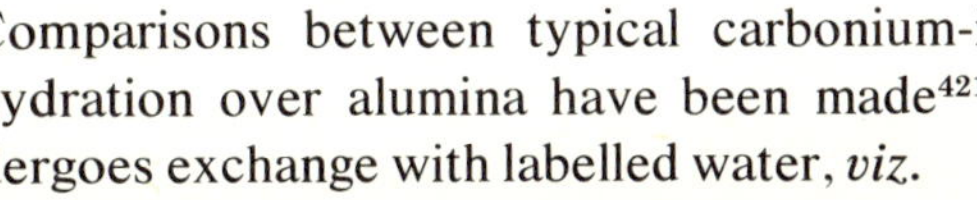

(167)

2.10.4 *Mechanism and orientation*

Comparisons between typical carbonium-ion processes in solution and dehydration over alumina have been made[421,424]. On alumina, ethyl alcohol undergoes exchange with labelled water, *viz*.

$$C_2H_5OH + H_2{}^{18}O \rightleftharpoons C_2H_5{}^{18}OH + H_2O \qquad (168)$$

at temperatures much lower than are required for dehydration[439] and on thoria (used to avoid catalyst poisoning by the sulphide) exchange between ethyl alcohol and hydrogen sulphide has been observed[440]. Although at sufficiently high temperatures, ethyl alcohol is intramolecularly dehydrated to ethylene[441], lower reaction temperatures (300°C) lead mainly to the condensation product, diethyl ether[442]. These results illustrate the greater lability of the C_α–OH bond than the C_β–H bond and a rapid absorption of the substrate on to the alumina surface, to give a complex or carbonium ion, followed by a slow proton transfer[443] would seem a reasonable description of the mechanism. Such a mechanism should be confirmed by deuterium isotope studies. Carbonium-ion intermediates are preferred for tertiary alcohols, but dehydrations of secondary and primary alcohols are considered to follow a concerted process leaning towards a carbonium ion. Even for a secondary alcohol, C_α–O separation may be insufficiently developed in the transition state to allow rearrangement to occur. The present knowledge of mechanism is dependent on product analysis rather than rate studies.

(*a*) *Primary alcohols*

The majority of investigations have been concerned with the dehydration of ethyl alcohol. Controversy has raged for many years over the possibility that diethyl ether is a precursor in the formation of ethylene. The inconsistency in the conclusions may arise from a failure of the earlier workers to appreciate the importance of the intrinsic acidity of the alumina. Whereas Balaceanu and Jungers[444] regard diethyl ether as the precursor of ethylene (169), Brey and Krieger[443] prefer simultaneous reactions (170), a mechanism favoured by Topchieva *et al.*[422] who further suggested a covalently bonded intermediate (171).

$$2\,C_2H_5OH \xrightarrow{-H_2O} C_2H_5OC_2H_5 \longrightarrow C_2H_5OH + H_2C{=}CH_2$$
$$C_2H_5OH \longrightarrow \tfrac{1}{2}\,C_2H_5OC_2H_5 + \tfrac{1}{2}\,H_2O \qquad (169)$$

$$2\,C_2H_5OH \longrightarrow H_2C{=}CH_2 + 2\,H_2O$$
$$2\,C_2H_5OH \longrightarrow C_2H_5OC_2H_5 + H_2O \qquad (170)$$

$$C_2H_5OH + \rangle AlOH \underset{H_2O}{\overset{-H_2O}{\rightleftharpoons}} \rangle AlOC_2H_5 \underset{-\frac{1}{2}H_2O}{\overset{\frac{1}{2}H_2O}{\rightleftharpoons}} \tfrac{1}{2}(C_2H_5)_2O + \rangle AlOH$$
$$\rangle AlOC_2H_5 \longrightarrow \rangle AlOH + H_2C{=}CH_2 \qquad (171)$$

The reversibility of ether formation has been demonstrated[445], but it is not known whether the ether is formed from the collision of gaseous and absorbed alcohol or from adsorbed alcohol molecules only. On alumina, ethyl alcohol and ether have similar stabilities but ether is much more stable on titanium dioxide[446] and is therefore not necessarily an intermediate in ethylene formation. The consecutive scheme is also inadequate for butanol dehydration[447]. Even the solution of the ether–ethylene problem may add little to the general understanding of the mechanism of dehydration as ethyl alcohol is rather an atypical substrate.

The low stability of primary carbonium ions makes their intermediacy highly improbable and any rearrangements have to be accounted for usually in terms of concerted mechanisms. The primary dehydration products of *n*-butanol are 1-butene (97.3%) and 2-butene in which the *cis* isomer predominates by a two-fold excess[435]. Whereas 1-butene possibly arises from an *anti*-elimination, the 2-butenes are best explained as arising by removal of the γ-hydrogen with a concerted migration of a beta hydrogen, *viz.*

(172)

Steric interactions between the alkyl substituents and the catalyst surface cause *trans*-olefin formation to be less favourable than *cis*. Dehydration of 2-methyl-1-propanol over a variety of alumina catalysts gives isobutylene (77–88%), similar amounts of 1-butene and *cis*-2-butene (4–10%) and smaller yields of *trans*-2-butene (2–4%)[448]. Non-classical rather than classical intermediates were suggested, as the least acidic alumina, which possessed the lowest dehydration activity, caused the greatest skeletal isomerisation (173).

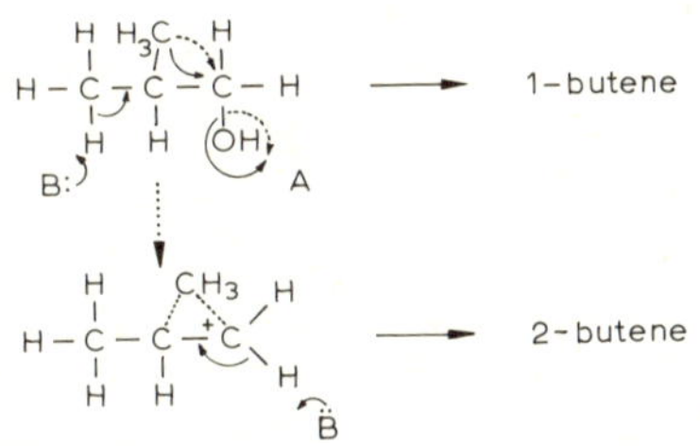

(173)

A more detailed study of the dehydration of 2-phenyl-1-propanol was reported and a radio tracer technique using ^{14}C was used to delineate the mechanism[448]. Of the reaction products, the allyl benzene formed was labelled only on the benzylic carbon and was thus produced by γ hydrogen abstraction and concerted phenyl migration. In the α-methylstyrene, the label was equally distributed between the 1 and 3 carbon atoms and this product arose either through a symmetrical cyclopropane intermediate or a tertiary carbonium ion, *viz.*

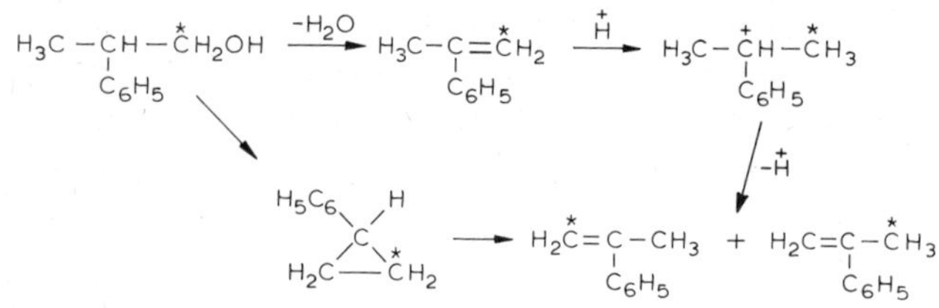

(174)

Phenyl migration occurs eight times more readily than methyl migration as indicated by the distribution of radioactivity in *trans*-β-methylstyrene. Aryl migration to the alpha carbon should increase with increasing electron donating substituents and the greater migration of a *p*-tolyl group (9%) than a phenyl group (6%) in the 2-arylethyl series supports this prediction.

The different nature of the intermediate in the dehydration of primary and tertiary alcohols, is substantiated by the contrasting dehydration behaviour of neopentyl and *t*-pentyl alcohols[449]. The carbonium ion mechanism for the tertiary alcohol leads only to 2-methyl-1-butene (57%) and 2-methyl-2-butene (42%), but neopentyl alcohol gives different proportions of these olefins (64–73%; 23–32%) along with small amounts of 1,1-dimethylcyclopropane. These latter facts are accommodated better in terms of a concerted mechanism for the dehydration of neopentyl alcohol, *viz*.

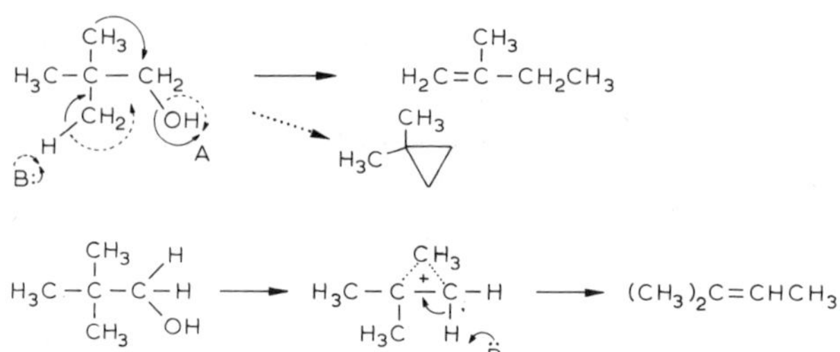

(175)

(*b*) *Secondary and tertiary alcohols*

Under conditions of kinetic control, the dehydration of 2-butanol follows the Saytzeff rule, but a greater yield of *cis*-2-butene than *trans*-2-butene is obtained. These observations have no parallel in acid- or base-catalysed or pyrolytic eliminations. However, the dehydration of 2,3-dimethyl-2-butanol gives 2,3-dimethyl-1-butene (88.4%) and 2,3-dimethyl-2-butene (9.9%) and is thus oriented towards the Hofmann rule despite being more probably a carbonium ion process. Under similar reaction conditions the quite distinctly different products arising from the secondary alcohol, 3,3-dimethyl-2-butanol [3,3-dimethyl-1-butene (70%), 2,3-dimethyl-1-butene (23.5%), 2,3-dimethyl-2-butene (3.9%), 1,1-dimethyl-2-methylcyclopropane (2.1%)] are accommodated in terms of concerted rather than a carbonium-ion mechanism.

A number of unexplained factors warrant mention. Orientation of elimination differs for secondary and tertiary structures. The peculiar predominance of *cis*- rather than *trans*-olefin may arise from the relative stabilities of the proton–olefin complexes[424], but a more certain conclusion would be possible if the stereochemistry of the dehydration in the acyclic series had been determined. Assumption of the *anti* stereospecificity known to be favoured by the cyclohexyl systems may be unsound especially in the light of the recent stereochemical findings in base-catalysed elimination reactions (Section 2.3.1(e)). The solution of the problem of the *cis/trans* ratios may lie in the duality of mechanism, namely the *syn-clinal/anti* complexity. Certainly recent results on the dehydration of *threo*- and *erythro*-2-methyl-4-deutero-3-pentanols on thoria show *syn-clinal* rather than *anti* stereospecificity as indicated by deuterium analysis of the *cis*- and *trans*-4-methyl-2-pentenes, but in these cases the *trans* isomer was formed in a three-fold excess over the *cis*-olefin[450]. Of course, the dehydration reactions on the less acidic thoria may not be good models for alumina but a knowledge of stereochemistry in the acyclic series might prove an invaluable aid in the elucidation of the mechanism. There is obviously plenty of scope for future kinetic investigations which at the moment sadly lag behind preparative studies.

2.11 DEHALOGENATIONS AND RELATED ELIMINATIONS

Dehalogenations are the most common of the olefin-forming beta-eliminations which do not involve hydrogen as one of the leaving groups. They can be induced (with the exception of fluorine) with a variety of metals, reducing agents and nucleophiles. Less frequently, only one of the leaving groups is a halogen and in the case of metal catalysis typical variations include OH, OR and OCOR, whereas with iodide ion, elimination has been achieved from β-halotosylates and β-ditosylates. The few available kinetic studies are concerned mainly with dehalogenation and interest has focussed on the stereochemistry rather than the transition-state structure. Often in synthesis, the need to protect an olefinic double bond from oxidation arises and addition of bromine, which can subsequently be easily removed, serves this purpose adequately. Vicinal dihalides are usually prepared in this way and consequently should not be regarded as readily available precursors for olefin synthesis.

2.11.1 *Iodide-induced elimination*

Dehalogenation of vicinal dibromides by sodium iodide in acetone or alcoholic solvents, *viz.*

References pp. 350–364

$$\mathrm{RCHBrCHRBr} + 3\,\bar{\mathrm{I}} \rightarrow \mathrm{RCH{=}CHR} + \bar{\mathrm{I}}_3 + 2\,\mathrm{Br}^- \qquad (176)$$

is first-order in each reactant, after an allowance for the iodide consumed by the liberated iodine has been made[451–454]. This observation excludes a mechanism which involves a di-iodo intermediate[456], but does not distinguish between an elimination and a rate-determining substitution to give a mono-iodo intermediate[457] which rapidly eliminates (177). The identity of the beta halogen has little influence

$$\mathrm{RCHBrCH_2Br} \xrightarrow[\text{slow}]{\bar{\mathrm{I}}} \mathrm{RCHBrCH_2I} \xrightarrow[\text{fast}]{\bar{\mathrm{I}}} \mathrm{RCH{=}CH_2} + \mathrm{BrI} \qquad (177)$$

on the rate of displacement of the alpha halogen by nucleophiles in the 1,2-dihaloethanes[46] and comparison of the overall rate of elimination and substitution by iodide ion indicates that the substitution–elimination mechanism occurs[451]. Stereochemical studies on *meso*-dideutero-dibromides substantiate this conclusion (178). When one of the halogens is situated on a primary carbon, the substitution–elimination mechanism is observed resulting in an apparent *syn*-elimination (178a)[458]. Steric considerations reduce the rate of the displacement reaction in secondary structures and 2,3-dibromobutane undergoes a direct *anti*-elimination[178b,454].

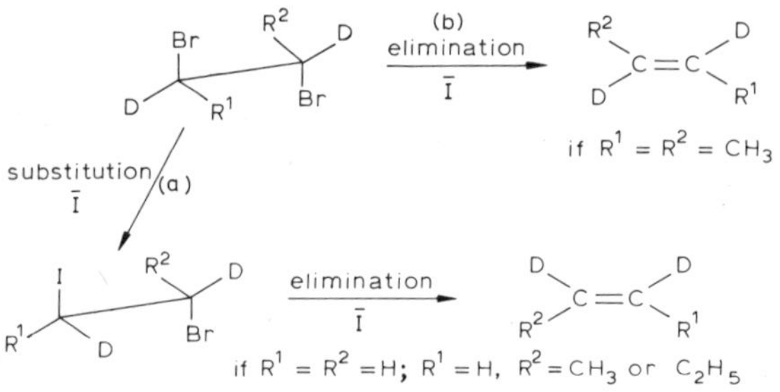

(178)

For the reaction of *sym*-tetrabromoethane with iodide ion in methanol a secondary isotope effect (k_H/k_D) of 1.28 at 80°C was observed[452]. On theoretical grounds, an inverse isotope effect ($k_H/k_D \simeq 0.7$) is predicted for a rate-determining displacement reaction and a normal isotope effect for an elimination reaction[459]. Methyl and bromine substituents have similar spatial requirements, so it is thus not surprising that the direct elimination mechanism, which is observed for 2,3-dibromobutane, is followed by *sym*-tetrabromoethane.

The need for coplanarity of the eliminating halogens in the transition state is confirmed by studies with cyclohexyl derivatives. In the rigid cumbersome triterpenoid series, 11,12-diequatorial dibromides are inert but the diaxial isomers undergo *anti*-elimination with iodide ion[460,461]. In the cyclohexyl system, *trans*-1,2-dihalides (or halide–sulphonate esters) undergo rapid elimination whereas the *cis* substrates react more slowly by the substitution–elimination route[462,463].

Although little is known concerning the identity of the transition state for these iodide induced reactions, the strict adherence to *anti* stereospecificity indicates the importance of π-orbital overlap and a high degree of double bond character seems probable. The 6-fold rate difference (almost the maximum predicted from eclipsing of methyl substituents) between the rates of formation of *trans*- and *cis*-2-butenes from 2,3-dibromobutane supports this hypothesis[453].

In the chalcone series

(179)

all substituents in the phenyl ring increase the rate of elimination[464]. Although this observation could reflect subtle changes in bond making and breaking in the transition state it is also possible that the bromine atom being abstracted by the iodide may change with a variation in the substituent. If, as has been suggested, a cyclic intermediate is involved, *viz.*[451]

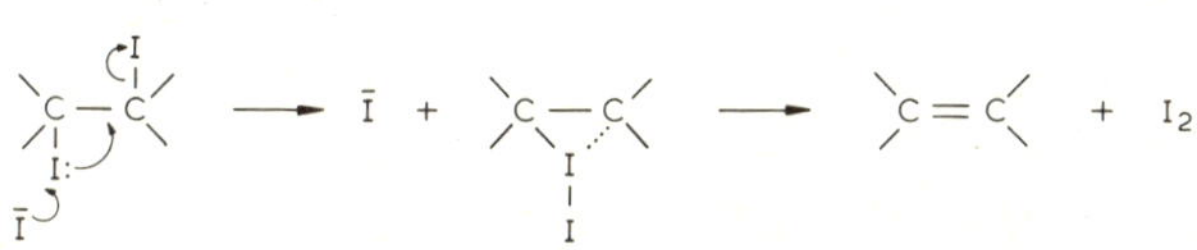
(180)

more fruitful results may arise from unsymmetrical dehalogenations in which a reasonable order of attack by iodide ion is I > Br > Cl in agreement with the tendencies to form halonium ions.

An E1cB reaction[465] or elimination from mono-[466] or di-iodo intermediates[467] are suggestions which have been proposed to account for the iodide-induced eliminations of toluenesulphonate esters of 1,2-diols in the carbohydrate, alicyclic and steroid fields. In the reaction of 1,2-ditosyloxy cyclohexanes with iodide ion, the rate of appearance of iodine is slower than that of sodium *p*-toluenesulphonate[468]. This observation excludes a direct elimination reaction on the substrates and the following schemes were proposed (181).

(a) *trans*-series

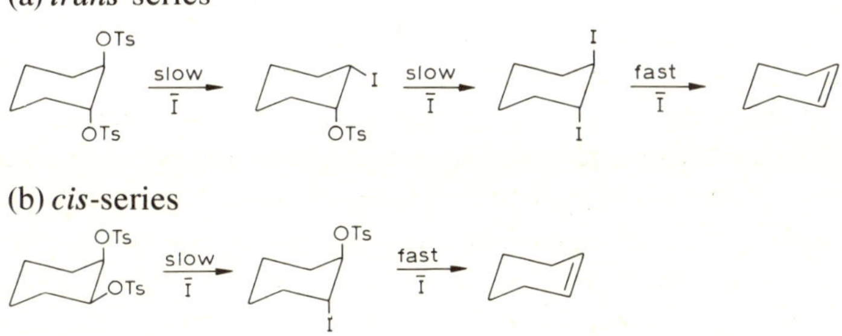

(b) *cis*-series
(181)

Depending on the stereochemistry, either a mono- or di-iodo intermediate is favoured and, although the *trans* isomer was too reactive under the reaction conditions, *cis*-2-iodocyclohexyl-*p*-toluenesulphonate was actually isolated.

Solvolytic deoxymercuration catalysed by iodide has been reported and the reaction shows both a first- and second-order component in iodide[469,470]. It seems unlikely that other halides will be as useful as iodide in dehalogenation as they are oxidised less easily. Rearrangement rather than gamma elimination is observed when 1,3-dihalosubstrates are treated with iodide ion[471].

Elimination from vicinal dihalides has also been accomplished with the reducing agent, lithium aluminium hydride[472] and the nucleophilic thiophenoxide[473] but mechanistic details other than *anti* stereospecificity are lacking.

2.11.2 *Metal-promoted eliminations*

(*a*) *Dehalogenations*

Heterogeneous dehalogenations of vicinal dihalides can be achieved smoothly with zinc in protic solvents or magnesium, lithium or sodium in ethereal solution. In general the elimination occurs at the surface of the metal and the reactivity of this surface is maintained by the presence of a solvent which redissolves any produced olefin and any metallic salt. However, in one case, catalysis by the zinc dihalide deposited on the metal surface during the elimination has been reported[474].

It seems unlikely that the influence of the metal is of an electrophilic nature, inducing a carbonium-ion mechanism, as the products of substitution and dehydrohalogenation are never found simultaneously. More plausibly the metal donates electrons, and carbanion or concerted processes with carbanion character rather than radical processes are usually favoured. Radical additions occur very easily to olefinic double bonds[475] and 1-bromo-1,2-diphenylethane undergoes radical bromination with *N*-bromosuccinimide without accompanying dehalogenation[476].

Elimination from *meso*-dideuterodibromoethane and the diastereoisomeric 2,3-dibromobutanes with zinc shows *anti* stereospecificity[458,477]. However, such stringent stereochemical behaviour is not shown in elimination from the higher homologues such as the dibromo-pentanes, -hexanes or -octanes[478]. In the cyclodecyl series a marked preference for *syn*-elimination is observed[479]. These findings have a parallel in base-catalysed Hofmann eliminations, and explanations in terms of steric interactions which cause preferential substrate conformations to be assumed adequately account for the experimental results. An alternative approach is to invoke carbanion intermediates in which rotation about the C_β–C_α bond is more rapid or of a similar rate to the ejection of bromide ion[478,480]. As products characteristic of a solvent interaction on a carbanion intermediate are not observed, the latter species must have a very transient existence. Steric and carbanion-stabilising influences have been suggested to account for the lack of stereospecificity of elimination from substrates possessing phenyl or ester functions[478,480].

The absence of solvolytic products can be explained if any intermediate carbanion is associated with the metal surface. An extension of this view is to consider these metal catalysed eliminations as proceeding through organometallic compounds ($BrMCR_2Br$; M = Mg or Zn) which break down by a concerted process[480].

Radical processes seem more probable with the monovalent alkali metals than with the divalent magnesium and zinc and complete non-stereospecificity is often observed[478,480]. Of course, initial radical formation could be followed by a rapid electron transfer to give a cabanion[10], which subsequently undergoes a unimolecular elimination to give the olefin.

(*b*) *Elimination involving only one halogen*

Elimination has been achieved with metallic catalysts for a number of halogen compounds possessing beta hydroxy, alkoxy or ester functions. In general, elimination catalyzed by zinc of bromo- and iodo-hydrins and their related alkyl ethers shows much less stereospecificity than is shown by the corresponding vicinal dihalides. Organo-metallic compounds, which decompose by a concerted *anti*-elimination when X is a good leaving group (*e.g.* halogen) but via a carbanion with consequent lower stereospecificity when X is a poor leaving group (*e.g.* OH,OCH_3) have been suggested as the probable intermediates[480].

Bromo- and iodo-hydrins have been converted into olefins by a variety of reagents in acetic acid. In view of the acidic medium and the inert nature of chlorohydrins, a mechanism involving a halonium ion intermediate was suggested, *viz.*[482]

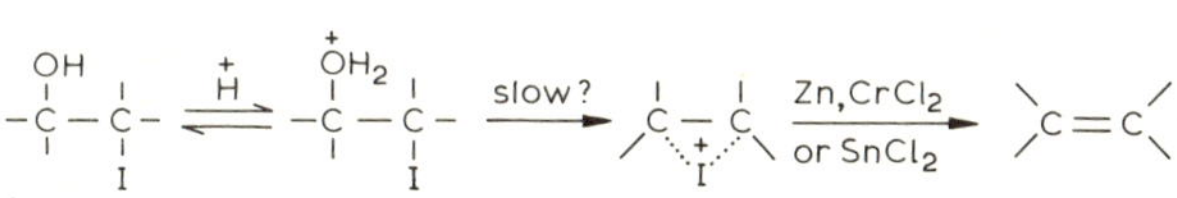

(182)

If the halonium ion mechanism is correct and the hydroxyl group is reversibly protonated by the acidic medium, then the rate of formation of the olefin should follow the acidity function H_R (formation of carbonium ions from carbinols[71]). Elimination from the iodo-hydrin of *trans*-3,4-epoxyoctane was completely non-stereospecific with zinc but gave 75–80% of *trans*-3-octene with chromous chloride and showed complete *anti*-stereospecificity with stannous chloride[482]. The change in stereospecificity with the variation in the metal cation is interesting. It is tempting to point out the difference in the valence states between chromous–chromic and stannous–stannic of one and two, respectively, as being the cause. Application of the Hammett equation to appropriate substrates might prove fruitful. For the chromous case (chromous sulphate in dimethyl formamide), radical intermediates have been suggested by Kray and Castro[483], but the oxidation of the metal is regarded as occurring after the initial formation of the radical intermediate, *viz.*

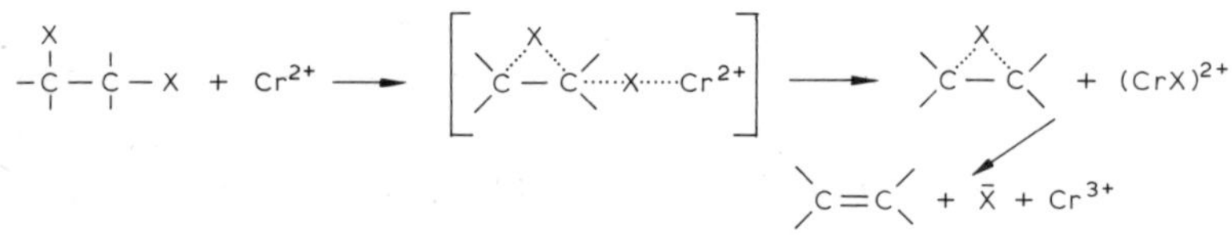

(183)

In the absence of a beta halogen, alkyl halides or 1,1-dihaloalkanes undergo coupling reactions indicative of radical intermediates. The presence of a beta halogen increases the rate of elimination by stabilising the forming radical[484].

It would be instructive to apply electron spin resonance spectroscopy in these various metal cation oxidations. Certainly the homogeneous nature of these reactions offers greater possibilities to the kineticist than is afforded by the heterogeneous nature of the metal-catalysed eliminations. Considering the recent discoveries of the synthetic utility of thallium in organic synthesis[485], thallous salts might prove to be extremely useful catalysts for dehalogenations. as under the reaction conditions the less stable olefins are rapidly isomerised.

2.12 DEHYDRATION OF ALCOHOLS

Alcohols are rarely dehydrated under strongly basic conditions, unless the beta hydrogen is activated by a carbonyl or carbon–carbon double-bond function[486]. As an alternative to solid surface catalysis (Section 2.10), dehydration of most alcohols can be achieved under homogeneous conditions with a number of Brönsted or Lewis acids. Strong mineral acids and iodine convert the hydroxyl function to a better leaving group (H_2O^+ or $I\overset{+}{O}H$) and give rise to carbonium ion mechanisms. Acids such as phosphoric or boric acid or potassium hydrogen sulphate can esterify the alcohol and the resulting ester decomposes by an E2, E1 or thermal process.

2.12.1 Using strong acids

In the presence of strong acids, most alcohols are dehydrated to give the thermodynamically more stable olefin. It is not safe to assume that such a strict adherence to the Saytzeff rule is indicative of a carbonium ion mechanism, as under the reaction conditions the less stable olefins are rapidly isomerised. Such thermodynamic control greatly restricts the synthetic utility of dehydration by strong acids, but the lability of the olefins to the reaction conditions can be put to a useful purpose from a kinetic standpoint. By studying the reverse reaction, namely the hydration of olefins, and applying the principle of microscopic reversibility, a more complete picture of the mechanism of dehydration can be obtained.

A number of different types of kinetic evidence have been used to show that the dehydration reactions proceed via carbonium-ion intermediates in an E1 process. Simple rate studies show that the order of reactivity decreases along the series tertiary alcohol > secondary > primary, the order of decreasing carbonium-ion stability. Skeletal rearrangements typical of carbonium ions have also been observed[487]. On a number of occasions, the rate of olefin formation has been shown to be a slower process than the exchange of the hydroxyl function with the reaction medium[488–492]. This can conveniently be demonstrated by studying the reaction in water enriched with $^{18}OH_2$ or, in the case of an optically active alcohol, by comparing the rates of racemisation and dehydration[493]. These observations indicate that exchange occurs before (184), and not simultaneously with, elimination.

$$\mathrm{ROH} \underset{k_{-1}}{\overset{k_1}{\rightleftharpoons}} \mathrm{ROH_2} \underset{k_{-2}}{\overset{k_2}{\rightleftharpoons}} \overset{+}{\mathrm{R}}\cdots\cdots\mathrm{OH_2} \xrightarrow{k_3} \text{olefin} \tag{184}$$

For tertiary alcohols, oxygen exchange occurs via the "loosely" solvated (encumbered) carbonium ion (k_{-2}) whereas for the less stable primary carbonium ions, exchange most plausibly occurs via a symmetrically solvated carbonium ion (185) or a direct displacement reaction (186), *viz*.

$$\mathrm{RCH_2OH} \underset{k_{-1}}{\overset{k_1}{\rightleftharpoons}} \mathrm{RCH_2\overset{+}{O}H_2} \underset{k_{-2}}{\overset{k_2}{\rightleftharpoons}} \mathrm{R{-}\overset{\overset{H_2O}{\vdots}}{\underset{\underset{H_2O}{\vdots}}{\overset{+}{C}}}H_2} \xrightarrow{k_3} \text{olefin} \tag{185}$$

$$\mathrm{RCH_2OH_2 + H_2{}^{18}O \rightleftharpoons RCH_2{}^{18}\overset{+}{O}H_2 + H_2O} \tag{186}$$

The nature of exchange mechanism with secondary alcohols will depend on the stability of the intermediate carbonium ion and possibly alcohols possessing an alpha-phenyl substituent simulate the reaction mode of tertiary alcohols, whereas the simple secondary alkyl alcohols possibly utilise a mechanism similar to that of the primary substrates. The extensive kinetic investigations of Manassen and Klein[488] on the dehydration of secondary butanol have been interpreted in terms of carbonium-ion intermediates which are symmetrically solvated. Supporting evidence for this idea is provided by the earlier studies on the rate of exchange of ^{18}O in optically active secondary butanol[492,494]. Racemisation occurs at twice the rate of exchange, which shows that each act of exchange is accompanied by an inversion of configuration. For secondary butanol $k_{exchange}/k_{elimination} = 2$, a much smaller figure than that of 27, which is observed for tertiary butyl alcohol[490,503]. To explain this difference it was pointed out that exchange from a solvated carbonium ion would require a greater activation energy than that from an encumbered or free tertiary carbonium ion, but that the activation energy for elimination from the secondary structure was lower than from the tertiary carbonium ion due to more effective participation by the protonated water molecule in aiding breaking of the C_β–H bond in the former case (187)[488].

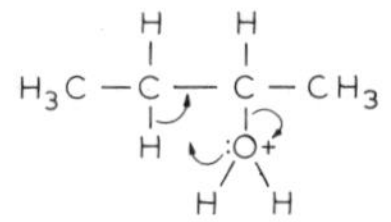

(187)

Whether such arguments are justified is debatable, as $k_{racemisation}/k_{elimination}$ varies widely for similar structures, being 58 for 1,2-diphenylethanol[493] but exceeding 100 for 1-phenylethanol[491,499]. For optically active 1-phenylethanol the ratio, $k_{exchange}/k_{racemisation}$, in aqueous sulphuric acid, was 0.82. This result indicates that exchange occurs at a partially shielded carbonium ion or alternatively by a simultaneous mixture of unimolecular and bimolecular mechanisms[491].

The activation energy for the acid-catalysed oxygen exchange of tertiary butyl alcohol of approximately 30 kcal.mole^{-1} and the difference between steps k_3 and k_{-2} (see 184, p. 295) is in close agreement with the values derived for E1 reactions of tertiary butyl substrates under solvolytic conditions[491]. The unimportance of E2 mechanisms involving $R\overset{+}{O}H_2$ has been ascribed to the lack of a suitable base in the aqueous acidic media[10]. Consequently, the relatively fast ionisation favoured from entropy considerations offsets the large activation energy and unimolecular rather than bimolecular elimination is observed.

If the rate-determining step in the acid-catalysed dehydration is the transfer of a proton from the carbonium ion to the reaction medium, then replacement of beta hydrogen by deuterium should reduce the rate of the reaction. In fact for dehydration of 2-deutero-1,2-diphenylethanol a primary isotope effect of 1.83 at 25°C in 5%-ethanol-95% aqueous sulphuric acid (% H_2SO_4, 53–59%) has recently been reported[493]. The *trans*-stilbene isolated retained 78.5% of the original deuterium and transformation of the product isotope effect into a kinetic isotope effect gives an expected value of 1.57. Remembering that the measured primary isotope effect incorporates a secondary isotope effect, the two values are in close agreement. The small size of this isotope effect indicates a highly unsymmetrical transition state for proton transfer. Considering the small activation energy for this process, a reactant-like transition state seems certain and proton transfer to the medium is presumably little developed. For the reverse reaction, considering the principle of microscopic reversibility, protonation of the olefin should be well developed in the transition state. The Brönsted component, $\alpha(=0.9)$, for protonation of isobutene by aqueous perchloric acid confirms this view[498]. From a subsequent measurement of the rate of exchange of *trans*-stilbene-α-d_1 in 40% sulphuric acid: after allowing for statistical and secondary and primary isotope effects, it was possible to estimate that at equilibrium only 0.31% of the 1,2-diphenylethanol remained. Combining these studies with investigations of the rate of isomerisation of *cis*-stilbene under the same reaction conditions[496], and the known difference in stabilities of *cis*- and *trans*-stilbenes[497], Noyce *et al.*[496] were able to construct a complete free energy profile for the isomerisation of *cis*-stilbene (Fig. 10).

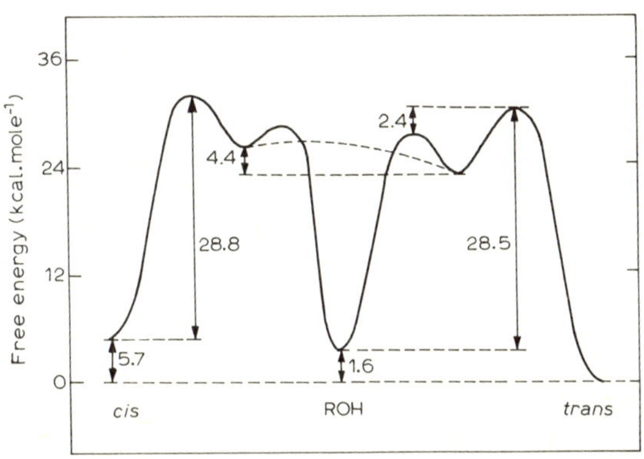

Fig. 10. Reaction coordinate diagram for the isomerisation of *cis*-stilbene. (Reproduced with permission from Noyce *et al.*[496].)

A difference in energy between the two conformers of the carbonium ion, *cis*-R^+ and *trans*-R^+ (188), of 4 kcal.mole^{-1} was estimated.

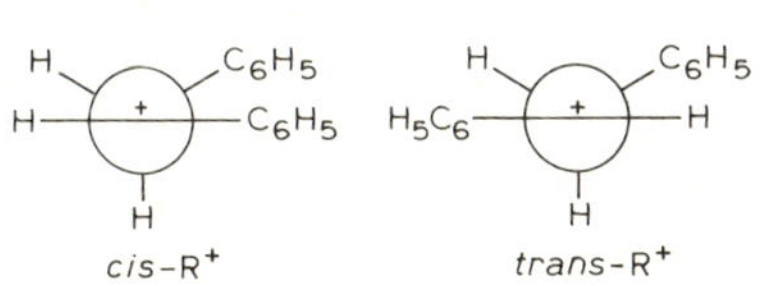

(188)

The investigations of Noyce *et al.*[493,496] can be used to demonstrate the mechanistic application of solvent isotope effects on rates of acid catalysed reactions. The dehydration of 1,2-diphenylethanol proceeds more rapidly in aqueous deutero-sulphuric than in aqueous sulphuric acid ($k_{D_2O}/k_{H_2O} = 1.62$ at 25°C in 5% ethanol 95% H_2O–H_2SO_4-d_2; 51% acid). In this case deuterium is not transferred during the rate-determining step and the influence of the isotopic substitution is to change the steady-state concentration of the reactive carbonium ions resulting from the pre-equilibria preceding the rate-determining step. As $D_3\overset{+}{O}$ is a stronger acid than H_3O, the substrate is protonated to a greater extent in the deuterated medium. In the isomerisation reaction of *cis*-stilbene under similar reaction conditions, $k_{H_2O}/k_{D_2O} = 2.59$. In this case the rate-determining step involves transfer of the isotopically substituted atom to the olefin and a normal isotope effect is observed.

In aqueous solution, the rates of acid-catalysed dehydrations increase more rapidly than the stoicheiometric acid concentration[493]. The gradual decrease in the concentration of water results in a shortage of solvating molecules and a concomitant increase in the activity of the acidic species (*e.g.* $\overset{+}{H}(H_2O)_4 \rightarrow \overset{+}{H}(H_2O)_3 \rightarrow \overset{+}{H}(H_2O)_2$ etc.). Medium acidity has to be measured from the extent of ionisation of weak bases[501] and the pioneer attempts were based on nitroaniline indicators (189). The difference in pK_a between two such indicators of similar basicity was found to be independent of the acidity of the medium, suggesting the ratio, $-\log\,(a_{H^+}\gamma_A/\gamma_{AH^+})$ was a property of the medium and not the indicator, being subsequently termed H_0 (190) and regarded as the tendency of the medium to protonate a neutral indicator molecule.

$$A + H^+ \xrightleftharpoons{Ka} AH^+$$

$$K_a = \frac{[AH^+]\gamma_{AH^+}}{[A]a_{H^+}\gamma_A}$$

where A is a nitroaniline, *a* refers to activities and γ refers to activity coefficients whence

$$pK_a = \log\frac{a_{H^+}\gamma_A}{\gamma_{AH^+}} + \log\frac{[A]}{[AH^+]} \tag{189}$$

for a second nitroaniline indicator, B,

$$\Delta pK = pK_a - pK_b = \log\frac{[A]}{[AH^+]} - \log\frac{[B]}{[BH^+]} + \log\frac{\gamma_A\gamma_{BH^+}}{\gamma_{AH^+}\gamma_B}$$

$$= \log\frac{[A]}{[AH^+]} - \log\frac{[B]}{[BH^+]}$$

if γ_A/γ_{AH^+} varies in the same way as γ_B/γ_{BH^+} with changes in the reaction medium

$$H_0 = -\log\frac{a_{H^+}\gamma_A}{\gamma_{AH^+}} \tag{190}$$

By anchoring the scale in the pH region in dilute aqueous solution (γ_A, $\gamma_{AH^+} \rightarrow 1$) it is possible to extend the H_0 function to the stronger acidic media by successive overlap of indicator pairs of similar but not identical basicity. Of particular relevance to the dehydration of alcohols is the protonation of carbinols and olefins. From the protonation and subsequent carbonium ion formation of triphenylmethanols the H_R function[71] (191) has been established and this rises more rapidly with increasing concentration of sulphuric acid in water than H_0. Thus, for

$$\mathrm{ROH + H^+ \xrightleftharpoons{K_R} R^+ + H_2O}$$

where ROH is a triphenylmethanol,

$$\log K_R = -\log a_{H^+} + \log \frac{[R^+]}{[ROH]} - \log \frac{\gamma_{ROH}}{a_{H_2O}\gamma_{R^+}} \tag{191}$$

and

$$H_R = -\log \frac{a_{H^+}\gamma_{ROH}}{a_{H_2O}\gamma_{R^+}}$$

If the activity coefficient ratios of these different types of indicator vary in an analogous way with changes in medium acidity, the H_R and the equilibrium constant for protonation of olefins[502] can be related to H_0 (see 192–194).

$$\mathrm{Ar_2C{=}CH_2 + \overset{+}{H} \xrightleftharpoons{K_0} Ar_2\overset{+}{C}CH_3}$$
$$\log K_0 = -\log (a_{H^+}\gamma_R/\gamma_{R^+}) \tag{192}$$

$$\log K_0 = H_0 + \log ([R^+]/[R]) \tag{193}$$

if γ_R/γ_{R^+} varies as does γ_A/γ_{AH^+} with changes in the reaction medium

$$H_R = H_0 + \log a_{H_2O} \tag{194}$$

These relationships however, are not observed and protonation of olefins defines a new acidity function, $H_{R'}$, *viz.*

$$H_{R'} = H_R + \log a_{H_2O} \tag{195}$$

These observations indicate that activity coefficients of the conjugate acids of aniline indicators reflect changes in the medium acidity in a different manner to highly delocalised carbonium ions. Explanations in terms of a greater tendency of the anilinium than carbonium ions to bond with the solvent have been proposed to account for the different behaviour[683].

It is with the limitation of being unable to assess the acidity of the medium independently of the type of indicator employed, that interpretation of the dependency of the rate of acid-catalysed dehydration of alcohols and hydration of olefins must be approached. As each of the various acidity functions run parallel to each other, a plot of the logarithm of the rate coefficient of an acid-catalysed reaction against an acidity function should give a linear correlation. The slope of such a plot, however, will only be unity if the ratio of activity coefficients of the substrate and its activated complex vary in the same way with changes in the reaction medium as the ratio of activity coefficients of the indicator molecule and its conjugate acid.

Although the rate coefficients for the dehydration of a number of tertiary alkyl alcohols (Table 26, 1–3) correlate with the H_0 function[503,504] a divergence is noted for 1,2-diphenylethanol[493]. For the former substrates, the activity coefficients of the nitroaniline indicators and their conjugate acids are appropriate models. However, the presence of an alpha phenyl substituent causes more effective delocalisation of the positive charge and there is a lower tendency for the activated complex arising from the carbonium ion to hydrogen bond with the solvent molecules of the medium. Consequently, with a decrease in the activity of the water, the free energy of solvation change suffers less in the case of 1,2-diphenylethanol than with the tertiary alkyl substrates and the rate of dehydration rises more steeply with increasing acidity. Making an allowance for a difference of one molecule of water between the activated

TABLE 26

THE RELATIONSHIP BETWEEN THE RATE COEFFICIENTS FOR DEHYDRATION, OXYGEN EXCHANGE AND RACEMISATION OF ALCOHOLS AND ISOMERISATION AND HYDRATION OF OLEFINS AT 25°C WITH HAMMETT ACIDITY FUNCTIONS

No.	*Reaction*	*Substrate*	*Solvent*	*Slope**	
1	Dehydration	*t*-BuOH	H_2O/H_2SO_4	1.00	503
2	Dehydration	*t*-AmOH	H_2O/H_2SO_4	1.00	503
3	Dehydration	1-Methylcyclohexanol	$CH_3COOH/H_2SO_4/H_2O$	1.00	504
4	Dehydration	1,2-Diphenylethanol	5% $C_2H_5OH/H_2O/H_2SO_4$	1.32	493
5	Dehydration	1,2-Diphenylethanol	5% $C_2H_5OH/H_2O/H_2SO_4$	1.02[a]	493
6	Racemisation	(+)β-Phenyl-β-hydroxy-propionic acid	H_2O/H_2SO_4	1.17	505
7	Exchange	*t*-BuOH	H_2O/H_2SO_4	1.20	503
8	Hydration	Trimethylethylene	H_2O/H_2SO_4	1.00	503
9	Hydration	Styrene	H_2O/H_2SO_4	1.26	506
10	Hydration	Styrene	H_2O/H_2SO_4	0.72[b]	506
11	Isomerisation	*cis*-Stilbene	H_2O/H_2SO_4	1.25	496
12	Isomerisation	*cis*-Cinnamic acid	H_2O/H_2SO_4	1.00	507

* Slope of a graph of log k_{obs} against $-H_0$, except [a] in which the abscissa was $-(H_0+\log a_{H_2O})$ and [b] $-H_R$

complex and the solvated conjugate acid of the nitroaniline appears to rectify the discrepancy and a plot of log k against $-(H_0+\log a_{H_2O})$ for the dehydration of 1,2-diphenylethanol has almost unit slope[493].

The variation in the magnitude of the response of the rate coefficients, for hydration or isomerisation of the olefins and the racemisation or oxygen exchange of the alcohols, to H_0 can also be explained by varying activity coefficient behaviour[506]. The indicators used to establish the H_R and $H_{R'}$ functions are rather atypical with respect to the dehydration and hydration reactions. Their conjugate acids are much more extensively delocalised than the relevant activated complexes. For instance the sensitivity of the rate of hydration of styrene to H_0 and $H_{R'}$ suggests that the activated complex is solvated to a degree intermediate between that of the conjugate acids of the nitroanilines and those of the 1,1-diaryl-olefins[506].

Although it has since been shown that slow proton transfers can follow H_0[508], the earlier disbelief in this view led Taft *et al.*[503] to suggest a mechanism in which the formation of a π-complex between the olefin and the solvent preceeded the subsequent rate-determining formation of the carbonium ion. Although not required by the principle of microscopic reversibility, the reverse step was also suggested to be rate-determining (196).

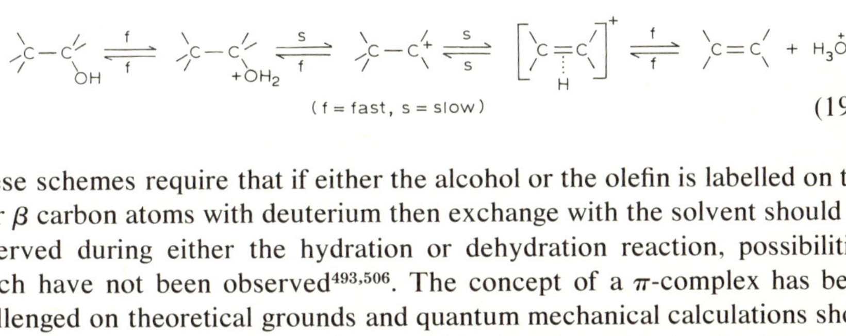

(f = fast, s = slow) (196)

These schemes require that if either the alcohol or the olefin is labelled on the α or β carbon atoms with deuterium then exchange with the solvent should be observed during either the hydration or dehydration reaction, possibilities which have not been observed[493,506]. The concept of a π-complex has been challenged on theoretical grounds and quantum mechanical calculations show that a π-complex should be much less stable than the carbonium ion[509].

A final comment is merited on the influence of substituents on the correlation of the rate coefficient with Hammett acidity functions[500]. For the series of substituted *cis*-stilbenes the response of rate coefficients of hydration to H_0 varied markedly and slopes (log k against $-H_0$) ranging from 0.88 to 1.27 were observed. The change in electron density, in going from reactant to transition state, at the heteroatom centres of the substituents obviously gives rise to further possible variations in activity coefficient behaviour.

There is a considerable amount of evidence opposing the intermediacy of π-complexes in the acid-catalysed dehydration of alcohols and in the hydration of olefins. In the 1,2-diphenylethanol system[493], if before or during the course of the rate-determining step the two central carbon atoms become equivalent then substituents in both phenyl groups should exert a similar influence on the rate of dehydration. However, if non-equivalence is maintained until after the rate-determining step, the influence of 1-aryl substituents on the rate coefficient should be predictable from the Hammett substituent constants, σ^{+}[510], whereas the influence of 2-aryl substituents should correlate with the normal Hammett substituent constants, σ. This latter circumstance is in fact observed. The 2-phenyl substituents exert only a modest effect on the rate ($\rho = -1$), while the 1-phenyl substituents, being adjacent and able to conjugate with the carbonium ion centre, exert a greater influence ($\rho^{+} = -3.77$). Further evidence against a π-complex is provided by the observation that 1-deutero-1,2-diphenylethanol gives *trans*-stilbene which retains all of the deuterium. This result shows that hydride rearrangements predicted if a π-complex is involved do not compete with elimination[500]. Isomerisation of 1,2-disubstituted *cis*-stilbenes shows the following correlation with the Hammett equation: $\log k_{XY} = -4.46 - 3.30 (\sigma_X^{+} + 0.29\sigma_Y)$, where X is potentially the greater electron-releasing substituent[496]. In the case of symmetrical stilbenes (X = Y), a correlation with the above equation was only observed if a statistical correction for the two identical sites of protonation was made. These results again demonstrate the non-equivalence of the two phenyl rings in the intermediate arising from or before the rate-determining step and accord with a carbonium ion rather than a π-complex.

There seems little doubt that π-complexes do not play a role in the rate-determining step of acid-catalysed dehydration of 1,2-diphenylethanols and related substrates. However, in simple alkyl structures in which carbonium ion stability is lower, π-complexes may offer an alternate pathway of lower energy. Certainly π-complexes appear to account for the greater preference for the formation of *cis*-but-2-ene rather than the thermodynamically more stable *trans* isomer from the dehydration of secondary butanol over solid catalysts[488]. Of course metal cations with a greater variety of orbitals do not really provide appropriate models for assessing the co-ordination properties of a proton. Considering all the available kinetic evidence the most plausible mechanism for dehydration can be described as (197), with route A for tertiary and secondary alcohols and route B for primary and secondary alcohols, *viz*.

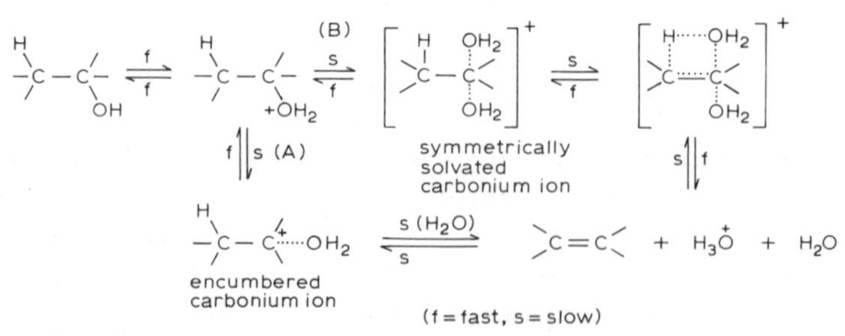

(197)

Of course, varying degrees of solvation other than illustrated in this scheme will be encountered and the desolvation of the carbonium ion centre and the loss of a beta proton may occur synchronously or in discrete steps.

2.12.2 Dehydration involving intermediate ester formation

For a number of acids, the intermediacy of an ester in the dehydration of an alcohol to an olefin has to be considered. Possibly the best-known example is that of ethyl hydrogensulphate, an intermediate which partitions to diethyl ether and ethylene, in the dehydration of ethyl alcohol by sulphuric acid. Should ester formation be more rapid than its subsequent decomposition, it should be possible to isolate the ester during the course of the reaction. An intermediate ester may well decompose thermally to give reaction products similar to those observed from the pyrolysis of the corresponding acetates or xanthates, provided that kinetic control is maintained. Of course some esters (*e.g.* *p*-toluene-sulphonates) ionise easily in polar solvents and would decompose by carbonium ion processes similar to those of the oxonium ions. Should ester formation be a slow process compared to its subsequent decomposition by a carbonium-ion mechanism, then it becomes more problematic to prove its intermediacy. In such circumstances, it is wisest to study separately the dehydration of both the alcohol and the ester and to combine the results with a knowledge of the rate of esterification of a similar structurally related alcohol which cannot undergo dehydration. It may, of course, be difficult to select an appropriately similar alcohol as stereochemical changes may be crucial. When treated with phthalic anhydride, borneol, which can form a carbonium ion by a synartetically assisted ionisation, undergoes dehydration to give a rearranged olefin, while the *endo* isomer is converted into a stable ester[10,511]. Oxonium ions cannot be formed with phthalic anhydride, and the phthalate esters of alcohols in which synartetic assistance is not possible decompose at elevated temperatures in accordance with the Saytzeff rule[512].

The acidity function criterion offers little prospect as a basis on which the reaction mechanism can be assigned as both 1-methylcyclohexanol and its acetate form olefins by unimolecular carbonium-ion processes which show a unit dependence on H_0 in acetic acid containing aqueous sulphuric acid[504]. The ten-fold greater reactivity of the ester was explained in terms of the greater electronegativity of the protonated ester function than the oxonium ion, which more than counterbalances the lower basicity of the acetate than the alcohol.

Kinetic studies concerning the intermediacy of esters in the dehydration of alcohols are few and evidence is less soundly based on product analysis. An interesting diversity of reaction products, suggestive of varying mechanisms, has been reported for the dehydration of α-terpineol with various acidic catalysts, *viz.*[513]

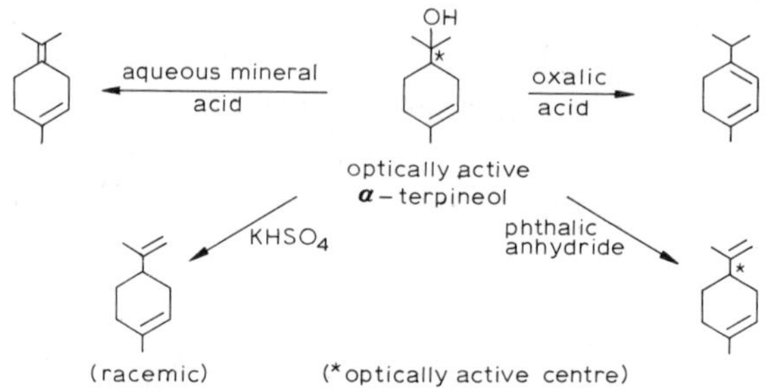

(198)

With aqueous mineral acids the unconjugated diene is the main product but with oxalic acid, in which solvation of the carbonium ion is less developed and hydrogen-bridging may assist ionisation, the conjugated diene is formed. With phthalic anhydride, decomposition occurs according to statistical and steric considerations and the optically active product is most plausibly formed from an intermediate ester pyrolysis. The elimination with potassium hydrogen sulphate is puzzling as a racemic product is obtained, but elimination again gives the least stable diene. The possible intermediacy of the 1,1-dimethyl ethylenic product has been suggested[514], but this seems surprising as any olefin rearrangement would surely result in a conjugated product.

For carbonium-ion processes, unless configuration is maintained by ion association, it is unsound to assume stereochemistry of elimination from product analysis, unless it is independently shown that racemisation at the carbon bonded to the hydroxy group is slower than olefin formation. Decompositions of ester intermediates should show a preference for *anti* stereospecificity if they are base-induced and *syn-clinal* stereospecificity when they occur under thermal conditions. Typical results on the dehydration of two stereoisomers in the rigid steroid series support the above generalisations (199)[515].

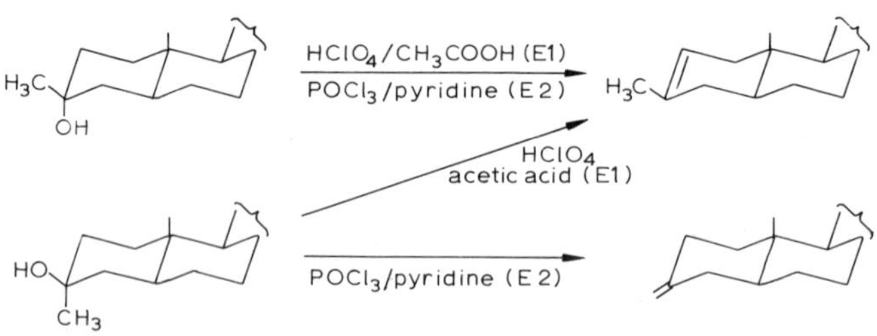

(199)

When dehydration is accomplished with perchloric acid in acetic acid, both alcohols give the *endo*-olefin as would be expected for carbonium ion processes. An E2 reaction of an oxonium ion, by analogy with the orientation observed for base-induced eliminations of 1-methylcycloalkyl trimethylammonium and dimethylsuphonium salts, would be expected to give the Hofmann product, especially in the case of the equatorial epimer which cannot attain *anti* stereospecificity for *endo*-olefin formation. Dehydration by phosphorus oxychloride in pyridine at 0°C occurs by *anti*-elimination with both isomers, the alcohols being converted into the esters ($ROPOCl_2$) which then undergo E2 reactions with the basic solvent.

2.13 DEAMINATION REACTIONS

The treatment of amines with nitrous acid results in an initial rate-determining nitrosation which is followed by diazotisation and subsequent deamination, *viz.*[516]

$$RNH_2 \xrightarrow{HNO_2} RNHNO \rightarrow RN{=}NOH \rightarrow R\overset{+}{N}_2 \rightarrow \overset{+}{R} \rightarrow \text{products} \quad (200)$$

For secondary amines, the reaction stops at the nitrosamine stage, but with primary aromatic amines fast reaction effectively ceases at the diazonium ion. However, for primary aliphatic amines the unstable diazonium ions rapidly lose nitrogen and the resulting carbonium ion decomposes, often yielding olefins as the minor reaction products[517], *e.g.* (201).

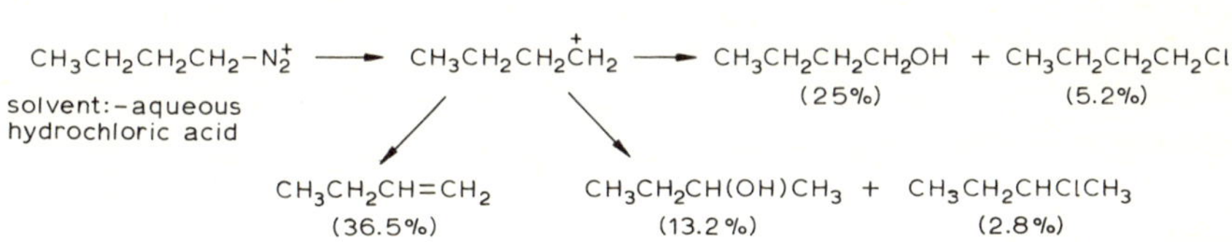

(201)

The multiplicity of the reaction products limits deamination as a useful synthetic method for the preparation of olefins but it is of great interest to compare the mode of reaction of the intermediate carbonium ion with those derived from solvolysis of alkyl halides and esters or acid-catalysed dehydration of alcohols. As nitrosation is the rate-determining step, evidence for carbonium-ion intermediates has to come from the nature and stereochemistry of the reaction products rather than from kinetic studies.

A number of features in deamination have their counterparts in solvolytic E1 reactions. Transannular 1,5 and 1,6 shifts have been observed in deaminations of eight- to twelve-membered alicyclic amines, *viz.*

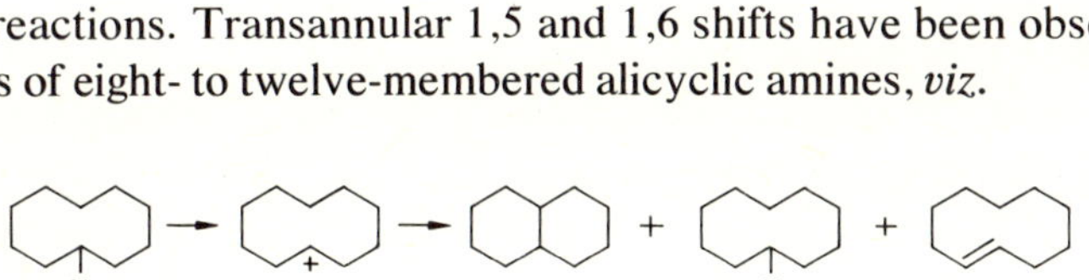

(202)

References pp. 350–364

and also in the solvolyses of the corresponding *p*-toluenesulphonate esters[10,518,519]. Studies with the 1-deutero-butyl-X system show that deamination and solvolysis of the *p*-toluenesulphonate involve similar amounts of racemisation[520]. Usually racemisation in deamination increases with the stability of the carbonium ion, indicating that for the substitution reaction, the less stable carbonium ions are still partly shielded by the departing nitrogen molecule[520]. As for other S_N1 reactions involving a neighbouring carboxyl group[521], deamination of α-amino acids occurs with retention of configuration[522].

The major difference between deamination and other processes involving carbonium ion mechanisms arises not from the substitution reactions but from the unimolecular eliminations and in particular the rearrangement reactions. Olefin-isomer distributions are generally different, and rearrangement reactions are seldom encountered in solvolytic processes except in the presence of non-nucleophilic reagents such as Friedel–Crafts catalysts[523] or under conditions of limiting solvolysis in a poorly ionising solvent of low nucleophilicity such as acetic acid[524]. In the latter case, the ionisation of the C–X bond is assisted by migration of an alkyl or aryl group or hydride transfer from the adjacent beta carbon. However, whereas a *p*-methoxy-phenyl migrates 500 times more rapidly than a phenyl group in the pinacol rearrangement, the difference in migratory aptitude is only 1.56 in pinacolic deamination[525]. In the acetolysis of *erythro*- and *threo*-1-methyl-2-phenylpropyl-*p*-toluenesulphonates, predominant migration is by the phenyl group and only a minor fraction involves hydride transfer. However, in deamination of the corresponding amines, the methyl group migrates most easily and hydrogen migration competes more favourably with phenyl migration for the *threo* isomer, whereas for the *erythro* isomer, phenyl migrates more rapidly than methyl by a factor of eight[526].

The above findings suggest that if a carbonium ion is involved in the decomposition of the diazonium ion, then its reactivity is modified in some way from that of carbonium ions generated by more conventional methods and the concept of a high energy or so-called "hot" carbonium ion has been developed[525–528]. The unstable aliphatic diazonium ion loses nitrogen so rapidly that neither neighbouring group participation nor solvent assistance, normally required in the ionisation of more stable carbonium ion precursors, is encountered. The resulting carbonium ion is an exceptionally reactive species as it possesses an unsolvated vacant *p* orbital and has no counter-ion in the immediate vicinity. Thus it is capable of promoting rearrangements of appropriately situated substituents on the adjacent carbon atom and rearrangement occurs more rapidly than molecular collision as hydride shifts occur intramolecularly, no tracer being incorporated when the deamination is performed in labelled solvents[529]. From a consideration of preferred orbital overlap, substituents situated *trans* with respect to the vacant orbital will migrate preferentially and the above-mentioned relative migratory aptitudes in deamination in the 1-methyl-2-phenylpropyl series can be explained in terms of the populations of the three conformations of the diazonium ions. A similar explanation has been used to account for the olefin distribution arising from the deamination of secondary butylamine, *viz.*

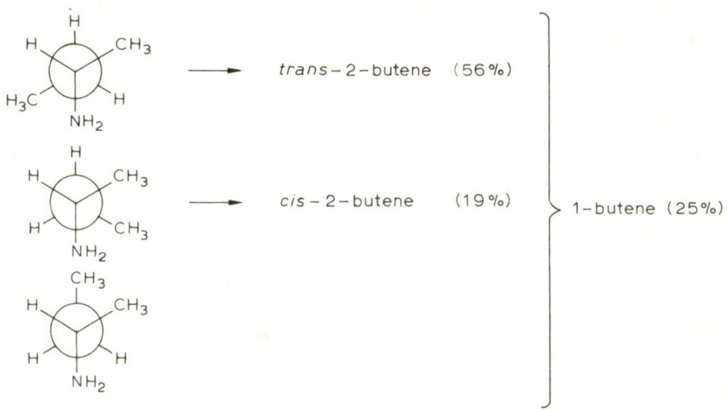

(203)

All three conformations can give rise to 1-butene and although this is not the major product it is formed in much greater abundance than from E1 reactions of 2-butyl halides or esters. Methyl rearrangement does not occur in this case as this would give rise to a less stable primary carbonium ion. A hydride shift is possible to give the same but essentially a new ion which subsequently breaks down, but this possibility can be excluded by the use of carbon-labelling experiments.

Ground-state control as suggested above depends on the barrier to rotation about a C–C bond being of greater energy than the barrier to the reaction of the carbonium ion. Considering that desolvation of the carbonium ion is not required before reaction, activation energies of less than 3–4 kcal.mole^{-1} seem likely. Of course an increase in the stability of the carbonium ion could cause bond rotation and carbonium-ion reaction to become more competitive and carbon-labelling studies on the pinacolic deamination of optically active 2-amino-1,1-diphenyl-propan-1-ol support this possibility (204).

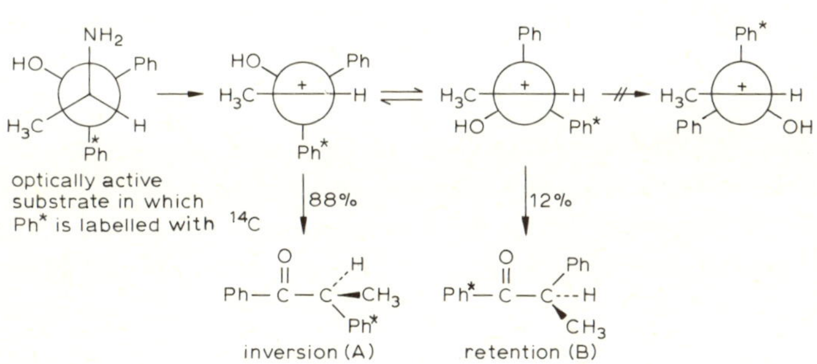

(204)

The major product (A) is formed with inversion of configuration[530] and migration of the labelled phenyl substituent[531]. However, the other ketone, formed with retention arose by migration of the unlabelled phenyl group, indicating that the life time of the carbonium ion is sufficiently long for some rotation to occur but is too short for complete rotation through 180° to be achieved.

As an alternative to the "hot" ion theory, the possible breakdown of the diazonium ion by a concerted rearrangement or bimolecular reaction with the solvent has been suggested, to account for the different behaviour from typical E1 reactions[520,532]. An E2 reaction has also been put forward, for alicyclic amines. The equatorial isomers of cyclic amines (*e.g. trans*-fused steroids[533,534], 1- and 2-decylamines[535] and heterocyclics[536]) deaminate to give predominantly the alcohol of retained configuration, while in the case of the axial isomer considerable olefin formation is observed and the alcohol with inverted configuration is obtained. These results can be accommodated on the "hot" ion theory in terms of ground-state control and inhibition to solvent approach from the "backside" by the ring system for an equatorial substituent. It is tempting to dismiss an E2 reaction of a diazonium ion in terms of failure to observe Hofmann orientation of elimination, but features which influence bimolecular reactions with more normal activation energies may not be applicable for low activation processes. Typical E1-type rearrangements are observed in deamination reactions of labile systems such as bornyl, fenchyl and isobornyl amines, and even apocamphylamine, which possesses the amino-group at a bridgehead forms an alcohol[537] (this could arise from decomposition of the diazo-hydroxide rather than the carbonium ion[538]), a conversion which cannot be achieved by more conventional unimolecular or bimolecular reactions.

Most of these observations accord with a carbonium-ion process, but in terms of energetics, if the potential energy well of an unsolvated carbonium ion has sides of lower energy than the energy of thermal vibrations, the carbonium ion may be by-passed and in effect a two-stage reaction degenerates into a single step, namely decomposition of the diazonium ion. With increasing carbonium-ion stability, the existence of an intermediate carbonium ion seems more realistic. The need to pay careful attention to reaction conditions when interpreting mechanisms of deamination has been stressed[10], and the possibility that decomposition to products occurs from the diazo-hydroxide rather than a diazonium ion or a carbonium ion cannot be ignored unless the solution pH is less than 2. Such intermediates may explain the apparent anomalies, that deamination of certain axial-amino steroids results in an alcohol of the same configuration and an olefin distribution biased considerably in favour of the Saytezeff rule (205)[539]. Recent results of Cohen and Daniewski[687] on the deamination of *trans*-2-decalylamines have challenged the *anti*-E2 hypothesis. Using specifically deuterated substrates they have shown that in acetic acid,

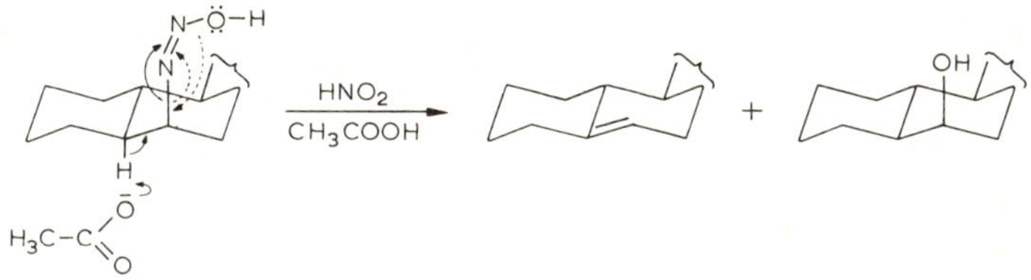

(205)

nitrous acid deamination of the axial amino substrate proceeds mainly (56–61%) by a *syn-clinal* elimination. This is assumed to occur via an intermediate ion pair, in which acetate is the anionic species (*cf.* Section 2.3.2). Whether such a phenomenon is general awaits studies with more ionising media.

3. Alkyne-forming beta-eliminations

Within the bounds of beta-elimination, alkynes are generally prepared from substituted olefins. A greater electronegativity is associated with sp^2- than sp^3-hybridised carbon and consequently olefinic carbon atoms accept positive charge less readily but negative charge more readily than the corresponding saturated carbon atoms[173]. Thus, intermediates are more likely to possess carbanion character than in the corresponding mechanisms in the saturated series and examples of carbonium-ion intermediates are rather less likely. As might be expected, the conversion of olefins into alkynes is generally base-catalysed, typical reactions including dehydrohalogenation and halogenative decarboxylation. A number of eliminations have also been successfully achieved with metals in organic solvents. A great multitude of synthetic approaches to alkynes and some of the salient features of the mechanisms have recently been reviewed by Köbrich[540].

3.1 BASE-CATALYSED ELIMINATION REACTIONS

The mechanisms of base-induced beta-eliminations leading to alkynes are highly dependent on the type of base employed, and for this reason the dehydrohalogenations induced by metal alkoxides are considered separately from those initiated by metal amides or metal alkyls.

3.1.1 Metal alkoxide-induced dehydrohalogenations

Kinetic studies indicate that carbanion mechanisms are more probable in elimination from alkenes than from saturated molecules. Hydrogen–deuterium exchange catalysed by sodium methoxide in methanol at 26°C occurs twenty five times more rapidly than *anti*-elimination from *cis*-dibromoethylene[541]. As might be expected, the primary kinetic isotope effect ($k_{C_2H_2Br_2}/k_{C_2D_2Br_2}$) for elimination is only 1.03 at 35°C and it seems that even *anti*-elimination in this case occurs by the E1cB mechanism (Section 2.1.3). The difference in the relative importance of the activation energy and entropy on the rate of elimination from *cis*- and the corresponding *trans*-dihaloethylenes had earlier been proposed to arise from a duality of mechanism, namely a carbanion *syn*-elimination but a concerted *anti*-elimination[542]. Considering the above results and also the observation of hydrogen exchange in *cis*-dichloroethylene and in all three *trans*-dihaloethylenes, there seems little reason to invoke such a duality[541].

Exchange occurred at a similar rate in *cis*-dibromo- and dichloro-ethylenes but was slower for the *trans* substrates following the order ClCH=CHCl > BrCH=CHBr > IHC=CHI. These results show that the exchange reaction is favoured by increased electronegativity of the alpha halogen and a beta halogen *trans* rather than *cis* to the ionising carbon–hydrogen bond. Earlier workers had observed hydrogen exchange in the absence of beta-elimination in the reaction of trichloroethylenes with calcium deuteroxide[543,544].

In alkoxide–alcohol media, hydrogen–deuterium exchange of olefinic hydrogen is not observed with hydrocarbon structures such as di- and tri-phenylethylenes[541] and, although exchange is exhibited by β-bromostyrene and its 4-nitro isomer, it is the hydrogen alpha to the halogen and not to the phenyl which is involved[545]. Carbanions derived by the ionisation of sp^2-hybridised carbon–hydrogen bonds appear to be stabilised more effectively by alpha halogen than by alpha phenyl substituents. Aryl substituents may have a greater influence on carbon–hydrogen acidity in saturated than in alkenyl systems. In the former case, the carbanion can be stabilised by an adjacent aryl group exerting both electron-withdrawing inductive and mesomeric influences. However, for carbanions arising from the ionisation of sp^2-hybridised carbon–hydrogen bonds, the stabilisation by an adjacent aryl group may be limited mainly to its inductive effect. For mesomeric stabilisation to assist carbanion formation, the conjugation between the π-electrons of the aryl group and those of the double bond must be overcome so that the aryl substituent can rotate until it is orthogonal with the developing electron pair of the ionising carbon–hydrogen bond. Thus, in attaining the transition state for carbanion formation, some of the resonance energy associated with the unionised olefin will be lost.

The observation of primary kinetic hydrogen isotope effects for elimination from β-bromostyrenes in alkoxide–alcohol media suggests these substrates follow an E2 rather than an E1cB mechanism[545,546]. Certainly greater carbanion character is associated with the transition states than those of the corresponding saturated halides. The magnitude of the isotope effect (Table 27) decreases

TABLE 27

PRIMARY KINETIC HYDROGEN ISOTOPE EFFECTS IN THE BASE-INDUCED ELIMINATION OF β-HALOGENOSTYRENES AT 30°C[545]

Substrate	$k_H/k_D(\bar{O}CH_3/CH_3OH)$	$k_H/k_D(\bar{O}Bu^t/t\text{-}BuOH)$
cis-$C_6H_5CD{=}CHBr$	2.9	4.6
cis-(*p*-$NO_2.C_6H_4$)$CD{=}CHBr$	2.2	4.3
cis-$C_6H_5CD{=}CHCl$	2.2	3.2
cis-(*p*-$NO_2.C_6H_4$)$CD{=}CHCl$	1.6	3.5
trans-(*p*-$NO_2.C_6H_4$)$CD{=}CHBr$		1.8
trans-(*p*-$NO_2.C_6H_4$)$CD{=}CHCl$		1.6

with (*i*) an increase in the acidity of the beta hydrogen, induced by the incorporation of the highly electron-withdrawing *p*-nitro substituent, (*ii*) a change to a poorer leaving group (Br → Cl), and (*iii*) a change from *anti*- to *syn*-orientation of the eliminating fragments. All of these changes produce a greater extent of proton transfer to the base in the transition state. As in the saturated systems, greater isotope effects, indicative of a more reactant-like transition state, are observed with an increase in the basic strength of the medium. However, if C–X bond breaking is retarded more than C–H bond breaking by a change from methanol to *t*-butyl alcohol, then greater carbanion character may be associated with the transition state in the latter solvent.

The stereochemical results are of great interest and, as with the saturated structures, changes in electron density appear to impose a greater influence on the leaving group when the latter is *anti*-related to the developing carbanion. Unlike the case of the saturated substrates, conformational considerations do not complicate the stereochemistry in alkenes and elimination occurs from either planar *anti* or *syn* orientation of the leaving fragments. The magnitude of the ratio, k_{anti}/k_{syn} varies widely, being influenced by both the substrate and the reaction conditions (Table 28). The differences in the free energies of activation are more meaningful than the rate ratios[10], as the relative importance of the enthalpy and entropy of activation depends greatly on the reaction conditions. However, as these activation parameters are seldom available, the present discussion has been based on the rate relationships. To explain the phenomenal rate difference between *anti*- and *syn*-elimination from β-bromostyrene, Cristol and Norris[547] proposed an E2 *anti*-elimination but a carbanion *syn*-elimination, in which inversion of configuration followed proton removal. However, for the styrenes the isotope studies mentioned above do not support a carbanion mechanism. Careful analysis shows that the k_{anti}/k_{syn} ratio decreases with the expected increasing carbanion character of the transition state. This observation suggests that the more developed is the electron pair of the carbanion at the beta carbon, the less selective it becomes on reaction and the relative stereochemical relationship of the leaving group becomes less important.

TABLE 28

RELATIVE RATES OF *anti*- AND *syn*-ELIMINATION FROM SOME HALOGENOALKENES

No.	*Substrate*	k_{anti}/k_{syn} [a]	*base/solvent*	*Ref.*
1	$C_6H_5CH{=}CHBr$	210,000	$\bar{O}H$, i-PrOH	547
2	p-$NO_2.C_6H_4CH{=}CHBr$	16,000	$\bar{O}H$, i-PrOH	547
3	$IHC{=}CHI$	12,000	$\bar{O}CH_3$, CH_3OH	542
4	p-$NO_2.C_6H_4CH{=}CHBr$	3,000	$\bar{O}Bu^t$, t-BuOH	545
5	$ClHC{=}CHCl$	3,000	$\bar{O}CH_3$, CH_3OH	542
6	p-$NO_2C_6H_4CH{=}CHCl$	700	$\bar{O}Bu^t$, t-BuOH	545
7	$C_6H_5CH{=}CClC_6H_5$	400	$\bar{O}H$, C_2H_5OH	548
8	$\bar{O}OC(Br)C{=}CHCO\bar{O}$	17	$\bar{O}H$, H_2O	547
9	$\bar{O}OC(Cl)C{=}CHCO\bar{O}$	10	$\bar{O}H$, H_2O	547
10	$ClHC{=}CHCl$	1.2	$LiCH_3$, ether	552
11	$C_6H_5CH{=}CHBr$	0.8	Li-piperidyl, ether	553
12	$C_6H_5CH{=}CBrC_6H_5$	0.6	Li-piperidyl, ether	553
13	$C_6H_5CH{=}CHBr$	0.5	LiC_6H_5, ether	554
14	$C_6H_5CH{=}CHBr$	0.2	LiC_6H_5, di-n-butyl ether	554
15	$C_6H_5CH{=}CClC_6H_5$	0.04	LiC_6H_5, ether	548

[a] *anti*-elimination from the *cis* isomer and *syn*-elimination from the *trans* isomer, at various reaction temperatures.

Release of steric interaction between bulky substituents related *cis* across the carbon–carbon double bond or unfavourable electronic interactions between the basic reagent and the *cis*-related negatively charged alpha substituent are suggestions which have been proposed to explain an increase or a reduction in the k_{anti}/k_{syn} ratio, respectively[10]. Certainly these two effects exert a major influence on the rate ratios.

Possibly the measurement of the chlorine isotope effect for a series of suitably substituted substrates may help to decide whether or not greater stereospecificity is associated with decreased carbanion character of the transition state for elimination from alkenes. Of course, such studies must be accompanied by labelling experiments to exclude the possibility that alkynes arise by an apparent alpha-elimination (Section 5.1) involving formation of a carbanion at the alpha carbon, which subsequently eliminates alpha halogen simultaneously with migration of a substituent from the beta carbon. Results with alcoholic media, however, tend to indicate this alternative mechanism is unimportant if a normal beta-elimination is possible.

3.1.2 Metal amide- and metal alkyl-induced dehydrohalogenations

Carbanions, which have been postulated as intermediates in dehydrohalogenations of halogenoalkenes in alkoxide–alcoholic media, have actually been prepared as their lithium salts in ethereal solvents. Chloroethylenes can be metallated with *n*-butyl-lithium in tetrahydrofuran at low temperatures and subsequent carboxylation and acidification yields the expected chloroacrylic acids, *viz.*

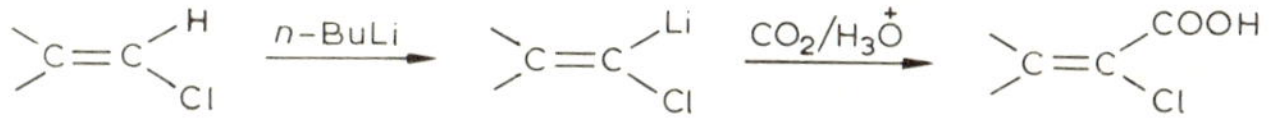

(206)

The configurational stability of alkenyl anions is obviously remarkably high compared with that of alkyl anions[549]. The order of stability of the carbanions of the "chloroethylenes", (207), reflects that *anti*-beta-elimination occurs more easily than *syn*-beta-elimination or alpha-elimination[550].

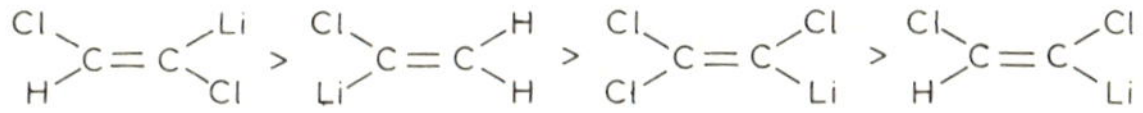

(207)

Metallated alkenyl anions of small ring systems generally possess greater thermodynamic stability than those of the corresponding open-chain compounds. The geometry of the ring systems greatly inhibits the formation of linear acetylenic bonds, and, even at room temperature, metallated derivatives of indenyl and acenaphthenyl substrates have been prepared[540]. Under more forcing conditions, the acetylenes formed in cyclic systems are so reactive that their existence is only transient but can be established by labelling or trapping experiments[540], *e.g.*

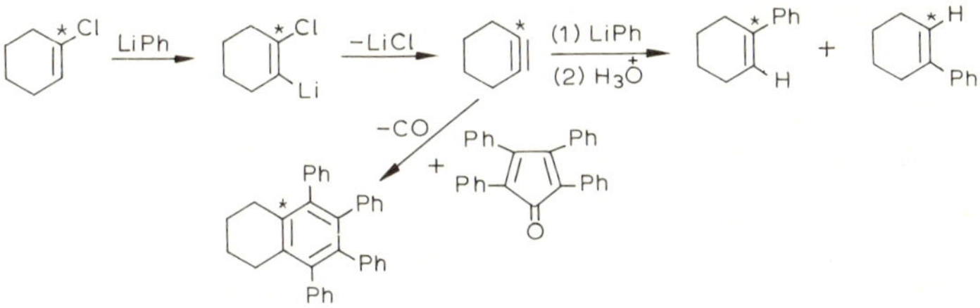

(208)

The most classic cyclic acetylene intermediate is undoubtedly benzyne[551].

Dehydrohalogenations of halogenoalkenes by metal amides or metal alkyls show a much lower stereochemical preference than those induced by metal alkoxides (Table 28, p. 312). In fact, *syn*-elimination generally occurs slightly more rapidly than *anti*-elimination, a preference explained in terms of ion pairing and competitive apparent alpha-elimination, which compete more favourably alongside *syn*- rather than *anti*-beta-elimination. These additional factors are easily demonstrated for β-bromostyrene. For the *trans* isomer, a cyclic *syn*-elimination involving an ion pair (209) is possible. The phenyl substituent is also favourably disposed to exert a nucleophilic displacement on the halogen from the carbanion intermediate (210). Whereas the *cis*-β-bromostyrene decomposes by an *anti*-beta-elimination, elimination can occur from the *trans*

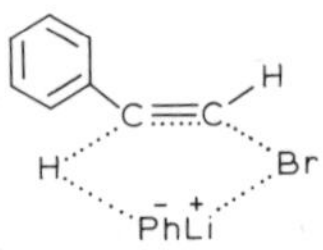

(209)

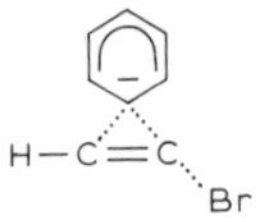

(210)

isomer by a combination of simultaneous processes, namely *syn*-beta-elimination, ion-associated cyclic elimination, and apparent alpha-elimination. That apparent alpha-eliminations are more competitive with beta-elimination when metal amides and alkyls are used as the bases is demonstrated by the work of Cristol *et al.*[548,554]. Metallation occurs more rapidly in β-bromostyrene than in the chlorostilbene series in which the acidic hydrogen is not alpha to the halogen substituent.

3.2 DEHALOGENATIVE DECARBOXYLATION

When heated at 120°C in concentrated aqueous alkaline solution, β-chloro-*cis*-cinnamic acid is converted into phenylacetylene. A dehydrohalogenation followed by a decarboxylation would explain this sequence, but more probably a direct dehalogenative decarboxylation occurs as such reactions are observed in the absence of vinylic hydrogen, *e.g.*

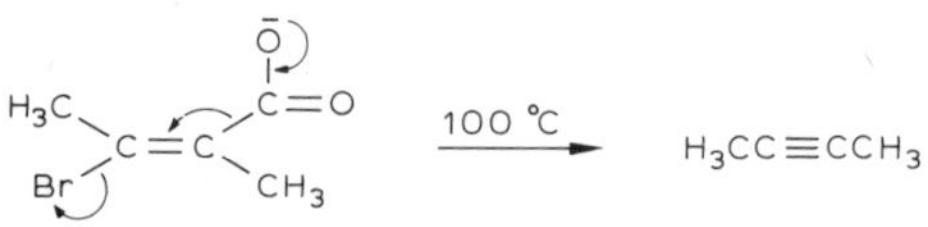

(211)

Of the two isomeric 2-bromo-3-carboxy-2-butenes, the substrate in which the halogen and the carboxyl function are *trans*-related yields dimethylacetylene while the other isomer is inert[555]. Really typical of fragmentation (Section 8.1) rather than elimination reactions, these decarboxylative processes show a general preference for an *anti* orientation of the eliminated fragments and have also been observed with sulphonate esters rather than halogens as the leaving groups[556]. Although carbanion intermediates may be involved, the stringent stereochemical requirement suggests a concerted elimination. In media of low basicity, fragmentation rather than dehalogenation or dehydrohalogenation is observed with the saturated structures such as *trans*-cinnamic acid dibromides. In ethanol, the major product is *cis*-β-bromostyrene which arises via formation of an intermediate beta-lactone, whereas, in water, initial ionisation to a zwitterion is followed by free rotation about the central carbon–carbon bond and a mixture of the isomeric β-bromostyrenes is formed[557,558], *viz.*

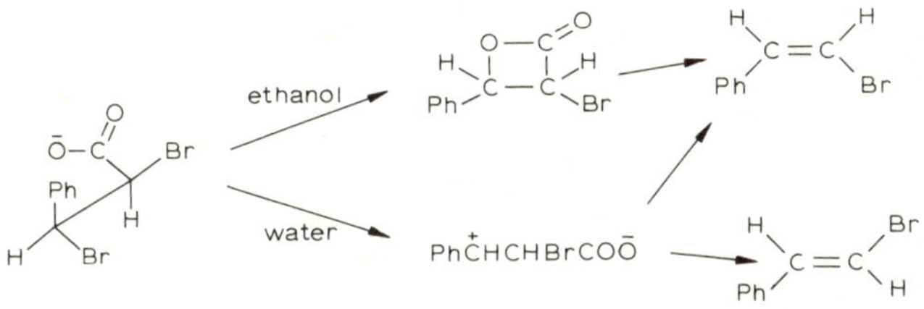

(212)

Electron-withdrawing substituents in the phenyl group promote the stereospecific route and electron-releasing substituents favour the carbonium ion mechanism (see Section 8.1)[559].

3.3 METAL-CATALYSED DEHALOGENATIONS

Dehalogenations occur smoothly when 1,2-dihalogenoalkenes are treated in suitable solvents with zinc, magnesium or sodium. Fluorine is not removed and this method can be used to prepare fluorinated alkynes[560]. Carbanion intermediates are usually suggested although formation of the Grignard reagent before elimination is observed with magnesium[561]. Elimination from β-halogenated vinyl ethers has also been achieved using sodium in dipolar aprotic solvents[540].

3.4 SOLVOLYTIC ELIMINATIONS

Although vinyl halides have a much lesser tendency than secondary alkyl halides to undergo solvolysis, some reaction products are most easily explained in terms of carbonium ion mechanisms. Facile fragmentation of β-bromo-*cis*-cinnamic acid to phenyl acetylene occurs in alkaline solution, but the *trans* isomer reacts much more slowly to give a mixture of phenyl acetylene and acetophenone[562]. The rate of fragmentation is increased by the presence of electron-releasing substituents in the phenyl ring, and a mechanism, involving initial formation of an intermediate zwitterion which partitions to the reaction products, has been suggested, *viz.*

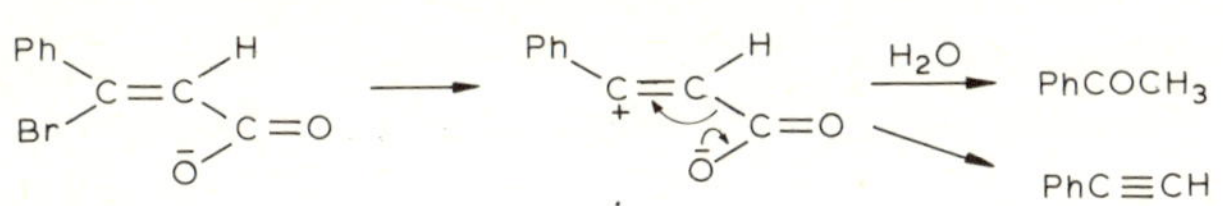

(213)

Phenyl acetylene and acetophenone are also the reaction products observed when β-bromostyrene is solvolysed in aqueous ethanol at 100°C[563]. This reaction is also strongly accelerated by electron-releasing aromatic substituents but electron-withdrawing substituents render the substrates almost inert. The possibility that the ketone arises from the addition of water to the olefin followed by an E_{CO} reaction (see Section 4.3) seems unlikely, as the rate of solvolysis is not influenced by varying the concentration of added triethylamine although this reagent induces bimolecular elimination from the strongly acidic *p*-nitro substrate. Moreover, as silver-ion catalysis is observed, it seems that the mechanism is of the E1 type, (214).

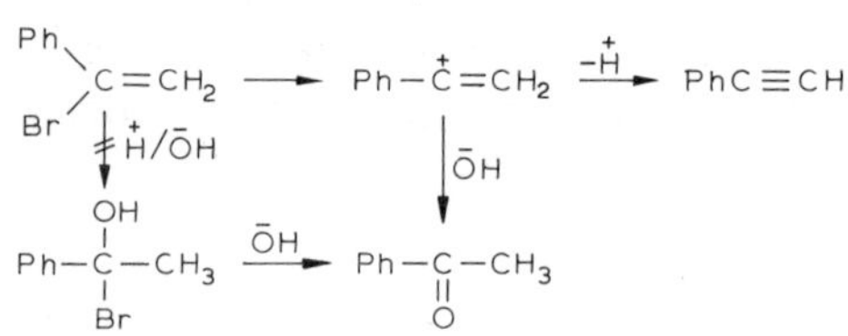

(214)

It would be interesting to investigate the analogous reactions of the α-chloro- and α-iodo-styrenes. Halogen-exchange experiments might also prove useful in establishing a carbonium-ion intermediate[563,564].

4. Beta-eliminations giving multiple bonds between carbon and other elements

4.1 COMPARISON WITH ALKENE- AND ALKYNE-FORMING ELIMINATIONS

Elimination reactions which give rise to multiple bonds between carbon and a heteroatom occur with particular facility and show many of the characteristics of olefin-forming processes. Most often one of the eliminating fragments is a hydrogen atom, which is easily removed from a heteroatom and the choice of mechanism, between an anion or a concerted process, will depend on the lability of the beta carbon–X bond, *e.g.*

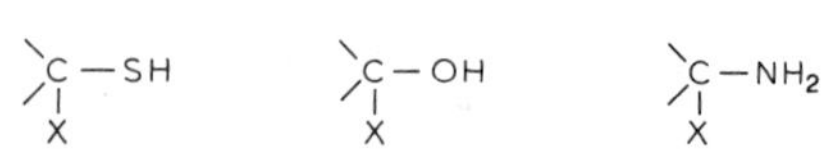

(215)

These types of reaction are more realistically considered as the reversal of the well-known addition reactions to multiple bonds and thus are not discussed in this chapter.

When hydrogen is removed from a carbon atom (216),

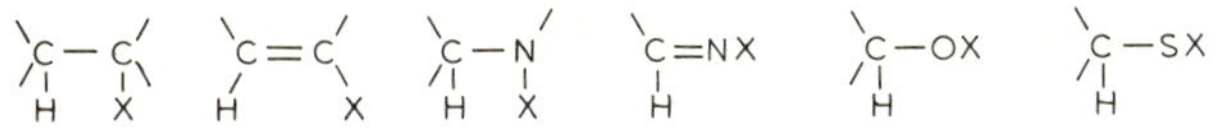

(216)

the beta-elimination of HX to form a multiple bond between carbon and a heteroatom occurs more easily than the formation of a carbon–carbon multiple bond. A number of factors may give rise to this difference in reactivity[7]. Usually X is an electronegative atom, or a group attached by an electronegative atom, or a group with strongly electron-attracting properties. The "group" X is thus more strongly bonded in the substrate when it is attached to carbon rather than to a heteroatom as the difference in the electronegativities of the constituent atoms of the C–X bond is greater than that of the heteroatom–X bond. In general, for beta-elimination, bond changes are usually well advanced in the transition state and consequently the transition states to a certain extent reflect the stability of the products. With increasing multiplicity of a bond, bond strength rises more rapidly when carbon is bonded to a heteroatom than to another carbon atom. Thus the activation energy for elimination leading to a carbon–heteroatom multiple bond is expected to be lower than that leading to the corresponding carbon–carbon multiple bond. The presence of the adjacent heteroatom rather than a carbon atom should also facilitate the reaction by imparting greater hydrogen acidity to the substrate.

There are numerous examples of reactions leading to multiple bonds between a carbon atom and a heteroatom but in the space available in this review it is only possible to mention a few of the interesting kinetic techniques which have led to an understanding of some of the reaction mechanisms involved.

4.2 FORMATION OF C=S

The second-order reaction of diphenylmethyl thiocyanate with isopropoxide gives thiobenzophenone, *viz.*[565,566].

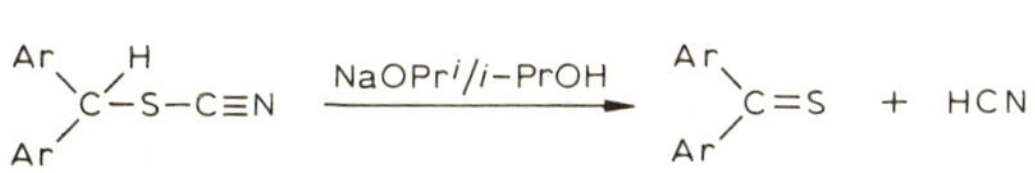

(217)

The presence of low-lying *d*-orbitals on the sulphur atom makes the intermediacy of a carbanion attractive. Racemisation, prior to elimination, is not detected when optically active 4-chlorodiphenylmethyl thiocyanate is allowed to react with isopropoxide in isopropanol. Recovered unreacted 4-phenyldiphenylmethyl thiocyanate is found not to have undergone hydrogen exchange with a labelled solvent. These findings accord better with a concerted rather than a stepwise carbanion mechanism for elimination. However, the transition state certainly possesses considerable carbanion character, the Hammett reaction constant being 3.5 and the primary kinetic hydrogen isotope effect, (k_H/k_D), only 3.0 at 20°C.

4.3 FORMATION OF C=O

Dehydrogenation of alcohols to carbonyl compounds is the simplest type of E_{CO} reaction. At elevated temperatures, metals or metal oxides catalyse both dehydration and dehydrogenation of ethyl alcohol. Dehydrogenation has a lower activation energy, but an increased reaction pressure favours dehydration. However, if the catalysts are heated to a high temperature, a process which causes "healing" of the irregularities in the crystal lattice, the dehydration activity is greatly reduced but the dehydrogenation activity is influenced to only a small extent[433]. These experimental facts suggest that dehydration of alcohols occurs within the pores of the catalyst while dehydrogenation is a "surface" reaction (see Section 2.10).

Suitable oxidising agents can be used to convert primary and secondary alcohols to carbonyl compounds. The chromic acid oxidation has been studied in great detail and in fact has been reviewed on a number of occasions[567–569]. The function of the various oxidation states of chromium has been previously discussed and only the changes in the organic molecules during the reaction are considered here.

Much of the kinetic work has been concerned with the oxidation of isopropyl alcohol to acetone in aqueous sulphuric or acetic acids. The rate-determining step definitely involves the breaking of a carbon–hydrogen bond, as isopropyl alcohol is oxidised 6–7 times faster than its 2-deutero derivative[570]. At all acidities, the rate of oxidation shows a first-order dependence on both the concentration of $HCrO_4^-$ and the alcohol, but the dependence on $[\overset{+}{H}]$ varies. In dilute acidic solution the rate is proportional to $[\overset{+}{H}]^2$, to $[\overset{+}{H}]$ in more concentrated solution[570] and to H_0 in 20–60% aqueous sulphuric acid[571]. These observations are consistent with two mechanisms: (*i*) a bimolecular electron transfer process (218)[571],

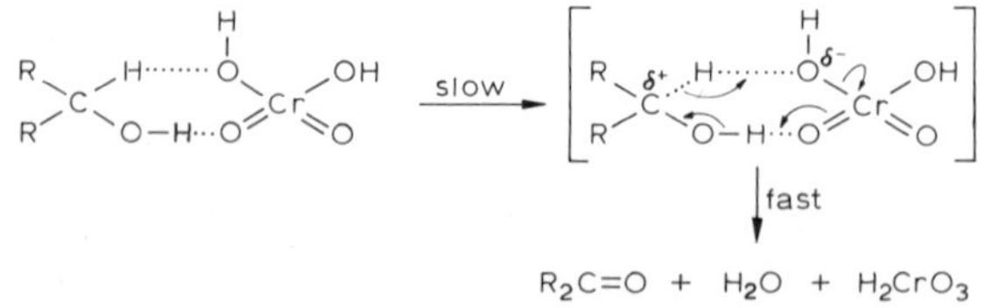

(218)

involving gradually more reactive chromic acid species with increasing acidity of the medium (*e.g.* $H_2CrO_4 \rightarrow H_3CrO_4^+ \rightarrow HCrO_3^+$), and (*ii*) a rapid pre-equilibrium giving a chromate ester, which subsequently decomposes by an E2 mechanism in which a water molecule acts as the base[570], *viz.*

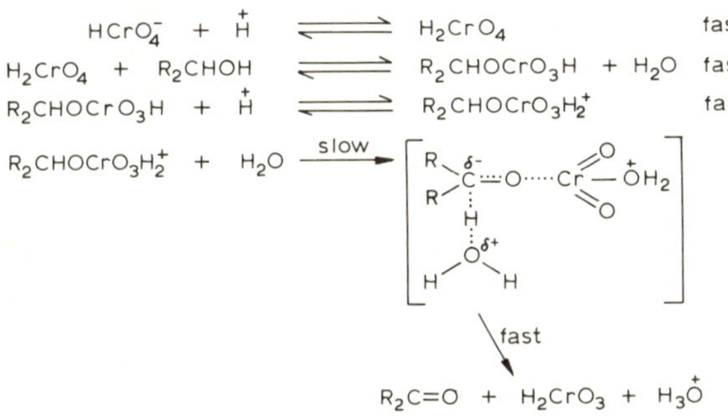

(219)

For a series of substituted aliphatic alcohols, there is a linear correlation between the logarithm of the rate coefficient for oxidation and the Hammett substituent constant, σ^*, the slope ρ^* being -1.06 ± 0.06 (ref. 572). Electron-releasing substituents in the aromatic ring of 1-phenylethanol enhance the rate of oxidation[573], but a linear correlation with the Hammett equation is only noted if the substituent constants σ^+ are used. Both of these observations show that the oxidation of an alcohol is encouraged by electron release towards the reaction centre and the σ^+ correlation implies that at some stage before or in the rate-determining step, direct conjugation between the substituent and a "carbonium-ion-type" centre is involved. These results are inconsistent with the chromate ester mechanism as initially designated (219) and suggest the rate-determining step involves a hydride rather than a proton transfer. Modification of scheme (219) to a unimolecular ester decomposition, analogous to the pyrolysis of other esters, accommodates the substituent influences (220).

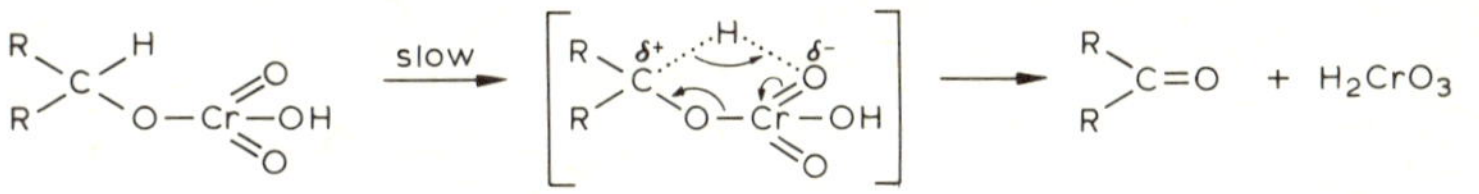

(220)

However, C–H bond breaking must be in advance of Cr–O bond breaking in the transition state[573].

The dependence of the logarithm of the rate coefficient for oxidation of isopropyl alcohol on the H_0 acidity function (slope 0.91) is difficult to accommodate by the scheme (219, 220), for which an abscissa of $-H_0 + \log a_{H_2O}$ would seem more appropriate. However, in view of the widely varying activity coefficient behaviour of weak bases in strongly acidic media, the slope of near unity for the above oxidation may be fortuitous and somewhat misleading. Clearcut evidence in favour of the ester–hydride transfer mechanism has recently been provided by kinetic isotope studies on the rate of oxidation of a highly sterically hindered alcohol (3β-28-diacetoxy-6β-hydroxy-18β-olean-12-ene)[574]. In mineral acids, chromate ester formation is rate-determining and no isotope effect is observed. However, in 60% aqueous acetic acid containing sodium acetate, an isotope effect (k_H/k_D) of 2 is found and ester formation and its subsequent decomposition contribute to the rate-determining step. The kinetics of oxidation of some α-hydroxycarboxylic acids have also been interpreted in terms of the ester–hydride transfer mechanism[575].

Nitrate and sulphonate esters can undergo carbonyl-forming eliminations along with substitution and olefin-forming elimination in basic solution, *e.g.*

$$\begin{aligned} RCH_2CH_2ONO_2 + \bar{B} &\xrightarrow{(i)} RCH_2CH_2B + NO_3^- && S_N2 \\ &\xrightarrow{(ii)} RCH_2CH_2\bar{O} + BNO_2 && S_N2 \\ &\xrightarrow{(iii)} RCH{=}CH_2 + NO_3^- + BH && E2 \\ &\xrightarrow{(iv)} RCH_2CH{=}O + NO_2^- + BH && E_{CO}2 \end{aligned} \qquad (221)$$

From a careful product analysis, it is possible to study the influence of structural variations on each of the reactions involved[576]. Carbonyl products are not obtained under solvolytic conditions only, but require the presence of a basic reagent[577]. The rate of such $E_{CO}2$ reactions is increased by the presence of electron-attracting substituents or by groups which can conjugate with the developing carbonyl bond[576]. In fact, for the carbonyl-forming elimination of benzyl nitrate, a Hammett reaction constant of 3.4 at 60°C in 90% aqueous ethanol was observed[576]. Although this is indicative of the development of extensive carbanion character at the benzylic carbon atom during the reaction, subsequent isotope studies confirm a concerted rather than a carbanion mechanism[578]. Unreacted benzyl nitrate does not exchange its benzylic hydrogen for deuterium from a labelled solvent. A primary kinetic hydrogen isotope effect (k_H/k_D) of 5 and a nitrogen isotope effect of 1.0196, indicating considerable bond breaking of the O–N bond in the transition state, are observed in ethanol containing sodium ethoxide at 60.2°C. For benzyl nitrate, the only side reaction is that of substitution and the deuterium isotope effect of only 1.16, and the lack of any nitrogen isotope effect, indicate this substitution occurs at carbon rather than nitrogen (221(i)). To measure the nitrogen isotope effects, it was necessary to separate the nitrate ion (substitution) and the nitrite ion (elimination) before nitrogen analysis by mass spectroscopy. Control experiments showed that no exchange between nitrate and nitrite ions occurred under the reaction conditions and that nitrite was rapidly reduced to ammonia by ferrous hydroxide while nitrate ion reacted only in the presence of silver ion.

The greater acidity of the fluorenyl than the benzyl group makes it a more attractive molecular system for investigations of possible carbanion mechanisms. However, isotope studies on the elimination from 9-fluorenyl nitrate with acetate ion in ethanol are consistent with a concerted reaction[579]. At 30°C, the primary kinetic hydrogen isotope effect is 4.3 and the nitrogen isotope effect is 1.009. These results are both lower than those obtained for benzyl nitrate and suggest, as might be expected, that greater carbanion character is associated with the transition state in the fluorenyl system. A change from ethanol to 85% aqueous ethanol as solvent causes no variation in the hydrogen isotope effect, but a greater nitrogen isotope effect (1.0131) is observed. Greater solvation by water aids the weakening of the O–N bond and causes a greater decrease in the composite force constant in the transition state.

The $E_{CO}2$ reaction of 9-fluorenyl toluene-*p*-sulphonate with sodium hydroxide in 90% methanolic acetone has a primary kinetic hydrogen isotope effect of only 2.3 at 25°C[580]. It is difficult to compare this result with that for the corresponding nitrate but if a transition state with considerable carbanion character is involved, then the indications are that nitrite is a better leaving group than toluene-*p*-sulphinate. It would be worth while to investigate the sulphur isotope effect for this latter reaction.

The base-catalysed transformation of α-phenylethyl *t*-butyl peroxide to substituted isomeric 2-chloro-2,7-dihydro-3,4:5,6-dibenzothiepin-1,1-dioxides, *viz.*

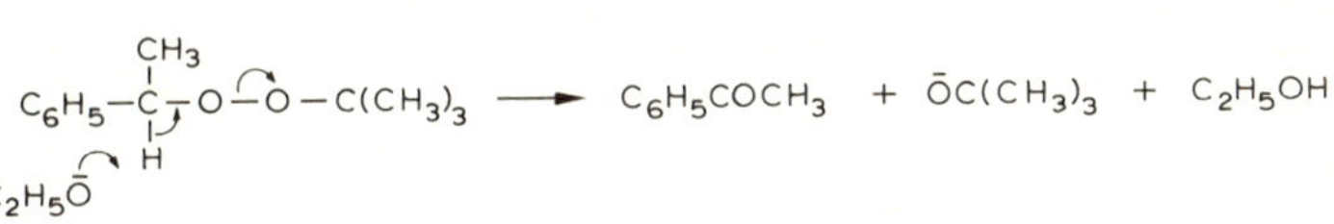

(222)

4.4 FORMATION OF C≡N AND C=N

There are comparatively very few kinetic studies of beta-eliminations giving imines or nitriles. Nitriles can be prepared by the acid-catalysed dehydration of amides or aldoximes or by the base-catalysed elimination from aldochlorimines or esters of aldoximes. Imines often result in the attempted preparation of *N*-chloramines, dehalogenation occurring in the weakly alkaline hypochlorite solution which is used as the chlorinating reagent.

Elimination from substituted α-aldochlorimines, (223),

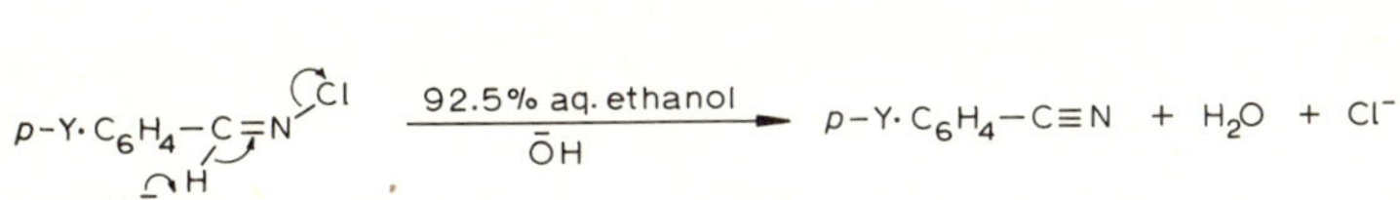

(223)

occurs more rapidly in the presence of electron-withdrawing groups and a

Hammett reaction constant of 2.24 is observed at 0°C[582]. Certainly this result indicates that the C–H bond breaking runs ahead of N–Cl bond breaking but as to whether these changes occur synchronously or in discrete steps cannot be inferred without further kinetic investigations.

The hydrolysis of *N*-acyl derivatives of substituted benzaldoximes in dilute aqueous alkaline solution gives rise to nitriles and benzaldoximes, *viz.*[583]

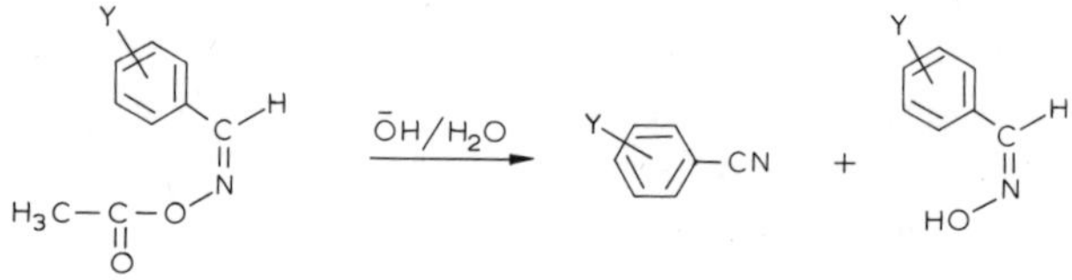

(224)

Electron withdrawing substituents in the benzene ring increase the rate and the competitive nature of the elimination reaction. The stereochemical arrangement of the eliminating fragments is of paramount importance, *anti*-elimination from the beta isomers competing favourably with the ester hydrolysis, while the alpha isomers, which must eliminate with *syn* stereospecificity, undergo only hydrolysis. In the alpha isomers, hydrogen bonding between the carbonyl function and the beta hydrogen will reduce the rate of the elimination, but exert little influence on the rate of the hydrolysis of the ester (225).

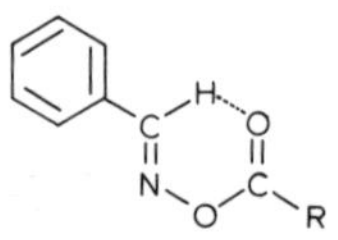

(225)

The influence of structural variations on the rates of pyrolysis of *N*-acyl derivatives of substituted α-benzaldoximes (226) has been made[584].

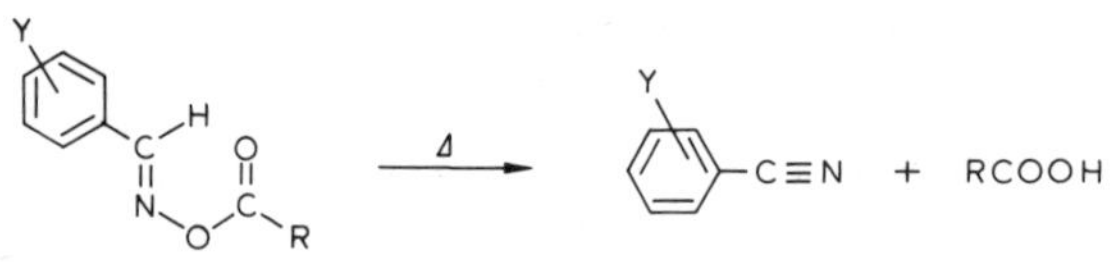

(226)

The rate of the reaction is increased by electron-releasing substituents in the benzene ring (in contrast to the mild enhancement by electron-withdrawing substituents in the pyrolysis of 1,2-diphenylacetates, (Section 2.9) and by electron-withdrawing substituents in the alkyl function, R (226). These facts are consistent with a cyclic mechanism in which N–O heterolysis is more advanced than C–H heterolysis or O–H bond formation in the transition state. The alpha isomers are more reactive than the beta isomers, but study of the latter is complicated as in the presence of traces of acid they are isomerised to the alpha form.

There is ample scope for further investigations of the mechanism of beta-eliminations yielding imines and nitriles. Some of the base-catalysed reactions may be examples of the carbanion mechanism and hydrogen isotope studies and the application of the Brönsted catalysis law to rates of elimination under various reaction conditions could prove informative.

5. Apparent alpha-eliminations

Earlier in this chapter, (Section 2.1.4) the possibility that some beta-eliminations might proceed by a base-catalysed alpha-elimination followed by hydrogen migration was considered. In this section, evidence is presented to demonstrate that a number of apparent alpha-eliminations do not actually involve carbenoid intermediates. Earlier reviews are available on this aspect of elimination[540,585].

5.1 ALKYNE-FORMING ELIMINATIONS

It has been known for many years that the treatment of 1-halo-2,2-diphenylethylenes with strong bases, such as alkoxides, metal amides or alkyls, produces diphenylacetylene[586]. With the alkoxide bases, exchange of alpha deuterium with the protic medium has been shown to occur more rapidly than product formation[587]. When a label is present in one of the aryl substituents, it is possible to demonstrate that the exchanged substrate has retained its original configuration even at moderately elevated temperatures. Moreover, making use of carbon tracers, it has been shown that it is the aryl substituent originally *trans* to the halogen in the substrate which undergoes migration[588], *viz.*

(227)

All the reactions are first-order with respect to each of the reactants and the facility of formation of the acetylene as the leaving group is changed, which follows the order Br > I ≫ Cl, can be explained in terms of the ease of heterolysis of the carbon–halogen bond and the acidity of the vinylic hydrogen atoms. These facts are consistent with a rapid equilibrium to give a carbanion which decomposes in the rate-determining step. Of the two mechanistic schemes which accommodate the results, the use of substituents in one of the aryl groups serves to support the first mechanism (228). For the alternative scheme (229), a rapid migration followed by a slow ejection of halide ion can be dismissed, as the second carbanion like the first should rapidly equilibrate with the solvent and it should be possible to isolate a bromostilbene intermediate. However, a slow migration followed by a rapid ejection of halide ion is not in disagreement with the above evidence.

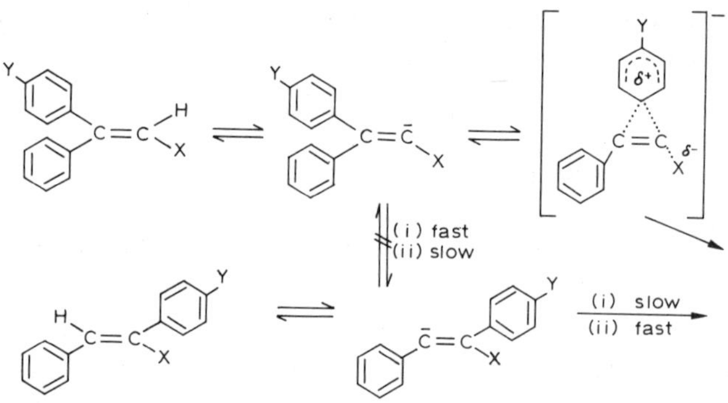

(228)

(229)

Substituents in one of the aryl groups can influence the rate of formation of the alkyne in two ways, (*i*) by changing the acidity of the α hydrogen and hence the steady-state concentration of the ionised substrate and (*ii*) by modifying the electron density in the migrating group. A comparison of the rate coefficients of reaction of *trans*- and *cis*-1-halo-2-Y-aryl-2-phenylethylenes (*trans* and *cis* refer to the stereochemistry of Y-aryl and the leaving group X) affords a means of assessing the latter influence alone. When Y is an electron-releasing substituent, the *trans* isomer rearranges more rapidly, but the *cis* isomer is more reactive when Y is an electron-withdrawing substituent. These results suggest that mechanism (228) is correct, the migrating aryl substituent possessing some carbonium character in the transition state[589].

When lithium alkyls are used as the base, it is found that, with diethyl ether as the solvent, the α-deuterated compound reacts more slowly than 1-chloro-2,2-diphenylethylene[590]. The rate-determining proton transfer may arise from the lower reactivity of an ion pair than a dissociated ionic base, for a change in the solvent to tetrahydrofuran causes proton removal to become a faster process than rearrangement. In this latter solvent, the halogenovinyllithium compounds can be prepared and characterised as their carboxylation products, when the metallation reactions are carried out at low temperatures[591]. Diphenyl acetylene is not the only product formed when 2,2-diphenyl-1-halogenoethylene is treated with lithium alkyls, as in addition to metal–hydrogen exchange, metal–halogen exchange also occurs, *viz.*[590]

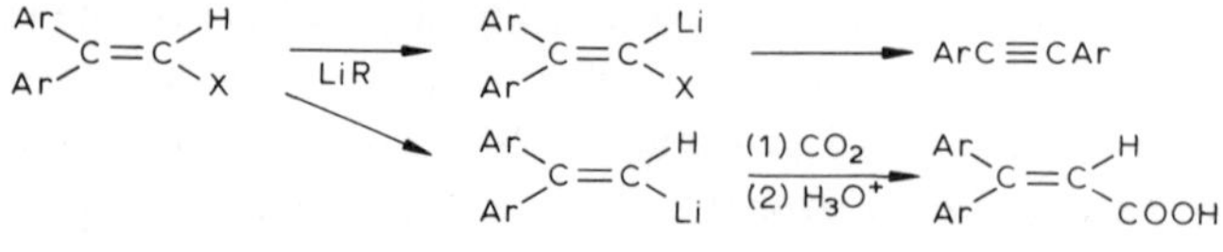

(230)

The competitive nature of these two reactions is determined primarily by the leaving group, 1-chloro-2,2-diphenylethylene yielding mainly the acetylene, whilst the bromo and iodo substrates afford mainly the carboxylic acid when the reaction mixture is treated with carbon dioxide.

5.2 DIMERIC OLEFIN-FORMING ELIMINATIONS

A number of organic substrates which do not possess hydrogen atoms beta to the leaving group or a migratable aryl substituent, undergo an apparent alpha-elimination, when treated with strongly basic reagents, to give dimeric-olefin products. Typical of this category of reaction are the conversions of benzyl halides and dimethylsulphonium ion to stilbenes[592,593], diphenylmethyl halides to tetraphenylethylenes[594], 9-halogenofluorenes to bifluorenylidene[595] and 9-halogenomethylenefluorene to a cumulene derivative[596]. To account for these transformations, two distinct reaction mechanisms have been suggested, (*i*) a carbene mechanism and (*ii*) a bimolecular displacement mechanism, *viz.*

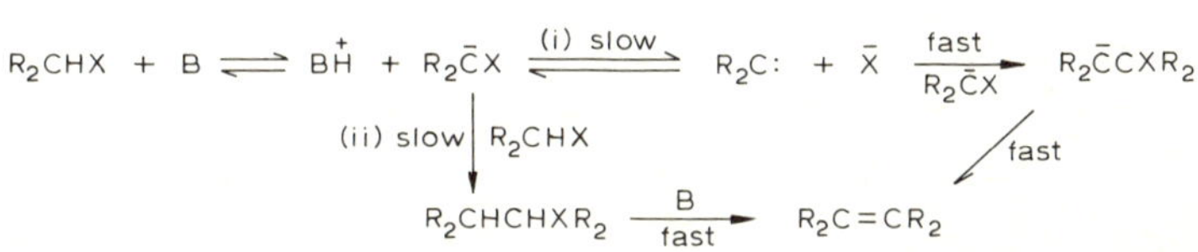

(231)

If the initial proton removal occurs in a rapid reversible equilibrium, then the determination of the reaction order in the substrate should suffice to distinguish between the two mechanisms.

Hauser *et al.*[594] were able to isolate the intermediate dimeric halide when 9-chlorofluorene was allowed to react with only one equivalent of sodium amide in liquid ammonia. In *t*-butyl alcohol containing its potassium salt or a dilute aqueous solution of benzyltrimethylammonium hydroxide, the rate of formation of bifluorenylidene is second-order in 9-bromofluorene and first-order in the basicity of the reaction medium as measured by the ionisation of nitroaniline indicators[595]. Under the same reaction conditions, protium exchange of 9-deutero-9-bromofluorene[595] and elimination from 9-bromo-9,9′-bifluorenyl[120] are much faster reactions than the conversion of 9-bromofluorene into bifluorenylidene. These facts are consistent with the displacement mechanism.

With sodium or potassium *t*-butoxides as the base, the rate of elimination of 9-halogenofluorenes sometimes shows a first-order dependence on both the medium basicity and the substrate, even though exchange of 9-deuterium is still a rapid process[597]. Factors which promote this change in kinetic order are the presence of powerful electron-withdrawing substituents in the fluorene nucleus, bromine rather than chlorine as the leaving group, and the addition of dipolar aprotic solvents such as dimethyl sulphoxide and tetramethylene sulphone.

Although these kinetic results are not inconsistent with a carbene mechanism, a number of additional factors are difficult to accommodate. Whatever the order of reaction in the substrate, in the absence of oxygen the yields of bifluorenylidenes are quantitative, even at spectroscopic concentrations. This observation requires that if a carbene intermediate is involved, then it is highly selective, always combining with the carbanion which is present in very low concentration, rather than undergoing insertion reactions with the solvent[598]. When the first-order law is followed, a change in the leaving group from bromine to chlorine produces only a small reduction in the rate coefficient (k_{Br}/k_{Cl} = 1.5 for 2-cyano-9-halogenofluorenes), whereas a larger change might be expected if the carbon–halogen bond is broken in the rate-determining step. Certainly, for substrates following the second-order law, a change of leaving group has a marked influence on the rate coefficient ($k_{Br}/k_{Cl} = 40$ for 9-halogenofluorenes).

In view of the above findings, rather than attribute a first-order dependence of rate on the substrate concentration to a rate-determining carbene formation, a modification of the displacement mechanism was suggested in which ion-pair dissociation to give more reactive solvent separated ions or free ions[599] becomes the slow step under certain conditions, *viz*.

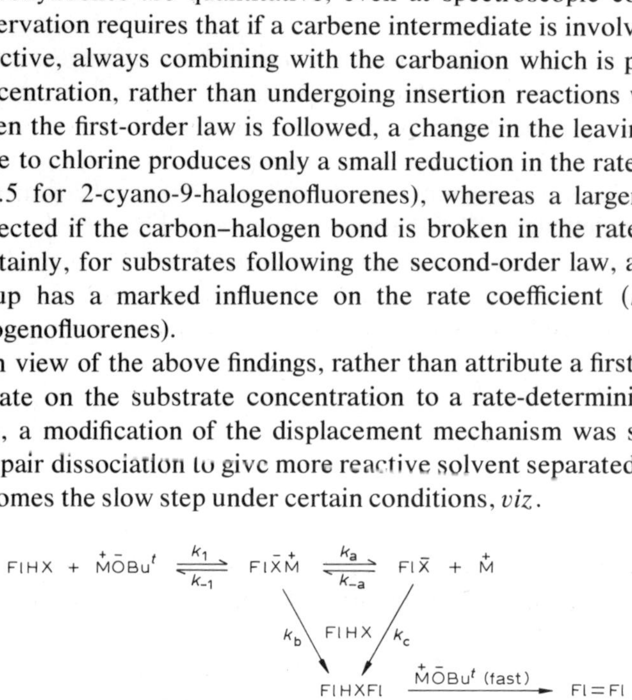

(232)

Fl = 9-fluorenyl– ; X = halogen; $\overset{+}{M}$ = Na$^+$ or K$^+$

From the steady state hypothesis,

$$\frac{d[Fl{=}Fl]}{dt} = \frac{k_1}{k_{-1}}\left\{k_b + \frac{k_a k_c}{k_{-a}\overset{+}{M} + k_c[FlHX]}\right. [FlHX]^2[\overset{+}{M}\overset{-}{O}Bu^t]$$

If $k_b \gg k_a k_c/(k_{-a}[\overset{+}{M}] + k_c[FlHX])$ or $k_{-a}[\overset{+}{M}] \gg k_c[FlHX]$, then the reaction will be second-order in the substrate, but if $k_b \ll k_a k_c/(k_{-a}[\overset{+}{M}] + k_c[FlHX])$ and $k_{-a}[\overset{+}{M}] \ll k_c[FlHX]$, a first-order dependence is expected. When the kinetic order in the 9-bromofluorene is unity, the addition of small amounts of sodium tetraphenylboron or sodium perchlorate decreases the rate coefficient only to the extent that the medium basicity is reduced. An increase in the concentration of the added salt causes the first-order dependence to change to a second-order dependence, attributable to $k_{-a}[\overset{+}{M}]$ becoming greater than k_c[FlHX]. Once a second-order law applies, the rate coefficient falls more sharply than the medium basicity with increasing concentration of the added salt. Added $[\overset{+}{M}]$ now not only influences the dissociation of the butoxide, but also the nucleophilicity of the carbanion with which it associates. The above observations only accord with the anticipated mechanism if the ions resulting from sodium tetraphenylboron or perchlorate are more dissociated than those derived from sodium *t*-butoxide. In support of this requirement, solutions of *t*-butyl alcohol containing the added salts show appreciable conductance but the conductivity of the butoxide solutions differs little from that of the pure solvent.

Despite earlier claims of a carbene mechanism[601], more recent kinetic studies indicate that the conversion of diphenylmethyl chloride into tetraphenylethylene by *t*-butoxide in dimethyl sulphoxide follows the displacement mechanism. The reaction is third-order, second in substrate and first in the medium basicity and α-hydrogen exchange in the substrate is a rapid process[602].

Unlike all the other substituted 9-bromofluorenes, the highly acidic 3-nitro-9-bromo isomer is converted extensively into its conjugate base by bases in *t*-butyl alcohol[600]. The rate of elimination can thus be followed by monitoring the absorption of the blue carbanion rather than the red olefinic product, as is otherwise employed. Extensive carbanion formation is also observed with 4-nitrobenzyl chloride and 4-nitrodiphenylmethyl chloride. For all three of these substrates, the rate of elimination shows a complex dependence on the substrate concentration, but an inverse dependence on the medium basicity. This observation is not consistent with a carbene mechanism but accords with the displacement scheme if the substrate is always more than half-ionised under the reaction conditions. To achieve a rapid successful preparative conversion to the dimeric olefin, it is necessary to keep the substrate in an excess of the base in order to maximise the concentration of unionised substrate on which the carbanion can exert a displacement.

In the less basic medium of aqueous acetone or aqueous dioxan containing hydroxide ion, a carbene mechanism has been suggested to explain the quantitative conversion of 4-nitrobenzyl chloride[495,603] and 4-nitrobenzyl dimethylsulphonium ion into *cis*- and *trans*-4,4′-dinitrostilbenes[604,605]. In both cases, the rate of elimination follows first-order kinetics in both the substrate and hydroxide ion, and deuterium exchange of labelled substrates occurs more rapidly than stilbene formation. In this aqueous medium, a rate-determining ion-pair dissociation is unlikely. However, again the selectivity of the carbene is astonishing and it is difficult to explain the lack of insertion products with the reaction medium or the failure to isolate cyclopropane adducts when the elimination reactions are performed in the presence of added olefins.

Recent investigations on the alkylations by ambident nucleophiles of aliphatic and aromatic compounds, and in particular of *ortho*- and *para*-nitrobenzyl chlorides, have shown that *C*-alkylation proceeds via radical anion intermediates and the rate of such a process can be reduced by the addition of radical inhibitors such as dinitrobenzenes[606,607]. It is worth noting that care was not taken to exclude light and oxygen from the reaction of the 4-nitrobenzyl substrates with hydroxide ion and stilbene formation may occur by a similar radical chain process, *viz*.

$$
\begin{aligned}
\mathrm{Ar'CH_2Cl + \bar{O}H} &\xrightleftharpoons{\text{fast}} \mathrm{Ar'\bar{C}HCl + H_2O} \\
\mathrm{Ar'\bar{C}HCl + Ar'CH_2Cl} &\longrightarrow \mathrm{Ar'\dot{C}HCl + (Ar'CH_2Cl)^{\bullet-}} \\
\mathrm{(Ar'CH_2Cl)^{\bullet-}} &\longrightarrow \mathrm{Ar'\dot{C}H_2 + Cl^-} \\
\mathrm{Ar'\dot{C}H_2 + Ar'\bar{C}HCl} &\longrightarrow \mathrm{(Ar'CH_2CHClAr')^{\bullet-}} \qquad (233) \\
\mathrm{(Ar'CH_2CHClAr')^{\bullet-} + Ar'CH_2Cl} &\longrightarrow \mathrm{(Ar'CH_2Cl)^{\bullet-} + (Ar'CH_2CHClAr')} \\
\mathrm{Ar'CH_2CHClAr'} &\xrightarrow[\bar{O}H]{\text{fast}} \mathrm{Ar'HC{=}CHAr'}
\end{aligned}
$$

$\mathrm{Ar'} = p\text{-}\mathrm{NO_2.C_6H_4}$-

Preliminary experiments show that *ortho*-dinitrobenzene almost completely prevents stilbene formation from 4-nitrobenzyl dimethylsulphonium ion with hydroxide and the major product, *p*-nitrobenzyl alcohol, is formed at a much slower rate than is the stilbene in the absence of the inhibitor[608].

There is obviously scope for further studies in this field and it would be worthwhile investigating these reactions in the absence of light and oxygen and subjecting all reactions to electron spin resonance studies.

6. Gamma-eliminations

6.1 NOMENCLATURE

Unlike their beta counterparts, gamma-eliminations, which most often give rise to three-membered rings, have received much less attention and it is only in recent years that more detailed mechanistic studies have been reported. Some of the general features of base-catalysed gamma-eliminations have been discussed in an earlier review[609]. In many cases the direct products of elimination are unstable under the reaction conditions and the intermediacy of a three-membered ring has to be established by carefully combining kinetic studies and subsequent product analysis. This approach is adequately illustrated by the well-known Neber, Ramberg-Bäcklund and Favorskii rearrangements.

In this text, a common system of nomenclature is used for all 1,3-eliminations. In a five-centre system,

$$Y\text{——}a^{\alpha}\text{——}b^{\beta}\text{——}c^{\gamma}\text{——}X \qquad (234)$$

X and Y are the leaving groups and the atoms *a*, *b* and *c* are labelled with respect to the more electronegative leaving group, Y. Most often, X is a hydrogen atom. Independent rotation about bonds *a–b* and *b–c* can give rise to numerous conformations and a greater variety of stereochemical arrangements of the leaving groups is available for concerted 1,3-eliminations than for the corresponding concerted beta-eliminations. Considering only staggered conformations for the substrate (or a reactant-like transition state), Nickon and Werstiuk[610] have proposed a concise nomenclature to describe the five possibilities, (235) below. When these conformations are modified to more product-like structures, only four arrangements are possible. Independent of the type of transition state (*i.e.* product- or reactant-like), a simple nonperspective notation and terminology was proposed to cater for the five possibilities. Of the substrate arrangements, only the W form can go directly to a product-like state without conformational rotation. For cases in which the leaving groups are situated on asymmetric carbon atoms, the stereochemical outcome of these latter atoms is also listed below (235).

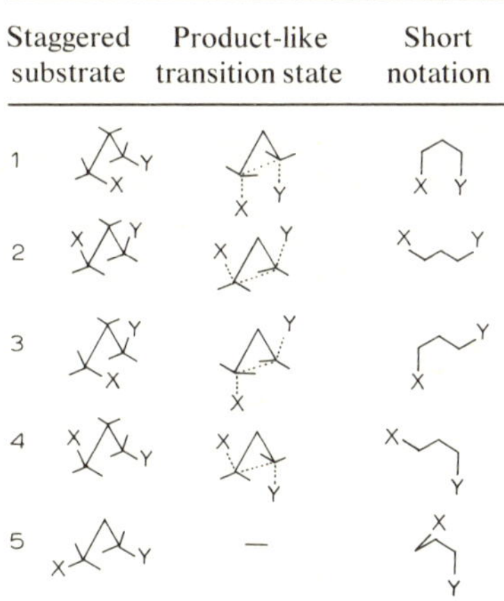

	Terminology	Stereochemical outcome
1	U	retention
2	W	inversion
3	*exo*-sickle	retention at *c*; inversion at *a*
4	*endo*-sickle	retention at *a*; inversion at *c*
5	*apo*-sickle	depends on which product-like transition state is adopted

(235)

The experimental evidence indicates that a number of 1,3-eliminations proceed via two-step reactions, analogous to the E1cB and E1 mechanisms for olefin formation. Nickon and Werstiuk have also proposed a terminology (236) to cater for rate-determining ring closure from the carbanion or carbonium ion in which the stereochemistry at atom *c* or *a*, respectively, has been lost, and the detectable stereochemical change is now confined to the remaining reaction centre.

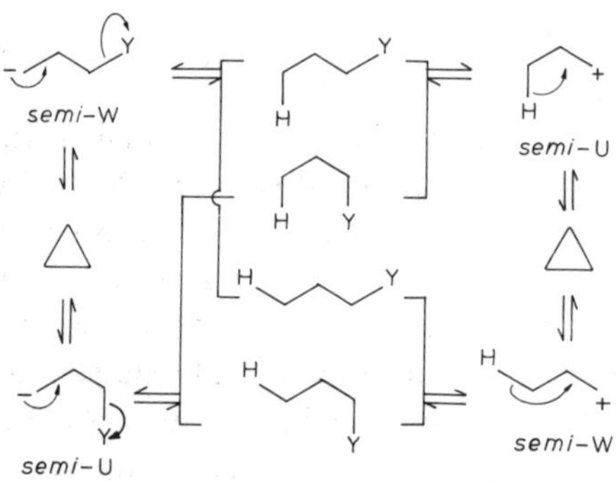

(236)

6.2 ELIMINATIONS YIELDING CYCLOPROPANE DERIVATIVES

6.2.1 *Base-catalysed and solvolytic eliminations*

The products of solvolysis of *exo*-norbornyl tosylate in *t*-butyl alcohol at 60°C are the *exo-t*-butyl ether (70%), norbornene (10%) and nortricyclene (20%)[611]. The solvolysis follows first-order kinetics and the rate is influenced little by low concentrations (0.04 *M*) of potassium *t*-butoxide. When the products arising from the 6-*endo*- and 6-*exo*-deuterated substrates are analysed, it is found that the two samples of the *exo-t*-butyl ether have identical infrared spectra and have retained all the deuterium originally present in the substrates. The nortricyclene samples have both lost the same fraction of deuterium. These facts are consistent with a rate-determining formation of a norbornyl cation (bridged or rapidly equilibrating) which subsequently partitions to the reaction products. In more concentrated basic solution, the relative yields of the elimination products are increased and a second-order component in the solvolytic rate coefficient is noted. In this case, the nortricyclene isolated from the two precursors contains appreciably different amounts of deuterium. That arising from the *exo*-deutero substrate has lost 33.4% of its original deuterium, whereas the nortricyclene obtained from the *endo*-deutero substrate has lost 52.9%. After allowance is made for the deuterium loss resulting from the competing unimolecular process, these results show that the *exo*-sickle arrangement is favoured over the W arrangement in the base-catalysed gamma-elimination of *exo*-norbornyl tosylate. A similar comparison of the deuterium content of nortricyclene isolated from the 6-deutero-*endo*-norbornyl tosylates indicates that substrate U geometry is preferred to the *endo*-sickle arrangement for the base-catalysed 1,3-elimination[612]. Of course, a change in the reaction temperature may be accompanied by variations in the extent of deuteration of the nortricyclene and generalisations concerning the apparent favoured stereochemistry for 1,3-elimination cannot be made with certainty until the activation parameters for the bimolecular eliminations have been measured.

A number of observations, dependent on product analysis, are available for the base-catalysed reactions of a number of 3-phenylpropyl derivatives. These substrates can undergo competitive beta- and gamma-eliminations, *viz*.

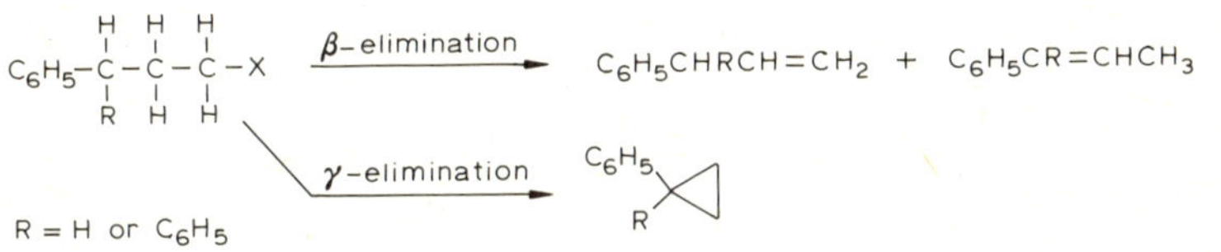

(237)

In liquid ammonia containing sodium amide, the bromide and chloride undergo mainly beta-elimination but the fluoride, tosylate and trimethylammonium ion all give mainly the cyclopropane[613,614]. With the poorer leaving groups, the carbanion character of the "displacing carbon atom" may be the decisive factor and the γ hydrogen atoms should be more acidic than their beta counterparts due to the adjacent phenyl substituent. Moderate steric inhibition to γ hydrogen abstraction has little influence on the reaction course and 3-(*o*-tolyl)-propyltrimethylammonium ion is converted in 87% yield into the cyclopropane. However, the more severe steric hindrance imposed by the mesityl substituent precludes gamma-elimination and only olefinic products are observed. Under the normal conditions of the Hofmann degradation, 3,3-diphenylpropyltrimethylammonium ion is converted into 1,1-diphenylpropene (72%) and 1,1-diphenylcyclopropane (28%), but 3-phenylpropyltrimethylammonium ion gives only products of beta-elimination. From all these results, it is obvious that cyclopropane rather than olefin formation is promoted by a poorer leaving group, a stronger base and a more acidic γ hydrogen atom. However, in the absence of more detailed kinetic investigations, it is not possible to decide whether the gamma-elimination is a concerted or a two-step process, or is in fact in some cases an alpha-elimination to give a carbene which undergoes an insertion reaction to give a cyclopropane. There is an obvious need for kinetic studies involving various deuterated substrates, and a comparison of the Hammett reaction constant for these gamma-eliminations with those already

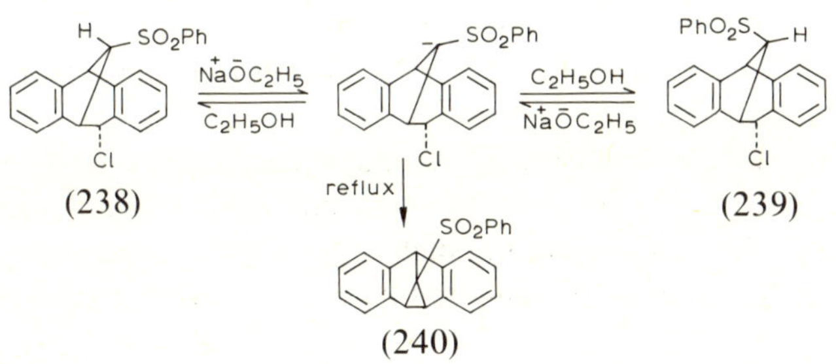

(238) (239) (240)

available for beta-eliminations in the 2-phenylethyl series would be of interest.

Using proton magnetic resonance spectroscopy, Cristol and Jarvis[615] showed that chloro-sulphone (238) epimerises in ethanol containing sodium ethoxide at room temperature, the equilibrium slightly favouring (239). When the solution is heated under reflux, a rapid gamma-elimination occurs to yield (240).

As epimerisation occurs much more rapidly than cyclopropane formation, the mechanism is most probably an E1cB-gamma-elimination. In this case, beta-elimination is precluded, due to the difficulty of forming a double bond at a bridgehead, in accordance with Bredt's rule. The *exo*-chloro-sulphone (241) rapidly epimerises to give almost completely the other isomer (242) in which

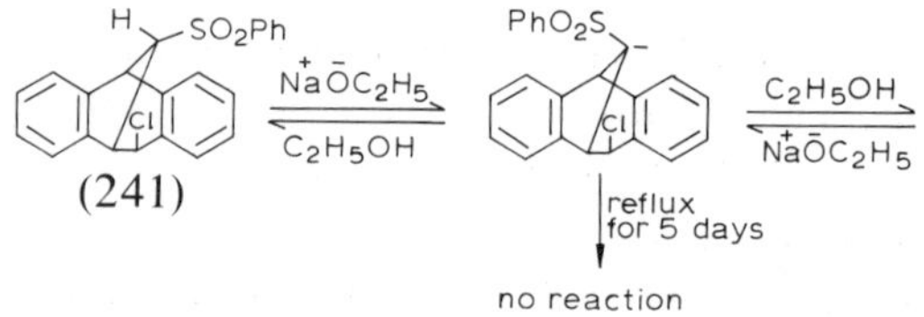

(241) (242)

the steric interaction between the chloro and sulphone substituents is reduced. However, even after heating under reflux for five days, no gamma-elimination was detected. In fact, the *exo*-chlorosulphones yield the cyclopropane only if a more basic reaction medium is used (*e.g.* $KOBu^t$ in dimethyl sulphoxide). In this case, the *endo*-chlorosubstrate is formed by epimerisation at the benzylic carbon and this subsequently then eliminates[616]. In an analogous case, 2-quadricyclylphenyl sulphone is formed in an E1cB-gamma-elimination from the *exo*-bromo but not the *endo*-bromo substrate (243)[617].

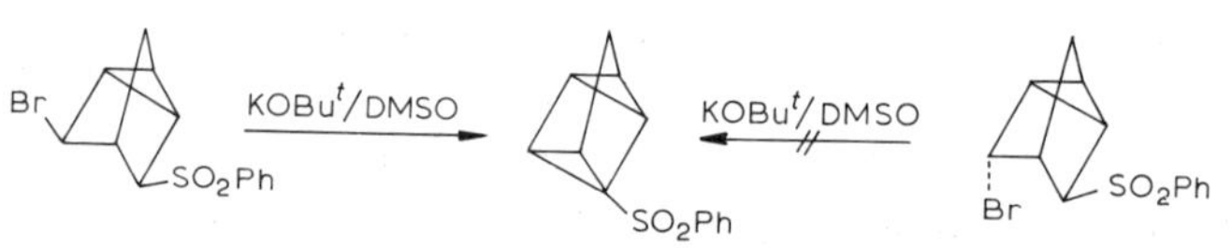

(243)

These results suggest that a 1,3-elimination can only occur if, in the carbanion, the leaving group is situated in a *semi*-W arrangement (see (236), p. 330) and not in a *semi*-U arrangement. This is in marked contrast to the ease of concerted 1,3-elimination from both the *endo*- and *exo*-norbornyl tosylates. Obviously more investigations are necessary and it would be of interest to study the behaviour of the above sulphones (238, 239, 241) in which the chlorine is replaced by a tosylate group.

The solvolysis of *exo*-norbornyl brosylate in acetic acid gives 4% elimination products, 98% of which is nortricyclene and only 2% is norbornene. The larger yield of the cyclopropane than the olefin was quoted as support for a non-classical rather than a rapidly equilibrating pair of classical carbonium ions[618].

6.3 DEAMINATION, DEOXIDEATION AND PYROLYSIS LEADING TO CYCLOPROPANES

The deamination of *n*-propylamine by nitrous acid and the deoxideation of *n*-propoxide both give propene and cyclopropane as the elimination products in a ratio of 9:1 (refs. 619, 620). A similar product ratio is obtained from the thermolysis of 1-diazopropane[621] and this raises the possibility that carbenoid intermediates are also involved in the former reactions. However, the products of deoxideation of 1,1-dideuteropropanol are consistent with a carbonium-ion mechanism involving a 1,3-elimination to give the cyclopropane, *viz.*[620]

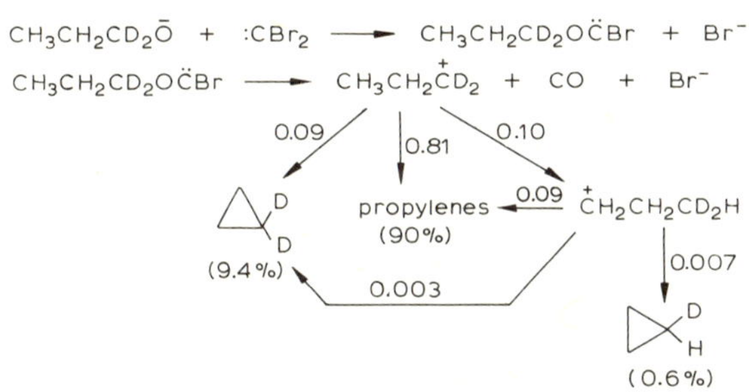

(244)

The small amount of monodeuterated cyclopropane could have arisen by a carbene insertion but a 1,3-hydride shift to give an isomeric intermediate carbonium ion, which then partitioned to elimination products, was preferred. As supporting evidence for this shift, rather than the normal 1,2-rearrangements, which are usually observed under solvolytic conditions, an earlier investigation of the deamination of carbon-labelled *n*-propylamine was cited[622]. Labelled 1-^{14}C-*n*-propylamine gives 1-^{14}C-*n*-propanol (92%) and 3-^{14}C-*n*-propanol (8%). This product distribution can be explained only by a 1,3-hydride shift, *viz.*

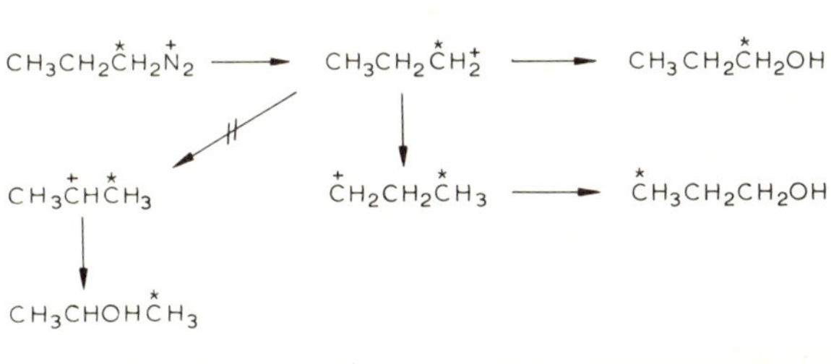

(245)

Two successive 1,2-hydride shifts can be ignored, as the initially formed secondary carbonium ion would give rise to some isopropanol. In addition, deamination of isopropylamine does not yield *n*-propanol. The intermediacy of a protonated cyclopropane is excluded as no label is found on the 2-carbon atom of the *n*-propanol. A similar deuterium labelling technique has been used to demonstrate the absence of carbenoid intermediates in a number of other deamination reactions leading to cyclopropanes[623,624]. Of course, care has to be taken to exclude the possibility that exchange of alpha deuterium has not occurred in the diazonium ion, but this seems unlikely as illustrated by the deamination of isobutylamine. The products of deamination, namely 1-butene, isobutylene and methylcyclopropane, had the same deuterium content when 1,1-dideutero-isobutylamine was deaminated in either deuterated or non-deuterated solvents[623].

The pyrolyses of *endo*-norbornyl acetate and *endo*-norbornyl methyl sulphone yield norbornene (7% and 16%) and nortricyclene (93% and 84%, respectively)[625]. Presumably, in harmony with pyrolyses giving the alkenes, the cyclopropanes are derived from *syn*-type eliminations which utilise a U arrangement of the leaving groups. This conclusion could be verified by pyrolysing the above substrates labelled with a hydrogen isotope in the 6 position. In the pyrolysis of bornyl chloride, gamma-elimination competes less favourably and the reaction products are tricyclene (55%), bornene (20%) and camphene (25%)[626]. The reaction is complicated as bornene and tricyclene are both prone to decompose under the reaction conditions, but an unchanged product analysis was obtained when the samples were isolated after between 15% and 50% reaction and the proportions are therefore meaningful. It would be interesting to study the stereochemistry of this latter cyclopropane-forming reaction and to test the possibility of carbonium-ion intermediates. Certainly, if a non-classical ion is formed, then the 6-*exo*-deutero-bornyl chloride will yield tricyclene with the same incorporation of deuterium as the 6-*endo* isomer, but if a concerted elimination with carbonium ion character is involved, a preference for a particular stereochemistry may be observed.

The "radical" induced 1,3-elimination of di-iodides to give cyclopropanes in almost quantitative yield, is only of preparative interest at present[627].

6.4 REACTIONS INVOLVING THREE-MEMBERED-RING INTERMEDIATES*

6.4.1 The Ramberg–Bäcklund reaction

When treated with base, α-halogenosulphones, possessing γ hydrogen atoms, undergo a 1,3-elimination to give an episulphone, which rapidly extrudes

* In this section, the leaving groups are regarded as situated on the α and γ carbon atoms, in line with the terminology outlined earlier, (234) p. 329, rather than on the α and α' carbons, as is more commonly used in connection with the Ramberg–Bäcklund, Favorskii and Neber rearrangements.

sulphur dioxide (see Introduction, p. 163) to give an olefin[628,629], *e.g.*

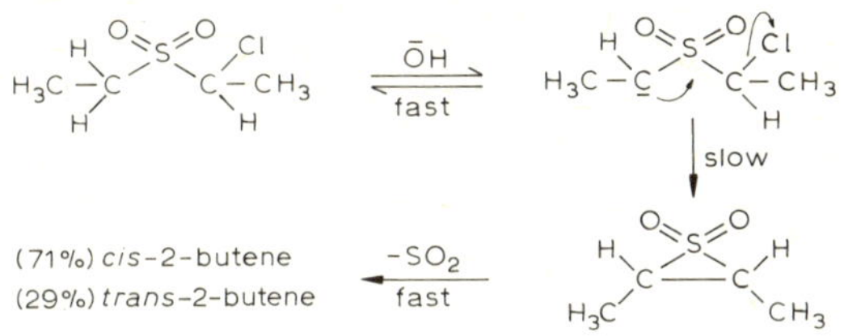

(246)

The release of halide ion follows first-order kinetics in both the substrate and in the base[630]. Deuterium exchange studies provide evidence in favour of a rapid reversible ionisation. If the reaction is performed in deuterated hydroxylic solvents, the olefins isolated are completely deuterated in the vinylic positions[631] and the substrate recovered in the early stages of the reaction is found to have undergone complete exchange of both the α and γ hydrogen atoms[632]. Carbenoid intermediates are not involved, as the reaction works equally well for both α,α-dihalogeno- and α,α,α-trihalogenosulphones[633], and some α-halogenosulphones which can undergo only alpha-elimination and not 1,3-elimination do not show the rearrangement[631]. Consequently, exchange of the acidic α hydrogen atoms is a side reaction as also might be exchange of the γ hydrogen atoms. However, a carbanion mechanism rather than a concerted elimination seems more probable, as the leaving group effect, k_{Br}/k_{Cl} (= 620 for 1,3-elimination from $C_6H_5CH_2SO_2CHXC_6H_5$ at 0°C), is considerably larger than that observed for concerted 1,2-eliminations in the 2-phenylethyl series[632]. Paquette[631a] has studied the stereochemistry of the cyclisation step using substituted isomeric 2-chloro-2,7-dihydro-3,4:5,6-dibenzothiepin-1,1-dioxides, *viz.*

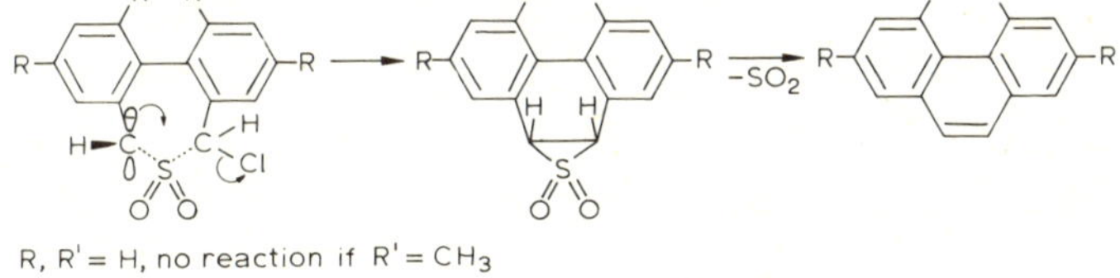

(247)

R, R' = H, no reaction if R' = CH_3

A transition state was envisaged in which all six carbon atoms of the heterocyclic seven-membered ring are mutually coplanar and the sulphone group is bent below this plane, so as to minimise the interaction between the large negative field of the sulphonyl oxygens and the site of displacement[629]. Apparently, it is necessary to achieve coplanarity of the two benzene rings to form the episulphone ring, as when R′ = CH_3 (247), coplanarity is prevented and the reaction does not occur. In agreement with the observations of Cristol *et al.*[615–617], it would seem reasonable to expect a *semi*-W conformation for 1,3-elimination from the carbanion. For substrates possessing an α hydrogen atom, this point cannot be verified as the stereochemistry at C_α is lost in the rapid exchange reaction, but it could be demonstrated for optically active tertiary (lacking hydrogen atoms) α-halogenosulphones or bromo-chlorosulphones possessing only one γ hydrogen atom which would thereby yield a bromo- or chloroalkene.

Using an alternative route, Bordwell *et al.*[632,634] have prepared a number of episulphones. These decompose much more rapidly than their precursor α-halogenosulphones in the Ramberg–Bäcklund reaction and unless very strongly basic conditions are used or the episulphones have particularly acidic hydrogen atoms, the geometrical arrangement of substituents in these three-membered rings is maintained in the olefinic products. The Ramberg–Bäcklund reaction usually gives a predominance of the *cis*-alkene and the stereochemistry is obviously determined at the cyclisation stage. Previous authors have explained this stereochemical preference in terms of greater London attractive forces between R and R′, (248)[631a] or greater ease of solvation in the transition state[609] leading to the *cis*-episulphone. An alternative approach adopted by Paquette[629] hinges on the greater ease of rotation in the carbanion precursor when attaining the *semi*-W conformation for formation of the *cis*-episulphone than for formation of the *trans*-episulphone (248). Consequently, the concentration of the conformer leading to *cis*-episulphone is greater than that leading to the *trans* compound.

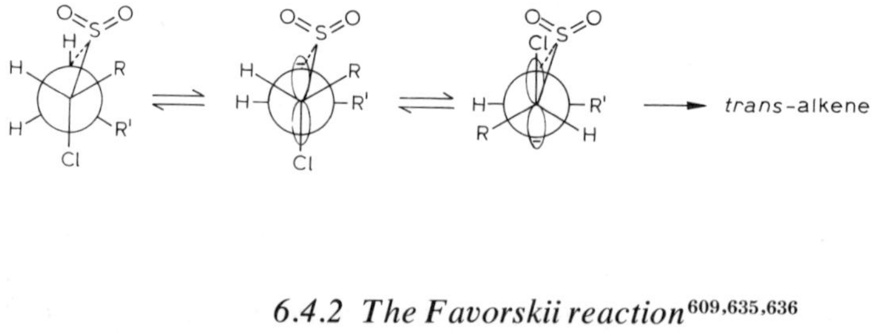

(248)

6.4.2 *The Favorskii reaction*[609,635,636]

The treatment of α-halogenoketones with nucleophilic bases gives re-arranged carboxylic acids or their derivatives. For substrates lacking γ hydrogen atoms, a *semi*-benzilic acid rearrangement is generally accepted

(249)[637], but for those possessing a γ hydrogen atom, a mechanism involving a cyclopropanone intermediate is usually suggested (250)[638].

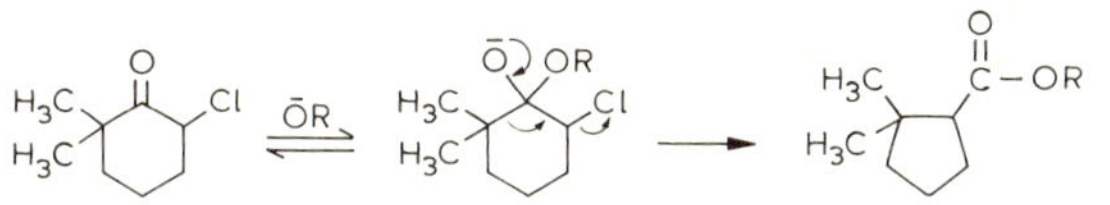

(249)

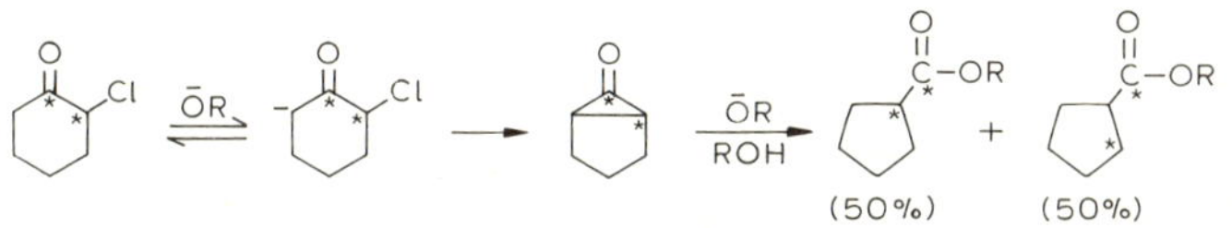

(250)

The cyclopropanone ring opens so as to give the more stable carbanion, which is rapidly protonated by the solvent[639,640]. Elimination of halide ion is first-order in both the substrate and the base.

The existence of a three-membered ring intermediate is suggested by a number of experimental observations. The isomeric α-halogenoketones, (251), are both converted into 3-phenylpropionic acid by hydroxide ion, indicating a common intermediate, *viz.*[641]

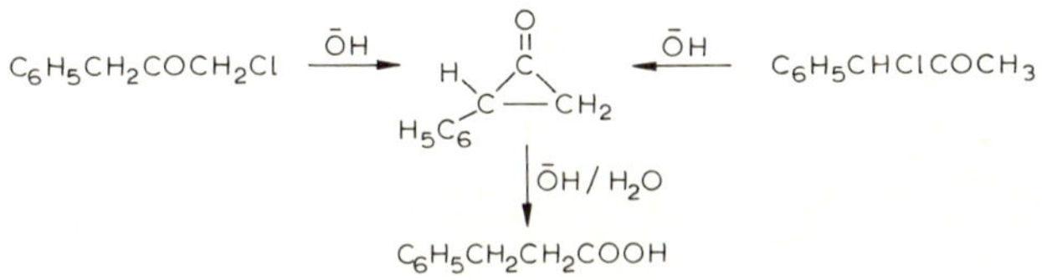

(251)

Treatment of 2-chlorocyclohexanone-1,2-^{14}C, in which the isotope is equally distributed between the carbon atoms, with less than one equivalent of base, yielded cyclopentane carboxylic acid. The isotope distribution in the recovered unreacted starting material was the same as in the original reactant. However, the label in the product was distributed 25% on both the alpha carbon and the two beta carbon atoms and 50% in the carboxyl group. These observations exclude halogen migration before rearrangement and are consistent with the product having arisen from an intermediate in which the alpha and gamma carbons of the substrate have become equivalent (250)[638,642]. Using secondary α,γ-dihalogenoketones, Breslow *et al.*[643] have isolated cyclopropenones, *viz.*

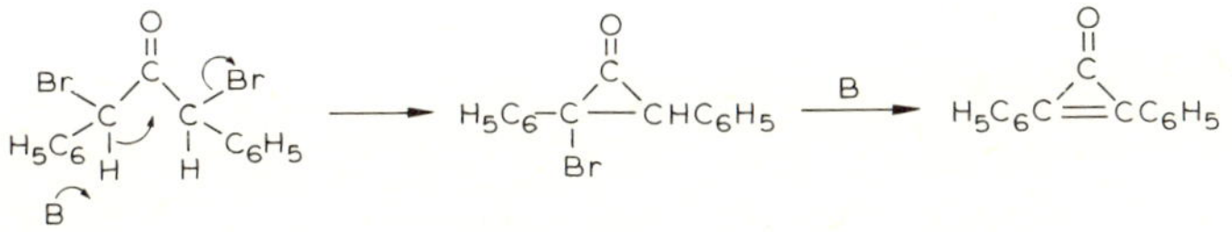

(252)

In the heterogeneous medium of sodium benzyloxide in ether, the stereochemistry of the Favorskii reaction has been shown to involve inversion of configuration at the α carbon (253).

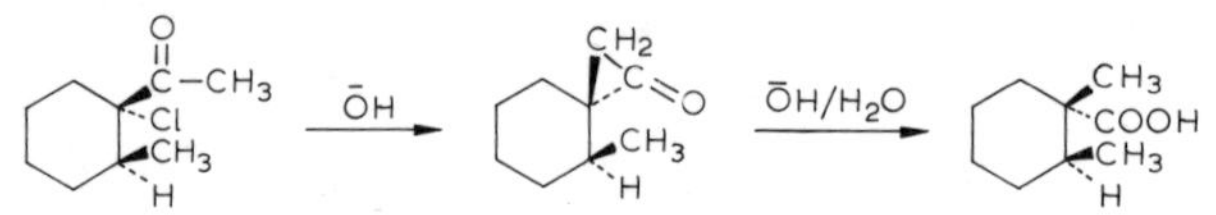

(253)

The other diastereoisomer gave the isomeric acid, again illustrating that inversion accompanies cyclopropanone formation[644].

The above results strongly support the postulation of a cyclopropanone intermediate, but other than implying inversion of configuration, they afford no information concerning the nature of the 1,3-elimination step. Recent reports suggest that the choice between a carbanion and essentially a concerted mechanism, is determined by the alpha halogen[645,646].

Mechanistic assignment was dependent on a comparison of the Favorskii reactions of corresponding α-chloro- and α-bromo-ketones, the primary kinetic hydrogen–deuterium isotope effect in the γ position, the extent of hydrogen exchange in the substrate in the early reaction stages, and on the incorporation or exchange of hydrogen isotopes in the isolated ester products (254). All the

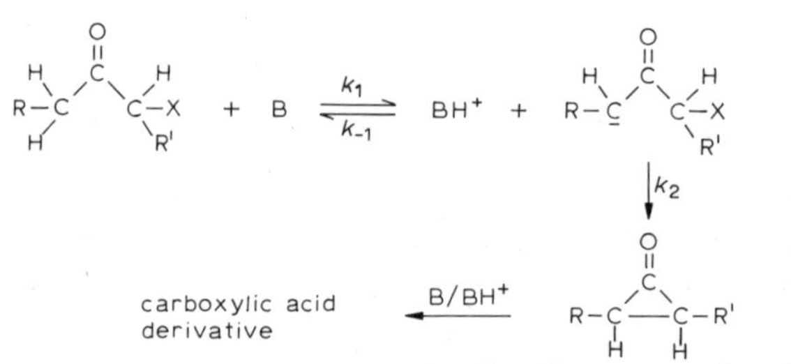

$$k_{obs} = \frac{k_1 k_2}{k_{-1}[BH^+] + k_2} \quad \equiv k_1 k_2/k_{-1}[BH^+] \text{ if } k_2 << k_{-1}[BH^+] \quad \equiv k_1 \text{ if } k_{-1}[BH^+] << k_2$$

(254)

*substrate**

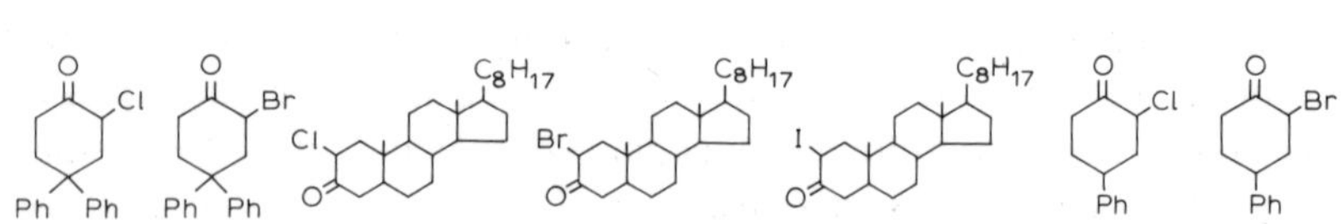

	1	2	3	4	5	6	7
k_H/k_D at 0°C	1.05	4.1	(4.0)	(4.2)	(5.1)	—	—
k_{Br}/k_{Cl} at 0°C	116(79)		36(5.7)		(6.8)**	52	
extent of exchange at C_γ	extensive	moderate	moderate	none	none	moderate	none

* Base; methanol/methoxide except values in parentheses for which the reaction medium was ethanol/ethoxide.
** k_I/k_{Cl}.

substrates showed complete exchange of the α hydrogen atoms, but for the chloro-substrates extensive exchange with the solvent also occurred, prior to the rate-determining step, at the γ position. The analyses of the esters were in agreement with the conclusion that extensive exchange at the γ carbon accompanies the reaction of the chloro compounds, but little occurs with the bromo (except 2-bromo-4,4-diphenylcyclohexanone) and the iodo compounds, which also exhibit larger kinetic isotope effects. These results accord with a carbanion mechanism in which the balance between k_{-1} and k_2 changes with the leaving group. For all the halogens, the acidity of the γ hydrogens should be virtually the same, but with a change to a more ionising substituent, k_2 will approach k_{-1}, and the observed rate coefficient thus approaches k_1. With the 4,4-diphenyl-2-halogenocyclohexanones, the leaving group ratio, k_{Br}/k_{Cl}, is large (= 116), and reflects to a certain extent, a rate-determining carbon–halogen bond cleavage. However, with the other substrates, in which exchange becomes less marked, the k_{Br}/k_{Cl} ratio diminishes and k_1 becomes rate-determining, certainly, for the bromine derivatives, and the reactions tend to become essentially more concerted in nature. An obvious approach to distinguish between the one- and two-step schemes is to correlate changes in rate coefficient with changes in the medium basicity as measured by an acidity function (Section 2.1.3).

There are many other interesting and important investigations concerning the Favorskii reaction but these are not concerned with the nature of the elimination step and the reader is recommended to consult earlier reviews[609,635,636].

Arguing that the geometry of the enolate π orbital is not suitable for effective S_N2-type overlap with the σ orbital of the α carbon atom, a number of authors have proposed that cyclopropanone formation is preceded by halide ion loss[647]. An intermediate zwitterion (255),

(255)

which is stabilised by mesomerism, was suggested, but as this has a plane of symmetry, unless it is asymmetrically solvated, it cannot be involved in the necessary stereospecific reaction. However, considerable dipolar character is associated with the transition state leading to cyclisation from the enolate anion as shown by the recent investigations of Bordwell and Scamehorn[684]. Under conditions in which exchange of γ hydrogen is a rapid process compared to halide-ion loss and a large leaving group effect (k_{Br}/k_{Cl}) is observed, electron-releasing aromatic substituents in α-chlorobenzylmethyl ketones facilitate the reaction. In fact, a linear correlation with the Hammett equation is only observed if substituent constants σ^+ are employed for those substituents which can conjugate with the developing positive charge at the α carbon atom. These facts, and a reaction constant of -2.37 at 0°C, are consistent with considerable C–X bond breaking and little C_α–C_γ bond formation in the transition state for cyclisation from the enolate anion. The authors suggest that dissociation occurs with the C–Cl bond parallel to the p orbitals of the adjacent carbon atom, and is aided by partial overlap of the orbital developing at C_α with that of the carbonyl carbon and oxygen atom as well as possible π overlap between the p orbitals on C_γ and C_α, *viz.*

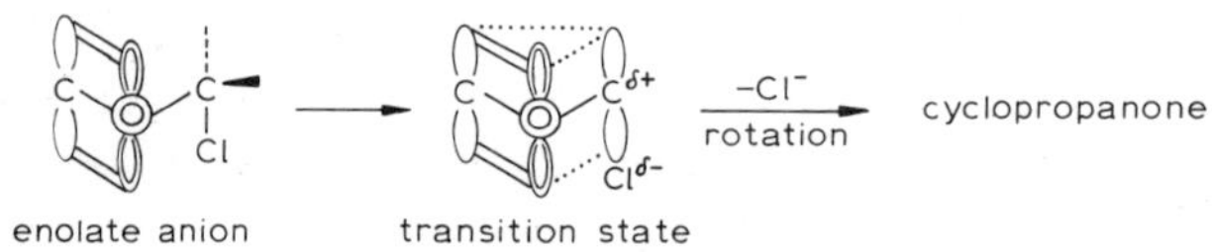

(256)

That the enolate anion does participate in aiding C–X bond breaking is suggested by the more rapid release of bromide ion from $PhCHBrCO\bar{C}H_2$ than from $PhCHBrCO\bar{O}$.

A similar application of the Hammett equation to the reaction of α-halogenobenzylmethyl sulphones with basic reagents, shows that the rearrangement is promoted by electron withdrawing substituents. In fact, for this particular Ramberg–Bäcklund reaction, a reaction constant of 0.8 indicates that although C–X bond breaking is extensive, it is less advanced in the transition state than C_γ–C_α bond formation[684,685]. A dipolar intermediate is less probable in the Ramberg–Bäcklund reaction than in the Favorskii reaction, as it would give rise to considerable electrostatic repulsion between the adjacent electrophilic sulphur and alpha-carbon atom (257).

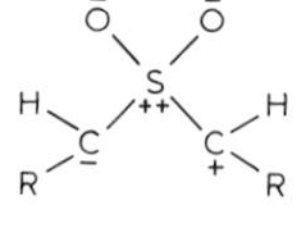

(257)

Of interest is a consideration of the change-over point from a cyclopropanone to a *semi*-benzilic acid mechanism. In cases in which gamma-hydrogen exchange does not occur before rearrangement, it should be possible to use a deuterium labelling technique to distinguish between these two mechanisms. In a basic deuterated hydroxylic solvent, substrates lacking α hydrogen and initially containing gamma hydrogen, should give a Favorskii product containing deuterium if the reaction follows the cyclopropanone mechanism, but unlabelled product if the *semi*-benzilic acid scheme operates. This approach has been used by Warnhoff *et al.*[670], who have shown recently that a delicate balance exists between the two mechanisms. The cyclopropanone route is promoted by an increase in the acidity of the γ hydrogen atom or an increase in the medium basicity.

6.4.3 The Neber rearrangement[609,648,649]

Fewer mechanistic details are available for the Neber rearrangement than for the closely related Favorskii and Ramberg–Bäcklund reactions. Oxime tosylates, when treated initially with base and then with water, yield α-amino-ketones, *viz.*

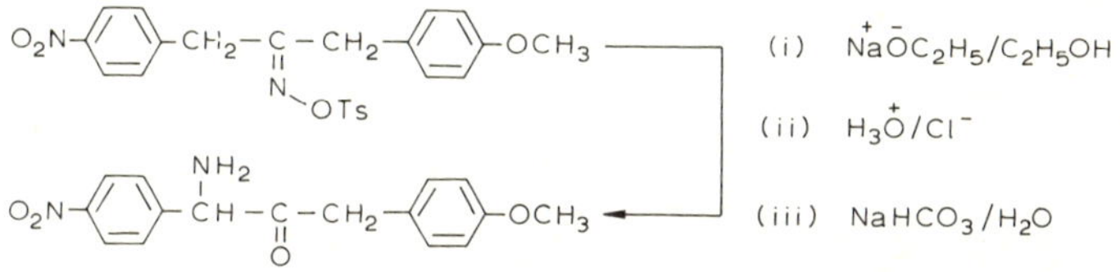

(258)

The presence of a gamma-hydrogen atom is essential and when more than one appropriate gamma carbon is available, the amino-group is substituted for the more acidic hydrogen, irrespective of the stereochemical relationship between the oxime tosylate and the hydrogen atom (258)[648,650,651]. In the only stereochemical study available, the amino group was found to occupy the equatorial conformation in the product (259), but at which stage this stereochemical orientation was introduced is not clear[652].

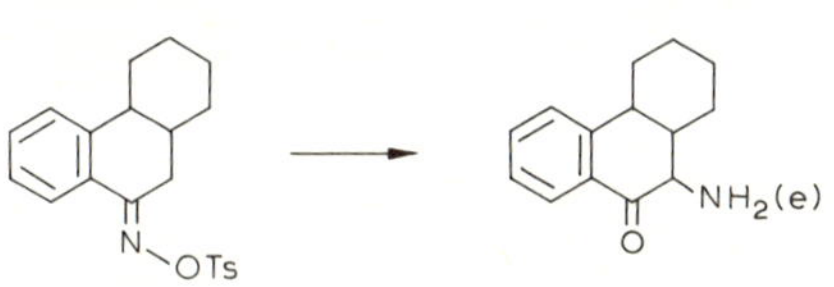

(259)

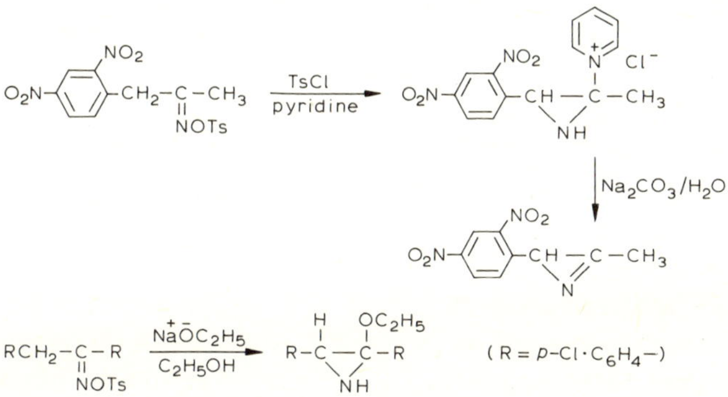

(260)

Under varying reaction conditions, a number of three-membered ring intermediates have been isolated (260)[650]. Bearing all these points in mind, two mechanisms, both involving a nitrene intermediate, have been suggested. Of these, the suggestion of House and Berkowitz[651] seems less probable as it requires that the intermediate nitrene is remarkably selective in reaction (*cf.* 258), and in fact attacks the less nucleophilic γ carbon atom, *viz.*

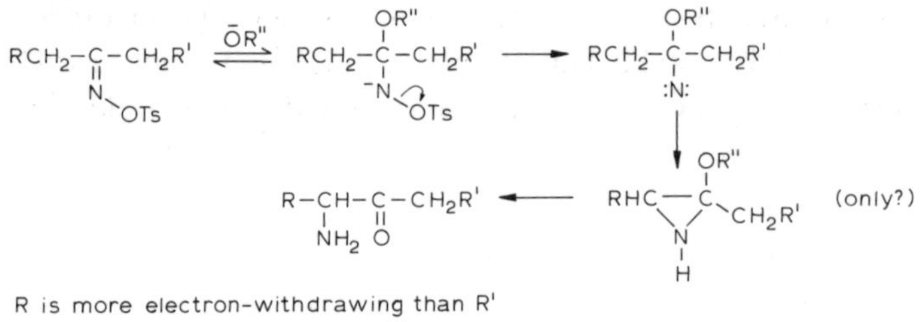

(261)

The alternative scheme of Hatch and Cram[650] seems more likely, the nitrene arising from a 1,3-elimination and being partially stabilised by the adjacent double bond, *viz.*

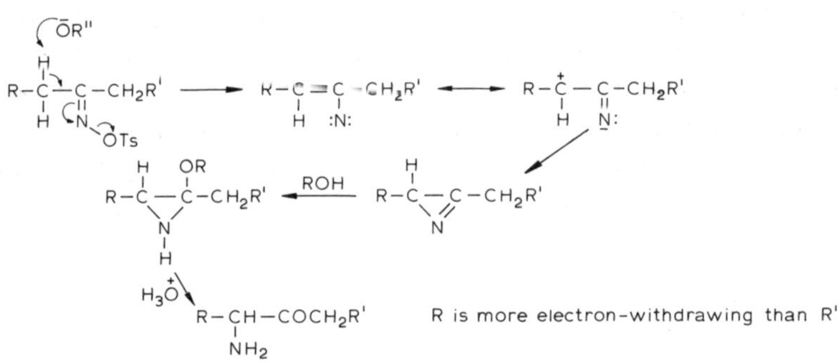

(262)

There is obviously great need for a more thorough mechanistic study. The nature of the nitrene (*e.g.* singlet or triplet), if one is in fact an intermediate, is unknown. The 1,3-elimination may be a stepwise process.

6.5 ELIMINATION GIVING THREE-MEMBERED HETEROCYCLIC RINGS

A wide variety of three-membered heterocyclic rings are formed by a 1,3-elimination of HX[653]. Such reactions occur with particular facility when the heteroatom carries the hydrogen atom and X is a halogen or an ester function. Typical examples are the conversions of 2-halogenothiols, 2-halogenoalcohols and 2-halogenoamines into cyclic sulphides, epoxides and aziridines, respectively. Although most reactions of this type require base catalysis, in some cases the nucleophilicity of the heteroatom is sufficient to affect the displacement reaction. In such cases, a simple law first-order in the substrate is followed, but in the presence of basic reagents, a second-order component in the rate is also noted (263).

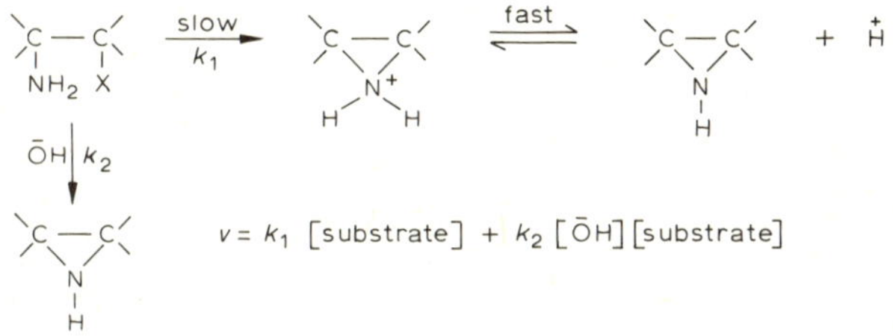

(263)

Either initial proton abstraction by the base or a concerted 1,3-elimination now competes with the internal displacement[654]. Cyclisation is accompanied by inversion of configuration at the carbon bearing the leaving group, as exemplified by α-lactam formation (264)[655].

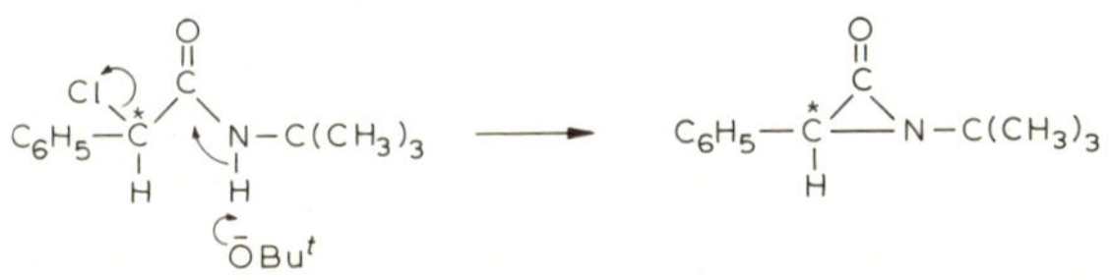

(264)

The greater separation of the leaving groups in 1,3-elimination than in 1,2-elimination increases the probability of carbanion relative to concerted reactions. It would seem a sound approach to utilise the response of some of the carbanion 1,3-eliminations to changes in the reaction conditions as models for estimating the carbanion character in the 1,2-processes.

7. Delta-eliminations

7.1 CLASSIFICATION

There are relatively few mechanistic studies of delta-eliminations. Characteristic of this class of reaction are not only the closures to 4-membered ring systems[653] but also some interesting reactions which involve the migration of a carbon–carbon double bond, *e.g.*[656]

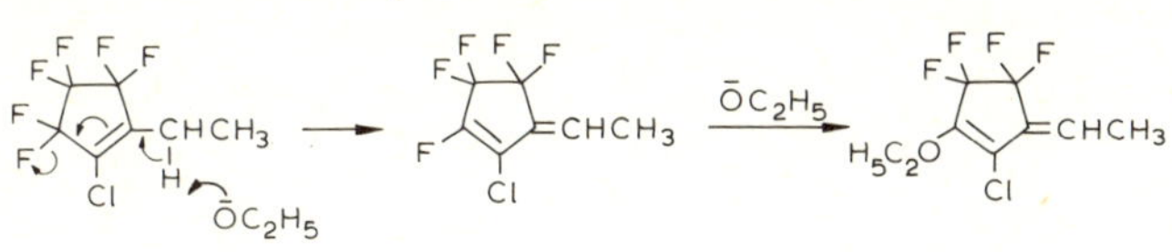

(265)

Under basic conditions, 1,4-elimination occurs up to 1200 times more rapidly from *trans*-9,10-dihydroanthracenes than from the *cis* substrates, (266)[657].

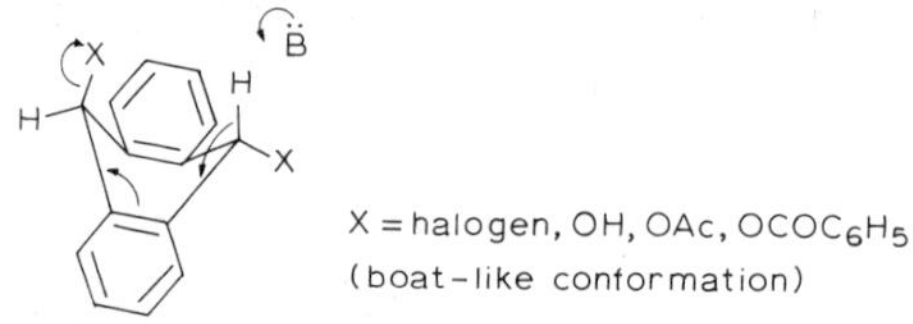

(266)

This marked preference for 1,4-*syn*-elimination is not surprising as it involves two simultaneous 1,2-*anti*-eliminations. On the other hand, 1,4-*anti*-elimination involves a 1,2-*anti*-elimination coupled with a 1,2-*syn-clinal*-elimination and consequently involves less favourable overlap of the developing *p* orbitals in the transition state.

7.2 COMPARISON OF 1,2- AND 1,4-ELIMINATION

From the analysis of the dichlorobenzenes produced in the alkaline dehydrochlorination of benzene tetrachlorides, Orloff and Kolka[658] predicted the relative ease of 1,2- to 1,4-elimination to be as follows: *anti*-1,2 > *anti* 1,4 > *syn*-1,2 > *syn*-1,4. This apparent reversal of the more favourable stereochemical arrangement for 1,4-elimination is explained by assuming that the first dehydrohalogenation always involves allylic hydrogen. Within this scheme, initial 1,4-elimination eventually leads to 1,2- and 1,3-dichlorobenzenes, whereas initial 1,2-elimination gives the 1,3- and 1,4-products. This interpretation has recently been challenged by de la Mare *et al.*[659] who have studied the dehydrohalogenations of the naphthalene tetrachlorides. Their rate data and product analyses accord with the reactivity order *anti*-1,2 > *syn*-1,2 > *syn*-1,4. Although it was not possible to place *anti*-1,4-elimination in the reactivity sequence, it was felt that this orientation was least likely. Initial abstraction of benzylic hydrogen is not necessary and when the benzylic hydrogen has mainly a "pseudo" equatorial conformation and the stereochemistry for 1,2-elimination is not *anti*, then neither of the leaving groups need be in the benzylic position (267).

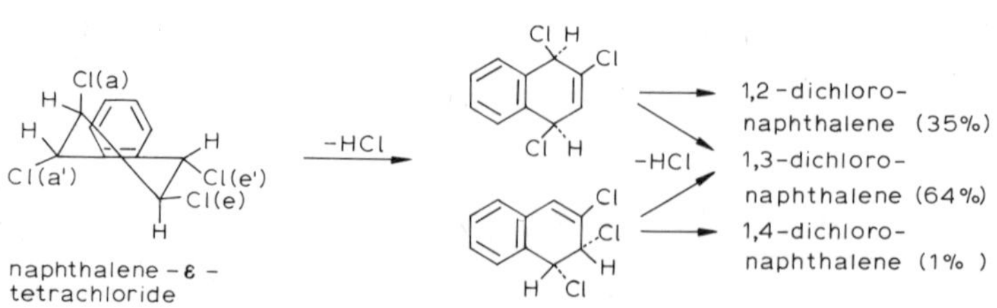

(267)

In Orloff and Kolka's original scheme, 1,4-elimination figured prominantly in the dehydrohalogenation of both α- and γ-benzene tetrachlorides, yet these substrates differ more than 40-fold in reactivity and have a similar stereochemistry for the second elimination. On the modified scheme, this rate difference is accommodated by consideration of the ease of 1,2-*anti*- to 1,2-*syn*-elimination.

The relative ease of 1,4-eliminations to the corresponding 1,2-eliminations has been suggested as a measure of transition state structure[660]. If considerable charge separation occurs, then 1,4-elimination should proceed much more easily, as its transition state, in resembling an allylic cation or anion, should be more stable than that of the 1,2-elimination. If the reaction is highly synchronous, then as the thermal resonance energy of butadiene is greater than that of an isolated double bond, a ratio of $k_{1,4}/k_{1,2}$ of greater than unity is predicted. Of course, in all cases, a greater loss in rotational entropy will result in attaining a planar transition state in 1,4-elimination as the freely rotating terminal groups become restricted. An additional complicating factor is that of substrate conformation, which has more influence for 1,2- than 1,4-eliminations due to the closer proximity of the substituents. As many factors will influence the population of the various conformations, a generalisation of this effect cannot be made with any certainty, but a general modification of the anticipated profile is expected. Thus, for transition states with extensive carbanion or carbonium ion character, $k_{1,4}/k_{1,2}$ should be large, whereas for a highly synchronous reaction, a small value or one of less than unity is predicted.

In the highly ionising medium of aqueous acetone, the 1,4-dimethanesulphonate underwent elimination with hydroxide very much faster than the corresponding 1,2-substrate[661]. A transition state with considerable carbonium character was thus proposed, (268) below. However, in ethanol containing ethoxide ion, dehydrohalogenation of 1,2-dichloroethane occurs slightly more rapidly than that of *trans*-1,4-dichlorobut-2-ene. A similar reactivity ratio is observed in the iodide-promoted dehalogenations of 1,2-diiodoethane and *trans*-1,4-diiodobut-2-ene. Both of these latter pairs of reactions would be expected to proceed via transition states in which double-bond character is extensively developed[660].

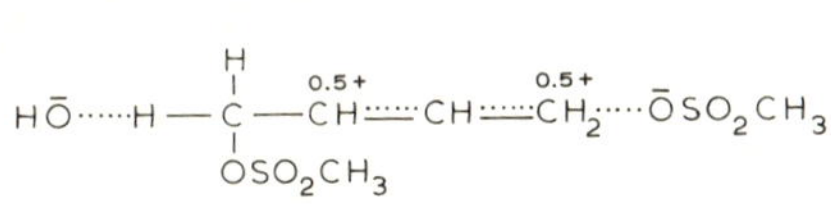
(268)

Before the $k_{1,4}/k_{1,2}$ ratio can be considered as a valid criterion for assessing transition state structure, it needs testing for many more model reactions. The examples selected were not ideal as the low reactivity towards elimination made the significance of solvolytic side reactions more important and measurement of the rate coefficients was not accurate enough for a meaningful calculation of the activation parameters to be made.

8. Fragmentation reactions

Just over a decade ago, fragmentation reactions were recognised as a new class of organic reactions by Grob and Bauman[662] and since that time Grob has written a number of excellent reviews on this subject[4,663–665]. There are now over sixty known examples of fragmentation reactions.

Many analogies between simple 1,2-eliminations and fragmentation reactions are apparent; see (269).

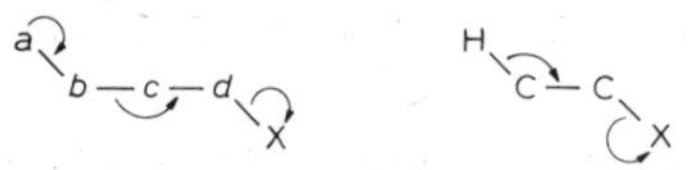

(269)

The leaving groups (X), generally termed the nucleofugal fragments, as they become one charge unit more negative on reaction, are typical of leaving groups usually found in 1,2-elimination. For example, X can be a halogen, an ester or an "onium" group. The electrofugal fragment (*a–b*), which donates an electron pair by which the central fragment (*c–d*) becomes more unsaturated, is analogous to the atoms, generally hydrogen, which are involved in beta-eliminations. Depending on the order in which the fragments are released, one- or two-step processes, equivalent to the E1, E2 and E1cB mechanisms, can be recognised in fragmentation reactions. However, the present knowledge of these processes is confined mainly to product analysis and stereochemical requirements, and mechanistic investigations have been relatively neglected to date. The terminology used in this text is that suggested by Grob and is illustrated by (270).

a———b	—c—d—	———X	→	$\overset{+}{a}$═b	+ c═d	+ $\bar{X}$	
electrofuge (or electrofugal group)	middle group	nucleofuge (or nucleofugal group)		electrofugal fragment	unsaturated fragment	nucleofugal fragment	(270)

8.1 OLEFIN-FORMING FRAGMENTATIONS

A number of γ-substituted alcohols undergo fragmentation reactions under the influence of acidic or basic catalysts. For instance, unsymmetrical 1,3-diols fragment in such a way that the nucleofugal fragment is formed from the hydroxyl group on the more highly substituted carbon, indicating the intermediacy of a carbonium ion, *e.g.*[666]

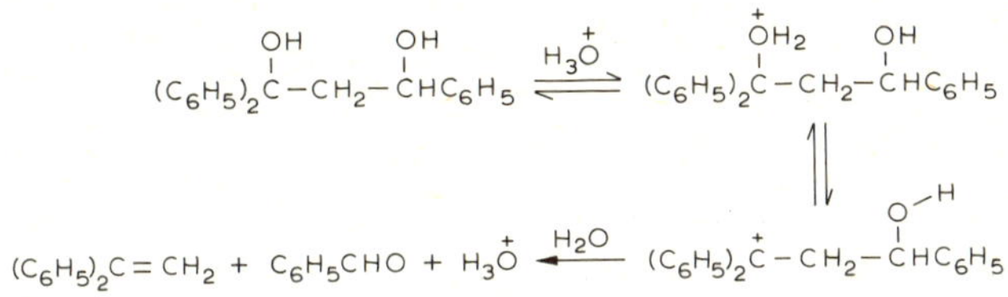

(271)

Similar fragmentations occur when γ-aminoalcohols are treated with nitrous acid[667]. If the nucleofuge is a neutral group, then fragmentation of hydroxyl compounds requires basic catalysis, *e.g.*[668]

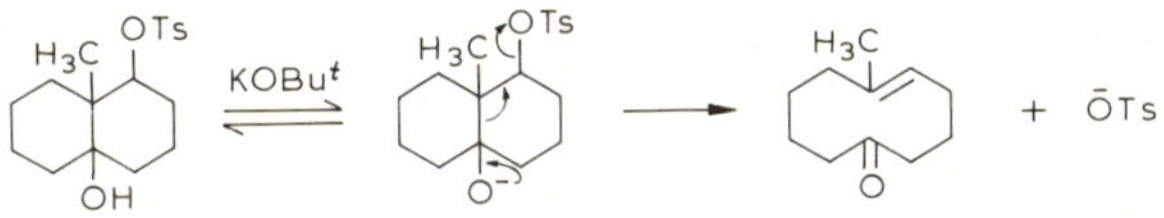

(272)

In a number of fragmentations, the electrofugal group is generated only after an initial addition of a nucleophilic species. Typical of this class of fragmentation are the reactions of β-halogenocarbonyl compounds, *e.g.*[669]

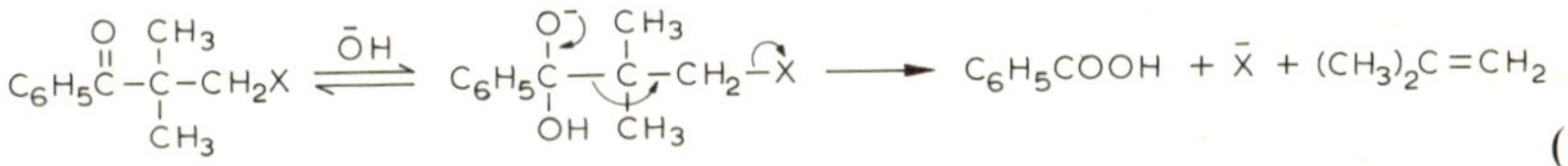

(273)

In weakly alkaline solution, dehalogenative decarboxylation of β-halocarboxylic acids occurs readily[669]. In the case of *trans*-cinnamic acid dibromides, the stereochemical fate of the β-bromostyrene is determined by the solvent and substituents in the benzene ring (274)[559].

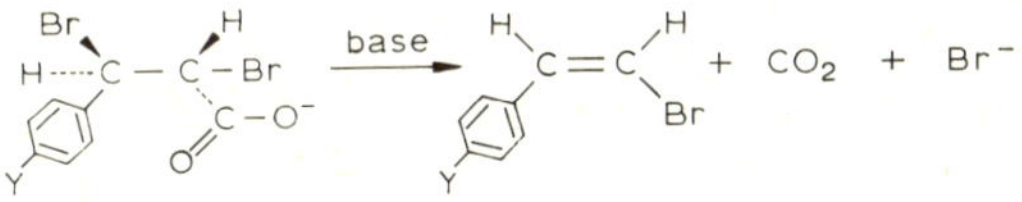

(274)

Y	H	H	H	OCH_3	OCH_3	NO_2	NO_2
solvent	H_2O	C_2H_5OH	acetone	H_2O	C_2H_5OH	H_2O	C_2H_5OH
% *cis*-alkene	22	87	99	100	100	67	100
% *trans*-alkene	78	13	—	—	—	23	—

The *cis*-alkene is formed alone in a stereospecific concerted reaction, analogous to an *anti*-E2 elimination, which is encouraged by electron-withdrawing substituents (Y) and a poorer ionising medium. However, electron-releasing substituents and a more ionising solvent promote benzylic carbonium ion character and a mixture of alkenes, in which the thermodynamically more stable *trans* product predominantes, is formed, *e.g.*

$$C_6H_5CHBrCHBrCOO^- \xrightarrow{\text{slow}} C_6H_5\overset{+}{C}HCHBrCOO^-$$
$$+Br^- \longrightarrow C_6H_5CH{=}CHBr+CO_2 \qquad (275)$$

As might be predicted for a reaction in which an increase in charge occurs in attaining the transition state, a positive salt effect on the rate coefficient is observed. In the zwitterion, rotation about the C_α–C_β bond is presumably more rapid than recombination of ions, as added bromide ion exerts only the same influence on the rate coefficient as perchlorate and acetate ions.

The intermediacy of a lactone (276) would seem reasonable, but the predominance of the *trans*-alkene is not explicable in this case. As long as the acid was completely ionised, the rate of dehalogenative decarboxylation was independent of pH (in aqueous solution). However, in more strongly alkaline solution (pH > 11), the E2 dehydrohalogenation becomes competitive, but the rate coefficient for the dehalogenative decarboxylation also increases[559,671]. An initial attack of hydroxide on the acid, giving (277) and a second-order term in the rate law, has been suggested.

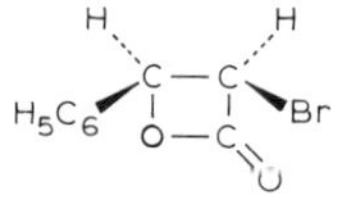

(276)

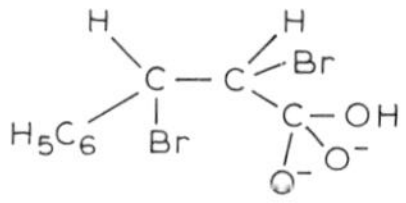

(277)

In a number of olefin-forming fragmentations, the electrofuge is the lone pair of an oxygen or nitrogen atom. One of the most well-known examples is the solvolysis of 4-bromoquinuclidine, *viz.*[672]

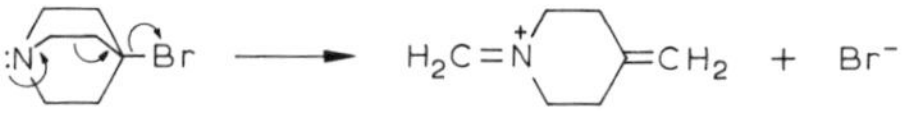

(278)

Only on rare occasions is the electrofugal fragment a carbonium ion, as most often appropriate structures undergo substitution or elimination reactions more easily. However, ions such as the very stable tropylium cation are exceptions (279)[673].

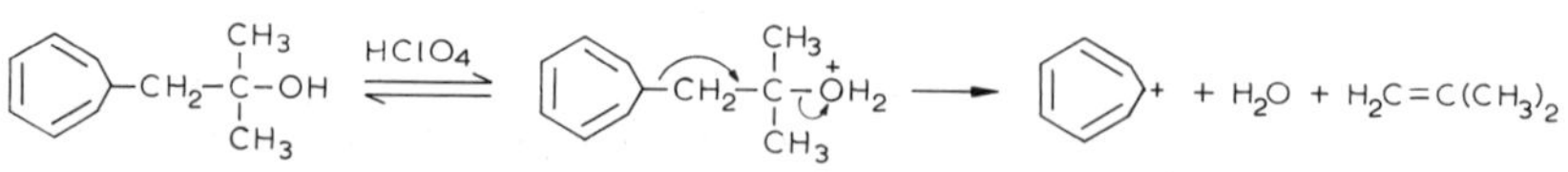

(279)

A large number of other fragmentation reactions leading to alkenes are known but these have not been characterised from a mechanistic standpoint[4].

8.2 ALKYNE-FORMING FRAGMENTATIONS

Fewer variations of fragmentations leading to alkynes are known than those giving alkenes, probably due to the attachment of the nucleofugal group to an unsaturated atom, which consequently reduces its reactivity[4]. Only the most active nucleofuges and electrofuges thus suffice for this series. Dehalogenative decarboxylation occurs easily, but as mentioned earlier, requires the *anti* orientation of the eliminating fragments. In the presence of a nucleophile β-halo-α,β-unsaturated carbonyl compounds fragment, but the intermediacy of an anion is assumed and not proven at present[674]. Examples of phosphonic acid as an electrofugal fragment are known[4].

8.3 FRAGMENTATIONS YIELDING MULTIPLE BONDS BETWEEN CARBON AND A HETEROATOM

Numerous examples of this variety of reaction have been summarised by Grob and Schiess[4]. The nucleofugal fragment can be attached to either the heteroatom or the carbon atom in the substrate. Typical of the former type are the side reaction in the Hofmann rearrangement when a substrate possesses a strongly electrofugal group (280)[675], a side reaction in the chromic acid oxidation of a secondary alcohol (281)[676], and an alkaline hydrogen peroxide oxidation of an epoxide (282)[677]. Cases in which the electrofuge resides on the heteroatom are the decomposition of *N*-*t*-butylimidoyl chloride (283)[678], and the rearrangement of an allophanyl chloride into a diisocyanate (284)[679]. In examples (281) and (283), the electrofugal fragment is a simple carbonium ion.

(280)

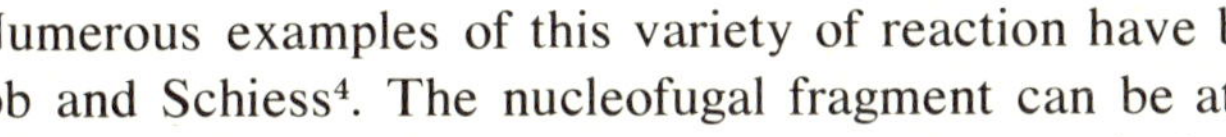

(281)

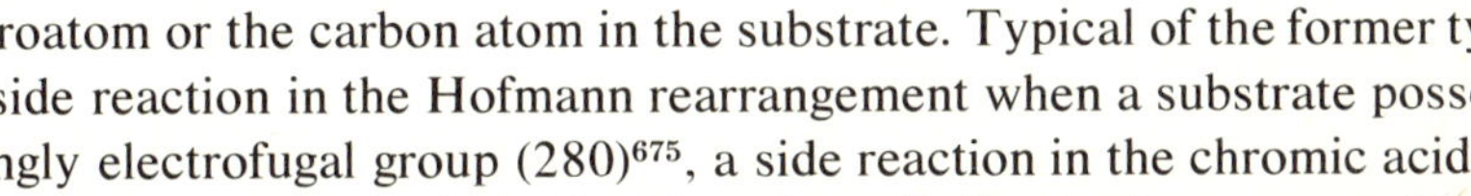

(282)

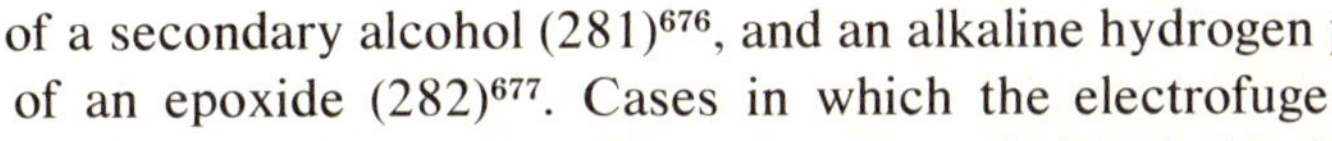

(283)

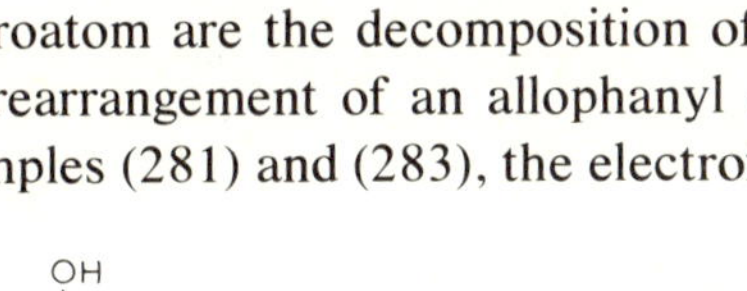

(284)

Fragmentation reactions which give rise to multiple bonding between nitrogen atoms are exemplified by decompositions of aromatic diazohydroxides in acidic media to give aryl cations, nitrogen and water. Such reactions generally occur in two steps and the intermediate diazonium ion is well characterised by its electrophilic properties.

The field of fragmentation reactions is obviously going to provide the physical organic chemist with abundant scope for investigation, since at the time of writing few mechanistic studies are available in this highly interesting field.

Acknowledgements

The author wishes to express his thanks to Professor W. H. Saunders, Jr., Dr. A. J. Boulton, Dr. C. D. Johnson and Dr. J. Banger for their helpful suggestions during the preparation of this chapter. Finally, he acknowledges the assistance of his wife, Sheila M. Cockerill, for the checking of the manuscript.

REFERENCES

1 *e.g.* D. J. Cram and G. S. Hammond, *Organic Chemistry*, 2nd edn., McGraw-Hill, London, 1964.
2 W. Hanhart and C. K. Ingold, *J. Chem. Soc.*, (1927) 997.
3 J. Hine and A. M. Dowell, *J. Am. Chem. Soc.*, 76 (1954) 2688.
4 C. A. Grob and P. W. Schiess, *Angew. Chem. Intern. Ed. Engl.*, 6 (1967) 1.
5 B. P. Stark and A. P. Duke, *Extrusion Reactions*, Pergamon, London, 1967.
6 C. K. Ingold, *Structure and Mechanism in Organic Chemistry*, Cornell University Press, Ithaca, New York, 1953, Chap. 8.
7 J. Hine, *Physical Organic Chemistry*, 2nd edn., McGraw-Hill, New York, 1962, Chap. 8.
8 D. J. Cram, in *Steric Effects in Organic Chemistry*, M. S. Newman (Ed.), J. Wiley, New York, 1956, Chap. 6.
9 E. S. Gould, *Mechanism and Structure in Organic Chemistry*, Holt, New York, 1959, Chap. 12.
10 D. V. Banthorpe, *Elimination Reactions*, Vol. 2, Elsevier, Amsterdam, 1963.
11 A. H. Corwin and M. M. Bursey, *Elements of Organic Chemistry*, Addison and Wesley, London, 1966, p. 511.
12 J. D. Roberts and M. C. Caesario, *Basic Principles of Organic Chemistry*, Benjamin, New York, 1965, p. 306.
13 Ref. 1, Chap. 16, p. 328.
14 W. H. Saunders, Jr., Elimination Reactions in Solution, in *The Chemistry of Alkenes*, S. Patai (Ed.), Interscience, Wiley, London, 1964, Chap. 2.
15 E. D. Hughes and C. K. Ingold, *J. Chem. Soc.*, (1933) 523.
16 L. Hey and C. K. Ingold, *J. Chem. Soc.*, (1933) 531; G. W. Fenton and C. K. Ingold, *J. Chem. Soc.*, (1939) 2342.
17 M. Zanger, C. A. Van der Werf and W. E. McEwen, *J. Am. Chem. Soc.*, 81 (1959) 3806.
18 (a) M. Grayson and P. T. Keough, *J. Am. Chem. Soc.*, 82 (1960) 3919.
(b) C. B. Parisek, W. E. McEwen and C. A. Van der Werf, *J. Am. Chem. Soc.*, 82 (1960) 5503.
19 C. K. Ingold, J. A. Jessop, K. I. Kuriyan and A. M. M. Mandour, *J. Chem. Soc.*, (1933) 533.

20 C. K. INGOLD AND J. A. JESSOP, *J. Chem. Soc.*, (1930) 708.
21 S. J. CRISTOL, *J. Am. Chem. Soc.* 67 (1945) 1494.
22 C. H. DEPUY AND D. H. FROEMSDORF, *J. Am. Chem. Soc.*, 79 (1957) 3710.
23 C. H. DEPUY AND C. A. BISHOP, *J. Am. Chem. Soc.*, 82 (1960) 2535.
24 N. B. CHAPMAN AND J. L. LEVY, *J. Chem. Soc.*, (1952) 1673.
25 D. J. CRAM, *J. Am. Chem. Soc.*, 74 (1952) 2149.
26 C. H. DEPUY AND C. A. BISHOP, *J. Am. Chem. Soc.*, 82 (1960) 2532.
27 D. Y. CURTIN AND D. B. KELLOM, *J. Am. Chem. Soc.*, 75 (1953) 6011.
28 E. D. HUGHES, *J. Am. Chem. Soc.*, 57 (1935) 708.
29 E. D. HUGHES AND B. J. MCNULTY, *J. Chem. Soc.*, (1937) 1283.
30 K. A. COOPER, E. D. HUGHES, C. K. INGOLD AND B. J. MCNULTY, *J. Chem. Soc.*, (1948) 2038.
31 W. HUCKEL, W. TAPPE AND G. LEGUTKE, *Ann. Chem.*, 543 (1940) 191.
32 E. D. HUGHES AND J. WILBY, *J. Chem. Soc.*, (1960) 4094.
33 E. D. HUGHES, C. K. INGOLD AND C. S. PATEL, *J. Chem. Soc.*, (1933) 526.
34 D. J. MCLENNAN, *Quart. Rev. (London)*, 21 (1967) 490.
35 E. M. HODNETT AND J. FLYNN, *J. Am. Chem. Soc.*, 79 (1957) 2300.
36 P. S. SKELL AND C. R. HAUSER, *J. Am. Chem. Soc.*, 67 (1945) 1661.
37 D. V. BANTHORPE AND J. H. RIDD, *Proc. Chem. Soc.*, (1964) 365.
38 C. R. HAUSER, D. G. HILL, B. STEWART, S. W. KANTOR AND W. A. JUDGE, *J. Am. Chem. Soc.*, 76 (1954) 5129.
39 S. M. LUCK, D. G. HILL, A. T. STEWART, JR. AND C. R. HAUSER, *J. Am. Chem. Soc.*, 81 (1959) 2784.
40 W. H. SAUNDERS, JR. AND D. H. EDISON, *J. Am. Chem. Soc.*, 82 (1960) 138.
41 V. J. SHINER, JR. AND M. L. SMITH, *J. Am. Chem. Soc.*, 80 (1958) 4095.
42 L. A. PAQUETTE AND L. D. WISE, *J. Org. Chem.*, 30 (1965) 228.
43 Y. ISKANDER AND Y. RIAD, *J. Chem. Soc.*, (1961) 223.
44 R. BRESLOW, *Tetrahedron Letters*, (1964) 399.
45 A. F. COCKERILL, S. ROTTSCHAEFER AND W. H. SAUNDERS, JR., *J. Am. Chem. Soc.*, 89 (1967) 901.
46 J. HINE, R. WEISBOECK AND O. B. RAMSAY, *J. Am. Chem. Soc.*, 83 (1961) 1222.
47 A. STREITWIESER, JR., A. P. MARCHAND AND A. H. PUDJAATMAKA, *J. Am. Chem. Soc.*, 89 (1967) 693.
48 J. HINE, R. WEISBOECK AND R. G. GHIRARDELLI, *J. Am. Chem. Soc.*, 83 (1961) 1219.
49 J. HINE, C. H. THOMAS AND S. J. EHRENSON, *J. Am. Chem. Soc.*, 77 (1955) 3886.
50 J. HINE, S. J. EHRENSON AND W. H. BRADER, *J. Am. Chem. Soc.*, 78 (1956) 2282.
51 W. T. MILLER, JR., E. W. FAGER AND P. H. GRISWALD, *J. Am. Chem. Soc.*, 70 (1948) 431.
52 J. HINE AND W. H. BRADER, *J. Am. Chem. Soc.*, 75 (1953) 3964.
53 W. H. SAUNDERS, JR. AND M. R. SCHREIBER, *Chem. Commun.*, (1966) 145.
54 R. P. BELL, *The Proton in Chemistry*, Cornell University Press, Ithaca, New York, 1959, p. 134.
55 A. A. FROST AND R. G. PEARSON, *Kinetics and Mechanism*, 2nd edn, Wiley, London, 1961, p. 213.
56 K. WIBERG, *Physical Organic Chemistry*, Wiley, London, 1964, p. 397.
57 B. D. ENGLAND AND D. J. MCLENNAN, *J. Chem. Soc. B*, (1966) 696.
58 R. F. HUDSON AND G. KLOPMAN, *J. Chem. Soc.*, (1964) 5.
59 R. E. LUTZ, D. F. HINCKLEY AND R. H. JORDAN, *J. Am. Chem. Soc.*, 73 (1951) 4647.
60 T. I. CROWELL, A. A. WALL, R. T. KEMP AND R. E. LUTZ, *J. Am. Chem. Soc.*, 85 (1963) 2521; 90 (1968) 4638.
61 S. PATAI, S. WEINSTEIN AND Z. RAPPOPORT, *J. Chem. Soc.*, (1962) 1741.
62 J. HINE AND L. A. KAPLAN, *J. Am. Chem. Soc.*, 82 (1960) 2915.
63 M. A. PAUL AND F. A. LONG, *Chem. Rev.*, 57 (1957) 1.
64 K. BOWDEN, *Chem. Rev.*, 66 (1966) 119.
65 C. H. ROCHESTER, *Quart. Rev. (London)*, 20 (1966) 511.
66 M. ANBAR, M. BOBTELSKY, D. SAMUEL, B. SILVER AND G. YAGIL, *J. Am. Chem. Soc.*, 85 (1963) 2380.

67 A. F. COCKERILL, *J. Chem. Soc. B*, (1967) 964.
68 R. STEWART AND J. P. O'DONNELL, *Can. J. Chem.*, 42 (1964) 1681, 1694.
69 R. STEWART AND D. DOLMAN, *Can. J. Chem.*, 45 (1967) 911.
70 M. J. JORGENSON AND D. R. HARTTER, *J. Am. Chem. Soc.*, 85 (1963) 878.
71 N. C. DENO, J. J. JARUZELSKI AND A. SCHREISHEIM, *J. Am. Chem. Soc.*, 77 (1955) 3044.
72 K. YATES, J. B. STEVENS AND A. R. KATRITZKY, *Can. J. Chem.*, 42 (1964) 1957.
73 E. M. ARNETT AND G. W. MACH, *J. Am. Chem. Soc.*, 86 (1964) 2671.
74 R. L. HINMAN AND J. LANG, *J. Am. Chem. Soc.*, 86 (1964) 3796.
75 K. BOWDEN AND R. STEWART, *Tetrahedron*, 21 (1965) 265.
76 K. BOWDEN AND A. F. COCKERILL, *Chem. Commun.*, (1967) 989.
77 J. H. RIDD AND R. A. MORE O'FERRALL, *J. Chem. Soc.*, (1963) 5030.
78 (a) A. ALLISON, C. BUMFORD AND J. H. RIDD, *Chem. Ind. (London)*, (1958) 718.
(b) J. H. RIDD AND R. A. MORE O'FERRALL, *J. Chem. Soc.*, (1963) 5035.
79 S. J. CRISTOL, *J. Am. Chem. Soc.*, 69 (1947) 338.
80 D. J. CRAM, D. A. SCOTT AND W. D. NIELSEN, *J. Am. Chem. Soc.*, 83 (1961) 3696.
81 C. J. M. STIRLING, *Chem. Ind. (London)*, (1960) 933.
82 P. S. SKELL AND J. H. MACNAMARA, *J. Am. Chem. Soc.*, 79 (1957) 85.
83 H. BOHME AND P. HELLER, *Chem. Ber.*, 86 (1953) 785.
84 H. BOHME AND P. HELLER, *Chem. Ber.*, 86 (1953) 443.
85 G. WITTIG AND R. POLSTER, *Ann. Chem.*, 599 (1956) 13.
86 J. RABIANT AND G. WITTIG, *Bull. Soc. Chim. France*, (1957) 798.
87 G. WITTIG AND R. POLSTER, *Ann. Chem.*, 612 (1958) 102.
88 G. WITTIG AND T. F. BURGER, *Ann. Chem.*, 632 (1960) 85.
89 A. C. COPE, N. A. LEBEL, P. T. MOORE AND W. R. MOORE, *J. Am. Chem. Soc.*, 83 (1961) 3861.
90 G. AYREY, E. BUNCEL AND A. N. BOURNS, *Proc. Chem. Soc.*, (1961) 458.
91 J. L. COKE AND M. P. COOKE, *J. Am. Chem. Soc.*, 89 (1967) 2779.
92 A. C. COPE AND A. S. MEHTA, *J. Am. Chem. Soc.*, 85 (1963) 1949.
93 S. OAE, W. TAGAKI AND A. OHNO, *J. Am. Chem. Soc.*, 83 (1961) 5036.
94 S. AŠPERGER, N. ILAKOVAC AND D. PAVLOVIĆ, *J. Am. Chem. Soc.*, 83 (1961) 5032.
95 A. N. BOURNS AND P. J. SMITH, *Proc. Chem. Soc.*, (1964) 366.
96 W. H. SAUNDERS, JR. AND D. PAVLOVIĆ, *Chem. Ind. (London)*, (1962) 180.
97 V. FRANZEN AND C. MERTZ, *Chem. Ber.*, 93 (1960) 2819.
98 V. FRANZEN AND H. J. SCHMIDT, *Chem. Ber.*, 94 (1961) 2937.
99 D. G. HILL, W. A. JUDGE, P. S. SKELL, S. W. KANTOR AND C. R. HAUSER, *J. Am. Chem. Soc.*, 74 (1952) 5599.
100 V. J. SHINER, JR. AND M. L. SMITH, *J. Am. Chem. Soc.*, 80 (1958) 4095.
101 L. FRIEDMAN AND J. G. BERGER, *J. Am. Chem. Soc.*, 83 (1961) 492.
102 W. KIRMSE AND W. VON E. DOERING, *Tetrahedron*, 11 (1960) 266.
103 V. FRANZEN, H. J. SCHMIDT AND C. MERTZ, *Chem. Ber.*, 94 (1961) 2942.
104 D. J. CRAM, C. H. DEPUY AND F. D. GREENE, *J. Am. Chem. Soc.*, 78 (1956) 790.
105 J. F. BUNNETT, *Angew, Chem. Intern. Ed. Engl.*, 1 (1962) 225.
106 D. V. BANTHORPE, in *Studies on Chemical Structure and Reactivity*, J. H. RIDD (Ed.), presented to Sir C. K. Ingold, Methuen, London, 1966, Chap. 3.
107 C. K. INGOLD, *Proc. Chem. Soc.*, (1962) 265.
108 W. H. SAUNDERS, JR., Kinetic Isotope Effects, in *Survey of Progress in Chemistry*, Academic Press, New York, 1966, Vol. 3, p. 109.
109 H. SIMON AND D. PALM, *Angew. Chem. Intern. Ed. Engl.*, 5 (1966) 920.
110 J. BIGELEISEN AND M. G. MAYER, *J. Chem. Phys.*, 15 (1947) 261.
111 J. BIGELEISEN AND M. WOLFSBERG, *Advan. Chem. Phys.*, 1 (1958) 15.
112 L. MELANDER, *Isotope Effects on Reaction Rates*, Ronald Press, New York, 1960.
113 C. J. COLLINS, *Advan. Phys. Org. Chem.*, 2 (1964) 60.
114 H. ZOLLINGER, *Advan. Phys. Org. Chem.*, 2 (1964) 163; *Angew. Chem.*, 70 (1958) 204.
115 E. A. HALEVI, *Progr. Phys. Org. Chem.*, 1 (1964) 109.
116 J. BIGELEISEN, *Pure Appl. Chem.*, 8 (1964) 217.

117 S. Glasstone, K. J. Laidler and H. Eyring, *The Theory of Rate Processes*, McGraw-Hill, New York, 1941.
118 K. Wiberg, *Chem. Rev.*, 55 (1955) 713.
119 V. J. Shiner, Jr. and M. L. Smith, *J. Am. Chem. Soc.*, 83 (1961) 593.
120 D. Bethell and A. F. Cockerill, *J. Chem. Soc. B*, (1966) 917.
121 F. H. Westheimer, *Chem. Rev.*, 61 (1961) 265.
122 R. P. Bell and J. E. Crooks, *Proc. Roy. Soc. (London), Ser. A*, 286 (1965) 285.
123 R. P. Bell and D. M. Goodall, *Proc. Roy. Soc. (London), Ser. A*, 294 (1966) 273.
124 F. A. Long and J. L. Longridge, *J. Am. Chem. Soc.*, 89 (1967) 1292.
125 A. V. Willi and M. Wolfsberg, *Chem. Ind. (London)*, (1964) 2097.
126 V. J. Shiner, Jr. and B. Martin, *Pure Appl. Chem.*, 8 (1964) 371.
127 G. Swain, E. C. Stivers, J. F. Reuwer, Jr. and L. J. Schaad, *J. Am. Chem. Soc.*, 80 (1958) 5885.
128 E. M. Hodnett and J. J. Sparapany, *Pure Appl. Chem.*, 8 (1964) 385.
129 J. Bigeleisen, *Science*, 110 (1949) 14.
130 Ref. 56, p. 355.
131 H. Simon and G. Mullhöfer, *Chem. Ber.*, 97 (1964) 2202.
132 H. Simon and G. Mullhöfer, *Pure Appl. Chem.*, 8 (1964) 379.
133 A. V. Willi, *Helv. Chim. Acta.*, 49 (1966) 1725.
134 J. F. Bunnett, G. T. Davis and H. Tanida, *J. Am. Chem. Soc.*, 84 (1962) 1606.
135 D. J. Cram and A. S. Winsgrove, *J. Am. Chem. Soc.*, 86 (1964) 5490.
136 W. H. Saunders, Jr., D. Bushmann and A. F. Cockerill, *J. Am. Chem. Soc.*, 90 (1968) 1775.
137 Ref. 54, Chap. 11.
138 E. S. Lewis and L. H. Funderburk, *J. Am. Chem. Soc.*, 89 (1967) 2322.
139 D. Bethell and A. F. Cockerill, *J. Chem. Soc. B*, (1966) 913.
140 A. V. Willi, *J. Phys. Chem.*, 70 (1966) 2705.
141 R. A. More O'Ferrall and J. Kouba, *J. Chem. Soc. B*, (1967) 985.
142 P. E. Yankwich, A. L. Promislow and R. F. Nystrom, *J. Am. Chem. Soc.*, 76 (1954) 5893.
143 A. F. Cockerill and W. H. Saunders, Jr., *J. Am. Chem. Soc.*, 89 (1967) 4985.
144 W. H. Stevens and R. W. Attree, *Can. J. Res.*, B 27 (1949) 807.
145 W. H. Saunders, Jr. and S. Ašperger, *J. Am. Chem. Soc.*, 79 (1957) 1612.
146 G. A. Ropp, *J. Am. Chem. Soc.*, 82 (1960) 842.
147 M. L. Bender and D. F. Hoeg, *J. Am. Chem. Soc.*, 79 (1957) 5649.
148 M. L. Bender and G. J. Buist, *J. Am. Chem. Soc.*, 80 (1958) 4304.
149 A. J. Kresge, N. N. Lechtin and K. N. Roa, *J. Am. Chem. Soc.*, 85 (1963) 1210.
150 W. H. Saunders, Jr. and S. E. Zimmerman, *J. Am. Chem. Soc.*, 86 (1964) 3789.
151 W. H. Saunders, Jr., A. F. Cockerill, S. Ašperger, L. Klasinc and D. Stefanović, *J. Am. Chem. Soc.*, 88 (1966) 848.
152 G. Ayrey, E. Buncel and A. N. Bourns, *Proc. Chem. Soc.*, (1961) 458.
153 G. Ayrey, A. N. Bourns and V. A. Vyas, *Can. J. Chem.*, 41 (1963) 1759.
154 A. Streitwieser, Jr., R. H. Jagow, R. C. Fahey and S. Suzuki, *J. Am. Chem. Soc.*, 80 (1958) 2326.
155 S. Ašperger, L. Klasinc and D. Pavlović, *Croat. Chem. Acta*, 36 (1964) 159.
156 L. S. Bartell, *J. Am. Chem. Soc.*, 83 (1961) 3567.
157 H. C. Brown and G. J. McDonald, *J. Am. Chem. Soc.*, 88 (1966) 2514.
158 G. J. Karabatsos, G. C. Sonnichsen, C. G. Papaioannon, S. E. Scheppele and R. L. Shone, *J. Amer. Chem. Soc.*, 89 (1967) 463.
159 E. A. Halevi, M. Nussim and A. Ron, *J. Chem. Soc.*, (1963) 866.
160 E. S. Lewis and C. E. Boozer, *J. Am. Chem. Soc.*, 74 (1952) 6306.
161 V. J. Shiner, Jr., *J. Amer. Chem. Soc.*, 75 (1953) 2925.
162 V. J. Shiner, Jr., *Tetrahedron*, 5 (1959) 243.
163 A. C. Cope, N. A. LeBel, H. H. Lee and W. R. Moore, *J. Am. Chem. Soc.*, 79 (1957) 4720.

164 W. H. SAUNDERS, JR., Investigation of Rates and Mechanisms of Reactions, in Part 1, of *Technique of Organic Chemistry*, S. L. FRIESS, E. S. LEWIS AND A. WEISSBERGER (Eds.), Wiley-Interscience, New York, 1961, Vol. 8.
165 L. J. STEFFA AND E. R. THORNTON, *J. Am. Chem. Soc.*, 89 (1967) 6149.
166 P. M. LAUGHTON AND R. E. ROBERTSON, *Can. J. Chem.*, 35 (1957) 1319.
167 L. P. HAMMETT, *Physical Organic Chemistry*, McGraw-Hill, New York, 1940, Chap. 7.
168 H. H. JAFFÉ, *Chem. Rev.*, 53 (1953) 191.
169 H. VAN BEKKUM, P. E. VERKADE AND B. M. WEPSTER, *Rec. Trav. Chim.*, 78 (1959) 815.
170 A. I. BIGGS AND R. A. ROBINSON, *J. Chem. Soc.*, (1961) 388.
171 G. S. HAMMOND, *J. Am. Chem. Soc.*, 77 (1955) 334.
172 J. ZÁVADA, J. KRUPIČKA AND J. SICHER, *Chem. Commun.*, (1967) 66.
173 C. A. COULSON, *Valence*, Oxford University Press, Oxford, England, 1952, p. 206.
174 M. SHIMA, D. N. BHATTACHARYYA, J. SMID AND M. SZWARC, *J. Am. Chem. Soc.*, 85 (1963) 1306.
175 A. STREITWIESER, JR. AND A. H. PUDJAATMAKA, unpublished results quoted by A. STREITWIESER, JR. AND J. H. HAMMONS, in *Progr. Phys. Org. Chem.*, 2 (1965) 57.
176 K. BOWDEN, A. F. COCKERILL AND J. R. GILBERT, *J. Chem. Soc. B*, (1970) 179.
177 L. F. BLACKWELL, A. FISCHER AND J. VAUGHAN, *J. Chem. Soc. B*, (1967) 1084.
178 C. H. DEPUY, G. F. MORRIS, J. S. SMITH AND R. J. SMAT, *J. Am. Chem. Soc.*, 87 (1965) 2421.
179 W. H. SAUNDERS, JR. AND R. A. WILLIAMS, *J. Am. Chem. Soc.*, 79 (1957) 3712.
180 J. N. BRÖNSTED AND K. PEDERSEN, *J. Physik. Chem.*, 108 (1924) 185.
181 Ref. 54, Chap. 10.
182 R. F. HUDSON, *Chimia (Aarau)*, 16 (1962) 173.
183 G. SCHWARZENBACH AND E. RUDIN, *Helv. Chim. Acta.*, 22 (1939) 360.
184 D. J. MCLENNAN, *J. Chem. Soc. B*, (1966) 709.
185 D. J. MCLENNAN, *J. Chem. Soc. B*, (1966) 705.
186 H. M. R. HOFFMANN, *J. Chem. Soc.*, (1965) 6753.
187 H. M. R. HOFFMANN, *J. Chem. Soc.*, (1965) 6762.
188 S. WINSTEIN, A. H. FAINBERG AND E. GRUNWALD, *J. Am. Chem. Soc.*, 79 (1957) 4151.
189 H. C. BROWN, The Transition State, *Chem. Soc., Spec. Publ. No. 16*, 1962, p. 140.
190 R. G. PEARSON AND J. SONGSTAD, *J. Am. Chem. Soc.*, 89 (1967) 1827.
191 G. M. FRASER AND H. M. R. HOFFMANN, *J. Chem. Soc. B*, (1967) 265.
192 H. M. R. HOFFMANN, *Tetrahedron Letters*, (1967) 4393.
193 R. F. HUDSON, personal discussion.
194 C. D. RITCHIE AND W. J. SAGER, *Progr. Phys. Org. Chem.*, 1 (1964) 323.
195 J. W. HILL AND A. FRY, *J. Am. Chem. Soc.*, 84 (1962) 2763.
196 J. ZÁVADA, M. SVOBODA AND J. SICHER, *Tetrahedron Letters*, (1966) 1627.
197 D. L. GRIFFITH AND B. SINGERMAN, *Chem. Commun.*, (1968) 438.
198 (a) W. H. SAUNDERS, JR. AND A. KATZ, *J. Am. Chem. Soc.*, 91 (1969) 4469; (b) W. H. SAUNDERS, JR. AND K. BROWN, unpublished results; (c) W. H. SAUNDERS, JR. AND D. S. BAILEY, *Chem. Commun.*, (1968) 1598.
199 A. F. COCKERILL, unpublished results.
200 W. KLYNE AND V. PRELOG, *Experientia*, 16 (1960) 521.
201 P. F. FRANKLAND, *J. Chem. Soc.*, (1912) 654.
202 E. L. ELIEL, N. L. ALLINGER, S. J. ANGYAL AND G. A. MORRISON, *Conformational Analysis*, Interscience, New York, 1965, p. 482.
203 G. H. STEWART AND H. EYRING, *J. Chem. Educ.*, 35 (1958) 550.
204 K. FUKUI AND H. FUJIMOTO, *Tetrahedron Letters*, (1965) 4303.
205 J. HINE, *J. Am. Chem. Soc.*, 88 (1966) 5525.
206 P. PFEIFFER, *Z. Physik. Chem.* (Leipzig), 48 (1904) 40.
207 F. G. BORDWELL AND P. S. LANDIS, *J. Am. Chem. Soc.*, 79 (1957) 1593.
208 J. ZAVADA AND J. SICHER, *Proc. Chem. Soc.*, (1963) 96.
209 J. ZÁVADA AND J. SICHER, *Collection Czech. Chem. Commun.*, 30 (1965) 438.
210 S. J. CRISTOL AND P. PAPPAS, *J. Org. Chem.*, 28 (1963) 2066.

211 E. D. Hughes and J. C. Maynard, *J. Chem. Soc.*, (1960) 4087.
212 R. D. Haworth, J. McKenna and R. G. Powell, *J. Chem. Soc.*, (1953) 1110.
213 S. Winstein and N. J. Holness, *J. Am. Chem. Soc.*, 77 (1955) 5562.
214 N. B. Chapman and R. E. Parker, *J. Chem. Soc.*, (1960) 3634.
215 B. Cross and G. H. Whitham, *J. Chem. Soc.*, (1960) 3892.
216 D. V. Stolow and W. Maya, *J. Am. Chem. Soc.*, 81 (1959) 330.
217 W. M. Jones, T. G. Squires and M. Lynn, *J. Am. Chem. Soc.*, 89 (1967) 318.
218 N. O. Brace, *J. Am. Chem. Soc.*, 86 (1964) 2428.
219 S. J. Cristol, *J. Am. Chem. Soc.*, 71 (1949) 1894.
220 E. D. Hughes, C. K. Ingold and R. Pasternak, *J. Chem. Soc.*, (1953) 3832.
221 S. J. Cristol, N. L. Hause and J. S. Meek, *J. Am. Chem. Soc.*, 73 (1951) 674.
222 S. J. Cristol and D. D. Fix, *J. Am. Chem. Soc.*, 75 (1953) 2647.
223 Ref. 8, p. 321.
224 J. Hine, R. D. Weimer, Jr., P. B. Langford and O. B. Ramsey, *J. Am. Chem. Soc.*, 88 (1966) 5522.
225 J. Hine and P. B. Langford, *J. Org. Chem.*, 27 (1962) 4149.
226 E. D. Hughes, C. K. Ingold and J. B. Rose, *J. Chem. Soc.*, (1953) 3839.
227 R. T. Arnold and P. N. Richardson, *J. Am. Chem. Soc.*, 76 (1954) 3649.
228 J. Weinstock and F. G. Bordwell, *J. Am. Chem. Soc.*, 77 (1955) 6706.
229 S. J. Cristol and F. R. Stermitz, *J. Am. Chem. Soc.*, 82 (1960) 4692.
230 J. Weinstock, R. G. Pearson and F. G. Bordwell, *J. Am. Chem. Soc.*, 76 (1954) 4748.
231 J. Weinstock, R. G. Pearson and F. G. Bordwell, *J. Am. Chem. Soc.*, 78 (1956) 3468, 3473.
232 J. Weinstock, J. L. Bernardi and F. G. Bordwell, *J. Am. Chem. Soc.*, 80 (1958) 4961.
233 F. G. Bordwell, R. L. Arnold and F. B. Biranowski, *J. Org. Chem.*, 28 (1963) 2496.
234 K. Jewers and J. McKenna, *J. Chem. Soc.*, (1958) 2209.
235 G. B. Kistiakowsky *et al.*, *J. Am. Chem. Soc.*, 58 (1936) 137; 59 (1937) 831.
236 H. C. Brown, R. S. Fletcher and R. B. Johanneson, *J. Am. Chem. Soc.*, 73 (1951) 212.
237 S. J. Cristol, F. R. Stermitz and P. S. Ramey, *J. Am. Chem. Soc.*, 78 (1956) 4939.
238 P. B. D. de la Mare, N. V. Klassen and R. Koenigsberger, *J. Chem. Soc.*, (1961) 5285.
239 S. J. Cristol and N. L. Hause, *J. Am. Chem. Soc.*, 74 (1952) 2193.
240 S. J. Cristol and R. P. Arganbright, *J. Am. Chem. Soc.*, 79 (1957) 3441.
241 N. A. LeBel, P. D. Beirne, E. R. Karger, J. C. Powers and P. M. Subramanian, *J. Am. Chem. Soc.*, 85 (1963) 3199.
242 N. A. LeBel, P. D. Beirne and P. M. Subramanian, *J. Am. Chem. Soc.*, 86 (1964) 4144.
243 H. Kwart, T. Takeshita and J. L. Nyce, *J. Am. Chem. Soc.*, 86 (1964) 2606.
244 J. K. Stille, F. M. Sonnenberg and T. H. Kinstle, *J. Am. Chem. Soc.*, 88 (1966) 4922.
245 S. J. Cristol and E. F. Hoegger, *J. Am. Chem. Soc.*, 79 (1957) 3438.
246 J. L. Coke and M. P. Cooke, Jr., *J. Am. Chem. Soc.*, 89 (1967) 6701.
247 C. W. Bird, R. C. Cookson, J. Hudec and R. O. Williams, *J. Chem. Soc.*, (1963) 410.
248 A. C. Cope, E. Ciganek and N. A. LeBel, *J. Am. Chem. Soc.*, 81 (1959) 2799.
249 F. S. Fawcett, *Chem. Rev.*, 47 (1950) 219.
250 W. von E. Doering and E. F. Schoenewaldt, *J. Am. Chem. Soc.*, 73 (1951) 2333.
251 Ref. 10, p. 132.
252 A. C. Cope and E. R. Trumbull, *Org. Reactions*, 11 (1961) 317.
253 K. Ziegler and H. Wilms, *Ann Chem.*, 567 (1950) 1.
254 A. T. Blomquist, L. H. Liu and J. C. Bohrer, *J. Am. Chem. Soc.*, 74 (1952) 3643.
255 A. C. Cope, P. T. Moore and W. R. Moore, *J. Am. Chem. Soc.*, 82 (1960) 1744.
256 J. Sicher, The Stereochemistry of Many-Membered Rings, *Progr. Stereochem.*, 3 (1962) 210.
257 H. C. Brown and K. Ichikawa, *Tetrahedron*, 1 (1957) 221.
258 J. Sicher, J. Závada and J. Krupička, *Tetrahedron Letters*, (1966) 1619.
259 D. J. Cram and J. E. McCarty, *J. Am. Chem. Soc.*, 76 (1954) 5740.
260 J. Sicher and J. Závada, *Collection Czech. Chem. Commun.*, 32 (1967) 2122.
261 J. Sicher, J. Krupicka and J. Závada, *Collection Czech. Chem. Commun.*, 31 (1966) 4273.
262 L. Schotsmans, P. J. C. Fierens and T. Verlie, *Bull. Soc. Chim. Belges*, 61 (1952) 609.

263 J. Závada, M. Svoboda and J. Sicher, *Tetrahedron Letters*, (1967) 1627.
264 M. Pánková, J. Sicher and J. Závada, *Chem. Commun.*, (1967) 394.
265 D. H. Froemsdorf, W. Dowd, W. A. Grifford and S. Meyerson, *Chem. Commun.*, (1968) 449.
266 J. L. Coke, M. P. Cooke, Jr. and M. C. Mourning, *Tetrahedron Letters*, (1968) 2247.
267 J. L. Coke and M. P. Cooke, Jr., *Tetrahedron Letters*, (1968) 2253.
268 J. Závada and J. Sicher, *Collection Czech. Chem. Commun.*, 32 (1967) 3701.
269 M. Cocivera and S. Winstein, *J. Am. Chem. Soc.*, 85 (1963) 1702.
270 A. Streitwieser, Jr. and T. D. Walsh, *Tetrahedron Letters*, (1963) 27.
271 P. S. Skell and W. L. Hall, *J. Am. Chem. Soc.*, 85 (1963) 2851.
272 D. J. Cram and M. R. V. Sahyun, *J. Am. Chem. Soc.*, 85 (1963) 1257.
273 M. Svoboda, J. Závada and J. Sicher, *Collection Czech. Chem. Commun.*, 32 (1967) 2104.
274 D. J. Cram and J. Tadanier, *J. Am. Chem. Soc.*, 81 (1959) 2737.
275 S. Winstein and J. Takahashi, *Tetrahedron*, 2 (1958) 316.
276 V. J. Shiner, Jr. and J. G. Jewett, *J. Am. Chem. Soc.*, 87 (1965) 1382.
277 V. J. Shiner, Jr. and J. G. Jewett, *J. Am. Chem. Soc.*, 87 (1965) 1383.
278 J. B. Hendrickson, *J. Am. Chem. Soc.*, 83 (1961) 4537.
279 W. H. Saunders, Jr. and K. T. Finlay, *J. Am. Chem. Soc.*, 87 (1965) 1384.
280 H. C. Brown and I. Moritani, *J. Am. Chem. Soc.*, 77 (1955) 3623.
281 V. J. Shiner, Jr., *J. Am. Chem. Soc.*, 74 (1952) 5285.
282 E. D. Hughes, C. K. Ingold, S. Masterman and B. J. MacNulty, *J. Chem. Soc.*, (1940) 899.
283 K. A. Cooper, E. D. Hughes, C. K. Ingold, B. J. MacNulty and L. I. Woolf, *J. Chem. Soc.*, (1948) 2043.
284 S. Winstein, D. Darwish and N. J. Holness, *J. Am. Chem. Soc.*, 78 (1956) 2915.
285 J. Hayami, N. Uno and A. Kaji, *Tetrahedron Letters*, (1968) 1385.
286 C. D. Ritchie and R. E. Uschold, *J. Am. Chem. Soc.*, 90 (1968) 2821.
287 T. E. Hogen-Esch and J. Smid, *J. Am. Chem. Soc.*, 87 (1965) 669; 89 (1967) 2764.
288 E. D. Hughes, C. K. Ingold and V. J. Shiner, Jr., *J. Chem. Soc.*, (1953) 3827.
289 H. C. Brown and R. S. Fletcher, *J. Am. Chem. Soc.*, 71 (1949) 1845; 72 (1950) 1223.
290 H. C. Brown and M. Nakagawa, *J. Am. Chem. Soc.*, 77 (1955) 3610, 3614.
291 H. C. Brown and Y. Okamoto, *J. Am. Chem. Soc.*, 77 (1955) 3619.
292 K. A. Cooper, E. D. Hughes, C. K. Ingold, G. A. Maw and B. J. MacNulty, *J. Chem. Soc.*, (1948) 2049.
293 E. D. Hughes, C. K. Ingold and L. I. Woolf, *J. Chem. Soc.*, (1948) 2084.
294 M. L. Dhar, E. D. Hughes and C. K. Ingold, *J. Chem. Soc.*, (1948) 2065.
295 E. L. Eliel and R. S. Ro, *Tetrahedron*, 2 (1958) 353.
296 E. L. Eliel and R. G. Haber, *J. Am. Chem. Soc.*, 81 (1959) 1249.
297 P. B. D. de la Mare and C. A. Vernon, *J. Chem. Soc.*, (1956) 41.
298 N. H. Cromwell and A. Hassner, *J. Am. Chem. Soc.*, 79 (1957) 230; 80 (1958) 901; D. N. Kevill, and N. H. Cromwell, *ibid* 83 (1961) 3812, 3815; D. N. Kevill, E. D. Weiler and N. H. Cromwell, *ibid*, 88 (1966) 4489.
299 H. M. R. Hoffmann and G. M. Fraser, *J. Chem. Soc. B*, (1967) 425.
300 J. Csapilla, *Chimia* (Aarau), 18 (1964) 37.
301 G. Biale, S. Winstein, A. J. Parker and M. Ruane, *Tetrahedron Letters*, (1968) 2113.
302 A. W. Hofmann, *Ann. Chem.*, 78 (1851) 253; 79 (1851) 11.
303 A. Saytzeff, *Ann. Chem.*, 179 (1875) 296.
304 P. A. S. Smith and S. Frank, *J. Am. Chem. Soc.*, 74 (1952) 509.
305 H. J. Lucas, R. T. Dillon and W. G. Young, *J. Am. Chem. Soc.*, 52 (1930) 1949.
306 N. L. McNiven and J. Read, *J. Chem. Soc.*, (1952) 153.
307 W. H. Jones, *Science*, 118 (1953) 387.
308 W. H. Saunders, Jr., S. R. Fahrenholtz, E. A. Caress, J. P. Lowe and M. R. Schreiber, *J. Am. Chem. Soc.*, 87 (1965) 3401.
309 H. C. Brown, I. Moritani and Y. Okamoto, *J. Am. Chem. Soc.*, 78 (1956) 2193.
310 H. C. Brown and I. Moritani, *J. Am. Chem. Soc.*, 77 (1955) 3607.

311 H. C. Brown, I. Moritani and M. Nakagawa, *J. Am. Chem. Soc.*, 78 (1956) 2190.
312 R. B. Turner, D. E. Nettleton and M. Perelman, *J. Am. Chem. Soc.*, 80 (1958) 1430.
313 D. V. Banthorpe, E. D. Hughes and C. K. Ingold, *J. Chem. Soc.*, (1960) 4054.
314 M. L. Dhar, E. D. Hughes, C. K. Ingold, A. M. M. Mandour, G. A. Maw and L. I. Woolf, *J. Chem. Soc.*, (1948) 2093.
315 E. D. Hughes, C. K. Ingold, G. A. Maw and L. I. Woolf, *J. Chem. Soc.*, (1948) 2077.
316 R. W. Taft, in *Steric Effects in Organic Chemistry*, M. S. Newman (Ed.), Wiley, New York, 1956, p. 619.
317 M. J. S. Dewar and H. S. Schmeising, *Tetrahedron*, 5 (1959) 166; 11 (1960) 96.
318 H. C. Brown and M. Nakagawa, *J. Am. Chem. Soc.*, 78 (1956) 2197.
319 H. C. Brown and O. H. Wheeler, *J. Am. Chem. Soc.*, 78 (1956) 2199.
320 H. C. Brown and I. Moritani, *J. Am. Chem. Soc.*, 78 (1956) 2203.
321 H. C. Brown and R. L. Klimisch, *J. Am. Chem. Soc.*, 88 (1966) 1425.
322 H. C. Brown and R. L. Klimisch, *J. Am. Chem. Soc.*, 88 (1966) 1430.
323 H. C. Brown, D. L. Griffith and D. L. Meges, *Chem. Commun.*, (1968) 90.
324 E. D. Hughes, C. K. Ingold and G. A. Maw, *J. Chem. Soc.*, (1948) 2072.
325 W. H. Saunders, Jr., S. R. Fahrenholtz and J. P. Lowe, *Tetrahedron Letters*, (1960) 1.
326 A. J. Berlin and F. R. Jansen, *Chem. Ind. (London)*, (1960) 998.
327 J. Hine and O. B. Ramsey, *J. Am. Chem. Soc.*, 84 (1962) 973.
328 R. A. Bartsch and J. F. Bunnett, *J. Am. Chem. Soc.*, 90 (1968) 408.
329 M. Calligaria, G. Illuminati and G. Marino, *J. Am. Chem. Soc.*, 89 (1967) 3518.
330 A. K. Colter and R. D. Johnson, *J. Am. Chem. Soc.*, 84 (1962) 3289.
331 A. K. Colter and D. R. McKelvey, *Can. J. Chem.*, 43 (1965) 1282.
332 A. Schriesheim and C. A. Rowe, *J. Am. Chem. Soc.*, 84 (1962) 3160.
333 D. H. Froemdorf and M. E. McCain, *J. Am. Chem. Soc.*, 87 (1965) 3983.
334 D. H. Froemsdorf, M. E. McCain and W. W. Wilkinson, *J. Am. Chem. Soc.*, 87 (1965) 3984.
335 D. H. Froemsdorf, W. Dowd and K. E. Leimer, *J. Am. Chem. Soc.*, 88 (1966) 2345.
336 D. H. Froemsdorf and M. D. Robbins, *J. Am. Chem. Soc.*, 89 (1967) 1737.
337 D. H. Froemsdorf, unpublished results.
338 H. C. Brown and R. L. Klimisch, *J. Am. Chem. Soc.*, 87 (1965) 5517.
339 W. H. Saunders, Jr. and I. N. Feit, *Chem. Commun.*, (1967) 610.
340 R. B. Turner and R. H. Garner, *J. Am. Chem. Soc.*, 80 (1958) 1424.
341 C. G. Swain and E. R. Thornton, *J. Am. Chem. Soc.*, 84 (1962) 817.
342 Ref. 10, p. 170.
343 H. R. Nace, *Org. Reactions*, 12 (1961) 57.
344 C. H. DePuy and R. W. King, *Chem. Rev.*, 60 (1960) 431.
345 J. L. Holmes and A. Maccoll, *Proc. Chem. Soc.*, (1957) 175.
346 J. L. Holmes and A. Maccoll, *J. Chem. Soc.*, (1963) 5919.
347 E. D. Hughes and C. K. Ingold, *Quart. Rev. (London)*, 6 (1952) 34.
348 C. G. Swain, R. F. Bader, R. M. Esteve, Jr. and R. N. Griffin, *J. Am. Chem. Soc.*, 83 (1961) 1951.
349 C. G. Swain, R. A. Wiles and R. F. Bader, *J. Am. Chem. Soc.*, 83 (1961) 1945.
350 D. J. Cram, *J. Am. Chem. Soc.*, 71 (1949) 3883.
351 E. R. Alexander and A. Mudrak, *J. Am. Chem. Soc.*, 72 (1950) 3194.
352 E. R. Alexander and A. Mudrak, *J. Am. Chem. Soc.*, 73 (1951) 59.
353 D. H. R. Barton, *J. Chem. Soc.*, (1949) 2174.
354 W. J. Bailey and R. A. Baylouny, *J. Am. Chem. Soc.*, 81 (1959) 2126.
355 F. G. Bordwell and P. S. Landis, *J. Am. Chem. Soc.*, 80 (1958) 2450.
356 F. G. Bordwell and P. S. Landis, *J. Am. Chem. Soc.*, 80 (1958) 6379.
357 E. R. Alexander and A. Mudrak, *J. Am. Chem. Soc.*, 72 (1950) 1810.
358 R. T. Arnold, G. G. Smith and R. M. Dodson, *J. Org. Chem.*, 15 (1950) 1256.
359 W. J. Bailey and L. Nicholas, *J. Org. Chem.*, 21 (1956) 854.
360 A. T. Blomquist and P. R. Taussig, *J. Am. Chem. Soc.*, 79 (1957) 3505.
361 F. G. Bordwell and P. S. Landis, *J. Am. Chem. Soc.*, 80 (1958) 6383.

362 W. J. Bailey and C. N. Bird, *A.C.S. Meeting, Miami, April, 1957*, 44–0.
363 A. C. Cope and N. A. LeBel, *J. Am. Chem. Soc.*, 82 (1960) 4656.
364 A. C. Cope, D. C. McLean and N. A. Nelson, *J. Am. Chem. Soc.*, 77 (1955) 1628.
365 D. H. R. Barton, A. J. Head and R. J. Williams, *J. Chem. Soc.*, (1952) 453.
366 T. Bamkole, *Ph.D. Thesis*, University of London, 1964.
367 R. C. Bicknell, *Ph.D. Thesis*, University of London, 1962.
368 D. H. Froemsdorf, C. H. Collins, G. S. Hammond and C. H. DePuy, *J. Am. Chem. Soc.*, 81 (1959) 643.
369 W. O. Haag and H. Pines, *J. Org. Chem.*, 24 (1959) 877.
370 J. P. W. Houtman, J. van Steenis and P. M. Heertjes, *Rec. Trav. Chim.*, 65 (1946) 781.
371 R. A. Benkeser, J. J. Hazdra and M. L. Burrous, *J. Am. Chem. Soc.*, 81 (1959) 5374.
372 W. J. Bailey and C. King, *J. Org. Chem.*, 21 (1956) 858.
373 C. H. DePuy and R. E. Leary, *J. Am. Chem. Soc.*, 79 (1957) 3705.
374 A. Maccoll and R. H. Stone, *J. Chem. Soc.*, (1961) 2756.
375 A. Maccoll, *J. Chem. Soc.*, (1958) 3398.
376 J. C. Sheer, E. C. Kooyman and F. L. J. Sixma, *Rec. Trav. Chim.*, 82 (1963) 1123.
377 P. J. Thomas, *J. Chem. Soc.*, (1959) 1192.
378 A. T. Blades, *Can. J. Chem.*, 36 (1958) 1043.
379 A. T. Blades and G. W. Murphy, *J. Am. Chem. Soc.*, 74 (1952) 6219.
380 A. Maccoll and P. J. Thomas, *J. Chem. Soc.*, (1958) 979.
381 G. D. Harden and A. Maccoll, *J. Chem. Soc.*, (1955) 2454.
382 D. H. R. Barton and K. E. Howlett, *J. Chem. Soc.*, (1949) 165.
383 A. Maccoll, in *Theoretical Organic Chemistry*, papers presented to the *Kekulé Symposium, London, 1958*, Butterworths Scientific Publications, London, 1959, p. 230.
384 D. H. R. Barton and A. J. Head, *Trans. Faraday. Soc.*, 46 (1950) 114.
385 D. H. R. Barton and P. F. Onyon, *Trans. Faraday. Soc.*, 45 (1949) 725.
386 G. D. Harden and A. Maccoll, *J. Chem. Soc.*, (1959) 1197.
387 D. H. R. Barton, A. J. Head and R. J. Williams, *J. Chem. Soc.*, (1951) 2039.
388 K. E. Howlett, *J. Chem. Soc.*, (1952) 4487.
389 A. Maccoll, *Chem. Soc. Spec. Publ., No. 16*, 1962, p. 158.
390 A. Maccoll, *Advan. Phys. Org. Chem.*, 3 (1965) 91.
391 A. Maccoll, *J. Chem. Soc.*, (1955) 965.
392 M. Szwarc, B. N. Ghosh and A. H. Sehon, *J. Chem. Phys.*, 18 (1950) 1142.
393 A. Maccoll, in *Theoretical Organic Chemistry*, Butterworths, London, 1959, p. 230.
394 G. T. Genaux, F. Kern and W. D. Walters, *Acta Chim. Acad. Sci. Hung.*, 75 (1963) 6191.
395 M. N. Das and W. D. Walters, *Z. Physik. Chem.*, *(Frankfurt)* 15 (1958) 22.
396 A. Maccoll and E. S. Swinbourne, *J. Chem. Soc.*, (1964) 149.
397 P. T. Good, *Ph. D. Thesis*, University of London, 1956.
398 A. T. Blades and P. W. Gilderson, *Can. J. Chem.*, 38 (1960) 1401.
399 A. T. Blades, P. W. Gilderson and M. G. H. Wallbridge, *Can. J. Chem.*, 40 (1962) 1527.
400 A. T. Blades, P. W. Gilderson and M. G. H. Wallbridge, *Can. J. Chem.* 40 (1962) 1533.
401 C. K. Ingold, *Proc. Chem. Soc.*, (1957) 279.
402 E. M. Bilger and H. Hibbert, *J. Am. Chem. Soc.*, 58 (1936) 823.
403 W. J. Bailey and W. F. Hale, *J. Am. Chem. Soc.*, 81 (1959) 647.
404 E. U. Emovan and A. Maccoll, *J. Chem. Soc.*, (1962) 335.
405 G. G. Smith and W. H. Wetzel, *J. Am. Chem. Soc.*, 73 (1959) 975.
406 E. U. Emovan, *J. Chem. Soc.*, (1963) 1246.
407 G. G. Smith, F. D. Bagley and R. Taylor, *J. Am. Chem. Soc.*, 83 (1961) 3647.
408 R. Taylor, G. G. Smith and W. H. Wetzel, *J. Am. Chem. Soc.*, 84 (1962) 4817.
409 G. G. Smith, D. A. K. Jones and D. F. Brown, *J. Org. Chem.*, 28 (1963) 403.
410 A. C. Cope and E. M. Acton, *J. Am. Chem. Soc.*, 80 (1958) 355.
411 J. Crosby and C. J. M. Stirling, *J. Am. Chem. Soc.*, 90 (1968) 6869.

412 P. G. Stevens and J. H. Richmond, *J. Am. Chem. Soc.*, 63 (1941) 3132.
413 P. D. Bartlett, *Organic Chemistry, an Advanced Treatise*, H. Gilman (Ed.), Wiley, New York, 1953, Vol. 3, p. 53.
414 E. Salomaa, *Ann. Acad. Sci. Fennicae, ser. A II*, 94 (1959) 1.
415 D. S. Tarbell and D. P. Harnish, *Chem. Rev.*, 49 (1951) 60.
416 R. F. Bader and A. N. Bourns, *Can. J. Chem.*, 39 (1961) 348.
417 G. L. O'Connor and H. R. Nace, *J. Am. Chem. Soc.*, 74 (1952) 5454; 75 (1953) 2118.
418 H. R. Nace, D. Manly and S. Fusco, *J. Org. Chem.*, 23 (1958) 687.
419 D. J. Cram, M. R. V. Sahyun and G. R. Knox, *J. Am. Chem. Soc.*, 84 (1962) 1734.
420 V. N. Ipatieff, *Catalytic Reactions at High Pressures and Temperatures*, Macmillan, New York, 1936, p. 60.
421 M. E. Winfield, in *Catalysis*, P. H. Emmett (Ed.), Reinhold, New York, 1960, Vol. VII, p. 93.
422 K. V. Topchieva, K. Yun Pin and J. V. Smirnova, *Advan. Catalysis*, 9 (1957) 799.
423 H. Pines and L. A. Schaap, *Advan. Catalysis*, 12 (1960) 117.
424 H. Pines and J. Manassen, *Advan. Catalysis*, 16 (1966) 49.
425 Y. Trambouze, *Compt. Rend.*, 236 (1953) 1261.
426 A. N. Webb, *Ind. Eng. Chem.*, 49 (1957) 261.
427 H. Pines and W. O. Haag, *J. Am. Chem. Soc.*, 82 (1960) 2471.
428 E. Eschigoya and T. Shiba, *Bull. Tokyo Inst. Technol. Ser. B*, 3 (1960) 133; *Chem. Abstr.*, 55 (1961) 20577.
429 G. Rienacker and K. Wencke, *Angew. Chem.*, 66 (1954) 479.
430 O. V. Krylov, S. Z. Roginskii and E. A. Fokina, *Dokl. Akad. Nauk SSSR*, 96 (1954) 1183.
431 J. M. Cowley, *Acta. Cryst.*, 6 (1953) 846.
432 J. H. de Boer and G. M. M. Houben, in *Proceedings of the International Symposium on the Reactivity of Solids*, Gothenburg, 1952, p. 237.
433 G. M. Schwab and E. Schwab-Agallidis, *J. Am. Chem. Soc.*, 71 (1949) 1806.
434 H. Pines and W. O. Haag, *J. Am. Chem. Soc.*, 83 (1961) 2847.
435 H. Pines and C. N. Pillai, *J. Am. Chem. Soc.*, 83 (1961) 3270.
436 K. Kochloefl, M. Kraus, C. S. Chon, L. Beranek and V. Bazant, *Collection Czech. Chem. Commun.*, 27 (1962) 1199.
437 R. C. Olberg, H. Pines and V. N. Ipatieff, *J. Am. Chem. Soc.*, 66 (1944) 1096.
438 J. Manassen and H. Pines, *Proc. 3rd Intern. Congr. Catalysis, Amsterdam, 1964*, North-Holland, Amsterdam, 1965, p. 845.
439 S. M. Karpacheva and A. M. Rozen, *Dokl. Akad. Nauk SSSR*, 81 (1951) 425.
440 P. Sabatier and A. Mailhe, *Compt. Rend.*, 150 (1910) 1569.
441 A. Grigorieff, *Bull. Soc. Chim. France*, 26 (1901) 612.
442 J. B. Senderens, *Comp. Rend.*, 146 (1908) 125.
443 W. S. Brey, Jr. and A. Krieger, *J. Am. Chem. Soc.*, 71 (1949) 3637.
444 J. C. Balaceanu and J. C. Jungers, *Bull. Soc. Chim. Belges*, 60 (1957) 476.
445 R. N. Pease and C. C. Yung, *J. Am. Chem. Soc.*, 46 (1924) 390.
446 F. Bischoff and H. Adkins, *J. Am. Chem. Soc.*, 47 (1925) 807.
447 H. Adkins and P. P. Perkins, *J. Am. Chem. Soc.*, 47 (1925) 1163.
448 H. Pines and J. Herling, unpublished results quoted in Ref. 424.
449 C. N. Pillai and H. Pines, *J. Am. Chem. Soc.*, 83 (1963) 3274.
450 A. J. Lundeen and R. Van Hoozer, *J. Org. Chem.*, 32 (1967) 3386.
451 J. Hine and W. H. Brader, Jr., *J. Am. Chem. Soc.*, 77 (1955) 361.
452 W. G. Lee and S. I. Miller, *J. Phys. Chem.*, 66 (1962) 655.
453 W. G. Young, D. Pressman and C. D. Coryell, *J. Am. Chem. Soc.*, 61 (1939) 1641.
454 S. Winstein, D. Pressman and W. G. Young, *J. Am. Chem. Soc.*, 61 (1939) 1645.
455 H. Booth, N. C. Franklin and G. C. Gidley, *J. Chem. Soc. C*, (1968) 1891.
456 C. F. Van Duin, *Rec. Trav. Chim.*, 43 (1924) 341; 45 (1926) 345; 47 (1928) 715.
457 E. Biilmann, *Rec. Trav. Chim.*, 36 (1917) 313.
458 W. M. Schubert, H. Steadly and B. S. Rabinovitch, *J. Am. Chem. Soc.*, 77 (1955) 5755.
459 S. I. Miller, *J. Phys. Chem.*, 66 (1962) 978.

460 D. H. R. Barton and W. J. Rosenfelder, *J. Am. Chem. Soc.*, 72 (1950) 1066.
461 D. H. R. Barton and W. J. Rosenfelder, *J. Chem. Soc.*, (1951) 1048.
462 H. L. Goering and H. H. Espy, *J. Am. Chem. Soc.*, 77 (1955) 5023.
463 S. J. Cristol, J. Q. Weber and M. C. Brindell, *J. Am. Chem. Soc.*, 78 (1956) 598.
464 T. L. Davis and R. F. Heggie, *J. Org. Chem.*, 2 (1938) 470.
465 D. R. James, R. W. Rees and C. W. Shoppee, *J. Chem. Soc.*, (1955) 1370.
466 A. B. Forster and W. G. Overend, *J. Chem. Soc.*, (1951) 3452.
467 P. Bladon and L. N. Owen, *J. Chem. Soc.*, (1950) 598.
468 S. J. Angyal and R. J. Young, *Australian J. Chem.*, 14 (1961) 8.
469 M. M. Kreevoy *et al.*, *J. Org. Chem.* 27 (1962) 726, 3421.
470 M. M. Kreevoy and M. A. Turner, *J. Org. Chem.*, 29 (1964) 1639.
471 W. M. Schubert and S. M. Leaky, Jr., *J. Am. Chem. Soc.*, 79 (1957) 381.
472 J. F. King and R. G. Pews, *Can. J. Chem.*, 42 (1964) 1294.
473 F. Weygand and H. G. Peine, *Rev. Chim. Acad. Rep. Populaire Roumaine*, 7 (1963) 1379; *Chem. Abstr.*, 61 (1964) 4208c.
474 T. Alfrey, H. C. Haos and C. W. Lewis, *J. Am. Chem. Soc.*, 74 (1952) 2097.
475 F. R. Mayo and C. Walling, *Chem. Rev.*, 27 (1940) 351.
476 F. D. Greene, W. A. Renes and J. W. Wilson, *J. Am. Chem. Soc.*, 79 (1957) 1416.
477 W. M. Schubert, B. S. Rabinovitch, N. R. Larson and V. Sims, *J. Am. Chem. Soc.*, 74 (1952) 4590.
478 W. G. Young, S. J. Cristol and T. Skei, *J. Am. Chem. Soc.*, 65 (1943) 2099.
479 J. Sicher, M. Havel and M. Svoboda, *Tetrahedron Letters*, (1968) 4269.
480 H. O. House and R. S. Ro, *J. Am. Chem. Soc.*, 80 (1958) 182.
481 M. R. Bridge, D. H. Davis, A. Maccoll, R. A. Ross, B. Stephenson and O. Banjoko, *J. Chem. Soc. B*, (1968) 805.
482 J. W. Cornforth, R. H. Cornforth and K. Mathew, *J. Chem. Soc.*, (1959) 112.
483 W. C. Kray, Jr. and C. E. Castro, *J. Am. Chem. Soc.*, 86 (1964) 4605.
484 C. E. Castro and W. C. Kray, Jr., *J. Am. Chem. Soc.*, 85 (1963) 2768.
485 A. McKillop, E. C. Taylor, G. H. Hawks, G. W. McLay and L. F. Elsom, *J. Am. Chem. Soc.*, 90 (1968) 2421, 2422, 2423.
486 G. Ohloff, *Ann. Chem.*, 627 (1959) 79.
487 F. C. Whitmore and K. C. Laughlin, *J. Am. Chem. Soc.*, 55 (1933) 3732.
488 J. Manassen and F. S. Klein, *J. Chem. Soc.*, (1960) 4203.
489 I. Dostrovsky and F. S. Klein, *J. Chem. Soc.*, (1955) 4401.
490 I. Dostrovsky and F. S. Klein, *J. Chem. Soc.*, (1955) 791.
491 E. Grunwald, A. Heller and F. S. Klein, *J. Chem. Soc.*, (1957) 2604.
492 C. A. Bunton and D. R. Llewellyn, *J. Chem. Soc.*, (1957) 3402.
493 D. S. Noyce, D. R. Hartter and R. M. Pollack, *J. Am. Chem. Soc.*, 90 (1968) 3791.
494 C. A. Bunton, A. Konasiewicz and D. R. Llewellyn, *J. Chem. Soc.*, (1955) 604.
495 S. B. Hanna, Y. Iskander and Y. Riad, *J. Chem. Soc.*, (1961) 217.
496 D. S. Noyce, D. R. Hartter and F. B. Miles, *J. Am. Chem. Soc.*, 90 (1968) 4633.
497 R. B. Williams, *J. Am. Chem. Soc.*, 64 (1942) 1395.
498 V. Gold and M. A. Kessick, *J. Chem. Soc.*, (1965) 6718.
499 W. M. Schubert and B. Lamm, *J. Am. Chem. Soc.*, 88 (1966) 120.
500 D. S. Noyce, D. R. Hartter and F. B. Miles, *J. Am. Chem. Soc.*, 90 (1968) 3794.
501 L. P. Hammett and A. J. Deyrup, *J. Am. Chem. Soc.*, 54 (1932) 2721.
502 N. C. Deno, P. T. Groves and G. Saines, *J. Am. Chem. Soc.*, 81 (1959) 5790.
503 R. H. Boyd, R. W. Taft, Jr., A. P. Wolf and D. R. Christman, *J. Am. Chem. Soc.*, 82 (1960) 4729 and references therein.
504 J. Roček, *Collection Czech. Chem. Commun.*, 25 (1960) 375.
505 D. S. Noyce and C. A. Lane, *J. Am. Chem. Soc.*, 84 (1962) 1635.
506 W. M. Schubert and B. Lamm, *J. Am. Chem. Soc.*, 88 (1966) 120.
507 D. S. Noyce, H. S. Avarbock and W. L. Reed, *J. Am. Chem. Soc.*, 84 (1962) 1647.
508 L. Melander and P. C. Mhyre, *Arkiv Kemi*, 13 (1959) 507.
509 I. I. Moiseev and Ju K. Syrkin, *Dokl. Akad. Nauk SSSR*, 115 (1957) 541.

510 H. C. Brown and Y. Okamoto, *J. Am. Chem. Soc.*, 80 (1958) 4979.
511 P. Ikeda and F. Fujita, *Bull. Inst. Chem. Res. Kyoto Univ.*, 7 (1928) 257.
512 G. M. Gutman and W. J. Hickinbotham, *J. Chem. Soc.*, (1951) 3344.
513 J. L. Simonsen, *The Terpenes*, Macmillan, New York, 1947, Vol. 1, p. 256.
514 Ref. 10, p. 151.
515 D. H. R. Barton, A. S. Campus-Neves and R. C. Cookson, *J. Chem. Soc.*, (1956) 3500.
516 J. H. Ridd, *Quart. Rev. (London)*, 15 (1961) 418.
517 F. C. Whitmore and D. P. Langlois, *J. Am. Chem. Soc.*, 54 (1932) 3441.
518 V. Prelog, *Angew. Chem.*, 70 (1958) 145.
519 R. A. Paphael, *Proc. Chem. Soc.*, (1962) 97.
520 A. Streitwieser, Jr. and W. D. Schaeffer, *J. Am. Chem. Soc.*, 79 (1957) 2888.
521 Ref. 6, p. 383.
522 P. Brewster, F. Hiron, E. D. Hughes, C. K. Ingold and P. A. D. S. Rao, *Nature*, 166 (1950) 179.
523 J. D. Roberts, R. E. McMahon and J. Hine, *J. Am. Chem. Soc.*, 72 (1950) 4237.
524 C. C. Lee, C. P. Slater and J. W. T. Spinks, *Can. J. Chem.*, 35 (1957) 1416.
525 D. Y. Curtin and M. C. Crew, *J. Am. Chem. Soc.*, 76 (1954) 3719.
526 D. J. Cram and J. E. McCarty, *J. Am. Chem. Soc.*, 79 (1957) 2866.
527 J. E. Burr and L. S. Ciereszko, *J. Am. Chem. Soc.*, 74 (1952) 5431.
528 J. D. Roberts *et al.*, *J. Am. Chem. Soc.*, 76 (1954) 4501; 78 (1956) 584.
529 L. G. Connell and R. W. Taft, Jr., *J. Am. Chem. Soc.*, 78 (1956) 5812.
530 H. I. Bernstein and F. C. Whitmore, *J. Am. Chem. Soc.*, 61 (1939) 1324.
531 B. M. Benjamin, H. J. Schaeffer and C. J. Collins, *J. Am. Chem. Soc.*, 79 (1957) 6160.
532 A. Streitwieser, Jr., *J. Org. Chem.*, 22 (1957) 861.
533 J. A. Mills, *J. Chem. Soc.*, (1953) 260.
534 A. K. Bose, *Experientia*, 9 (1953) 256.
535 W. G. Dauben, R. C. Tweit and C. Mannerskantz, *J. Am. Chem. Soc.*, 76 (1954) 4420; W. G. Dauben and J. Jiu, *ibid*, 4426; W. G. Dauben, R. C. Tweit and R. L. MacLean, *ibid*, 77 (1955) 48.
536 A. Stoll, Th. Petrzilka, J. Rutschmann, A. Hofmann and Hs. H. Günthard, *Helv. Chim. Acta.*, 37 (1954) 2039.
537 P. D. Bartlett and L. H. Knox, *J. Am. Chem. Soc.*, 61 (1939) 3184.
538 P. D. Bartlett, in *Organic Chemistry*, H. Gilman (Ed.), Wiley, New York, 1953, Vol. 3, p. 45.
539 C. W. Shoppee, D. E. Evans and G. H. R. Summers, *J. Chem. Soc.*, (1957) 97; C. W. Shoppee, R. J. W. Gembyn, D. E. Evans and G. H. R. Summers, *ibid*, 4364; C. W. Shoppee and J. C. P. Sly, *ibid*, (1959) 345; C. W. Shoppee, S. K. Roy and B. S. Goodrich, *ibid*, (1961) 1583.
540 G. Köbrich, *Angew Chem. Intern. Ed. Engl.*, 4 (1965) 49.
541 S. I. Miller and W. G. Lee, *J. Am. Chem. Soc.*, 81 (1959) 6313.
542 S. I. Miller and R. M. Noyes, *J. Am. Chem. Soc.*, 74 (1952) 629.
543 L. C. Leitch and H. J. Bernstein, *Can. J. Res.*, 28B (1950) 35.
544 T. J. Houser, R. B. Bernstein, R. G. Miekka and J. C. Angus, *J. Am. Chem. Soc.*, 77 (1955) 6201.
545 G. Marchese, F. Naso and G. Modena, *J. Chem. Soc. B*, (1968) 958; G. Marchese, F. Naso, G. Modena and N. Tangari, *Boll. Sci. Fac. Ind. Bologna*, 26 (1968) 209.
546 G. Marchese, G. Modena and F. Naso, *Chem. Commun.*, (1966) 492.
547 S. J. Cristol and W. P. Norris, *J. Am. Chem. Soc.*, 76 (1954) 3005.
548 S. J. Cristol and R. S. Bly, *J. Am. Chem. Soc.*, 83 (1961) 4027.
549 D. Y. Curtin and W. J. Koehl, *J. Am. Chem. Soc.*, 84 (1962) 1967.
550 G. Köbrich and K. Flory, *Tetrahedron Letters*, (1964) 1137.
551 H. Heaney, *Chem. Rev.*, 62 (1962) 81.
552 H. G. Viehe, *Chem. Ber.*, 82 (1959) 1950.
553 V. Mack and K. Herbig, unpublished results cited by R. Huisgen and J. Sauer, *Angew. Chem.*, 72 (1960) 91.

554 S. J. Cristol and R. F. Helmreich, *J. Am. Chem. Soc.*, 77 (1955) 5034.
555 J. Wislicenus and M. Henze, *Ann. Chem.*, 313 (1900) 243.
556 J. Fleming and J. Harley-Mason, *Proc. Chem. Soc.*, (1961) 245; *J. Chem. Soc.*, (1963) 4771.
557 S. J. Cristol and W. P. Norris, *J. Am. Chem. Soc.*, 75 (1953) 2645.
558 E. Grovenstein and D. E. Lee, *J. Am. Chem. Soc.*, 75 (1953) 2639.
559 E. R. Trumbull, R. T. Finn, K. M. Ibno-Rosa and C. K. Sauers, *J. Org. Chem.*, 27 (1962) 2339.
560 H. G. Vieke and E. Franchimant, *Chem. Ber.*, 95 (1962) 319.
561 H. Normant, *Advan. Org. Chem.*, II (1960) 1.
562 C. A. Grob, *Gazz. Chim. Ital.*, 92 (1962) 902.
563 C. A. Grob and G. Cesh, *Helv. Chim. Acta*, 47 (1964) 194.
564 C. A. Grob, J. Csapilla and G. Cesh, *Helv. Chim. Acta*, 47 (1964) 1590.
565 A. Ceccon, U. Tonellato and U. Miotti, *Chem. Commun.*, (1966) 586.
566 U. Tonellato, U. Miotti, M. Padovan and A. Ceccon, unpublished results.
567 F. H. Westheimer, *Chem. Rev.*, 45 (1949) 419.
568 W. A. Waters, *Quart. Rev. (London)*, 12 (1958) 277.
569 N. Venkatasubramanian, *J. Sci. Ind. Res. (India)*, 22 (1963) 397.
570 F. H. Westheimer and N. Nicolaides, *J. Am. Chem. Soc.*, 71 (1949) 25.
571 J. Roček and J. Krupička, *Collection Czech. Chem. Commun.*, 23 (1958) 2068.
572 J. Roček, *Collection Czech. Chem. Commun.*, 25 (1960) 1052.
573 H. Kwart and P. S. Francis, *J. Am. Chem. Soc.*, 77 (1955) 4907.
574 J. Roček, F. H. Westheimer, A. Eschenmoser, L. Moldoványi and J. Schreiber, *Helv. Chim. Acta*, 45 (1962) 2554.
575 G. V. Bakore and D. Narain, *J. Chem. Soc.*, (1963) 3419.
576 J. W. Baker and T. G. Heggs, *J. Chem. Soc.*, (1955) 616.
577 J. W. Baker and D. M. Easty, *J. Chem. Soc.*, (1952) 1193, 1208.
578 A. N. Bourns and E. Buncel, *Can. J. Chem.*, 38 (1960) 2457.
579 A. N. Bourns and P. J. Smith, *Can. J. Chem.*, 44 (1966) 2553.
580 A. Ledwith, G. W. Cowell and D. G. Morris, *J. Chem. Soc. B*, (1967) 697.
581 N. Kornblum and H. E. De La Mare, *J. Am. Chem. Soc.*, 73 (1951) 880.
582 C. R. Hauser, J. W. LeMaistre and A. E. Rainsford, *J. Am. Chem. Soc.*, 57 (1935) 1056.
583 M. Benger and O. L. Brady, *J. Chem. Soc.*, (1950) 1221.
584 D. Ambrose and O. L. Brady, *J. Chem. Soc.*, (1950) 1243.
585 W. Kirmse, *Angew. Chem. Intern. Ed. Engl.*, 4 (1965) 1.
586 P. Fritsche, *Ann. Chem.*, 279 (1894) 319; W. P. Buttenburg, *ibid*, 279 (1894) 324; H. Wiechell, *ibid*, 279 (1894) 337.
587 J. G. Pritchard and A. A. Bothner-by, *J. Phys. Chem.*, 64 (1960) 1271.
588 A. A. Bothner-by, *J. Am. Chem. Soc.*, 77 (1955) 3293.
589 W. M. Jones and R. Damico, *J. Am. Chem. Soc.*, 85 (1963) 2273.
590 D. Y. Curtin and E. W. Flynn, *J. Am. Chem. Soc.*, 81 (1959) 4714.
591 G. Köbrich and H. Trapp, *Z. Naturforsch.*, 18b (1963) 1125.
592 C. R. Hauser, W. R. Brasen, P. S. Skell, S. W. Kantor and A. E. Brodhag, *J. Am. Chem. Soc.*, 79 (1957) 397.
593 M. S. Kharasch, W. Nudenberg and E. K. Fields, *J. Am. Chem. Soc.*, 66 (1944) 1276.
594 C. R. Hauser, S. W. Kantor, P. S. Skell and W. R. Brasen, *J. Am. Chem. Soc.*, 78 (1956) 1653.
595 D. Bethell, *J. Chem. Soc.*, (1963) 666.
596 R. Kuhn and G. Platzer, *Chem. Ber.*, 73 (1940) 1410.
597 D. Bethell, A. F. Cockerill and D. B. Frankham, *J. Chem. Soc. B*, (1967) 1287.
598 P. Miginiac, *Bull. Soc. Chim. France*, (1962) 2000.
599 T. Shimomura, K. J. Tolle, J. Smid and M. Szwarc, *J. Am. Chem. Soc.*, 89 (1967) 796.
600 A. F. Cockerill, *Ph. D. Thesis*, The University of Liverpool, 1965; A. F. Cockerill and R. T. Hargreaves, unpublished results.
601 A. Ledwith and Yang-Shih Lin, *Chem. Ind. (London)*, (1964) 1867.
602 S. B. Hanna and L. G. Wideman, *Chem. Ind. (London)*, (1968) 486.

603 D. M. DOLEIB AND Y. ISKANDER, *J. Chem. Soc. B*, (1967) 1154, 1159.
604 C. G. SWAIN AND E. R. THORNTON, *J. Am. Chem. Soc.*, 83 (1961) 4033.
605 I. ROTHBERG AND E. R. THORNTON, *J. Am. Chem. Soc.*, 86 (1964) 3296.
606 G. A. RUSSELL AND W. C. DANEN, *J. Am. Chem. Soc.*, 90 (1968) 347.
607 R. C. KERBER, G. W. URRY, AND N. KORNBLUM, *J. Am. Chem. Soc.*, 87 (1965) 4520; N. KORNBLUM, R. E. MICHEL AND R. C. KERBER, *ibid*, 88 (1966) 5660, 5662; N. KORNBLUM, T. M. DAVIES, G. W. EARL, N. L. HOLY, R. C. KERBER, M. T. MUSSER AND D. H. SNOW, *ibid*, 89 (1967) 725.
608 E. R. THORNTON, personal communication of unpublished observations.
609 D. J. CRAM, *Fundamentals of Carbanion Chemistry*, Academic Press, New York, 1965, pp. 243–256.
610 A. NICKON AND N. H. WERSTIUCK, *J. Am. Chem. Soc.*, 89 (1967) 3914.
611 A. NICKON AND N. H. WERSTIUCK, *J. Am. Chem. Soc.*, 89 (1967) 3915.
612 A. NICKON AND N. H. WERSTIUCK, *J. Am. Chem. Soc.*, 89 (1967) 3917.
613 C. L. BUMGARDNER, *Chem. Commun.*, (1965) 374.
614 C. L. BUMGARDNER, *J. Am. Chem. Soc.*, 83 (1961) 4420, 4423.
615 S. J. CRISTOL AND B. B. JARVIS, *J. Am. Chem. Soc.*, 88 (1966) 3095.
616 S. J. CRISTOL AND B. B. JARVIS, *J. Am. Chem. Soc.*, 89 (1967) 401.
617 S. J. CRISTOL, J. K. HARRINGTON AND M. S. SINGER, *J. Am. Chem. Soc.*, 88 (1966) 1529.
618 S. WINSTEIN, E. CLIPPINGER, R. HOWE AND E. VOGELFANGER, *J. Am. Chem. Soc.*, 87 (1965) 376.
619 P. S. SKELL AND I. STARER, *J. Am. Chem. Soc.*, 82 (1960) 2971.
620 P. S. SKELL AND I. STARER, *J. Am. Chem. Soc.*, 84 (1962) 3962.
621 L. FRIEDMAN AND H. SHECHTER, *J. Am. Chem. Soc.*, 81 (1959) 5512.
622 O. A. REUTOV AND T. N. SHATKINA, *Tetrahedron*, 18 (1962) 237.
623 A. T. JUREWICZ AND L. FRIEDMAN, *J. Am. Chem. Soc.*, 89 (1967) 149.
624 G. J. KARABATSOS, R. A. MOUNT, D. O. RICKTER AND S. MEYERSON, *J. Am. Chem. Soc.*, 88 (1966) 5651.
625 F. M. SONNENBERG AND J. K. STILLE, *J. Org. Chem.*, 31 (1966) 3441.
626 R. C. BICKNELL AND A. MACCOLL, *Chem. Ind., (London)*, (1961) 1912.
627 L. KAPLAN, *J. Am. Chem. Soc.*, 89 (1967) 1753.
628 L. RAMBERG and B. BÄCKLUND, *Arkiv Kemi Mineral. Geol.*, 13A (1940) No. 27: *Chem. Abstr.*, 34 (1940) 4725.
629 For a recent review see L. A. PAQUETTE, *Accounts Chem. Res.*, 1 (1968) 209.
630 F. G. BORDWELL AND G. D. COOPER, *J. Am. Chem. Soc.*, 73 (1951) 5187.
631 (a) L. A. PAQUETTE, *J. Am. Chem. Soc.*, 86 (1964) 4085.
(b) N. P. NEUREITER AND F. G. BORDWELL, *J. Am. Chem. Soc.*, 85 (1963) 1209.
(c) N. P. NEUREITER, *J. Amer. Chem. Soc.*, 88 (1966) 558.
632 F. G. BORDWELL AND J. M. WILLIAMS, JR., *J. Am. Chem. Soc.*, 90 (1968) 435.
633 L. A. PAQUETTE, L. S. WITTENBROOK AND V. W. KANE, *J. Am. Chem. Soc.*, 89 (1967) 4487 and earlier references.
634 F. G. BORDWELL, J. M. WILLIAMS, JR., E. B. HOYT, JR. AND B. B. JARVIS, *J. Am. Chem. Soc.*, 90 (1968) 429.
635 R. JACQUIER, *Bull. Soc. Chim. France*, 17 [5] (1950) D35.
636 A. S. KENDE, *Organic Reactions*, Wiley, New York, 1960, Vol. 11, p. 261.
637 J. M. CONIA AND J. SALAUN, *Bull. Soc. Chim. France*, (1964) 1957.
638 R. B. LOFTFIELD, *J. Am. Chem. Soc.*, 73 (1951) 4707.
639 W. B. HAMMOND AND N. J. TURRO, *J. Am. Chem. Soc.*, 88 (1966) 2880.
640 C. RAPPE AND L. KNUTSSON, *Acta Chem. Scand.*, 21 (1967) 2205.
641 W. D. MCPHEE AND E. KLINGSBERG, *J. Am. Chem. Soc.*, 66 (1944) 1132.
642 R. B. LOFTFIELD, *J. Am. Chem. Soc.*, 73 (1951) 4707.
643 R. BRESLOW, J. POSNER AND A. KREBS, *J. Am. Chem. Soc.*, 85 (1963) 234.
644 G. STORK AND I. BOROWITZ, *J. Am. Chem. Soc.*, 82 (1960) 4307.
645 F. G. BORDWELL, R. R. FRAME, R. G. SCAMEHORN, J. G. STRONG AND S. MEYERSON, *J. Am. Chem. Soc.*, 89 (1967) 6706.
646 H. R. NACE AND B. A. OLSEN, *J. Org. Chem.*, 32 (1967) 3438.

647 (a) A. A. SACKS AND J. G. ASTON, *J. Am. Chem. Soc.*, 73 (1951) 3902.
(b) J. G. BURR AND M. J. S. DEWAR, *J. Chem. Soc.*, (1954) 1201.
(c) A. W. FORT, *J. Am. Chem. Soc.*, 84 (1962) 2620, 2625.
648 P. W. NEBER, A. BURGARD AND W. THIER, *Ann. Chem.*, 526 (1936) 277 and earlier papers.
649 C. O'BRIEN, *Chem. Rev.*, 64 (1964) 81.
650 M. J. HATCH AND D. J. CRAM, *J. Am. Chem. Soc.*, 75 (1953) 33, 38.
651 H. O. HOUSE AND W. F. BERKOWITZ, *J. Org. Chem.*, 28 (1963) 307, 2271.
652 G. DREFAHL AND D. MARTIN, *Chem. Ber.*, 93 (1960) 2497.
653 For a number of examples see *Heterocyclic Compounds with Three- and Four-membered Rings*, A. WEISSBERGER (Ed.), Wiley, London, 1964, Vol. 19.
654 H. E. BAUMGARTEN, J. F. FUERHOLZER, R. D. CLARK AND R. D. THOMPSON, *J. Am. Chem. Soc.*, 85 (1963) 3303.
655 H. E. BAUMGARTEN, R. L. ZEY AND U. KROLLS, *J. Am. Chem. Soc.*, 83 (1961) 4469.
656 J. D. PARK AND R. J. MCMURTRY, *J. Org. Chem.*, 32 (1967) 2397.
657 S. J. CRISTOL, N. L. HAUSE AND J. S. MEEK, *J. Am. Chem. Soc.*, 73 (1951) 674; S. J. CRISTOL and D. D. FIX, *J. Am. Chem. Soc.*, 75 (1953) 2647.
658 H. D. ORLOFF AND A. J. KOLKA, *J. Am. Chem. Soc.*, 76 (1954) 5484.
659 P. B. D. DE LA MARE, R. KOENIGSBERGER AND J. S. LOMAS, *J. Chem. Soc. B*, (1967) 834.
660 R. F. HUDSON AND R. J. WITHEY, *J. Chem. Soc. B*, (1966) 237.
661 R. F. HUDSON, J. ARENDT AND A. MANCUSO, *J. Chem. Soc. B*, (1967) 1069.
662 C. A. GROB AND W. BAUMAN, *Helv. Chim. Acta*, 38 (1955) 594.
663 C. A. GROB, *Experientia*, 13 (1957) 126.
664 C. A. GROB, *Theoretical Organic Chemistry, Report on the Kekulé Symposium*, Butterworths, London, 1958, p. 114.
665 C. A. GROB AND F. OSTERMAYER, *Helv. Chim. Acta*, 45 (1962) 1119.
666 J. ENGLISH AND F. V. BRUTCHER, *J. Am. Chem. Soc.*, 74 (1952) 4279.
667 J. ENGLISH AND A. D. BLISS, *J. Am. Chem. Soc.*, 78 (1956) 4057.
668 P. S. WHARTON AND G. A. HIEGEL, *J. Org. Chem.*, 30 (1965) 3254.
669 F. NERDEL, D. FRANK AND H. J. LENGERT, *Chem. Ber.*, 98 (1965) 728.
670 E. M. WARNHOFF, C. M. WONG AND W. T. TAI, *J. Am. Chem. Soc.*, 90 (1968) 514.
671 W. R. VAUGHAN AND R. L. CRAVEN, *J. Am. Chem. Soc.*, 77 (1955) 3795.
672 P. BRENNEISEN, C. A. GROB, R. A. JACKSON AND M. OHTA, *Helv. Chim. Acta*, 48 (1965) 146.
673 K. CONROW, *J. Am. Chem. Soc.*, 81 (1959) 5461.
674 K. BODENDORF AND R. MAYER, *Chem. Ber.*, 98 (1965) 3554.
675 C. L. ARCUS AND B. S. PRYDAL, *J. Chem. Soc.*, (1954) 4018.
676 P. T. LANSBURY, V. A. PATTISON AND J. W. DIEHL, *Chem. Ind. (London)*, (1962) 653.
677 J. HOFFMANN, *J. Am. Chem. Soc.*, 79 (1957) 504.
678 C. A. GROB, H. P. FISCHER, W. RANDENBUSCH AND J. ZERGENYI, *Helv. Chim. Acta*, 47 (1964) 1003.
679 A. A. R. SAYIGH, J. N. TILLEY AND H. ULRICH, *J. Org. Chem.*, 29 (1964) 3344.
680 J. ZÁVADA, M. PÁNKOVÁ AND J. SICHER, *Chem. Commun.*, (1968) 1145.
681 M. PÁNKOVÁ, J. ZÁVADA AND J. SICHER, *Chem. Commun.*, (1968) 1142.
682 J. SICHER, J. ZÁVADA AND M. PÁNKOVÁ, *Chem. Commun.*, (1968) 1147.
683 N. C. DENO, in *Survey of Progress in Chemistry*, A. F. SCOTT (Ed.), Interscience, New York, 1963, Vol. 1, p. 155.
684 F. G. BORDWELL AND R. G. SCAMEHORN, *J. Am. Chem. Soc.*, 90 (1968) 6751.
685 M. D. WOLFINGER, *Ph.D. Dissertation*, North Western University, U.S.A., June, 1968, quoted in ref. 684.
686 A. F. COCKERILL, *Tetrahedron Letters*, (1969) 4913.
687 T. COHEN AND A. R. DANIEWSKI, *J. Am. Chem. Soc.*, 91 (1969) 533.

Appendix

Since this chapter was submitted, a number of interesting papers on elimination reactions has appeared in the literature. Some are listed and briefly described below.

A.1. Alkene-forming eliminations

A.1.1 BASE- AND SOLVENT-CATALYSED ELIMINATIONS

A number of examples of the E1cB mechanism has been reported (*cf.* Section 2.1.3)[1-7].

Syn-elimination (*cf.* Section 2.3.1) has been shown to be a much more common occurrence in base-catalysed E2 reactions than was previously thought to be the case. The stereochemistry of E2 reactions has been reviewed by Sicher[8] and several examples of *syn*-elimination in both the acyclic and cyclic series have been demonstrated. *Syn*-elimination is promoted by solvents which cause ion association[9,10], and by bulky leaving groups such as trialkylammonium ions with sterically hindered bases in encumbered substrates[11,12]. In the elimination from cyclopentyltrimethylammonium ions, smaller isotope effects (k_H/k_D) are associated with *syn-* than are associated with *anti*-elimination[13]. This observation is in agreement with earlier results which indicated that *syn*-elimination involves a transition state with greater carbanion character than is involved in *anti*-elimination. *Anti*-elimination has been confirmed as the sole stereochemical path in the elimination reactions of 2-phenylethyl[14,15] and 2-butyl derivatives[16]. A more detailed analysis of the principle of least motion as applied to the stereochemistry of elimination reactions has appeared[17] (*cf.* Section 2.3.1).

The theory of the variable transition state for base-catalysed E2 reactions has been reviewed[18,19] and has been used to explain the Hammett reaction constants[20-26], kinetic isotope effects[26-29], and orientational features[29-32] of some base-catalysed elimination reactions (*cf.* Section 2.2).

The Brönsted equation has been used to confirm concerted rather than stepwise base-catalysed eliminations[33,34]. However, parallel Brönsted plots were found for primary, secondary, and tertiary amines. Bordwell *et al.* have reported values of the Brönsted component, β, which are greater than unity for the ionisation of psuedo acids, such as nitroalkanes, for which proton removal by a base is not diffusion-controlled[35,36]. Values greater than unity are not to be expected for the ionisation of N–H and O–H bonds[37], but these large values for the nitroalkanes do indicate limitations in the use of Brönsted components in the assignment of mechanism of organic reactions which involve C–H ionisation (*cf.* Section 2.2.2.(b)).

An ion-pair mechanism has been proposed for the elimination of 1-phenylethyl bromide with ethoxide ion as the second-order rate coefficient decreases with increasing base concentration[38,39]. However, this variation in rate can be attributed to association of the base in the more concentrated solution[40].

Details of the theoretical calculations of isotope effects (both k_H/k_D and k_{32_S}/k_{34_S}) for various model transition states in the elimination of 2-phenylethyldimethylsulphonium ion have appeared[41] (*cf.* Section 2.2.3).

Parker *et al.*[42] and Betrame *et al.*[43] have presented evidence to support transition states with features which are closely related to those suggested previously for the merged mechanism of substitution and elimination (*cf.* Section 2.4). However, they regard bimolecular elimination and substitution as occurring through different transition states. The so-called E2C mechanism is preferred for bases with poor hydrogen basicity but good carbon nucleophilicity. On the other hand, the normal E2 carbanion-type transition state, in which direct interaction between the base and the alpha carbon is lacking, is preferred for elimination reactions with more basic reagents. This latter transition state is now termed as E2H. Eck and Bunnett[44] have presented evidence which is not consistent with the E2C mechanism and they regard the E2C transition state as an unnecessary addition to the terminology at present used for the mechanisms of elimination.

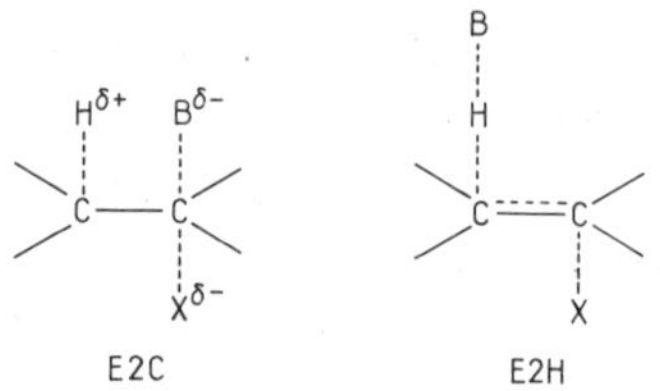

The factors which may influence the *trans*/*cis* olefin ratios arising from base-induced elimination reactions have been discussed and are summarised in a recent paper by Feit and Gitlin[45]. These factors include, in addition to eclipsing effects, steric effects of unsymmetrical leaving groups and bases, and the incursion of *syn*-elimination (*cf.* Section 2.2.2.(d)).

A.1.2 PYROLYTIC ELIMINATIONS IN THE GAS PHASE

Maccoll[46] has written a comprehensive review on the pyrolysis of alkyl halides. A description of the experimental techniques which are used in the measurement of the kinetics of gas-phase reactions is included. Results obtained by different experimental procedures are generally in good agreement. Heterolytic pyrolytic eliminations are regarded as occurring through transition states with considerable ionic character. The kinetics of a number of dehydrofluorinations have also been measured[47,48].

A number of non-stereospecific acetate and xanthate pyrolyses have been observed[49–54]. *Anti*-elimination is rationalised by invoking neighbouring-group participation by an appropriately situated methyl ester group in one case[52]. The array of products arising from one of the xanthate pyrolyses can be accounted for in terms of competing heterolytic and homolytic eliminations[53]. The pyrolyses of isothiocyanates and thiocyanates bear a close resemblance to the ester eliminations[55,56].

A.1.3 DEHALOGENATIONS (*cf.* Section 2.11)

The mechanisms of dehalogenations have been reviewed by Miller[57] and in a series of papers[58], the stereoselectivity of the dehalogenation of the stilbene dibromides with a wide variety of reagents has been discussed. The *meso*-stilbene dibromide always eliminates to give the thermodynamically more stable alkene, namely *trans*-stilbene which is product of apparent *anti*-elimination. However, the *dl*-stilbene dibromide gives both *cis*- and *trans*-stilbenes, and the ratio of these products can provide useful mechanistic information. One-electron reductants, such as chromous ion, give rise to intermediate radical formation in which rotation about the C_α–C_β bond allows thermodynamic control of the reaction. Two-electron reductants, such as iodide ion in dimethyl formamide, induce highly stereoselective *anti*-elimination. In protic solvents, carbonium ion intermediates were proposed to explain the trend towards thermodynamic control. Miller has proposed a reaction mechanism which embraces elimination, substitution, and electrophilic addition to alkenes.

A.1.4 DEHYDRATION OF ALCOHOLS (*cf.* Sections 2.10 and 2.12)

Two reviews have appeared which are concerned particularly with the role of solid catalysts in dehydration reactions[59,60]. Kieboom and van Bekkum[61] have reported details of the dehydration of 2-aryl-3-methyl-2-butanols by sulphuric acid and by dimethyl sulphoxide. Surprisingly, for both reactions, the kinetic product is the thermodynamically less stable 2-aryl-3-methyl-1-butene, which subsequently isomerises to 2-aryl-3-methyl-2-butene. Steric interactions were proposed to account for the initial formation of the less stable alkene in the reaction of the carbonium ion. Significant proportions of the Hoffmann

References pp. 370–372.

alkene have also been reported in the eliminations of 2,3-diphenyl-2-propanol under a variety of conditions[62].

The mechanism of dehydration of alcohols by hexamethylphosphortriamide[63], phosphorus oxychloride and thionyl chloride in the presence of bases[64], and sulfur tetrafluoride[65] have been studied. Kirk and Shaw[64] have shown that it is unwise to assume *anti*-stereospecificity in dehydration with phosphorus oxychloride or thionyl chloride with basic catalysts. The stereoselectivity is greatly dependent upon the basicity and steric environment of the base. Various steroidal alcohols undergo stereoselective *syn*-elimination upon treatment with methyl (carboxysulfamoyl) triethylammonium hydroxide inner salt[66].

A.1.5 DEAMINATION REACTIONS (*cf.* Section 2.13)

Two extensive reviews concerned with the mechanisms of deamination reactions have appeared[67,68]. White and Woodcock[68] have compared the deamination by nitrous acid to those of the closely related nitrosoamide and triazene decompositions. Ion-pair intermediates were invoked to account for the stereochemistry of the substitution reactions.

Lee[69] has written a review on the role of protonated cyclopropanes in solvolytic reactions, and considerable attention has been focussed on these possible intermediates in deamination reactions. Combinations of carbon and deuterium labelling experiments have been used to reveal the identity of the intermediates[70–78]. Lee and Wan[71] have compared the products of deamination and of the closely related deoxidation reactions. They concluded that similar intermediates were involved in the two reactions.

Diazo alkanes are sometimes involved as intermediates in deamination, especially in aprotic solvents[79,80]. The counterion[78] and micellar formation[81] are other factors which can determine the products of deamination.

A.2. Eliminations giving alkynes (*cf.* Section 3)

Rappoport[82] has presented a detailed outline of the mechanisms of the reactions of vinyl halides with nucleophilic reagents. Modena *et al.*[83,84] have provided further evidence in support of a spectrum of transition states for elimination from activated vinyl halides induced by alkoxide bases. Cristol and Whittemore[85] have shown that the stereoselectivity of elimination from vinyl halides is largely determined by the choice of basic reagent; alkoxide bases encourage *anti*-elimination, whereas *syn*-elimination and alpha-elimination become dominant with lithium alkyls.

An (E1cB) ion-pair mechanism has been invoked for the elimination from *cis*-1,2-dibromoethylene with triethylamine in dimethyl formamide[86]. There have been numerable examples of vinyl cation intermediates in the solvolytic eliminations of vinyl halides[87–95] and Hanack[96] has reviewed the subject.

A.3. Eliminations giving imines, nitriles, thiocarbonyls and carbonyls (*cf.* Sections 4.2, 4.3, and 4.4)

There is further evidence supporting a continuum of transition states for bimolecular thiocarbonyl eliminations[97–99]. Compared to alkene eliminations, the transition states have extensive carbanion character, but little double-bond character, attributable to the low effectiveness of π-electron overlap involving the $3p$ orbitals of sulphur.

Braumann and Hill[100] have studied the dehydrofluorinations giving imine and nitrile products. These dehydrofluorinations occur under considerably milder conditions than the corresponding eliminations leading to alkenes, in line with the weaker nature of the N–F than the C–F bond.

A.4. Alpha-eliminations (*cf.* Section 5)

Bethell[101] has reviewed the chemical properties of carbenes and has discussed the mechanisms of the reactions in which they are formed.

A.5. Gamma-eliminations

Bordwell *et al.*[102–105] have continued their investigations into the mechanisms of the Ramberg–Bäcklund and Favorskii reactions. A change in mechanism from rate-determining C_α–X ionisation to C_γ–H proton abstraction for the Favorskii reaction using methoxide in methanol is brought about[106] by a change in structure from $PhCH_2COCH_2Cl$ to $PhCH_2COCHClCH_3$. Evidence in favour of a dipolar intermediate in the cyclisation step of the Favorskii reaction is provided[107,108] by the presence of an indanone among the reaction products in the decompositions of $Ph_2CClCOCH_3$ and $Ph_2CHCOCH_2Cl$ (*cf.* Section 6.4).

Steric control in the fission of crowded cyclopropanones intermediates has been demonstrated[109]. Hassner *et al.*[110] have reported details of γ-elimination leading to aziridines from β-iodoazides.

A.6. Fragmentation reactions

Grob[111] has reviewed the mechanisms of fragmentation of γ-amino derivatives. Examples of unimolecular, bimolecular, and E1cB-type fragmentations are discussed (*cf.* Section 8).

References pp. 370–372

REFERENCES TO APPENDIX

1 R. A. More O'Ferrall and S. Slae, *J. Chem. Soc. B*, (1970) 260, 268.
2 L. R. Fedor, *J. Am. Chem. Soc.*, 91 (1969) 908, 913.
3 L. R. Fedor and R. C. Cevestri, *J. Am. Chem. Soc.*, 92 (1970) 4610.
4 F. G. Bordwell, M. M. Vestling and K. C. Yee, *J. Am. Chem. Soc.*, 92 (1970) 5950.
5 F. G. Bordwell, K. C. Yee and A. C. Knipe, *J. Am. Chem. Soc.*, 92 (1970) 5945.
6 J. Crosby and C. J. M. Stirling, *J. Chem. Soc. B*, (1970) 671.
7 J. Crosby and C. J. M. Stirling, *J. Chem. Soc. B*, (1970) 679.
8 J. Sicher, *Angew. Chem. Intern. Ed. Engl.*, 11 (1972) 200.
9 W. F. Bayne and E. I. Snyder, *Tetrahedron Letters*, (1971) 571.
10 H. C. Brown and K.-T. Liu, *J. Am. Chem. Soc.*, 92 (1970) 200.
11 D. S. Bailey and W. H. Saunders, Jr., *J. Am. Chem. Soc.*, 92 (1970) 6904.
12 D. S. Bailey, F. C. Montgomery, G. W. Chodak and W. H. Saunders, Jr., *J. Am. Chem. Soc.*, 92 (1970) 6911.
13 H. C. Brown and W. H. Saunders, Jr., *J. Am. Chem. Soc.*, 92 (1970) 4292.
14 A. N. Bourns and A. C. Frosst, *Can. J. Chem.*, 48 (1970) 133.
15 A. N. Bourns and P. J. Smith, *Can. J. Chem.*, 48 (1970) 125.
16 R. Bartsch, *Tetrahedron Letters*, (1970) 297.
17 O. S. Tee, *J. Am. Chem. Soc.*, 91 (1969) 7144.
18 R. A. More O'Ferrall, *J. Chem. Soc. B*, (1970) 274.
19 J. F. Bunnett, *Surv. Progr. Chem.*, (1969) 5.
20 J. Banger, A. F. Cockerill and G. L. O. Davies, *J. Chem. Soc. B*, (1971) 498.
21 Y. Yano and S. Oae, *Tetrahedron*, 26 (1970) 27.
22 S. Oae and Y. Yano, *Tetrahedron*, 24 (1968) 5721.
23 Y. Yano and S. Oae, *Tetrahedron*, 26 (1970) 67.
24 R. Baker and M. J. Spillett, *J. Chem. Soc. B*, (1969) 482.
25 D. J. McLennan and R. J. Wong, *J. Chem. Soc. Perkin Trans. II*, (1972) 72.
26 P. J. Smith and S. K. Tsui, *Tetrahedron Letters*, (1972) 917.
27 J. I. Hayami, N. Ono and A. Kaji, *Bull. Soc. Chem. Japan*, 44 (1971) 1628.
28 T. A. Spencer, M. C. R. Kendall and I. D. Reingold, *J. Am. Chem. Soc.*, 94 (1972) 1250.
29 W. H. Saunders, Jr. and A. T. Ashe, *J. Am. Chem. Soc.*, 91 (1969) 4473.
30 W. H. Saunders, Jr. and I. N. Feit, *J. Am. Chem. Soc.*, 92 (1970) 5615.
31 R. A. Bartsch and J. F. Bunnett, *J. Am. Chem. Soc.*, 91 (1969) 1376.
32 R. A. Bartsch and J. F. Bunnett, *J. Am. Chem. Soc.*, 91 (1969) 1382.
33 T. A. Spencer, M. R. C. Kendall and I. D. Reingold, *J. Am. Chem. Soc.*, 94 (1972) 1250.
34 D. J. Hope, M. R. C. Kendall and T. A. Spencer, *J. Am. Chem. Soc.*, 94 (1972) 1254.
35 F. G. Bordwell, W. J. Boyle, Jr., J. A. Hantala and K. C. Yee, *J. Am. Chem. Soc.*, 91 (1969) 4002.
36 F. G. Bordwell and W. J. Boyle, Jr., *J. Am. Chem. Soc.*, 93 (1971) 511.
37 R. A. Marcus, *J. Am. Chem. Soc.*, 91 (1969) 7224.
38 R. A. Sneen and H. M. Robbins, *J. Am. Chem. Soc.*, 91 (1969) 3100.
39 R. A. Sneen and J. W. Larson, *J. Am. Chem. Soc.*, 91 (1969) 362.
40 D. J. McLennan, *Tetrahedron Letters*, (1971) 2317.
41 A. M. Katz and W. H. Saunders, Jr., *J. Am. Chem. Soc.*, 91 (1969) 4469.
42 (a) G. Biale, A. J. Parker, S. G. Smith, I. D. R. Stevens and S. Winstein, *J. Am. Chem. Soc.*, 92 (1970) 115. (b) G. Biale, A. J. Parker and S. Winstein, *J. Am. Chem. Soc.*, 94 (1972) 2235. (c) A. J. Parker, M. Ruane and S. Winstein, *J. Am. Chem. Soc.*, 94 (1972) 2228. (d) D. Cook and A. J. Parker, *Tetrahedron Letters*, (1969) 4901. (e) G. Biale, D. Cook, D. J. Lloyd, A. J. Parker, I. D. R. Stevens, J. Takahashi and S. Winstein, *J. Am. Chem. Soc.*, 93 (1971) 4735.
43 P. Beltrame, G. Biale and S. Winstein, *J. Am. Chem. Soc.*, 94 (1972) 2240.

44 D. ECK AND J. F. BUNNETT, *J. Am. Chem. Soc.*, 91 (1969) 3099.
45 I. N. FEIT AND L. F. GITLIN, *J. Chem. Soc. Chem. Commun.*, (1972) 561.
46 A. MACCOLL, *Chem. Rev.*, 69 (1969) 33.
47 P. CADMAN, M. DAY, A. W. KIRK AND A. F. TROTMAN-DICKINSON, *Chem. Commun.*, (1970) 203.
48 G. E. MILLWARD, R. HARTIG AND E. TSCHNIKOW-ROUX, *Chem. Commun.*, (1971) 465.
49 H. KWART AND H. G. LING, *Chem. Commun.*, (1969) 302.
50 E. S. LEWIS AND E. R. NEWMAN, *J. Am. Chem. Soc.*, 91 (1969) 7455.
51 J. C. LEFFINGWELL AND R. E. SHACKELFORD, *Tetrahedron Letters*, (1970) 2003.
52 E. E. SMISSMAN, J. P. LI AND M. W. CREESE, *J. Org. Chem.*, 35 (1970) 1352.
53 R. E. GILMAN *et al.*, *Can. J. Chem.*, 48 (1970) 970.
54 C. DJERASSI AND W. S. BRIGGS, *J. Org. Chem.*, 33 (1968) 1625.
55 N. BARROETA, A. MACCOLL AND A. FAVA, *J. Chem. Soc. B*, (1969) 347.
56 N. BARROETA, A. MACCOLL, M. CAVAZZA, L. CONGIU AND A. FAVA, *J. Chem. Soc. B*, (1971) 1264, 1267.
57 S. I. MILLER, *Advan. Phys. Org. Chem.*, 6 (1968) 185.
58 I. M. MATHAI, K. SCHUG AND S. I. MILLER, *J. Org. Chem.*, 35 (1970) 1733; I. M. MATHAI AND S. I. MILLER, *J. Org. Chem.*, 35 (1970) 3416; W. K. KWOK, I. M. MATHAI AND S. I. MILLER, *J. Org. Chem.*, 35 (1970) 3420; W. K. KWOK AND S. I. MILLER, *J. Am. Chem. Soc.*, 92 (1970) 4599; C. S. TSAI LEE, I. M. MATHAI AND S. I. MILLER, *J. Am. Chem. Soc.*, 92 (1970) 4602.
59 H. KNOZINGER, in *The Chemistry of the Hydroxyl Group*, S. PATAI (Ed.), Interscience, London, 1971, p. 642.
60 H. NOLLER, P. ANDRÉU AND M. HUNGER, *Angew. Chem. Intern. Ed. Engl.*, 10 (1971) 172.
61 A. P. G. KIEBOOM AND H. VAN BEKKUM, *Rec. Trav. Chim.*, 88 (1969) 1424.
62 I. HO AND J. G. SMITH, *Tetrahedron*, 26 (1970) 4277.
63 A. MONSON, *Tetrahedron Letters*, (1971) 567.
64 D. N. KIRK AND P. M. SHAW, *J. Chem. Soc. C*, (1970) 182.
65 R. E. A. DEAR, E. E. GILBERT AND J. J. MURRAY, *Tetrahedron*, 27 (1971) 3345.
66 P. CRABBÉ AND C. LEON, *J. Org. Chem.*, 35 (1970) 2594.
67 D. V. BANTHORPE, Rearrangements involving amino groups, in *The Chemistry of the Amino Group*, S. PATAI (Ed.), Interscience, London, 1968.
68 E. H. WHITE AND D. J. WOODCOCK, Cleavage of the carbon–nitrogen bond, in *The Chemistry of the Amino Group*, S. PATAI (Ed.), Interscience, London, 1968.
69 C. C. LEE, *Progr. Phys. Org. Chem.*, 7 (1970) 129.
70 C. C. LEE, J. E. KRUGER AND E. W. C. WONG, *J. Am. Chem. Soc.*, 87 (1965) 3985.
71 C. C. LEE AND K. M. WAN, *J. Am. Chem. Soc.*, 91 (1969) 6416.
72 E. I. SNYDER, *J. Am. Chem. Soc.*, 91 (1969) 5118.
73 L. FRIEDMAN AND A. T. JUREWICZ, *J. Am. Chem. Soc.*, 91 (1969) 1800, 1803.
74 G. J. KARABATSOS, M. ANAND, D. O. RICKTER AND S. MEYERSON, *J. Am. Chem. Soc.*, 92 (1970) 1254.
75 G. J. KARABATSOS, R. A. MOUNT, D. O. RICKTER AND S. MEYERSON, *J. Am. Chem. Soc.*, 92 (1970) 1248.
76 G. J. KARABATSOS, N. HSI AND S. MEYERSON, *J. Am. Chem. Soc.*, 92 (1970) 621.
77 G. J. KARABATSOS, C. E. ORZECH, JR., J. L. FRY AND S. MEYERSON, *J. Am. Chem. Soc.*, 92 (1970) 606.
78 C. J. COLLINS AND B. M. BENJAMIN, *J. Am. Chem. Soc.*, 92 (1970) 3182, 3183.
79 L. FRIEDMAN, A. T. JUREWICZ AND J. H. BAYLESS, *J. Am. Chem. Soc.*, 91 (1969) 1795.
80 L. FRIEDMAN AND J. H. BAYLESS, *J. Am. Chem. Soc.*, 91 (1969) 1790.
81 R. A. MOSS AND D. W. REGER, *J. Am. Chem. Soc.*, 91 (1969) 7539.
82 Z. RAPPOPORT, *Advan. Phys. Org. Chem.*, 7 (1969) 1.
83 D. LANDINI, F. MONTANARI, G. MODENA AND F. NASO, *J. Chem. Soc. B*, (1969) 243.
84 G. MARCHESE, F. NASO, N. TANGARI AND G. MODENA, *J. Chem. Soc. B*, (1970) 1196.
85 S. J. CRISTOL AND C. A. WHITTEMORE, *J. Org. Chem.*, 34 (1969) 705.

86 W. K. Kwok, W. G. Lee and S. I. Miller, *J. Am. Chem. Soc.*, 91 (1969) 468.
87 C. A. Grob and R. Spaar, *Tetrahedron Letters*, (1969) 1439.
88 G. Modena and U. Tonnellato, *J. Chem. Soc. B*, (1971) 374, 381.
89 G. Capazzi, G. Melloni, G. Modena and U. Tonnellate, *Chem. Commun.*, (1969) 1520.
90 G. Capazzi, G. Melloni and G. Modena, *J. Chem. Soc. B*, (1970) 2625.
91 Z. Rappoport and A. Gul, *J. Am. Chem. Soc.*, 91 (1969) 5246.
92 Z. Rappoport and Y. Apeloig, *J. Am. Chem. Soc.*, 91 (1969) 6734.
93 P. J. Stang and R. Summerville, *J. Am. Chem. Soc.*, 91 (1969) 4600.
94 S. A. Sherrod and R. G. Bergmann, *J. Am. Chem. Soc.*, 91 (1969) 2115.
95 M. Hanack and T. Bassler, *J. Am. Chem. Soc.*, 91 (1969) 2117.
96 M. Hanack, *Accounts Chem. Res.*, 3 (1970) 209.
97 U. Miotti and A. Ceccon, *Ric. Sci.*, 38 (1968) 824.
98 A. Ceccon, U. Miotti, U. Tonnellato and M. Padovan, *J. Chem. Soc. B*, (1969) 1084.
99 U. Miotti, U. Tonellato and A. Ceccon, *J. Chem. Soc. B*, (1970) 325.
100 S. K. Brauman and M. E. Hill, *J. Org. Chem.*, 34 (1969) 3381; *J. Am. Chem. Soc.*, 89 (1967) 2127, 2131.
101 D. Bethell, *Advan. Phys. Org. Chem.*, 7 (1969) 153.
102 F. G. Bordwell and M. W. Carlson, *J. Am. Chem. Soc.*, 91 (1969) 3951.
103 F. G. Bordwell and S. C. Crooks, *J. Am. Chem. Soc.*, 91 (1969) 2084.
104 F. G. Bordwell, E. Doomes and P. W. R. Corfield, *J. Am. Chem. Soc.*, 92 (1970) 2581.
105 F. G. Bordwell, R. G. Scamehorn and W. R. Springer, *J. Am. Chem. Soc.*, 91 (1969) 2087.
106 F. G. Bordwell and M. W. Carlson, *J. Am. Chem. Soc.*, 92 (1970) 3370, 3377.
107 F. G. Bordwell, R. G. Scamehorn and A. C. Knipe, *J. Am. Chem. Soc.*, 92 (1970) 2172.
108 R. G. Scamehorn and F. G. Bordwell, *J. Am. Chem. Soc.*, 93 (1971) 3410.
109 C. Rappe, L. Knutsson, N. J. Turro and R. B. Gagasian, *J. Am. Chem. Soc.*, 92 (1970) 2032.
110 A. Hassner, G. J. Matthews and F. W. Fowler, *J. Am. Chem. Soc.*, 91 (1969) 5046.
111 C. A. Grob, *Angew. Chem. Intern. Ed. Engl.*, 8 (1969) 535.

Chapter 4

Carbene and Carbenoid Formation and Reactions

W. KIRMSE

1. Introduction

The common overall reaction in carbene chemistry is methylene transfer. A (substituted) methylene fragment is eliminated from the "carbene precursor" and inserted into the "carbene acceptor". Kinetic methods have been widely employed to determine whether divalent carbon intermediates are involved in these reactions. Two major lines of investigation may be distinguished, (*i*) kinetic studies of the decomposition of carbene precursors, and (*ii*) determination of the relative rates of product formation. The mechanistic possibilities range from the intervention of "free" carbenes to fully concerted methylene transfer reactions.

2. Decomposition of carbene precursors

2.1 PHOTOLYSIS OF DIAZO COMPOUNDS

Diazo compounds show a weak absorption at longer wavelengths in the visible region, and a somewhat stronger absorption at shorter wavelengths. An $n \rightarrow \pi^*$ transition may be responsible for the weak absorption bands, but this is not clear. The most important primary photodissociative process is presumed to be reaction (1), *viz.*

$$R_2C{=}N_2 + h\nu \rightarrow R_2C: + N_2 \tag{1}$$

The question has been raised[1] whether photoexcited states of diazo compounds, rather than or in addition to carbenes, participate in product formation.

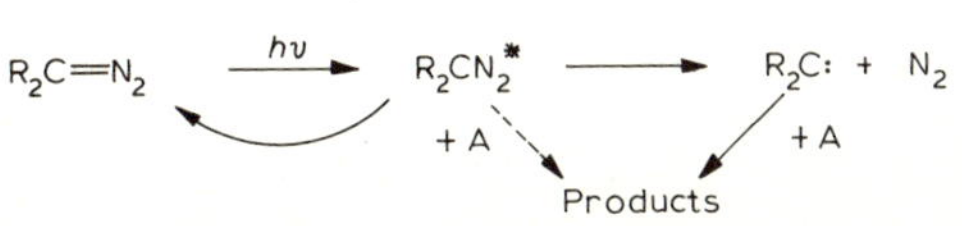

(2)

If so, the quantum yield of the photolysis should depend on the nature and concentration of the reactant(s). The fraction of excited molecules returning to the ground state will be lowered by diverting them to other reaction paths, as indicated in (2). This argument applies regardless of whether $R_2CN_2^*$ represents an electronically excited state or a vibrationally excited ground state. The quantum yield of the photolysis of diazomethane is not precisely known; the reported value, $\phi_{N_2} \sim 4$ (gas phase), may be in error because of the extinction coefficient used[2,3]. The initial rates of photolysis were found to be identical, however, in cyclohexane and cyclohexene[4]. Methylcyclohexane is the only product obtained from cyclohexane whereas cyclohexene produces a mixture of norcarane and methylcyclohexenes. Similarly, the photolysis of diphenyldiazomethane ($\phi_{N_2} = 0.78$ in methanol at $\lambda_{max} = 2880$ Å)[5] proceeded with identical initial rates in a variety of solvents which give rise to entirely different products[6].

The quantum yields of the photolysis of diazocarbonyl compounds are solvent-dependent (Tables 1 and 2). These data do not necessarily imply solvent participation in the photodissociative step. The variation of solvent produces

TABLE 1

QUANTUM YIELDS (ϕ_{N_2}) OF ETHYL DIAZOACETATE PHOTOLYSIS[7]

	Wavelength (A)				
Solvent	*2600*	*2804*	*3130*	*3650*	*4200*
Heptane	1.11	1.02	0.54	0.31	0.20
Ethanol	1.35	1.1	0.34	0.15	0.12
Methanol	1.42	1.1	0.37	0.14	0.12
Water	2.8		0.36	0.12	

TABLE 2

QUANTUM YIELDS OF THE PHOTOLYSIS OF SUBSTITUTED DIAZOACETOPHENONES[5,8], *viz.*

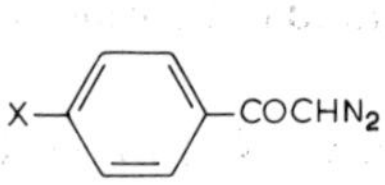

	n-Hexane		*Methanol*		*Ethanol–Water (20:80)*	
X	λ_{max} *(mμ)*	ϕ_{N_2}	λ_{max} *(mμ)*	ϕ_{N_2}	λ_{max} *(mμ)*	ϕ_{N_2}
H	(277)	0.55	294	0.46	298	0.40
CH_3	(286)	0.62	296	0.41	301	0.30
CH_3O	283	0.50	304	0.34		
Cl	(287)	0.62	299	0.41		
NO_2	299	0.31	307	0.18		

spectral shifts which can be explained as substantially due to mesomeric effects. A varying extent of conjugative phenomena in the photoexcited state may cause an excitation energy redistribution more or less favorable for dissociation.

2.2 THERMOLYSIS OF DIAZO COMPOUNDS

The kinetics of the pyrolysis of diazomethane have been investigated by several research groups[9–11]. Above 250°C the decomposition appeared to be homogeneous and first-order. Shantarovitch[9], studying the decomposition of diazomethane in the presence of nitrogen in a flow system, obtained $k = 8\times10^{10}\exp(-31{,}750/RT)\ \text{sec}^{-1}$. Rabinovitch and Setser[10] pyrolyzed diazomethane in the presence of hydrocarbons, and found $k = 1.2\times10^{12}\exp(-34{,}000/RT)\ \text{sec}^{-1}$. The products were the same as those formed by photolysis at room temperature. Kinetic studies over the range 293–393°C in the presence of excess hydrogen[11] afforded $k = 9\times10^{11}\exp(-32{,}000/RT)\ \text{sec}^{-1}$.

Schenck *et al.*[12] have argued that, on thermochemical grounds, the rate data reported above cannot refer to the unimolecular dissocation $CH_2N_2 \rightarrow CH_2{:} + N_2$. Recent estimates[13,14] of $\Delta H_f^\circ(CH_2)$ are 92 kcal.mole^{-1} and 94 kcal.mole^{-1}. While there have been a number of arguments[15–17] advocating various values of $\Delta H_f^\circ(CH_2N_2)$ ranging from 49 to 103 kcal.mole^{-1}, no agreement on the magnitude of this quantity has been reached. Schenck *et al.*[12] have combined $\Delta H_f^\circ(^1CH_2) = 96$ kcal.mole^{-1} and $\Delta H_f^\circ(CH_2N_2) = 49$ kcal.mole^{-1} to make the dissociation of CH_2N_2 endothermic by 47 kcal.mole^{-1}. This is more than the observed energy of activation. However, one should not expect a reliable estimate by combining a very high value of $\Delta H_f^\circ(CH_2)$ with the lowest value of $\Delta H_f^\circ(CH_2N_2)$. It should be remembered that $\Delta H_f^\circ(CH_2N_2) =$ 49 kcal.mole^{-1} is obtained from appearance potential data which, on the other hand, lead to an unreasonably low value of 66.4 kcal.mole^{-1} for ΔH_f°-(CH_2)[18]. Furthermore, the close agreement of the rate coefficients obtained in the presence of different reactants is hardly consistent with a bimolecular decomposition of diazomethane.

The thermal decomposition of diphenyldiazomethane in aprotic (hydrocarbon) solvents gives benzophenone azine and tetraphenylethane as the principal identifiable products[19], while in hydroxylic solvents the appropriate derivative of benzhydrol accompanies the azine[20]. Kinetic studies of the decomposition of diphenyldiazomethane in xylene[21] and in 1-methylnaphthalene[22] at about 100°C have shown that the reaction is first-order in diazo compound: this has been interpreted as supporting the intermediate formation of diphenylcarbene, *viz.*

References pp. 411–415

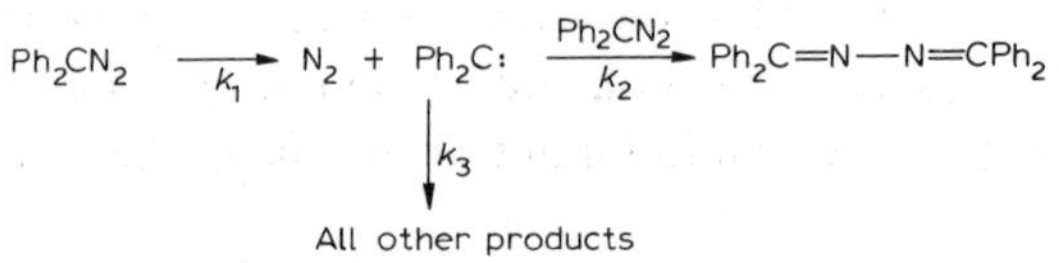

(3)

According to (3), the rate of disappearance of the diazo compound is given by

$$v = -\mathrm{d}[\mathrm{Ph_2CN_2}]/\mathrm{d}t = k_1[\mathrm{Ph_2CN_2}] + k_2[\mathrm{Ph_2C:}][\mathrm{Ph_2CN_2}] \tag{4}$$

and, assuming the establishment of a steady-state concentration of the intermediate carbene, this becomes

$$v = k_1[\mathrm{Ph_2CN_2}]\left(1 + \frac{k_2[\mathrm{Ph_2CN_2}]}{k_2[\mathrm{Ph_2CN_2}] + k_3}\right) \tag{5}$$

Integration and application of limiting conditions yields the expression

$$\ln\left\{\frac{[\mathrm{Ph_2CN_2}]}{[\mathrm{Ph_2CN_2}]_0}\left(\frac{(2k_2/k_3)[\mathrm{Ph_2CN_2}]_0 + 1}{(2k_2/k_3)[\mathrm{Ph_2CN_2}] + 1}\right)^{1/2}\right\} = -k_1 t \tag{6}$$

Results for acetonitrile, benzene, and toluene solutions of diphenyldiazomethane[23] are given in Table 3. k_2/k_3 has been evaluated from the yield of ketazine, and k_1 has been derived from k_{obs} by means of eqn. (6). Values of k_1 do not differ significantly in the two aromatic solvents. The somewhat larger value of k_1 in acetonitrile indicates a small medium effect. Clearly, the nature of the solvent has a marked influence on the partitioning of the intermediate though not on its formation.

In aqueous acetonitrile, the disappearance of diphenyldiazomethane followed first-order kinetics at all water concentrations[23]. The observed velocity

TABLE 3

THERMAL DECOMPOSITION OF DIPHENYLDIAZOMETHANE AT 85°C[23]

Solvent	$[Ph_2CN_2]_0$ *(M)*	10^4k_{obs} *(sec^{-1})*	10^4k_1 *(sec^{-1})*	*Ketazine (%)*
CH_3CN[a]	0.0533	1.73	1.55	19.7
	0.0148	1.60	1.58	7.0
Benzene	0.0538	1.60	0.92	61.5
	0.0182	1.30	1.10	36.5
Toluene	0.0504	0.90	0.82	10.6
	0.0252	0.97	0.85	6.9

[a] $\Delta H^\ddagger = 27.2 \pm 0.2$ kcal.mole^{-1}; $\Delta S^\ddagger = 0.0 \pm 0.4$ eu, from data at 65, 75, 85, and 95°C.

constants varied little in the range 0–10 *M* water, and, moreover, *decreased* in that range where the yields of benzhydrol rise from zero to > 90% of the total product. Eqns. (3)–(6) are adapted to these experiments by using k'_3[ROH] instead of k_3. k_2/k_3 values with ethanolic acetonitrile were not significantly different from those with aqueous acetonitrile for the same concentration of ROH. The value of k_2/k_3 did not change significantly when D_2O replaced H_2O. The reaction of diphenylcarbene with water and alcohol would thus appear to involve attack on the oxygen atom rather than protonation to give a diphenylcarbonium ion[6].

The thermal decomposition of phenyldiazomethane in anhydrous acetonitrile is of intermediate order in diazo compound between first and second[24]. Benzalazine, *cis*- and *trans*-stilbene are primary products. It has been assumed that the reaction proceeds by two paths, a unimolecular reaction in which phenylcarbene is formed, and a bimolecular reaction involving two molecules of phenyldiazomethane, *viz.*

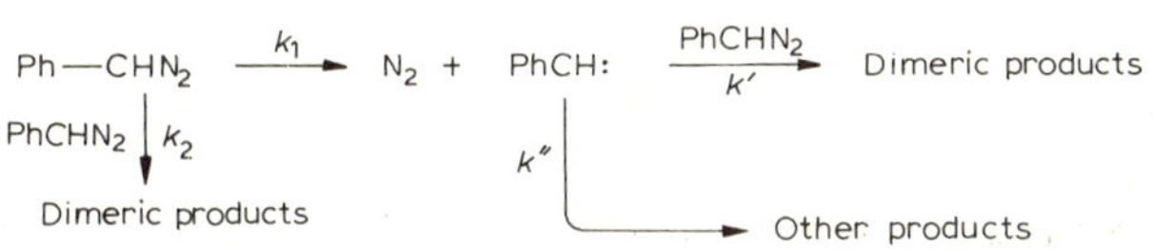

(7)

An approximate treatment, assuming $k' \gg k''$, yields $k_1 = 7.8 \times 10^{-6}$ sec^{-1} and $k_2 = 3.3 \times 10^{-4}$ l.mole^{-1}.sec^{-1} at 85°C. The rate of the thermal decomposition of phenyldiazomethane in acetonitrile increases rapidly with added water. This affords a contrast with diphenyldiazomethane, and suggests that a considerable fraction of the diazo compound reacts directly with water.

A large variety of compounds is susceptible to attack both by the carbene

TABLE 4

DECOMPOSITION OF DIAZOFLUORENE (4×10^{-3} MOLE) IN XYLENE (10.0 ML)[25]

Reactant Y (4×10^{-3} *mole*)	*Temp.* (°C)	$10^2 k_2$ (*l.mole*$^{-1}$.*sec*$^{-1}$)	$10^6 k_1$ (*sec*$^{-1}$)
Dibenzyl azodicarboxylate	60	3.3	0.8
Azodibenzoyl	60	1.3	0.8
Diethyl azodicarboxylate	80	2.1	3.3
Azobenzene	80		3.3
Ethyl phenylazocarboxylate	80		3.3
Phenylazocarboxamide	80		3.3
ω,ω'-Azotoluene	80		3.3
p-Nitrosodimethylaniline	80	0.3	3.3
Ethyl cyanoacetate	110		50
p-Toluonitrile	110		50
Nitromethane	110		50

and the diazo compound. In these cases, reactions of the second-order may compete with the unimolecular decomposition of the diazo compound ((8) and Table 4).

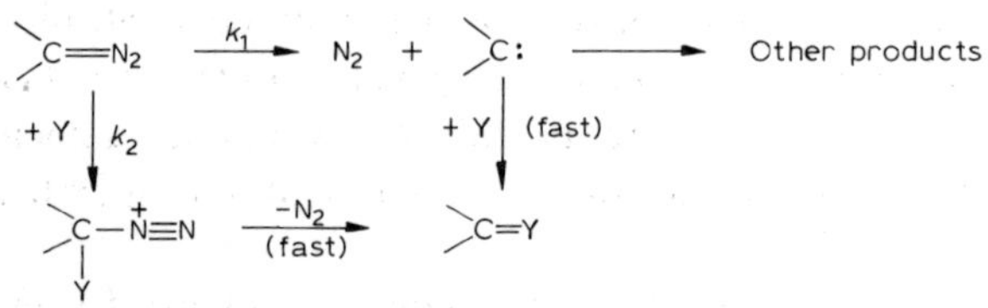

(8)

In contrast to the results in Table 4, two bimolecular processes have been claimed to compete in the reaction of 9-diazofluorene with triarylphosphines[26], *viz.*

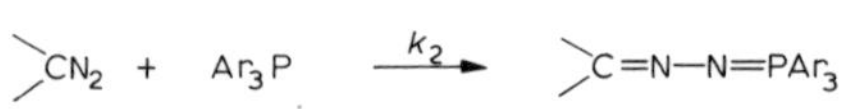

(9)

(9′)

k_2' is independent of the structure of the phosphine [$10^3 k_2' = 0.080$ (35°C), 0.15 (40°C), 0.20 (45°C) l.mole^{-1}.sec^{-1}] whereas k_2 increases upon introduction of donor substituents in the *p*-position of Ar. It has been suggested that the observed reaction (9′) may be catalytic rather than purely thermal[25].

Frequently, volumetric methods have been used in studying the thermal decomposition of diazo compounds. The results of such studies should be interpreted with caution since labile, nitrogen-containing intermediates (*e.g.* pyrazolines) may be formed in the first step of reaction (10).

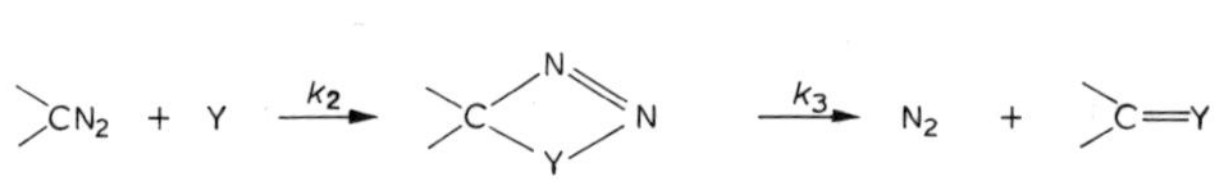

(10)

If $k_2 \gg k_3$, reactions of the first order are observed which are by no means indicative of a carbene mechanism. Thus, the decomposition of 9-diazofluorene was of apparent first order in the presence of diethyl fumarate, acrylonitrile, and other reactants containing polar double bonds[27].

2.3 PHOTOLYSIS OF KETENE

Earlier work has indicated a marked difference in the photolytic behavior of ketene at wavelengths around 2700 A from that at wavelengths around 3700 A. When ketene is irradiated with light of 2700 A, the primary dissociative yield is unity ($\phi_{CO} = 2.0$ within experimental error) and is not diminished by addition of oxygen[28]. At 3650 A dissociation is pressure- and temperature-dependent[28,29], and the quantum yield is markedly reduced by addition of oxygen[29]. Plots of twice the quantum yield of carbon monoxide against ketene concentration are linear over a wide range of concentration and show no signs that ϕ_{CO} reaches a limiting value at high concentrations[28,30].

At short wavelengths, the methylene formed is almost entirely in its singlet state, whereas at long wavelengths the photolysis of ketene produces predominantly triplet methylene[31] (methods for the estimation of singlet and triplet methylene fractions are discussed on p. 393). The intervention of a triplet state of ketene is indicated by quenching experiments with biacetyl[32] and, more conclusively[33], *cis*-butene-2.

Addition of *cis*-butene-2 to the photolysis of ketene causes a reduction in the rate of formation of carbon monoxide. Ethylene production diminishes more rapidly and is approaching zero while the yield of carbon monoxide is still finite. The decline in production of ethylene is accompanied by a steady increase in isomerisation of *cis*-butene-2 to the *trans* isomer. This behavior is ascribed to an energy transfer involving triplet-state ketene molecules[33].

A mechanism of ketene photolysis recently proposed by Strachan and Thornton[34] accounts for most of these observations.

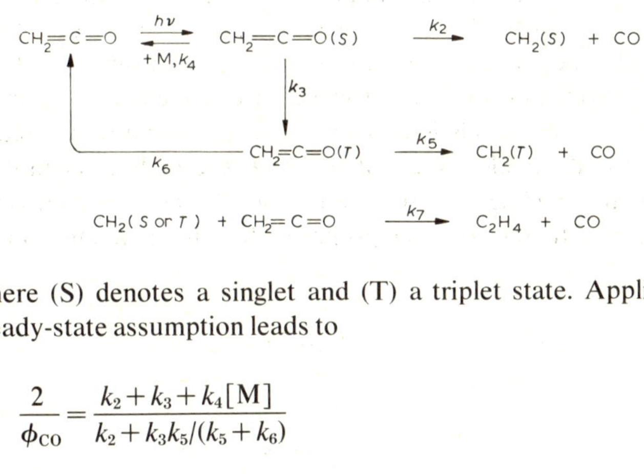

(11)

where (S) denotes a singlet and (T) a triplet state. Application of the usual steady-state assumption leads to

$$\frac{2}{\phi_{CO}} = \frac{k_2 + k_3 + k_4[\mathrm{M}]}{k_2 + k_3 k_5/(k_5 + k_6)} \tag{12}$$

and

$$\alpha = \frac{k_2}{k_2 + k_3 k_5/(k_5 + k_6)} \tag{13}$$

where α is the fraction of singlet methylene. The rate coefficient k_2 is assumed to have the form, $k_2 = \nu(1 - E_0/E)^{s-1}$ (Slater Theory). At short wavelengths, when the photon energy is large, k_2 will be larger than k_3 and $k_4[\mathrm{M}]$ at ordinary pressures, in which case $2/\phi_{CO} = 1$ and $\alpha = 1$. At long wavelengths, k_2 will become either zero or at least much smaller than k_3 so that $\alpha = 0$ and

$$\frac{2}{\phi_{CO}} = \frac{k_5 + k_6}{k_5} + \frac{(k_5 + k_6)k_4}{k_5 k_3}[M] \tag{14}$$

The slope and intercept of $2/\phi_{CO}$ *versus* [M] at four temperatures between 37 and 300°C (3660 A) were consistent with

$$k_4 = 9.61 \times 10^9 T^{1/2}\ \text{l.mole}^{-1}.\text{sec}^{-1}$$

$$k_3 = 1.66 \times 10^9 \exp(-1500/RT)\ \text{sec}^{-1}$$

$$k_5/k_6 = 20.7 \exp(-3500/RT)$$

Sulfur hexafluoride and octafluorocyclobutane were found to be equally as efficient as ketene at deactivating the excited singlet-state molecules, a result which strengthens the assumption that deactivation occurs on every collision.

2.4 Simmons–Smith reaction

Methylene iodide and a zinc–copper couple react in ether solution to give bis-(iodomethyl)zinc · zinc iodide[35]. Closely related reagents are accessible by the reaction of zinc halides with diazomethane[36]. These reagents have been widely used in the synthesis of cyclopropanes from olefins. Norcarane formation from cyclohexene and bis(iodomethyl)zinc was found to be second order, $k = (6.3 \pm 0.5) \times 10^{-3}$l.mole^{-1}.sec^{-1} at 0°C[37]. These findings cannot be reconciled with the intermediacy of free methylene. A one-step methylene transfer mechanism has therefore been proposed for the Simmons–Smith reaction, *viz*.

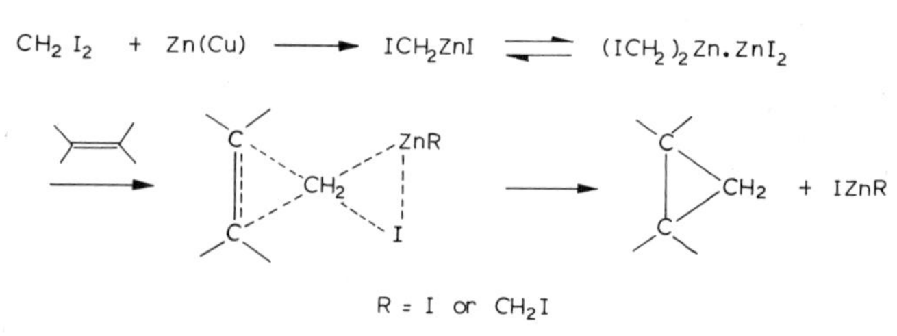

(15)

2.5 Decomposition of alkylcarbenoids

α-Elimination of hydrogen chloride from 1-chloroalkanes and α-elimination of halogen from *gem*-dihaloalkanes produce cyclopropanes by intramolecular insertion. The intermediates of α-elimination reactions are more selective than those produced in the photolysis or thermolysis of diazo compounds. Discrimination in favor of tertiary and secondary C–H bonds, as compared to primary C–H bonds, is more pronounced in the α-elimination reactions[38]. The deuterium isotope effect of the carbenoid intramolecular insertion depends on the nature of the halogen[39], and is appreciably larger than that observed in the decomposition of diazoalkanes[40].

α-Iodoneopentyllithium has been prepared in ether solution at $-116°C$, and characterized by protolysis which gives $> 91\%$ of neopentyl iodide. Goldstein and Dolbier[39] have tried to determine on the basis of isotope effects whether the formation of 1,1-dimethylcyclopropane at $-90°C$ ($97 \pm 3\%$ yield) from α-iodoneopentyllithium proceeds in one step or by way of an intermediate B, *viz.*

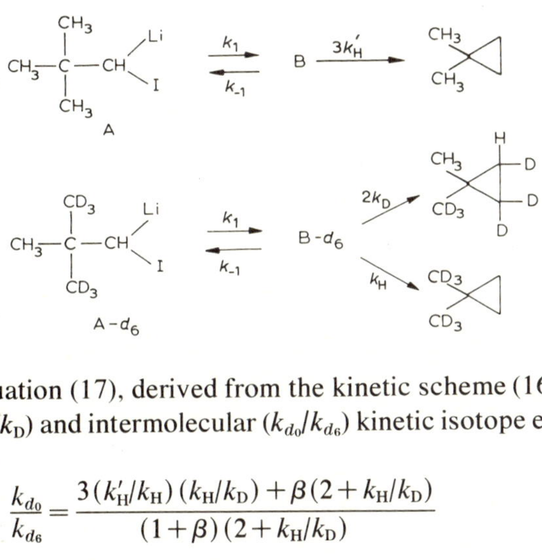

(16)

Equation (17), derived from the kinetic scheme (16), relates the intramolecular (k_H/k_D) and intermolecular (k_{d_0}/k_{d_6}) kinetic isotope effects.

$$\frac{k_{d_0}}{k_{d_6}} = \frac{3(k'_H/k_H)(k_H/k_D) + \beta(2 + k_H/k_D)}{(1+\beta)(2 + k_H/k_D)} \tag{17}$$

where $\beta = 3k'_H/k_{-1}$. $k_H/k_D = 2.76 \pm 0.09$ has been found by NMR analysis of the cyclopropanes obtained from $A - d_6$, and $k_{d_0}/k_{d_6} = 1.50 \pm 0.12$ by mass spectrometry of the cyclopropanes after $3.6 - 16\%$ conversion of mixtures of A and $A - d_6$. If B is absent in the carbenoid insertion, $\beta = 0$. $k_{d_0}/k_{d_6} = 1.74$ is calculated from eqn. (17) assuming $\beta = 0$ and $k'_H/k_H = 1$. The calculated k_{d_0}/k_{d_6} is somewhat larger than the experimental value (1.50). The agreement may be improved by assuming $k'_H/k_H < 1$, and has been interpreted by Goldstein to indicate the absence of B. However, β also approaches zero in the case of a fully established pre-equilibrium ($k_{-1} \gg k'_H$). Furthermore, eqn. (17) is rather sensitive to experimental error ($\beta = 0.48 \pm 0.30$ from $k'_H/k_H = 1$ and the experimental values of k_H/k_D and k_{d_0}/k_{d_6}). Thus, the outcome of this elegantly designed experiment is less definitive than one might have hoped.

References pp. 411–415

2.6 α-ELIMINATION OF BENZYLIC HALIDES

A number of benzylic halides and onium salts under the influence of base undergo an apparent α-elimination to yield "dimeric" olefins, *viz.*

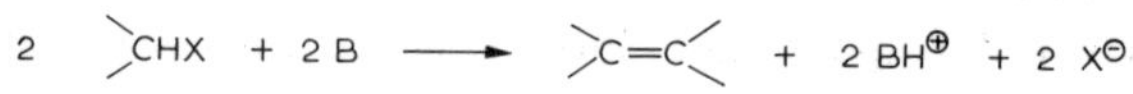

(18)

Two types of mechanism have been suggested.

(*i*) Carbene mechanism

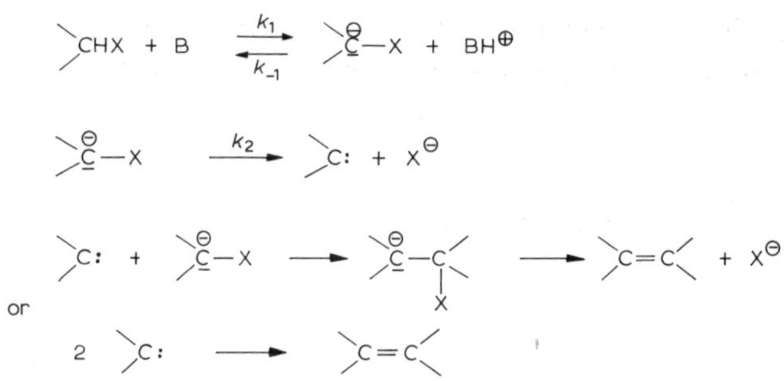

(19a)

(19b)

(19c)

(19d)

(*ii*) Bimolecular displacement mechanism

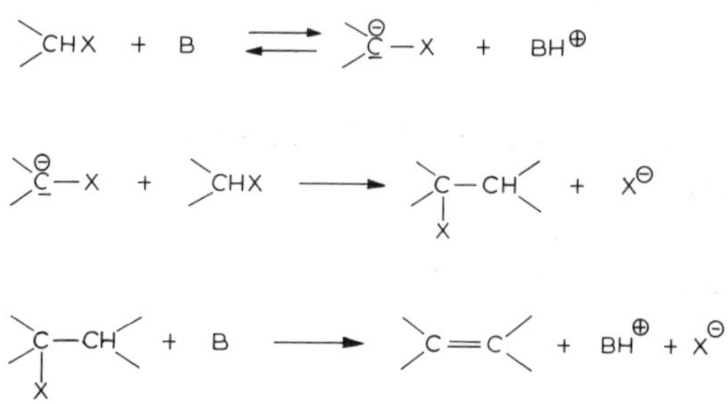

(20a)

(20b)

(20c)

With sodamide in liquid ammonia, "dimeric halides" (20b) were isolated in good yield[41]. Experiments with optically active α-phenylethyl chloride indicated that the displacement reaction (20b) proceeds with Walden inversion[42]. Clearly, the bimolecular displacement mechanism operates in the benzyl and benzhydryl series at low temperatures.

At higher temperatures, and with aqueous or alcoholic base, isolation of "dimeric halides" is not possible. Provided that proton transfers occur rapidly, the slow steps for the two mechanisms would be (19b) and (20b), steps which are respectively uni- and bimolecular. Determination of the kinetic order thus provides a method of distinguishing the operative mechanism.

The reaction of 4-nitrobenzyl chloride in the presence of alkali to give 4,4′-

dinitrostilbene is first order in each reactant[43]. 4-Nitrobenzylchloride reacts faster (60 times in acetone, and 35 times in dioxan) than benzyl chloride (which produces benzyl alcohol). The considerable reactivity of 4-nitrobenzyl chloride suggests that the first stage in its reaction with alkali is removal of a proton. 4-Nitrobenzyl chloride, recovered after partial reaction in dioxan–D_2O, contains deuterium. The reversibility of deprotonation, combined with first-order kinetics, suggests a carbene mechanism (scheme 19). If $k_{-1} \gg k_2$ no deuterium isotope effect should be observed with α-deutero-4-nitrobenzyl chloride. The observation of a small effect ($k_H/k_D = 1.28$) indicates that k_{-1} is of similar magnitude as k_2.

The product of reaction of sodium hydroxide with *p*-nitrobenzyldimethylsulfonium tosylate in water at 60°C is a mixture of *cis*- and *trans*-4,4′-dinitrostilbenes in 99% yield[44]. Again, the reaction is first order in each reactant. After a short reaction time in DO^-/D_2O the benzylic protons were at least 97% exchanged for deuterium[45]. This eliminates the carbanion displacement mechanism. The over-all sulfur isotope effect k_{32}/k_{34} is 1.0066 ± 0.0008, lower than that of the first-order solvolysis of *t*-butyl dimethylsulfonium iodide (1.0177 ± 0.0014)[44]. With a series of sulfonium salts, the electrical nature of the substituent on the leaving group appeared to affect the yield of 4,4′-dinitrostilbene only slightly[46]. In contrast, 3-nitrobenzyldimethylsulfonium bromide, 4-nitrobenzyldimethylammonium and -phosphonium bromides afforded little, if any 4,4′-dinitrostilbene[45].

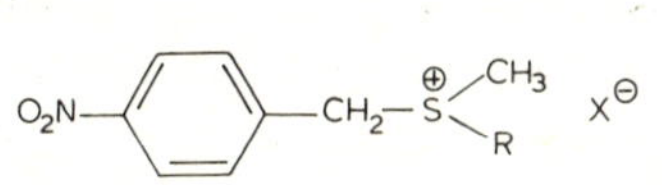

9-Halogenofluorenes also undergo this type of elimination, forming 9,9′-bifluorenylidene. A thorough kinetic study of this system has been made, using benzyltrimethylammonium hydroxide in aqueous *t*-butanol and potassium *t*-butoxide in *t*-butanol as the base[47]; under both conditions the conversion of 9-bromofluorene into bifluorenylidene is much slower than the proton transfer to and from the 9-carbon. The initial rate of formation of bifluorenylidene is proportional to the square of the concentration of 9-bromofluorene. This result suggests strongly the operation of the bimolecular displacement mechanism. Consequently, the observed rate coefficients (Table 5) refer to reaction (20b). This conclusion has been checked by estimating the rate of bifluorenylidene formation from 9-bromo-9,9′-bifluorenyl (the "dimeric" halide); this was found to be 30–50 times that of bifluorenylidene formation from 9-bromofluorene[48]. Thus, after the first 2–3% of reaction, the rate of appearance of bifluorenylidene from 9-bromofluorene is indeed the rate of formation of 9-bromo-9,9′-bifluorenyl.

A variety of substituted 9-bromofluorenes has also been investigated

TABLE 5

RATE COEFFICIENTS[a] FOR THE DISAPPEARANCE OF SUBSTITUTED 9-BROMOFLUORENES IN *t*-BUTANOL AT 30°C[49]

	$C_6H_5CH_2\overset{+}{N}(CH_3)_3\overset{-}{OH}$		*t-BuOK*	
Substituent	*Order*	k^b	*Order*	k^b
	2	7.3	2	0.37
3-CH_3	2	1.6		
2-OCH_3	2	22	2	1.35
3-Br	2	1.4×10^3	2	1.8×10^2
2-Cl	2	3.0×10^3	1	1.3×10^{-2}
2-Br	2	7.7×10^3	1	1.8×10^{-2}
4-CN	2	8.9×10^6	1	0.85
2-NO_2	2	2.4×10^7	1	1.32

[a] Referring to media for which 4-O_2N–C_6H_4-NH-/4-O_2N-C_6H_4-NH_2 = 0.89

[b] $l.mole^{-1}.sec^{-1}$ or sec^{-1}, as appropriate

(Table 5)[49]. In *t*-butanol containing benzyltrimethylammonium hydroxide and water, all compounds are converted into the corresponding substituted bi-fluorenylidenes according to a second-order kinetic law. With potassium *t*-butoxide as base, however, compounds containing powerful electron-with-drawing substituents give first-order kinetics. Dipolar aprotic additives (*e.g.* dimethyl sulfoxide, tetramethylenesulfone) also promote the change in kinetic form. By a detailed examination of the kinetics of these reactions in the presence of an added electrolyte it has been shown that, for systems obeying the first-order kinetic law, the rate-determining process is the dissociation of a rapidly formed ion pair consisting of a 9-halofluorenyl anion and an alkali metal cation[49a]. Carbene formation does not appear to play any part in these reactions.

Although a similar effect of ion pairing would not be expected in aqueous solution, first-order kinetics are no longer considered as a sufficient criterion of a carbene mechanism.

2.7 SOLVOLYSIS OF ETHYNYLCARBINYL HALIDES

The solvolysis of 3-halo-3-alkyl-1-butynes (Table 6) is very slow as compared to the solvolysis of *t*-alkyl halides, but is strongly accelerated by base. Under these conditions the reactions are first-order in both halide and base concentrations, and proceed predominantly with displacement of the halogen (only 3% elimination was found under basic conditions, as compared to 35% in the neutral solvolysis)[50–52].

TABLE 6

SOLVOLYSIS OF ETHYNYLCARBINYL HALIDES, $RR'C(X)-C\equiv CH$, IN 80% ETHANOL AT 25°C

R	*R'*	*X*	*Neutral solvent* 10^7k_1 (sec^{-1})	*Basic solvent* 10^4k_2 ($l.mole^{-1}.sec^{-1}$)	*Ref.*
CH_3	CH_3	Cl	2.37	7.1	53
			2.3	4.52	56
			2.50	3.53	52
CH_3	C_2H_5	Cl	3.45	3.17	52
CH_3	$i\text{-}C_4H_9$	Cl	9.78	7.38	52
CH_3	CH_3	Br	46	115	54, 55
CD_3	CD_3	Br	25	88	55

Replacement of the acetylenic hydrogen by an alkyl group speeds the rate of neutral solvolysis by a factor of 10^3, but base catalysis is no longer observed[53].

The base-catalyzed exchange of the acetylenic protons in C_2H_5OD/D_2O is appreciably faster than the base-promoted solvolysis[54]. Although the rate of exchange was not measured, a lower limit of about sixteen times the solvolysis rate was established by analysis of the recovered halide after partial conversion. This result eliminates rate-determining formation of the conjugate base and suggests S_N1 ionization of the conjugate base to give a zwitterion-carbene, *viz.*

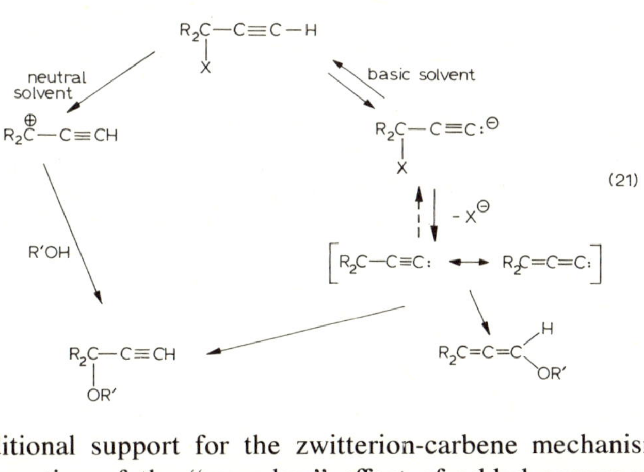

(21)

Additional support for the zwitterion-carbene mechanism comes from the observation of the "mass-law" effect of added common ion salts[54]. In the presence of 0.22 *M* sodium bromide, the initial rate of the second-order reaction of 3-bromo-3-methyl-1-butyne is depressed by about 40%. The same concentration of either sodium perchlorate or sodium nitrate depresses the rate by only about 18%.

Hexadeutero-3-bromo-3-methyl-1-butyne undergoes the neutral solvolysis 1.84 times slower than the protonated analog (Table 6). The second-order hydroxide ion-promoted solvolysis is slowed by this deuterium substitution by only a factor of 1.31 (ref. 55). This is reasonably explained in terms of differing demand for stabilization of positive charge in the two transition states.

Similar conclusions have been drawn from the effects of hydrostatic pressure on the solvolysis of 3-chloro-3-methyl-1-butyne[56]. The activation volume of the neutral solvolysis is -15 cm^3.mole^{-1}, in close agreement with the values observed for *t*-alkyl halides. The net effect of pressure on the base-promoted reaction is remarkably small ($\Delta V^* = +2$ cm^3.mole^{-1}). The most reasonable interpretation is that the departure of chloride ion is accompanied by the creation of another pair of charges, *i.e.* the zwitterion-carbene.

The ambident zwitterion-carbene intermediate can react with nucleophiles to give products originating from both α and γ attack. Hydroxide and ethoxide ions prefer the propargyl to the allenyl position exclusively, while azide prefers it by a factor of 8, thiophenoxide by a factor of 1.6, and bromide attacks the two positions almost equally[57]. The reaction with trimethylamine produces quaternary ammonium halides; these have propargylic structure when $R = R' = CH_3$ or $-(CH_2)_5-$. When R and R′ are larger than CH_3 the products are allenic[58]. Diethyl sodiomethylmalonate affords propargylic and allenic products in the ratio of 2.6:1 (ref. 59). Under aprotic conditions the zwitterion-carbene adds on to olefins with formation of alkenylidencyclopropanes[60] (*cf.* p. 407).

2.8 BASIC HYDROLYSIS OF HALOFORMS

Although relevant observations[61] had been made as early as 1862, it was not until 1950 that the mechanism of the basic hydrolysis of haloforms was elucidated by Hine and his co-workers. Today the reactions of haloforms with base are the most thoroughly investigated examples of α-elimination.

2.8.1 *Two-step (carbanion) mechanism*

The basic hydrolysis of haloforms is first-order with respect to both haloform and hydroxide ion[62,63]. Several arguments eliminate nucleophilic attack of hydroxide ion on haloform, S_N2 displacement (22), as a possible mechanism.

$$\begin{aligned} &CHX_3 + HO^{\ominus} \xrightarrow{\text{slow}} CHX_2OH + X^{\ominus} \\ &CHX_2OH \xrightarrow[HO^{\ominus}, H_2O]{\text{fast}} CO, HCO_2^{\ominus} \end{aligned} \qquad (22)$$

TABLE 7

KINETIC PARAMETERS FOR THE BASIC HYDROLYSIS OF HALOFORMS AND RELATED COMPOUNDS IN 66.7% DIOXAN–WATER[64]

	10^4k (l.mole^{-1}.sec^{-1})			$\Delta H^{\ddagger}$	$\Delta S^{\ddagger}$
	0°C	*25.3°C*	*35.7°C*	*(kcal.mole^{-1})*	*(eu)*
$CHCl_3$	0.0104	0.602	3.19	26.3	+10.5
$CHBrCl_2$	0.352	16.0	86.5	25.6	+14.8
$CHBr_2Cl$	0.137	8.01	39.8	26.0	+14.4
$CHBr_3$		3.20	15.4	25.0	+9.4
$CHCl_2I$	0.118		39.9	27.2	+18.8
CHBrClF	191	2150		25.8	+28.4
CH_2Br_2			0.0061		
CH_2BrCl			0.025		

The replacement of an α hydrogen of an alkyl halide by halogen decreases S_N2 reactivity. Chloroform, however, is about one thousandfold more reactive in basic hydrolysis than methylene chloride[62]. Every bromine-containing haloform studied (Table 7) is at least 600 times as reactive toward hydroxide ions in 66.7% aqueous dioxan as methylene bromide[64]. Toward weakly basic nucleophiles, such as thiophenoxide ion, the predicted reactivity order is obeyed; haloforms have been found to be less reactive than the corresponding methylene halides[65]. The reaction of haloforms with sodium thiophenoxide is strongly accelerated, however, by the presence of hydroxide ions[62,64]. These observations are quite unexplainable in terms of scheme (22).

Investigation of the base-catalyzed hydrogen–deuterium exchange of haloforms shows that trihalocarbanions are formed in a fast and reversible reaction. Hine *et al.*[66–70] have measured the rates of the hydroxide-catalyzed transformations of deuterated haloforms to the corresponding protonated compounds in homogeneous aqueous solution. With non-fluorine-containing haloforms the rate coefficient for the basic hydrolysis was only 0.1–0.001% of that for deuterium exchange, *i.e.* the exchange reaction was pseudo-unimolecular. With haloforms such as $CHCl_2F$, CHBrClF, and $CHBr_2F$ the rates of hydrolysis and deuterium exchange were comparable, and the kinetic treatment was more complicated. (For difluorinated compounds see p. 391.)

Comparison of the resultant data reveals that the α halogen substituents facilitate carbanion formation in the order I ~ Br > Cl > F, which is almost the reverse of that expected from the inductive effect. *d*-Orbital resonance and polarizability increase in the order F < Cl < Br < I, and it is difficult to assess their relative importance in carbanion stabilization[68].

The kinetic isotope effect of proton abstraction has been measured directly with the rapidly hydrolyzable haloforms $CHCl_2F$ ($k_H/k_D = 1.76$ at 0°C and 1.52

at 20°C)[67] and CHBrClF ($k_H/k_D = 1.74$ at 0°C and 1.69 at 15°C)[68]. With chloroform the rate of exchange of $CDCl_3$ in alkaline H_2O has been compared to the rate of exchange of $CHCl_3$ in alkaline D_2O; the resultant estimate of k_H/k_D was 1.48 at 20°C[66,67]. The small isotope effects suggest a transition state with little stretching of the carbon–hydrogen bond.

Having established that there is a pre-equilibrium between haloforms and trihalocarbanions, the rate-determining step of the over-all hydrolysis remains to be located among the subsequent reactions of the trihalocarbanions. Two mechanisms are compatible with the observed kinetics; *viz.*

$$X_3C^{\ominus} \xrightarrow{\text{slow}} X_2C: + X^{\ominus}$$

$$X_2C: \xrightarrow[\text{HO}^{\ominus},\ H_2O]{\text{fast}} CO, HCO_2^{\ominus} \tag{23}$$

$$X_2C: + Y^{\ominus} \longrightarrow X_2YC^{\ominus}$$

$$X_3C^{\ominus} + H_2O \xrightarrow{\text{slow}} H_2\overset{\oplus}{O}-\overset{\ominus}{C}X_2 + X^{\ominus}$$

$$H_2\overset{\oplus}{O}-\overset{\ominus}{C}X_2 \xrightarrow[\text{HO}^{\ominus},\ H_2O]{\text{fast}} CO, HCO_2^{\ominus} \tag{24}$$

$$X_2C^{\ominus} + Y^{\ominus} \longrightarrow X_2YC^{\ominus} + Y^{\ominus}$$

Mechanism (23) is a S_N1 reaction of the trihalocarbanion, proceeding via the dihalocarbene intermediate, whereas mechanism (24) involves S_N2 displacement at the trihalocarbanion. It is impossible to distinguish between (23) and (24) if water, present in large excess, is the only nucleophile available. The two mechanisms lead to different predictions, however, concerning the effect of additional nucleophilic anions. The case of chloride ion, added to the basic hydrolysis of chloroform, is particularly instructive. No kinetic effect at all should result from the operation of mechanism (24). With mechanism (23), however, the combination of $Cl^{\ominus}$ with $:CCl_2$ to give $Cl_3C^{\ominus}$ is a reversal of the rate-determining step. Therefore the over-all rate will be decreased by a "mass-law" effect.

The rate of the basic hydrolysis of chloroform in the presence of 0.08 and 0.16 *N* sodium chloride was shown to be 92.9 and 86.6%, respectively, of that observed in the presence of an equal amount of sodium perchlorate, nitrate, or fluoride[63]. A specific salt effect appears improbable as chloride ions slow the basic hydrolysis of chloroform as effectively in the presence of 1.5 *N* sodium perchlorate as in its absence.

Additional support for mechanism (23) is provided by the success of the following quantitative treatment of the decrease in rate produced by added halide ions[63]. Dichlorocarbene may react with any of the three reagents, water (rate coefficient, k_w), hydroxide ion (k_h), and halide ion $X^{\ominus}$ (k_x). If k represents the initial rate coefficient in the presence of a given concentration of sodium perchlorate and k' represents that found in the presence of the same concentration of halide ion, then

$$\frac{k'}{k} = \frac{k_w[H_2O] + k_h[HO^{\ominus}]}{k_w[H_2O] + k_h[HO^{\ominus}] + k_x[X^{\ominus}]} \tag{25}$$

where $[H_2O]$, $[HO^{\ominus}]$, and $[X^{\ominus}]$ are initial concentrations in the run for which k' is the initial rate coefficient. Solution of (25) for the competition factor k_x/k_w yields

$$\frac{k_x}{k_w} = \frac{\left([H_2O] + \frac{k_h}{k_w}[HO^{\ominus}]\right)\left(\frac{k}{k'} - 1\right)}{[X^{\ominus}]} \tag{26}$$

A plot of log (k_x/k_w) against the nucleophilic constants n_x of Swain and Scott[71] gives a straight line. In calculating the values of k_x/k_w (Table 8) from k/k' by means of eqn. (26), $k_h/k_w = 229$ was *chosen* in accord with the nucleophilic constant of hydroxide ion. The linear relationship for the various halide ions shows that the relative amounts by which these ions decrease the rate of the basic hydrolysis of chloroform are proportional to their relative nucleophilicities.

The knowledge of the activation volume (ΔV^*) has also been helpful in distinguishing between mechanisms (23) and (24)[56,72]. Since in the equilibrium

$$CHCl_3 + HO^{\ominus} \rightleftharpoons Cl_3C^{\ominus} + H_2O$$

neither the number of ionic charges nor the number of molecules changes, ΔV is likely to be very small. Any participation of water in the rate-determining

TABLE 8

COMPETITION FACTORS TOWARD DICHLOROCARBENE ($0.08\ N$ NaOH)[63]

Ion	*Concn.* *(N)*	k'/k	k_x/k_w	n_x
Chloride	0.08	0.929	56.8	3.04
Chloride	0.16	0.866	61.1	
Chloride	0.20	0.778	79.1	
Bromide	0.08	0.819	164	3.89
Bromide	0.20	0.571	158	
Iodide	0.08	0.608	480	5.04
Iodide	0.16	0.426	532	

step (scheme 24) is expected to lead to a negative activation volume, whereas simple dissociation of $Cl_3C^{\ominus}$ (scheme 23) would be characterized by a volume expansion. The observed value, $\Delta V^* = +16\ \text{cm}^3.\text{mole}^{-1}$, suggests that the transition state leads to free dichlorocarbene.

Accepting the carbene mechanism (23) as generally valid, the over-all rate coefficients for the basic hydrolysis of haloforms may be expressed in terms of the individual steps as follows.

$$\text{CHXYZ} + \text{HO}^{\ominus} \underset{k_{-1}}{\overset{k_1}{\rightleftharpoons}} \text{XYZC}^{\ominus} + \text{H}_2\text{O} \xrightarrow{k_2} \text{XYC:} + \text{Z}^{\ominus}$$

$$k = \frac{k_1(k_2/k_{-1})}{1+(k_2/k_{-1})} \tag{27}$$

The values of k_1 are known or may be closely approximated from data obtained on deuterated haloforms by use of deuterium kinetic isotope effects. A correlation of the values of k_2/k_{-1} has been attempted by means of eqn. (28)[73].

$$\log\frac{(k_2/k_{-1})_{\text{CHXYZ}}}{(k_2/k_{-1})_{\text{CHCl3}}} = M_\text{X}+M_\text{Y}+N_\text{Z}+\log(n_z/3)+(c-d)\log\frac{(k_1)_{\text{CHXYZ}}}{(k_1)_{\text{CHCl}_3}} \tag{28}$$

This equation contains parameters for (*1*) the relative abilities of the various halogens to stabilize dihalocarbenes (M), (*2*) the relative ease with which halogens separate as anions from the intermediate trihalocarbanions (N), and (*3*) the relative carbanion character of the transition states for protonation and carbene formation ($c-d$). Log $(n_z/3)$ is a statistical correction for the number of Z atoms present in the haloform. The parameters derived from measurements on 12 haloforms[74] are given in Table 9. The k_2/k_{-1} values calculated with the aid of these parameters approximate the experimental values with a maximum deviation of 12%. According to the M parameters of Table 9, halogens facilitate carbene formation in the order F $\gg$ Cl $>$ Br $>$ I. Assuming a singlet state of the dihalocarbenes, this order may be attributed to the relative abilities of the halogens to supply electrons to the carbon atom by mesomeric effects, *viz.*

TABLE 9

M AND N PARAMETERS OF EQN. (28)[74]

M_F	3.071	N_Br	1.090
M_Cl	0.000 (standard)	N_I	0.327
M_Br	−1.164	N_Cl	0.000 (standard)
M_I	−1.696	$c-d = 0.114$	

$$\mathrm{X{-}C{-}X} \leftrightarrow \overset{\oplus}{\mathrm{X}}{=}\overset{\ominus}{\mathrm{C}}{-}\mathrm{X} \leftrightarrow \mathrm{X}{-}\overset{\ominus}{\mathrm{C}}{=}\overset{\oplus}{\mathrm{X}}$$

The N parameters indicate the unusual order Br > I > Cl for the abilities of the halogens to depart as halide ions from the trihalocarbanions. An investigation of the hydrolysis of bromochloroiodomethane in the presence of sodium iodide has shown that chloroiodocarbene is in fact the major intermediate[74].

2.8.2 *One-step (concerted) mechanism*

As mentioned above, the replacement of bromine or chlorine in haloforms by fluorine *decreases* the rate coefficient of carbanion formation, k_1. The over-all rate of basic hydrolysis, however, *increases* because fluorine stabilizes the carbene intermediate. Because of the increase in k_2, the ratio k_2/k_{-1} of haloforms CHX_2F is in the order of 2–10.

One should expect carbanion formation to become rate-controlling upon introduction of a second fluorine atom, with a corresponding *decrease* of the over-all rate of hydrolysis. Haloforms $CHXF_2$ were found, however, to hydrolyse much faster than they could conceivably form carbanions[69,75] (Table 10). The basic hydrolysis of deuterobromodifluoromethane was shown to proceed with negligible isotopic exchange[69]. The lack of isotopic exchange would be consistent with irreversible carbanion formation, but this assumption does not account for the large rate coefficients. Hine and Langford[69] have, therefore, suggested that the removal of a proton and the elimination of a halide ion are concerted processes, *viz.*

$$\mathrm{HO^{\ominus} + HCF_2X \rightarrow \overset{\ominus}{H}O{-}{-}{-}H{-}{-}{-}CF_2{-}{-}{-}\overset{\ominus}{X} \rightarrow H_2O + {:}CF_2 + X^{\ominus}} \qquad (29)$$

TABLE 10

KINETIC DATA FOR THE BASIC HYDROLYSIS OF HALOFORMS $CHXF_2$ IN AQUEOUS SOLUTION[69,75]

	10^4k(l.mole^{-1}.sec^{-1})		$\Delta H^\ddagger$ (kcal.mole^{-1})	$\Delta S^\ddagger$ (eu)
$CHClF_2$	0.17 (0°C)	8.0 (25°C)	24.4	9.0
$CHBrF_2$	20.8 (0°C)	374 (20.1°C)	22.3	11.1
$CDBrF_2$	10.3 (0°C)			
CHF_2I	96 (0°C)	600 (21.2°C)		

"The loss of halide ion appears to add driving force to the proton removal.... Thus it is possible to go directly to what is probably the most stable of the dihalocarbenes and by-pass what would be a relatively unstable trihalomethyl anion"[69]. The kinetic isotope effect observed with $CDBrF_2$ ($k_H/k_D = 2.02 \pm$

0.14) exceeds the isotope effects found in trihalocarbanion formation (p. 387), a fact which is also consistent with a concerted mechanism.

2.9 DECOMPOSITION OF PHENYL(BROMODICHLOROMETHYL)-MERCURY

Seyferth *et al.*[76] have studied the kinetics of the reaction of phenyl(bromodichloromethyl)mercury with olefins, in which CCl_2 is transferred to the olefins with formation of *gem*-dichlorocyclopropanes in high yield. The *initial* rates are essentially independent of olefin concentration and proportional to the first power of the mercurial concentration. They do, however, show a small dependence on the nature of the olefin, the observed reaction rates decreasing as the nucleophilicity of the olefin decreases (Table 11).

TABLE 11

KINETICS OF OLEFIN–$PhHgCCl_2Br$ REACTIONS AT 39.0°C IN BENZENE SOLUTION[76]

Olefin (0.0202 M)	*PhHgBr (M)*	*10^5k (sec^{-1})*
Cyclooctene	0	1.18 ± 0.03
	0.0037	1.00 ± 0.04
2,3-Dimethyl-2-pentene	0	1.48 ± 0.03
	0.0037	1.47 ± 0.02
1-Heptene	0	0.54 ± 0.01
	0.0037	0.37 ± 0.01

The initial rates are retarded by added phenylmercuric bromide. This effect is greatest for 1-heptene, less for cyclooctene, and not perceptible for 2,3-dimethyl-2-pentene. These observations provide strong evidence that the decomposition of $PhHgCCl_2Br$ is reversible and suggest the following mechanism.

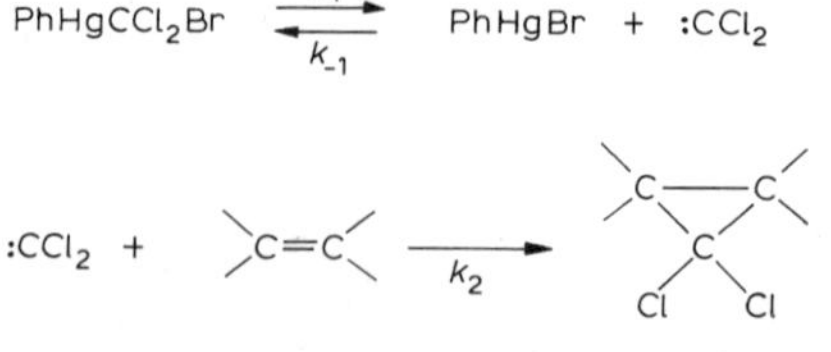

(30)

$$v = \frac{k_1\,[PhHgCCl_2Br]}{1 + \dfrac{k_{-1}\,[PhHgBr]}{k_2\,[olefin]}} \tag{31}$$

The rate expression (31) may be derived from the mechanism, and has been used to calculate rate *versus* time plots which were in good agreement with experimentally determined curves. In the early stages of the reaction [olefin] $\gg$ [PhHgBr]. Therefore, the rate coefficients k (Table 11) obtained from *initial* rates closely approximate k_1. The k_{-1}/k_2 ratio can be calculated from the initial rates in the presence and absence of PhHgBr. It is found to be ~1.0 for the cyclooctene case, and 2.5 for the 1-heptene case. The error is large in this estimation, which does not agree very well with the relative reactivities of the olefins toward dichlorocarbene as determined in competitive experiments, k (1-heptene) = 0.184 k (cyclooctene). The high reactivity of 2,3-dimethyl-2-pentene toward dichlorocarbene (18.4 as compared to cyclooctene) is responsible for the negligible effect of added PhHgBr in this system (Table 11).

The initial rate of the mercurial + 2,3-dimethyl-2-pentene reaction was also measured in 1,2-dimethoxyethane, and found to be slower by a factor of 0.56 than the reaction in benzene. The small solvent effect is inconsistent with a strongly solvent-complexed CCl_2 intermediate.

It should be noted that CH_2 transfer to olefins using monohalomethylmercury compounds proceeds by a concerted mechanism similar to that of the Simmons–Smith reaction[76a]. Bis(bromomethyl)mercury is stable in refluxing benzene; only in the presence of a CH_2 acceptor does CH_2 extrusion from the mercurial take place. The nature of the olefin has a large effect on the rate of the reaction (2,3-dimethyl-2-pentene, 26; 3-ethyl-2-pentene, 4.2; cyclohexene, 1.0; 1-heptene, 0.225).

The reaction of $Hg(CH_2Br)_2$ with a number of organosilicon hydrides to give methylsilanes also appears to involve direct transfer of CH_2. A Hammett plot showed a good correlation between $\log k_{rel}$ and σ^0 of substituted phenyldimethylsilanes[76b].

3. Relative reactivities of carbenes and carbenoids

3.1 RELATIVE RATES OF METHYLENE REACTIONS

3.1.1 Determination of singlet and triplet methylene fractions

The concern with the chemical behavior of singlet and triplet methylene was instigated by the spectroscopic findings of Herzberg and Shoosmith[77,78]. Their results indicated that the flash photolysis of diazomethane yields methylene molecules in excited singlet states which are subsequently energetically degraded to the triplet ground state through collision with inert gas molecules. The spectroscopic data do not exclude the direct formation of triplet methylene, together with the single species, in the photolytic process. Consequently, triplet methylene may be produced in photolysis systems by two paths. (*i*) Intersystem crossing of an excited singlet precursor to an excited triplet state which dissociates to give triplet methylene. In the vapor phase photochemistry of many organic carbonyl compounds, excited triplet-state molecules are formed, and ketene (diazomethane) might exhibit similar behavior. (*ii*) Spin inversion of initially formed singlet methylene through collisional deactivation.

Information concerning electronic configurations of methylene during the course of chemical reactions can only be adduced by an interplay of intuitive reasoning and experiment. The most widely employed criterion for arriving at the electronic multiplicity of a carbene has been advanced by Skell and Woodworth[79] and concerns the stereochemistry of addition of such a fragment to olefinic bonds. It states that singlet carbenes should add in one step to yield cyclopropanes in a stereospecific fashion (32). Triplet carbenes ought to react by way of a diradical intermediate which, being essentially a free rotator, leads to nonstereospecific ring closure. Rearrangements of the intermediate diradical, as indicated in reaction (33), may compete with cyclopropane formation.

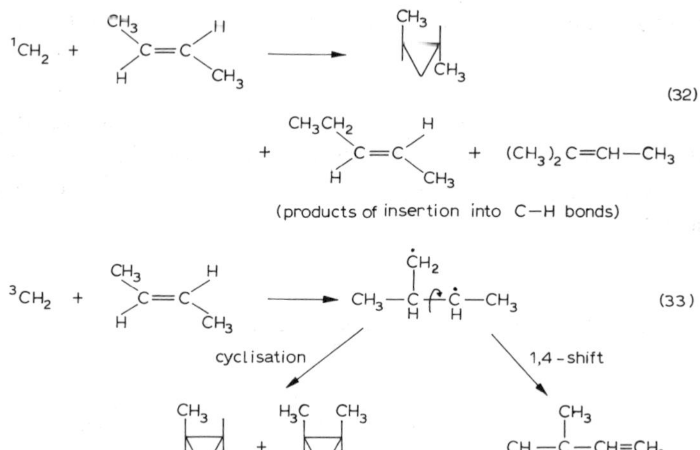

(32)

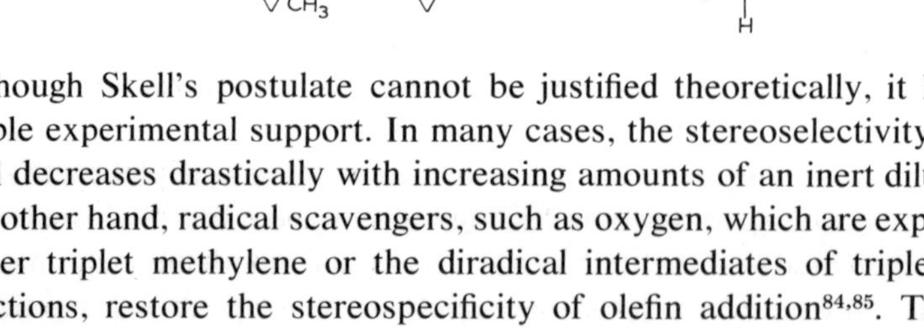

(33)

Although Skell's postulate cannot be justified theoretically, it has received ample experimental support. In many cases, the stereoselectivity of the addition decreases drastically with increasing amounts of an inert diluent[80–83]. On the other hand, radical scavengers, such as oxygen, which are expected to trap either triplet methylene or the diradical intermediates of triplet methylene reactions, restore the stereospecificity of olefin addition[84,85]. The mercury-photosensitized decomposition of ketene[86] and the benzophenone-sensitized photolysis of diazomethane[87] also lead to non-stereospecific addition of methylene. Both reactions involve energy transfer from the excited triplet state of the sensitizer.

For a further differentiation of singlet from triplet methylene it has been assumed that singlet methylene undergoes mainly insertion, reaction (34), and triplet methylene undergoes mainly abstraction, reaction (35)[82,85,87].

$$^{1}CH_2 + R{-}H \longrightarrow R{-}CH_3 \qquad (34)$$

$$\begin{array}{ccccccc} ^{3}CH_2 + R{-}H & \longrightarrow & CH_3\cdot & + & R\cdot & \longrightarrow & R{-}CH_3 \\ & & \downarrow & & \downarrow & & \\ & & CH_3{-}CH_3 & & R{-}R & & \end{array} \qquad (35)$$

The validity of this assumption has been questioned[88]; the total amount of insertion products $R{-}CH_3$ was found to exceed the possible contribution from radical recombination, as estimated from the yields of R–R and $CH_3{-}CH_3$. Ring and Rabinovitch[88] have, therefore, suggested that methylene both in the singlet and triplet states inserts into carbon–hydrogen bonds. The calculations of Ring and Rabinovitch were based on the preposition that under their experimental conditions (800-fold excess of nitrogen, approximately 1.6 atm) methylene was entirely in its triplet state. It should be pointed out, however, that similar conditions did not afford the same relative amounts of stereoisomeric cyclopropanes starting from either *cis*- or *trans*-2-butene[81].

The application of both criteria to gas-phase reactions is complicated further by the formation of vibrationally excited products. Both the insertion and addition reactions of methylene are exothermic by approximately 93 kcal. mole^{-1} (based on recent estimates[89] of ΔH_f° $(CH_2) = 94$ kcal.mole^{-1}). Vibrationally excited alkanes and alkenes may dissociate into free radicals, and excited cyclopropanes may undergo structural and geometrical isomerizations unless collisionally stabilized[90]. The occurrence of "hot molecule" reactions excludes any reasonable estimation of singlet and triplet methylene fractions. The data presented in the following paragraphs have been taken from experiments at "high-pressures", which are thought to ensure complete collisional deactivation of excited reaction products.

For a methylene–olefin system, both the addition and abstraction reaction of triplet methylene should be taken into account. The photolysis of ketene[84,85] and diazomethane[85] in the presence of *trans*-2-butene has been studied in some detail. The fraction of triplet methylene formed can be determined by taking the ratio of the sum of triplet products to total products, where

$$\text{sum of triplet products} = \text{cyclopropanes from triplet } CH_2 + \text{3-methyl-1-butene} + 2\,C_2H_6 + CH_4 \qquad (36)$$

cf. reaction (33) for the formation of cyclopropanes and 3-methyl-1-butene and (35) for the formation of ethane. Methane is thought to arise through further hydrogen abstraction by methyl radicals. *cis*-2-pentene has also been found in small yields and was added to the triplet products[85]. It may also result, however, from singlet methylene insertion into *cis*-2-butene formed by ketene-photosensitized isomerization of *trans*-2-butene[33]. The assignment of "cyclopropanes from triplet CH_2" requires some estimate of the *trans*/*cis* ratio of 1,2-dimethylcyclopropane formed in reaction (33), *e.g.* by a comparison of the amounts of *trans*- and *cis*-1,2-dimethylcyclopropane relative to *trans*-2-pentene both in the presence and absence of oxygen[84]. Such calculations yield a *trans*/*cis* ratio in the range 1.25–1.65. The mercury-photosensitized decomposition of ketene in the presence of *trans*-2-butene[86] leads to much higher *trans*/*cis* ratios (3–4) which are inconsistent with other data of the direct photolysis. These comments may serve to indicate the uncertainties involved in the determination of triplet methylene fractions.

On the basis of eqn. (36), the yields of triplet methylene from ketene are found to be roughly 15% at 2800 A, 15% at 3130 A, 30% at 3340 A and 40% at 3660 A[84]. Equation (36) neglects the formation of "singlet-type" products from triplet methylene via radical recombination (35). If the reduction in yield of *trans*-2-pentene and 2-methyl-2-butene by added oxygen is also considered to be the result of triplet CH_2 reactions, a recalculation of triplet methylene fractions yields 20% triplet at 2800 A, 55% at 3340 A and 60% at 3660 A for ketene photolysis[91]. Assuming that 10% added oxygen suppresses all triplet products and measuring product yields relative to an internal standard, Eder and Carr[91a] arrived at $29 \pm 3\%$ triplet CH_2 at 3130 A, and $87 \pm 2\%$ at 3660 A. For the diazomethane–*trans*-2-butene system, the fraction of triplet methylene was found to be 12% at 3550–4000 A[85].

In a methylene–alkane system, the determination of the electronic multiplicity of CH_2 rests solely on the relative rates of insertion and abstraction processes. In the ketene–propane system[85] the interesting feature is the wavelength dependence of the ratios of *n*-butane to isobutane. At short wavelengths the ratio was found to be 1.93 while at long wavelengths it was about 0.37. This effect is reasonably attributed to the wavelength dependence of the triplet methylene fraction. The relevant reactions are

$$
\begin{aligned}
{}^1CH_2 + CH_3CH_2CH_3 &\xrightarrow{k_1} CH_3CH_2CH_2CH_3 \\
&\xrightarrow{k_2} (CH_3)_2CH \\
{}^3CH_2 + CH_3CH_2CH_3 &\xrightarrow{k_3} CH_3\cdot + CH_3CH_2CH_2\cdot \xrightarrow{k_{3,3}} n\text{-hexane} \\
&\qquad\qquad \searrow \xrightarrow{k_{3,4}} \text{2-methylpentane}
\end{aligned} \tag{37}
$$

$$\xrightarrow{k_4} CH_3\cdot + CH_3\dot{C}HCH_3 \xrightarrow{k_{4,4}} \text{2,3-dimethylbutane}$$

$$CH_3\cdot + CH_3CH_2CH_2\cdot \xrightarrow{k_5} CH_3CH_2CH_2CH_3$$

$$CH_3\cdot + CH_3\dot{C}HCH_3 \xrightarrow{k_6} (CH_3)_3CH$$

$$2CH_3\cdot \xrightarrow{k_7} CH_3\text{–}CH_3$$

Among the C_6 hydrocarbons, the yield of *n*-hexane was very small as compared to that of 2,3-dimethylbutane. It was concluded that the concentration of *n*-propyl radicals in this system is small, and the formation of *n*-butane by radical recombination (k_5) may be neglected in a first-order approximation. The fraction of triplet methylene then corresponds to $2C_2H_6$ + isobutane from triplet CH_2. The latter quantity may be derived from a steady-state treatment by use of the known relative rate coefficients of radical recombinations, *viz.*

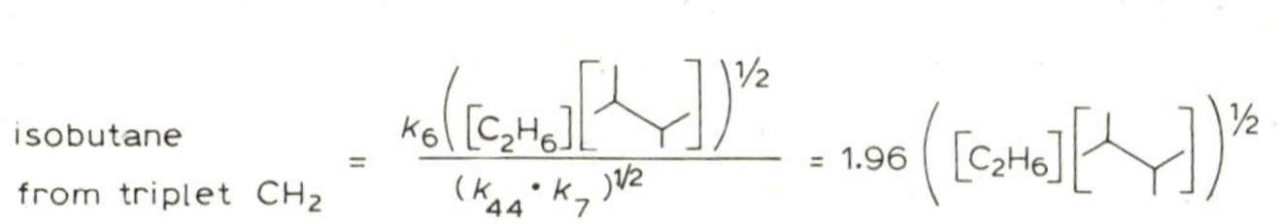

(38)

The fraction of triplet methylene calculated on the basis of eqn. (38) is 20% at 2800 A and 77% at 3800 A[85], slightly higher than the values obtained for the ketene–*trans*-2-butene system.

A somewhat different method has been employed for the diazomethane–isopentane system[91]. The formation of pentenes is attributed to the disproportionation of pentyl radicals and, therefore, indicative of abstraction by triplet methylene. The assumption that all of the products eliminated by added oxygen arise through reactions involving triplet methylene leads to the expression

$$\text{fraction of triplet } CH_2 = \frac{[\text{pentenes}] + [\text{hexanes}] - [\text{hexanes}]_{O_2}}{[\text{pentenes}] + [\text{hexanes}]} \tag{39}$$

where subscript O_2 indicates experiments with added oxygen (which produce exclusively hexanes). This method, unlike those mentioned above, requires the determination of absolute yields by means of an internal standard. The proportion of the methylenes that are triplet was calculated to be at least 15%. The actual fraction of triplet methylene is probably closer to 20% as eqn. (39) omits pentyls (P.) participating in 2 P. → P–P, 2 P. → pentanes + pentenes, and P. + C_2H_5 → C_7H_{16}. (Ethyl radicals arise in diazomethane–alkane systems through the reaction $CH_3\cdot + CH_2N_2 \rightarrow C_2H_5\cdot + N_2$.) The invariance of pentene yields over the range of wavelengths 3660–4358 A indicates that triplet methylene formation is independent of the excitation energy in diazomethane, in contrast with ketene where the fraction of triplet methylene shows a marked increase at longer wavelengths.

No typical "triplet products" are observed when methylene reacts in the liquid phase. The addition to olefins is stereospecific, and oxygen does not affect the product ratios. These findings do not necessarily imply the absence of triplet methylene. A "solvent-cage" effect may be responsible for inhibiting all radical–radical reactions other than methyl–alkyl recombination. A greater selectivity in the liquid phase than in the gas phase with added oxygen[91] may be indicative of triplet participation.

3.1.2 Intramolecular competition

The measurement of the relative rates of insertion of methylene into carbon–hydrogen bonds in hydrocarbons can, in principle, be accomplished by measuring the relative rates of formation of the stable hydrocarbon products. However, there are two complications. (*a*) The products are vibrationally excited (*cf.* p. 395) and may dissociate into free radicals. The subsequent recombination of the free radicals so formed may produce hydrocarbons in proportions quite different from those resulting from the insertion reaction. Such effects are eliminated at high pressures which ensure complete collisional deactivation of excited products. (*b*) Free radicals may also arise through abstraction reactions by triplet methylene. As outlined in the preceeding section, photolysis of diazomethane at all wavelengths, and photolysis of ketene at < 2800 A produce only minor amounts of triplet CH_2 (< 20%). Radical scavengers may be added to inhibit the recombination of radicals eventually formed in these systems.

Early work on the gas-phase interaction of methylene with carbon–hydrogen bonds[92,93] showed selective reactivity in the order, tertiary C–H > secondary C–H > primary C–H. Methylene from diazomethane[92] was found to be less selective than methylene from ketone[93]. The differences in the relative reactivity of methylene from the two sources have been attributed to excess translational energy of methylene from CH_2N_2[94,95]. Recently, however, it has been shown that the photolysis of ketene produces more triplet CH_2 than the photolysis of diazomethane (*cf.* preceeding section). Part of the reported selectivity[92,93] must be attributed to triplet-methylene effects.

On the basis of recent work with radical scavengers (Table 12) it appears that there is no real difference in the reactivity of singlet methylene from diazomethane and that from ketene towards C–H bonds. Very probably, methylene is translationally thermalized before reacting. This assumption is consistent with the lack of wavelength dependence for singlet methylene

TABLE 12

RELATIVE RATES OF SINGLET METHYLENE REACTIONS WITH ALKANES

Alkane	*Methylene precursor*	*Conditions*	*Relative rates of insertion "per bond"* Primary C–H	Secondary C–H	Tertiary C–H	*Ref.*
Propane	CH_2N_2	Gas phase, O_2 added	1.00	1.11		96
	CH_2CO	Gas phase, 2800 A	1.00	1.15[a]		85
	CH_2CO	Gas phase, 3130 A, NO added	1.00	1.30		97
n-Butane	CH_2CO	Gas phase, 3130 A, NO added	1.00	1.31		97
Isobutane	CH_2CO	Gas phase, 3130 A, NO added	1.00		1.2	97
n-Pentane	CH_2N_2	Liquid, unfiltered radiation, 15°C	1.00	1.04		98
Isopentane	CH_2N_2	Gas phase, unfiltered radiation, O_2 added	1.00[b]	1.22	1.39	91
	CH_2CO	Gas phase, unfiltered radiation, O_2 added	1.00[b]	1.22	1.42	91
	CH_2N_2	Liquid, 4358 A	1.00	1.22	1.51	91
2,3-Dimethyl-butane	CH_2N_2	Liquid, unfiltered radiation, 15°C	1.00		1.2	98

[a] Calculated from the experimental value 1.55, taking into account 20% of triplet CH_2 and the ratio *n*-butane/isobutane = 0.22 for butane formed from triplet CH_2.

[b] 2-Methylpentane and 3-methylpentane were formed in "per bond" ratios of 1.03 (CH_2N_2) and 1.06 (CH_2CO).

insertion[91], and is also consistent with evidence that the collision yield for singlet insertion is much less than that for addition to double bonds (Table 13). Obviously, insertion does not occur on every collision of methylene with hydrocarbons, and only a few collisions would be needed for translational thermalization.

Whereas a small but significant selectivity of methylene in gas-phase reactions is well established, the evidence for liquid-phase reactions is less definitive. Studies using diazomethane[98,99], phenylcyclopropane[100], or 9,10-dihydro-9,10-methanophenanthrene[100] as the CH_2 precursor have been interpreted as indicating random attack on primary, secondary and tertiary C–H bonds. However, some results for the liquid phase indicate slight selectivity[98], and recent liquid-phase experiments with isopentane[91] show even greater selectivity at the tertiary position than does singlet-methylene insertion in the gas phase.

Comparatively few data are available for methylene–olefin systems with added radical scavengers. Since the triplet fraction in the photolysis of diazomethane is known to be small ($<20\%$) and independent of the wavelength (p. 397), some diazomethane reactions without added scavengers have also been entered in Table 13. No reliable data on the relative rates of insertion and addition exist for the simplest olefin, ethylene. Diazomethane–ethylene reactions were studied from the point of view of "hot molecule" chemistry, and were not extended to sufficiently high pressures[101]. The results of the photolysis of ketene in the presence of ethylene with light of 3100 and 3600 A extrapolate to a 85% yield of cyclopropane at 1600–2400 torr[102], *i.e.* the relative rate of insertion into a carbon–hydrogen bond compared with addition to the double bond is 0.044. An even smaller rate ratio of 0.028 has been observed with ketene–ethylene mixtures at 7 atm total pressure[103]. Irradiation was performed in pyrex bulbs with the full light of a medium pressure mercury arc. Obviously only the longer wavelengths are effective under these conditions, and a large fraction of triplet methylene should be expected. Correction for 60% of triplet methylene (*cf.* p. 397) gives a rate ratio of approximately 0.07 which agrees with those reported in Table 13.

The data collected in Table 13 show a fair constancy for the rates of insertion per bond. The slight upward trend with increasing alkyl substitution scarcely exceeds the limits of experimental error. The presence of a trifluoromethyl group, however, decreases the rate of methylene insertion relative to addition significantly[104]. The deactivation becomes as great as a factor of 7 to 8 for the rate of allylic C–H insertion to addition for 4,4,4-trifluoro-1-butene, as compared to the corresponding rate ratio for 1-butene.

The oxygen atom of aliphatic ethers has a small promoting effect on methylene insertion into neighboring carbon–hydrogen bonds[109–111]. By analogy with the behavior of alkanes, discrimination appears to be more pronounced in the

TABLE 13

RELATIVE RATES OF SINGLET METHYLENE REACTIONS WITH OLEFINS

Reactant	Methylene precursor	Conditions (photolysis)	Relative rates of insertion (addition 1.00)			Ref.
			Vinylic C–H	Allylic C–H	Non-allylic C–H	
Propene	CH_2CO	Gas phase, 3200 A, O_2 added	0.070	0.097		104
	CH_2N_2	Gas phase, unfiltered radiation[a]	0.09	0.11		95, 105
3,3,3-Trifluoro-propene	CH_2CO	Gas phase, 3200 A, O_2 added	0.033 (H-1)			104
			0.030 (H-2)			104
1-Butene	CH_2CO	Gas phase, 3200 A, O_2 added	0.070	0.125	0.12	104
4,4,4-Trifluoro-1-butene	CH_2CO	Gas phase, 3200 A, O_2 added	0.050 (H-1)	0.015		104
			0.023 (H-2)			
trans-2-Butene	CH_2CO	Gas phase, O_2 added, 2800 A	0.088	0.13		84
		3130 A	0.082	0.13		84
		3340 A	0.092	0.125		84
	CH_2N_2	Gas phase, 4358 A[a]	0.11	0.15		106
Isobutene	CH_2N_2	Gas phase, unfiltered radiation[a]	0.09	0.11		94
Cyclohexene	CH_2N_2	Liquid phase, unfiltered radiation[a]	0.125	0.155[b]		98
			0.12	0.16[b]		87
Allene	CH_2N_2	Gas phase, 4358 A[a]	0.11			107
1,3-Butadiene	CH_2N_2	Gas phase, 4358 A[a]	0.085			108

[a] Participation of some triplet methylene (< 20%) in these experiments appears likely (*cf.* p. 397).

[b] 3-Methylcyclohexene and 4-methylcyclohexene have not been separated; *IR* analysis indicates 1 : 1 composition.

gas phase than in the liquid phase (Table 14). Insertion of methylene into C–H bonds is accompanied by a small amount of elimination which is thought to proceed by way of an oxonium ylide[110,112]. Little, if any, carbon–oxygen insertion occurs in the liquid phase. In the case of ethyl ether, the absence of C–O insertion has been demonstrated with the aid of $^{14}CH_2N_2$; no ^{14}C was found in a position adjacent to oxygen[112]. Less than 2% of tetrahydropyran was obtained from tetrahydrofuran in the liquid phase[109,111], but in the gas phase 11% and 21% of tetrahydropyran were formed with diazomethane and ketene, respectively, as the methylene precursor. Added oxygen did not reduce the yield of tetrahydropyran; triplet methylene produced by the mercury-photosensitized decomposition of ketene did not yield tetrahydropyran. Thus, C–O insertion must be due to singlet methylene, and the phase dependence of this reaction remains obscure.

TABLE 14

RELATIVE RATES OF METHYLENE INSERTION INTO ETHERS

(Photolysis of Diazomethane)

Ether	*Conditions*	α	β	γ	α'	*Ref.*
$CH_3OCH_2CH_3$	Gas, O_2	1.49	1.00		1.13	111a, b
	Liquid	1.30	1.00		1.00	
$(CH_3CH_2)_2O$	Liquid	1.23	1.00			109
$CH_3OCH_2CH_2CH_3$	Gas	1.29	1.00	0.83	1.02	110
	Gas, O_2	1.12	1.00	0.95	1.00	111a
	Liquid	1.00	1.00	0.98	0.98	111a
$CH_3OCH(CH_3)_2$	Gas, O_2	1.73	1.00		1.39	111a
	Liquid	1.21	1.00		1.13	
$CH_3OC(CH_3)_3$	Gas, O_2		1.00		1.54	111a
	Liquid		1.00		1.25	
Tetrahydrofuran	Gas	1.62	1.00			111
	Gas, O_2	1.35	1.00			111a
	Liquid	1.26	1.00			109

3.1.3 Intermolecular competition

Methylene, produced by photolysis of diazomethane in mixtures of liquid alkanes, does not discriminate between the components to any extent which exceeds intramolecular discrimination[99]. Deviations from random insertion were within experimental error.

Relative rates of reaction of a number of olefins with singlet and triplet methylene have been determined in the gas phase[113]. Singlet methylene was generated by direct photolysis of ketene at ~2600 A and the triplet by mercury-photosensitized decomposition of ketene. The evaluation of relative rate coefficients for two olefins (Ol_1, Ol_2) has been based on the competition of the two olefins for methylene. α in eqn. (40) represents the respective fractions of methylene consumed in formation of the particular products P. Values of α were determined in separate experiments with the olefins taken singly, *viz.*

$$\begin{aligned} CH_2 + Ol_1 &\rightarrow \alpha_1 P_1 + \cdots \\ CH_2 + Ol_1 &\rightarrow \alpha_1 P_2 + \cdots \end{aligned} \quad \text{where } \alpha = \frac{[P]}{[CO]} \tag{40}$$

At very low conversions of the olefins, the following expressions can be used alternatively for the determination of the relative rate coefficients (k_2/k_1).

$$\frac{k_2}{k_1} = \frac{[P_2]\alpha_1[Ol_1]}{[P_1]\alpha_2[Ol_2]} \tag{41a}$$

$$\frac{k_2}{k_1} = \left(\frac{\alpha_1[CO]}{[P_1]} - 1\right)\frac{[Ol_1]}{[Ol_2]} \tag{41b}$$

Where values of k_2/k_1 from (41a) and (41b) were compared, the differences between the two sets were quite small. The k_2/k_1 values from eqn. (41b), based on the determination of only one product, are listed in Table 15. The values in column 2 of Table 15 are affected only slightly if allowance is made for 10% of triplet methylene which may be produced in the direct photolysis of ketene at

TABLE 15

RELATIVE RATES OF REACTIONS OF $^3CH_2$ AND $^1CH_2$ WITH OLEFINS IN THE GAS PHASE AT 24°C

	$^3CH_2$	$^1CH_2$		
	k_2/k_1	$(k_2/k_1)_{f=0}$[a]	$(k_2/k_1)_{f=0.1}$[b]	$(k_2/k_1)_{add}$[c]
Ethylene	0.35	0.51	0.54	0.73
Propene	0.35	0.65	0.71	0.88
1-Butene	0.56	0.33	0.89	0.91
Isobutene	1.00[d]	1.00[d]	1.00[d]	1.00[d]
cis-2-Butene	0.33	0.70	0.78	0.67
trans-2-Butene	0.31	0.71	0.79	0.76
2-Methyl-2-butene	0.64	1.08	1.17	1.13
2,3-Dimethyl-2-butene	0.96	1.10	1.13	0.79
1,3-Butadiene	6.64	2.01	1.73	2.86

[a] Fraction of triplet methylene assumed to be zero.
[b] Fraction of triplet methylene assumed to be 0.1 (10%).
[c] Calculated from the values in column 3, $f = 0.1$.
[d] Taken as standard for comparison.

~ 2600 A along with the singlet methylene (*cf.* p. 402). The only significant change is found for 1,3-butadiene which reacts much faster with triplet than with singlet methylene. The values of k_2/k_1 in columns 1–3 of Table 15 represent the relative rates of *overall* attack on olefins by CH_2. There are strong indications that triplet methylene interacts predominantly with the double bond of olefins. Thus, the k_2/k_1 values of column 1 of Table 15 represent, to a good approximation, the relative rates of addition of $^3CH_2$. Recent studies[113a,b] have confirmed the existence of an appreciable concentration of singlet methylene in the photolysis of mercury–ketene systems. These results may imply small changes in the reactivity of triplet methylene (Table 15, column 1). The values in column 2 and 3 of Table 15 should be corrected for the competing insertion processes to give $(k_2/k_1)_{add}$ of $^1CH_2$ (column 4 of Table 15). This correction has been achieved with the aid of data given in Table 13. For some olefins not listed in Table 13 allowance was made for C–H insertion by analogy with the other olefins.

The data in Table 15 show that both $^3CH_2$ and $^1CH_2$ exhibit relatively little discrimination in their interaction with double bonds. The somewhat smaller discriminating power of $^1CH_2$ might be expected in view of its higher energy content. It is of interest that especially in the case of $^3CH_2$ the attack on 1,3-butadiene occurs appreciably more rapidly than on the monoolefins. There is little justification from Table 15 for discerning an electrophilic reactivity trend of singlet methylene.

3.2 RELATIVE RATES OF INSERTION REACTIONS

Various carbenes are capable of insertion into C–H bonds but are more discriminating than methylene. Pertinent data are listed in Table 16. Resonance stabilization by donor substituents in expected to lower the reactivity of the divalent carbon; diarylcarbenes, dihalocarbenes, alkoxycarbenes, etc. do not insert at all. The increase in selectivity due to acceptor groups, such as $-CO_2R$ and –CN, must be the result of other effects. It has been suggested[117] that polar resonance structures contribute to the transition state of the insertion reaction. The "weight" of a resonance structure with negative charge on the carbene carbon will increase if the charge can be dispersed over $-CO_2R$ or –CN groups. On the other hand, inductive and hyperconjugative effects stabilize the positive charge on the alkane in the order tertiary > secondary > primary carbon. Any polarization of the transition state, as follows,

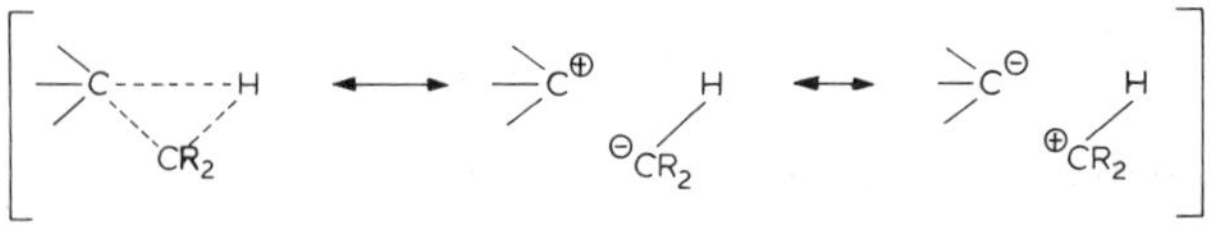

TABLE 16

RELATIVE RATES OF CARBENE INSERTION INTO C–H BONDS (LIQUID PHASE)

Carbene	*Carbene precursor**	*Alkane*	*Secondary C–H / Primary C–H*	*Tertiary C–H / Primary C–H*	*Ref.*
$:CH_2$	CH_2N_2, $h\nu$	Isopentane	1.22	1.51	91
		2,3-Dimethylbutane		1.2	98
Cyclopenta-dienylidene	Diazocyclopentadiene, $h\nu$	2,3-Dimethylbutane		7.3	114
PhCH:	$PhCHN_2$, $h\nu$	*n*-Pentane	6.3		115
	$PhCHN_2$, $h\nu$	*n*-Pentane	8.3		116
	Phenyloxiranes, $h\nu$	*n*-Pentane	8.4		116
	Phenylcyclopropanes, $h\nu$	*n*-Pentane	8.6		116
$:CH\text{–}CO_2CH_3$	$N_2CH\text{–}CO_2CH_3$, $h\nu$	*n*-Pentane	2.3		117
		Isobutane		3.1	117
		2,3-Dimethylbutane		2.9	117
$:C(CO_2C_2H_5)_2$	$N_2C(CO_2C_2H_5)_2$, $h\nu$	*n*-Pentane	8.4		117
		Isobutane		21.0	117
		2,3-Dimethylbutane		12.5	117
$:C(CN)_2$	$N_2C(CN)_2$, Δ, 80°C	*n*-Butane	4.6		118
		Isobutane		12.0	118
:CHCl	N_2CHCl, Δ, −10°C; $h\nu$, −50°C	*n*-Pentane	20		119
:CHBr	N_2CHBr, Δ, −10°C; $h\nu$, −50°C	*n*-Pentane	25		119

* $h\nu$ indicates photolysis and Δ thermal decomposition.

accentuates the discrimination in favor of secondary and tertiary C–H bonds.

The intramolecular insertion of alkylcarbenes appears to involve an eclipsed conformation the energy of which depends on the number of alkyl–alkyl interactions. As a consequence of steric effects, highly-branched alkylcarbenes display an apparent selectivity in favor of primary C–H bonds whereas less-branched structures insert preferentially into seconday C–H bonds[120a]. For the example given in scheme (42),

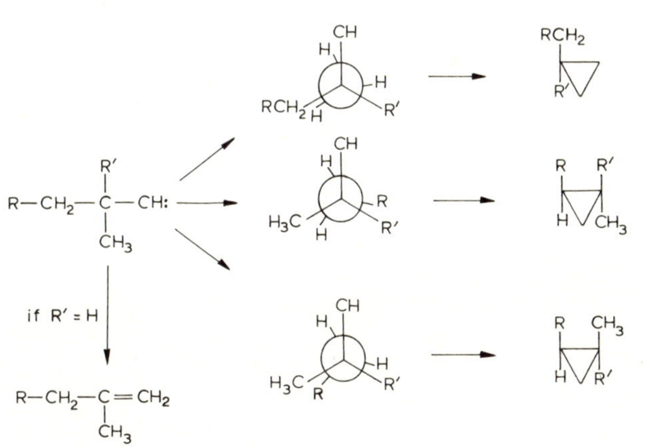

(42)

the rate of insertion into secondary C–H bonds as compared to primary C–H bonds is 1.4 if R′ = H and 0.78 if R′ = CH_3 (R = CH_3). A C–H bond α to an electron-donating substituent (R = Ph, OCH_3 or $N(CH_3)_2$) is strongly deactivated with respect to intramolecular carbene insertion (Table 17)[120b]. This was discussed in terms of inductive effects. A polar transition state with charge separation is clearly excluded.

Carbenoids, *i.e.* species intermediate in α-elimination reactions and catalytic decomposition of diazoalkanes, have so far failed to give products of intermolecular insertion. In contrast, alkylcarbenoids do insert into β and γ C–H bonds. They show a greater selectivity in favor of secondary and tertiary C–H bonds than to "free" alkylcarbenes (Table 18)[120c,d] (*cf.* p. 381 for isotope effects).

TABLE 17

RELATIVE RATES OF INTRAMOLECULAR INSERTION OF ALKYLCARBENES, *viz.*

$$R{-}CH_2{-}\overset{CH_3}{\underset{CH_3}{C}}{-}CH{:}$$

R	*Secondary C–H* / *Primary C–H*	*R*	*Secondary C–H* / *Primary C–H*
CH_3	0.78	$CH{=}CH_2$	0.43
C_2H_5	0.85	Ph	0.50
$CH(CH_3)_2$	0.95	$N(CH_3)_2$	0.21
$CH_2CH_2CH_3$	0.95	OCH_3	0.06

TABLE 18

RELATIVE RATES OF INTRAMOLECULAR INSERTION OF ALKYLCARBENES AND ALKYLCARBENOIDS

	Secondary C–H / *Primary C–H* $R{=}CH_3CH_2C(CH_3)_2{-}$	*Tertiary C–H* / *Primary C–H* $R{=}(CH_3)_2CHC(CH_3)_2{-}$
$R{-}CHN_2$, Pyrolysis (160°C)	0.78	0.38
$R{-}CHN_2$, Photolysis	0.81	0.63
$R{-}CHN_2$ + CuI	4.8	1.5
$R{-}CHN_2$ + AgI	2.5	2.1
$R{-}CH_2Cl$ + $NaNH_2$	1.59	1.06
$R{-}CHI_2$ + Na	1.44	0.74
$R{-}CHI_2$ + Mg	2.18	1.74
$R{-}CHI_2$ + Zn	6.4	4.0

3.3 RELATIVE RATES OF ADDITION REACTIONS

The data collected in Table 19 have been arranged according to the type

TABLE 19

RELATIVE RATES OF CARBENE AND CARBENOID ADDITION TO ALKENES (LIQUID PHASE)

Carbene *[Carbenoid]*	*Precursor**	R–CH=CH$_2$	RHC=CHR (*trans*)	RHC=CHR (*cis*)	cyclohexene	$R_2C=CH_2$	$R_2C=CHR$	$R_2C=CR_2$	*Ref.*
[:CH$_2$]	CH$_2$I$_2$ + Zn(Cu) in (C$_2$H$_5$)$_2$O	0.14[a]	0.17[b]	0.33[c]	0.40	1.00[d]	0.86[e]	0.51[f]	37
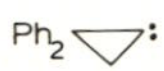		0.22[g]	0.42[h]	1.15[i]	1.24	1.00[k]		0.41[f]	121
	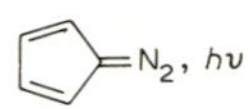	1.25[a]			1.33		1.00[e]	0.99[f]	114
[(CH$_3$)$_2$C=C=C:]	(CH$_3$)$_2$C(Cl)–C≡CH + KOC(CH$_3$)$_3$	0.05[a]			0.25	1.00[d]	0.94[e]	3.56[f]	122
PhCH:	PhCHN$_2$, *hν*; *syn*	0.56[l]	1.10[h]	1.01[i]		1.00[k]	1.9[e]		123
	anti	0.56		0.93			1.8[e]		
[PhCH:]	PhCHBr$_2$ + RLi; *syn*	0.65[l]	0.59[h]	1.29[i]		1.00[k]	1.2[e]		123
	anti	0.30		0.53			0.82		
[:CH–CO$_2$C$_2$H$_5$]	N$_2$CH–CO$_2$C$_2$H$_5$ + CuSO$_4$	0.56[a]			0.62		1.00[e]	1.0[f]	124
:C(CO$_2$CH$_3$)$_2$	N$_2$C(CO$_2$CH$_3$)$_2$, *hν*	0.47[g]	0.23[m]	0.55[n]			1.00[e]	0.88[f]	125
:CF$_2$	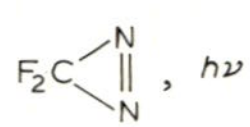			0.08[i]		1.00[k]			126
[:CFCl]	Cl$_2$FC–CO–CCl$_2$F + KOC(CH$_3$)$_3$, −10°C	0.009[l]	0.097[h]	0.14[i]		1.00[k]	6.5[e]	31[f]	127
:CHCl	N$_2$CHCl, Δ, −30°C	0.74[l]	1.09[h]	0.99[i]		1.00[k]	1.18[e]	1.20[f]	119
[:CHCl]	CH$_2$Cl$_2$ + RLi *syn* −35°C	0.35[g]	0.45[h]	1.55[i]	0.91	1.00[k]	2.2[e]	2.8[f]	128
	anti	0.10		0.28	0.28		1.38		
:CHBr	N$_2$CHBr, Δ, −30°C	0.75[l]	1.10[h]	1.02[i]		1.00[k]		1.18[f]	119
[:CCl$_2$]	CHCl$_3$ + KOC(CH$_3$)$_3$ −10°C	0.011[l]	0.15[h]	0.23[i]	0.13	1.00[k]	2.8[e]	6.5[f]	127

TABLE 19 (*Contd.*)

Carbene [Carbenoid]	Precursor*	R–CH=CH$_2$	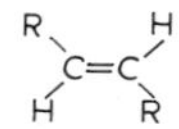	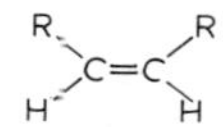		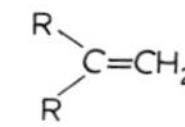	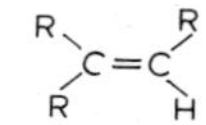	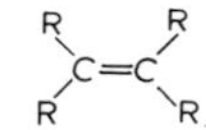	Ref.
[:CBr$_2$]	CHBr$_3$ + KOC(CH$_3$)$_3$	0.045[a]			0.27	1.00[k]	2.0[e]	1.85[f]	129
Ph–C–Br	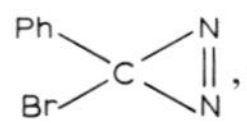, *hν* *syn*-Br		0.25[h]	0.65[i]		1.00[k]	2.9[e]	4.44[f]	130
	anti-Br			0.42			2.2		
[Ph–C–Br]	PhCHBr$_2$ + KOC(CH$_3$)$_3$ *syn*-Br		0.14[h]	0.33[i]		1.00[k]	1.45[e]	1.65[f]	131
	anti-Br			0.25			1.13		131

* *hν* indicates photolysis and Δ thermal decomposition.

[a] 1-Hexene; [b] *trans*-3-Hexene; [c] *cis*-3-Hexene; [d] 2-Methyl-1-butene; [e] 2-Methyl-2-butene; [f] 2,3-Dimethyl-2-butene; [g] 1-Pentene; [h] *trans*-2-Butene; [i] *cis*-2-Butene; [k] Isobutene; [l] 1-Butene; [m] *trans*-3-Methyl-2-pentene; [n] *cis*-3-Methyl-2-pentene.

TABLE 20

RELATIVE RATES OF DICHLOROCARBENE ADDITION TO OLEFINS[134]

	PhHgCCl$_3$ + NaI in dimethoxyethane at 80°C	*PhHgCCl$_2$Br in dimethoxyethane at 80°C*	*Cl$_3$C–CO$_2$Na in dimethoxyethane at 80°C*	*PhHgCCl$_2$Br + NaI, −15°C*	*CHCl$_3$ + KOC(CH$_3$)$_3$ −15°C*
2,3-Dimethyl-2-butene				43.6	53.7
2,3-Dimethyl-2-pentene	23.2	22.5	24.8	22.7	
3-Ethyl-2-pentene	3.13	3.54	3.52	4.73	
2-Ethyl-1-pentene	2.30	2.31			
Cyclohexene	1.00	1.00	1.00	1.00	1.00
cis-3-Heptene	0.835	0.83	0.80	0.81	0.89
trans-3-Heptene	0.537	0.52	0.52	0.435	0.435
1-Heptene	0.218	0.24	0.22	0.11	

of substitution of the olefin. It is clear from Table 19 that virtually all carbenes and carbenoids behave as electrophiles toward olefinic bonds; that is, the reactivity of the olefin increases with increased substitution of the double bond with alkyl groups. If the divalent intermediate displays sizable steric requirements in the transition state, a reversal of reactivity with highly substituted olefins may be observed. The Simmons–Smith reagent – probably not a simple monomeric species – exhibits such behavior[37]. The relative rates of cyclopropane formation using the Simmons–Smith reagent reagent have been determined for a number of cyclic olefins not listed in Table 19[132]. The normal sequence cyclopentene (1.60) > cycloheptene (1.18) > cyclohexene (1.00) is followed. Among the substituted cyclohexenes, 1-methylcyclohexene (2.14) and 1-phenylcyclohexene (0.30) represent the extremes in reactivity. The unusually low factors associated with alkyl and aryl substitution suggest a balance between small inductive and steric effects. Norbornene reacts only 1.70 times faster than cyclohexene[132] to give the *exo*-cyclopropane product exclusively[133].

2,2-Diphenylcyclopropylidene, generated thermally (spontaneously) from the diazo compound, shows a similar reversal in reactivity with highly substituted olefins[121]. In this case the authors have attributed the reversal to steric hindrance during addition of the free carbene. In contrast, steric assistance may operate in the addition of dihalocarbenes to highly substituted olefins. The preferred *syn*-addition of several halocarbenes and halocarbenoids (Table 19) indicates favorable non-bonded interactions between halogen and alkyl groups.

The more reactive carbenes have little ability to distinguish among differently substituted olefins. Thus, the response of relative rates of chlorocarbene and bromocarbene addition to changes in the nucleophilicity of the olefins is small, and no distinct difference in the discriminating abilities of these carbenes exists[119]. The halocarbenes insert into C–H bonds more selectively than does phenylcarbene (Table 16), but the reverse holds for addition to olefins (Table 19). Despite such inconsistencies in the reactivity order, it is obvious that the indiscriminate species of Table 19 do insert into C–H bonds, and the highly selective ones do not.

Much emphasis has been placed on relative rates as a means of distinguishing between carbenes and carbenoids. Seyferth *et al.*[134] have studied the relative reactivities of olefins toward "dichlorocarbene" from various precursors (Table 20). The values obtained with $PhHgCCl_3 + NaI$ at 80°C were virtually identical with those obtained at the same temperature using $PhHgCCl_2Br$ in benzene or dimethoxyethane, and $Cl_3C{-}CO_2Na$ in dimethoxyethane. The values obtained at −15°C with $PhHgCCl_2Br + NaI$ were virtually identical with those obtained at this temperature using chloroform and potassium *t*-butoxide as the dichlorocarbene source. This evidence, together with the first-order kinetics of the decomposition of $PhHgCCl_2Br$ (p. 392), suggests that dichlorocarbene generated in the various elimination reactions is rather weakly complexed with either leaving groups or solvent.

Inspection of Table 19 shows three carbene–carbenoid pairs, namely, :CHPh, :CHCl, and Ph–C–Br. The most significant discrepancies exist in the reactivities of the intermediates formed from chlorodiazomethane and α-elimination on methylene chloride[119]. The pronounced stereoselectivity observed in the α-elimination reaction giving rise to relatively large *syn/anti* ratios is reduced to random addition when the halocarbenes are generated from the diazo precursors. The competition data with different olefins indicate that a considerably less reactive intermediate is formed in the α-elimination.

In the phenylcarbene (carbenoid) system, *syn/anti* ratios differ substantially from unity in the addition reactions involving organolithium reagents whereas only small stereoselectivities are observed in the photolysis. The influence of aryl substituents on both reactions can be summarized by the following sequence of diminishing stereoselectivity: p-OCH_3 > p-CH_3 > m-Cl > p-Cl > H[124]. The *syn/anti* ratios are always greater in pentane as compared to ether, and differ slightly with benzal bromide and benzal iodide as starting materials[135]. Clearly, the transition state of the carbenoid addition involves lithium halide. The relative rate data, on the other hand, show relatively minor variations with the degree of olefin substitution in both reaction types (Table 19).

Phenylbromocarbene, generated by diazirine photolysis, is apparently more selective than the corresponding carbenoid from benzal bromide and potassium *t*-butoxide[131]. The electrophilic behavior of the carbenoid may be obscured by steric factors in the reaction with highly substituted olefins, as discussed above. In principle, however, complexation of a singlet carbene may increase or decrease its electrophilicity, depending on whether the complexing agent supplies electrons to the vacant *p* orbital (Lewis base) or withdraws electrons from the occupied sp^2 orbital (Lewis acid). More data on this problem is clearly warranted.

Knowledge of triplet carbenes in solution is still inadequate. Neither the work of Hammond *et al.*[87] on methylene nor that of Jones and Rettig[136] on fluorenylidene presents a detailed description of the properties of both spin states. In the former case the process of photosensitization apparently does not produce triplet methylene cleanly, in the latter the properties of singlet fluorenylidene are obscured by ready intersystem crossing to the more stable triplet. Jones *et al.*[137] have compared the direct and benzophenone-sensitized photolysis of methyl diazomalonate. The direct photolysis yields a species which adds stereoselectively to *cis*- and *trans*-4-methyl-2-pentene (< 10% of the "wrong" cyclopropane). By contrast, the photosensitized decomposition gives about 88% of *trans*-cyclopropane starting from either *cis*- or *trans*- olefin. The rate data in Table 21 are, therefore, thought to represent the behavior of

TABLE 21

RELATIVE RATES OF ADDITION OF SINGLET AND TRIPLET BIS-CARBOMETHOXYCARBENE[137]

	$N_2C(CO_2CH_3)_2$	
Olefin	*Direct photolysis*	*Benzophenone-sensitized photolysis*
2,3-Dimethyl-2-butene	0.88	0.33
2-Methyl-2-butene	1.00	1.00
1-Pentene	0.47	0.46
3,3-Dimethyl-1-butene	0.48	0.48
cis-4-Methyl-2-pentene	0.55	0.15
trans-4-Methyl-2-pentene	0.23	0.13
2,3-Dimethyl-1,3-butadiene	1.3	4.4
1,3-Butadiene		4.5

TABLE 22

RELATIVE RATES OF CARBENOID ADDITION TO UNSATURATED SILANES

		Relative rates	
n in $(CH_3)_3Si\text{–}(CH_2)_nCH{=}CH_2$	*Dipole moment 25°C*	$N_2CH\text{–}CO_2C_2H_5$ $+ CuSO_4$	$CHCl_3$ $+ KOC(CH_3)_3$
0	0.33	0.52	0.047
1	0.59	2.94	4.97
2	0.477	1.44	1.20
1-Heptene		1.00	1.00

singlet and triplet bis-carbomethoxycarbene to a good approximation. They correlate reasonably well with the behavior of singlet and triplet methylene (Table 15, p. 403). In both cases, a large rate factor of triplet addition is observed on going from monoolefins to 1,3-dienes.

The effect of substituents other than alkyl on the reactivity of olefins has received little attention so far. The variation with n in the compounds $(CH_3)_3Si(CH_2)_nCH{=}CH_2$ of the dipole moment[138], the rates of reaction with ethyl diazoacetate ($CuSO_4$-catalyzed)[139], and the rates of dichlorocarbene addition[140] are listed in Table 22. The maximum at $n = 1$ reflects the opposite effects of $p_\pi\text{–}d_\pi$ bonding from the double bond to silicon and of the inductive effect of the silyl group.

REFERENCES

1 G. O. Schenck and R. Steinmetz, *Ann. Chem.*, 668 (1963) 19.
2 F. W. Kirkbridge and R. G. W. Norrish, *J. Chem. Soc.*, (1933) 119.
3 R. K. Brinton and D. H. Volman, *J. Chem. Phys.*, 19 (1951) 1394.
4 E. Müller, H. Renner and W. Rundel, *Z. Naturforsch.*, 21b (1966) 751.

5 W. Kirmse and L. Horner, *Ann. Chem.*, 625 (1959) 34.
6 W. Kirmse, *Ann. Chem.*, 666 (1963) 9.
7 E. Wolf, *Z. Physik. Chem.*, B17 (1932) 46.
8 U. Mazzucato, G. Cauzzo and A. Foffani, *Tetrahedron Letters*, (1963) 1525.
9 P. S. Shantarovitch, *Dokl. Akad. Nauk SSSR*, 116 (1957) 255.
10 B. S. Rabinovitch and D. W. Setser, *J. Am. Chem. Soc.*, 83 (1961) 750.
11 W. J. Dunning and C. C. McCain, *J. Chem. Soc. B*, (1966) 68.
12 G. v. Bünau, P. Potzinger and G. O. Schenck, *Tetrahedron*, 21 (1965) 1293.
13 W. A. Chupka and C. Lifshitz, *J. Chem. Phys.*, 48 (1968) 1109.
14 V. H. Dibeler, M. Krauss, R. M. Reese and N. F. Harlee, *J. Chem. Phys.*, 42 (1965) 3791.
15 G. S. Paulett and R. Ettinger, *J. Chem. Phys.*, 39 (1963) 825, 3534; 41 (1964) 2557.
16 J. M. Bell, *J. Chem. Phys.*, 41 (1964) 2556.
17 J. C. Hassler and D. W. Setser, *J. Am. Chem. Soc.*, 87 (1965) 3793.
18 A. Langer, J. A. Hipple and D. P. Stevenson, *J. Chem. Phys.*, 22 (1954) 1836.
19 W. E. Parham and W. R. Hasek, *J. Am. Chem. Soc.*, 76 (1954) 937.
20 H. Staudinger, E. Anthes and F. Pfenninger, *Chem. Ber.*, 49 (1916) 1928.
21 G. Murgulescu and T. Oncescu, *J. Chim. Phys.*, 58 (1961) 508.
22 H. Reimlinger, *Chem. Ber.*, 97 (1964) 339.
23 D. Bethell, D. Whittaker and J. D. Callister, *J. Chem. Soc.*, (1965) 2466.
24 D. Bethell and D. Whittaker, *J. Chem. Soc. B*, (1966) 778.
25 E. Fahr, K. H. Keil, H. Lind and F. Scheckenbach, *Z. Naturforsch.*, 20b (1965) 526.
26 H. Goetz and H. Juds, *Ann. Chem.*, 678 (1964) 1.
27 L. Horner and E. Lingnau, *Ann. Chem.*, 591 (1955) 21.
28 A. N. Strachan and W. A. Noyes, Jr., *J. Am. Chem. Soc.*, 76 (1954) 3258.
29 G. B. Porter, *J. Am. Chem. Soc.*, 79 (1957) 827.
30 G. A. Traylor and G. B. Porter, *J. Chem. Phys.*, 36 (1962) 1353.
31 S. Ho, I. Unger and W. A. Noyes, Jr., *J. Am. Chem. Soc.*, 87 (1965) 2297.
32 W. A. Noyes, Jr. and I. Unger, *Pure Appl. Chem.*, 9 (1964) 461.
33 R. B. Cundall and A. S. Davies, *J. Am. Chem. Soc.*, 88 (1966) 1329.
34 A. N. Strachan and D. E. Thornton, *J. Phys. Chem.*, 70 (1966) 952.
35 H. E. Simmons and R. D. Smith, *J. Am. Chem. Soc.*, 80 (1958) 5323; 81 (1959) 4256.
36 G. Wittig and K. Schwarzenbach, *Angew. Chem.*, 71 (1959) 652; *Ann. Chem.*, 650 (1961) 1.
37 E. P. Blanchard and H. E. Simmons, *J. Am. Chem. Soc.*, 86 (1964) 1337.
38 W. Kirmse and G. Wächtershäuser, *Tetrahedron*, 22 (1966) 73.
39 M. J. Goldstein and W. R. Dolbier, Jr., *J. Am. Chem. Soc.*, 87 (1965) 2293.
40 W. Kirmse, H.-D. v. Scholz and H. Arold, *Ann. Chem.*, 711 (1968) 22.
41 C. R. Hauser, W. R. Brasen, P. S. Skell, S. W. Kantor and A. E. Brodhag, *J. Am. Chem. Soc.*, 78 (1956) 1653.
42 W. R. Brasen, S. W. Kantor, P. S. Skell and C. R. Hauser, *J. Am. Chem. Soc.*, 79 (1957) 397.
43 S. B. Hanna, Y. Iskander and Y. Riad, *J. Chem. Soc.*, (1961) 217.
44 C. G. Swain and E. R. Thornton, *J. Am. Chem. Soc.*, 83 (1961) 4033.
45 I. Rothberg and E. R. Thornton, *J. Am. Chem. Soc.*, 85 (1963) 1704; 86 (1964) 3296.
46 I. Rothberg and E. R. Thornton, *J. Am. Chem. Soc.*, 86 (1964) 3302.
47 D. Bethell, *J. Chem. Soc.*, (1963) 666.
48 D. Bethell and A. F. Cockerill, *J. Chem. Soc. B*, (1966) 917.
49 D. Bethell and A. F. Cockerill, *Proc. Chem. Soc.*, (1964) 283.
49a D. Bethell, A. F. Cockerill and D. B. Frankham, *J. Chem. Soc. B*, (1967) 1287.
50 G. F. Hennion and D. E. Maloney, *J. Am. Chem. Soc.*, 73 (1951) 4735.
51 G. F. Hennion and E. G. Teach, *J. Am. Chem. Soc.*, 75 (1953) 1653.
52 G. F. Hennion and K. W. Nelson, *J. Am. Chem. Soc.*, 79 (1957) 2142.
53 A. Burawoy and E. Spinner, *J. Chem. Soc.*, (1954) 3752.
54 V. J. Shiner, Jr. and J. W. Wilson, *J. Am. Chem. Soc.*, 84 (1962) 2402.

55 V. J. Shiner, Jr., J. W. Wilson, G. Heinemann and N. Solliday, *J. Am. Chem. Soc.*, 84 (1962) 2408.
56 W. J. LeNoble, *J. Am. Chem. Soc.*, 87 (1965) 2434.
57 V. J. Shiner, Jr. and J. S. Humphrey, Jr., *J. Am. Chem. Soc.*, 89 (1967) 622.
58 G. F. Hennion and C. V. DiGiovanna, *J. Org. Chem.*, 30 (1965) 3696.
59 A. F. Bramwell, L. Crombie and M. H. Knight, *Chem. Ind. (London)*, (1965) 1265.
60 H. D. Hartzler, *J. Am. Chem. Soc.*, 81 (1959) 2024; 83 (1961) 4990, 4997.
61 A. Geuther, *Ann. Chem.*, 123 (1862) 121.
62 J. Hine, *J. Am. Chem. Soc.*, 72 (1950) 2438.
63 J. Hine and A. M. Dowell, Jr., *J. Am. Chem. Soc.*, 76 (1954) 2688.
64 J. Hine, A. M. Dowell, Jr. and J. E. Singley, Jr., *J. Am. Chem. Soc.*, 78 (1956) 479.
65 J. Hine, C. H. Thomas and S. J. Ehrenson, *J. Am. Chem. Soc.*, 77 (1955) 3886.
66 J. Hine, R. C. Peek, Jr. and B. D. Oakes, *J. Am. Chem. Soc.*, 76 (1954) 827.
67 J. Hine and N. W. Burske, *J. Am. Chem. Soc.*, 78 (1956) 3337.
68 J. Hine, N. W. Burske, M. Hine and P. B. Langford, *J. Am. Chem. Soc.*, 79 (1957) 1406.
69 J. Hine and P. B. Langford, *J. Am. Chem. Soc.*, 79 (1957) 5497.
70 J. Hine, R. Butterworth and P. B. Langford, *J. Am. Chem. Soc.*, 80 (1958) 819.
71 C. G. Swain and C. B. Scott, *J. Am. Chem. Soc.*, 75 (1953) 141.
72 W. J. LeNoble and M. Duffy, *J. Am. Chem. Soc.*, 86 (1964) 4512.
73 J. Hine and S. J. Ehrenson, *J. Am. Chem. Soc.*, 80 (1958) 824.
74 J. Hine and F. P. Prosser, *J. Am. Chem. Soc.*, 80 (1958) 4282.
75 J. Hine and A. D. Ketley, *J. Org. Chem.*, 25 (1960) 606.
76 D. Seyferth, J. Y.-P. Mui and J. M. Burlitch, *J. Am. Chem. Soc.*, 89 (1967) 4953.
76a D. Seyferth, R. M. Turkel, M. A. Eisert and L. J. Todd, *J. Am. Chem. Soc.*, 91 (1969) 5027.
76b D. Seyferth, R. Damrauer, R. M. Turkel and L. J. Todd, *J. Organometal. Chem.*, 17 (1969) 367.
77 G. Herzberg and J. Shoosmith, *Nature*, 183 (1959) 1801.
78 G. Herzberg, *Proc. Roy. Soc. (London), Ser. A*, 262 (1961) 291; *Can. J. Phys.*, 39 (1961) 1511.
79 P. S. Skell and R. C. Woodworth, *J. Am. Chem. Soc.*, 78 (1956) 4496, 6427; 81 (1959) 3383.
80 F. A. L. Anet, R. F. W. Bader, and A. M. van der Auwera, *J. Am. Chem. Soc.*, 82 (1960) 3217.
81 H. M. Frey, *J. Am. Chem. Soc.*, 82 (1960) 5947.
82 R. F. W. Bader and J. I. Generosa, *Can. J. Chem.*, 43 (1965) 1631.
83 B. S. Rabinovitch, K. W. Watkins and D. F. Ring, *J. Am. Chem. Soc.*, 87 (1965) 4960.
84 R. W. Carr, Jr. and G. B. Kistiakowsky, *J. Phys. Chem.*, 70 (1966) 118.
85 S. -Y. Ho and W. A. Noyes, Jr., *J. Am. Chem. Soc.*, 89 (1967) 5091.
86 F. J. Duncan and R. J. Cvetanovic, *J. Am. Chem. Soc.*, 84 (1962) 3593.
87 K. R. Kopecky, G. S. Hammond and P. A. Leermakers, *J. Am. Chem. Soc.*, 84 (1962) 1015.
88 D. F. Ring and B. S. Rabinovitch, *J. Am. Chem. Soc.*, 88 (1966) 4285; *Can. J. Chem.*, 46 (1968) 2435.
89 V. H. Dibeler, M. Krauss, R. M. Reese and F. N. Harlee, *J. Chem. Phys.*, 42 (1965) 3791.
90 H. M. Frey, in *Carbene Chemistry* W. Kirmse (Ed.), Academic Press, New York, 1964, p. 213.
91 B. M. Herzog and R. W. Carr, Jr., *J. Phys. Chem.*, 71 (1967) 2688.
91a T. W. Eder and R. W. Carr, Jr., *J. Phys. Chem.*, 73 (1969) 2074.
92 H. M. Frey, *J. Am. Chem. Soc.*, 80 (1958) 5005.
93 H. M. Frey and G. B. Kistiakowsky, *J. Am. Chem. Soc.*, 79 (1957) 6373.
94 H. M. Frey, *Proc. Roy. Soc. (London), Ser.* A, 250 (1959) 409.
95 J. N. Butler and G. B. Kistaikowsky, *J. Am. Chem. Soc.*, 83 (1961) 1324.
96 G. Z. Whitten and B. S. Rabinovitch, *J. Phys. Chem.*, 69 (1965) 4348.

97 M. L. Halberstadt and J. R. McNesby, *J. Am. Chem. Soc.*, 89 (1967) 3417.
98 W. v. E. Doering, R. G. Buttery, R. G. Laughlin and N. Chaudhuri, *J. Am. Chem. Soc.*, 78 (1956) 3224.
99 D. B. Richardson, M. C. Simmons and I. Dvoretzky, *J. Am. Chem. Soc.*, 82 (1960) 5001; 83 (1961) 1934.
100 D. B. Richardson, L. R. Durett, I. M. Martin, S. C. Slaymaker and I. Dvoretzky, *J. Am. Chem. Soc.*, 87 (1965) 2763.
101 J. W. Simons, B. S. Rabinovitch and F. H. Dorer, *J. Phys. Chem.*, 70 (1966) 1076.
102 B. S. Rabinovitch, E. Tschuikow-Roux and E. W. Schlag, *J. Am. Chem. Soc.*, 81 (1959) 1083.
103 F. Casas, J. A. Kerr and A. F. Trotman-Dickenson, *J. Chem. Soc.*, (1964) 3655; (1965) 1141.
104 F. H. Dorer and B. S. Rabinovitch, *J. Phys. Chem.*, 69 (1965) 1964.
105 J. N. Butler and G. B. Kistiakowsky, *J. Am. Chem. Soc.*, 82 (1960) 759.
106 H. M. Frey, *Proc. Roy. Soc. (London), Ser.* A, 251 (1959) 575.
107 H. M. Frey, *Trans. Faraday Soc.*, 57 (1961) 951.
108 H. M. Frey, *Trans. Faraday Soc.*, 58 (1962) 516.
109 W. v. E. Doering, L. H. Knox and M. Jones, Jr., *J. Org. Chem.*, 24 (1959) 136.
110 H. M. Frey, *Rec. Trav. Chim.*, 83 (1964) 117.
111 H. M. Frey and M. A. Voisey, *Chem. Commun.*, (1966) 454.
111a H. M. Frey and M. A. Voisey, *Trans. Faraday Soc.*, 64 (1968) 954.
111b M. A. Voisey, *Trans. Faraday Soc.*, 64 (1968) 3058.
112 V. Franzen and L. Fikentscher, *Ann. Chem.*, 617 (1958) 1.
113 S. Krzyzanowski and R. J. Cvetanović, *Can. J. Chem.*, 45 (1967) 665.
113a D. C. Montague and F. S. Rowland, *J. Phys. Chem.*, 72 (1968) 3705.
113b H. M. Frey and R. Walsh, *Chem. Commun.*, (1969) 158.
114 R. A. Moss, *J. Org. Chem.*, 31 (1966) 3296.
115 C. D. Gutsche, G. L. Bachman and R. S. Coffey, *Tetrahedron*, 18 (1962) 617.
116 H. Dietrich, G. W. Griffin and R. C. Petterson, *Tetrahedron Letters*, (1968) 153.
117 W. v. E. Doering and L. H. Knox, *J. Am. Chem. Soc.*, 78 (1956) 4947; 83 (1961) 1989.
118 E. Ciganek, *J. Am. Chem. Soc.*, 88 (1966) 1979.
119 G. L. Closs and J. J. Coyle, *J. Am. Chem. Soc.*, 87 (1965) 4270.
120 (a) W. Kirmse and G. Wächtershäuser, *Tetrahedron*, 22 (1966) 63.
(b) W. Kirmse, H. J. Schladetsch and H. W. Bücking, *Chem. Ber.*, 99 (1966) 2579.
(c) W. Kirmse and G. Wächtershäuser, *Tetrahedron*, 22 (1966) 73.
(d) W. Kirmse and K. Horn, *Chem. Ber.*, 100 (1967) 2698.
121 W. M. Jones, M. H. Grasley and W. S. Brey, Jr., *J. Am. Chem. Soc.*, 85 (1963) 2754.
122 H. D. Hartzler, *J. Am. Chem. Soc.*, 83 (1961) 4997.
123 G. L. Closs and R. A. Moss, *J. Am. Chem. Soc.*, 86 (1964) 4042.
124 P. S. Skell and R. M. Etter, *Chem. Ind. (London)*, (1958) 624.
125 M. Jones, Jr., A. Kulczycki and K. F. Hummel, *Tetrahedron Letters*, (1967) 183.
126 R. A. Mitsch, *J. Am. Chem. Soc.*, 87 (1965) 758.
127 R. A. Moss and R. Gerstl, *J. Org. Chem.*, 32 (1967) 2268; *Tetrahedron*, 23 (1967) 2549.
128 G. L. Closs and G. M. Schwartz, *J. Am. Chem. Soc.*, 82 (1960) 5729.
129 W. v. E. Doering and W. A. Henderson, *J. Am. Chem. Soc.*, 80 (1958) 5274.
130 P. S. Skell and A. Y. Garner, *J. Am. Chem. Soc.*, 78 (1956) 5430.
131 R. A. Moss and R. Gerstl, *Tetrahedron*, 22 (1966) 2637; *Tetrahedron Letters*, (1967) 4905.
132 B. Rickborn and J. H-H. Chan, *J. Org. Chem.*, 32 (1967) 3576.
133 H. E. Simmons, E. P. Blanchard and R. D. Smith, *J. Am. Chem. Soc.*, 86 (1964) 1347.
134 D. Seyferth, M. E. Gordon, J. Y.-P. Mui and J. M. Burlitch, *J. Am. Chem. Soc.*, 89 (1967) 959.
135 R. A. Moss, *J. Org. Chem.*, 30 (1965) 3261.
136 M. Jones, Jr. and K. R. Rettig, *J. Am. Chem. Soc.*, 87 (1965) 4013, 4015.

137 M. Jones, Jr., W. Ando and A. Kulczycki, Jr., *Tetrahedron Letters*, (1967) 1391.
138 J. Nagy, S. Ferenczi-Gresz and V. F. Mironov, *Z. Anorg. Allgem. Chem.*, 347 (1966) 191.
139 I. A. D'yakonov, G. V. Golodnikov and I. B. Repinskaja, *Zh. Obschch. Khim.*, 35 (1965) 2181.
140 J. Cudlin and V. Chvalovsky, *Collection Czech. Chem. Commun.*, 27 (1962) 1658.

Chapter 5

Kinetics of Alkene Isomerizations

ROBERT H. DeWOLFE

1. Introduction

A number of interesting and useful organic reactions involve isomerizations of substances having one or more carbon–carbon double bonds. This chapter deals with the kinetics of reactions of alkenes, cycloalkenes and substituted alkenes which involve migration of carbon–carbon double bonds, with or without structural alteration of the carbon skeleton of the starting materials. These reactions include prototropic and anionotropic rearrangements, several concerted unimolecular isomerizations such as the Cope and Claisen rearrangements, and a number of non-concerted thermal isomerization reactions.

2. Anionotropic isomerizations

Anionotropic rearrangements involve migrations of electronegative groups. They comprise two mechanistically related groups of reactions: substitutions and isomerizations. Allylic substitution reactions in which skeletal rearrangement accompanies nucleophilic substitution, *viz.*

(1)

are reviewed elsewhere (refs. 96, 98 and 99 and Vol. 12A of this series, Chapter 1). Many allylic anionotropic isomerization reactions are special cases of allylic substitution in which the entering and leaving groups are identical. Some allylic isomerizations, however, involve intramolecular migration of the functional group. The net result in either case is the transfer of a functional group from one end of a three-carbon allylic system to the other, *viz.*

(2)

For purposes of discussion, allylic isomerizations are divided into two groups:

References pp. 481–487

those which occur under non-solvolytic conditions and those which accompany solvoloysis reactions.

2.1 NON-SOLVOLYTIC ISOMERIZATIONS

2.1.1 Isomerizations of allylic esters

Qualitative studies of the effects of allylic substituents, acyl substituents, and reaction conditions on rates of isomerization of allylic esters played a prominent role in the development of mechanistic theories of anionotropic rearrangements[40,207]. Burton and Ingold[64–66] observed that α-phenylallyl p-nitrobenzoate isomerizes more rapidly than α-phenylallyl acetate, that electron-releasing α-substituents facilitate isomerization of α-substituted allyl acetates, and that the rate of isomerization of α-phenylallyl p-nitrobenzoate increases as the dielectric constant of the solvent increases. All of these observations support the S_N1' mechanism, which involves dissociation of the allylic ester into an allylic carbonium carboxylate ion pair, which reassociates to the starting material and its allylic isomer, *viz.*

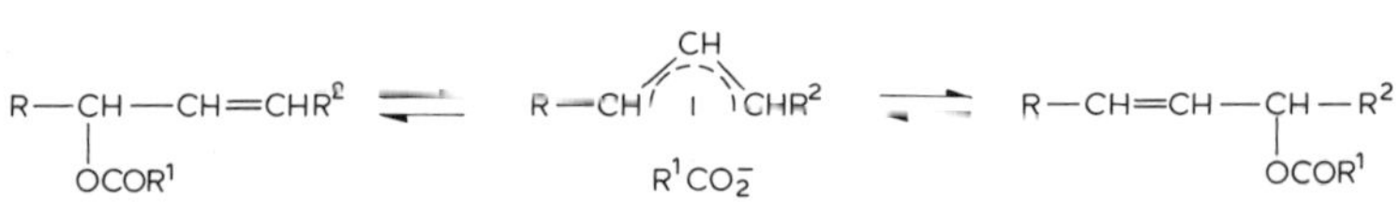 (3)

In non-polar solvents recombination of the ion pair exhibits considerable stereospecificity. The carbonyl oxygen of the starting material tends to become the alkyl oxygen of the product (see below), and the carboxylate ion tends to re-bond to the same side of the planar allylic system as that from which it departed[159,210].

Meisenheimer *et al.*[239] were the first to study the kinetics of allylic ester isomerizations. They followed the isomerization of α-phenylallyl p-nitrobenzoate and α-phenylallyl 3,4,5-tribromobenzoate to the corresponding cinnamyl benzoates by a tedious analytical procedure involving use of melting point–composition diagrams. They observed that isomerization of the molten esters at 137°C is autocatalytic due to partial slow decomposition of the esters to substituted benzoic acids and unidentified products, and demonstrated that the reaction is catalyzed by added benzoic acids. The rate law for these reactions is

$$k_i = k_1 + k_2[\text{ArCO}_2\text{H}]$$

where k_i is the observed first-order rate coefficient of isomerization. The rates of isomerization of solutions of the α-phenylallyl esters in aprotic solvents increased with increasing dielectric constant of the solvent. Meisenheimer and Beutter[238] observed only a moderate salt effect upon the isomerization of α-phenylallyl acetate in acetic acid solutions containing added acetate ion, and Catchpole and Hughes[71] observed a similar small acceleration of the isomerization of α-phenylallyl p-nitrobenzoate in acetonitrile by added lithium p-nitrobenzoate. These salt effects were considered too small to be due to the incursion of abnormal bimolecular substitution (S_N2'). The early experimental studies of allylic ester isomerizations were critically discussed by Ingold and co-workers[72] who concluded that most of the reactions studied involved the S_N1' mechanism illustrated by (3).

Braude[39,40] later pointed out that mechanistic conclusions based on these early studies of allylic ester isomerization are of doubtful validity, since they do not account for the sensitivity of these reactions to adventitious acidic impurities. Braude *et al.*[53–56] and Pocker[254–256] studied the kinetics of isomerization of α-phenylallyl esters [(3), $R = C_6H_5$, $R^1 = YC_6H_4$, $R^2 = H$] in the weakly polar, aprotic solvent chlorobenzene by ultraviolet spectrophotometry. Isomerization of α-phenylallyl p-nitrobenzoate in chlorobenzene at 130°C is strongly catalyzed by p-nitrobenzoic acid, but uncatalyzed isomerization also occurs[55,56]. At 130°C the apparent first-order rate coefficient is given by $[k_1 = \{1.3 \times 10^{-6} + 2.34 \times 10^{-4}[p\text{-}O_2NC_6H_4CO_2H]\}\ \text{sec}^{-1}$.

When the reaction is carried out in solutions containing $p\text{-}O_2NC_6H_4{}^{14}CO_2H$, isomerization is considerably faster than exchange of p-nitrobenzoyl groups at low p-nitrobenzoic acid concentrations ($< \sim 0.01\ M$). At sufficiently high p-nitrobenzoic acid concentrations ($\sim 0.1\ M$), the rate of exchange actually exceeds the rate of isomerization, and exchange without rearrangement is about twice as fast as exchange with rearrangement[54,55]. Isomerization of carbonyl-^{18}O labeled α-phenylallyl p-nitrobenzoate in chlorobenzene at 130°C was found to occur with considerable stereospecificity: in the uncatalyzed isomerization, the carbonyl oxygen atom of the α-phenylallyl ester becomes the alkyl oxygen of the cinnamyl ester[53], *viz.*

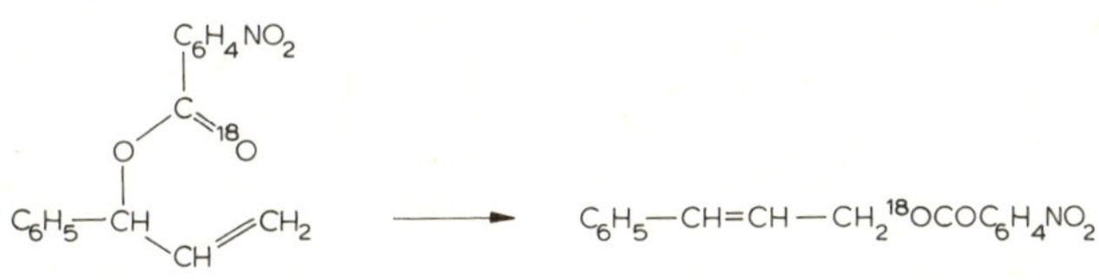

(4)

For isomerization of α-phenylallyl p-nitrobenzoate in chlorobenzene at 155°C, $k_i = \{1.16 \times 10^{-5} + 6.62 \times 10^{-4}[O_2NC_6H_4CO_2H]\}\ \text{sec}^{-1}$. The reaction rate is not significantly changed by addition of $3.6 \times 10^{-3}\ M$ pyridine or $4.5 \times 10^{-3}\ M$ tetrabutylammonium p-nitrobenzoate to the reaction mixture[254]. α-Phenylallyl benzoate isomerizes more slowly than the p-nitrobenzoate ester under these conditions, as expected[255], $k_i = \{1.0 \times 10^{-6} + 3.4 \times 10^{-5} [C_6H_5CO_2H]\}$ sec^{-1}. In the presence of $C_6H_5{}^{14}CO_2H$, isotopic exchange occurred somewhat more slowly than rearrangement. Cinnamyl benzoate did not undergo benzoate exchange under the same experimental conditions. In chlorobenzene solutions containing either isotopically labeled benzoic acid (0.06 *M*) or isotopically labeled tetrabutylammonium benzoate (0.098 *M*), about 70% of the rearrangement occurs with exchange and about 25% proceeds intramolecularly. α-Phenylallyl benzoate isomerizes about 25% faster in chlorobenzene at 155°C in the presence of 0.09 *M* tetrabutylammonium benzoate-^{18}O than in solutions containing no added salt and undergoes isotope exchange almost as fast as it isomerizes.

All of these experimental observations are consistent with a mechanism involving phenylallyl carbonium ion as an intermediate. The stereospecificity of the uncatalyzed reaction suggests a tight ion-pair intermediate which collapses to products more rapidly than it dissociates and also more rapidly than the carboxylate ion rotates relative to the carbonium ion. Exchange of benzoate ions between the ion pair and dissolved benzoate salt prior to collapse of the ion pair to a mixture of allylic esters would account for the observed isotopic exchange.

The stereospecific intramolecular isomerization of α-phenylallyl esters in aprotic solvents might conceivably be a concerted S_Ni' reaction with a cyclic transition state (I) instead of an S_N1' reaction involving a tight ion pair intermediate. If so, structural and solvent effects on reactivity require that the

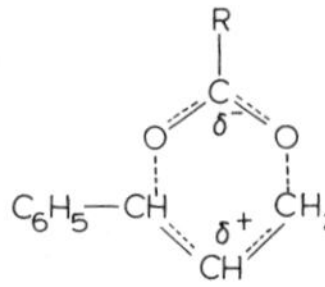

I

transition state involve significant charge separation. Electron-withdrawing acyl substituents, which increase the stability of carboxylate ions, increase the rate of uncatalyzed allylic ester isomerizations. For example, the approximate rate coefficients of isomerization of α-phenylallyl acetate, benzoate and *p*-nitrobenzoate in chlorobenzene at 130°C are 6.3×10^{-8} sec^{-1}, 2.3×10^{-7} sec^{-1}, 1.3×10^{-6} sec^{-1}, respectively[56]. In the same solvent at 155°C, α-phenylallyl *p*-nitrobenzoate isomerizes about twelve times as rapidly as the corresponding benzoate[254,255]. The observation of Meisenheimer *et al.*[239] that α-phenylallyl 3,4,5-tribromobenzoate isomerizes 50–100 times faster in polar aprotic solvents such as acetonitrile and nitromethane than in non-polar solvents such as benzene or pentane also suggests that the reaction involves ionic intermediates. Further support for this conclusion is provided by data of Addy and

Noyes[1], from which it can be estimated that α-phenylallyl p-chlorobenzoate isomerizes almost as rapidly in the polar solvent N-methylpyrrolidone ($k_i = 3.8 \times 10^{-6}$ sec^{-1} at 140°C) as does α-phenylallyl p-nitrobenzoate in chlorobenzene. In the absence of solvent effects, the p-chlorobenzoate ester should be less reactive than the p-nitrobenzoate ester.

In summary, the limited data available permit no clear-cut decision between the S_Ni' mechanism with charge separation in the transition state, and the S_N1' mechanism with tight ion-pair intermediates. The fact that allylic ester isomerizations almost certainly involve ionic intermediates under solvolytic conditions (see below) suggests that the reactions under non-solvolytic conditions probably also occur by an ionic mechanism.

Lewis and co-workers studied the kinetics of vapor-phase isomerization of allylic esters by means of a flow reactor technique[225,226]. At 245°C, α- and γ-methylallyl chloroformates isomerize[226] at approximately the same rate ($k = 9 \times 10^{-2}$ sec^{-1}). Effects of allylic and acyl substituents on rates of isomerization of allylic acetates, trifluoroacetates and carbonates were also studied[225]. At 325°C, methyl substituents on the allylic system of the ester increase reactivity, and trifluoromethyl substituents decrease reactivity, relative to the unsubstituted ester. α- and γ-methyl allyl trifluoroacetates isomerize somewhat more rapidly than the analogous acetate esters. The energies and entropies of activation of these reactions indicate that not all of the allylic esters investigated isomerize by the same mechanism. Methyl-substituted allylic trifluoroacetates and dimethyl-substituted allylic acetates isomerize at 325°C with small positive or small negative entropies of activation, while α- and γ-methylallyl acetates and α- and γ-trifluoromethylallyl trifluoroacetate isomerizations have substantial negative entropies of activation. It is reasonable to suppose that the reactions with entropies of activation near zero have ion-pair-like transition states, while those having negative entropies of activation are more concerted in character, with less charge separation in the transition state.

2.1.2 Isomerization of allylic halides

Isomerization reactions of allylic halides ((2), X = halogen) are well known (ref. 99, p. 720), and occur in both aprotic and protic solvents. Isomerizations which accompany solvolysis reactions of allylic chlorides are discussed in Section 2.2.2.

Allylic chlorides isomerize considerably less readily than the corresponding bromides and iodides, and monosubstituted allylic chlorides such as α- and γ-methyl-allyl chloride are reasonably stable. An α-aryl substituent or two alkyl substituents increases the lability of allylic chlorides considerably, and the kinetics of isomerization of a limited number of reactive allylic chlorides have been studied.

The kinetics of the essentially irreversible isomerization of α-phenylallyl chloride to *trans*-cinnamyl chloride ((5), Ar = C_6H_5)

$$\text{ArCHClCH=CH}_2 \rightarrow \text{ArCH=CHCH}_2\text{Cl} \quad (5)$$

were studied spectrophotometrically by Waight and co-workers[57,287,303] and by Noyes[262]. The isomerization is first order in allylic chloride in chlorobenzene at 90°C[57]. The reaction proceeds slowly in the absence of catalysts, but is catlyzed by chloroacetic, dichloroacetic, and trichloroacetic acids. The catalytic coefficients of these acids parallel their acid strengths. Rates of isomerization of α-phenylallyl chloride at 40°C in chlorobenzene, dimethylformamide, 2-methyl-2-butanol, 2-propanol, ethanol, methanol and aqueous dioxane correlate well with the ionizing power of the solvents[303]. The solvent effect is reflected in the energy of activation rather than in the entropy of activation: $\Delta S^{\ddagger}$ has values near −20 e.u. in chlorobenzene, dimethylformamide and aqueous 85% dioxane, while E_a varies from 22 kcal mole^{-1} (chlorobenzene) to 16 kcal mole^{-1} (aq. dioxane). The isomerization in ethanol and dimethylformamide is accelerated by added lithium chloride, lithium bromide, lithium azide, tetraethylammonium chloride, tetraethylammonium nitrate and tetraethylammonium perchlorate[262,303]. The linear variation of reaction rate with salt concentration, and the observed catalytic effect of salts having non-nucleophilic anions, suggest that the catalytic effects are salt effects and not caused by the intervention of competing nucleophilic substitutions. The catalytic effectiveness of added salts in dimethylformamide solutions is practically independent of the nature of the cation, but depends on the size of the anion. Large anions are less effective in stabilizing the rate-limiting transition state than smaller ones[262]. Similar salt effects have been observed for solvolysis reactions[266,304].

The solvent and salt effects on α-phenylallyl chloride isomerization are consistent with a rate-limiting transition state in which there is substantial charge separation, but do not permit a choice between a concerted S_Ni' process and a two-step S_N1' reaction involving an intimate ion-pair intermediate. Loss of optical activity by resolved α-phenylallyl chloride in dimethylformamide solutions occurs faster, and is more sensitive to the ionic strength of the medium than isomerization to cinnamyl chloride[287,312]. This observation suggests that the racemization reaction proceeds via an ion-pair intermediate, whereas the isomerization occurs by an intra-molecular (S_Ni') mechanism. A transition state having considerable carbonium ion character is required by the solvent and salt effects referred to above, and by the fact[303] that rates of isomerization of α-phenylallyl, α-4-fluorophenylallyl and α-*m*-tolyallyl chlorides ((5), Ar = C_6H_5, *p*-FC_6H_4, *m*-$CH_3C_6H_4$) correlate with Brown's σ^+ substituent constants[62] rather than with Hammett's σ substituent constants. (ref. 182, p. 188).

The mechanism of the acid catalysis of α-phenylallyl chloride isomerization has not been investigated in detail. In aprotic solvents such as chlorobenzene, Brönsted acids probably catalyze the reaction by hydrogen bonding with chloride, which should stabilize the S_Ni' transition state.

It has long been recognized that Lewis acids catalyze the isomerization of allylic chlorides (ref. 99, p. 721). The isomerization of 4-chloro-1,2-butadiene to 2-chloro-1,3-butadiene in hexane solutions is first order in allylic chloride and first order in cuprous chloride[73].

Isomerization of *cis*- and *trans*-5-methyl-2-cyclohexenyl chlorides in acetonitrile solutions containing tetraethylammonium chloride was studied by Noyes and co-workers[223]. These chlorides undergo simultaneous *cis–trans* isomerization, mutarotation, racemization and chloride exchange, *e.g.*

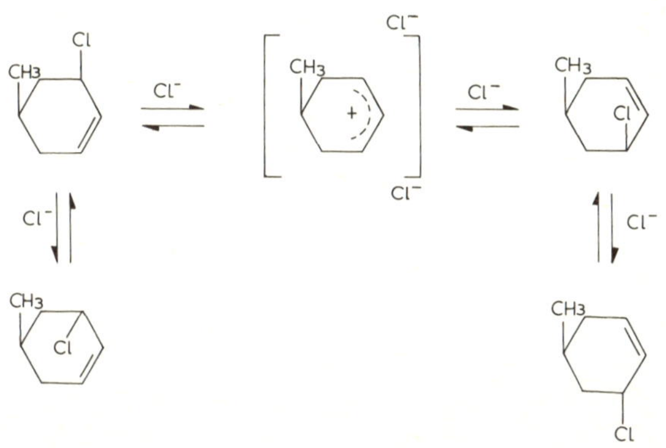

(6)

All of these processes are first order in allylic chloride and in tetraethylammonium chloride. At 65°C, the following rate coefficients were measured: for *cis*-RCl → *trans*-RCl, $k = 5.3 \times 10^{-3}$ l mole^{-1} sec^{-1}; for *trans*-RCl → *cis*-RCl, $k = 2.20 \times 10^{-3}$ l mole^{-1} sec^{-1}; for exchange of *cis*-RCl with $(C_2H_5)_4N^+Cl^-$, $k = 4.9 \times 10^{-3}$ l mole^{-1} sec^{-1}; for exchange of *trans*-RCl with $(C_2H_5)_4N^+Cl^-$, $k = 2.4 \times 10^{-3}$ l mole^{-1} sec^{-1}; for mutarotation of optically active *cis*- or *trans*-RCl, $k = 10.3 \times 10^{-3}$ l mole^{-1} sec^{-1}; and for racemization of the equilibrium mixture from mutarotation, $k = 0.40 \times 10^{-3}$ l mole^{-1} sec^{-1}.

The interpretation of these rate coefficients involves a complex line of reasoning based on several plausible assumptions. Briefly, the authors' conclusions were as follows: *cis–trans* isomerization and mutarotation occur at appreciable rates only in the presence of ionic chloride. Racemization may occur very slowly in the absence of ionic chloride, although the data do not permit a definite decision on this point. The most important reaction involves simultaneous *cis–trans* isomerization (with reversal of sign of optical rotation) and exchange with ionic chloride, probably by an S_N2 process. Since most events leading to racemization do not result in *cis–trans* isomerization, substitution by a *trans*-S_N2' process is unimportant. Further, since within experimental error *cis–trans* isomerization and exchange occur at the same rate, *cis*-S_N2' processes (which would result in exchange without isomerization) are also unimportant. A mechanism for the chloride ion catalyzed racemization involving a symmetrical transition state having two chloride ions partially bonded to opposite sides of the allylic system is proposed by the authors, who argue on steric grounds that such a transition state should be more readily formed from the *cis* chloride and should usually revert to racemic *cis* chloride. Both of these assumptions are compatible with the experimental results. A reaction scheme incorporating the above mechanistic features is outlined in (6).

The isomerization of 2,3,3-trichloro-1,1,3-trifluoropropene to 1,2,3-trichlor-1,3,3-trifluoropropene in chloroform solutions containing tetraethylammonium chloride apparently occurs by an S_N2' mechanism[178], *viz.*

$$CF_2{=}CClCCl_2F \xrightarrow{(C_2H_5)_4N^+Cl^-} ClCF_2CCl{=}CClF \qquad (7)$$

At 25°C, the reaction is first order in both olefin and tetraethylammonium chloride ($k = 3.5 \times 10^{-3}$ l mole^{-1} sec^{-1}). This abnormal replacement reaction may be attributable to the fact that fluorine substituents increase the susceptibility of olefinic carbon to nucleophilic attack.

Allylic bromides are much more labile than the corresponding chlorides, and isomerize to equilibrium mixtures unless special precautions are taken. The isomerization reactions are catalyzed by salts, Lewis acids and free radicals, and may occur in the absence of catalysts. Hatch *et al.*[186] found that isomerization of 3,4-dibromo-1-butene to *trans*-1,4-dibromo-2-butene in *n*-decane, *viz.*

$$CH_2BrCHBrCH{=}CH_2 \rightarrow CH_2BrCH{=}CHCH_2Br \qquad (8)$$

follows first-order kinetics in the region 60–100°C, with $E_a = 18.6$ kcal.mole^{-1}, $\log A = 10.71$, and $\Delta S^{\ddagger} = -12.6$ eu. This reaction is presumably uncatalyzed. The negative entropy of activation, together with the low polarity of the solvent used, suggests that the reaction occurs via a cyclic transition state in which the migrating bromine atom is partially bonded to both the α and γ carbons of the allylic system.

Lithium bromide catalyzes the isomerization of α- and γ-methylallyl bromides in acetone solution[124,125].

$$CH_3CHBrCH{=}CH_2 \underset{}{\overset{LiBr}{\rightleftharpoons}} CH_3CH{=}CHCH_2Br \qquad (9)$$

The kinetics of this reaction were studied using a method of analysis based on the rapid exchange reaction of γ-methylallyl bromide with radioactive lithium bromide. In acetone at 50°C, the isomerization of α-methylallyl bromide to γ-methylallyl bromide is approximately 0.75-order in lithium bromide. The activation parameters were $E_a = 19.4$ kcal.mole^{-1}, log $A = 9.40$, and $\Delta S^{\ddagger} = -18$ eu at 50°C. The rate of the reverse reaction could not be determined precisely but was about one third that of the forward reaction. The isomerization reaction was attributed to S_N2' substitution by bromide ion attacking the γ carbon of the allylic system. However, a mechanism involving lithium ion-catalyzed ion-pair formation (S_N1') is not excluded by the available data.

The lability of allylic iodides and their sensitivity to free radical-induced isomerization, polymerization, and decomposition reactions has discouraged kinetic studies of their substitution and isomerization reactions.

One isomerization reaction whose kinetics were studied[68] occurs when allyl iodide-1-^{14}C is treated with iodine in hexane solution in the dark, *viz.*

$$CH_2{=}CH^{14}CH_2I \overset{I_2}{\rightleftharpoons} CH_2ICH{=}^{14}CH_2 \qquad (10)$$

At 30°C, the rate of isomerization is about 20% of the rate of iodine exchange. It was not possible to obtain reproducible kinetic data, but the rate coefficient for isomerization is roughly 10^{-7} l.mole^{-1}.sec^{-1} under the conditions used. Formation of a π-complex intermediate between allyl iodide and molecular iodine was proposed to rationalize the experimental results.

2.2 ANIONOTROPIC ISOMERIZATIONS ACCOMPANYING SOLVOLYSIS REACTIONS

Various aspects of these rearrangements have been reviewed previously[96,98,99,158].

2.2.1 *Isomerizations of esters*

Isomerization may accompany nucleophilic substitution during the solvolysis of carboxylate and arenesulfonate esters of unsaturated alcohols. Most of the substrates for which such isomerization reactions have been observed are esters of allylic alcohols. If such an ester is optically active by virtue of having an asymmetric α-carbon atom, three experimentally observable processes occur simultaneously: racemization, solvolysis, and allylic isomerization. All three of these processes involve formation of planar, achiral allylic carbonium ions as intermediates, *viz.*

References pp. 481–487

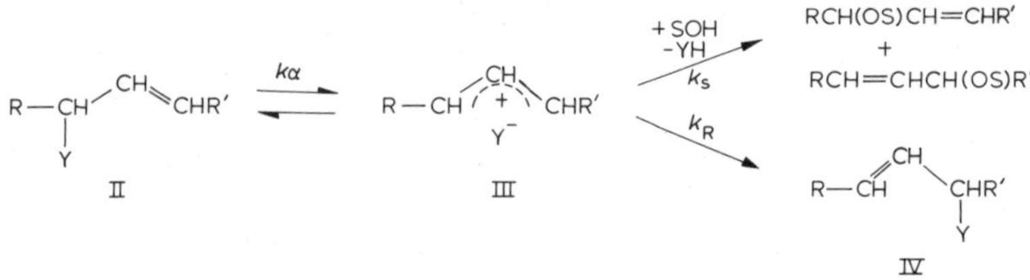

(11)

In general, the polarimetrically measured rate of racemization exceeds the rate of formation of solvolysis products and rearranged ester, as would be expected if an initially formed ion pair can revert to racemic starting material as well as produce solvolysis products and isomeric ester.

For unsymmetrically substituted allylic esters, in which R and R′ of II are different, the rate of isomerization is measured by multiplying the rate coefficient for disappearance of the starting material by the fraction of isomeric ester in the reaction products (applying a correction, where necessary, for solvolysis of the rearranged ester). All of the unsymmetrically substituted esters whose isomerizations have been studied kinetically under solvolytic conditions are α-arylallyl esters, whose isomerizations are essentially irreversible and yield γ-arylallyl esters which solvolyze much more slowly than the starting material.

Sneen and Rosenberg[293,294] studied the kinetics of solvolysis and isomerization of racemic α-aryl-γ-methylallyl p-nitrobenzoates in aqueous dioxane solutions [(11), $R = C_6H_5$, $p\text{-}CH_3C_6H_4$; $R' = CH_3$; $Y = p\text{-}O_2NC_6H_4CO_2$; $SOH = H_2O$]. In 60–80% dioxane–water at 25° and 50°C, the rate of isomerization is approximately equal to the rate of solvolysis. For isomerization of the α-phenyl ester in 70% dioxane, $\Delta H^{\ddagger} = 21$ kcal.mole^{-1} and $\Delta S^{\ddagger} = -12$ eu. In a given solvent, the α-p-tolyl ester is about ten times as reactive as the α-phenyl ester. Both isomerization and solvolysis rates are insensitive to added azide ion, although allylic azides are among the reaction products when azide ion is present in the reaction mixture. The m value[323] is approximately 0.7 for rearrangement and 0.85 for solvolysis. All of these observations are compatible with a mechanism involving dissociation of the allylic ester into a tight ion pair, which can either collapse to the starting material and its allylic isomer, or dissociate irreversibly to a carbonium ion which yields solvolysis products. When α-methyl-γ-phenylallyl p-nitrobenzoate is solvolyzed in aqueous acetone, however, all observable ion pair returns regenerate starting material[228].

If the allylic ester is symmetrically substituted [*i.e.* R = R′ in (11)] allylic isomerization can be observed only in the solvolysis of optically active esters. If it is assumed that the carboxylate group of the tight ion-pair intermediate III of scheme (11) remains on the same side of the planar allylic carbonium ion, ion pair collapse either regenerates the original molecule II or its enantiomer (allylic isomer) IV. If k_α = rate coefficient for loss of optical activity and k_t = the titrimetrically determined rate coefficient of formation of solvolysis products, it can be shown that the rate coefficient for isomerization is given by $k_i = 1/2(k_\alpha - k_t)$. This treatment has been applied to solvolysis and rearrangement reactions of *trans*-α,γ-dimethylallyl *p*-nitrobenzoate and *cis*- and *trans*-5-methyl-2-cyclohexenyl hydrogen phthalates and *p*-nitrobenzoates by Goering *et al.*[163,164,166,171–174]. Their observations are summarized below.

In aqueous 90% acetone at 99.6°C, k_i for *trans*-α,γ-dimethylallyl *p*-nitrobenzoate is 2.6×10^{-6} sec^{-1} [(11), R = R′ = CH_3; Y = *p*-$O_2NC_6H_4CO_2$; SOH = H_2O]. The rate is not affected by addition of small amounts of lithium *p*-nitrobenzoate or *p*-nitrobenzoic acid. In aqueous 60% acetone at 60°C, $k_i = 1.5 \times 10^{-6}$ sec^{-1}. By using carbonyl-^{18}O-labeled ester, it was shown that ^{18}O scrambling is about twice as fast as allylic rearrangement, which implies that the carboxyl group of the *p*-nitrobenzoate ion in the tight ion pair rotates relative to the allylic carbonium ion in some instances before the ions recombine to form enantiomeric esters[171,172]. This and similar observations on other allylic esters (see below) rules out the concerted S_Ni' mechanism as the only mechanism for allylic ester isomerization under solvolytic conditions.

α,γ-Dimethylallyl hydrogen phthalate rearranges about ten times faster than the *p*-nitrobenzoate ester in aqueous dioxane. In aqueous 90% acetone, activation parameters for isomerization of this ester are: E_a = 30 kcal.mole^{-1} and $\Delta S^\ddagger = -0.8 \pm 2.0$ eu. Variation of the rate of rearrangement in aqueous acetone at 80°C with solvent composition yields an *m* value of 0.43 for k_α (the rate of ion pair formation)[166].

The cyclic, symmetrically substituted 5-methyl-2-cyclohexenyl system yields more information on the detailed mechanism of allylic isomerizations than does the open-chain α,γ-dimethylallyl system. With the cyclohexenyl esters it is possible to determine whether the carboxylate ion migrates to the opposite side of the allylic system from that to which it was bonded in the starting material, since such migration results in *cis–trans* isomerization. In general it does not: racemization is not accompanied by *cis–trans* isomerization of the unsolvolyzed ester [(12), Y = $ArCO_2$][163,164,166,173,174].

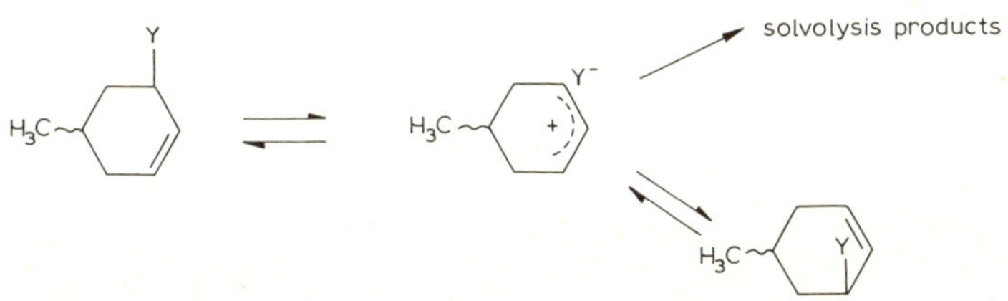

(12)

The sensitivity of the rate of isomerization of *cis*-5-methyl-2-cyclohexenyl hydrogen phthalate to solvent polarity in aqueous 50–90% acetone is clearly consistent with a carbonium ion mechanism for the racemization and solvolysis reactions[173]. In 60% dioxane, activation parameters for the rearrangement reaction are $\Delta H^{\ddagger} = 27.6$ kcal.mole^{-1} and $\Delta S^{\ddagger} = -1.2$ e.u. In aqueous 90% acetone at 100°C, the *cis*-hydrogen phthalate ester rearranges faster ($k_i = 7.5 \times 10^{-6}$ sec^{-1}) than the *trans* isomer ($k_i = 3.8 \times 10^{-6}$ sec^{-1}). Both of these esters rearrange somewhat faster than the corresponding *p*-nitrobenzoates under comparable conditions[174]. Isomerization and solvolysis of the hydrogen phthalate esters is accelerated by added phthalic acid.

Cis- and *trans*-5-methyl-2-cyclohexenyl *p*-nitrobenzoates also undergo racemization without *cis–trans* isomerization of the unsolvolyzed ester in aqueous acetone[163,164,174]. For the carbonyl-^{18}O-labeled *cis-p*-nitrobenzoate, racemization and oxygen scrambling occur at the same rate[164] within experimental error. This implies that, although the carboxyl ion remains on the same side of the allylic system from which it originated, on average it rotates relative to the allylic system more rapidly than it recombines with the carbonium ion to form racemic ester. In contrast, during rearrangement of the carbonyl-^{18}O-labeled *trans-p*-nitrobenzoate under the same conditions, oxygen scrambling is about twice as fast as racemization[163]. This implies that for this ester there are at least two conformations of the tight ion-pair intermediate, one of which is more likely to collapse to the starting material than to its enantiomer.

Most of the anionotropic ester isomerizations whose kinetics have been studied involve simple allylic compounds. A few kinetic investigations of solvolysis and isomerization reactions of doubly unsaturated allylic esters and non-allylic unsaturated esters have been reported. Two examples of such reactions are mentioned here. Spirodienyl *p*-nitrobenzoate (V) undergoes simultaneous allylic isomerization to VI, *viz.*

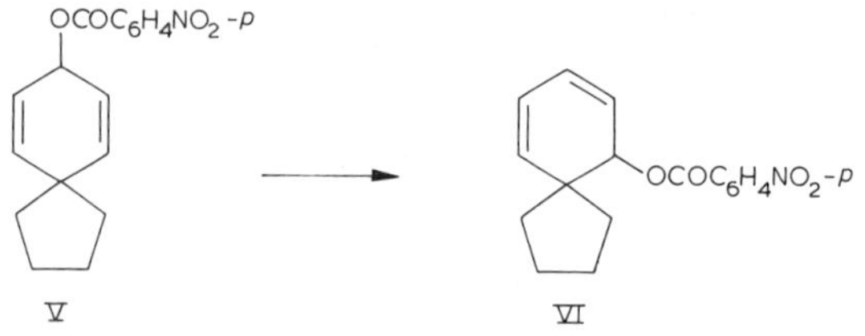

(13)

and aromatization and rearrangement to tetralin (presumably via a common ion-pair intermediate)[324] when solvolyzed in aqueous acetone[152,322]. *Cis*- and *trans*-5-cyclodecen-1-yl *p*-nitrobenzoates and tosylates undergo simultaneous solvolysis and isomerization to decalin derivatives in aqueous acetone[160,165,169], *viz.*

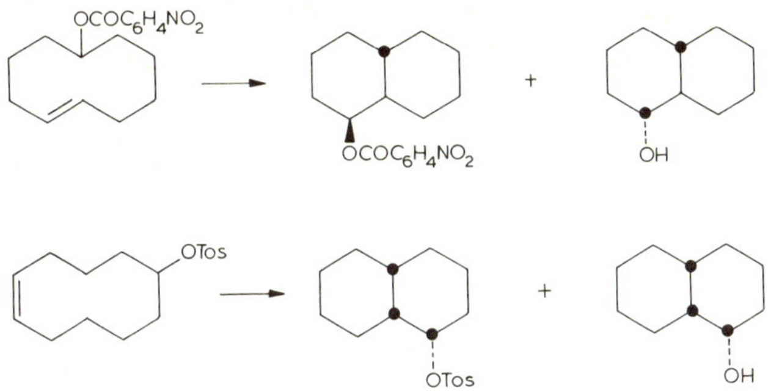

(14)

Acetolysis of *exo*-bicyclo[2.1.1]hex-2-en-5-yl methoxyacetate is accompanied by isomerization to *exo*-bicyclo[3.1.0]hex-2-en-6-yl methoxyacetate[237].

2.2.2 *Isomerizations of allylic halides during solvolyses*

Young *et al.*[328] reported in 1951 that the solvolysis of α,α-dimethylallyl chloride in acetic acid buffered with potassium or lithium acetate is accompanied by partial isomerization to γ,γ-dimethylallyl chloride, which solvolyzes only about an eighth as fast as its tertiary isomer. It was possible to calculate from the titrimetric data rate coefficients for the isomerization reaction as well as for the two solvolysis reactions. Since neither the reaction rates nor the product compositions were influenced by added potassium chloride, it was concluded that the isomerization is intramolecular. This is the first documented example of internal return from a carbonium ion–anion pair, *viz.*

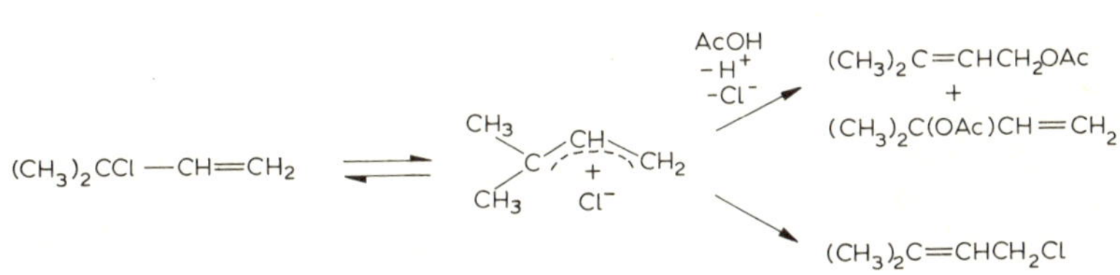

(15)

The solvolysis and isomerization of α,α-dimethylallyl chloride in aqueous 75% ethanol is more complex[97]. In this solvent, the rate of isomerization at 25°C is doubled by the addition of molar LiCl. In the presence of 0.68 *M* $Li^{36}Cl$, about 63% of the chlorine incorporated into the primary chloride during rearrangement came from radioactive chloride in solution. Thus, under these conditions isomerization involves external return of chloride from solution as well as internal return from a tight ion pair intermediate.

α-Phenylallyl chloride also undergoes simultaneous solvolysis and allylic isomerization[304,312]. In aqueous dioxane and in alcohols, the effects of added salts and of solvent ionizing power on spectrophotometrically determined rates of solvolysis and isomerization to cinnamyl chloride indicate that both reactions proceed through the same ion-pair intermediate. As is true of allylic ester solvolyses, the rate of isomerization is somewhat less sensitive to solvent ionizing power than the rate of solvolysis. The fact that the rate of disappearance of optical activity during solvolysis in aqueous dioxane is greater than the rates of isomerization and solvolysis demonstrates that the achiral ion pair collapses to give both cinnamyl chloride and racemic α-phenylallyl chloride.

The kinetics of solvolysis of optically active *cis*- and *trans*-5-methyl-2-cyclohexenyl chlorides provides additional insight into the ion-pair internal return mechanism of allylic isomerization[170]. Solvolysis of both of these chlorides in ethanol and acetic acid is accompanied by loss of optical activity without *cis–trans* isomerization of the unreacted chloride [(12), Y=Cl]. The fact that rate of loss of optical activity exceeds rate of formation of solvolysis products provides evidence for internal return from carbonium chloride ion pairs, unaccompanied by geometrical isomerization. This conclusion is supported by the observation that the ratio of racemization rate to solvolysis rate is insensitive to added ionic chloride.

Since ethanol and acetic acid have similar ionizing powers, it is not surprising that racemization of each optically active chloride occurs at about the same rate in the two solvents. The more nucleophilic solvent, ethanol, is much more efficient than acetic acid in reacting with the allylic carbonium ion, with the result that isomerization is faster, and solvolysis slower, in acetic acid than in ethanol. That is, the partitioning of the ion pair between solvolysis products and racemic chloride is more in favor of solvolysis products in the more nucleophilic solvent.

One allylic fluoride isomerization has been studied kinetically. Solvolysis of 1,3,5,7,7,7-hexafluoro-1-heptene in 97.3% formic acid, *viz.*

$$CF_3CH_2CHFCH_2CHFCH{=}CHF \xrightarrow{HCO_2H/H_2O} CF_3CH_2CHFCH_2CH{=}CHCHO + CF_3CH_2CHFCH_2CH{=}CHCHF_2 \quad (16)$$

is accompanied by some allylic isomerization[108]. Isomerization is only about a fifth as fast as solvolysis. The solvolysis reaction is acid-catalyzed, while the rate of isomerization is relatively independent of the acidity of the solvent. The isomerization reaction presumably involves recombination of an ion-pair intermediate.

3. Oxotropic isomerizations of allylic compounds

Allylic alcohols, esters, and ethers undergo acid-catalyzed isomerizations, and allylic alcohols and ethers isomerize at appreciable rates only in the presence of acids. The migrating groups in these reactions are uncharged molecules derived from the oxonium ion conjugate acids of the starting materials. For this reason, and because they are mechanistically related to anionotropic isomerizations, Braude[39,40] named these reactions oxotropic rearrangements.

A number of mechanistic criteria, most of which involve kinetic data, show that these reactions generally involve dissociation of the conjugate acid of the starting material to an allylic carbonium ion and a water, carboxylic acid, or alcohol molecule, followed by recombination of the carbonium ion with an uncharged molecule from its solvation shell, *viz.*

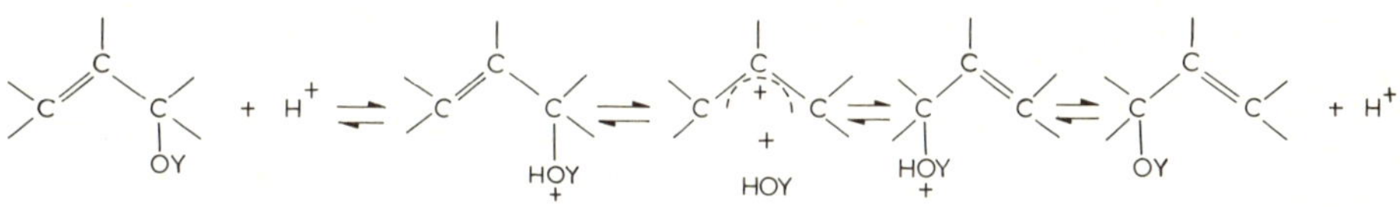

(Y = H, R, RCO) (17)

The kinetic data which lead to this conclusion are summarized in the discussion which follows.

3.1 ACID-CATALYZED ISOMERIZATIONS OF ALLYLIC ALCOHOLS

The most thoroughly investigated group of oxotropic isomerizations are those of allylic alcohols. Braude, who conducted extensive studies of allylic alcohol isomerizations, reviewed them[39,40] in 1950.

Allylic alcohol isomerizations are catalyzed by proton acids and by Lewis acids such as boron trifluoride[44], and rarely occur at measurable rates in the absence of acids. Allylic alcohols having a single alkyl substituent in the α or γ positions isomerize at appreciable rates only in the presence of moderately concentrated acids at elevated temperatures, conditions which usually cause simultaneous formation of ethers, dienes, and carbonyl compounds. Because of the problem of competing reactions, only one kinetic study of isomerizations of monoalkyl-substituted allylic alcohols has been published. Young and Franklin[327], using gas chromatography as an analytical technique, studied the kinetics of the reversible isomerizations of *cis*- and *trans*-γ-methylallyl alcohols to α-methylallyl alcohol in aqueous 70% dioxane–2.9 *M* H_2SO_4. The kinetic results, including the fact that optically active α-methylallyl alcohol racemizes more than five times faster than it isomerizes, are compatible with a carbonium-ion mechanism for the isomerizations.

Most of the kinetic studies of allylic alcohol isomerizations used alcohols having unsaturated α-substituents such as aryl, vinyl, or ethynyl groups, which greatly increase the rate of isomerization. Further, the increase in the extent of conjugated unsaturation which accompanies isomerization makes it feasible to follow these reactions spectrophotometrically in solutions of such low alcohol concentration that competing ether formation is negligible, and the greater thermodynamic stability of the conjugated isomerization product renders the reaction effectively irreversible.

Braude *et al.*[10,37,38,41–52,58] carried out extensive studies of the effects of substituents, solvent composition, temperature and acid concentration on rates of isomerization of α-ethynyl, α-vinyl, and α-aryl substituted allylic alcohols. They found that α-, β-, and γ-methyl substituents on the allylic system increase isomerization rates by factors of the order of 10^2, 10^1, and 10^3 respectively, and that alkyl substitution increases isomerization rate primarily by lowering the energy of activation. The effects of multiple substituents on activation energy are approximately additive in the absence of steric hindrance. Unsaturated α substituents greatly facilitate the isomerization of allylic alcohols. An α-phenyl substituent has a rate accelerating effect similar to that of two α-alkyl groups, and an α-vinyl group is only slightly less activating than an α-phenyl group. Probably because of its electronegativity, an α-ethynyl group is much less activating than an α-vinyl or α-phenyl group.

The activating effects of unsaturated α-substituents are probably due to π-orbital overlap with the allylic system in the transition state for carbonium ion formation. This hypothesis is supported by the fact that α- and β-methyl substituents which sterically hinder attainment of coplanarity of the allylic system and the unsaturated substituent produce abnormally small rate accelerations, and may even decrease isomerization rate. The activating effects of alkyl substituents probably also arise because of their stabilizing effect on the transition state for carbonium ion formation. Selected data (mostly from Braude's publications) which illustrate substituent effects on reactivity are collected in Table 1.

Aryl substituents which can release electrons to the allylic system increase the ease of isomerization of α-arylallyl alcohols. Aryl substituent effects support the hypothesis that the rate-limiting transition state has considerable carbonium-ion character. The specific rate coefficients for acid catalyzed isomerization of α-aryl-γ-methylallyl alcohols in aqueous 60% dioxane correlate poorly with Hammett's σ-substituent constants (see ref. 182, p. 184) whereas a plot of log k_{H^+} *versus* Brown's σ^+-substituent constants[62] is linear, with a slope of -2.7.

The rate of isomerization of two different allylic alcohols has been found

TABLE 1

SUBSTITUENT EFFECTS ON RATES OF ISOMERIZATION OF ALLYLIC ALCOHOLS IN AQUEOUS 60% DIOXANE AT 30°C

R^1	R^2	R^3	R^4	*Relative rate*	k_{H^+}(*l.mole*$^{-1}$.*sec*$^{-1}$)
$HC{\equiv}C{-}CR^1(OH){-}CR^2{=}CR^3R^4 \xrightarrow{H^+} HC{\equiv}C{-}CR^1{=}CR^2{-}CR^3R^4OH$					
H	H	H	H	1.0	1.7×10^{-7}
CH_3	H	H	H	20	
H	CH_3	H	H	5	
H	H	H	CH_3	60	
CH_3	H	H	CH_3	9000	
CH_3	H	CH_3	CH_3	4×10^6	
$C_6H_5{-}CR^1(OH){-}CR^2{=}CR^3R^4 \xrightarrow{H^+} C_6H_5{-}CR^1{=}CR^2{-}CR^3R^4OH$					
H	H	H	H	1.0	6.3×10^{-5}
CH_3	H	H	H	0.7	
H	CH_3	H	CH_3	0.6	
H	H	H	CH_3	460	
H	H	CH_3	CH_3	21,000	
CH_3	H	CH_3	CH_3	2100	
CH_3	CH_3	CH_3	CH_3	2600	
H	H	H	p-$C_6H_4NO_2$	1.3	
H	H	H	α-$C_{10}H_7$	3300	
H	H	H	p-$C_6H_4OCH_3$	22,000	
$RCH(OH)CH{=}CHCH_3 \xrightarrow{H^+} RCH{=}CHCH(OH)CH_3$					
$HC{\equiv}C{-}$					6×10^{-6}
$H_2C{=}CH{-}$					2.9×10^{-2}
C_6H_5					3.0×10^{-2}

to parallel Hammett's acidity function (ref. 182, Chap. 9), rather than hydrogen-ion concentration. For isomerization of α-ethynyl-γ-methylallyl alcohol in aqueous 20% ethanol solutions of hydrochloric or sulfuric acids, plots of log k_1 *versus* $-H_0$ were linear[46] with slopes of 1.09. Similar linear plots of slopes 1.3 and 1.25 were obtained for isomerization of α-phenylallyl alcohol in aqueous 60% dioxane[67] and in water[218]. This type of acidity dependence indicates that water does not function as a nucleophile in the rate-limiting step of the reaction.

Braude *et al.* studied the dependence of rates of acid-catalyzed isomerization of α-ethynyl-γ-methylallyl alcohol[37] and α-phenyl-γ-methylallyl alcohol[49,51] on the composition of mixed solvents such as aqueous ethanol and aqueous dioxane. If allowance is made for the effect of solvent on the concentration of the reactive conjugate acid of the alcohol, these data show that the rate of isomerization of the conjugate acid is nearly independent of solvent composition[181], as would be expected for a unimolecular rate-limiting dissociation of the conjugate acid to a solvated carbonium ion. Kwart and Herbig[218] reached a similar conclusion by applying their solvent composition criterion to Braude's data.

Observed Arrhenius activation parameters for allylic alcohol isomerizations are also consistent with a unimolecular rate-limiting step. In aqueous ethanol and aqueous dioxane, differences in reactivity due to structural changes in the allylic system are reflected primarily in the energy of activation. Values of entropy of activation are generally near zero[10,38,42,43,45–48,50,58,67,258]. Harris and Weale[185] found that the rate of isomerization of α-phenylallyl alcohol in acidic aqueous acetone and acidic aqueous dioxane increases slightly as the pressure is increased to 5000 atm. From the rate data, small negative volumes of activation ($\Delta V^{\ddagger} = -1$ to -5 $cm^3 mole^{-1}$) were calculated. These data indicate that the rate-limiting transition state has a slightly smaller volume than the starting materials. This result is not inconsistent with the carbonium ion mechanism since water is less-strongly hydrogen bonded to the solvated carbonium ion than to the hydronium ion and alcohol from which it was formed.

Kinetic studies of racemization of optically active allylic alcohols, of isotope exchange of ^{18}O-labeled allylic alcohols, and of *cis–trans* isomerization of suitably substitued allylic alcohols have contributed to knowledge of the detailed mechanism and stereochemistry of allylic alcohol isomerization.

In aqueous 40% dioxane–hydrochloric acid solutions, optically active α-phenylallyl alcohol racemizes about 2.5 times faster than it isomerizes to cinnamyl alcohol[161]. Under the same conditions, ^{18}O-labeled α-phenylallyl alcohol undergoes nearly complete oxygen exchange with the solvent during isomerization. Racemization of the unrearranged alcohol is slightly faster than oxygen exchange by the unrearranged alcohol[162]. Cinnamyl alcohol undergoes oxygen exchange under the same conditions much more slowly than α-phenylallyl alcohol[67]. These results, together with other data, exclude the S_Ni' and S_N2' mechanisms for this isomerization reaction and indicate that racemization, exchange, and isomerization involve a common phenylallylcarbonium ion intermediate which reacts with water reversibly to form α-phenylallyl alcohol and essentially irreversibly to form cinnamyl alcohol.

α-Phenyl-γ-methylallyl alcohol behaves somewhat differently. In acidic aqueous 40% dioxane this alcohol racemizes at the same rate that it undergoes allylic isomerization[152]. Further, the α-methyl-γ-phenylallyl alcohol produced by isomerization of the optically active α-phenylallyl alcohol is completely racemic. Optically active α-methyl-γ-phenylallyl alcohol racemizes about 1/56th as fast as its allylic isomer under these conditions. Formation of racemic α-methyl-γ-phenylallyl alcohol from optically active α-phenyl-γ-methylallyl alcohol excludes stereospecific S_Ni' or S_N2' mechanisms for this reaction, which apparently proceeds via a carbonium-ion intermediate which reacts with water nearly exclusively at the methyl-substituted carbon.

The acid-catalyzed racemization, *cis–trans* isomerization and oxygen exchange of *cis*- and *trans*-5-methyl-2-cyclohexenols were studied by Goering *et al.*[167,168,175]. In aqueous 35% acetone–perchloric acid solutions, these alcohols undergo simultaneous racemization (allylic isomerization) and *cis–trans* isomerization. With each isomer, the rate of racemization is several times the rate of *cis–trans* isomerization. By using ^{18}O-labeled alcohols, it was shown that with the *cis* alcohol most of the racemization is intramolecular (*i.e.* occurs without oxygen exchange), while with the *trans* alcohol, interconversion of enantiomers results in nearly complete oxygen exchange.

These reactions almost certainly involve carbonium-ion intermediates, as do nearly all other interconversions in the 5-methyl-2-cyclohexenyl system[304]. The kinetic and stereochemical results, however, lead to the surprising conclusion that the conjugate acids of the *cis* and *trans* alcohols do not dissociate to the same carbonium ion. Goering *et al.* rationalized the experimental data by assuming that the isomeric alcohols react mainly via quasi-axial conformations which give rise to carbonium ions differing conformationally and in the manner in which they are solvated. These carbonium ion collapse to products more rapidly than they are interconverted.

The bicyclic carbonium ion derived from *exo*- and *endo*-bicyclo[3.2.1]oct-3-ene-2-ol in aqueous acetone reacts with water predominantly at the *exo* side. Thus, the *endo* alcohol racemized and underwent *exo–endo* isomerization at the same rate, while the *exo* alcohol racemized 260 times faster than it underwent geometrical isomerization at 20°C[306].

The carbonium-ion intermediates in oxotropic isomerizations of allylic alcohols usually react with water at both ends of the allylic system. Reaction at the original α carbon regenerates the starting material and is detectable in the case of optically active or isotopically labeled alcohols. It has already been mentioned that both allylic carbons of the phenylallylcarbonium ion react with water at comparable rates[161,162], while the phenylmethylallylcarbonium ion reacts with water predominately at the methyl-substituted carbon[152]. The pentad carbonium ion derived from α-vinyl-γ-methylallyl alcohol reacts with water about ten times faster at the secondary methyl-substituted carbon than at the primary carbon[286]. The two allylic carbons of a phenylarylcarbonium ion react with water at about the same rate, regardless of the nature of the aryl substituent[21,45,58]. Thus, although there is a rate difference of some five powers of ten for the isomerization of α-phenyl-γ-*p*-nitrophenylallyl alcohol and α-phenyl-γ-*p*-methoxyphenylallyl alcohol, for each alcohol the rates of the forward and reverse reactions are approximately the same[21].

When isomerization of an allylic alcohol can lead to geometrically isomeric products, both geometrical isomers are generally among the kinetically controlled products. This has been demonstrated for isomerizations of α-methylallyl alcohol[327], the 5-methyl-2-cyclohexenols[165,168,175], α-ethynyl-γ-methylallyl alcohol[12], and α-ethynyl-α-methylallyl alcohol[250] and is probably generally

References pp. 481–487

true. The ratio of geometrical isomers formed presumably depends on the relative free energies of the conformations of the conjugate acids which dissociate to yield the geometrically isomeric allylic carbonium ions and on the energy barrier to interconversion of the *cis* and *trans* carbonium ions.

3.2 ACID-CATALYZED ISOMERIZATION OF ALLYLIC ESTERS

Allylic esters undergo both uncatalyzed and acid-catalyzed heterolysis. There is an important difference between the two processes: uncatalyzed alkyl carbon–oxygen bond fission produces a carbonium carboxylate ion pair, while acid-catalyzed alkyl carbon–oxygen fission results in formation of an allylic carbonium ion and an uncharged carboxylic acid molecule.

In aprotic solvents, the allylic carbonium ion and carboxylic acid molecule produced by acid-catalyzed heterolysis of an allylic ester may recombine to form the starting material and its allylic isomer. Kinetic studies show that α-phenylallyl *p*-nitrobenzoate undergoes concurrent uncatalyzed and *p*-nitrobenzoic acid-catalyzed isomerization to cinnamyl benzoate in chlorobenzene solutions containing *p*-nitrobenzoic acid[56,254]. Similarly, benzoic acid catalyzes the isomerization of α-phenylallyl benzoate in chlorobenzene[255]. When this reaction was carried out using ^{18}O-labeled benzoic acid, ^{18}O appeared in both the starting material and its allylic isomer: under the conditions used, only about a quarter of the isomerization reaction is intramolecular.

In protic solvents, the allylic carbonium ion formed by acid-catalyzed alkyl carbon–oxygen bond fission can recombine either with the carboxylic acid molecule or with a solvent molecule. The electrostatic attraction between the carbonium and carboxylate ions, which is a major factor in isomerization of allylic esters by ion-pair internal return during solvolysis, is absent in the acid-catalyzed reaction. The more numerous, usually more nucleophilic, solvent molecules in the solvation shell of the carbonium ion should compete effectively with the departed carboxylic acid molecule and solvolysis rather than isomerization should be the predominant reaction. For example, in the presence of 0.05 *M* perchloric acid, solvolyses of *cis*- and *trans*-5-methyl-2-cyclohexenyl *p*-nitrobenzoates are not only very much faster than in the absence of the acid, but polarimetric and titrimetric rates of solvolysis of optically-active esters were identical within experimental error[174]. For these esters, the acid-catalyzed solvolysis was not accompanied by a detectable amount of isomerization. Braude[38] reported, on the basis of indirect evidence, that isomerization accompanies acid-catalyzed hydrolysis of α-ethynyl-γ-methylallyl acetate in aqueous dioxane. It was shown that, under some experimental conditions, the spectrophotometrically determined rate of appearance of the rearranged 1-yne-3-ene chromophore exceeds the titrimetrically determined rate of hydrolysis, but rearranged ester was not actually isolated. The data show that acid-catalyzed heterolysis of the acetate ester occurs at a rate comparable with acid-catalyzed isomerization of the allylic alcohol from which it is derived.

In acetic acid, solvolysis and intramolecular rearrangement of an allylic acetate would both result in allylic isomerization. Pocker[256] found that the concurrent racemization, isomerization and isotopic exchange reactions of (−)α-phenylallyl-^{14}C-acetate are catalyzed by perchloric acid, and that plots of log (reaction rate) *versus* $-H_0$ are linear with slopes of 1.1. Isotope exchange with solvent acetic acid and racemization occurred at the same rate, which was about 50% faster than isomerization. Cinnamyl acetate recovered after complete isomerization contained no isotopic label, although labeled cinnamyl acetate was shown to undergo exchange at a rate slower than either exchange of isomerization of α-phenylallyl acetate. These results preclude isomerization by an intramolecular S_Ni' mechanism and support a mechanism involving rate-limiting formation of phenylallylcarbonium ion which reacts with acetic acid to give both cinnamyl acetate and racemic α-phenylallyl acetate.

3.3 ISOMERIZATION OF AN ALLYLIC ETHER

The only kinetic study of acid-catalyzed isomerization of a simple allylic ether was published by Braude and Gore[44] who found that methyl α-phenylallyl ether isomerizes to methyl cinnamyl ether in anhydrous dioxane containing either boron trifluoride or hydrogen chloride. At 25°C, the ether is a quarter to an eighth as reactive as the corresponding alcohol.

4. Alkene isomerizations involving homogeneous, catalyzed hydrogen migrations

This section concerns alkene isomerizations in which hydrogen, rather than an anion or uncharged hydroxylic molecule, is the migrating entity. Only isomerizations involving migration of carbon–carbon double bonds under the influence of acidic, basic, or other catalysts in homogeneous systems are considered.

4.1 ACID-CATALYZED ALKENE ISOMERIZATIONS

Few kinetic studies of acid-catalyzed alkene isomerizations have been reported. This is probably due in part to the complexity of these reactions and the number of competing processes in which the carbonium ion intermediates are involved. In general, these reactions involve initial reaction with a proton or Lewis acid to form a carbonium ion, which can revert to the starting material or lose a proton to form an isomeric alkene, *viz.*

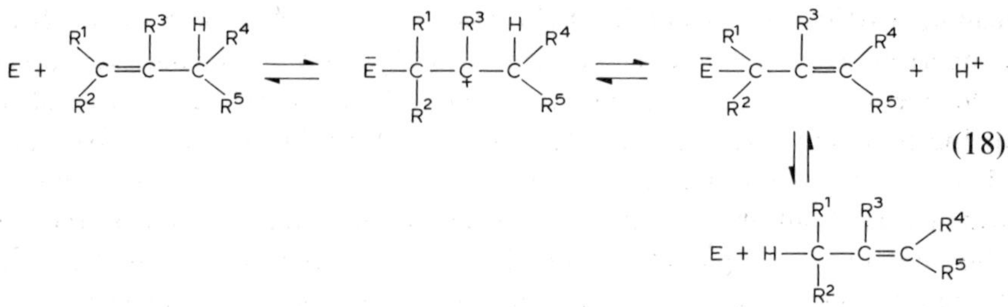

(18)

When the catalyst is a proton acid, the reaction is complete after the second step.

Cram and Sahyun[85] studied the reversible *p*-toluenesulfonic acid-catalyzed isomerization of *cis*- and *trans*-2-phenyl-2-butenes and 2-phenyl-1-butene in acetic acid solutions, *viz.*

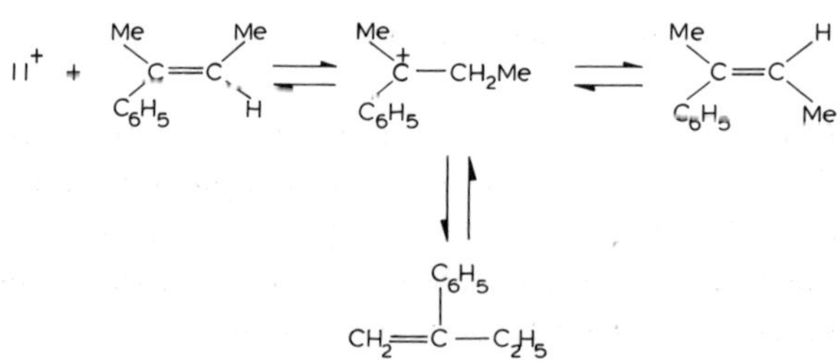

(19)

Initial rates of disappearance of each of the isomeric phenylbutenes are first order in alkene and first order in *p*-toluenesulfonic acid. The kinetic and equilibrium data are accounted for by assuming that all three alkenes form the same carbonium ion, which is deprotonated to form 9% *cis*-2-phenyl-2-butene, 1% *trans*-2-phenyl-2-butene and 90% 2-phenyl-1-butene in glacial acetic acid – 0.01 *M* *p*-toluenesulfonic acid at 50°C.

In aqueous 98.7% dioxane, perchloric acid catalyzes the isomerization of 3-methylenecholestane to 3-methyl-2-cholestene[78],

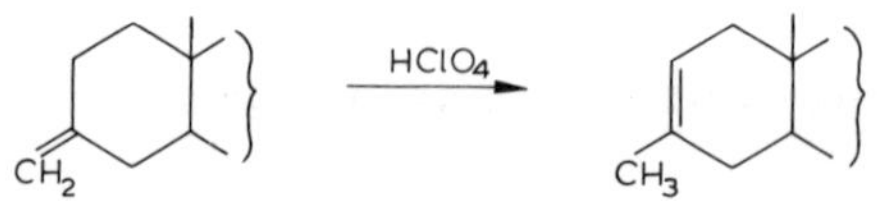

(20)

Experiments with 2β-D-, 2α-D-, and 2,2,4,4-D_4-3-methylenecholestene showed that at 30°C the unlabeled and deuterium-labeled compounds isomerize at the same rate, but that an equatorial 2 α-deuterium atom is lost more rapidly than an axial 2β-deuterium atom. For the isomerization of the unlabeled alkene, $E_a = 24$ kcal.mole^{-1} and $\Delta S^{\ddagger} = +0.7$ eu. The results are consistent with a mechanism involving rate-limiting proton transfer to methylene carbon, followed by proton loss from C-2.

Powell and Whiting[260] made a kinetic study of the isomerization of *trans*-Δ^1-octalin and $\Delta^{1,9}$-octalin in 33% benzene–67% acetic acid and 33% benzene–67% sulfolane solutions at 30°C, *viz.*

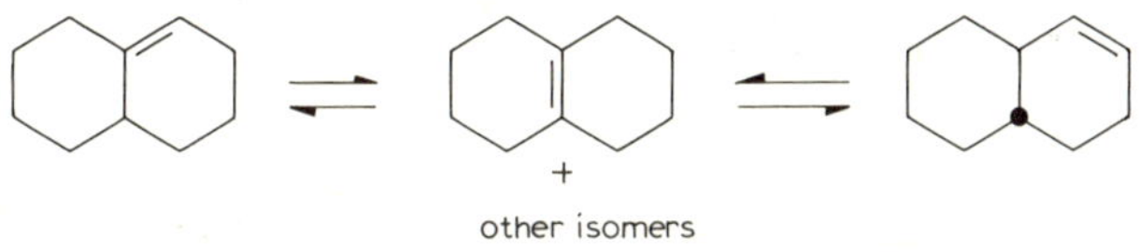

(21)

Catalytic effectiveness of various acids in 0.25 *M* solution in benzene–acetic acid for isomerization of $\Delta^{1,9}$-octalin increased in the order: $BF_3 \cdot HOAc$, 1; *p*-$CH_3C_6H_4SO_3H$, 6; H_2SO_4, 20; HBF_4, 200; $HClO_4$, 500. In benzene–sulfolane solutions, catalytic effectiveness of 0.25 *M* solutions of various acids for the same reaction increased in the order: CF_3CO_2H, < 1; H_2SO_4, 15; $BF_3 \cdot (C_2H_5)_2O$, 200; BF_3, 3000; HBF_4, > 3000. The outstanding catalytic efficiency of fluoroboric acid in benzene–sulfolane solutions was ascribed to the high dielectric constant of the solvent, the absence of nucleophilic species stronger than benzene or fluoride ion, and the absence of electrophiles other than hydrogen ion.

33% Methyl bromide–67% sulfolane was used as solvent in a study of the kinetics and products of the fluoroboric acid-catalyzed isomerization of 1-octene, *trans*-3-octene and *trans*-4-octene[69]. At 0°C, 0.067 *M* HBF_4 caused the three octenes to isomerize at comparable rates to equilibrium mixtures of *cis*- and *trans*-2,3, and 4-octenes in which *trans*-3-octene predominated. Fluoroboric acid was about 10^7 times more effective as a catalyst than *p*-toluenesulfonic acid in the solvent used.

Clayton and Eastham[76] found that in 1,2-dichloroethane at 20°C, the rate of boron trifluoride–acetic acid-catalyzed isomerization of 1-butene to *cis*- and *trans*-2-butenes is proportional to the concentration of both free boron trifluoride and of boron trifluoride–acetic acid complex. The kinetic results are evidence for pre-equilibrium formation of an adduct between the butene and the boron trifluoride–acetic acid complex.

The kinetics of gas-phase hydrogen bromide-catalyzed isomerization of butenes[231] lead to the conclusion that radical chain reactions are not involved in the interconversion of 1-butene and *cis*- and *trans*-2-butenes at 300°C. Activation parameters for the disappearance of 1-butene, calculated from second-order rate coefficients, are $E_a = 26.3$ kcal.mole^{-1}, $\log A = 11.8$, and

References pp. 481–487

$\Delta S^{\ddagger} = -6$ eu. A bimolecular mechanism involving a cyclic transition state was suggested for this reaction.

Carbonium ions are Lewis acids, but most are too reactive to permit study of their effectiveness as catalysts for olefin isomerizations. An exception is tropylium ion. The kinetics of tropylium perchlorate-catalyzed isomerization of 7-methyl-1,3,5-cycloheptatriene to a mixture of 1-, 2-, and 3-methyl-1,3,5-cycloheptatrienes in acetonitrile solutions at 25°C were studied[77]. Isomerization rate was proportional to 7-methyltropylidene concentration and tropylium-ion concentration. The reaction probably involves reversible hydride transfer tropylium ion and the methyltropylidenes.

4.2 BASE-CATALYZED ALKENE ISOMERIZATIONS

Allylic protons of alkenes undergo base-catalyzed exchange much more rapidly than either vinylic or saturated aliphatic protons[288]. The enhanced kinetic acidity of such protons is due to the fact that their removal results in formation of resonance-stabilized allylic carbanions. These carbanions react with proton donors to form the original alkenes and sometimes one or more isomeric alkenes, *viz.*

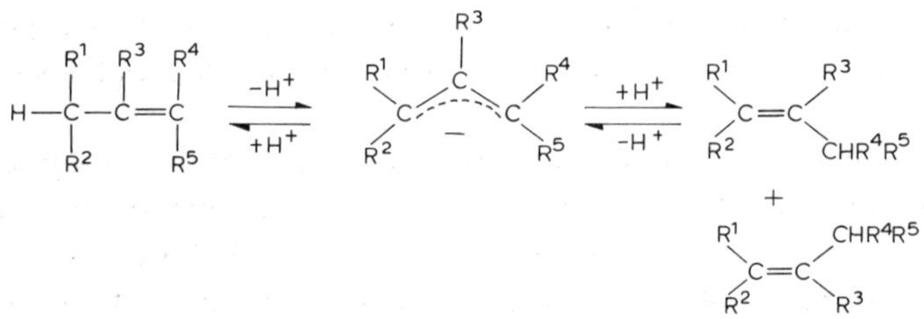

(22)

The mechanistic and stereochemical options available to allylic carbanions are fully as complex as those available to allylic carbonium ions. Much that is known concerning the mechanisms of base-catalyzed alkene isomerizations was obtained from stereochemical, isotope exchange, and product composition studies. The present discussion is limited to alkene isomerizations whose kinetics have been studied in homogeneous systems. Isomerizations involving allylic carbanion intermediates have been reviewed by Cram[84].

Cram *et al.*[86,113,114,204] studied the kinetics and stereochemistry of isomerizations of several phenyl-substituted alkenes. 3-Phenyl-1-butene isomerizes in the presence of bases to a mixture of about 98% *cis*-2-phenyl-2-butene and 2% *trans*-2-phenyl-2-butene[86]

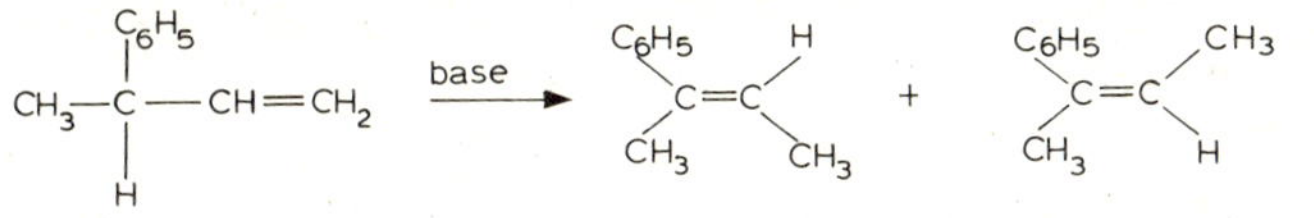

(23)

Rates of isomerization were measured in triethyl carbinol, ethylene glycol dimethyl ether, *tert.*-butyl alcohol, ethylene glycol, and dimethyl sulfoxide, using potassium alkoxides and tetramethylammonium hydroxide as catalysts. Solvent and substrate deuterium isotope effects on isomerization rate were studied, along with rates of racemization of optically active 3-phenyl-1-butene and rates of isotope exchange without isomerization. The kinetic and stereochemical results show that the isomerization involves both intra-molecular and intermolecular hydrogen migration, and provide support for a mechanism involving a carbanion intermediate hydrogen bonded at both ends of the allylic system to an alcohol, which undergoes alcohol exchange with the solvent.

The kinetics and stereochemistry of isomerization of diphenylbutenes VII–X, *viz.*

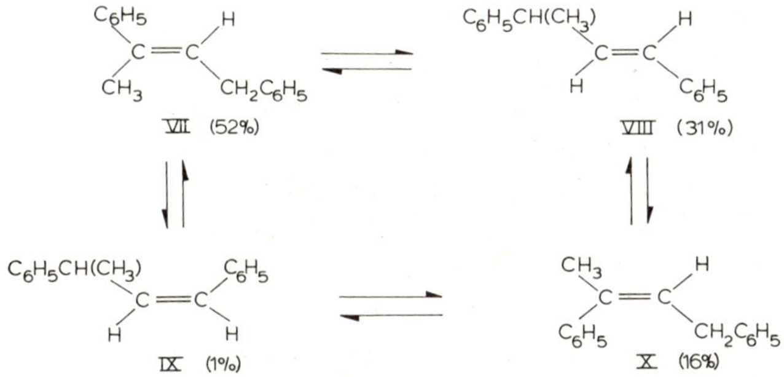

(24)

in *tert.*-butyl alcohol containing potassium *tert.*-butoxide were studied[113]. At 40°C, the four olefins equilibrated to a mixture containing the amounts listed under their formulas. The initial rates of isomerization decreased in the order VIII > VII > X > IX, which correlates with the predicted order of decreasing stability of the allylic carbanion intermediates involved. VII and VIII are interconverted much more rapidly than they form IX and X. Isotope exchange experiments showed that the carbanion intermediate for interconversion of VII and VIII collapses to VII much more rapidly than it collapses to VIII. The kinetic and stereochemical results are accounted for if it is assumed that different conformations of each olefin give rise to two isomeric configurationally discrete allylic carbanions, whose relative stabilities are determined by the magnitude of 1,3-steric interactions of substituents *cis* to one another in the plane of the anion.

In *tert.*-butyl alcohol containing potassium *tert.*-butoxide, allylbenzene isomerizes at 25°C to a mixture of 93.95% *trans*-propenylbenzene, 5.73% *cis*-propenylbenzene and 0.32% allylbenzene[114]. The reaction was 2.3 times faster in $(CH_3)_3COD$ than in $(CH_3)_3COH$. From the kinetic, equilibrium, and isotope exchange data it was calculated that *trans*-phenylallylcarbanion undergoes proton capture to give *trans*-propenylbenzene about three times faster than to give allylbenzene, while *cis*-phenylallylcarbanion is converted to *cis*-propenylbenzene ten to fifteen times faster than it is converted to allylbenzene.

The kinetics of potassium *tert.*-butoxide-catalyzed interconversion of *cis*- and *trans*-α-methylstilbene and α-benzylstyrene was studied in *tert.*-butyl alcohol solutions[204], *viz.*

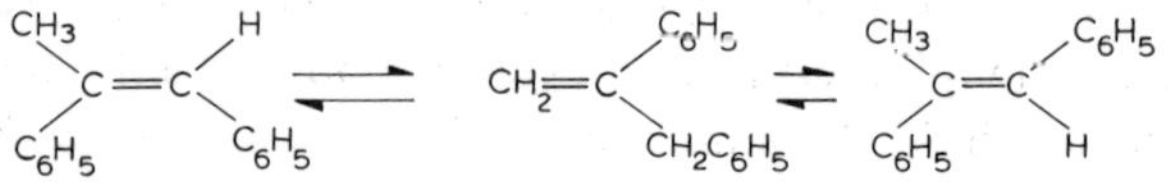

(25)

The kinetic data together with isotope exchange and solvent isotope effect data are consistent with a mechanism involving non-interconverting *cis* and *trans* allylic carbanion intermediates, with partially intramolecular proton transfer.

Schriesheim *et al.*[6–9,275–280] measured the rates of potassium *tert.*-butoxide-catalyzed isomerizations of a number of alkenes and cycloalkenes in dimethyl sulfoxide solutions. Isomerization rates increase with increasing potassium *tert.*-butoxide concentration in dilute solutions, but are nearly independent of catalyst concentration above about 0.2 molar[276]. The principal kinetically controlled products of isomerization of 1-alkenes are *cis*-2-alkenes. This stereospecificity is presumably due to the greater stability of *cis*-allylic carbanions than their *trans* isomers. $C_3H_7CH{=}CH_2$ isomerizes 4.4 times faster than $C_3D_7CD{=}CD_2$; the isotope effect is due to different energies of activation, which indicates that it reflects the difference in zero-point energies of the protio and deuterio olefins[6]. The observed k_H/k_D ratio compares with a calculated primary isotope effect of 5.2, and provides support for the assumptions that C–H bond cleavage is rate-limiting and that the C–H bond is largely broken in the transition state for carbanion formation.

Labeling experiments showed that most of the isomerization does not involve exchange: that is, the prototropic shift is largely intramolecular. This suggests that the protonated base formed simultaneously with the allylic carbanion is not free to exchange, but is, instead, rather tightly held by the anion XI. Proton exchange occurs by a dissociated carbanion intermediate XII, *viz.*

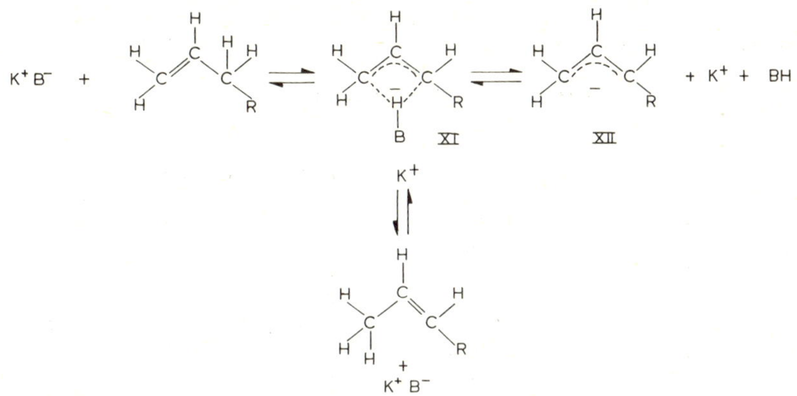

(26)

This mechanism accounts for the observed primary isotope effect, the slow rate of exchange, and the large negative entropies of activation. It receives additional support from the observation that, for potassium *tert.*-butoxide catalyzed-isomerization of 1-butene in dipolar aprotic solvents, $\log k_i$ is proportional to the reciprocal of the dielectric constant of the solvent[275].

The cation of the metal *tert.*-butoxide strongly influences the rates but not the products of alkene isomerizations. For isomerization of 1-butene in dimethyl sulfoxide solutions at 55°C, the relative catalytic effectiveness of alkali *tert.*-butoxides increase in the order[9]: NaOBu^t, 1.0; KOBu^t, 116; CsOBu^t, 284, RbOBu^t, 447. This is probably attributable to the fact that large cations are more weakly bonded to the alkoxide ion than smaller cations[296]. The anion of the alkoxide also strongly influences its catalytic effectiveness. Potassium *tert.*-butoxide is 126 times as effective a catalyst for 1-butene isomerization as potassium methoxide in dimethyl sulfoxide at 55°C[9]. The rate of potassium *tert.*-butoxide-catalyzed 1-butene isomerization in DMSO is strongly retarded by addition of *tert.*-butyl alcohol to the solvent, probably due to hydrogen bonding between the alkoxide and the alcohol[279].

The most effective solvent–catalyst systems for alkene isomerizations are hexamethylphosphoramide solutions of alkali alkylamides. For example, sodium dimethylamide in hexamethylphosphoramide is about 300,000 times as effective in catalyzing the isomerization of 2,4,4-trimethyl-1-pentene as is potassium *tert.*-butoxide in dimethylsulfoxide[275].

Rates of potassium *tert.*-butoxide-catalyzed isomerization of 1-alkenes and methylenecycloalkanes in dimethylsulfoxide at 55°C are summarized in Table 2. For the acyclic as well as exocyclic alkenes, isomerization rates parallel rates of base-catalyzed enolization of structurally analogous ketones. This is further evidence that the rate-limiting step in the alkene isomerizations is proton abstraction to form the allylic carbanion.

TABLE 2

RATES OF ISOMERIZATION OF 1-ALKENES IN DMSO–0.77 *M* KOBu^t AT 55°C[a]

$CH_2{=}CR^1{-}CHR^2R^3 \rightarrow CH_3CR^1{=}CR^2R^3$

R^1	R^2	R^3	k_{rel}[b]	$\Delta H^{\ddagger}$ (kcal.mole^{-1})	$\Delta S^{\ddagger}$ (eu)
H	H	CH_3	1.0[c]	16.6	−22
CH_3	H	CH_3	0.15		
C_2H_5	H	CH_3	0.091		
H	H	C_2H_5	0.57		
H	H	$CH(CH_3)_2$	0.18	21.6	−17
H	H	$C(CH_3)_3$	0.0072	20.3	−22
H	CH_3	CH_3	0.24		
H	CH_3	C_2H_5	0.13		
H	CH_3	$CH(CH_3)_2$	0.036		
CH_3	H	C_2H_5	0.078		
CH_3	H	$CH(CH_3)_2$	0.036		
CH_3	H	$C(CH_3)_3$	0.0013		
H	H	$CH{=}CH_2$	$\sim 20 \times 10^5$		
H	H	C_6H_5	$\sim 6 \times 10^5$		
	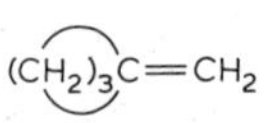		2.73	13.3	−27
	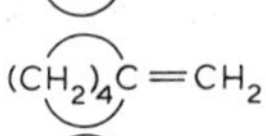		1.16	17.3	−17
	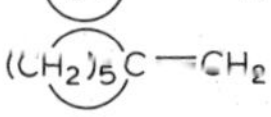		0.0026	27.1	+0.7
	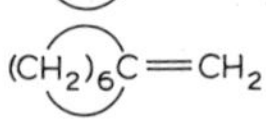		0.015	19.7	18
	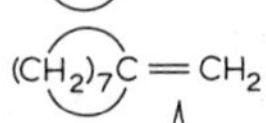		0.044	16.8	−25
			0.0032	20.5	−20
	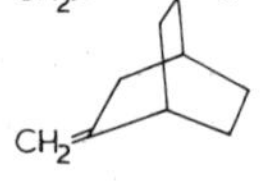		0.010	22	−17
	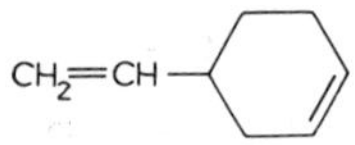		0.45		

[a] Data from refs. 8, 277, 280.

[b] Relative rates are statistically corrected for number of allylic protons.

[c] $k_1 = 8.40 \times 10^{-4}$ sec^{-1}.

1,4-Pentadiene and allylbenzene are about 10^6 times as reactive as 1-butene. The effect on reactivity of an alkyl group, relative to a methyl group at the same position, is nearly independent of whether the alkyl group is bonded to the 2, 3, or 4 position of the 1-alkene chain. The effects of unbranched alkyl substituents or reactivity correlate well with Taft's polar substituent constants[297], but isopropyl and tertiary butyl groups have much larger deactivating effects than predicted by their polar substituent constants. These bulky groups apparently influence reactivity not only by their polar effects, but also by steric hindrance to carbanion formation and steric inhibition of carbanion solvation. 1-Alkene substituents fall in the following order with respect to their effect on isomerization rates: $CH_2{=}CH \sim C_6H_5 \gg H > CH_3 > C_2H_5 \sim n\text{-}C_3H_7 > CH(CH_3)_2 > C(CH_3)_3$.

Rates of isomerization of methylenecycloalkanes vary over a range of three powers of ten, with methylenecyclobutane being the most reactive and methylenecyclohexane the least reactive[277]. The observed reactivities and activation parameters suggest that the transition states for proton abstraction structurally resemble allylic carbanions. Methylenecyclobutane and methylenecyclopentane, which can most readily assume the planar geometry of the allylic carbanion, and in which eclipsed hydrogen–hydrogen interactions are relieved on passing to the transition state, are most reactive. Torsional (eclipsing) effects appear to outweigh angle strain effects in determining reactivities of methylenecycloalkanes, while the reverse is true for isomerization of 1-methylcycloalkenes[7].

Most of the published kinetic data on base-catalyzed alkene isomerizations are for reactions of 1-alkenes. Alkyl substitution stabilizes alkenes and products of isomerization of terminal alkenes isomerize at slower rates than their precursors. Rates of isomerization of methylcycloalkenes can be estimated from kinetic and equilibrium data[7] or measured directly[301]. These compounds isomerize at slower rates than the corresponding methylenecycloalkanes[277], as expected.

Bergson *et al.*[14–18,247,248,314], using NMR spectrometry and polarimetry as analytical techniques, made a detailed study of the kinetics and mechanisms of base-catalyzed isomerization, racemization, and mutarotation reactions of 1-substituted and 1,3-disubstituted indenes. The prototropic rearrangements are completely stereospecific and intramolecular when aliphatic amines are used as catalysts in solvents of low or moderate polarity—the proton becomes re-attached to the same side of the ring system from which it departed under these conditions[247]. With alcohols as solvents and alkoxides as catalysts, extensive racemization occurs indicating that indenes form symmetrically solvated carbanions in alcohol solvents[314]. In dimethyl sulfoxide, isomerization was only partly stereospecific even when aliphatic amines were used as catalysts. These, and other observations, are rationalized by the reaction scheme

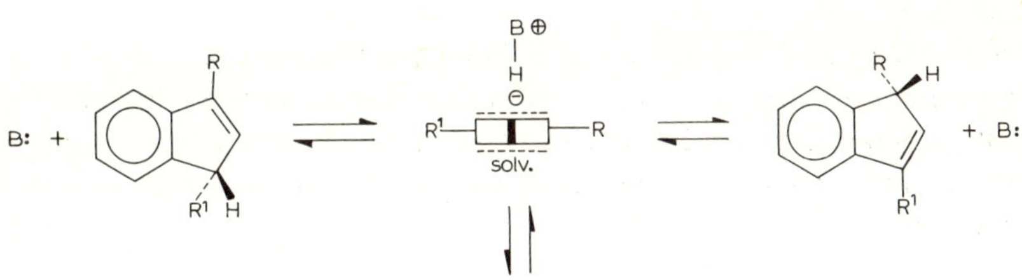

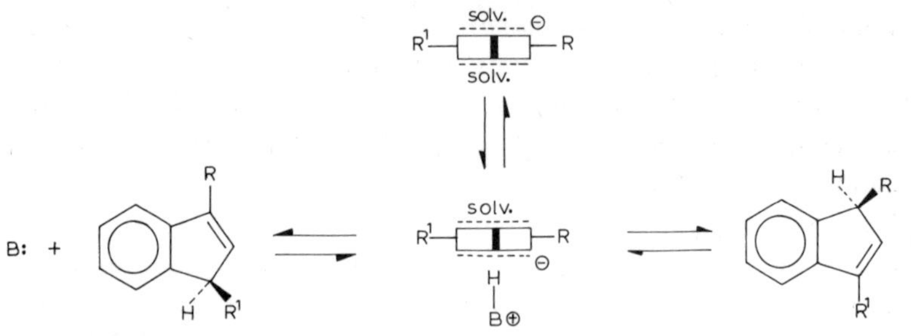

(27)

Intramolecular proton migration results in isomerization with mutarotation, whereas formation and protonation of a symmetrically solvated carbanion results in simultaneous racemization and isomerization. In all cases studied, the kinetic data support the assumption that proton abstraction by the basic catalyst is rate-limiting.

Alkyl substitution at position 3 slows the rate of diazabicyclooctane (DABCO)-catalyzed isomerization of 1-methyl-3-substituted indenes in pyridine solution. At 30°C, the relative rates of isomerization of 1-methyl-3-R-indenes decreased in the order R = H, 70; R = Pr^i, 1.25; R = CD_3, 1.11; R = Bu^t, 1.0[247,314].

In pyridine solutions, the statistically corrected relative catalytic coefficients of tertiary amines for 1-methylindene isomerization decreased in the order[248,314]: quinuclidine, 80; DABCO, 10; triethylamine, 1. The smaller catalytic effectiveness of DABCO than quinuclidine is attributable to its weaker basicity; $\Delta S^\ddagger$ is −30 eu for each of these bicyclic bases. On the other hand, triethylamine is about as basic as quinuclidine, but must lose considerable rotational freedom in the rate-limiting proton transfer. This is reflected in the more negative entropy of activation (−39 eu) for the triethylamine-catalyzed reaction. In pyridine solution, there is a close correlation between pK_a of the catalyzing base and $\Delta H^\ddagger$ for 1-methylindene isomerization. Asymmetric catalysis was demonstrated in the quinine-catalyzed isomerization of optically active 1-methylindene in pyridine at 25°C: the dextrorotatory indene isomerized nearly twice as fast as its enantiometer[247].

Isomerization of 1-deutero-1,3-dimethylindine at 30°C in pyridine, benzene, and methanol, using DABCO as catalyst, occurs with relative rates of 6.5, 1.14, and 1.0 respectively[248]. The low relative rate in methanol is probably due to hydrogen bonding between the solvent and catalyst. The relative rates in the two aprotic solvents decrease in order of the decreasing dielectric constants of the solvents.

In the rearrangement of 1-protio- and 1-deuterio-1,3-dimethylindene, the isotope effect tended to increase with increasing base strength of the catalyst. At 30°C, potassium *tert.*-butoxide in *tert.*-butyl alcohol, sodium methoxide in methanol and DABCO in methanol gave k_H/k_D ratios of 8.1, 6.7, and 4.8, respectively[248].

A number of base-catalyzed prototropic isomerizations have been the subject of less extensive kinetic studies than those discussed above. These are summarized briefly.

4-Phenyl-1-butene isomerizes to 1-phenyl-1-butene in ethylene glycol dimethyl ether and tetraethylene glycol dimethyl ether at rates which are inversely proportional to the square root of the concentration of the potassium *tert.*-butoxide catalyst. This is the concentration dependence expected if potassium *tert.*-butoxide ion pairs dissociate only slightly in the solvents used, and if only the dissociated alkoxide ion catalyzes the reaction[302].

The effects of molecular weight and alkali metal cation on rates of double bond migration in allyloxypropylene glycol polyether alkoxides [CH_2=$CHCH_2(OC_3H_6)_nO^-M^+$] in tetrahydrofuran showed that the triproyplene glycol ether ($n=3$) is more reactive than its higher or lower homologues, presumably due to chelation of the cation by the polyether chain[296].

Potassium polyethyleneglycolates, $\overset{+}{K}\overset{-}{O}(CH_2CH_2O)_nCH_3$, were shown to be much more effective catalysts for prototropic isomerizations of butenylbenzenes than potassium *tert.*-butoxide, in ether and dioxane solutions[265]. When $n = 7$–8, these polyether alkoxides were several hundred times more effective than potassium *tert.*-butoxide, presumably due to "self solvation" of the potassium ion by oxygen in the polyether anion.

The kinetics of the reversible potassium *tert.*-butoxide-catalyzed isomerization of cyclopentenyl- and cyclohexenyl-*n*-butylsulfones in *tert.*-butyl alcohol at 35°C showed that there is little resonance interaction between an α-ethylenic group and a sulfone function[244].

The kinetics and equilibria of isomerization of 1,3-phenylarylpropenes (aryl = naphthyl, anthryl, and phenanthryl) were studied in methanolic sodium methoxide[267].

The relative rates of lithium 2-aminoethylamide-catalyzed isomerization of *trans*-4-octene, *cis*-4-octene, and 1-octene in ethylene diamine solutions at 25°C are[69] 1, 3.5, and 90. *Cis*-2-octene was the major product from 1-octene, while *trans*- and *cis*-3-octene were the principal initial products from *cis*- and *trans*-4-octene, respectively.

The effects of catalyst structure (secondary and tertiary aliphatic and heterocyclic amines), solvent, temperature, and substrate structure on rates and products of prototropic isomerizations of β,γ-unsaturated nitriles and esters were studied by Mialhe and Vessière[240].

The kinetics of interconversion of α- and β-ionones in acetic acid solutions of triethylamine were studied[211]. Under these conditions, the catalyst is present mostly as triethylammonium acetate, and the rates of isomerization of the unsaturated ketones are proportional to the square of the stoichiometric concentration of triethylamine. This is the concentration dependence expected if the only effective catalyst is the free amine.

One example of an intramolecular 1,5-proton transfer has been studied.

The kinetics and products of isomerization of methyl 2-[(4-diphenylmethylene)-2,5-cyclohexadienyl]-2-methylpropanoate to methyl *p*-diphenylmethyl-α,α-dimethylphenylacetate in *tert.*-butyl alcohol solutions of DABCO and methanol solutions of potassium methoxide, *viz.*,

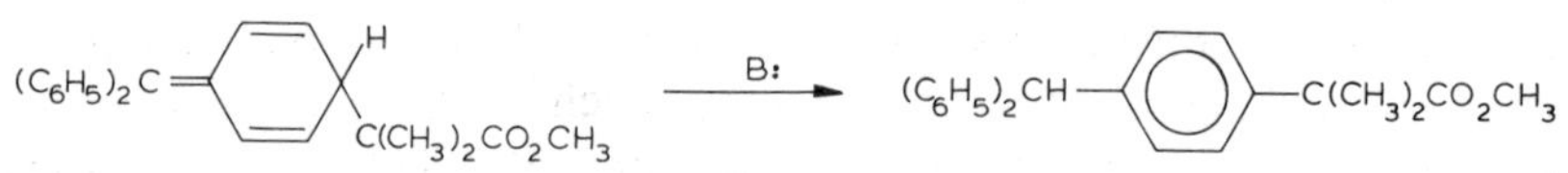

(28)

indicate that the proton shift is largely intramolecular and occurs by a 1,5-proton shift rather than by two consecutive 1,3-proton shifts[87].

4.3 IODINE-CATALYZED ALKENE ISOMERIZATIONS

Benson *et al.*[13,111,112] studied the kinetics of iodine atom-catalyzed isomerizations of butenes and pentadienes in order to obtain data for estimating the resonance energies of butenyl and pentadienyl radicals. The isomerization of gaseous 1-butene in the presence of iodine vapor

$$CH_3CH_2CH{=}CH_2 + I \underset{k_{-1}}{\overset{k_1}{\rightleftharpoons}} \underset{(cis+trans)}{CH_3CH{=}CHCH_3} + I \tag{29}$$

follows the rate law

$$k_1 = \{[K/(1+K)]/K_{I_2}{}^{1/2}[I_2]^{1/2}t\}\ln\{[B_1]/[B_1]_0 - 1/(1+K)\}[(1+K)/K]$$

where $K = k_1/k_{-1}$, K_{I_2} is the equilibrium constant for dissociation of iodine into atoms, and $[B_1]$ is the concentration of 1-butene. Activation parameters for k_1 are $E_a = 12.4$ kcal.mole^{-1} and $\log A = 9.0$. From these and other data, the resonance energy of the butenyl radical is estimated to be 12.6 ± 0.8 kcal.mole^{-1}. A similar study of the iodine-catalyzed isomerization of 1,4-pentadiene to 1,3-pentadiene yielded $E_a = 9.6$ kcal.mole^{-1} and $\log A = 9.15$, from which the resonance energy of the pentadienyl radical is estimated[111] to be 15.4 kcal.mole^{-1}.

4.4 ALKENE ISOMERIZATIONS CATALYZED BY TRANSITION METAL COMPLEXES

Several kinetic studies of alkene isomerizations catalyzed by transition-metal coordination complexes in homogeneous solutions have been reported.

Tetrachloropalladate(II) ion catalyzes the interconversion of 1- and 2-butenes in aqueous solutions containing chloride and hydronium ions[242,253]. Sodium tetrachloropalladate(II) catalyzes the conversion of allylbenzene to propenylbenzene in acetic acid solutions[93]. Tetrakis(ethylene)μ,μ'-dichlororhodium(I) catalyzes butene isomerization in methanolic hydrogen chloride solutions[88]. Cyclooctadienes isomerize in benzene–methanol solutions of dichlorobis-(triphenylphosphine)platinum(II) and stannous chloride[299]. Chloroplatinic acid–stannous chloride catalyzes the isomerization of pentenes[25]. Coordination complexes of zero-valent nickel with tris(2-biphenylyl)phosphite or triphenylphosphine catalyze the isomerization of *cis*-1,2-divinylcyclobutane to a mixture of *cis,cis*-1,5-cyclooctadiene and 4-vinylcyclohexene[189]. Detailed discussions of reaction kinetics and mechanisms appear in the papers cited.

5. Unimolecular thermal isomerization reactions

There is a large group of isomerization reactions in which carbon–carbon double bond migrations are achieved simply by heating the reactant. Some of these reactions, such as the Claisen and Cope rearrangements, involve concerted intramolecular electronic reorganizations and are examples of sigmatropic rearrangements[197,326]. Others involve transient diradical intermediates. All of them are unimolecular and occur without catalysts.

5.1 THERMAL ISOMERIZATIONS OF ALLYLIC COMPOUNDS

A number of allylic compounds, including allyl vinyl ethers, allylic azides, allylic thiocyanates, and allylic sulfinates, undergo unimolecular thermal isomerization reactions. Most of these reactions are concerted, four-center processes and in most of them the allylic substituent, as well as the allylic carbon skeleton, undergoes alteration.

5.1.1 Vinyl allyl ether isomerizations

Vinyl allyl ethers and allyl aryl ethers are converted by heating to γ,δ-unsaturated carbonyl compounds and allyl phenols, respectively. The two groups of reactions are mechanistically related, and are referred to as Claisen rearrangements. Claisen rearrangements of allyl aryl ethers have been reviewed[208,264,289,298] and their kinetics are discussed in Chapter 3 Volume 13 of this series. The present discussion is limited to thermal isomerizations of vinyl allyl ethers.

When a vinyl allyl ether is isomerized, a σ bond forms between the β carbon of the vinyl group and the γ carbon of the allyl group. Simultaneously, π bonds are formed between the vinyl α carbon and oxygen and between the α and β carbons of the allyl system, and the original π bonds become σ bonds, *viz.*

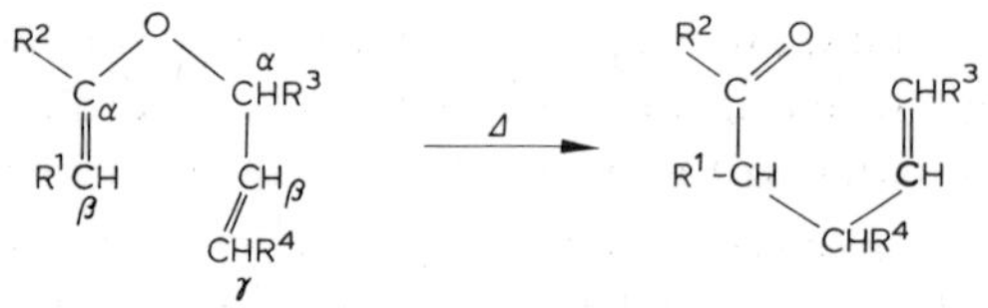

(30)

The reaction is highly stereospecific[193] and is always accompanied by inversion of the structure of the allyl group. This inversion is experimentally detectable when the α carbon[140,257] or γ carbon[261] bears an alkyl substituent and is detectable in the case of allyl vinyl ether itself if the α or γ carbons of the allyl group are labeled with ^{14}C (ref. 257). That the α carbon of the vinyl group of the ether becomes the carbonyl carbon of the product is demonstrated by the isomerization of isopropenyl allyl ether to hex-5-ene-2-one[295], and by the isomerization of $CH_2{=}CDOC_3H_5$ to $CH_2{=}CHCH_2CH_2CDO$[257].

The kinetics of isomerizations of vinyl allyl ethers have been studied in the gas phase, in solution, and in the neat liquid ether. In the gas phase, isomerization rate is independent of pressure down to about 1 torr, below which the rate diminishes with decreasing pressure in the manner characteristic of unimolecular reactions[257,295]. The rate of isomerization is not influenced significantly by glass surfaces[295], but is faster in solution in either non-polar or moderately polar solvents than in the gas phase. Reactivity is increased by methyl substitution of either the vinyl or the allyl group of the ether. All of the isomerizations thus far studied have negative entropies of activation, with values ranging from -8 to -13 eu. Isomerization of vinyl allyl ether (neat liquid) at 130°C exhibits a volume of activation[61] of -18 $cm^3.mole^{-1}$. Published kinetic data are summarized in Table 3.

The first-order kinetics, lack of dependence of rate upon pressure in the gas phase above 1 torr, absence of catalysis, and stereospecificity all show that vinyl allyl ether isomerizations are unimolecular reactions. The effects of substrate structure and solvents on reactivity indicate that the rate-limiting transition state does not resemble an ion pair. Methyl substituents on the α and γ carbons of the allylic group increase reactivity by only about 10- and 2.5-fold respectively, which is very much less than substituent effects on ionic allylic reactions. While the isomerization of vinyl α-methylallyl ether is about ten times faster in organic solvents than in the vapor phase, the solvent effect is small and does not correlate with solvent polarity.

TABLE 3

THERMAL ISOMERIZATIONS OF VINYL ALLYL ETHERS

Compound	*Reaction medium*	E_a *(kcal.mole^{-1})*	*log A*	$\Delta S^{\ddagger}_{278}$ *(eu)*	k_{rel}[a]	*Ref.*
CH_2=$CHOCH_2CH$=CH_2	Vapor	30.6	11.70	−7	1[b]	284
CH_2=$C(CH_3)OCH_2CH$=CH_2	Vapor	29.3	11.73	−7	5.2	295
CH_2=$CHOCH_2C(CH_3)$=CH_2	Vapor	29.1	11.15	−9	1.7	143
CH_2=$CHOCH(CH_3)CH$=CH_2	Vapor	27.9	11.32	−9	10	140
CH_2=$CHOCH(CH_3)CH$=CH_2	1-Decene	25.7	10.78	−11	130	140
CH_2=$CHOCH(CH_3)CH$=CH_2	Benzonitrile	25.3	10.91	−10	83	140
$C_4H_9O_2CCH$=$C(CH_3)OCH_2CH$=$CHCH_3$	Neat liquid	28.0	11.72	−7	23	261
$C_3H_5O_2CCH$=$C(CH_3)OCH_2CH$=CH_2	Neat liquid	26.3	10.43	−13	9	261

[a] Relative rates at 430°K, calculated from activation parameters.
[b] Calculated $k_1 = 1.2 \times 10^{-4}$ sec^{-1} at 430°K.

The kinetic and other evidence strongly supports a concerted mechanism having a cyclic 6-membered transition state, *viz.*

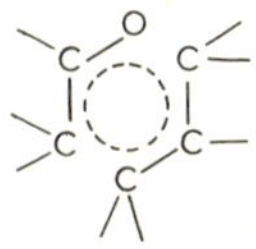

The loss of rotational freedom when the cyclic transition state is formed from the acyclic starting material accounts for the observed negative entropies and volumes of activation. It has been established that the transition state of the closely related Cope rearrangement (see below) resembles the chair conformation of cyclohexane. An analogous transition state structure for the Claisen rearrangement would account for the fact that isomerization of vinyl α-methylallyl ether yields 97% *trans*-4-hexenal and 3% *cis*-4-hexenal, since there should be a preference for an equatorial orientation of the methyl group in a cyclohexane chair-like transition state[140].

5.1.2 Isomerizations of allylic azides

Isomerizations of allylic azides also appear to involve concerted, non-ionic rearrangements. Butenyl and pentenyl azides isomerize slowly at room temperature in a variety of solvents[153], *viz.*

$$CH_3\text{–}CR{=}CH\text{–}CH_2\text{–}N_3 \rightleftharpoons CH_3\text{–}CR(N_3)\text{–}CH{=}CH_2 \quad (31)$$

where R = H or CH_3.

TABLE 4

ISOMERIZATION OF ALLYLIC AZIDES IN ACETONE AT 25 °C[a]

$$R^3R^2C{=}CR^1CH_2N_3 \underset{k_{-1}}{\overset{k_1}{\rightleftharpoons}} N_3CR^2R^3CR^1{=}CH_2$$

R^1	R^2	R^3	10^5k_1 (sec^{-1})	10^5k_{-1} (sec^{-1})
H	H	CH_3	3.8	7.2
H	CH_3	CH_3	7.1	24
H	H	CH_2N_3	0.85	1.2
CH_3	CH_3	CH_2N_3	5.3	2.3

[a] Data from refs. 153 and 305.

The rate of allylic azide isomerization is remarkably insensitive to substituent effects and solvent ionizing power. An α- or γ-methyl group increases reactivity by only a factor of 2–4, while an electron-withdrawing azido group exerts a slight rate-retarding effect (see Table 4). Rates of isomerization of butenyl and pentenyl azides are changed by only about a factor of ten when the solvent is changed from pentane to 70% aqueous acetone[153]. In contrast, isomerization of the analogous allylic chlorides is many powers of ten faster in aqueous acetone than in pentane.

Allylic azide isomerizations resemble other concerted thermal isomerizations of allylic compounds in their low sensitivity to solvent ionizing power and allylic substituents and also in having negative entropies and volumes of activation. Entropies of activation for isomerization of pentenyl azides average -10 to -11 eu. in pentane, ether, and acetone solutions[153]. In methylene chloride at 20°C, the volumes of activation for isomerization of α-methylallyl azide and γ-methylallyl azide are -9.5 cm^3.mole^{-1} and -7.9 cm^3.mole^{-1}, respectively[221].

The insensitivity of allylic azide isomerization rates to substituent effects and solvent ionizing power, and the relatively small value of the negative volume of activation[221,222] make it very unlikely that ion pair intermediates are involved in these reactions. The negative entropy and volume of activation are compatible with a concerted process involving a cyclic transition state resembling a 1,2,3-triazine. However, the known ability of alkyl and aryl azides to add to olefins[2,33] suggests the possibility of an intramolecular addition–elimination mechanism,[153] *viz.*

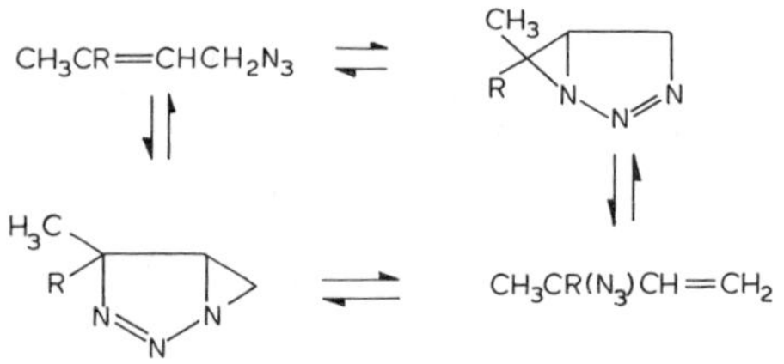

(32)

5.1.3 Isomerizations of allylic thiocyanates

Allylic thiocyanates isomerize to allylic isothiocyanates

$$RR^1C{=}CR^2{-}CH_2{-}SCN \rightleftharpoons RR^1C(NCS){-}CR^2{=}CH_2 \tag{33}$$

in a reaction which involves structural inversion of both the allylic system and the functional group. While the isomerization of allyl thiocyanate proceeds nearly to completion in non-polar media, it is measurably reversible in acetonitrile[206]. γ-Methyl and γ,γ-dimethylallyl thiocyanates isomerize to equilibrium mixtures containing appreciable amounts of unrearranged thiocyanates under all conditions studied[206].

Allyl thiocyanate isomerizes at similar rates in toluene, nitrobenzene, dimethylformamide, cyclohexane, and acetonitrile[206,291]. The reaction is slowest in the two most polar solvents, dimethylformamide and acetonitrile, and its rate is not appreciably affected by dissolved potassium thiocyanate.

Allyl and β-methylallyl thiocyanates isomerize at almost the same rate[206]. In cyclohexane solution, γ-methylallyl thiocyanate isomerizes about 10 times faster, and γ,γ-dimethylallyl thiocyanate about 30 times faster than allyl thiocyanate[291]. The effect of γ-methyl substitution on isomerization rate is somewhat larger in acetonitrile than in cyclohexane solutions. The substituent effects are several orders of magnitude smaller than would be expected for S_N1' anionotropic isomerizations.

The insensitivity of isomerization rate to solvent and structural changes suggests that isomerizations of most allylic thiocyanates are intramolecular and involve cyclic transition states in which there is little charge separation. This view is supported by the fact that isomerizations of allyl and β-methylallyl thiocyanates in toluene solution have substantial negative entropies of activation (−9.4 eu and −8.7 eu, respectively)[291].

5.1.4 Isomerization of allylic thionbenzoates

Another thermal isomerization which results in structural inversion of both the allylic system and the functional group is the apparently irreversible rearrangement of allylic thionbenzoates to allylic thiolbenzoates.

(34)

The rate of this isomerization is increased by α- or γ-alkyl substituents on the allylic system. The relative rates of isomerization of allyl, γ-methylallyl and α-methylallyl thionbenzoates at 100°C are 1, 3.1, and 35 in cyclohexane and 1, 5.5, and 52 in acetonitrile[292]. These effects are some two powers of ten smaller than substituent effects on ionization of allylic chlorides (ref. 98, p. 786).

For the same allylic thionbenzoates, isomerization rates increase with ionizing power of the solvent. Again, the effects are small: the isomerization reactions are only 5–7 times faster in acetic acid and acetonitrile than in cyclohexane[292]. The dependence of rate on solvent ionizing power corresponds to an m value[323] of about 0.13, which is about the same as the m value for allylic azide isomerization and is much smaller than the values for reactions which involve ion pair formation[292].

The entropy of activation for isomerization of allyl thionbenzoate, -10 eu, is similar to entropies of activation for other cyclic, concerted allylic isomerizations[234]. Additional support for a concerted mechanism is provided by secondary deuterium isotope effects on the rate of isomerization of allyl thionbenzoate. Each α-deuterium atom lowers the isomerization rate by about 6%, while each γ-deuterium atom increases the rate by about 3%[234]. Since allylic reactions which proceed by carbonium ion, carbanion, and radical pathways are known to be retarded by about twice as much for each α-deuteron, while an allylic chloride solvolysis exhibited a γ-deuterium isotope effect of zero, the observed secondary isotope effects provide support for a concerted process whose transition state has little carbonium ion, carbanion, or radical character.

5.1.5 Isomerizations of allylic sulfinates, sulfenates and sulfoxides

Braverman and Stabinsky[60] reported that allylic arenesulfinates isomerize to allylic sulfones

$$CH_2{=}CH{-}CHR{-}O{-}SO{-}Ar \rightarrow Ar{-}SO_2{-}CH_2{-}CH{=}CHR \qquad (35)$$

with structural inversion of the allylic group. The low sensitivity of these reactions to allylic substituents and solvent polarity suggests that they have a concerted intramolecular mechanism.

Attempts to prepare allylic sulfenates yield allylic sulfoxides instead[23,60,263]. This fact prompted Mislow *et al.*[23,263] to suggest that racemization of optically active allyl aryl sulfoxides involves their reversible isomerization to thermodynamically less stable achiral allyl arenesulfonates. The enthalpies of activation for racemization of allyl sulfoxides are some 20 kcal.mole^{-1} smaller than for racemization of non-allylic sulfoxides, and the entropies of activation are substantially more negative. Further, racemization of sulfoxides having deuterium-labeled allyl groups results in no α,γ-deuterium scrambling. Racemization is faster in solvents of low polarity than in solvents of high polarity (for example, allyl *p*-tolyl sulfoxide racemizes 30 times faster in methylcyclohexane than in ethanol), as would be expected if the sulfoxide is stabilized more by solvation than is the transition state for formation of the hypothetical sulfenate. All of these observations are accounted for by a mechanism involving reversible, concerted, intramolecular isomerization of the sulfoxides to sulfenate esters, *viz*.

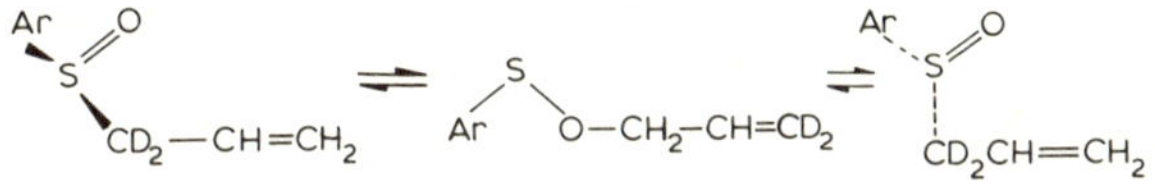

(36)

5.1.6 *Isomerization of allylic phosphinates*

Allyl diphenylphosphinate isomerizes when heated to allyldiphenylphosphine oxide[192].

$$(C_6H_5)_2P\text{-}OCH_2CH{=}CHR \rightarrow (C_6H_5)_2P(O)\text{-}CHRCH{=}CH_2 \qquad (37)$$

In benzene solution the entropy of activation is -14 eu. Isomerization of deuterium-labeled allyl diphenylphosphinate and α- and γ-methylallyl diphenylphosphinates showed that the reaction proceeds with complete inversion of allylic structure and is effectively irreversible. The activation parameters and stereospecificity of this reaction are consistent with a concerted, cyclic mechanism.

5.2 THERMAL ISOMERIZATIONS OF ALKENES, DIENES AND POLYENES

In this section are discussed a number of thermal isomerization reactions of substances having one or more carbon–carbon double bonds in which skeletal reorganizations occur without cleavage of bonds to substituents which may be bonded to the carbon skeleton.

5.2.1 *Isomerizations of 1,5-dienes (the Cope rearrangement)*

The Cope rearrangement of 1,5-dienes is the all-carbon analog of the Claisen rearrangement. It was discovered by Cope and Hardy[81], who found that ethyl(1-methylisopropenyl)allylcyanoacetate isomerizes when heated at 150°C to ethyl 2-cyano-3,4-dimethyl-2,6-heptadienoate, *viz.*

(38)

Similar isomerizations of allylalkenyl malonitriles, malonic esters, and cyanoacetic esters were studied by Cope *et al.*[80,82,83,127,224,318], and subsequently many examples of Cope rearrangements of unsubstituted 1,5-dienes were reported.

Cope rearrangements have been reviewed previously[105,151,264]. The present discussion is principally concerned with the kinetics of these isomerizations.

The bonding changes which occur during a Cope rearrangement are strictly analogous to those which occur in the Claisen rearrangement of vinyl allyl ethers. There are other similarities as well. Cope rearrangements occur at similar rates in the gas phase and in polar and non-polar solvents. They are kinetically first-order, highly stereospecific, and are unaffected by catalysts. The products they form, their negative entropies of activation, and their low activation energies (which exclude formation of allylic radical intermediates) all suggest that Cope rearrangements are intramolecular concerted processes.

In the transition state of a Cope rearrangement the two allylic groups are held together in a cyclic complex in which breaking of old bonds is concerted with formation of new ones. The allylic groups in this complex might be held together by bonding interactions between all three pairs of atoms, or by bonding forces between only the terminal pairs of atoms. In the first case the transition state would resemble a cyclohexane boat conformation, and in the second it would resemble a cyclohexane chair conformation. Experiments by Doering and Roth[103] and by Hill and Gilman[195] demonstrated that the chair-like arrangement is the preferred transition state geometry whenever the structure of the diene permits it. A rationalization of the observed preferred transition state geometries in terms of orbital symmetries was given by Hoffmann and Woodward[196].

Secondary deuterium isotope effects for Cope rearrangements of $CH_3CH{=}C(C_2H_5)C(CN)_2CH_2CH{=}CD_2$ ($k_H/k_D = 0.94$ at 85°C) and $CH_3CH{=}C(C_2H_5)$-$C(CN)_2CD_2CH{=}CH_2$ ($k_H/k_D = 1.19$ at 85°C) suggest that the transition state more closely resembles reactants than products, as expected from their relative thermodynamic stabilities[203].

The rates of Cope rearrangements are nearly independent of the reaction medium. They occur readily in the gas phase and in two instances occurred at similar rates in a variety of solvents. The rate of isomerization of diethyl allylisopropenylmalonate at 145°C is nearly the same in the neat liquid and in solutions in 1,2,3-trichloropropane, bromobenzene, anisole, and *n*-decane[3,20], and the rate of isomerization of ethyl allyl(1-ethylpropenyl)cyanoacetate varies only slightly in solutions in 1,4-dibromobutane, 1-octanol, dibutyl cellosolve, decalin, and glycerol[313].

Cope rearrangements typically have entropies of activation of about −10 eu. (see Table 5) and isomerization of ethyl allyl(1-ethyl-propenyl) cyanoacetate at 119°C has volumes of activation of about -7 $cm^3.mole^{-1}$ in a variety of solvents[313]. The negative entropies and volumes of activation are consistent with rate-limiting formation of a cyclic transition state which has fewer rotational degrees of freedom than the ground state[290,313].

Table 5 summarizes kinetic data for Cope rearrangements whose rates have been measured at several temperatures. The relative reactivities, calculated from activation parameters, are only approximate since the reactions were carried out under a variety of conditions.

The activation energies for isomerization of unsubstituted and methyl-substituted 1,5-hexadienes are about 35 $kcal{:}mole^{-1}$, some 5 $kcal.mole^{-1}$ less than the estimated activation energy for cleavage of 1,5-hexadiene to two allyl radicals[105]. Methyl substitution at C-3 has little effect on reactivity. *Cis*- and *trans*-1,5-heptadienes, and 6-methyl-1,5-heptadiene, in which there are one or two methyl groups on the terminal carbon of one of the allylic systems, are less reactive than the hexadienes due to higher energies and more negative entropies of activation. A *cis* methyl group on C-6 is more deactivating than a *trans* methyl group.

Unsaturated substituents on C-4 of the 1,5-diene system, such as cyano and carbethoxy groups, strongly activate the compound toward Cope rearrangements by lowering the activation energy. Such substituents presumably stabilize the transition state for isomerization by resonance interaction with the delocalized π system of the allylic group to which they are attached. The second double bond of the allenic system of 1,5,6-trienes also facilitates Cope rearrangements, again probably due to resonance stabilization of the transition states.

Fusion of a strained ring system to the 3,4 positions of a 1,5-hexadiene also activates the diene toward isomerization provided that the *vic* vinyl groups of the divinylcycloalkane are *cis* to one another. *Cis*-1,2-divinylcyclobutane isomerizes about a million times faster than 1,5-hexadiene, and all attempts to synthesize *cis*-1,2-divinylcyclopropane yielded its isomerization product, 1,4-cycloheptadiene, instead[307].

Trans-1,2-divinylcyclopropanes and cyclobutanes can form neither chair-like nor boat-like Cope transition states without preliminary carbon–carbon bond fission, and these compounds rearrange much more slowly than their *cis* isomers. *Trans*-divinylcyclopropane isomerizes readily only at tempera-

TABLE 5

THERMAL ISOMERIZATIONS OF 1,5-DIENES

Reaction	*Reaction medium*	E_a (*kcal. mole*$^{-1}$)	*log A*	$\Delta S^{\ddagger}_{298}$ (*eu*)	k_{rel}[a]	*Refs.*
	Vapor	34.2	10.55	−12	1.00[b]	5, 147
	Vapor	35.7	10.54	−12	0.14	5, 147
		35.5	11.1	−10	0.7	102, 235, 249
	Vapor	34.6	11.13	−9	2.2	150
	Vapor	35.3	10.39	−13	0.17	5, 147
	Vapor	36.7	10.66	−12	5.7×10^{-2}	5, 147
	Vapor	36.9	10.39	−13	2.2×10^{-2}	150
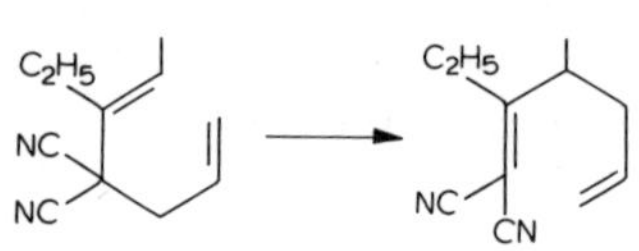	Neat	25.8	11.0	−10	1.2×10^{5}	83, 127
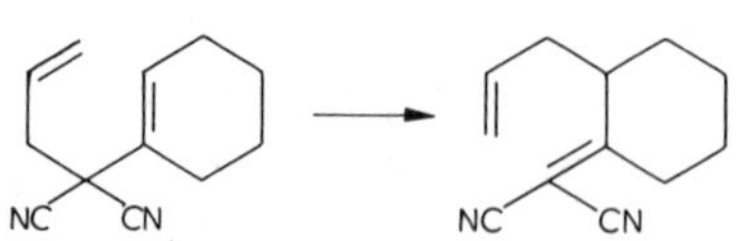	Neat	26.2	10.8	−11	5.1×10^{4}	83, 127
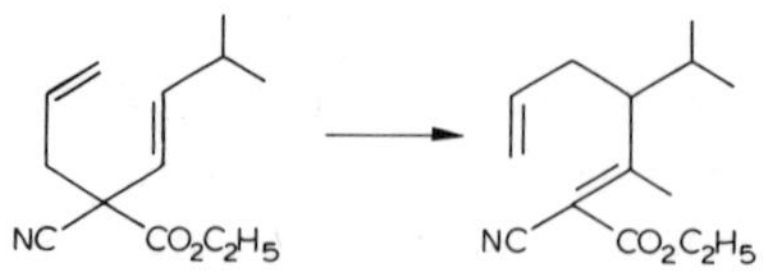	Neat	28.6	10.3	−13	6.6×10^{2}	83, 127
	Neat	28.4	10.45	−13	7.1×10^{2}	3

TABLE 5 (*Contd.*)

Reaction	*Reaction medium*	E_a *(kcal. mole^{-1})*	*log A*	$\Delta S^{\ddagger}_{298}$ *(eu)*	k_{rel}[a]	*Refs.*
$CH_2{=}C{=}CH(CH_2)_2CH{=}CH_2$ 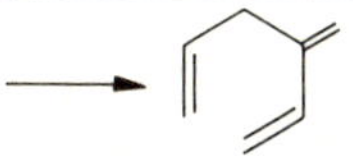	Vapor	28.5	9.97	−15	4.1×10^2	136
	Vapor	31.7	12.47	−3	1.8×10^3	135
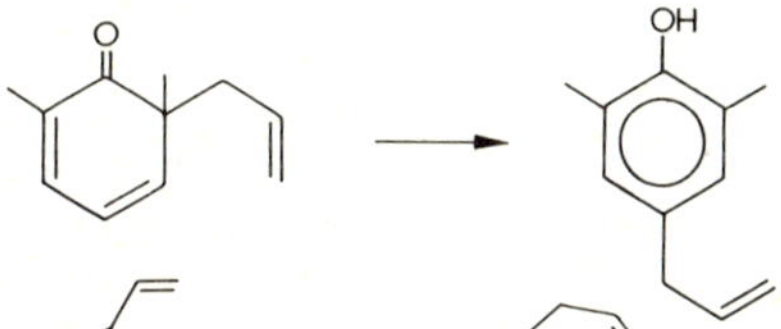	Neat	26.8	11.81	−7	2.0×10^4	94
		32.1	12.09	−5	5.4×10^3	310
	Neat	23.8	10.77	−11	7.8×10^5	183
	Neat	34.9	13.17	0	1.7×10^2	183
	C_9H_{20}- o-$Cl_2C_6H_4$	32.2	14.36	+5	8×10^4	183, 300
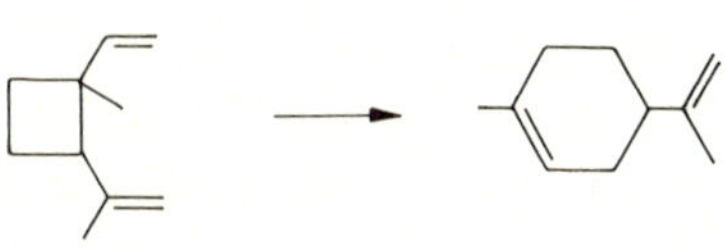	Neat	35.0	13.82	+3	5.3×10^2	183
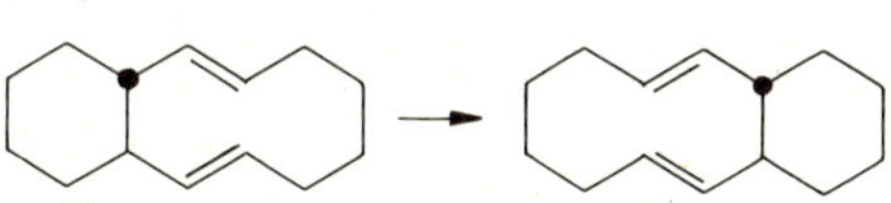	C_8H_{16}	25	11.2	−9	4.6×10^5	317
	C_6H_{12}	22.9	11.67	−7	1.8×10^7	63
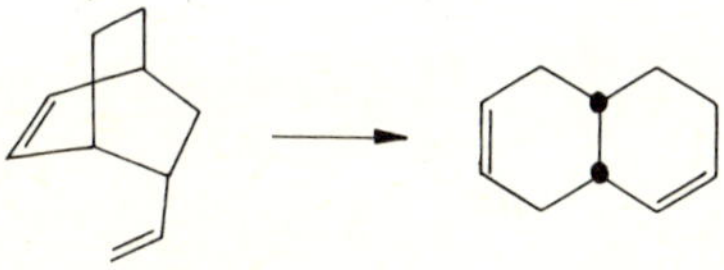	Vapor	44.2	12.8	−2	5.3×10^{-3}	22

[a] Relative rates at 400°K, calculated from activation parameters.
[b] Calculated $k_1 = 7.6 \times 10^{-9}$ sec^{-1} at 400°K.
[c] Major product.

tures above 200°C[307], and *trans*-1,2-divinylcyclobutane isomerizes less than a thousandth as fast as its *cis* isomer. Further, isomerizations of *trans*-1,2-divinylcyclobutanes have positive entropies of activation and yield mixtures of products. These reactions probably are not concerted but proceed via diradical intermediates[183].

Favorable pre-orientation of the 1,5-diene system by bicyclic ring structures also enhances reactivity. For example, isomerization of *trans,trans*-2,8-*trans*-bicyclo[8.4.0]tetradodecadiene is about half a million times faster than isomerization of 1,5-hexadiene.

5.2.2 *Structurally degenerate Cope rearrangements*

If the vinyl groups of *cis*-1,2-divinylcyclopropane are joined by a methylene bridge, Cope rearrangement of the resulting homotropylidene yields a product identical with the starting material.

 (39)

This reaction is said to be a structurally degenerate Cope rearrangement[105].

Non-reactions of the sort exemplified by the valence tautomerism of 3,4-homotropylidene obviously cannot be studied by conventional methods which involve measuring the rate of disappearance of a starting material or rate of formation of a product. However, many of these degenerate rearrangements occur so rapidly that they cause averaging of some or all of the NMR frequencies of protons in the substrate at temperatures which are experimentally accessible. The rates of these processes can be estimated by use of temperature-dependent line broadening or spin-echo NMR techniques.

The Cope "rearrangement" of 3,4-homotropylidene is facilitated by both pre-orientation and steric strain effects, and should occur quite rapidly. In fact, NMR spectroscopy shows that a molecule of 3,4-homotropylidene rearranges about a thousand times per second at 180°C, and about once per second at −50°C[104,105].

4,8-Bridged derivatives of homotropylidene also undergo extremely facile degenerate Cope rearrangements. Perhaps the most intriguing of these reactions involves the substance tricyclo[3.3.2.0^{4,6}]deca-2,7,9-triene, or bullvalene (XIII). Bullvalene can have its ten carbon atoms arranged in more than

XIII

TABLE 6

DEGENERATE COPE REARRANGEMENTS OF 4,8-BRIDGED BICYCLO 5.1.0 OCTA-2,5-DIENES

Y	E_a *(kcal.mole^{-1})*	*log A*	$\Delta S^{\ddagger}_{298}$ *(eu)*	k_{rel}[a]	*Refs.*
$-CH_2-$	10.4	15.7	+12	60,000	281, 283
$-CCl_2-$	13.6	16.0	+13	25	251, 281, 282
$-C(=O)-$	8.1	11.2	−9	500	219
$-CH_2-CH_2-$	12.6	15.0	+8	50	251, 281–283
$-CH=CH-$	11.5	12.1	−5	1[b]	4, 273, 281, 283
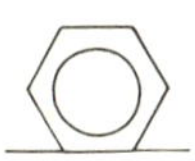	12.9	15.6	+11	70	251, 281, 282

[a] Relative rates at 200°K, calculated from activation parameters.
[b] Calculated rate coefficient at 200°K = 4×10^{-1} sec^{-1}.

1.2 million different ways, all of which are structurally identical and interconvertible by means of degenerate Cope rearrangements[105].

Kinetic data for degenerate Cope rearrangements of a number of bridged derivatives of homotropylidene are summarized in Table 6. All of these bridged homotropylidenes undergo very rapid valence tautomerism at ordinary temperatures. Interestingly, bullvalene is the least reactive compound of the series. There is a clear tendency for reactivity to increase as the length of the bridge decreases.

Schroeder and co-workers studied Cope rearrangements of monosubstituted bullvalenes. These reactions are not degenerate, since there are four isomeric forms of monosubstituted bullvalene. However, the isomers are interconverted so rapidly that it is possible only to measure average rates of rearrangement of the isomer mixtures. The relative average isomerization rates at 80°C were[251,281]: substituent = H, 100; C_6H_5, 45; Br, 40; $(CH_3)_3CO$, 14.5; $(CH_3)_2CHO$ 6.9; C_2H_5O, 5.4; CH_3O, 4.2. The fluorobullvalenes isomerize slowly on the NMR time scale[252].

5.2.3 *Intramolecular electrocyclic isomerizations of alkenes and cycloalkenes*

Electrocyclic transformations involve formation of single bonds between the terminal carbons of conjugated systems of double bonds, and the reverse of these processes[154,325]. These reactions, like Cope rearrangements, are concerted, intramolecular, and stereospecific.

(a) Electrocyclic ring closures

Electrocyclic ring closures occur in the gas phase and in solution, are relatively insensitive to solvent polarity[199,201] and do not require catalysis. Typically, they are also stereospecific: bond formation between terminal carbons of triene systems occurs in a disrotatory manner, and bond formation between termini of conjugated diene and tetraene systems occurs in a conrotatory manner[325]. Kinetic data on electrocyclic ring closures are summarized in Table 7.

TABLE 7

ELECTROCYCLIC CYCLIZATIONS OF DIENES, TRIENES, AND TETRAENES

Reaction	*Reaction medium*	E_a *(kcal.mole⁻¹)*	*log A*	$\Delta S^{\ddagger}_{298}$ *(eu)*	*Refs.*
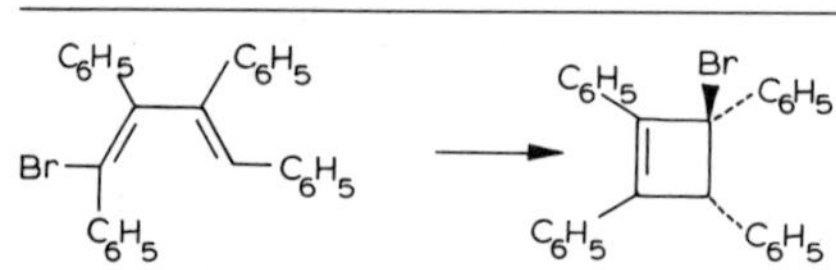	$CDCl_3$	19.6	9.7	−16	106
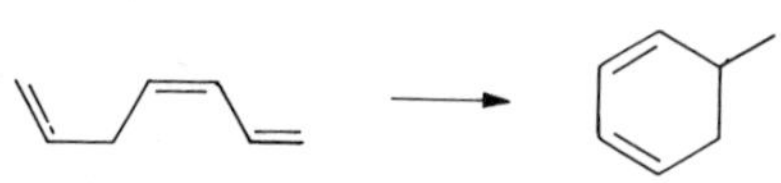	Vapor	29.9	11.85	−6	227, 274
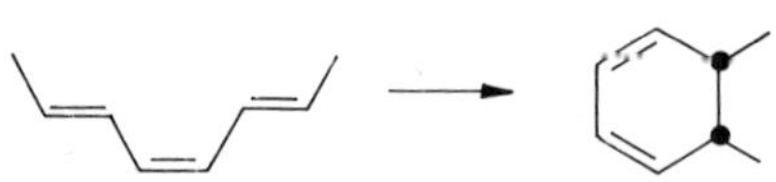	Pentane	29.4	11.5	−8	236
	Vapor	31	12	−5	308
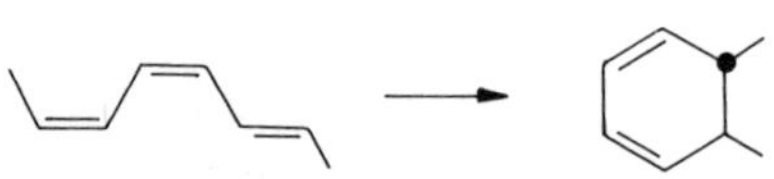		33	10.8	−11	274
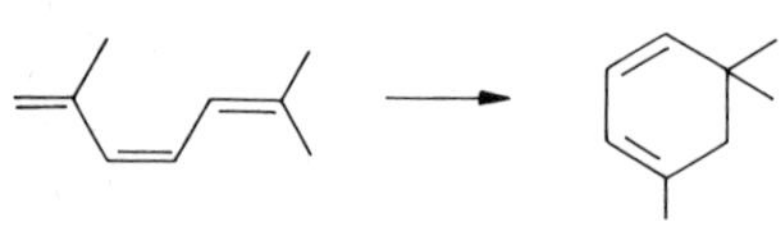	Vapor	31.8	10.5	−12	110
	CCl_4	7	12	−5	180
		26	13	−1	156
		21[a]			198
	C_6H_{12}	23.6	12.2	−5	156
	CCl_4	20.7	10.5	−13	308

TABLE 7 (*Contd.*)

Reaction	*Reaction medium*	E_a *(kcal.mole⁻¹)*	*log A*	$\Delta S^{\ddagger}_{298}$ *(eu)*	*Refs.*
		28	13	−1	156
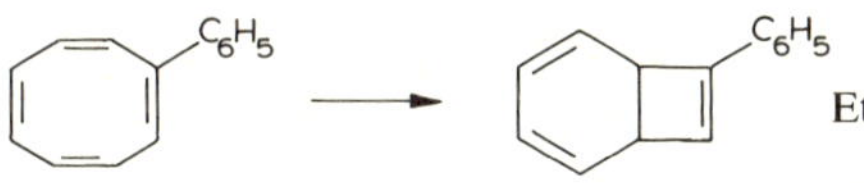	EtOAc	25.6	12.7	−3	201
	C_6H_{14}	17.0	11.04	−10	177
	$CDCl_3$	22.3	10.5	−13	199
	$CDCl_3$	15·6	9	−19	200
	$CDCl_3$	18.5	9.5	−17	200

[a] Estimated, assuming $\log A = 13$.

Electrocyclic ring closures are reversible, and only those which yield significant amounts of cyclic products at equilibrium have been studied. Most butadiene–cyclobutene equilibria greatly favor the butadienes (see below). An exception is the cyclization of *trans*, *cis*-1,2,3,4-tetraphenyl-1-bromo-1,3-butadiene, which undergoes the predicted conrotatory cyclization to a significant extent[106].

Electrocyclic ring closures of acyclic trienes and tetraenes have substantial negative entropies of activation, as anticipated for reactions in which rotational freedom is lost on assembling the cyclic transition states. Conversions of the more rigid cyclic trienes and tetraenes to bicyclic hydrocarbons usually have less negative entropies of activation.

Activation energies for conversion of acyclic conjugated trienes to cyclohexadienes are about 12 kcal.mole^{-1} larger than activation energies for conversion of acyclic conjugated tetraenes to cyclooctatrienes. In both classes of reactions, *trans* methyl groups on terminal unsaturated carbons have little effect on activation energies, while *cis* methyl groups increase activation energies for cyclization by 3–4 kcal.mole^{-1}. This is probably due to hindrance to overlap between the terminal carbons of the conjugated systems by the *cis* methyl groups.

Activation energies for conversion of cyclic trienes to bicyclic dienes are smaller than those for ring closures of acyclic trienes and activation energies for formation of bicyclo[4.3.0] systems are smaller than those for formation of bicyclo[4.2.0] systems. The only data for formation of a bicyclo[4.1.0] system are for cyclization of 1-oxa-2,4,6-cycloheptatriene, which has a very low energy of activation. The equilibrium for cyclization of cycloheptatriene to bicyclo[4.1.0]hepta-2,3-diene (norcaradiene) is too far toward cycloheptatriene to permit study of this interesting reaction.

(*b*) *Electrocyclic ring openings of cyclobutenes*

The only electrocyclic ring cleavages which consistently yield products more stable than the starting materials are those of cyclobutenes. Most of the kinetic data on electrocyclic ring openings, therefore, involve isomerizations of cyclobutenes to conjugated dienes. These reactions have been reviewed in part by Frey and Walsh[132,151] and by Criegee[89].

Most isomerizations of cyclobutenes to butadienes are first-order reactions which are unaffected by glass surfaces and free radicals. In the gas phase, their rates are independent of pressure above about 10 torr. The rates of gas-phase isomerization of cyclobutene[187], 1-methylcyclobutene[209], and 3-methylcyclobutene[138] decrease as the pressure is lowered below 10 torr in a fashion which is in reasonable agreement with theoretical calculations[117]. In solution, rates of cyclobutene isomerizations are insensitive to solvent polarity[35,92,141]. The isotope effect, k_H/k_D, for isomerization of cyclobutene and hexadeuterocyclobutene is 1.4 at 175°C[70,281].

Orbital symmetry considerations[154,325] lead to the prediction that cyclobutene isomerizations should involve conrotatory ring opening. The stereospecificity of these reactions and the products they form show that this is indeed the case whenever conrotatory opening is sterically possible. For example, *cis*-3,4-disubstituted cyclobutenes yield *cis, trans*-dienes, and *trans*-3,4-disubstituted cyclobutenes yield *trans,trans*-dienes, the products of conrotatory ring opening. The small entropies of activation and energies of activation of these reactions support the view that they are concerted processes involving simultaneous twisting of the cyclobutene ring and stretching of the 3,4 carbon bond. Isomerizations of cyclobutenes whose geometry prevents conrotatory opening have substantial positive entropies of activation and large activation energies.

Kinetic data for isomerizations of cyclobutenes and bicyclic and tricyclic hydrocarbons having cyclobutene rings are summarized in Table 8. A number of conclusions concerning the effects of structure on reactivity can be drawn from these data. Entropies of activation for isomerization of most monocyclic cyclobutenes fall in the range −2 to +3 eu. Alkyl substituents at C-1 increase the energy of activation by about 2–3 kcal.mole^{-1}. Two alkyl substituents at C-1 and C-2 increase the energy of activation by about 4 kcal.mole^{-1}, while

TABLE 8

ELECTROCYCLIC RING OPENINGS OF CYCLOBUTENES AND CYCLOHEXADIENES

Reaction	*Reaction medium*	E_a *(kcal.mole^{-1})*	*log A*	$\Delta S^{\ddagger}_{298}$ *(eu)*	*Refs.*
	Vapor	32.2	13.04	−1	70, 79, 187
	Vapor	33.8	13.66	+2	144
	Vapor	35.1	13.79	+3	129
	Vapor	34.8	13.76	+3	146
	Vapor	34.5	13.55	+2	100
	Vapor	34.7	13.55	+2	100
	Vapor	34.2	13.48	+2	100
	Vapor	34.1	13.48	+2	100
	Vapor	31.6	13.53	+2	131
	Vapor	36.0	13.84	+3	130
	C_6H_5CN	34.4	13.23	0	141
	$C_{12}H_{24}$	34.6	13.29	0	141
	o-$C_6H_4(CO_2CH_3)_2$	35.2	13.64	+2	
	Vapor	33.0	13.65	+2	139
	Vapor	33.4	13.52	+2	139

References pp. 481–487

TABLE 8 (*Contd.*)

Reaction	*Reaction medium*	E_a (*kcal.mole*$^{-1}$)	*log A*	$\Delta S^{\ddagger}_{298}$ (*eu*)	*Refs.*
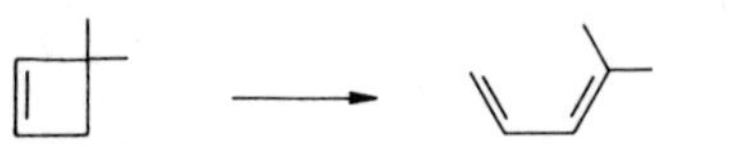	Vapor	36.1	13.93	+4	145
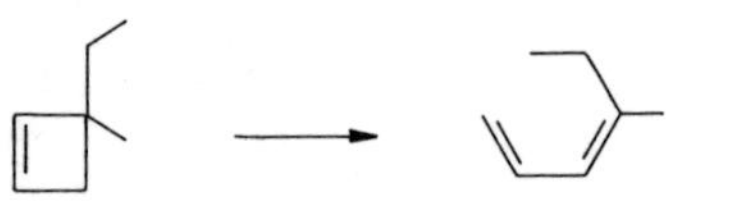	Vapor	35.2	13.50	+2	148
	Vapor	35.9	13.53	+2	148
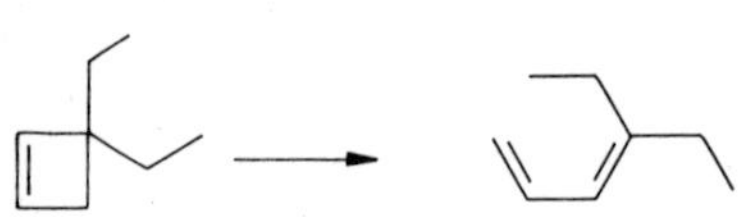	Vapor	34.7	13.53	+2	148
	Vapor	37.0	13.90	+3	145
	Vapor $C_6H_5CH_3$	37.4 35.8	14.10 13.4	+4 +1	36 92
	Vapor $C_6H_5CH_3$	33.6 32.0	13.85 13.3	+3 +1	36 92
	$C_6H_5CH_3$	34.0	13.0	−1	92
	$C_6H_5CH_3$	40	15.2	+9	92
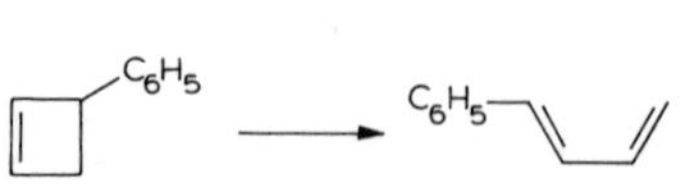	C_9H_{20}	26	12.4	−4	259
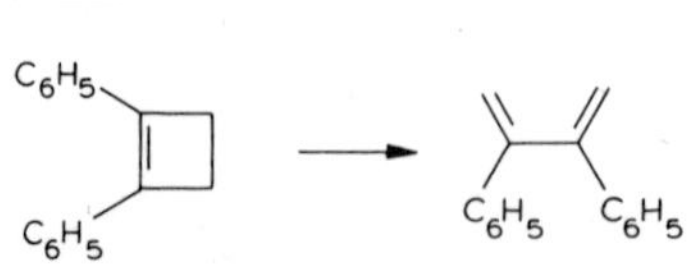	Decalin	32	12.8	−2	11

TABLE 8 (*Contd.*)

Reaction	*Reaction medium*	E_a *(kcal.mole^{-1})*	*log A*	$\Delta S^{\ddagger}_{298}$ *(eu)*	*Refs.*
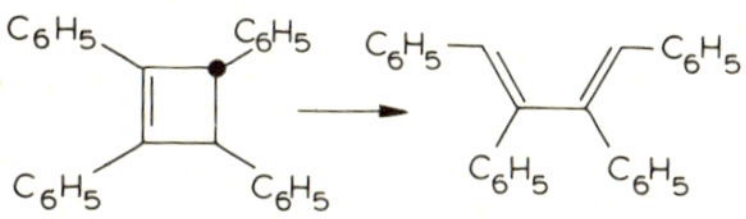	THF	21	11.1	−10	128
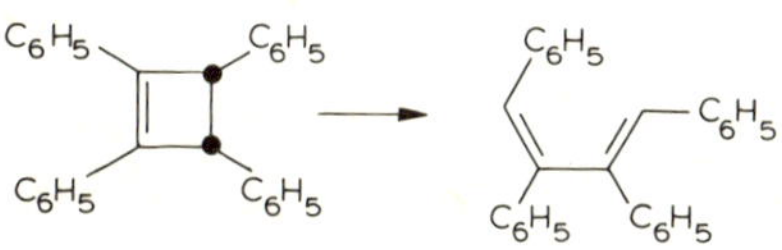	THF	25	12.8	−2	128
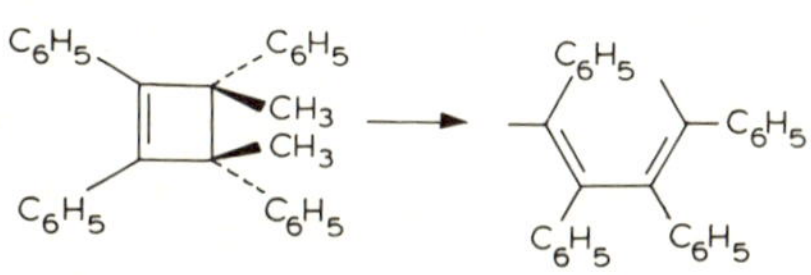	CCl_4/C_5H_5N	24.7	13.2	0	107
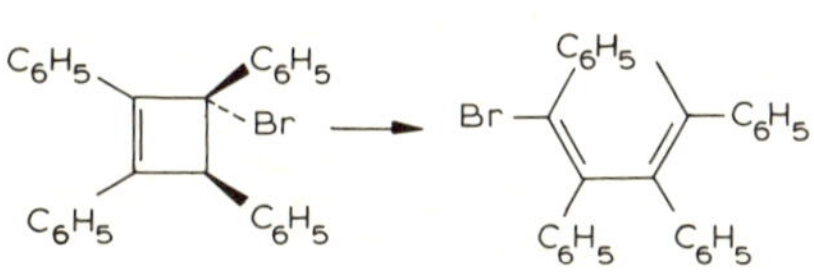	CD_2Cl_2	25	13	−1	106
	Vapor	26.9	14.2	+5	59, 176
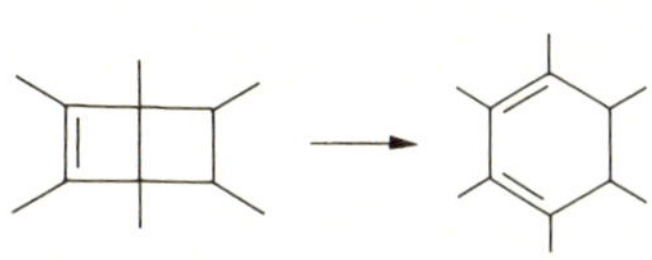		42	15.6	+11	89
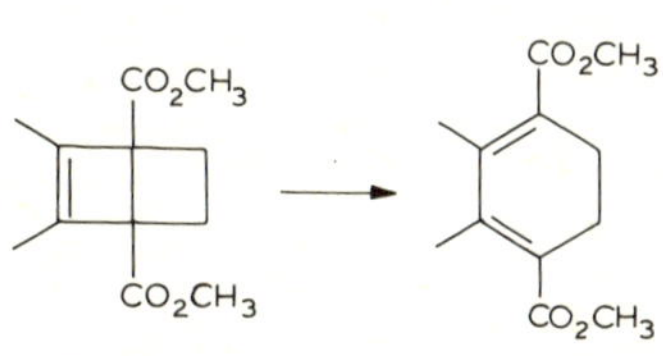	C_6H_{14}	30.6	14.6	+6	285
		26[a]			
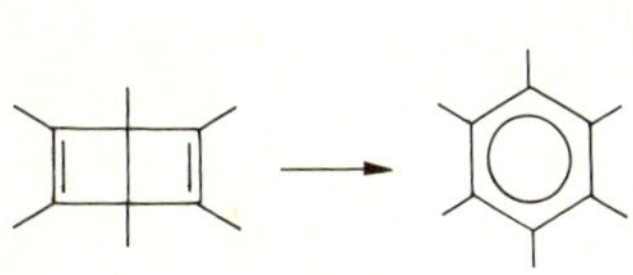		31.1	12	−6	311

References pp. 481–487

TABLE 8 (*Contd.*)

Reaction	*Reaction medium*	E_a (*kcal.mole*$^{-1}$)	*log A*	$\Delta S^{\ddagger}_{298}$ (*eu*)	*Refs.*
	Vapor o-$C_6H_4(CO_2CH_3)_2$	45.5 45.9	14.31 14.65	+5 +7	35 35
		39.5	14	+4	321
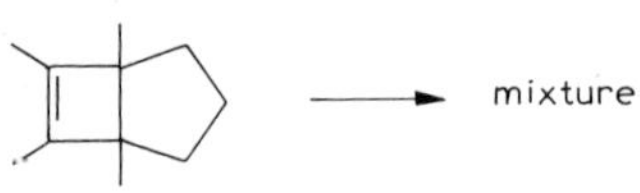		49[a]			90
	Vapor	43.2	14.13	+4	36
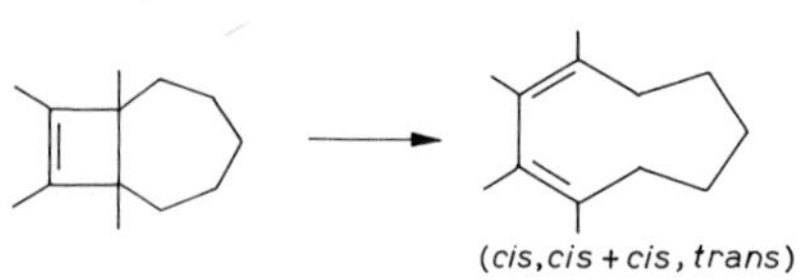		45[a]			90
		18.7	11.96	−6	308
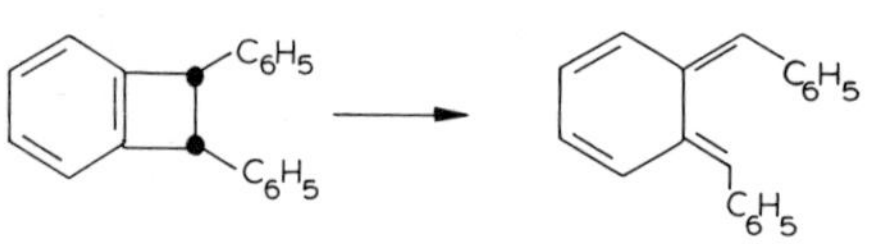	EtOAc	25.2	13.2	0	202
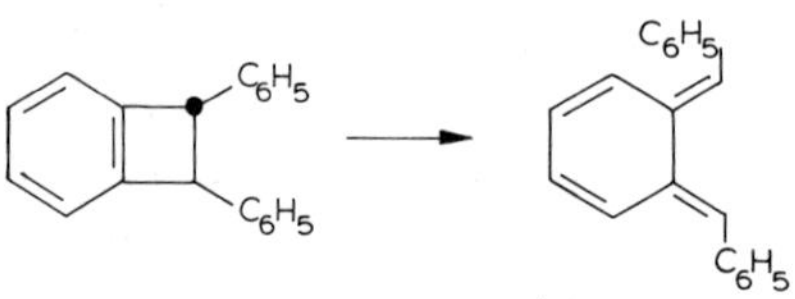	EtOAc	20.6	11.7	−7	202
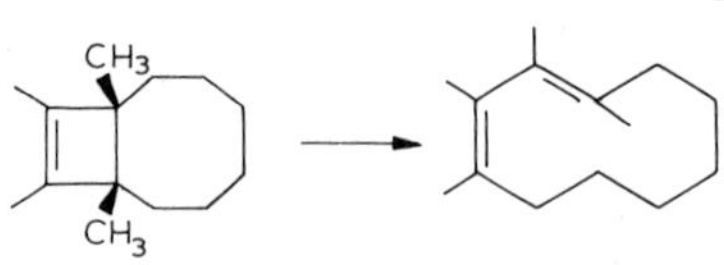		36	13.3	0	90
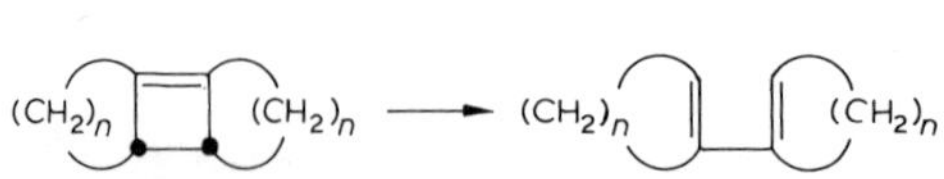	C_7H_{16} $n = 4$ $n = 5$	41.6 45	13.0 14	−1 +3	91 91
	C_7H_{14} $n = 4$ $n = 5$ $n = 6$	29.2 27.2 27.6	13.4 11.6 11.4	+1 −7 −8	91 91 91

[a] Calculated, assuming $\log A = 13$.

two phenyl substituents at the same positions have little effect on E_a. A 3-alkyl substituent lowers the energy of activation by 1–2 kcal.mole^{-1}, and a 3-phenyl substituent lowers it by about 6 kcal.mole^{-1}; two *trans*-3,4-phenyl groups lower E_a by about 11 kcal.mole^{-1}. These effects may be due to stabilization of the transition state by hyperconjugation and resonance interactions of the 3 and 4 substituents with the developing diene system.

If conrotatory ring opening forces a substituent on C-3 into close proximity with C-4, formation of the transition state for cyclobutene cleavage is sterically hindered, and the energy of activation is increased. For this reason, two alkyl substituents on C-3 raise the energy of activation by 2–4 kcal.mole^{-1} and cleavages of *cis*-3,4-disubstituted cyclobutenes have higher activation energies than those of their *trans*-3,4-disubstituted isomers.

Activation energies for isomerizations of bicyclic and tricyclic hydrocarbons having fused cyclobutenes systems yield information on the extent to which symmetry-allowed conrotatory ring opening is favored over disrotatory opening. If a polymethylene bridge is *cis* fused to C-3 and C-4 of the cyclobutene system, conrotatory opening of the cyclobutene ring results in formation of a highly strained *cis,trans*-cycloalkadiene. This strain is so severe in the case of small and medium ring dienes that conrotatory opening does not occur. Instead, the *cis*-fused bicyclic compounds isomerize either by concerted disrotatory opening or via a diradical intermediate to form *cis,cis*-1,3-cycloalkadienes. With the exception of bicyclo[2.1.0]pent-2-ene, in which cyclobutene ring opening is facilitated by the strain energy of the cyclopropane ring, these compounds isomerize with substantially larger than normal energies and entropies of activation. Bicyclo[3.2.0]hept-6-ene and bicyclo[4.2.0]oct-7-ene isomerizations have activation energies of 45.5 and 43.2 kcal.mole^{-1}. These activation energies indicate that the free energy of activation for conrotatory cyclobutene ring opening is at least 10–12 kcal.mole^{-1} less than that for disrotatory opening, in agreement with theoretical predictions[229]. A similar conclusion is reached from the energetics of isomerization of tricyclic hydrocarbons of types XIV and XV, which rearrange to 1,1′-bis(cycloalkenyls),

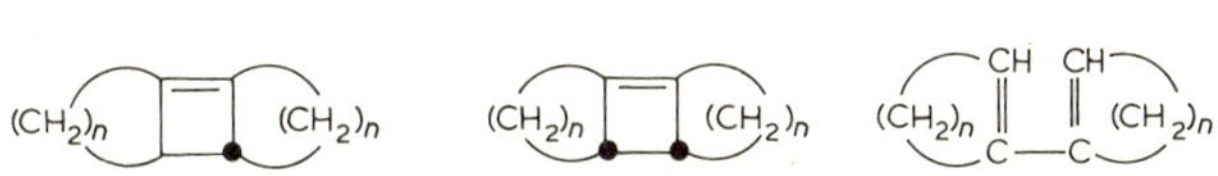

XIV (n = 4,5,6) (n = 4,5) XV XVI

XVI. Conrotatory opening of *cis* isomer XV would result in formation of XVI having one *trans* cycloalkene ring, whereas conrotatory opening of the trans isomer XIV yields XVI with two *cis* cycloalkene rings. Both XIV and XV ($n = 4,5$) yield *cis,cis*-bis(cycloalkenyls), but the energies of activation for their formation from *trans* isomer XIV are 12–15 kcal.mole^{-1} lower than for their formation from the *cis* isomer XV.

References pp. 481–487

If a *cis*-fused 3,4-polymethylene bridge is sufficiently large, isomerization yields the symmetry-allowed *cis,trans*-1,3-cycloalkadiene. Thus, isomerization of *cis*-1,8,9,10-tetramethylbicyclo[6.2.0]dec-9-ene yields *cis,trans*-1,2,3,4-tetramethyl-1,3-cyclodecadiene[90].

Carbon–carbon double bonds in *cis*-3,4-fused ring systems generally lower the energy of activation for isomerization relative to energies of activation for analogous isomerizations of cyclobutenes having saturated *cis*-3,4-polymethylene bridges. Thus, E_a for hexamethylbicyclo[2.2.0]hexadiene isomerization to hexamethylbenzene is 31.1 kcal.mole^{-1}, compared with 42 kcal.mole^{-1} for isomerization of hexamethylbicyclo[2.2.0]hexene, and the energy of activation for isomerization of bicyclo[3.2.0]hepta-2,6-diene is about 6 kcal.mole^{-1} less than for isomerization of bicyclo[3.2.0]-hept-6-ene. The energy of activation for isomerization of bicyclo[4.2.0]-octa-2,4,7-triene to cyclooctatetraene is some 25 kcal.mole^{-1} less than that for isomerization of bicyclo[4.2.0]oct-7-ene to cyclooctadiene, indicating that the first of these reactions involves a symmetry-allowed concerted disrotatory opening of the cyclohexadiene ring.

5.2.4 *Thermal isomerizations of vinylcycloalkanes*

(*a*) *Vinylcyclopropane isomerizations*

Thermal isomerizations of vinylcyclopropanes usually yield cyclopentenes as the principal products, although acyclic dienes are sometimes among the products and in some instances are the major products. These reactions have been reviewed in part by Frey and Walsh[132,151] and by Sarel *et al.*[272].

Isomerizations of most vinylcyclopropanes are first order and are unaffected by glass surfaces and isomerization products. In two instances they were shown to be insensitive to added nitric oxide[315] or oxygen[214]. In the gas phase, their rates are independent of pressure above about 5 torr but fall off with diminishing pressure at lower pressures[315].

Most vinylcyclopropane isomerizations have energies of activation near 50 kcal.mole^{-1} and entropies of activation near zero. Activation parameters and reaction products of vinyl cyclopropane isomerizations are summarized in Table 9.

These rearrangements may involve formation of diradicals as discrete intermediates, or they may be concerted processes in which bond breaking and bond making occur simultaneously. Their insensitivity to nitric oxide and oxygen and the absence of dimeric products suggest that long-lived radicals are not intermediates. The effects of substituents on reactivity can be accounted for in terms of either a concerted or a diradical mechanism. However, the

TABLE 9

ISOMERIZATIONS OF VINYLCYCLOPROPANES

Reaction	Reaction medium	E_a (kcal.mole^{-1})	log A	$\Delta S^{\ddagger}_{298}$ (eu)	Refs.
	Vapor	50.0	13.7	+2	126, 315
	Vapor	50.9	13.89	+3	137
	Vapor	51.1	14.29	+5	34
	Vapor	50.0	13.79	+3	120
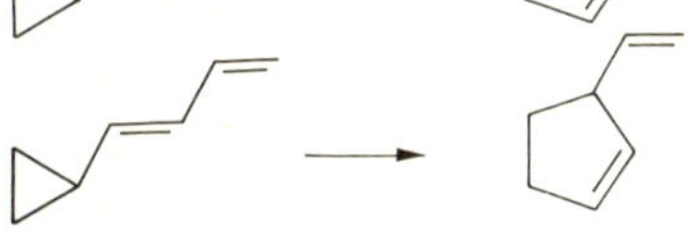	Vapor	44.7	13.44	+1	134
	Vapor	54.6	14.00	+4	116
	Vapor	56.5	14.61	+6	116
	Vapor	53.0	13.33	+1	116
	Vapor	52.1	13.25	0	116
	Vapor	51.3	14.01	+4	34
 (X = H, F, CH_3, CH_3O)	Neat	52	13.4	+1	19, 213
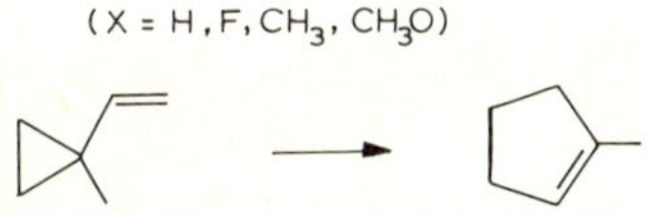	Vapor	49.4	14.11	+4	119

TABLE 9 (*Contd.*)

Reaction	*Reaction medium*	E_a *(kcal.mole^{-1})*	*log A*	$\Delta S^{\ddagger}_{298}$ *(eu)*	*Refs.*
	Vapor	50.5	14.16	+4	115
	Vapor	31.2	11.03	−10	121, 122, 270
	Vapor	48.6	13.67	+2	121, 270
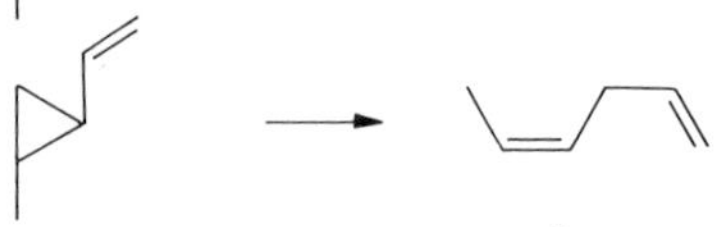	Vapor	48.6	14.74	+7	121, 270
	Vapor	50.2	14.28	+5	123
	Vapor	50.2	14.09	+4	123
	Vapor	39	13.4	+1	179
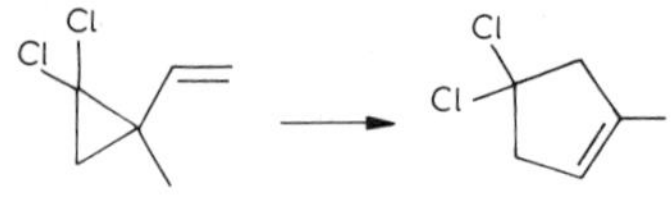	Neat	29	9	−18	214

stereochemistry of isomerization of deuterium-labeled vinylcyclopropanes requires a diradical mechanism for these reactions[117,209,319,320].

According to this mechanism, the initial step in typical vinylcyclopropane isomerizations is cleavage of the cyclopropane ring to form an allylic diradical, which undergoes intramolecular reorganization more rapidly than it enters into bimolecular reactions with other radicals, *viz.*

$$+ CH_2{=}CHCH{=}CHCH_3 + CH_2{=}CHCH_2CH{=}CH_2 \quad (40)$$

The deuterium labeling experiments referred to above indicate that the diradical cyclizes to starting material considerably more rapidly than it is converted to products.

The energies of activation for vinylcyclopropane isomerizations are 12–13 kcal.mole^{-1} less than those for geometrical isomerizations of 1,2-disubstituted cyclopropanes. The formation of a diradical intermediate which is stabilized by allylic resonance accounts for this difference in activation energies very nicely, since the difference is approximately equal to the resonance energy of an allylic radical. The diradical mechanism also accounts for the approximately zero entropies of activation, since the decrease in rotational freedom accompanying formation of the planar allylic radical is approximately compensated for by the increase in rotational freedom accompanying cleavage of the cyclopropane ring.

1-Alkyl, 1-aryl, and *trans*-2-alkyl substituents on the ethylenic system of 1-cyclopropylethylenes have little effect on the activation energies for their isomerizations. Ethylenic substituents *cis* to the cyclopropyl group raise the energy of activation for isomerization by about 4 kcal.mole^{-1}. This may be due to steric hindrance to formation of the transition state for the product-forming step of the reaction, since a cyclopentene-like transition state cannot be assembled from the *cis*-substituted diradical without deforming the planar geometry of the allylic system and consequent loss of resonance stabilization. A *trans* vinyl substituent on the ethylenic system lowers the activation energy for isomerization by about 5 kcal.mole^{-1}, presumably due to greater delocalization of the unpaired allylic electron in the diradical intermediate. Rates of isomerization of 1-aryl-1-cyclopropylethylenes are almost independent of the nature of the aryl substituent[19,212,213,215].

The alkyl substituents in 1-alkyl-1-vinyl and *trans*-1-alkyl-2-vinyl-cyclopropanes have little effect on reactivity. A *cis* alkyl substituent may have a profound effect: for example, the energy of activation is nearly 20 kcal.mole^{-1} smaller and the entropy of activation is 10 eu more negative, for isomerization of *cis*-1-methyl-2-vinylcyclopropane than for typical vinylcyclopropane rearrangements. Further, the sole product of this reaction is *cis*-1,4-hexadiene. This and related reactions, which occur by a concerted mechanism involving 1,5-hydrogen migration, are discussed in the next section.

Bicyclic cyclopropanes such as bicyclo[3.1.0]hex-2-ene and bicyclo[5.1.0]-oct-2-ene rearrange with normal activation energies[123,179]. Apparently formation of the transition state for 1,5-hydrogen migration is sterically hindered in the case of these compounds, and isomerization involves cleavage to cyclic diradicals followed by 1,2-hydrogen shifts. Halogen substituents on the cyclopropane ring, which should stabilize diradical intermediates, activate vinylcyclopropanes for isomerization[214].

(*b*) *Vinylcyclobutane isomerizations*

The kinetics of thermal isomerizations of vinylcyclobutanes have received little attention. Isomerization of isopropenylcyclobutane to 1-methylcyclohexene has an activation energy of 51 kcal.mole^{-1} and an entropy of activation of 8 eu[118]. These values are consistent with a diradical mechanism analogous to that proposed for vinylcyclopropane rearrangements.

5.2.5 *Isomerizations involving thermal sigmatropic migrations*

Thermal sigmatropic shifts involve the uncatalyzed intramolecular migration of a σ bond, flanked by one or more π-electron systems, to a new position within the molecule, *viz.*

(41)

Most examples of this class of reaction involve hydrogen shifts, but reactions are known in which alkyl, aryl, and other groups undergo sigmatropic migrations[154].

(a) 1,3-Sigmatropic shifts

Thermal 1,3-sigmatropic hydrogen shifts are required by orbital symmetry to occur antarafacially, which usually is sterically impossible. Carbon, which can migrate as a quasi-π radical, can undergo concerted suprafacial 1,3 migration with inversion of configuration at the migrating center[326]. One reaction of this type has been studied kinetically. Isomerization of *exo*-5-acetoxybicyclo[2.1.1]hex-2-ene to *exo*-6-acetoxybicyclo[3.1.0]hex-2-ene

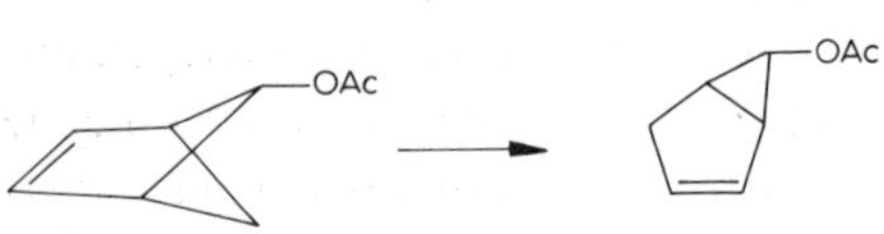

(42)

in dodecane solutions has an activation energy of about 28 kcal.mole^{-1} and an entropy of activation of about -2 eu[154]. This reaction is about 50% faster in dodecane solution than in acetic acid solution.

(b) Dienyl and homodienyl sigmatropic 1,5-hydrogen shifts

Most of the known thermal sigmatropic rearrangements involve 1,5-shifts of the migrating atom or group. Like other concerted intramolecular rearrangements, 1,5-hydrogen shifts occur at comparable rates in the gas phase and in solution and are insensitive to solvent polarity[26,27]. These migrations occur suprafacially and the required transition state geometry frequently can be attained without prohibitive steric interactions. Many 1,5-sigmatropic shifts involve migrations of hydrogen in *cis*-1,3-diene systems. A closely related group of reactions involve homodienyl systems. These reactions, which have been reviewed in part[151,155], are generalized as

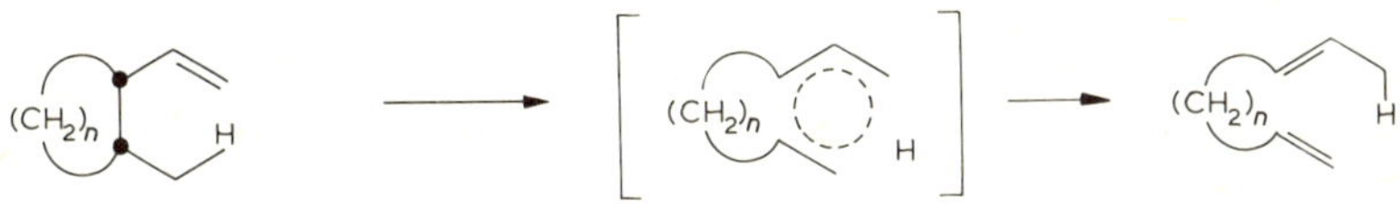

(43)

In dienyl 1,5-shifts, $n = 0$; in homodienyl shifts, $n = 1$. There is one example of the reverse of reaction (43) in which $n = 3$. Table 10 lists activation parameters for 1,5-hydrogen shifts which have been studied at several temperatures.

1,5-Hydrogen shifts in acyclic dienes have six-membered cyclic transition states and their entropies of activation (about −10 eu) are similar to those of Cope rearrangements. In the preferred transition-state geometry, the five carbons of the rearranging system are coplanar or nearly so, with C-1 and C-5 bridged by the migrating hydrogen atom

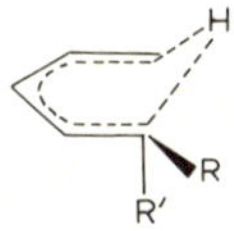

Only *cis*-1,3-dienes can attain this geometry, and only *cis* dienes isomerize by this mechanism. If R and R′ differ in size, the most stable transition state is that in which the bulkier group is oriented equatorially. This accounts for the observation that isomerization of *cis*-1,3-hexadiene gives predominantly *cis,trans*-2,4-hexadiene[142].

The transition-state geometry in homodienyl 1,5-hydrogen shifts is probably non-planar, *viz.*

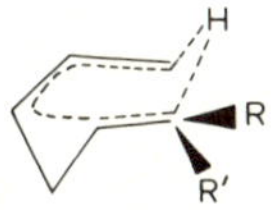

This geometry should be favorable for overlap of the π orbital of the vinyl group with the developing p orbitals of the cyclopropyl carbons and the C–H bond[155]. In a transition state of this geometry, the steric environments of groups R and R′ are similar. This probably accounts for the observation that isomerization of 1,1-diethyl-2-vinyl-cyclopropane yields approximately equal amounts of *cis,cis*- and *cis,trans*-3-ethyl-2,5-heptadiene.

The large isotope effect for migrating hydrogen relative to migrating deuter-

TABLE 10

ALKENE ISOMERIZATIONS INVOLVING 1,5-HYDROGEN SHIFTS

Reaction	*Reaction medium*	E_a *(kcal.mole^{-1})*	*log A*	$\Delta S^{\ddagger}_{298}$ *(eu)*	*Refs.*
	Vapor	36.3	11.45	−8	271
	Vapor	37.7	11.38	−8	271
	Vapor	32.5	10.80	−11	142
	Vapor	32.8	11.24	−9	133
	Vapor	31.6	10.77	−11	149
	Vapor	34.5	11.03	−10	149
	CCl_4	24.3	12.11	−5	269
	Neat	20.4	11	−10	232, 233
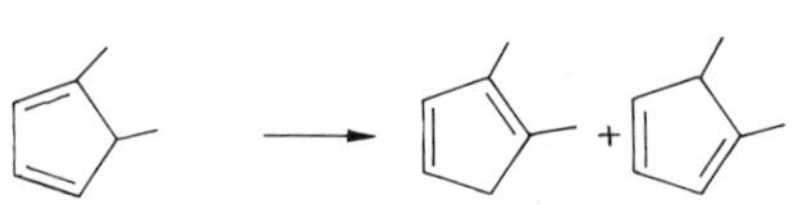	Neat	23	12	−4	232, 233
		34	11.7	−7	236, 274
		36	12.5	−3	236, 274
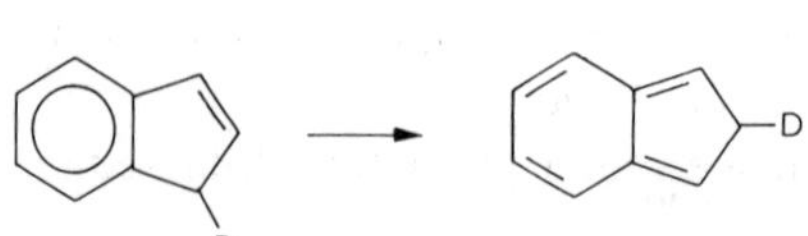		36[a]			269

TABLE 10 (*Contd.*)

action	Reaction medium	E_a (kcal.mole^{-1})	log A	$\Delta S^{\ddagger}_{298}$ (eu)	Refs.
	Heptane	26.5	11.5	−8	27
	Heptane	26.9	11.5	−8	27
		31	11	−10	29, 30
	Vapor	33.3	12.6	−3	32, 109
	Neat	26.4	10.04	−15	32, 245, 316
	Decane	27.6	10.8	−11	26, 28, 31, 32
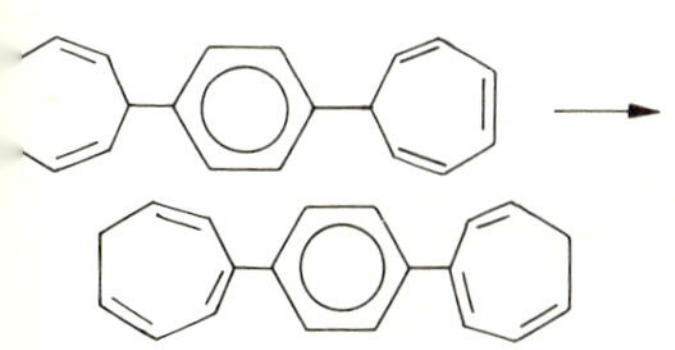	CCl_4	30.6	12.5	−3	243
	Neat	30	11	−10	155
	Neat	28.9	11.74	−7	268
	Vapor	31.2	11.03	−10	121, 122, 270
	Vapor	33.5	11.41	−8	151
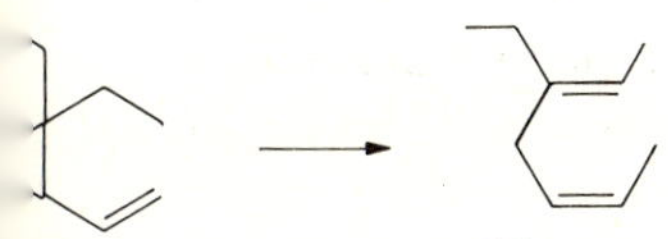	Vapor	33.7	11.32	−9	151
	Vapor	33.6	11.27	−9	151

erences pp. 481–487

TABLE 10 (*Contd.*)

Reaction	*Reaction medium*	E_a (*kcal.mole*$^{-1}$)	*log A*	$\Delta S^{\ddagger}_{298}$ (*eu*)	*Refs.*
		28	9.5	−17	246
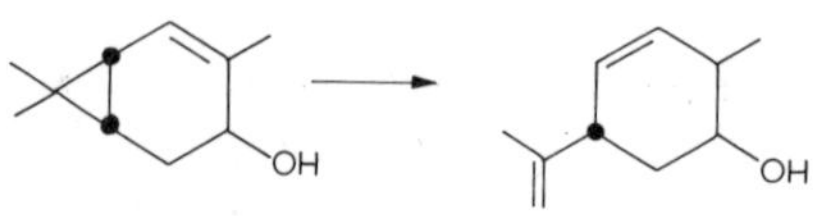		34	11.2	−9	246
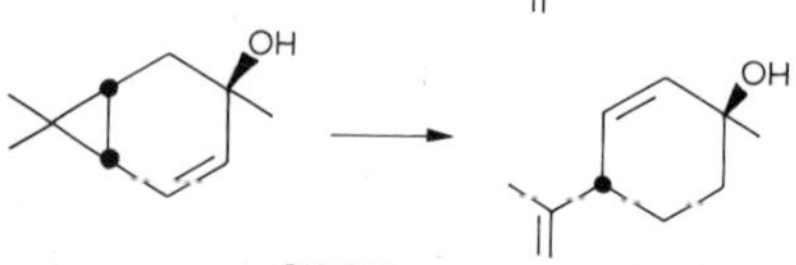		29	11.7	−7	246
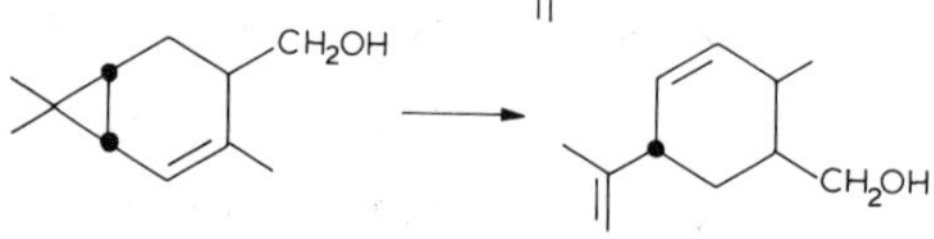		29	11.7	−7	246
	Vapor	38.6	12.76	−2	179
	Neat	32	12.0	−6	155
		28	11.4	−8	157
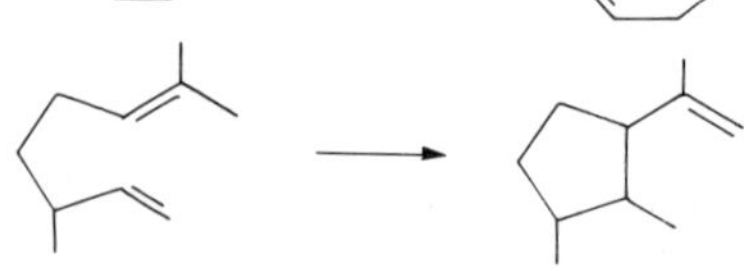	Vapor	35.2	9.55	−17	205

[a] Calculated, assuming log $A = 12$.

ium indicates that the hydrogen is roughly equally bonded to both terminal carbons in the transition state. Thus, $k_H/k_D = 12$ at 25°C for isomerization of *cis*-1,3-pentadiene[271] and $k_H/k_D = 5.0$ for isomerization of 1,3,6-cyclooctatriene to 1,3,5-cyclooctatriene at 120°C[216]. The secondary deuterium isotope effect on the latter reaction was negligible.

The activation energy for 1,5-hydrogen shifts in *cis*-1,3-pentadiene is about 36 kcal.mole^{-1}. Alkyl substitution on carbons 2, 4, or 5 of the rearranging system lowers the activation energy by 2–3 kcal.mole^{-1}. Dienyl 1,5-hydrogen shifts in cyclic dienes have rates and activation energies comparable to those for acyclic dienes, whether they occur in six-, seven-, eight- or ten-membered rings[155]. 1,5-Hydrogen shifts in cyclopentadienes have energies of activation nearly 10 kcal.mole^{-1} lower than those for other cyclic dienes. This is probably due in part to the absence of interactions between hydrogens on C-1 and C-5 of the pentadienyl system, and in part to the unique molecular orbitals of the cyclopentadienyl system[155].

The conformations required for homodienyl 1,5-hydrogen shifts can be achieved in open-chain, six-membered ring, and eight-membered ring cases, and these reactions have similar activation energies. However, there are severe hydrogen–hydrogen interactions in the saddle-shaped transition state for homodienyl 1,5-hydrogen shift in bicyclo[5.1.0]oct-2-ene, and the activation energy for this reaction is 7–8 kcal.mole^{-1} larger than that for 1,5-hydrogen shifts in cycloheptadienes. The energy of activation for 1,5-hydrogen shifts in bicyclo-[3.1.0]hex-2-enes is larger than 44 kcal.mole^{-1}, probably due to the inability of the bicyclohexene system to achieve the required transition state[194].

(*c*) *Dienyl 1,5-shifts of other groups*

Kinetics of a few reactions in which groups other than hydrogen undergo 1,5-shifts have been reported. Kinetics of isomerizations of substituted indenes show that, in shifts from C-1 to C-2 of the indene system, hydrogen migrates much faster than phenyl, which in turn migrates in preference to methyl[241]. In the gas phase, methyl groups migrate in 1,5,5- and 2,5,5-trimethylcyclopentadienes with activation energies of about 45 kcal.mole^{-1} and small positive entropies of activation[95,191]. Ciganek[75] studied the kinetics of reactions which involve 1,5-shifts of cyano groups in dicyanocycloheptadiene systems.

(*d*) *Thermal 1,7-hydrogen shifts*

The only thermal trienyl 1,7-hydrogen shifts whose kinetics have been studied are reversible isomerizations of D vitamins and some of their structural analogues

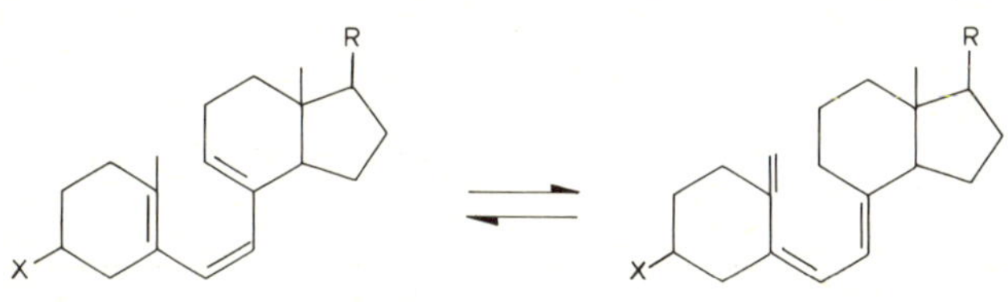

(44)

where X = OH, R = $CH(CH_3)(CH_2)_3CH(CH_3)_2$ (ref. 184); X = p-$O_2NC_6H_4$-CO_2, R = $CH(CH_3)CH{=}CHCH(CH_3)CH(CH_3)_2$ (ref. 220) and

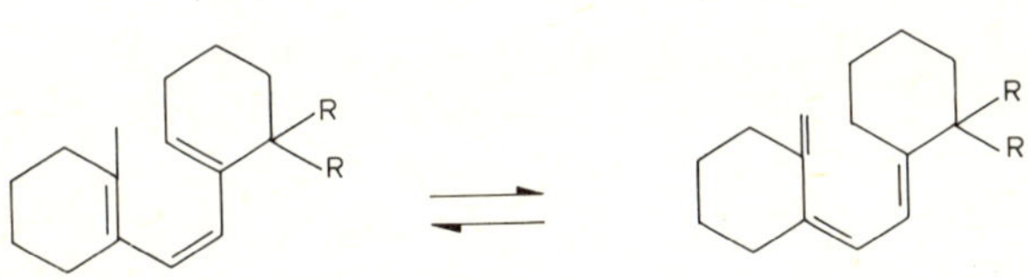

(45)

where R = H or CH_3 (ref. 188).

References pp. 481–487

These reactions are independent of light, solvent, surfaces, acids, and bases and are not affected by free radical inhibitors[188]. They have activation energies near 20 kcal.mole^{-1} and small to substantial negative entropies of activation. These isomerizations presumably involve antarafacial sigmatropic 1,7-hydrogen shifts. It is interesting that their activation energies are about 10 kcal.mole^{-1} smaller than those for typical 1,5-hydrogen shifts.

5.2.6 *Thermal isomerizations of methylenecycloalkanes*

Methylenecyclopropanes and methylenecyclobutanes undergo reversible thermal isomerizations. These reactions probably involve initial cleavage of the cycloalkane rings to allylic diradicals, which revert both to starting materials and products. Activation parameters for three of these reactions are listed in Table 11.

TABLE 11

GAS-PHASE ISOMERIZATIONS OF METHYLENECYCLOALKANES

Reactions	E_a *(kcal.mole^{-1})*	*log A*	$\Delta S^{\ddagger}$ *(eu)*	*Ref.*
	40.4	~13	~0	74
	49.5	14.47	+6	102
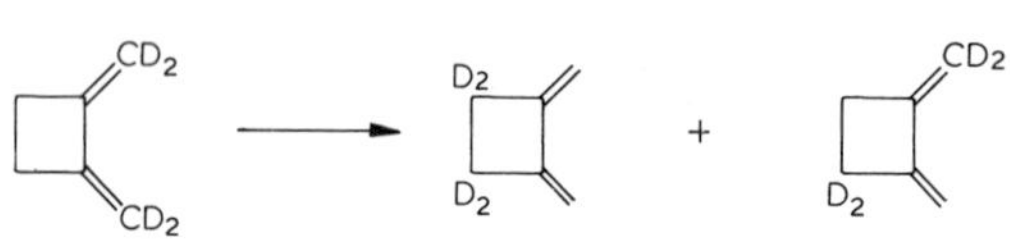	46.8	14.15	+4	101

5.2.7 *Thermal isomerizations of norbornadienes*

7-Substituted norbornadienes isomerize at 170°C to substituted cycloheptatrienes[230], *viz.*

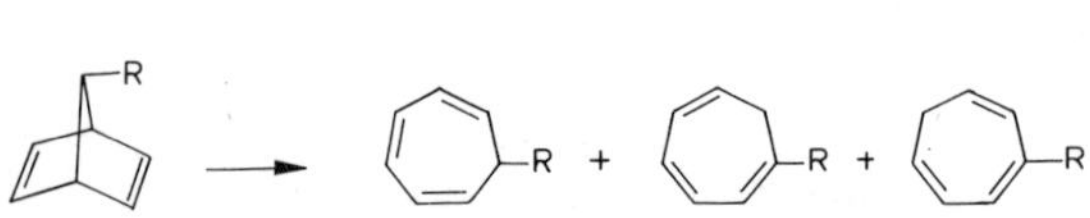

(46)

When R = $(CH_3)_3CO$, the energy of activation in decane solution is 35 kcal. mole^{-1}, and the entropy of activation is approximately zero. Relative reactivities in decane at 170°C are: R = $(CH_3)_3CO$, 14.3; R = CH_3O, 1.8; R = C_6H_5, 2.6.

The energy of activation for isomerization of norbornadiene itself is considerably larger[24,190]. The reaction is mechanistically complex, and yields toluene and decomposition products as well as cycloheptatriene. Under the conditions used, part of the toluene is formed by isomerization of the cycloheptatriene[190,217], possibly via norcaradiene. In the gas phase the energy of activation for formation of cycloheptatriene from norbornadiene is about 51 kcal.mole^{-1}, and log A is about 14.8 (refs. 24, 190). Activation parameters for the formation of toluene directly from norbornadiene in the gas phase[190] are E_a = 53 kcal.mole^{-1} and log A = 14.2. These reactions probably involve initial cleavage of the C-1, C-7 bond in norbornadiene to yield an allylic diradical which can cyclize to norcaradiene (a precursor for both cycloheptatriene and toluene) as well as undergo other reactions.

REFERENCES

1 J. K. Addy and R. M. Noyes, *J. Chem. Soc.*, (1964) 2473.
2 K. Alder, *Ann. Chem.*, 485 (1931) 211.
3 G. R. Aldridge and G. W. Murphy, *J. Am. Chem. Soc.*, 73 (1951) 1158.
4 A. Allerhand and H. S. Gutowsky, *J. Am. Chem. Soc.*, 87 (1965) 4092.
5 A. Amano and M. Uchiyama, *J. Phys. Chem.*, 69 (1965) 1278.
6 S. Bank, C. A. Rowe and A. Schriesheim, *J. Am. Chem. Soc.*, 85 (1963) 2115.
7 S. Bank, C. A. Rowe, A. Schriesheim and L. A. Naslund, *J. Am. Chem. Soc.*, 89 (1967) 6897.
8 S. Bank, C. A. Rowe, A. Schriesheim and L. A. Naslund, *J. Org. Chem.*, 33 (1968) 221.
9 S. Bank, A. Schriesheim and C. A. Rowe, *J. Am. Chem. Soc.*, 87 (1965) 3244.
10 H. C. Barany, E. A. Braude and J. A. Coles, *J. Chem. Soc.*, (1951) 2093.
11 M. A. Battiste and M. E. Burns, *Tetrahedron Letters*, (1968) 523.
12 I. Bell, E. R. H. Jones and M. C. Whiting, *J. Chem. Soc.*, (1957) 2597.
13 S. W. Benson, A. N. Bose and P. Nangia, *J. Am. Chem. Soc.*, 85 (1963) 1388.
14 G. Bergson, *Acta Chem. Scand.*, 17 (1963) 2691.
15 G. Bergson, *Acta Chem. Scand.*, 18 (1964) 2003.
16 G. Bergson and A. M. Weidler, *Acta Chem. Scand.*, 17 (1963) 862.
17 G. Bergson and A. M. Weidler, *Acta Chem. Scand.*, 17 (1963) 1798.
18 G. Bergson and A. M. Weidler, *Acta Chem. Scand.*, 18 (1964) 1498.
19 A. J. Berlin, L. P. Fisher and A. D. Ketley, *Chem. Ind. (London)*, (1965) 509.
20 D. C. Berndt, *J. Chem. Eng. Data*, 14 (1969) 112.
21 S. C. Bernstein, *J. Org. Chem.*, 33 (1968) 3486.
22 J. A. Berson and E. J. Walsh, *J. Am. Chem. Soc.*, 90 (1968) 4731.
23 P. Bickart, F. W. Carson, J. Jacobus, E. G. Miller and K. Mislow, *J. Am. Chem. Soc.*, 90 (1968) 4869.
24 J. H. Birely and J. P. Chesick, *J. Phys. Chem.*, 66 (1962) 568.
25 G. C. Bond and M. Hellier, *J. Catalysis*, 7 (1967) 217.
26 A. P. ter Borg and H. Kloosterziel, *Rec. Trav. Chim.*, 82 (1963) 741.
27 A. P. ter Borg and H. Kloosterziel, *Rec. Trav. Chim.*, 82 (1963) 1189.

28 A. P. TER BORG AND H. KLOOSTERZIEL, *Rec. Trav. Chim.*, 84 (1965) 245.
29 A. P. TER BORG, H. KLOOSTERZIEL AND N. VAN MEURS, *Proc. Chem. Soc.*, (1962) 359.
30 A. P. TER BORG, H. KLOOSTERZIEL AND N. VAN MEURS, *Rec. Trav. Chim.*, 82 (1963) 717.
31 A. P. TER BORG, H. KLOOSTERZIEL AND Y. L. WESTPHAL, *Rec. Trav. Chim.*, 86 (1967) 474.
32 A. P. TER BORG, E. RAZENBERG AND H. KLOOSTERZIEL, *Rec. Trav. Chim.*, 84 (1965) 1230.
33 J. H. BOYER AND F. C. CANTER, *Chem. Rev.*, 54 (1954) 1.
34 G. R. BRANTON AND H. M. FREY, *J. Chem. Soc. A*, (1966) 1342.
35 G. R. BRANTON, H. M. FREY AND D. C. MONTAGUE, *Trans. Faraday Soc.*, 62 (1966) 659.
36 G. R. BRANTON, H. M. FREY AND R. F. SKINNER, *Trans. Faraday Soc.*, 62 (1966) 1546.
37 E. A. BRAUDE, *J. Chem. Soc.*, (1944) 443.
38 E. A. BRAUDE, *J. Chem. Soc.*, (1948) 794.
39 E. A. BRAUDE, *Ann. Rept. Chem. Soc.*, 46 (1949) 125.
40 E. A. BRAUDE, *Quart. Rev.*, 4 (1950) 404.
41 E. A. BRAUDE, J. A. COLES, E. A. EVANS AND C. J. TIMMONS, *Nature*, 177 (1956) 1167.
42 E. A. BRAUDE AND J. S. FAWCETT, *J. Chem. Soc.*, (1950) 800.
43 E. A. BRAUDE AND J. S. FAWCETT, *J. Chem. Soc.*, (1952) 4158.
44 E. A. BRAUDE AND P. H. GORE, *Nature*, 173 (1954) 1091.
45 E. A. BRAUDE AND P. H. GORE, *J. Chem. Soc.*, (1959) 41.
46 E. A. BRAUDE AND E. R. H. JONES, *J. Chem. Soc.*, (1944) 436.
47 E. A. BRAUDE AND E. R. H. JONES, *J. Chem. Soc.*, (1946) 122.
48 E. A. BRAUDE AND E. R. H. JONES, *J. Chem. Soc.*, (1946) 128.
49 E. A. BRAUDE, E. R. H. JONES AND E. S. STERN, *J. Chem. Soc.*, (1946) 396.
50 E. A. BRAUDE AND E. S. STERN, *J. Chem. Soc.*, (1947) 1096.
51 E. A. BRAUDE AND E. S. STERN, *J. Chem. Soc.*, (1948) 1982.
52 E. A. BRAUDE AND C. J. TIMMONS, *J. Chem. Soc.*, (1953) 3138.
53 E. A. BRAUDE AND D. W. TURNER, *Chem. Ind. (London)*, (1955) 1223.
54 E. A. BRAUDE AND D. W. TURNER, *J. Chem. Soc.*, (1958) 2404.
55 E. A. BRAUDE, D. W. TURNER AND E. S. WAIGHT, *Nature*, 173 (1954) 863.
56 E. A. BRAUDE, D. W. TURNER AND E. S. WAIGHT, *J. Chem. Soc.*, (1958) 2396.
57 E. A. BRAUDE, G. VALKANAS AND E. S. WAIGHT, *Chem. Ind. (London)*, (1956) 314.
58 E. A. BRAUDE AND E. S. WAIGHT, *J. Chem. Soc.*, (1953) 419.
59 J. T. BRAUMAN AND D. M. GOLDEN, *J. Am. Chem. Soc.*, 90 (1968) 1920.
60 S. BRAVERMAN AND Y. STABINSKY, *Chem. Commun.*, (1967) 270.
61 K. R. BROWER, *J. Am. Chem. Soc.*, 83 (1961) 4370.
62 H. C. BROWN AND Y. OKAMOTO, *J. Am. Chem. Soc.*, 80 (1958) 4979.
63 J. M. BROWN, *Chem. Commun.*, (1965) 226.
64 H. BURTON, *J. Chem. Soc.*, (1928) 1650.
65 H. BURTON, *J. Chem. Soc.*, (1934) 1268.
66 H. BURTON AND C. K. INGOLD, *J. Chem. Soc.*, (1928) 904.
67 C. A. BUNTON, Y. POCKER AND H. DAHN, *Chem. Ind. (London)*, (1958) 1516.
68 W. P. CAIN AND R. M. NOYES, *J. Am. Chem. Soc.*, 81 (1959) 2031.
69 M. D. CARR, J. R. P. CLARKE AND M. C. WHITING, *Proc. Chem. Soc.*, (1963) 333.
70 R. W. CARR AND W. D. WALTERS, *J. Phys. Chem.*, 69 (1965) 1073.
71 A. G. CATCHPOLE AND E. D. HUGHES, *J. Chem. Soc.*, (1948) 1.
72 A. G. CATCHPOLE, E. D. HUGHES AND C. K. INGOLD, *J. Chem. Soc.*, (1948) 8.
73 P. CHANG, HUA HSUEH HSUEH PAO, 26 (3) (1960) 117; *Chem. Abstr.*, 59 (1963) 398.
74 J. P. CHESICK, *J. Am. Chem. Soc.*, 85 (1963) 2720.
75 E. CIGANEK, *J. Am. Chem. Soc.*, 89 (1967) 1458.
76 J. M. CLAYTON AND A. M. EASTHAM, *J. Chem. Soc.*, (1963) 1636.
77 K. CONROW, *J. Am. Chem. Soc.*, 83 (1961) 2343.
78 R. C. COOKSON, D. P. G. HAMON AND R. E. PARKER, *J. Chem. Soc.*, (1962) 5014.
79 W. COOPER AND W. D. WALTERS, *J. Am. Chem. Soc.*, 80 (1958) 4220.
80 A. C. COPE AND L. FIELD, *J. Am. Chem. Soc.*, 71 (1949) 1589.
81 A. C. COPE AND E. M. HARDY, *J. Am. Chem. Soc.*, 62 (1940) 441.
82 A. C. COPE, C. M. HOFMANN AND E. M. HARDY, *J. Am. Chem. Soc.*, 63 (1941) 1852.

83 A. C. Cope, K. E. Hoyle and D. Heyl, *J. Am. Chem. Soc.*, 63 (1941) 1843.
84 D. J. Cram, *Fundamentals of Carbanion Chemistry*, Academic Press, New York, 1968, Chap. 5.
85 D. J. Cram and M. R. V. Sahyun, *J. Am. Chem. Soc.*, 85 (1963) 1257.
86 D. J. Cram and R. T. Uyeda, *J. Am. Chem. Soc.*, 86 (1964) 5466.
87 D. J. Cram, F. Willey, H. P. Fischer, H. M. Relles and D. A. Scott, *J. Am. Chem. Soc.*, 88 (1966) 2759.
88 R. Cramer, *J. Am. Chem. Soc.*, 88 (1966) 2272.
89 R. Criegee, *Angew. Chem. Intern. Ed. Engl.*, 7 (1968) 559.
90 R. Criegee, G. Bolz and R. Askani, *Chem. Ber.*, 102 (1969) 275.
91 R. Criegee and H. G. Reinhart, *Chem. Ber.*, 101 (1968) 102.
92 R. Criegee, D. Seebach, R. E. Winter, B. Boerretzen and H. A. Brune, *Chem. Ber.*, 98 (1965) 2339.
93 B. Cruickshank and N. R. Davis, *Australian J. Chem.*, 19 (1966) 815.
94 D. Y. Curtin and R. J. Crawford, *J. Am. Chem. Soc.*, 79 (1957) 3156.
95 J. W. de Haan and H. Kloosterziel, *Rec. Trav. Chim.*, 84 (1965) 1594.
96 P. B. D. de la Mare, in *Molecular Rearrangements*, P. de Mayo (Ed.), Vol. 1, Interscience, New York, 1963, Chap. 2.
97 P. B. D. de la Mare and C. A. Vernon, *J. Chem. Soc.*, (1954) 2504.
98 R. H. DeWolfe and W. G. Young, *Chem. Rev.*, 56 (1956) 753.
99 R. H. DeWolfe and W. G. Young, in *The Chemistry of Alkenes*, Saul Patai (Ed.), Interscience, New York, 1964, p. 681.
100 D. Dickens, H. M. Frey and R. F. Skinner, *Trans. Faraday Soc.*, 65 (1969) 453.
101 W. von E. Doering and W. R. Dolbier, *J. Am. Chem. Soc.*, 89 (1967) 4533.
102 W. von E. Doering and J. Gilbert, *Tetrahedron Suppl.*, 7 (1966) 397.
103 W. von E. Doering and W. R. Roth, *Tetrahedron*, 18 (1962) 67.
104 W. von E. Doering and W. R. Roth, *Tetrahedron*, 19 (1963) 715.
105 W. von E. Doering and W. R. Roth, *Angew. Chem. Intern. Ed. Engl.*, 2 (1963) 115.
106 G. A. Doorakian and H. H. Freedman, *J. Am. Chem. Soc.*, 90 (1968) 3582.
107 G. A. Doorakian and H. H. Freedman, *J. Am. Chem. Soc.*, 90 (1968) 5310.
108 T. J. Dougherty, *J. Am. Chem. Soc.*, 86 (1964) 2236.
109 K. W. Egger, *J. Am. Chem. Soc.*, 89 (1967) 3688.
110 K. W. Egger, *Helv. Chim. Acta*, 51 (1968) 422.
111 K. W. Egger and S. W. Benson, *J. Am. Chem. Soc.*, 88 (1966) 241.
112 K. W. Egger, D. M. Golden and S. W. Benson, *J. Am. Chem. Soc.*, 86 (1964) 5420.
113 S. W. Ela and D. J. Cram, *J. Am. Chem. Soc.*, 88 (1966) 5777.
114 S. W. Ela and D. J. Cram, *J. Am. Chem. Soc.*, 88 (1966) 5791.
115 C. S. Elliott and H. M. Frey, *J. Chem. Soc.*, (1965) 4289.
116 C. S. Elliott and H. M. Frey, *J. Chem. Soc.*, (1965) 345.
117 C. S. Elliott and H. M. Frey, *Trans. Faraday Soc.*, 62 (1969) 895.
118 R. J. Ellis and H. M. Frey, *Trans. Faraday Soc.*, 59 (1963) 2076.
119 R. J. Ellis and H. M. Frey, *J. Chem. Soc.*, (1964) 959.
120 R. J. Ellis and H. M. Frey, *J. Chem. Soc.*, (1964) 4188.
121 R. J. Ellis and H. M. Frey, *J. Chem. Soc.*, (1964) 5578.
122 R. J. Ellis and H. M. Frey, *Proc. Chem. Soc.*, (1964) 221.
123 R. J. Ellis and H. M. Frey, *J. Chem. Soc. A*, (1966) 553.
124 B. D. England, *J. Chem. Soc.*, (1955) 1615.
125 B. D. England and E. D. Hughes, *Nature*, 168 (1951) 1002.
126 M. C. Flowers and H. M. Frey, *J. Chem. Soc.*, (1961) 3547.
127 E. G. Foster, A. C. Cope and F. Daniels, *J. Am. Chem. Soc.*, 69 (1947) 1893.
128 H. H. Freedman, G. A. Doorakian and V. R. Sandel, *J. Am. Chem. Soc.*, 87 (1965) 3019.
129 H. M. Frey, *Trans. Faraday Soc.*, 58 (1962) 957.
130 H. M. Frey, *Trans. Faraday Soc.*, 59 (1963) 1619.
131 H. M. Frey, *Trans. Faraday Soc.*, 60 (1964) 83.

132 H. M. Frey, *Advan. Phys. Org. Chem.*, **4** (1966) 148.
133 H. M. Frey and R. J. Ellis, *J. Chem. Soc.*, (1965) 4770.
134 H. M. Frey and A. Krantz, *J. Chem. Soc. A*, (1969) 1159.
135 H. M. Frey and A. M. Lamont, *J. Chem. Soc. A*, (1969) 1592.
136 H. M. Frey and D. H. Lister, *J. Chem. Soc. A*, (1967) 26.
137 H. M. Frey and D. C. Marshall, *J. Chem. Soc.* (1962) 3981.
138 H. M. Frey and D. C. Marshall, *Trans. Faraday Soc.*, 61 (1965) 1715.
139 H. M. Frey, D. C. Marshall and R. F. Skinner, *Trans. Faraday Soc.*, 61 (1965) 861.
140 H. M. Frey and D. C. Montague, *Trans. Faraday Soc.*, 64 (1968) 2369.
141 H. M. Frey, D. C. Montague and I. D. R. Stevens, *Trans. Faraday Soc.*, 63 (1967) 372.
142 H. M. Frey and B. M. Pope, *J. Chem. Soc. A*, (1966) 1701.
143 H. M. Frey and B. M. Pope, *J. Chem. Soc. B*, (1966) 209.
144 H. M. Frey and B. M. Pope, *Trans. Faraday Soc.*, 65 (1969) 441.
145 H. M. Frey, B. M. Pope and R. F. Skinner, *Trans. Faraday Soc.*, 63 (1967) 1166.
146 H. M. Frey and R. F. Skinner, *Trans. Faraday Soc.*, 61 (1965) 1918.
147 H. M. Frey and R. K. Solly, *Trans. Faraday Soc.*, 64 (1968) 1858.
148 H. M. Frey and R. K. Solly, *Trans. Faraday Soc.*, 65 (1969) 448.
149 H. M. Frey and R. K. Solly, *J. Chem. Soc. A*, (1969) 733.
150 H. M. Frey and R. K. Solly, *Trans. Faraday Soc.*, 65 (1969) 1372.
151 H. M. Frey and R. Walsh, *Chem. Rev.*, 69 (1969) 103.
152 E. C. Friedrichs and S. Winstein, *Tetrahedron Letters*, (1962) 475.
153 A. Gagneux, S. Winstein and W. G. Young, *J. Am. Chem. Soc.*, 82 (1960) 5956.
154. G. B. Gill, *Quart. Rev.*, 22 (1968) 338.
155 D. S. Glass, R. S. Boikiss and S. Winstein, *Tetrahedron Letters*, (1966) 999.
156 D. S. Glass, J. W. H. Whatthey and S. Winstein, *Tetrahedron Letters*, (1965) 377.
157 D. S. Glass, J. Zirner and S. Winstein, *Proc. Chem. Soc.*, (1963) 276.
158 H. L. Goering, *Record Chem. Progr. (Kresge-Hooker Sci. Lib.)*, 21 (1960) 109.
159 H. L. Goering, J. P. Blanchard and E. F. Silversmith, *J. Am. Chem. Soc.*, 76 (1954) 5409.
160 H. L. Goering and W. D. Closson, *J. Am. Chem. Soc.*, 83 (1961) 3511.
161 H. L. Goering and R. E. Dilgren, *J. Am. Chem. Soc.*, 81 (1959) 2556.
162 H. L. Goering and R. E. Dilgren, *J. Am. Chem. Soc.*, 82 (1960) 5744.
163 H. L. Goering and J. T. Doi, *J. Am. Chem. Soc.*, 82 (1960) 5850.
164 H. L. Goering, J. T. Doi and K. D. McMichael, *J. Am. Chem. Soc.*, 86 (1964) 1951.
165 H. L. Goering, H. H. Espy and W. D. Closson, *J. Am. Chem. Soc.*, 81 (1959) 329.
166 H. L. Goering and R. W. Greiner, *J. Am. Chem. Soc.*, 79 (1957) 3464.
167 H. L. Goering and R. R. Josephson, *J. Am. Chem. Soc.*, 83 (1961) 2585.
168 H. L. Goering and R. R. Josephson, *J. Am. Chem. Soc.*, 84 (1962) 2779.
169 H. L. Goering and R. F. Myers, *J. Am. Chem. Soc.*, 91 (1969) 3386.
170 H. L. Goering, T. D. Nevitt and E. F. Silversmith, *J. Am. Chem. Soc.*, 77 (1955) 5026.
171 H. L. Goering and M. M. Pombo, *J. Am. Chem. Soc.*, 82 (1960) 2515.
172 H. L. Goering, M. M. Pombo and K. D. McMichael, *J. Am. Chem. Soc.*, 85 (1963) 965.
173 H. L. Goering and E. F. Silversmith, *J. Am. Chem. Soc.*, 77 (1955) 1129.
174 H. L. Goering and E. F. Silversmith, *J. Am. Chem. Soc.*, 77 (1955) 6249.
175 H. L. Goering and E. F. Silversmith, *J. Am. Chem. Soc.*, 79 (1957) 348.
176 D. M. Golden and J. K. Brauman, *Trans. Faraday Soc.*, 65 (1969) 464.
177 T. D. Goldfarb and L. Lindqvist, *J. Am. Chem. Soc.*, 89 (1967) 4588.
178 H. Goldwhite and C. M. Valdez, *Chem. Commun.* (1969) 7.
179 W. Grimme, *Chem. Ber.*, 98 (1965) 756.
180 H. Gunther, *Tetrahedron Letters*, (1965) 4085.
181 B. Gutbezahl and E. Grunwald, *J. Am. Chem. Soc.*, 75 (1953) 572.
182 L. P. Hammett, *Physical Organic Chemistry*, McGraw-Hill, New York, 1940.
183 G. S. Hammond and C. D. De Boer, *J. Am. Chem. Soc.*, 86 (1964) 899.
184 K. H. Hanewald, M. P. Rappoldt and J. R. Roburgh, *Rec. Trav. Chim.*, 80 (1961) 1003.
185 R. T. Harris and K. E. Weale, *J. Chem. Soc.*, (1956) 953.

186 L. F. Hatch, P. D. Gardner and R. E. Gilbert, *J. Am. Chem. Soc.*, 81 (1959) 5943.
187 W. P. Hauser and W. D. Walters, *J. Phys. Chem.*, 67 (1963) 1328.
188 E. Havinga and J. L. M. A. Schlatmann, *Tetrahedron*, 16 (1961) 146.
189 P. Heimbach and W. Brenner, *Angew. Chem. Intern. Ed. Engl.*, 6 (1967) 800.
190 W. C. Herndon and L. L. Lowry, *J. Am. Chem. Soc.*, 86 (1964) 1922.
191 W. C. Herndon and J. M. Mannion, *J. Org. Chem.*, 33 (1968) 4504.
192 A. W. Herriott and K. Mislow, *Tetrahedron Letters*, (1968) 3013.
193 R. K. Hill, *Tetrahedron Letters*, (1964) 3239.
194 R. K. Hill and N. W. Gilman, *Tetrahedron Letters*, (1967) 1421.
195 R. K. Hill and N. W. Gilman, *Chem. Commun.*, (1967) 619.
196 R. Hoffmann and R. B. Woodward, *J. Am. Chem. Soc.*, 87 (1965) 4389.
197 R. Hoffmann and R. B. Woodward, *Accounts Chem. Res.*, 1 (1968) 17.
198 R. Huisgen and G. Boche, *Tetrahedron Letters*, (1965) 1769.
199 R. Huisgen, A. Dahmen and H. Huber, *J. Am. Chem. Soc.*, 89 (1967) 7130.
200 R. Huisgen, A. Dahmen and H. Huber, *Tetrahedron Letters*, (1969) 1461.
201 R. Huisgen and F. Mietzsch, *Angew. Chem. Intern. Ed. Engl.*, 3 (1964) 83.
202 R. Huisgen and H. Seidel, *Tetrahedron Letters*, (1964) 3381.
203 K. Humski, T. Strelkov, S. Borcic and D. E. Sunko, *Chem. Commun.*, (1969) 693.
204 D. H. Hunter and D. J. Cram, *J. Am. Chem. Soc.*, 86 (1964) 5478.
205 W. T. Huntsman and T. H. Curry, *J. Am. Chem. Soc.*, 80 (1958) 2252.
206 A. Iliceto, A. Fava and U. Mazzucato, *Tetrahedron Letters*, No. 11, (1960) 27.
207 C. K. Ingold, *Structure and Mechanism in Organic Chemistry*, Cornell University Press, Ithaca, New York, 1953, p. 586.
208 A. Jefferson and F. Scheinmann, *Quart. Rev.*, 22 (1968) 391.
209 M. A. Karlsson, *Ph.D. Thesis*, University of Rochester, 1967.
210 J. Kenyon, S. M. Partridge and H. Phillips, *J. Chem. Soc.*, (1937) 207.
211 A. Kergomard, F. Pigeret and M. Renard, *Bull. Soc. Chim. France*, (1962) 1648.
212 A. D. Ketley, *Tetrahedron Letters*, (1964) 1687.
213 A. D. Ketley, A. J. Berlin and L. P. Fisher, *J. Org. Chem.*, 31 (1966) 2648.
214 A. D. Ketley, A. J. Berlin, E. Gorman and L. P. Fisher, *J. Org. Chem.*, 31 (1966) 305.
215 A. D. Ketley and J. L. McClanahan, *J. Org. Chem.*, 30 (1965) 942.
216 H. Kloosterziel and A. P. ter Borg, *Rec. Trav. Chim.*, 84 (1965) 1305.
217 K. N. Klump and J. P. Chesick, *J. Am. Chem. Soc.*, 85 (1963) 130.
218 H. Kwart and J. Herbig, *J. Am. Chem. Soc.*, 85 (1963) 226.
219 J. B. Lambert, *Tetrahedron Letters*, (1963) 1901.
220 M. Legrand and J. Mathieu, *Compt. Rend.*, 245 (1957) 2502.
221 W. J. LeNoble, *J. Phys. Chem.*, 67 (1963) 2451.
222 W. J. LeNoble, *Progr. Phys. Org. Chem.*, 5 (1967) 207.
223 D. G. Lesnini, P. D. Buckley and R. M. Noyes, *J. Am. Chem. Soc.*, 90 (1968) 668.
224 H. Levy and A. C. Cope, *J. Am. Chem. Soc.*, 66 (1944) 1684.
225 E. S. Lewis, J. T. Hill and E. R. Newman, *J. Am. Chem. Soc.*, 90 (1968) 662.
226 E. S. Lewis and K. Witte, *J. Chem. Soc. B*, (1968) 1198.
227 K. E. Lewis and H. S. Steiner, *J. Chem. Soc.*, (1964) 3080.
228 E. C. Lindsay, *Ph.D. Dissertation*, University of Wisconsin, 1968; *Dissertation Abstr.*, 29 (B) (1968) 1983.
229 E. C. Lupton, *Tetrahedron Letters*, (1968) 4209.
230 R. J. Lustgarten and H. J. Richey, *Tetrahedron Letters*, (1966) 4655.
231 A. Maccoll and R. A. Ross, *J. Am. Chem. Soc.*, 87 (1965) 4997.
232 S. McLean and P. Haynes, *Tetrahedron Letters*, (1964) 2385.
233 S. McLean and P. Haynes, *Tetrahedron*, 21 (1965) 2329.
234 K. D. McMichael, *J. Am. Chem. Soc.*, 89 (1967) 2943.
235 R. Malojcic, K. Humski, S. Borcic and D. E. Sunko, *Tetrahedron Letters*, (1969) 2003.
236 E. N. Marvell, G. Caple and B. Schatz, *Tetrahedron Letters*, (1965) 385.
237 S. Masamune, S. Takada, N. Nakatsuka, R. Vukov and E. N. Cain, *J. Am. Chem. Soc.*, 91 (1969) 4322.

238 J. MEISENHEIMER AND G. BEUTTER, *Ann. Chem.*, 508 (1934) 58.
239 J. MEISENHEIMER, W. SCHMIDT AND G. SCHAFER, *Ann. Chem.*, 501 (1933) 131.
240 Y. MIALHE AND R. VESSIÈRE, *Bull. Soc. Chim. France*, (1968) 4181.
241 L. L. MILLER, R. GREISINGER AND R. F. BOYER, *J. Am. Chem. Soc.*, 91 (1969) 1579.
242 I. I. MOISEEV, S. V. PESTRIKOV AND L. M. SVERZH, *Izv. Akad. Nauk SSSR, Ser. Khim.*, (1966) 1866.
243 R. W. MURRAY AND M. L. KAPLAN, *J. Am. Chem. Soc.*, 88 (1966) 3527.
244 K. MUZIKA, M. PROCHAZKA AND M. PALECEK, *Collection Czech. Chem. Commun.*, 34 (1969) 635.
245 T. NOZOE AND K. TAKAHASHI, *Bull. Chem. Soc. Japan*, 38 (1965) 665.
246 G. OHLOFF, *Tetrahedron Letters*, (1965) 3795.
247 L. OHLSSON, I. WALLMARK AND G. BERGSON, *Acta Chem. Scand.*, 20 (1966) 750.
248 L. OHLSSON, S. WOLD AND G. BERGSON, *Arkiv. Kemi*, 29 (1968) 351.
249 H. E. O'NEAL AND S. W. BENSON, *J. Chem. Phys.*, 71 (1967) 2903.
250 W. OROSHNIK, *J. Am. Chem. Soc.*, 78 (1956) 2651.
251 J. F. M. OTH, R. MERENYI, J. NIELSEN AND G. SCHROEDER, *Chem. Ber.*, 98 (1965) 3385.
252 J. F. M. OTH, R. MERENYI, H. ROETTELE AND G. SCHROEDER, *Tetrahedron Letters*, (1968) 3941.
253 S. V. PESTRIKOV, I. I. MOISEEV AND L. M. SVERZH, *Kinetika i Kataliz*, 10 (1969) 74; *Chem. Abstr.*, 70 (1969) 105673.
254 Y. POCKER, *J. Chem. Soc.*, (1958) 4318.
255 Y. POCKER, *J. Chem. Soc.*, (1958) 4323.
256 Y. POCKER, *Proc. Chem. Soc.*, (1959) 226.
257 Y. POCKER, *Proc. Chem. Soc.*, (1961) 141.
258 Y. POCKER AND M. J. HILL, *J. Am. Chem. Soc.*, 91 (1969) 3243.
259 M. POMERANTZ AND P. H. HARTMAN, *Tetrahedron Letters*, (1968) 991.
260 J. W. POWELL AND M. C. WHITING, *Proc. Chem. Soc.*, (1960) 412.
261 J. W. RALLS, R. E. LUNDIN AND G. F. BAILEY, *J. Org. Chem.*, 28 (1963) 3521.
262 D. J. RAWLINSON AND R. M. NOYES, *J. Chem. Soc.*, (1963) 1793.
263 D. R. RAYNER, E. G. MILLER, P. BICKART, A. J. GORDON AND K. MISLOW, *J. Am. Chem. Soc.*, 88 (1966) 3138.
264 S. J. RHOADS, in *Molecular Rearrangements*, P. DE MAYO (Ed.), Vol. 1, Interscience, New York, 1963, Chap. 11.
265 O. A. ROKSTAD, J. UGELSTAD AND H. LID, *Acta Chem. Scand.*, 23 (1969) 782.
266 S. D. ROSS AND M. M. LABES, *J. Am. Chem. Soc.*, 79 (1957) 4155.
267 R. ROSS AND E. S. WAIGHT, *J. Chem. Soc.*, (1965) 6710.
268 W. R. ROTH, *Ann. Chem.*, 671 (1964) 25.
269 W. R. ROTH, *Tetrahedron Letters*, (1964) 1009.
270 W. R. ROTH AND J. KOENIG, *Ann. Chem.*, 688 (1965) 28.
271 W. R. ROTH AND J. KOENIG, *Ann. Chem.*, 699 (1966) 24.
272 S. SAREL, J. YOVELL AND M. SAREL-IMBER, *Angew. Chem. Intern. Ed. Engl.*, 7 (1968) 577.
273 M. SAUNDERS, *Tetrahedron Letters*, (1963) 1699.
274 B. S. SCHATZ, *Dissertation Abstr.*, 27 (B) (1967) 3870.
275 A. SCHRIESHEIM, *Trans. N.Y. Acad. Sci.*, 31 (1969) 97.
276 A. SCHRIESHEIM, J. E. HOFMANN AND C. A. ROWE, *J. Am. Chem. Soc.*, 83 (1961) 3731.
277 A. SCHRIESHEIM, R. J. MULLER AND C. A. ROWE, *J. Am. Chem. Soc.*, 84 (1962) 3164.
278 A. SCHRIESHEIM AND C. A. ROWE, *Tetrahedron Letters*, (1962) 405.
279 A. SCHRIESHEIM AND C. A. ROWE, *J. Am. Chem. Soc.*, 84 (1962) 3160.
280 A. SCHRIESHEIM, C. A. ROWE AND L. NASLUND, *J. Am. Chem. Soc.*, 85 (1963) 2111.
281 G. SCHROEDER, *Angew. Chem. Intern. Ed. Engl.*, 4 (1965) 752.
282 G. SCHROEDER, R. MERENYI AND J. F. M. OTH, *Tetrahedron Letters*, (1964) 773.
283 G. SCHROEDER AND J. F. M. OTH, *Angew. Chem. Intern. Ed. Engl.*, 6 (1967) 414.
284 F. W. SCHULER AND G. W. MURPHY, *J. Am. Chem. Soc.*, 72 (1950) 3155.
285 D. SEEBACH, *Chem. Ber.*, 97 (1964) 2953.
286 J. M. SHACKELFORD AND L. H. SCHWARTZMAN, *J. Org. Chem.*, 27 (1962) 1047.
287 M. Y. SHANDALA, E. S. WAIGHT AND M. WEINSTOCK, *J. Chem. Soc. B*, (1966) 590.

288 A. I. SHATENSHTEIN, *Advan. Phys. Org. Chem.*, 1 (1963) 177.
289 H. J. SHINE, *Aromatic Rearrangements*, Elsevier, Amsterdam, 1967, p. 89.
290 M. SIMONETTA, G. FAVINI, C. MARIANI AND P. GRAMACCIONI, *J. Am. Chem. Soc.*, 90 (1968) 1280.
291 P. A. S. SMITH AND D. W. EMERSON, *J. Am. Chem. Soc.*, 82 (1960) 3076.
292 S. G. SMITH, *J. Am. Chem. Soc.*, 83 (1961) 4285.
293 R. A. SNEEN, *J. Am. Chem. Soc.*, 82 (1960) 4261.
294 R. A. SNEEN AND A. M. ROSENBERG, *J. Am. Chem. Soc.*, 83 (1961) 900.
295 L. STEIN AND G. W. MURPHY, *J. Am. Chem. Soc.*, 74 (1952) 1041.
296 E. C. STEINER AND R. O. TRUCKS, *J. Am. Chem. Soc.*, 90 (1968) 720.
297 R. W. TAFT, in *Steric Effects in Organic Chemistry*, M. S. NEWMAN (Ed.), Wiley, New York, 1956, Chap. 13.
298 D. S. TARBELL, *Org. Reactions*, 2 (1944) 1.
299 H. A. TAYIM AND J. C. BALAIR, *J. Am. Chem. Soc.*, 89 (1967) 3420.
300 D. J. TRECKER AND J. P. HENRY, *J. Am. Chem. Soc.*, 86 (1964) 902.
301 H. TRUKSMA, H. STEINBERG AND T. J. DEBOER, *Rec. Trav. Chim.*, 82 (1963) 1057.
302 J. UGELSTAD AND O. A. ROKSTAD, *Acta Chem. Scand.*, 18 (1964) 474.
303 G. VALKANAS AND E. S. WAIGHT, *J. Chem. Soc.*, (1959) 2720.
304 G. VALKANAS, E. S. WAIGHT AND M. WEINSTOCK, *J. Chem. Soc.*, (1963) 4248.
305 C. A. VANDER WERF AND V. L. HEASLEY, *J. Org. Chem.*, 31 (1966) 3534.
306 J. C. VLANZY, *Dissertation Abstr.*, 29 (B) (1968) 1992.
307 E. VOGEL, *Angew. Chem. Intern. Ed. Engl.*, 2 (1963) 1.
308 E. VOGEL, W. GRIMME AND E. DINNE, *Tetrahedron Letters*, (1965) 391.
309 E. VOGEL, H. KIEFER AND W. R. ROTH, *Angew. Chem.*, 76 (1964) 432.
310 E. VOGEL AND R. SUNDERMAN, quoted in ref. 132, p. 163.
311 H. C. VOLGER AND H. HOGEVEEN, *Rec. Trav. Chim.*, 86 (1967) 830.
312 E. S. WAIGHT AND M. WEINSTOCK, *Proc. Chem. Soc.*, (1961) 334.
313 C. WALLING AND M. NAIMAN, *J. Am. Chem. Soc.*, 84 (1962) 2628.
314 A. M. WEIDLER AND G. BERGSON, *Acta Chem. Scand.*, 18 (1964) 1487.
315 C. A. WELLINGTON, *J. Phys. Chem.*, 66 (1962) 1671.
316 E. WETH AND A. S. DREIDING, *Proc. Chem. Soc.*, (1964) 59.
317 P. S. WHARTON AND R. A. KRETCHMAR, *J. Org. Chem.*, 33 (1968) 4258.
318 D. E. WHYTE AND A. C. COPE, *J. Am. Chem. Soc.*, 65 (1943) 1999.
319 M. R. WILCOTT AND V. H. CARGLE, *J. Am. Chem. Soc.*, 91 (1969) 4310.
320 M. R. WILCOTT AND V. H. CARGLE, *J. Am. Chem. Soc.*, 89 (1967) 723.
321 M. R. WILCOTT AND E. GOERLAND, *Tetrahedron Letters*, (1966) 6341.
322 S. WINSTEIN, E. C. FRIEDRICH AND S. SMITH, *J. Am. Chem. Soc.*, 86 (1964) 306.
323 S. WINSTEIN, E. GRUNWALD AND H. W. JONES, *J. Am. Chem. Soc.*, 73 (1951) 2700.
324 S. WINSTEIN, S. SMITH AND D. DARWISH, *Tetrahedron Letters*, No. 16 (1959) 24.
325 R. B. WOODWARD AND R. HOFFMANN, *J. Am. Chem. Soc.*, 87 (1965) 395.
326 R. B. WOODWARD AND R. HOFFMANN, *J. Am. Chem. Soc.*, 87 (1965) 2511.
327 W. G. YOUNG AND J. S. FRANKLIN, *J. Am. Chem. Soc.*, 88 (1966) 785.
328 W. G. YOUNG, S. WINSTEIN AND H. L. GOERING, *J. Am. Chem. Soc.*, 73 (1951) 1958.

INDEX

C

F

N

T